Zhilin Li Lubin Vulkov
Jerzy Waśniewski (Eds.)

Numerical Analysis and Its Applications

Third International Conference, NAA 2004
Rousse, Bulgaria, June 29 - July 3, 2004
Revised Selected Papers

Springer

Volume Editors

Zhilin Li
North Carolina State University
Center for Research in Scientific Computation, Department of Mathematics
Raleigh, NC 27695-8205, USA
E-mail: zhilin@math.ncsu.edu

Lubin Vulkov
University of Rousse, Department of Mathematics
Studentska str. 8, 7017 Rousse, Bulgaria
E-mail: vulkov@ami.ru.acad.bg

Jerzy Waśniewski
DTU, Danish IT Centre for Education and Research
2800 Lyngby, Denmark
E-mail: jerzy.wasniewski@uni-c.dk

Library of Congress Control Number: 2005920590

CR Subject Classification (1998): G.1, F.2.1, G.4, I.6, G.2, J.2

ISSN 0302-9743
ISBN 3-540-24937-0 Springer Berlin Heidelberg New York

Springer is a part of Springer Science+Business Media

springeronline.com

Printed in Germany

Typesetting: Camera-ready by author, data conversion by Scientific Publishing Services, Chennai, India
Printed on acid-free paper SPIN: 11392989 06/3142 5 4 3 2 1 0

Preface

This volume of the Lecture Notes in Computer Science series contains the proceedings of the 3rd Conference on Numerical Analysis and Its Applications, which was held at the University of Rousse, Bulgaria, June 29–July 3, 2004. The conference was organized by the Department of Numerical Analysis and Statistics at the University of Rousse with the support of the Department of Mathematics of North Carolina State University.

This conference continued the tradition of the two previous meetings (1996, 2000 in Rousse) as a forum where scientists from leading research groups from the "East" and "West" are provided with the opportunity to meet and exchange ideas and establish research cooperations. More than 100 scientists from 28 countries participated in the conference.

A wide range of problems concerning recent achievements in numerical analysis and its applications in physics, chemistry, engineering, and economics were discussed. An extensive exchange of ideas between scientists who develop and study numerical methods and researchers who use them for solving real-life problems took place during the conference.

We thank the plenary lecturers, Profs. R. Lazarov and V. Thome, and the key lecturers and the organizers of the minisymposia, T. Boyadjiev, T. Donchev, E. Farkhi, M. Van Gijzen, S. Nicaise, and M. Todorov, for their contributions. We recognize the effort required to prepare these key lectures and organize the minisymposia. We appreciate your sharing your knowledge of modern high-performance computing numerical methods with the conference participants. We also thank I. Brayanov for the help in putting together the book.

The 4th Conference on Numerical Analysis and Its Applications will take place in 2008.

October 2004

Zhilin Li
Lubin Vulkov
Jerzy Waśniewski

Table of Contents

Invited Papers

Research Papers

Uniform Convergence of a Monotone Iterative Method for a Nonlinear Reaction-Diffusion Problem

Igor Boglaev

Institute of Fundamental Sciences, Massey University,
Private Bag 11-222, Palmerston North, New Zealand
I.Boglaev@massey.ac.nz

Abstract. This paper deals with a monotone iterative method for solving a nonlinear parabolic reaction-diffusion problem. The monotone iterative method based on the method of upper and lower solutions is constructed. A rate of convergence of the method is estimated. Uniform convergence properties of the monotone iterative method are studied. Numerical experiments are presented.

1 Introduction

We are interested in a monotone iterative method for solving the nonlinear reaction-diffusion problem

$$-\mu^2\left(\frac{\partial^2 u}{\partial x^2}+\frac{\partial^2 u}{\partial y^2}\right)+\frac{\partial u}{\partial t}=-f(P,t,u), \tag{1}$$

$$P=(x,y),\quad (P,t)\in Q=\Omega\times(0,t_F],\quad \Omega=\{0<x<1,0<y<1\},$$

$$f_u(P,t,u)\geq 0,\quad (P,t,u)\in\overline{Q}\times(-\infty,\infty),\quad (f_u\equiv\partial f/\partial u),$$

where μ is a small positive parameter. The initial-boundary conditions are defined by

$$u(P,t)=g(P,t),\ (P,t)\in\partial\Omega\times(0,t_F],\quad u(P,0)=u^0(P),\ P\in\overline{\Omega},$$

where $\partial\Omega$ is the boundary of Ω. The functions $f(P,t,u)$, $g(P,t)$ and $u^0(P)$ are sufficiently smooth. Under suitable continuity and compatibility conditions on the data, a unique solution $u(P,t)$ of (1) exists (see [6] for details). For $\mu\ll 1$, problem (1) is singularly perturbed and characterized by the boundary layers of width $O(\mu|\ln\mu|)$ at the boundary $\partial\Omega$ (see [2] for details).

In the study of numerical solutions of nonlinear singularly perturbed problems by the finite difference method, the corresponding discrete problem is usually formulated as a system of nonlinear algebraic equations. A major point about this system is to obtain reliable and efficient computational algorithms for computing the solution. In the case of the parabolic problem (1), the implicit method is

Z. Li et al. (Eds.): NAA 2004, LNCS 3401, pp. 1–13, 2005.

usually in use. On each time level, this method leads to a nonlinear system which requires some kind of iterative scheme for the computation of numerical solutions. A fruitful method for the treatment of these nonlinear systems is the method of upper and lower solutions and its associated monotone iterations (in the case of "unperturbed" problems see [8], [9] and references therein). Since the initial iteration in the monotone iterative method is either upper or lower solutions, which can be constructed directly from the difference equation without any knowledge of the exact solution, this method eliminates the search for the initial iteration as is often needed in Newton's method. This elimination gives a practical advantage in the computation of numerical solutions.

In [4], for solving nonlinear singularly perturbed reaction-diffusion problems of elliptic type, we proposed the discrete monotone iterative method based on the method of upper and lower solutions. In this paper, we extand the monotone approach from [4] for solving the nonlinear parabolic reaction-diffusion problem (1).

The structure of the paper is as follows. In Section 2, we construct a monotone iterative method for solving the implicit difference scheme which approximates the nonlinear problem (1) and study convergence properties of the proposed method. Section 3 is devoted to the investigation of uniform convergence of the monotone iterative method based on Shishkin- and Bakhvalov-type meshes. The final Section 4 presents results of numerical experiments.

2 Monotone Iterative Method

On $\overline{Q}$ introduce a rectangular mesh $\overline{\Omega}^h \times \overline{\Omega}^\tau$, $\overline{\Omega}^h = \overline{\Omega}^{hx} \times \overline{\Omega}^{hy}$:

$$\overline{\Omega}^{hx} = \{x_i,\ 0 \le i \le N_x;\ x_0 = 0,\ x_{N_x} = 1;\ h_{xi} = x_{i+1} - x_i\},$$

$$\overline{\Omega}^{hy} = \{y_j,\ 0 \le j \le N_y;\ y_0 = 0,\ y_{N_y} = 1;\ h_{yj} = y_{j+1} - y_j\},$$

$$\overline{\Omega}^{\tau} = \{t_k = k\tau,\ 0 \le k \le N_\tau,\ N_\tau \tau = t_F\}.$$

For the approximation of the continuous problem (1), we use the implicit difference scheme

$$\mathcal{L}^h U(P,t) + \frac{1}{\tau}\left[U(P,t) - U(P,t-\tau)\right] = -f(P,t,U), \tag{2}$$

$$U(P,t) = g(P,t),\ (P,t) \in \partial\Omega^h \times \Omega^\tau, \quad U(P,0) = u^0(P),\ P \in \overline{\Omega}^h,$$

where $\mathcal{L}^h U(P,t)$ is defined by

$$\mathcal{L}^h U = -\mu^2 \left(\mathcal{D}_x^2 + \mathcal{D}_y^2\right) U.$$

$\mathcal{D}_x^2 U(P,t)$, $\mathcal{D}_y^2 U(P,t)$ are the central difference approximations to the second derivatives

$$\mathcal{D}_x^2 U_{ij}^k = (\hbar_{xi})^{-1}\left[\left(U_{i+1,j}^k - U_{ij}^k\right)(h_{xi})^{-1} - \left(U_{ij}^k - U_{i-1,j}^k\right)(h_{xi-1})^{-1}\right],$$

$$\mathcal{D}_y^2 U_{ij}^k = (\hbar_{yj})^{-1} \left[\left(U_{i,j+1}^k - U_{ij}^k \right) \left(h_{yj}^k \right)^{-1} - \left(U_{ij}^k - U_{i,j-1}^k \right) \left(h_{yj-1} \right)^{-1} \right],$$

$$\hbar_{xi} = 2^{-1} \left(h_{xi-1} + h_{xi} \right), \quad \hbar_{yj} = 2^{-1} \left(h_{yj-1} + h_{yj} \right),$$

where $U_{ij}^k \equiv U(x_i, y_j, t_k)$.

2.1 Statement of the Monotone Iterative Method

Now, we construct an iterative method for solving the nonlinear difference scheme (2) which possesses the monotone convergence. This method is based on the method of upper and lower solutions from [2]. Represent the difference equation from (2) in the equivalent form

$$\mathcal{L}U(P,t) = -f(P,t,U) + \tau^{-1} U(P,t-\tau), \quad \mathcal{L}U(P,t) \equiv \left(\mathcal{L}^h + \tau^{-1} \right) U(P,t).$$

We say that on a time level $t \in \Omega^\tau$, $\overline{V}(P,t)$ is an upper solution with a given function $V(P,t-\tau)$, if it satisfies

$$\mathcal{L}\overline{V}(P,t) + f\left(P,t,\overline{V}\right) - \tau^{-1} V(P,t-\tau) \geq 0, \ P \in \Omega^h,$$

$$\overline{V}(P,t) = g(P,t), \ P \in \partial\Omega^h.$$

Similarly, $\underline{V}(P,t)$ is called a lower solution on a time level $t \in \Omega^\tau$ with a given function $V(P,t-\tau)$, if it satisfies the reversed inequality and the boundary condition.

Additionally, we assume that $f(P,t,u)$ from (1) satisfies the two-sided constraints

$$0 \leq f_u \leq c^*, \quad c^* = \text{const}. \tag{3}$$

The iterative solution $V(P,t)$ to (2) is constructed in the following way. On each time level $t \in \Omega^\tau$, we calculate n_* iterates $V^{(n)}(P,t), \ P \in \overline{\Omega}^h, n = 1, \ldots, n_*$ using the recurrence formulas

$$\begin{aligned}\left(\mathcal{L} + c^*\right) Z^{(n+1)}(P,t) = - & \left[\mathcal{L}V^{(n)}(P,t) + f\left(P,t,V^{(n)}\right) \right. \\ & \left. -\tau^{-1} V(P,t-\tau) \right], \quad P \in \Omega^h,\end{aligned} \tag{4}$$

$$Z^{(n+1)}(P,t) = 0, \ P \in \partial\Omega^h, \quad n = 0, \ldots, n_* - 1,$$

$$V^{(n+1)}(P,t) = V^{(n)}(P,t) + Z^{(n+1)}(P,t), \ P \in \overline{\Omega}^h,$$

$$V(P,t) \equiv V^{(n_*)}(P,t), \ P \in \overline{\Omega}^h, \quad V(P,0) = u^0(P), \ P \in \overline{\Omega}^h,$$

where an initial guess $V^{(0)}(P,t)$ satisfies the boundary condition

$$V^{(0)}(P,t) = g(P,t), \ P \in \partial\Omega^h.$$

2.2 Convergence of the Monotone Iterative Method

On $\overline{\Omega}^h = \overline{\Omega}^{hx} \times \overline{\Omega}^{hy}$, we represent a difference scheme in the following canonical form

$$d(P)W(P) = \sum_{P' \in S(P)} e(P,P')\,W(P') + F(P), \; P \in \Omega^h, \tag{5}$$

$$W(P) = W^0(P), \; P \in \partial\Omega^h,$$

and suppose that

$$d(P) > 0, \; e(P,P') \geq 0, \; c(P) = d(P) - \sum_{P' \in S'(P)} e(P,P') > 0, \; P \in \Omega^h,$$

where $S'(P) = S(P) \setminus \{P\}$, $S(P)$ is a stencil of the difference scheme. Now, we formulate a discrete maximum principle and give an estimate on the solution to (5).

Lemma 1. *Let the positive property of the coefficients of the difference scheme (5) be satisfied.*

(i) If $W(P)$ satisfies the conditions

$$d(P)W(P) - \sum_{P' \in S(P)} e(P,P')\,W(P') - F(P) \geq 0 (\leq 0), \; P \in \Omega^h,$$

$$W(P) \geq 0 (\leq 0), \; P \in \partial\Omega^h,$$

then $W(P) \geq 0 (\leq 0), \; P \in \overline{\Omega}^h$.

(ii) The following estimate on the solution to (5) holds true

$$\|W\|_{\overline{\Omega}^h} \leq \max\left[\|W^0\|_{\partial\Omega^h}; \|F/c\|_{\Omega^h}\right], \tag{6}$$

where

$$\|W\|_{\overline{\Omega}^h} = \max_{P \in \overline{\Omega}^h} |W(P)|, \quad \|W^0\|_{\partial\Omega^h} = \max_{P \in \partial\Omega^h} |W^0(P)|.$$

The proof of the lemma can be found in [11].

Theorem 1. *Let $V(P,t-\tau)$ be given and $\overline{V}^{(0)}(P,t)$, $\underline{V}^{(0)}(P,t)$ be upper and lower solutions corresponding to $V(P,t-\tau)$. Suppose that $f(P,t,u)$ satisfies (3). Then the upper sequence $\left\{\overline{V}^{(n)}(P,t)\right\}$ generated by (4) converges monotonically from above to the unique solution $\mathcal{V}(P,t)$ of the problem*

$$\mathcal{L}V(P,t) + f(P,t,V) - \tau^{-1}V(P,t-\tau) = 0, \; P \in \Omega^h, \tag{7}$$

$$V(P,t) = g(P,t), \; P \in \partial\Omega^h,$$

the lower sequence $\left\{\underline{V}^{(n)}(P,t)\right\}$ generated by (4) converges monotonically from below to $V(P,t)$:

$$\mathcal{V}(P,t) \leq \overline{V}^{(n+1)}(P,t) \leq \overline{V}^{(n)}(P,t) \leq \overline{V}^{(0)}(P,t), \; P \in \overline{\Omega}^h,$$

$$\underline{V}^{(0)}(P,t) \le \underline{V}^{(n)}(P,t) \le \underline{V}^{(n+1)}(P,t) \le \mathcal{V}(P,t), \quad P \in \overline{\Omega}^h,$$

and the sequences converge with the linear rate $\rho = c^*/(c^* + \tau^{-1})$.

Proof. We consider only the case of the upper sequence. If $\overline{V}^{(0)}(P,t)$ is an upper solution, then from (4) we conclude that

$$(\mathcal{L} + c^*)\, Z^{(1)}(P,t) \le 0,\ P \in \Omega^h, \quad Z^{(1)}(P,t) = 0,\ P \in \partial\Omega^h.$$

From Lemma 1, by the maximum principle for the difference operator $\mathcal{L} + c^*$, it follows that $Z^{(1)}(P,t) \le 0, P \in \bar{\Omega}^h$. Using the mean-value theorem and the equation for $Z^{(1)}$, we have

$$\mathcal{L}\overline{V}^{(1)}(P,t) + f\left(P,t,\overline{V}^{(1)}\right) - \frac{V(P,t-\tau)}{\tau} = -\left[c^* - f_u^{(1)}(P,t)\right] Z^{(1)}(P,t), \quad (8)$$

where $f_u^{(1)}(P,t) \equiv f_u\left[P,t,\overline{V}^{(0)}(P,t) + \theta^{(1)}(P,t)Z^{(1)}(P,t)\right], 0 < \theta^{(1)}(P,t) < 1$. Since the mesh function $Z^{(1)}(P,t)$ is nonpositive on Ω^h and taking into account (3), we conclude that $\overline{V}^{(1)}(P,t)$ is an upper solution. By induction we obtain that $Z^{(n)}(P,t) \le 0, P \in \overline{\Omega}^h, n = 1,2,\ldots$, and prove that $\left\{\overline{V}^{(n)}(P,t)\right\}$ is a monotonically decreasing sequence of upper solutions.

Now we shall prove that the monotone sequence $\left\{\overline{V}^{(n)}(P,t)\right\}$ converges to the solution of (7). Similar to (8), we obtain

$$\mathcal{L}\overline{V}^{(n)}(P,t) + f\left(P,t,\overline{V}^{(n)}\right) - \frac{V(P,t-\tau)}{\tau} = -\left[c^* - f_u^{(n)}(P,t)\right] Z^{(n)}(P,t), \quad (9)$$

and from (4), it follows that $Z^{(n+1)}(P,t)$ satisfies the difference equation

$$(\mathcal{L} + c^*)\, Z^{(n+1)}(P,t) = \left(c^* - f_u^{(n)}\right) Z^{(n)}(P,t), \ P \in \Omega^h.$$

Using (6) and (3), we conclude

$$\left\| Z^{(n+1)}(t) \right\|_{\overline{\Omega}^h} \le \rho^n \left\| Z^{(1)}(t) \right\|_{\overline{\Omega}^h}, \quad \rho = \frac{c^*}{c^* + \tau^{-1}}. \quad (10)$$

This proves convergence of the upper sequence to the solution $\mathcal{V}$ of (7) with the linear rate ρ. In view of $\lim \overline{V}^{(n)} = \mathcal{V}$ as $n \to \infty$, we conclude that $\mathcal{V} \le \overline{V}^{(n+1)} \le \overline{V}^{(n)}$.

The uniqueness of the solution to (7) follows from estimate (6). Indeed, if by contradiction, we assume that there exist two solutions $\mathcal{V}_1$ and $\mathcal{V}_2$ to (7), then by the mean-value theorem, the difference $\delta\mathcal{V} = \mathcal{V}_1 - \mathcal{V}_2$ satisfies the following difference problem

$$(\mathcal{L} + f_u)\, \delta\mathcal{V}(P,t) = 0,\ P \in \Omega^h, \quad \delta\mathcal{V}(P,t) = 0,\ P \in \partial\Omega^h.$$

By (6), this leads to the uniqueness of the solution to (7).

Theorem 2. *Let $V^{(0)}(P,t)$ be an upper or a lower solution in the iterative method (4), and let $f(P,t,u)$ satisfy (3). Suppose that on each time level the number of iterates n_* satisfies $n_* \geq 2$. Then the following estimate on convergence rate holds*

$$\max_{t\in\Omega^\tau} \|V(t)-U(t)\|_{\overline{\Omega}^h} \leq C(\rho)^{n_*-1},$$

where $U(P,t)$ is the solution to (2) and constant C is independent of τ. Furthermore, on each time level the sequence $\{V^{(n)}(P,t)\}$ converges monotonically.

Proof. Introduce the notation

$$W(P,t) = U(P,t) - V(P,t),$$

where $V(P,t) \equiv V^{(n_*)}(P,t)$. Using the mean-value theorem, from (2), (9), conclude that $W(P,\tau)$ satisfies

$$(\mathcal{L} + f_u(P,\tau))\, W(P,\tau) = \left[c^* - f_u^{(n_*)}(P,\tau)\right] Z^{(n_*)}(P,\tau), \;\; P \in \Omega^h,$$

$$W(P,\tau) = 0, \;\; P \in \partial\Omega^h,$$

where $f_u(P,\tau) \equiv f_u\left[P,\tau,U(P,\tau)+\theta(P,\tau)W(P,\tau)\right],\; 0 < \theta(P,\tau) < 1$, and we have taken into account that $V(P,0) = U(P,0)$. By (6), (3) and (10),

$$\|W(\tau)\|_{\overline{\Omega}^h} \leq c^*\tau\rho^{n_*-1} \left\|Z^{(1)}(\tau)\right\|_{\overline{\Omega}^h}.$$

Estimate $Z^{(1)}(P,\tau)$ from (4) by (6),

$$\left\|Z^{(1)}(\tau)\right\|_{\overline{\Omega}^h} \leq \tau \left\|\mathcal{L}V^{(0)}(\tau) + f\left(V^{(0)}\right) - \tau^{-1}u^0\right\|_{\overline{\Omega}^h} \leq C_1,$$

where C_1 is independent of τ. Thus,

$$\|W(\tau)\|_{\overline{\Omega}^h} \leq \tilde{C}_1\tau\rho^{n_*-1}, \quad \tilde{C}_1 = c^*C_1, \tag{11}$$

where $\tilde{C}_1$ is independent of τ. Similarly, from (2), (9), it follows that

$$\begin{aligned}(\mathcal{L} + f_u(P,2\tau))\, W(P,2\tau) &= \tau^{-1}W(P,\tau)\\ &\quad + \left[c^* - f_u^{(n_*)}(P,2\tau)\right] Z^{(n_*)}(P,2\tau).\end{aligned}$$

By (6),

$$\|W(2\tau)\|_{\overline{\Omega}^h} \leq \|W(\tau)\|_{\overline{\Omega}^h} + c^*\tau\rho^{n_*-1}\left\|Z^{(1)}(2\tau)\right\|_{\overline{\Omega}^h}.$$

Estimate $Z^{(1)}(P,2\tau)$ from (4) by (6),

$$\left\|Z^{(1)}(2\tau)\right\|_{\overline{\Omega}^h} \leq \tau \left\|\mathcal{L}V^{(0)}(2\tau) + f\left(V^{(0)}\right) - \tau^{-1}U(\tau)\right\|_{\overline{\Omega}^h} \leq C_2,$$

where C_2 is independent of τ. From here and (11), we conclude

$$\|W(2\tau)\|_{\overline{\Omega}^h} \leq \left(\tilde{C}_1 + \tilde{C}_2\right)\tau\rho^{n_*-1}, \; \tilde{C}_2 = c^*C_2.$$

By induction, we prove

$$\|W(t_k)\|_{\overline{\Omega}^h} \leq \left(\sum_{l=1}^{k} \tilde{C}_l \right) \tau \rho^{n_*-1}, \; k = 1, \ldots, N_\tau, \tag{12}$$

where all constants $\tilde{C}_l$ are independent of τ. Denoting

$$C_0 = \max_{1 \leq l \leq N_\tau} \tilde{C}_l,$$

and taking into account that $N_\tau \tau = t_F$, we prove the estimate in the theorem with $C = t_F C_0$.

Remark 1. Consider the following approach for constructing initial upper and lower solutions $\overline{V}^{(0)}(P,t)$ and $\underline{V}^{(0)}(P,t)$. Suppose that for t fixed, a mesh function $R(P,t)$ is defined on $\overline{\Omega}^h$ and satisfies the boundary condition $R(P,t) = g(P,t)$ on $\partial\Omega^h$. Introduce the following difference problems

$$\mathcal{L}Z_q^{(0)}(P,t) = q\left|\mathcal{L}R(P,t) + f(P,t,R) - \tau^{-1}V(P,t-\tau)\right|, \; P \in \Omega^h, \tag{13}$$

$$Z_q^{(0)}(P,t) = 0, \; P \in \partial\Omega^h, \quad q = 1, -1.$$

Then the functions $\overline{V}^{(0)}(P,t) = R(P,t) + Z_1^{(0)}(P,t)$, $\underline{V}^{(0)}(P,t) = R(P,t) + Z_{-1}^{(0)}(P,t)$ are upper and lower solutions, respectively.

We check only that $\overline{V}^{(0)}(P,t)$ is an upper solution. From the maximum principle, it follows that $Z_1^{(0)}(P,t) \geq 0$ on $\bar{\Omega}^h$. Now using the difference equation for $Z_1^{(0)}$, we have

$$\mathcal{L}\left(R + Z_1^{(0)}\right) + f\left(R + Z_1^{(0)}\right) - \tau^{-1}V(P,t-\tau) = F(P,t) + |F(P,t)| + f_u^{(0)} Z_1^{(0)},$$

$$F(P,t) \equiv \left|\mathcal{L}R(P,t) + f(P,t,R) - \tau^{-1}V(P,t-\tau)\right|.$$

Since $f_u^{(0)} \geq 0$ and $Z_1^{(0)}$ is nonnegative, we conclude that $\overline{V}^{(0)}(P,t)$ is an upper solution.

Remark 2. Since the initial iteration in the monotone iterative method (4) is either an upper or a lower solution, which can be constructed directly from the difference equation without any knowledge of the solution as we have suggested in the previous remark, this algorithm eliminates the search for the initial iteration as is often needed in Newton's method. This elimination gives a practical advantage in the computation of numerical solutions.

Remark 3. The implicit two-level difference scheme (2) is of the first order accuracy with respect to τ. From here and since $\rho \leq c^*\tau$, one may choose $n_* = 2$ to keep the global error of the monotone iterative method (4) consistent with the global error of the difference scheme (2).

3 Uniform Convergence of the Monotone Iterative Method (4)

Here we analyse a convergence rate of the monotone iterative method (4) defined on meshes of the general type introduced in [10]. On these meshes, the implicit difference scheme (2) converges μ-uniformly to the solution of (1).

A mesh of this type is formed in the following manner. We divide each of the intervals $\overline{\Omega}^x = [0,1]$ and $\overline{\Omega}^y = [0,1]$ into three parts $[0,\sigma_x]$, $[\sigma_x, 1-\sigma_x]$, $[1-\sigma_x, 1]$, and $[0,\sigma_y]$, $[\sigma_y, 1-\sigma_y]$, $[1-\sigma_y, 1]$, respectively. Assuming that N_x, N_y are divisible by 4, in the parts $[0,\sigma_x]$, $[1-\sigma_x,1]$ and $[0,\sigma_y]$, $[1-\sigma_y,1]$ we allocate $N_x/4+1$ and $N_y/4+1$ mesh points, respectively, and in the parts $[\sigma_x, 1-\sigma_x]$ and $[\sigma_y, 1-\sigma_y]$ we allocate $N_x/2+1$ and $N_y/2+1$ mesh points, respectively. Points σ_x, $(1-\sigma_x)$ and σ_y, $(1-\sigma_y)$ correspond to transition to the boundary layers. We consider meshes $\overline{\Omega}^{hx}$ and $\overline{\Omega}^{hy}$ which are equidistant in $\left[x_{N_x/4}, x_{3N_x/4}\right]$ and $\left[y_{N_y/4}, y_{3N_y/4}\right]$ but graded in $\left[0, x_{N_x/4}\right]$, $\left[x_{3N_x/4}, 1\right]$ and $\left[0, y_{N_y/4}\right]$, $\left[y_{3N_y/4}, 1\right]$. On $\left[0, x_{N_x/4}\right]$, $\left[x_{3N_x/4}, 1\right]$ and $\left[0, y_{N_y/4}\right]$, $\left[y_{3N_y/4}, 1\right]$ let our mesh be given by a mesh generating function ϕ with $\phi(0)=0$ and $\phi(1/4)=1$ which is supposed to be continuous, monotonically increasing, and piecewise continuously differentiable. Then our mesh is defined by

$$x_i = \begin{cases} \sigma_x \phi\left(\xi_i\right), & \xi_i = i/N_x,\ i = 0,\ldots,N_x/4; \\ ih_x, & i = N_x/4+1,\ldots,3N_x/4-1; \\ 1-\sigma_x\left(1-\phi\left(\xi_i\right)\right), & \xi_i = \left(i-3N_x/4\right)/N_x,\ i = 3N_x/4+1,\ldots,N_x, \end{cases}$$

$$y_j = \begin{cases} \sigma_y \phi\left(\xi_j\right), & \xi_j = j/N_y,\ j = 0,\ldots,N_y/4; \\ jh_y, & j = N_y/4+1,\ldots,3N_y/4-1; \\ 1-\sigma_y\left(1-\phi\left(\xi_j\right)\right), & \xi_j = \left(j-3N_y/4\right)/N_y,\ j = 3N_y/4+1,\ldots,N_y, \end{cases}$$

$$h_x = 2\left(1-2\sigma_x\right)N_x^{-1}, \quad h_y = 2\left(1-2\sigma_y\right)N_y^{-1}.$$

We also assume that $d\phi/d\xi$ does not decrease. This condition implies that

$$h_{xi} \le h_{x,i+1},\ i = 1,\ldots,N_x/4-1, \quad h_{xi} \ge h_{x,i+1},\ i = 3N_x/4+1,\ldots,N_x-1,$$

$$h_{yj} \le h_{y,j+1},\ j = 1,\ldots,N_y/4-1, \quad h_{yj} \ge h_{y,j+1},\ j = 3N_y/4+1,\ldots,N_y-1.$$

3.1 Shishkin-Type Mesh

We choose the transition points σ_x, $(1-\sigma_x)$ and σ_y, $(1-\sigma_y)$ in Shishkin's sense (see [7] for details), i.e.

$$\sigma_x = \min\left\{4^{-1}, \upsilon_1 \mu \ln N_x\right\}, \quad \sigma_y = \min\left\{4^{-1}, \upsilon_2 \mu \ln N_y\right\},$$

where υ_1 and υ_2 are positive constants. If $\sigma_{x,y} = 1/4$, then $N_{x,y}$ are very large compared to $1/\mu$ which means that the difference scheme (2) can be analysed using standard techniques. We therefore assume that

$$\sigma_x = \upsilon_1 \mu \ln N_x, \quad \sigma_y = \upsilon_2 \mu \ln N_y.$$

Consider the mesh generating function ϕ in the form

$$\phi(\xi) = 4\xi.$$

In this case the meshes $\overline{\Omega}^{hx}$ and $\overline{\Omega}^{hy}$ are piecewise equidistant with the step sizes

$$N_x^{-1} < h_x < 2N_x^{-1}, \quad h_{x\mu} = 4\upsilon_1 \mu N_x^{-1} \ln N_x, \tag{14}$$

$$N_y^{-1} < h_y < 2N_y^{-1}, \quad h_{y\mu} = 4\upsilon_2 \mu N_y^{-1} \ln N_y.$$

The implicit difference scheme (2) on the piecewise uniform mesh (14) converges μ-uniformly to the solution of (1):

$$\max_{(P,t)\in\overline{\Omega}^h\times\overline{\Omega}^\tau} |U(P,t) - u(P,t)| \le M\left((\ln N/N)^2 + \tau\right), \quad N = \min\{N_x; N_y\},$$

where constant M is independent of μ, N and τ. The proof of this result can be found in [7]. From here and Theorem 2, we conclude

$$\max_{(P,t)\in\overline{\Omega}^h\times\overline{\Omega}^\tau} |V(P,t) - u(P,t)| \le M\left((\ln N/N)^2 + \tau\right) + C(\rho)^{n_*-1}, \tag{15}$$

where constant C is independent of τ.

Without loss of generality, we assume that the boundary condition in (1) is zero, i.e. $g(P,t) = 0$. This assumption can always be obtained via a change of variables. Let the initial function $V^{(0)}(P,t)$ be chosen on each time level in the form of (13), i.e. $V^{(0)}(P,t)$ is the solution of the following difference problem

$$\mathcal{L}V^{(0)}(P,t) = q\left|f(P,t,0) - \tau^{-1}V(P,t-\tau)\right|, \quad P \in \Omega^h, \tag{16}$$

$$V^{(0)}(P,t) = 0, \quad P \in \partial\Omega^h, \quad q = 1, -1,$$

where $R(P,t) = 0$. Then the functions $\overline{V}^{(0)}(P,t)$, $\underline{V}^{(0)}(P,t)$ corresponding to $q = 1$ and $q = -1$ are upper and lower solutions, respectively.

Theorem 3. *Suppose that the boundary condition in (1) is zero, and an initial upper or a lower solution $V^{(0)}(P,t)$ is chosen by (16). Let $f(P,t,u)$ satisfy (3). If on each time level the number of iterates of the monotone iterative method (4) satisfies $n_* \ge 2$, then the solution of the monotone iterative method (4) on the piecewise uniform mesh (14) converges μ-uniformly to the solution of the problem (1). The estimate (15) holds true with $\rho = c^*/\left(c^* + \tau^{-1}\right)$ and constants C and M are independent of μ, N and τ.*

Proof. By (6),

$$\left\|V^{(0)}(t)\right\|_{\overline{\Omega}^h} \le \tau\left\|f(P,t,0) - \tau^{-1}V(t-\tau)\right\|_{\overline{\Omega}^h}. \tag{17}$$

Using the mean-value theorem, from (3) and (12), it follows that

$$\begin{aligned}\left\|Z^{(1)}(t_l)\right\|_{\overline{\Omega}^h} &\leq \tau\Big[\left\|\mathcal{L}^{h\tau}V^{(0)}(t_l)\right\|_{\overline{\Omega}^h} + c^*\left\|V^{(0)}(t_l)\right\|_{\overline{\Omega}^h} \\ &\quad + \left\|f(P,t_l,0) - \tau^{-1}V(t_l-\tau)\right\|_{\overline{\Omega}^h}\Big] \\ &\leq (2\tau + c^*\tau^2)\left\|f(P,t_l,0) - \tau^{-1}V(t_l-\tau)\right\|_{\overline{\Omega}^h} \\ &\leq (2\tau + c^*\tau^2)\left\|f(P,t_l,0)\right\|_{\overline{\Omega}^h} + (2+c^*\tau)\left\|V(t_l-\tau)\right\|_{\overline{\Omega}^h} \\ &\leq C_l.\end{aligned}$$

To prove that all constants C_l are independent of the small parameter μ, we have to prove that $\|V(t_l-\tau)\|_{\overline{\Omega}^h}$ is μ-uniformly bounded. For $l=1$, $V(P,0) = u^0(P)$, where u^0 is the initial condition in the differential problem (1), and, hence, C_1 is independent of μ, τ and N, where we assume $\tau \leq \tau_0$. For $l=2$, we have

$$\left\|Z^{(1)}(t_2)\right\|_{\overline{\Omega}^h} \leq (2\tau + c^*\tau^2)\left\|f(P,t_2,0)\right\|_{\overline{\Omega}^h} + (2+c^*\tau)\left\|V(t_1)\right\|_{\overline{\Omega}^h} \leq C_2,$$

where $V(P,t_1) = V^{(n_*)}(P,t_1)$. As follows from Theorem 1, the monotone sequences $\left\{\overline{V}^{(n)}(P,t_1)\right\}$ and $\left\{\underline{V}^{(n)}(P,t_1)\right\}$ are μ-uniformly bounded from above by $\overline{V}^{(0)}(P,t_1)$ and from below by $\underline{V}^{(0)}(P,t_1)$. From (17) at $t=t_1$, we have

$$\left\|V^{(0)}(t_1)\right\|_{\overline{\Omega}^h} \leq \tau\left\|f(P,t_1,0) - \tau^{-1}u^0(P)\right\|_{\overline{\Omega}^h} \leq K_1,$$

where constant K_1 is independent of μ, τ and N. Thus, we prove that C_2 is independent of μ, τ and N. Now by induction on l, we prove that all constants C_l in (12) are independent of μ, and, hence, constant $C = t_F c^* \max_{1\leq l\leq N_\tau} C_l$ in Theorem 2 is independent of μ, τ and N. From here and (15), we conclude that the monotone iterative method (4) converges μ-uniformly to the solution of the differential problem (1).

3.2 Bakhvalov-Type Mesh

We choose the transition points σ_x, $(1-\sigma_x)$ and σ_y, $(1-\sigma_y)$ in Bakhvalov's sense (see [3] for details), i.e.

$$\sigma_x = \upsilon_1\mu\ln(1/\mu), \quad \sigma_y = \upsilon_2\mu\ln(1/\mu), \tag{18}$$

$$\phi(\xi) = \frac{\ln[1-4(1-\mu)\xi]}{\ln\mu}.$$

The difference scheme (2) on the Bakhvalov-type mesh converges μ-uniformly to the solution of (1):

$$\max_{(P,t)\in\overline{\Omega}^h\times\overline{\Omega}^\tau} |U(P,t) - u(P,t)| \leq M\left(N^{-1}+\tau\right), \quad N = \min\{N_x, N_y\},$$

where constant M is independent of μ, N and τ. The proof of this result can be found in [3].

For the monotone iterative method (4) on the Bakhvalov-type mesh, Theorem 3 holds true with the following error estimate

$$\max_{(P,t)\in\overline{\Omega}^h\times\overline{\Omega}^\tau} |V(P,t) - u(P,t)| \leq M\left(N^{-1} + \tau\right) + C(\rho)^{n_*-1},$$

where constants C and M are independent of μ, N and τ.

4 Numerical Experiments

Consider problem (1) with $f(P,t,u) = (u-4)/(5-u)$, $g(P,t) = 1$ and $u^0(P) = 1$, which models the biological Michaelis-Menton process without inhibition [5]. This problem gives $c^* = 1$.

On each time level t_k, the stopping criterion is chosen in the form

$$\left\| V^{(n)}(t_k) - V^{(n-1)}(t_k) \right\|_{\overline{\Omega}^h} \leq \delta,$$

where $\delta = 10^{-5}$. All the discrete linear systems are solved by $GMRES$-solver [1].

It is found that in all the numerical experiments the basic feature of monotone convergence of the upper and lower sequences is observed. In fact, the monotone property of the sequences holds at every mesh point in the domain. This is, of course, to be expected from the analytical consideration.

Consider the monotone iterative method (4) on the Shishkin-type mesh (14) with $N_x = N_y$. Since for our data set we allow $\sigma_x > 0.25$, the step size $h_{x\mu}$ is calculated as

$$h_{x\mu} = \frac{4\min\{0.25, \sigma_x\}}{N_x}.$$

In Table 1, for $\tau_1 = 10^{-1}$, $\tau_2 = 5\times10^{-2}$ and $\tau_3 = 10^{-2}$ and for various values of μ and N_x, we give the average (over ten time levels) numbers of iterations n_{τ_1}, n_{τ_2}, n_{τ_3} required to satisfy the stopping criterion. From the data, we conclude that for N_x fixed, the numbers of iterations are independent of the perturbation parameter μ and the number of iterations n_τ as a function of τ is monotone increasing. These numerical results confirm our theoretical results stated in Theorems 2 and 3.

Now, consider the monotone iterative method (4) on the Bakhvalov-type mesh (18) with $N_x = N_y$. In this case, for all values of μ and N_x presented in Table 1, we have $n_{\tau_1} = 5$, $n_{\tau_2} = 4$, $n_{\tau_3} = 3$. Thus, the main features of the monotone iterative method (4) on the Shishkin-type mesh highlighted from Table 1 hold true for the algorithm on the Bakhvalov-type mesh.

Table 2 presents the numerical experiments corresponding to ones in Table 1, when on each time level the Newton iterative method on the Bakhvalov-type mesh is in use. We denote by (k^*) if on time level t_k more than 100 iterations

Table 1. Average numbers of iterations on the Shishkin-type mesh

μ	n_{T_1}; n_{T_2}; n_{T_3}		
10^{-1}	5; 4; 3	5; 4; 3	5; 4; 3
10^{-2}	4; 3.8; 2.8	4.9; 4; 3	5; 4; 3
10^{-3}	4; 3.8; 2.8	4.9; 4; 3	5; 4; 3
10^{-4}	4; 3.8; 2.8	4.9; 4; 3	5; 4; 3
10^{-5}	4; 3.8; 2.8	4.9; 4; 3	5; 4; 3
$N_x/4$	16	32	64, 128, 256

Table 2. Average numbers of iterations for the Newton's method on the Bakhvalov-type mesh

μ	n_{T_1}; n_{T_2}; n_{T_3}			
10^{-1}	(9*); 8.1; 4.1	(9*); 8.1; 4.1	(9*); 8.1; 4.1	(8*); 8.1; 4.1
10^{-2}	18.4; 8.1; 4.1	17.1; 8.1; 4.1	(10*); 8.1; 4.1	(8*); 8.1; 4.1
10^{-3}	18.3; 8.1; 4.1	17.5; 8.1; 4.1	(9*); 8.1; 4.1	(8*); 8.1; 4.1
10^{-4}	18.8; 8.1; 4.1	17.8; 8.1; 4.1	(10*); 8.1; 4.1	(9*); 8.1; 4.1
10^{-5}	17.2; 8.1; 4.1	17.5; 8.1; 4.1	19.3; 8.1; 4.1	(9*); 8.1; 4.1
$N_x/4$	16	32	64	128

is needed to satisfy the stopping criterion, or if the method diverges. The experimental results show that, in general, the Newton method cannot be used successfully for this test problem.

References

1. R. Barrett, M. Berry, T. F. Chan, J. Demmel, J. Donato, J. Dongarra, V. Eijkhout, R. Pozo, C. Romine and H. Van der Vorst, Templates for the Solution of Linear Systems: Building Blocks for Iterative Methods, SIAM, Philadelphia, 1994.
2. I. Boglaev, Numerical solution of a quasi-linear parabolic equation with a boundary layer, USSR Comput. Maths. Math. Phys. 30 (1990) 55-63.
3. I. Boglaev, Finite difference domain decomposition algorithms for a parabolic problem with boundary layers, Comput. Math. Applic. 36 (1998) 25-40.
4. I. Boglaev, On monotone iterative methods for a nonlinear singularly perturbed reaction-diffusion problem, J. Comput. Appl. Math. 162 (2004) 445-466.
5. E. Bohl, Finite Modelle Gewöhnlicker Randwertaufgaben, Teubner, Stuttgart, 1981.
6. O.A. Ladyženskaja, V.A. Solonnikov and N.N. Ural'ceva, Linear and Quasi-Linear Equations of Parabolic Type, Academic Press, New York, 1968.
7. J.J.H Miller, E. O'Riordan and G.I. Shishkin, Fitted Numerical Methods for Singular Perturbation Problems, World Scientific, Singapore, 1996.
8. C.V. Pao, Monotone iterative methods for finite difference system of reaction-diffusion equations, Numer. Math. 46 (1985) 571-586.

9. C.V. Pao, Finite difference reaction diffusion equations with nonlinear boundary conditions, Numer. Methods Partial Diff. Eqs. 11 (1995) 355-374.
10. H.-G. Roos and T. Linss, Sufficient conditions for uniform convergence on layer adapted grids, Computing 64 (1999) 27-45.
11. A. Samarskii, The Theory of Difference Schemes, Marcel Dekker Inc., New York-Basel, 2001.

The Dynamics of Matrix Coupling with an Application to Krylov Methods

Françoise Chaitin-Chatelin

Université Toulouse 1 and CERFACS,
42 avenue G. Coriolis 31057 Toulouse Cedex 1, France
chatelin@cerfacs.fr,
http//www.cerfacs.fr/~chatelin

Abstract. Given the matrices A and E in $\mathbb{C}^{n\times n}$, we consider, for the family $A(t) = A + tE$, $t \in \mathbb{C}$, questions such as i) existence and analyticity of $t \mapsto R(t,z) = (A(t) - zI)^{-1}$, and ii) limit as $|t| \to \infty$ of $\sigma(A(t))$, the spectrum of $A(t)$. The answer depends on the Jordan structure of $0 \in \sigma(E)$, more precisely on the existence of trivial Jordan blocks (of size 1). The results of the theory of Homotopic Deviation are then used to analyse the convergence of Krylov methods in finite precision.

Keywords: Sherman-Morrison formula, Schur complement, Jordan structure, frontier point, critical point, Ritz value, eigenprojection, analyticity, singularity, backward analysis, Krylov method.

1 Introduction

A and E are given matrices in $\mathbb{C}^{n\times n}$, which are coupled by the complex parameter t to form $A(t) = A + tE$. $\sigma(A)$ (resp. $re(A) = \mathbb{C} - \sigma(A)$) denotes the spectrum (resp. resolvent set) of A. We study the two maps:

$$t \in \mathbb{C} \mapsto R(t,z) = (A(t) - zI)^{-1},$$

for z given in $re(A)$, and

$$t \in \mathbb{C} \mapsto \sigma(A(t)).$$

Such a framework is useful to perform a **backward analysis** for computational methods which are **inexact**: one has access to properties of $A(t)$ by means of the resolvent matrix $R(0,z) = (A - zI)^{-1}$, $z \in re(A)$, only. In this context, the question of the behavior of $R(t,z)$ and $\sigma(A(t))$ as $|t| \to \infty$ arises naturally [6]. Such a study is also of interest for engineering when the parameter t has a physical meaning and can be naturally unbounded [10].

Various approaches are useful, ranging from analytic/algebraic spectral theory [1,2,3,6,10] to linear control systems theory [12]. The theory surveyed here is **Homotopic Deviation** [4,5,11] which specifically looks beyond analyticity for $|t|$ large. The case of interest corresponds to a singular matrix E. The tools are elementary linear algebra based on the Sherman-Morrison formula, the Schur complement and on the Jordan structure of $0 \in \sigma(E)$. It also relies on Lidskii's perturbation theory [18].

Z. Li et al. (Eds.): NAA 2004, LNCS 3401, pp. 14–24, 2005.

1.1 Presentation of the Paper

The paper is organized as follows. The mathematical setting is given in the rest of Section 1. Then Section 2 analyses the convergence rates for the two analytic developments for $R(t,z)$ around 0 and ∞. A similar analysis for $\sigma(A(t))$ is performed in Section 3. This results in a complete homotopic **backward analysis** for the eigenproblem for A, in terms of $t \in \mathbb{C}$, the homotopy parameter. The theory is used in Section 4 to explain the extreme robustness of inexact Krylov methods to very large perturbations [5, 15].

Due to space limitation, most proofs have been omitted. A more comprehensive development of the theory can be found in [17] and [11].

1.2 Notation

We set

$$F_z = -E(A - zI)^{-1},\ z \in re(A)$$

Formally

$$R(t,z) = R(0,z)(I - tF_z)^{-1}.$$

exists for $t \neq \frac{1}{\mu_z}$, $0 \neq \mu_z \in \sigma(F_z)$ and is computable as

$$R(t,z) = R(0,z)\sum_{k=0}^{\infty}(tF_z)^k \text{ for } |t| < \frac{1}{\rho(F_z)},\ \rho(F_z) = \max|\mu_z|.$$

When Rank $E = n$, $0 \notin \sigma(F_z)$, and the eigenvalues of F_z are denoted by μ_{iz}, $i = 1, \cdots, n$. Therefore $R(t,z)$ is defined for almost all $t \in \mathbb{C}$, $t \neq t_i$, with $t_i = \frac{1}{\mu_{iz}}$, $i = 1, \cdots, n$. Consequently z is an eigenvalue of the n matrices $A(t_i)$, $i = 1, \cdots, n$. What happens in the limit $|t| \to \infty$? It is easy to check that under the assumption that E is *regular* and $|t| \to \infty$, the limit of the resolvent matrix $R(t,z)$ (resp. the spectrum $\sigma(A(t))$ is 0 (resp. at ∞). To get a richer situation where the limit resolvent may be nonzero, and eigenvalues may stay at finite distance, we assume that $E \neq 0$ is *singular*, or *rank deficient*, $r =$ Rank E, $1 \leq r < n$. We set $\hat{\mathbb{C}} = \mathbb{C} \cup \{\infty\}$; card $\hat{\mathbb{C}} =$ card $\mathbb{C} = c$ denotes the cardinal of the (complex) continuum.

1.3 $E = UV^H$ with $U,\ V \in \mathbb{C}^{n \times r}$ of Rank r, $1 \leq r < n$

Any singular matrix $E \neq 0$ of Rank r can be written under the form

$$E = UV^H, \text{ with } U, V \in \mathbb{C}^{n \times r} \text{ of Rank } r,\ 1 \leq r < n,$$

where U, V of Rank r represent a basis for Im E, Im E^H respectively [12]. F_z has now Rank r, so that at most r eigenvalues μ_{iz}, $i = 1, \cdots, r$ are nonzero. They are the r eigenvalues of

$$M_z = -V^H(A - zI)^{-1}U \in \mathbb{C}^{r \times r},\ z \in re(A).$$

By applying the *Sherman-Morrison formula* [12] we have that

$$R(t,z) = R(0,z)[I_n - tU(I_r - tM_z)^{-1}V^H R(0,z)] \tag{1}$$

exists for $t \neq \frac{1}{\mu_z}$, $0 \neq \mu_z \in \sigma(M_z)$. For $z \in re(A)$, $R(t,z)$ is not defined when $t \in \mathbb{C}$ satisfies $t\mu_z = 1$, $0 \neq \mu_z \in \sigma(M_z)$. If M_z is regular, this is equivalent to $t \in \sigma(M_z^{-1})$.

Therefore $z \in re(A)$ is an eigenvalue of $A+tE$ iff $t\mu_z = 1$. This means that any z in $re(A)$ is an inexact value for A at homotopic distance $|t|$, that is z is an exact eigenvalue of the r matrices $A(t_i) = A + t_i E$ with $t_i = \frac{1}{\mu_{iz}} \in \mathbb{C}$, $i = 1, \cdots, r$, when M_z is of Rank r.

When $r > 1$, the homotopic distance is not uniquely defined.

The matrix M_z of order $r < n$ will play the key role in the analysis of our problem, similar to the role of the transfer matrix in linear control theory [12].

1.4 The Limit of $R(t,z)$ When $|t| \to \infty$, for $z \in re(A)$

We suppose that $|t| > 1/\min|\mu_z|$ for M_z of Rank r.

Proposition 1. *For* $1 \leq r < n$, z *given in* $re(A)$ *such that Rank* $M_z = r$, $\lim_{|t|\to\infty} R(t,z)$ *exists and is given by*

$$R(\infty,z) = R(0,z)[I_n + UM_z^{-1}V^H R(0,z)]$$

Proof. Straightforward [11, 7].

When M_z^{-1} exists, the asymptotic resolvent $R(\infty,z)$ exists and is computable in *closed form.* This shows the dual role played by the two quantities $|t_1| = 1/\max|\mu_z| = 1/\rho(M_z)$ and $|t_r| = 1/\min|\mu_z| = \rho(M_z^{-1})$.

1) $|t_1|$ defines the largest analyticity disk for $R(t,z)$: it rules the convergence of the initial analytic development

$$R(t,z) = R(0,z)[I_n - tU\sum_{k=0}^{\infty}(tM_z)^k V^H R(0,z)] \tag{2}$$

based on M_z and valid for $|t| < |t_1|$ (around 0).

The series expansion(2) becomes *finite* when M_z is nilpotent ($\rho(M_z) = 0$).

2) $|t_r|$ defines the smallest value for $|t|$ beyond which $R(t,z)$ is analytic in $s = 1/t$: it rules the convergence of the asymptotic analytic development:

$$\begin{aligned} R(t,z) &= R(0,z)[I_n + UM_z^{-1}\textstyle\sum_{k=0}^{\infty}(sM_z^{-1})^k V^H R(0,z)] \\ &= R(\infty,z) + R(0,z)UM_z^{-1}\textstyle\sum_{k=1}^{\infty}(tM_z)^{-k} V^H R(0,z), \end{aligned} \tag{3}$$

based on M_z^{-1} and valid for $|t| > |t_r|$, $s = 1/t$, (around ∞).

Observe that M_z^{-1} cannot be nilpotent (because it is invertible).

1.5 Frontier of Existence for $R(\infty, z) = \lim_{|t| \to \infty} R(t, z)$

In general, $\lim_{z \to \lambda} \rho(M_z) = \infty$ for $\lambda \in \sigma(A)$. If $\lambda \in \sigma(A)$ is such that $\lim_{z \to \lambda} M_z = M_\lambda$ is defined (hence $\rho(M_\lambda) < \infty$, see [3]) we say that λ is *unobservable* by the deviation process (A, E) [11], see also § 3.3.

Definition 1. *The frontier points form the set* $F(A, E) = \{z \in re(A);\ 0 \in \sigma(M_z)\}$ *of points in* $re(A)$ *for which* $R(\infty, z)$ *does not exist. The critical points form the set* $\mathcal{C}(A, E)$ *of frontier points such that* $\rho(M_z) = 0$.

The inclusion $\mathcal{C}(A, E) \subset F(A, E))$ becomes an equality when $r = 1$. In general, $F(A, E)$ is a finite set of isolated points in $re(A) \subset \mathbb{C}$. We shall prove below that when $0 \in \sigma(E)$ is semi-simple, then card $F(A, E) \leq n - r$.

An exceptional case when card $F(A, E) = c$ or 0 is provided by the particular matrix $A = \lambda I$, which entails $M_z = \frac{1}{z - \lambda} V^H U$. Clearly, M_z is regular (resp. singular) for $z \neq \lambda$ when 0 is semi simple (resp. defective).

Similarly, it will be shown that $\mathcal{C}(A, E)$ is a finite set of at most $n - 1$ points, unless the map $t \mapsto \sigma(A(t))$ is constant for $t \in \mathbb{C}$, and $\mathcal{C}(A, E) = F(A, E) = re(A)$. This situation requires E to be nilpotent.

2 Convergence Rates for the Two Analytic Developments for $R(t, z)$

As z varies in $re(A)$, the convergence rate for (2) (resp. (3) is described by the map : $\varphi_1 : z \mapsto \rho(M_z)$ (resp. $\varphi_2 : z \mapsto \rho(M_z^{-1})$)

2.1 The Spectral Portrait φ_1

The map φ_1 is the homotopic analogue of the popular normwise spectral portrait map : $z \mapsto \|(A - zI)^{-1}\|$, [6]. In φ_1, the matrix $(A - zI)^{-1}$ of order n is replaced by M_z of order $r < n$, and $\|\cdot\|$ by $\rho(\cdot)$.

An important consequence is that φ_1 can localize the critical points ($\rho = 0$) when they are isolated, whereas the normwise spectral portrait cannot, see specifically the paragraph 2.3.

The map $\varphi_1 : z \mapsto \rho(M_z)$ is subharmonic with singularities at the observable eigenvalues of A ($\rho = \infty$) and the critical points ($\rho = 0$). We assume that there exist observable eigenvalues. Subharmonicity in $\mathbb{C}$ is the 2D-analogue of monotonicity in $\mathbb{R}$. It allows to order the ε−level sets, $\varepsilon > 0$ by inclusion. As z varies outside the disk $\{z;\ |z| \leq \rho(A)\}$, $\rho(M_z)$ decreases from $+\infty$ to 0 ($\rho(M_z) \to 0$ as $|z| \to \infty$). Therefore the set $\Gamma_0^\alpha = \{z \in \mathbb{C};\ \rho(M_z) = \alpha\}$ consists of a finite number of closed curves. For α small enough, there exists one single exterior curve around all the others which enclose local minima or isolated critical points.

The associated domain of convergence for(2) is the unbounded region outside the outer curve and inside the inner curves. See Figure 1, a) on the left. See also [7, 8].

2.2 The Frontier Portrait φ_2

The map $\varphi_2 : z \mapsto \rho(M_z^{-1}) = \rho_2$ is also subharmonic with singularities ($\rho = \infty$) at points in $F(A, E)$. We assume that $A \neq \lambda I$, and that $F(A, E)$ is a non empty finite set. When $|z|$ increases away from $F(A, E)$, $\rho(M_z^{-1})$ decreases to a local minimum to increase again ($\rho(M_z^{-1}) \to \infty$ as $|z| \to \infty$). For $\beta \geq \beta_\star > 0$, the set $\Gamma_\infty^\beta = \{z \in \mathbb{C};\ \rho(M_z^{-1}) = \beta\}$ consists of a finite number of closed curves. And for β large enough, there exists one single exterior curve around the others which enclose the points in $F(A, E)$. We observe that in exact arithmetic, it is conceivable that $\rho(M_z^{-1})$ can be 0 at observable eigenvalue of A, for which M_z is not defined, hence $\mu_{min} = \infty(\frac{1}{\mu_{min}} = 0)$ where μ_{min} is an eigenvalue for M_z of minimal modulus.

The associated domain of convergence for (3) is the bounded region inside the outer curve and outside the inner ones. See Figure 1, b) on the right and [10]. The shaded areas represent the respective analyticity domains for $R(t, z)$ around 0 ($|t| < \frac{1}{\alpha}$) and ∞ ($|t| > \beta$), with α small or β large, $\alpha \leq \beta$.

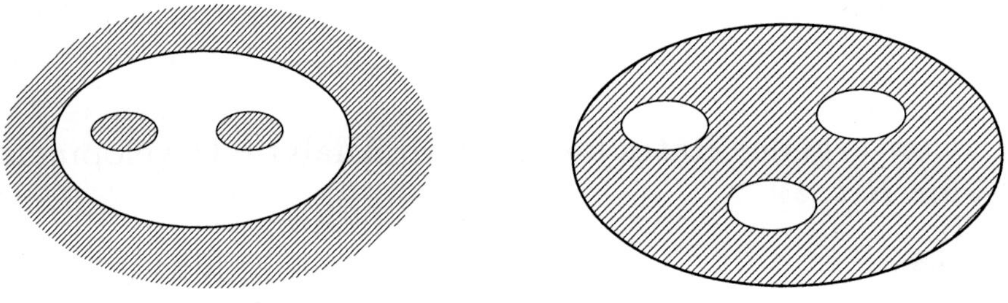

a) Γ_0^α for (2) b) Γ_∞^β for (3)

Fig. 1. Analytic representations for $R(t, z)$, $\alpha \leq \beta$

2.3 The Critical Points

When they exist, the critical points in $\mathcal{C}(A, E) \subset F(A, E)$ are singularities for φ_1 (at 0) and for φ_2 (at ∞).

At an isolated critical point, there is an abrupt change in the representation of $R(t, z)$. The symmetry of the dual analytic representation, valid locally for $|t|$ small (around 0) or large (around ∞) is broken in favour of 0.

The finite representation:

$$R(t, z) = R(0, z)[I_n - tU \sum_{k=0}^{r-1} (tM_z)^k V^H R(0, z)] \tag{4}$$

as a polynomial in t of degree $\leq r$, is valid for t everywhere in $\mathbb{C}$. The limit as $|t| \to \infty$ is not defined.

If M_z is nilpotent for any z in $re(A)$, $\sigma(A)$ is unobservable but $R(0, z)$ is *not* defined for $z \in \sigma(A)$.

2.4 The Case $r = 1$

The matrix M_z of order r reduces to the scalar μ_z. And $\mu_z \mu_z^{-1} = 1$. Therefore $\mathcal{C}(A, E) = F(A, E)$, and we can choose $\alpha = \beta = 1$. The unique set $\Gamma_0^1 = \Gamma_\infty^1$ reduces to the set Γ studied in [7].

There are at most $n - 1$ critical points [4] unless $\sigma(A(t))$ is invariant under $t \in \mathbb{C}$. In this case $\mathcal{C}(A, E) = re(A)$ and can be extended to $\mathbb{C}$ by continuity of $z \mapsto \rho(M_z) = 0$.

The symmetry between 0 and ∞ expressed by $s = 1/t$ is also carried by $\rho(M_z^{-1}) = 1/\rho(M_z)$. Convergence at 0 (resp. ∞) for (2) is equivalent to divergence at 0 (resp. ∞) for (3) for any z not critical ($\rho(M_z) > 0$). Such an exact symmetry does not hold for $r > 1$ since any z in $re(A)$, which is not a frontier point, is simultaneously an eigenvalue for r matrices $A(t)$, instead of just one. We shall continue this analysis in Section 3, after the comparison of the normwise versus homotopic level sets to follow.

2.5 Normwise Versus Homotopic Level Sets for $||\cdot||$, ρ_1, ρ_2.

A normwise backward analysis yields the well-known identity for $\varepsilon > 0$:

$$R_\varepsilon^N = \{z \in re(A);\ ||(A - zI)^{-1}|| \geq \frac{1}{\varepsilon}\} = \{z \in \sigma(A+E) \cap re(A),\ ||E|| \leq \varepsilon\} = S_\varepsilon^N,$$

where the sets cannot be empty [6]. N stands for normwise.

The homotopic analogue of R_ε^N is given by $R_\varepsilon = \{z \in re(A);\ \rho(M_z) \geq \frac{1}{\varepsilon}\}$ which can be empty for $\varepsilon > 0$ if all the eigenvalues of A are unobservable by (A, E). Such a situation corresponds to $\rho(M_z) = 0$ for any $z \in re(A)$.

The analogue of S_ε^N consists of the z in $re(A)$ which are eigenvalues of $A + tE$, at distance $|t| \leq \varepsilon$. Because there can be r such matrices for any given z in $re(A)$, the homotopic distance is not uniquely defined.

For example, one can choose a distance which is a) minimal or b) maximal. This corresponds to :

a) $|t| = \dfrac{1}{|\mu_{max}|}$: $A(t)$ is the *closest* matrix having z as its eigenvalue. Then $S_\varepsilon^a = R_\varepsilon$ [9]. This is the only possibility when $r = 1$.

b) $|t| = \dfrac{1}{|\mu_{min}|}$: $A(t)$ is the *farthest* matrix, then $S_\varepsilon^b \subset R_\varepsilon$. The maximal distance induces the level set for $\varphi_2 : \rho(M_z^{-1}) \leq \varepsilon$ [10].

3 The Spectrum $\sigma(A(t))$ as $|t| \to \infty$

Because E is singular, it is possible that some eigenvalues $\lambda(t)$ of $A(t)$ remain at finite distance when $|t| \to \infty$ [4].

Observing the evolution $t \mapsto \lambda(t)$ as $t \in \mathbb{C}$ leads to the distinction between invariant and evolving eigenvalues for A, according to the :

$E = e_{k+1}e_k^T : B(h) = B + hE = H_{k+1}$. E is nilpotent ($E^2 = 0$) with Rank 1, and $\sigma(E) = \{(0^1)^{k-1}, (0^2)\}$. For k fixed, $1 < k < n$, we set $H^- = H_{k-1}$, $H = H_k$, $H^+ = H_{k+1}$: these are the three successive Hessenberg matrices constructed by the Arnoldi decomposition, of order $k-1$, k and $k+1$. We define $u = (\tilde{u}^T, u_k)^T$. When $u_k \neq 0$, we set $\Omega = H^- - \frac{h_{k\,k-1}}{u_k}\tilde{u}e_{k-1}^T$.

We assume that $H_k = H$ is *irreducible*, therefore $\sigma(H^-) \cap \sigma(H) = \emptyset$ and $h_{k\,k-1} \neq 0$ in particular. $\sigma(B) = \sigma(H) \cup \{a\}$.

With the notation of Section 3, $0 \in \sigma(E)$ has the multiplicities $g' = k - 1 < g = k < m = k + 1$. Therefore $g' \geq 1$ for $k \geq 2$. The eigenspace K' is $K' = \text{lin}\,(e_1, \cdots, e_{k-1})$, and P' is the orthogonal projection on K', $P = I_{k+1}$. Thus $\Pi' = H_{k-1} = H^-$. The matrix M_z reduces to the scalar $\mu_z = -e_k^T(B - zI_{k+1})^{-1}e_{k+1}$, for $z \notin \sigma(B)$. Finally, because $r = 1$, $\mathcal{C}(B, E) = F(B, E)$ in $re(B)$. We survey the results established in [5].

1) Theory tells us that in the generic case where $u_k \neq 0$, Lim $= \sigma(\Omega)$, and Lim $\cap\, re(B) = \mathcal{C}(B, E)$ contains at most $k-1$ critical points in $re(B)$. Exactly two eigenvalues of H^+ escape to ∞ as $|h| \to \infty$.

2) The resolvent $R(h, z) = (H^+ - zI_{k+1})^{-1}$, defined for $z \notin \sigma(H^+)$ (i.e $h\mu_z \neq 1$) is a modification of $R(0, z) = (B - zI_{k+1})^{-1}$ by a rank 1 matrix depending on h through the multiplicative scalar $\frac{h}{h\mu_z - 1}$ only. The dependance is rational (resp. linear) when $\mu_z \neq 0$ (resp. $\mu_z = 0$, i.e z critical in Lim $\cap\, re(B)$).

Let (ξ, p) be an exact eigenpair for $H^- : H^- p = \xi p$ for $p \in \mathbb{C}^{k-1}$. We consider the augmented vector $\hat{\psi}_l = (p^T, 0)^T$ in $\mathbb{C}^l$, $l \geq k$, and define $h^- = h_{k\,k-1}$, $p_{k-1} = e_{k-1}^T p$.

The pair $(\xi, \hat{\psi}_l)$ is a pseudo eigenpair for H_l, $l \geq k$ corresponding to the residual vector $(h^- p_{k-1})e_k$ in $\mathbb{C}^l$. The pair $(\xi, \hat{\psi}_l)$ *cannot be improved* by inverse iteration using the Hessenberg form H_l, for $l \geq k+1$ (see [13] for a numerical illustration). The only solution is to *restart* with an improved starting vector v_1.

In exact arithmetic, the algorithmic analysis of the inner loop is easy under the assumption of irreducibility : either v_1 is an invariant vector for A and the algorithm stops exactly (with $h = 0$) for $k < n$, or v_1 is not invariant and the algorithm has to be run to completion ($k = n$).

In finite precision, the analysis is more delicate, since the mathematical analysis for convergence ($h \to 0$) is valid only when round-off can be ignored. And it is well known that round-off cannot be ignored when "convergence" takes place [13, 14, 16].

"Convergence" in finite precision means "near-reducibility", and this can happen with $|h|$ **large**, although this seems numerically counter-infinitive at first sight.

The algorithmic dynamics for "convergence" entails that there exist points in $\sigma(H^-)$, $\sigma(H)$ and $\sigma(H^+)$ which are very close, in spite of the fact that an exact coincidence is ruled out by the assumption of irreducibility for A.

The dynamics expressed in finite precision makes it possible that a value $z \in \sigma(H^+)$ which is close to $\sigma(\Omega)$ corresponds to a large h : z can be *nearly critical.* To the three above mentionned spectra, it may be necessary to add the fourth spectrum $\sigma(\Omega)$. This is certainly the case when $||\Omega - H^{-}|| = |h^{-}|\frac{||\tilde{u}||}{|u_k|}$ is small.

Observe that $\frac{||\tilde{u}||}{|u_k|} = \tan\psi$, where ψ is the acute angle between the direction spanned by u_k and e_k.

Therefore a complete explanation for the "convergence" of Krylov methods in finite precision requires to complement the classical point of view of exact *convergence* ($h \to 0$), valid when the arithmetic can be regarded as **exact**, by the novel notion of *criticality* ($|h| \to \infty$) which takes care of the effect of **finite precision** when they cannot be ignored.

The reader is refered to [5] to see precisely how this new notion clarifies the finite precision behaviour of such key aspects of Krylov methods as the Arnoldi residual, an algorithmic justification for restart and the extreme robustness to very large perturbations [15]. The notion of criticality offers therefore a theoretical justification for the highly successful heuristics which are commonly in use. It also shows why $|h_{k+1\,k}|$ small can be a misleading indicator for the nearness to exact reducibility.

References

1. F. Chatelin. **Spectral approximation of linear operators**. Academic Press, New York, 1983.
2. F. Chatelin. **Valeurs propres de matrices**. Masson, Paris, 1988.
3. F. Chatelin. **Eigenvalues of matrices**. Wiley, Chichester, 1993. Enlarged Translation of the French Publication with Masson.
4. F. Chaitin-Chatelin. About Singularities in Inexact Computing. Technical Report TR/PA/02/106, CERFACS, Toulouse, France, 2002.
5. F. Chaitin-Chatelin. The Arnoldi method in the light of Homotopic Deviation theory. Technical Report TR/PA/03/15, CERFACS, Toulouse, France, 2003.
6. F. Chaitin-Chatelin and V. Frayssé. **Lectures on Finite Precision Computation**. Publ., Philadelphia, 1996.
7. F. Chaitin-Chatelin and E. Traviesas. Homotopic perturbation - Unfolding the field of singularities of a matrix by a complex parameter: a global geometric approach. Technical Report TR/PA/01/84, CERFACS, 2001.
8. F. Chaitin-Chatelin and E. Traviesas. Qualitative Computing. Technical Report TR/PA/02/58, CERFACS, Toulouse, France, 2002. To appear in **Handbook of Computation**, B. Einarsson ed. , SIAM Philadelphia.
9. F. Chaitin-Chatelin, V. Toumazou, E. Traviesas. *Accuracy assessment for eigencomputations: variety of backward errors and pseudospectra.* Lin. Alg. App. 309, 73-83, 2000. Also available as Cerfacs Rep. TR/PA/99/03.
10. F. Chaitin-Chatelin and M.B. van Gijzen. Homotopic Deviation theory with an application to computational acoustics. Technical Report TR/PA/04/05, CERFACS, Toulouse, France, 2004.
11. F. Chaitin-Chatelin. Computing beyond analyticity. Matrix Algorithms in Inexact and Uncertain Computing. Technical Report TR/PA/03/110, CERFACS, Toulouse, France, 2003.

12. P. Lancaster, M. Tismenetsky. **Theory of Matrices**. Academic Press, New York, 1987.
13. F. Chaitin-Chatelin, E. Traviesas and L. Plantié. *Understanding Krylov methods in finite precision* in Numerical Analysis and Applications, NNA 2000 (L. Vulkov, J. Wasviewski, P. Yalamov eds.), Springer Verlag Lectures Notes in CS, vol. 1988, pp. 187-197, 2000. Also available as Cerfacs Rep. TR/PA/00/40.
14. F. Chaitin-Chatelin. Comprendre les méthodes de Krylov en précision finie : le programme du Groupe Qualitative Computing au CERFACS. Technical Report TR/PA/00/11, CERFACS, Toulouse, France, 2000.
15. F. Chaitin-Chatelin and T. Meškauskas. Inner-outer iterations for mode solver in structural mechanics: application to the Code-Aster. Contract Report FR/PA/02/56, CERFACS, Toulouse, France, 2002.
16. B.N. Parlett. **The Symmetric Eigenvalue Problem.** Prentice Hall, Englewood Cliffs, 1980.
17. F. Chaitin-Chatelin. The dynamics of matrix coupling with an application to Krylov methods. Technical Report TR/PA/04/29, CERFACS, Toulouse, France, 2004.
18. J. Moro, J.V. Burke and M.L. Overton, *On the Lidskii-Lyusternik-Vishik Perturbation Theory for Eigenvalues with Arbitrary Jordan Structure.* SIAM J. Matrix Anal. Appl. 18 (1997), pp. 793-817.

All Cerfacs Reports are available from:
`http://www.cerfacs.fr/algor/reports/index.html`

High Precision Method for Calculating the Energy Values of the Hydrogen Atom in a Strong Magnetic Field

M.G. Dimova and M.S. Kaschiev

[1] Institute of Mathematics and Informatics, Sofia, Bulgaria
mkoleva@math.bas.bg
[2] Present address: South West University, Blagoevgrad, Bulgaria
Permanent address: Institute of Mathematics and Informatics,
Sofia, Bulgaria
kaschiev@math.bas.bg

Abstract. The new method for calculating the energy values of the hydrogen atom in a strong magnetic field ($0 \leq B \leq 10^{13}G$) with high degree of accuracy is developed in this paper. The proposed method is based on the Kantorovich method for solving eigenvalue problems. This method successfully reduces the given two dimensional spectral problem for the Schrödinger equation to the spectral problems for one one-dimensional equation and the system of ordinary second-order differential equations. The rate of convergence is examined numerically and is illustrated in the table. The results are in good agreement with the best one up to now and show that 12 significant digits are computed in the obtaining energy values.

1 Introduction

A growing interest in the problem of the hydrogen atom in strong magnetic fields is motivated by its various applications in different branches of physics – astrophysics, solid state physics, atomic spectroscopy.

In recent years, several new techniques were carried out to treat the whole region of a magnetic field. Very accurate numerical results have been obtained. These include Hartree–Fock–like calculations [13], [15], variational calculations [14], [6], [7], [19], [8], finite–element analysis [11], [16], [17], finite–different methods [9]. The most precise calculations reported to date are presented in [12], where the solution is expressed as a power series in the radial variables with coefficients being polynomials in the sine of the polar angle.

A goal of this paper is to show the peculiarities of a modern implementation of the Kantorovich method to numerical solution of the multi-dimensional eigenvalue problems and also to point out some prospects of its application to the calculations of the low-energy spectrum of the hydrogen atom.

Z. Li et al. (Eds.): NAA 2004, LNCS 3401, pp. 25–34, 2005.

The Schrödinger equation for the hydrogen atom, written in spherical coordinates, has the following form

$$-\left(\frac{1}{r^2}\frac{\partial}{\partial r}r^2\frac{\partial}{\partial r}+\frac{1}{r^2\sin\theta}\frac{\partial}{\partial\theta}\sin\theta\frac{\partial}{\partial\theta}\right)\Psi(r,\theta)+V(r,\theta)\Psi(r,\theta)=\epsilon\Psi(r,\theta), \quad (1)$$

$0 \le r < \infty$ and $0 \le \theta \le \pi$. The potential function $V(r,\theta)$ is given by the formula

$$V(r,\theta)=\frac{m^2}{r^2\sin^2\theta}-\frac{2}{r}+\gamma m+\frac{1}{4}\gamma^2 r^2\sin^2\theta,$$

where $m = 0, \pm 1, \ldots$ is a magnetic quantum number, $\gamma = B/B_0$, $B_0 \cong 4.70 \times 10^9 G$ is a dimensionless parameter, which determines the strength of the field. In these expressions ϵ is the energy of the bound state and Ψ is the corresponding wave function. This function satisfies the following boundary conditions

$$\begin{aligned} &\lim_{\theta\to 0}\sin\theta\frac{\partial\Psi}{\partial\theta}=0,\ \frac{\partial\Psi}{\partial\theta}(r,\frac{\pi}{2})=0,\ 0\le r<\infty, \quad \text{for the even parity state,}\\ &\Psi(r,\frac{\pi}{2})=0,\ 0\le r<\infty, \quad \text{for the odd parity state,}\end{aligned} \quad (2)$$

$$\lim_{r\to 0} r^2\frac{\partial\Psi}{\partial r}=0, \qquad \lim_{r\to\infty} r^2\Psi=0. \quad (3)$$

The wave function is normalized as

$$\int_0^\infty\int_0^{\pi/2} r^2\sin\theta\Psi^2(r,\theta)d\theta dr=1. \quad (4)$$

2 Reduction of the 2D Problem by the Kantorovich Method

Consider a formal expansion of the solution of (1)–(3) using the finite set of one-dimensional basis functions $\{\Phi_i(\theta;r)\}_{i=1}^{n_{max}}$

$$\Psi(r,\theta)=\sum_{i=1}^{n_{max}}\chi_i(r)\Phi_i(\theta;r). \quad (5)$$

In (5), functions $\chi(r)^T = (\chi_1(r), \chi_2(r), \ldots, \chi_{n_{max}}(r))$ are unknown, and surface functions $\Phi(\theta;r)^T = (\Phi_1(\theta;r), \Phi_2(\theta;r), \ldots, \Phi_{n_{max}}(\theta;r))$ form an orthonormal basis for each value of radius r which is treated here as a parameter.

In the Kantorovich approach [10] functions $\Phi_i(\theta;r)$ are determined as solutions of the following one-dimensional parametric eigenvalue problem

$$\begin{aligned} -\frac{\partial}{\partial\theta}\sin\theta\frac{\partial\Phi(\theta;r)}{\partial\theta}+\hat{V}(r,\theta)\Phi(\theta;r)&=E(r)\sin\theta\Phi(\theta;r),\\ \hat{V}(r,\theta)&=r^2\sin\theta V(r,\theta)\end{aligned} \quad (6)$$

with boundary conditions derived from (2)

$$\lim_{\theta\to 0}\sin\theta\frac{\partial\Phi}{\partial\theta}=0,\ \frac{\partial\Phi}{\partial\theta}(\frac{\pi}{2};r)=0,\ 0\le r<\infty,\quad \text{for the even parity state,}$$
$$\Phi(\frac{\pi}{2};r)=0,\ 0\le r<\infty,\quad \text{for the odd parity state.} \tag{7}$$

Since the operator in the left side of (6) is self-adjoint, its eigenfunctions are orthonormal

$$\int_0^{\pi/2}\sin\theta\Phi_i(\theta;r)\Phi_j(\theta;r)d\theta=\delta_{ij}, \tag{8}$$

δ_{ij} is Kroneker's δ–symbol.The problem (6)–(7) is solved for each value of $r\in\omega_r$, where $\omega_r=(r_1,r_2,\dots,r_k,\dots)$ is a given set of r.

The Galerkin weak form of the problem (1)–(3) is

$$\int_0^\infty\int_0^{\pi/2}\left(r^2\sin\theta\left(\frac{\partial\Psi}{\partial r}\right)^2+\sin\theta\left(\frac{\partial\Psi}{\partial\theta}\right)^2+\hat{V}\Psi^2\right)d\theta dr=\epsilon\int_0^\infty\int_0^{\pi/2}r^2\sin\theta\Psi^2 d\theta\ dr. \tag{9}$$

After substitution of expansion (5) into (9), using (6), (7), (8) and obtained from them identity

$$\int_0^{\pi/2}\left(\sin\theta\frac{\partial\Phi_i(\theta;r)}{\partial\theta}\frac{\partial\Phi_j(\theta;r)}{\partial\theta}+\hat{V}(r,\theta)\Phi_i(r,\theta)\Phi_j(\theta;r)\right)d\theta=\frac{E_i(r)+E_j(r)}{2}\delta_{ij}, \tag{10}$$

the solution of problem (1)–(3) is transformed to a solution of an eigenvalue problem for a system of n_{max} ordinary second-order differential equations for determining the energy ϵ and coefficients (radial wave functions) $\chi(r)^T=(\chi_1(r),\chi_2(r),\dots,\chi_{n_{max}}(r))$ of expansion (5)

$$-\mathbf{I}\frac{1}{r^2}\frac{d}{dr}r^2\frac{d\chi}{dr}+\frac{\mathbf{U(r)}}{r^2}\chi+\mathbf{Q}(r)\frac{d\chi}{dr}+\frac{1}{r^2}\frac{dr^2\mathbf{Q}(r)\chi}{dr}=\epsilon\mathbf{I}\chi, \tag{11}$$

$$\lim_{r\to 0}r^2\frac{\partial\chi}{\partial r}=0,\qquad \lim_{r\to\infty}r^2\chi=0. \tag{12}$$

Here $\mathbf{I}$, $\mathbf{U}(r)$, and $\mathbf{Q}(r)$ are finite $n_{max}\times n_{max}$ matrices, elements of which are given by relations

$$U_{ij}(r)=\frac{E_i(r)+E_j(r)}{2}\delta_{ij}+r^2H_{ij}(r),\quad H_{ij}(r)=H_{ji}(r)=\int_0^{\pi/2}\sin\theta\frac{\partial\Phi_i}{\partial r}\frac{\partial\Phi_j}{\partial r}d\theta,$$
$$Q_{ij}(r)=-Q_{ji}(r)=-\int_0^{\pi/2}\sin\theta\Phi_i\frac{\partial\Phi_j}{\partial r}d\theta,\quad I_{ij}=\delta_{ij},\quad i,j=1,2,\dots,n_{max}. \tag{13}$$

Thus, the application of the Kantorovich approach makes the problem (1)–(3) equivalent to the following problems:

- Calculation of potential curves $E_i(r)$ and eigenfunctions $\Phi_i(\theta;r)$ of the spectral problem (6)–(7) for a given set of $r \in \omega_r$.
- Calculation of the derivatives $\dfrac{\partial \Phi}{\partial r}$ and computation of the corresponding integrals (see (13)) necessary for obtaining matrix elements of radial coupling $U_{ij}(r)$ and $Q_{ij}(r)$.
- Calculation of energies ϵ and radial wave functions $\chi(r)$ as eigensolutions of one-dimensional eigenvalue problem (11)–(12) and examination of the convergence of obtained eigensolutions as a function of number of channels n_{max}.

3 Computation of the Matrix Elements of Radial Coupling

Calculation of potential matrices $\mathbf{U}(r)$ and $\mathbf{Q}(r)$ ([4]) with sufficiently high accuracy is a very important step of solving the system of radial equations (11). This is a very difficult problem for most of numerical methods usually used in this case. To obtain the desired energies and wave functions with required precision, the derivatives $\dfrac{\partial \Phi}{\partial r}$ should be computed with the highest possible accuracy.

An effective method, allowing to calculate derivative $\dfrac{\partial \Phi}{\partial r}$ with the same accuracy as achieved for eigenfunctions of (6), has been developed in [3]. For completeness here we outline it briefly. Taking a derivative of (6) with respect to r, we get that $\dfrac{\partial \Phi}{\partial r}$ can be obtained as a solution of the following right hand side boundary problem

$$
\begin{aligned}
-\frac{\partial}{\partial \theta}\sin\theta\frac{\partial}{\partial \theta}\frac{\partial \Phi(\theta;r)}{\partial r} + \hat{V}(r,\theta)\frac{\partial \Phi(\theta;r)}{\partial r} - E(r)\sin\theta\frac{\partial \Phi(\theta;r)}{\partial r} &= \\
E'(r)\sin\theta\Phi(\theta;r) - \frac{\partial \hat{V}(r,\theta)}{\partial r}\Phi(\theta;r)&,
\end{aligned}
\tag{14}
$$

where the boundary conditions for function $\dfrac{\partial \Phi}{\partial r}$ are the same as for function Φ. Taking into account that $E(r)$ is an eigenvalue of the operator, defined by(6), problem (14) will have a solution *if and only if the right hand side term is orthogonal to the eigenfunction* Φ. From this condition we find

$$
E'(r) = \int_0^{\pi/2} \frac{\partial \hat{V}(r,\theta)}{\partial r}\Phi^2(\theta;r)d\theta. \tag{15}
$$

Now problem (14) has a solution, but it is not unique. From the normalization condition (8) we obtain the required additional condition

$$
\int_0^{\pi/2} \sin\theta\Phi(\theta;r)\frac{\partial \Phi(\theta;r)}{\partial r}d\theta = 0, \tag{16}
$$

guaranteeing the uniqueness of the solution.

4 Matrix Representations of the Eigenvalue Problems

For numerical solution of one-dimensional eigenvalue problems (6)–(7), (11)–(12) and boundary value problem (14)–(16), the high-order approximations of the finite element method ([18], [5]) elaborated in the papers [1], [2] have been used. One-dimensional finite elements of order $k = 1, 2, \ldots, 10$ have been implemented. Using the standard finite element procedures [5], problems (6)–(7) and (11)–(12) are approximated by the generalized algebraic eigenvalue problem

$$\mathbf{A}\mathbf{F}^h = E^h \mathbf{B}\mathbf{F}^h. \tag{17}$$

$\mathbf{A}$ and $\mathbf{B}$ are the finite element matrices, corresponding to problems (6)–(7) or (11)–(12) (see [1],[2]), E^h is the corresponding eigenvalue and $\mathbf{F}^h$ is the vector approximating solutions of (6)–(7) or (11)–(12) on the finite-element grid.

Let us consider a numerical algorithm for the computation of derivative $\dfrac{\partial \Phi}{\partial r}$, proposed in [3]. It follows from (17) that we should solve the following linear system of algebraic equations

$$\mathbf{K}\mathbf{u} \equiv (\mathbf{A} - E^h \mathbf{B})\mathbf{u}^h = \mathbf{b}, \quad \mathbf{u}^h = \frac{\partial \mathbf{F}^h}{\partial r}, \tag{18}$$

The finite element matrices $\mathbf{A}$ and $\mathbf{B}$ are defined as

$$\begin{aligned} A_{ij} &= \int_0^{\pi/2} \left(\sin\theta \frac{\partial \varphi_i(\theta)}{\partial \theta} \frac{\partial \varphi_j(\theta)}{\partial \theta} + \hat{V}(r,\theta)\varphi_i(\theta)\varphi_j(\theta) \right) d\theta, \\ B_{i,j} &= \int_0^{\pi/2} \sin\theta \varphi_i(\theta)\varphi_j(\theta) d\theta, \end{aligned} \tag{19}$$

where $\varphi_i(\theta), \varphi_j(\theta)$ are the finite element polynomials.

Now we determinate the derivative $(E^h)'$ and the right hand side $\mathbf{b}$ using the formulas

$$(E^h)' = (\mathbf{F}^h)^T \mathbf{P}\mathbf{F}^h, \quad \mathbf{b} = ((E^h)'\mathbf{B} - \mathbf{P})\mathbf{F}^h,$$

$$P_{ij} = \int_0^{\pi/2} \frac{\partial \hat{V}(r,\theta)}{\partial r} \varphi_i(\theta)\varphi_j(\theta) d\theta.$$

Since E^h is an eigenvalue of (14), matrix $\mathbf{K}$ in (18) is degenerate. The algorithm for solving (18) can be written in three steps as follows:

Step 1. The additional condition (16) has the form

$$(\mathbf{u}^h)^T \mathbf{B}\mathbf{F}^h = 0.$$

Denote by s a number determined by the condition

$$|\mathbf{B}\mathbf{F}^h|_s = \max_{1 \le i \le N} |\mathbf{B}\mathbf{F}^h|_i, \quad D_s = (\mathbf{B}\mathbf{F}^h)_s,$$

where N is the order of matrices above.

Step 2. Solve two systems of algebraic equations

$$\overline{\mathbf{K}}\overline{v} = \overline{\mathbf{b}}, \quad \overline{\mathbf{K}}\overline{w} = \mathbf{d},$$

where

$$\mathbf{d}^T = (K_{1s}, K_{2s}, \ldots, K_{Ns}), \quad d_s = 0, \quad \overline{b}_i = b_i, \quad \overline{b}_s = 0,$$

$$\overline{\mathbf{K}}_{ij} = \mathbf{K}_{ij}, \quad i \neq s, \; j \neq s, \quad \overline{\mathbf{K}}_{is} = 0, \; i \neq s, \quad \overline{\mathbf{K}}_{sj} = 0, \; j \neq s, \quad \overline{\mathbf{K}}_{ss} = 1.$$

In this way we have $\overline{v}_s = 0$ and $\overline{w}_s = 0$.

Step 3. Find constants γ, γ_1 and γ_2 as

$$\gamma_1 = \overline{\mathbf{v}}^T \mathbf{B} \mathbf{F}^h, \quad \gamma_2 = \overline{\mathbf{w}}^T \mathbf{B} \mathbf{F}^h, \quad \gamma = -\frac{\gamma_1}{D_k - \gamma_2}.$$

After that derivative $\mathbf{u}^h = \dfrac{\partial \mathbf{F}^h}{\partial r}$ is obtained using formula

$$u_i^h = \overline{v}_i - \gamma \overline{w}_i, \quad i \neq s, \quad u_s^h = \gamma.$$

Let E_l, Φ_l and $\dfrac{\partial \Phi_l}{\partial r}$ be the exact solutions of (6)–(7), (11)–(12) or (14)–(16) and $E_l^h, \mathbf{F}_l^h, \mathbf{u}_l^h$ be the corresponding numerical solutions. Then the following estimates are valid [18]

$$|E_l - E_l^h| \leq c_1(E_l) h^{2k}, \quad ||\Phi_l - \mathbf{F}_l^h||_0 \leq c_2(E_l) h^{k+1},$$
$$||\frac{\partial \Phi_l}{\partial r} - \mathbf{u}_l^h||_0 \leq c_3 h^{k+1}, \quad c_1 > 0, \quad c_2 > 0, \quad c_3 > 0,$$

where h is the grid step, k is the order of finite elements, l is the number of the corresponding eigensolution, and constants c_1, c_2 and c_3 do not depend on step h. ¿From the consideration above it is evident, that the derivative computed has the same accuracy as the calculated eigenfunction.

5 Numerical Results

In this section we present our numerical results for the energy spectrum of hydrogen atom in strong magnetic fields ($0 \leq \gamma \leq 2000$). Ten eigensolutions ($n_{max} = 10$) of the problem (6)–(7) are calculated that corresponding to get 10 equations of system (11). Finite elements of order $k = 4$ are implemented. Problem (6)–(7) has been solved using a grid with 157 finite elements in the interval $[0, \pi/2]$ and consists of 527 nodes. The finite element grid of r has been chosen as follows, 0 (100) 3 (70) 20 (80) 100 (number in parentheses denotes the number of finite elements of order $k = 4$ in each intervals [0,3], [3,20], [20,100]). This grid is composed of 999 nodes. The maximum number of unknowns of the system (11) ($n_{max} = 10$) is 9990. To calculate the finite element matrices in each finite element $[\theta_{i-1}, \theta_i]$ or $[r_{j-1}, r_j]$ the Gaussian quadrature with $k+1$ nodes have

Table 1. Convergence of the energy states $\epsilon(1s_0)$ and $\epsilon(2s_0)$ (in a.u.) with the number of coupled channels n_{max} for $\gamma = 4$

n_{max}	$\epsilon(1s_0)$	$\epsilon(2s_0)$
1	**1**.613 354 906 00	**3.6**72 561 204 50
2	**1**.450 371 395 55	**3.6**27 227 848 92
3	**1.4**38 964 768 53	**3.62**3 088 870 19
4	**1.438** 433 429 69	**3.62**2 352 872 53
5	**1.438** 405 323 92	**3.622 3**11 478 09
6	**1.438 404** 033 39	**3.622 3**07 326 59
7	**1.438 403** 971 08	**3.622 307** 091 84
8	**1.438 403** 968 24	**3.622 307** 073 74
9	**1.438 403 968** 11	**3.622 307 072** 71
10	**1.438 403 968** 10	**3.622 307 072** 64

been used. All matrix elements $U_{ij}(r)$ and $Q_{ij}(r)$ are calculated in the Gaussian nodes with respect to r.

The algebraic problem (17) has been solved by the subspace iteration method [5], which allows us to find simultaneously first several low eigensolutions. All eigenvalues and the corresponding matrix elements of the problem (6)–(7) are calculated with relative accuracy of 10^{-12}.

A study of the convergence of the states energy of $\epsilon(1s_0)$ and $\epsilon(2s_0)$ with the number of radial equations (n_{max}) is demonstrated in Table 1 for $\gamma = 4$. One can see that the energy values converge monotonically from above, with the 10-channel values being $\epsilon(1s_0) = 1.43840396810$ a.u. and $\epsilon(2s_0) = 3.62230707264$ a.u. Fig.1 and Fig.2 show the first four eigenfunctions of problem (6)–(7) for $\gamma = 4$, $r = 5$ and $r = 15$. As can been seen in these figures the new orthogonal basis, obtained as eigenfunctions of problem (6)–(7) are localized around the point $\theta = 0$ when the radius r grows up.

The bounding energies of the states $\epsilon(1s_0)$, $\epsilon(2s_0)$, $\epsilon(2p_0)$, $\epsilon(3p_0)$ are calculated for a great number of values of γ, $0 \leq \gamma \leq 2000$. The Table 2 compares the computing energies $\mathcal{E} = (\gamma - \epsilon)/2$ with the energies obtained in [12] for some values of γ.

6 Conclusions

In the present work the bound states of hydrogen atom in strong magnetic field are calculated using the spherical coordinates. The reduction of the two-dimensional problem to the one-dimensional one has been performed using the

9. Ivanov, M.V.: The hydrogen atom in a magnetic field of intermediate strength. J. Phys. B **21** (1988) 447–462
10. Kantorovich, L.V., Krilov, V.I.: *Approximate Methods of Higher Analysis* (Wiley, New York, 1964) (Gostekhteorizdat, Moscow, 1952)
11. Kaschiev, M.S., Vinitsky, S.I., Vukajlovic F.R.: Hydrogen atom H and H_2^+ molecule in strong magnetic fields. Phys. Rev. A **22** (1980) 557–559
12. Kravchenko, Yu.P., Liberman, M.A., .Johanson, B.: Exact solution for a hydrogen atom in a magnetic field of arbitrary strength. Phys. Rev. A **54** (1996) 287–305
13. Rösner, W., Wunner, G., Herold, H., Ruder, H.: Hydrogen atoms in arbitrary magnetic fields. I. Energy levels and wavefunctions. J. Phys. B **17** (1986) 29–52
14. Rech, P.C., Gallas, M.R., Gallas, J.A. C.,: Zeeman diamagnetism in hydrogen at arbitrary field strengths. J. Phys. B **19** (1986) L215–L219
15. H. Ruder, G. Wunner, H. Herold, and F. Geyer, *Atoms in Strong Magnetic Fields* (Springer–Verlag, Berlin, 1994)
16. Shertzer, J.: Finite-element analysis of hydrogen in superstrong magnetic fields. Phys. Rev. A **39** (1989) 3833–3835
17. Shertzer, J., Ram–Mohan, L.R., Dossa, D.: Finite-element calculation of low-lying states of hydrogen in a superstrong magnetic field. Phys. Rev. A **40** (1989) 4777–4780
18. G. Strang and G.J. Fix, *An Analysis of the Finite Element Method* (Prentice Hall, Englewood Cliffs, New York, 1973)
19. Jinhua Xi, Xinghong He and Baiwen Li: Energy levels of the hydrogen atom in arbitrary magnetic fields obtained by using B-spline basis sets. Phyz. Rev. A **46** (1992) 5806–5811

Splitting Methods and Their Application to the Abstract Cauchy Problems*

I. Faragó

Eotvos Lorand University, Department of Applied Analysis,
1117 Budapest, Pazmany P. setany 1/c, Hungary
faragois@cs.elte.hu

Abstract. In this paper we consider the interaction of the operator splitting method and applied numerical method to the solution of the different sub-processes. We show that the well-known fully-discretized numerical models (like Crank-Nicolson method, Yanenko method, sequential alternating Marchuk method, parallel alternating method, etc.), elaborated to the numerical solution of the abstract Cauchy problem can be interpreted in this manner. Moreover, on the base of this unified approach a sequence of the new methods can be defined and investigated.

1 Introduction

In the modelling of complex time-depending physical phenomena the simultaneous effect of several different sub-processes has to be described. The operators describing the sub-processes are as a rule simpler than the whole spatial differential operator. The operator splitting method is one of the most powerful methods to solve such problems. For simplicity, in this paper we assume that there are only two sub-processes. The generalization to more sub-processes is straightforward.

The mathematical models of the above problems can be written in the form of abstract Cauchy problems (ACP) of special structure. Namely, let $\mathbf{S}$ denote some normed space and we consider the initial value problem in the Banach space $\mathbf{S}$ of the form

$$\left.\begin{aligned} \frac{dw(t)}{dt} &= (A+B)w(t), \quad t \in (0,T) \\ w(0) &= w_0, \end{aligned}\right\} \tag{1}$$

where $w : [0,T) \to \mathbf{S}$ is the unknown (abstract) function, $w_0 \in \mathbf{S}$ is a given element, A and B are given operators of type $\mathbf{S} \to \mathbf{S}$. Then defining the splitting step by $\tau > 0$, $\tau << T$ we consider the sequence of initial value problems of the form

* Supported by Hungarian National Research Founds (OTKA) under grant N. T043765

Z. Li et al. (Eds.): NAA 2004, LNCS 3401, pp. 35–45, 2005.

$$\frac{dw_1^n}{dt}(t) = Aw_1^n(t), \qquad (n-1)\tau < t \le n\tau,$$
$$w_1^n((n-1)\tau) = w_2^{n-1}((n-1)\tau), \tag{2}$$

and

$$\frac{dw_2^n}{dt}(t) = Bw_2^n(t), \qquad (n-1)\tau < t \le n\tau,$$
$$w_2^n((n-1)\tau) = w_1^n(n\tau), \tag{3}$$

for $n = 1, 2, \ldots N$, where N denotes the supremum of the integers N_1 such that $N_1\tau \le T$.

Here $w_2^0(0) = w_0$ and this splitting is called *sequential splitting.* The function $w_{sp}^N(n\tau) = w_2^n(n\tau)$, defined at the points $t_n = n\tau$ is called splitting solution of the problem.

We can define other types of splittings, too. The most popular and widely used one is the so-called *Strang splitting (or Strang-Marchuk splitting)*, defined as follows [6], [10]:

$$\frac{dw_1^n}{dt}(t) = Aw_1^n(t), \qquad (n-1)\tau < t \le (n-0.5)\tau,$$
$$w_1^n((n-1)\tau) = w_3^{n-1}((n-1)\tau), \tag{4}$$

$$\frac{dw_2^n}{dt}(t) = Bw_2^n(t), \qquad (n-1)\tau < t \le n\tau,$$
$$w_2^n((n-1)\tau) = w_1^n((n-0.5)\tau), \tag{5}$$

and

$$\frac{dw_3^n}{dt}(t) = Aw_3^n(t), \qquad (n-0.5)\tau < t \le n\tau,$$
$$w_3^n((n-0.5)\tau) = w_2^n(n\tau), \tag{6}$$

for $n = 1, 2, \ldots N$.

Here $w_3^0(0) = w_0,$ and the function $w_{sp}^N(n\tau) = w_3^n(n\tau)$, defined at the points $t_n = n\tau$ is the corresponding splitting solution of the problem.

Another alternative is *the symmetrically weighted sequential splitting* (SWS splitting) [3], [5] which means the following. We execute two sequential splittings in different ordering (which are independent tasks) and at the end of each step the splitting approximation is computed as the arithmetical mean of the results. This means that the algorithm of the computation is as follows:

$$\frac{du_1^n(t)}{dt} = Au_1^n(t), \qquad (n-1) < t \le n\tau,$$
$$u_1^n((n-1)\tau) = u_2^{n-1}((n-1)\tau); \tag{7}$$

$$\frac{du_2^n(t)}{dt} = Bu_2^n(t), \qquad (n-1) < t \le n\tau,$$
$$u_2^n((n-1)\tau) = u_1^n(n\tau); \tag{8}$$

and

$$\frac{dv_1^n(t)}{dt} = Bv_1^n(t), \qquad (n-1) < t \le n\tau,$$
$$v_1^n((n-1)\tau) = v_2^{n-1}((n-1)\tau); \tag{9}$$

$$\frac{dv_2^n(t)}{dt} = Av_2^n(t), \qquad (n-1) < t \le n\tau,$$
$$v_2^n((n-1)\tau) = v_1^n(n\tau); \tag{10}$$

and

$$w_{sp}^N(n\tau) := \frac{u_2^n(n\tau) + v_2^n(n\tau)}{2} \tag{11}$$

for $n = 1, 2, \ldots N$, where $u_2^0(0) = v_2^0(0) = w_0$.

Typically, the splitting solution is different from the exact solution, i.e., there is some error, called *local splitting error*, defined as

$$Err_{sp}(\tau) := w(\tau) - w_{sp}^N(\tau). \tag{12}$$

E.g., provided that $A, B \in \mathcal{B}(\mathbf{S})$ (the set of bounded operators), for the sequential splitting this error is

$$Err_{sp}(\tau) = [\exp(A+B)\tau - \exp(B\tau)\exp(A\tau)]w_0. \tag{13}$$

For the other splittings the local splitting error for bounded operators can be defined similarly.

Definition 1. *A splitting is called consistent if*

$$\|Err_{sp}(\tau)\| = \mathcal{O}(\tau^{p+1}) \tag{14}$$

with $p > 0$ and p is called order of the splitting.

As an easy computation shows, the sequential splitting is of first order, the Strang splitting and the SWS splitting are of second order for bounded operators [4].

2 Splitting as a Discretization Process

We could see that the use of any splitting to problem (1) in fact results in a discretization process: while the true solution $w(t)$ of (1) is defined on the time interval $[0, T]$, the splitting solution $w_{sp}^N(n\tau)$ is defined on the mesh $\Omega_\tau := \{n \cdot \tau,\, n = 0, 1 \ldots N\}$.

Remark 1. Let us compare the splitting discretizaton method with the well-known one-step numerical methods applied to ordinary differential equations. (See e.g. [1].) Considering the one-step numerical method of Runge-Kutta type of the form

$$y_{n+1} = y_n + \tau\Phi(\tau, y_n, y_{n+1}), \tag{15}$$

the local approximation error is defined as follows

$$\hat{l} = y(\tau) - y_1. \tag{16}$$

The RK-method is called consistent if $\hat{l} = \mathcal{O}(\tau^{p+1})$ with $p > 0$ and p is called the order of the numerical method. This implies that the notions of consistency and order coincide for the splitting methods and the numerical methods for the ODE's.

The splitting discretization process can be written in the form

$$\begin{aligned} w^{n+1} &= C(\tau)w^n, \\ w^0 &= w_0 \end{aligned} \tag{17}$$

where, for simplicity we have used the notation $w^n = w_{sp}^N(n\tau)$. Here $\{C(\tau)\}_{\tau>0}$ is a one-parameter family of operators $\mathbf{S} \to \mathbf{S}$ and they are defined by the applied splitting. For bounded operators we can define $C(\tau)$ directly for the different splittings introduced above:

– for the sequential splitting

$$C_{ss}(\tau) = \exp(\tau B)\exp(\tau A); \tag{18}$$

– for the Strang splitting

$$C_{St}(\tau) = \exp(\frac{\tau}{2}A)\exp(\tau B)\exp(\frac{\tau}{2}A), \tag{19}$$

– and for the SWS splitting

$$C_{SWS}(\tau) = \frac{1}{2}\left[\exp(\tau B)\exp(\tau A) + \exp(\tau A)\exp(\tau B)\right]. \tag{20}$$

In virtue of the general theory of abstract numerical methods, we can consider a splitting method as an abstract numerical method, defined in some Banach space $\mathbf{S}$. For such methods we have the following [8]

Definition 1. *The numerical method $C(\tau)$ which is applied to the well-posed ACP problem (1), is called consistent if for all fixed $t \in [0, T)$ the relation*

$$\lim_{\tau\to 0}\left\|\left(\frac{C(\tau) - I}{\tau} - (A+B)\right) w(t)\right\| = 0 \tag{21}$$

holds for the solutions $w(t)$ which correspond to all possible choices of the initial elements w_0.

Since the abstract function $w(t)$ is the solution of the ACP, therefore for all $t \in (0,T)$ there exists the limit

$$\lim_{\tau \to 0} \left\| \frac{w(t+\tau)-w(t)}{\tau} - (A+B)w(t) \right\| = 0. \tag{22}$$

Consequently, (21) is satisfied if and only if the relation

$$\lim_{\tau \to 0} \left\| \frac{w(t+\tau)-C(\tau)w(t)}{\tau} \right\| = 0 \tag{23}$$

holds, that is, the numerical method $C(\tau)$ is consistent if and only if (23) holds. Substituting $t=0$ into the formula, we see that the expression in norm (which is called local approximation error of the numerical method) is the local splitting error. Hence, the sequential, the Strang and the SWS splittings as abstract discretization methods are consistent.

However, as it follows from the abstract theory of numerical methods, the consistency itself doesn't yield the convergence. Our aim is to use the Lax theorem. Therefore we introduce the following

Definition 2. *We say that a splitting discretization with the operator $C(\tau)$ is stable if for all fixed $t \in [0,T)$ there exists a constant $K>0$ such that the relation*

$$\|C(\tau)^n\| \leq K \tag{24}$$

holds for all $n \in \mathbb{N}$ such that $n\tau \leq t$.

Hence the condition

$$\|C(\tau)\| \leq 1 + K\tau \tag{25}$$

(where $K \geq 0$ is a constant) is a sufficient condition for the stability. The contractivity of the operators $\exp(\tau A)$ and $\exp(\tau B)$, i.e.,

$$||\exp(\tau A)|| \leq 1; \quad ||\exp(\tau B)|| \leq 1 \tag{26}$$

in some norm is clearly a sufficient condition for the stability (24).

Remark 2. The sequential splitting is unconditionally stable for bounded operators. This follows from the relation

$$\begin{aligned} &\|C_{ss}(\tau)^n\| \leq \|C_{ss}(\tau)\|^n \leq [\|\exp(\tau B)\|]^n \, [\|\exp(\tau A)\|]^n \leq \\ &\leq [\exp(\tau\|B\|)]^n \, [\exp(\tau\|A\|)]^n = \exp(n\tau\|B\|)\exp(n\tau\|A\|) = \\ &= \exp\left[n\tau(\|A\|+\|B\|)\right] \leq \exp\left[t(\|A\|+\|B\|)\right] \end{aligned} \tag{27}$$

which holds for all $n \in \mathbb{N}$ such that $n\tau \leq t$. Hence the splitting is stable with the constant $K = \exp\left[t(\|A\|+\|B\|)\right]$.

To prove the similar statements for the Strang and SWS splittings, as an easy exercise, is left to the reader.

The Lax equivalence theorem says that for a well-posed ACP a consistent numerical method is convergent if and only if it is stable [8]. Hence, we have

Theorem 1. *Assume that $A, B \in \mathcal{B}(\mathbf{S})$. Then the sequential, Strang and SWS splittings are convergent for the well-posed ACP (1).*

3 Application of Numerical Methods in the Split Sub-problems

In Section 2 we analyzed the convergence under the assumption that the split sub-problems are solved exactly. Usually, this cannot be done and we apply some numerical method to their solution with the discretization parameter Δt. Clearly, the condition $\Delta t \leq \tau$ should be satisfied. Aiming at finishing the numerical solving process at the endpoint of the time intervals, where the sub-problems are posed, we select $\Delta t = \tau/K$, where K is some integer.

In this manner, we can define a new numerical discretization method, based on the splitting. Obviously, for the different sub-problems different numerical methods can be chosen with different discretization parameters. Hence, assuming the use of the same numerical method on each splitting step, the total discretization operator depends on the choice of the splitting, the splitting step, the applied numerical methods and their step sizes. For instance, for the sequential splitting, by the choice of some numerical method NM1 with the step-size $\Delta t_1 = \tau/K_1$ for the first sub-problem, and a numerical method NM2 with the step-size $\Delta t_2 = \tau/K_2$ for the second sub-problem, the total numerical discretization operator can be defined as $C_{tot} = C_{tot}(\tau, NM1, K_1, NM2, K_2)$. (For simplicity, we will use the notation $C(\tau)$ when $NM1, K_1, NM2$ and K_2 are fixed.)

Example 1. Let us consider the sequential splitting applied to the ACP (1). We solve each problem with the explicit Euler (EE) method and we choose $\Delta t = \tau$ for both sub-problems. (I.e., $C_{tot} = C_{tot}(\tau, EE, 1, EE, 1)$). If we denote by y_1^n and y_2^n the approximations to $w_1^N(n\tau)$ and $w_2^N(n\tau)$, respectively, the numerical schemes are

$$\frac{y_1^{n+1} - y_1^n}{\tau} = Ay_1^n, \quad \frac{y_2^{n+1} - y_2^n}{\tau} = By_2^n \tag{28}$$

and $y_2^n = y_1^{n+1}$. Hence

$$y_2^{n+1} = (I + \tau B)(I + \tau A)y_1^n. \tag{29}$$

Consequently,

$$C(\tau) = (I + \tau B)(I + \tau A). \tag{30}$$

Example 2. Let us consider again the sequential splitting for the ACP (1) and apply the explicit Euler (EE) method with $\Delta t = \tau/K$ for both sub-problems. (I.e., $C_{tot} = C_{tot}(\tau, EE, K, EE, K)$). Then

$$C(\tau) = (I + \frac{\tau}{K}B)^K (I + \frac{\tau}{K}A)^K. \tag{31}$$

Obviously, in order to prove the convergence of the combined numerical discretization, we can apply the Lax theorem with the choice $C(\tau) = C_{tot}$. For illustration, we prove a simple case.

Theorem 1. *Assume that $A, B \in \mathcal{B}(\mathbf{S})$. Then the sequential splitting with the explicit Euler method with the choice $\Delta t = \tau$ is convergent for the well-posed ACP (1).*

Proof. By the use of (30), we get

$$\frac{C(\tau) - I}{\tau} - (A + B) = \tau A \cdot B, \tag{32}$$

which means that combined method is consistent for the bounded operators. As for the stability, for (30) we get the relation

$$\begin{aligned} \|C(\tau)^n\| = \| \left((I + \tau B)(I + \tau A)\right)^n \| \le (1 + \tau\|B\|)^n (1 + \tau\|A\|)^n \le \\ \le \exp(t\|A\|) \exp(t\|B\|) = \exp\left(t(\|A\| + \|B\|)\right), \end{aligned} \tag{33}$$

which proves the stability.

4 Some Combined Schemes

In Section 3 we defined the combined discretization methods. In this section we show the relation of these methods to some well-known discretization methods.

4.1 Crank-Nicolson Method

Let us consider the Cauchy problem

$$\begin{aligned} \frac{dw}{dt} &= Aw(t); \quad 0 < t \le T, \\ w(0) &= w_0. \end{aligned} \tag{34}$$

If we use the trivial splitting $A = \frac{1}{2}A + \frac{1}{2}A$, then (34) can be split into two sub-problems via sequential splitting. (For simplicity, we write only the first step.)

$$\begin{aligned} \frac{dw_1^1(t)}{dt} &= \frac{1}{2}Aw_1^1(t); \quad 0 < t \le \tau, \\ w_1^1(0) &= w_0. \end{aligned} \tag{35}$$

$$\frac{dw_2^1(t)}{dt} = \frac{1}{2}Aw_2^1(t), \quad 0 < t \leq \tau$$
$$w_2^1(0) = w_1^1(\tau) \tag{36}$$

Applying the explicit Euler method for the sub-problem (35) and the implicit Euler (IE) method for (36) with $\Delta t = \tau$, we obtain

$$\frac{y_1^1 - y_1^0}{\tau} = \frac{1}{2}Ay_1^0; \quad y_1^0 = w_0,$$

$$\frac{y_2^1 - y_2^0}{\tau} = \frac{1}{2}Ay_2^1; \quad y_2^0 = y_1^1.$$

This implies that for the above special decomposition the discretization operator for the sequential splitting is $C_{tot} = C_{tot}(\tau, EE, 1, IE, 1)$ and has the form

$$C(\tau) = (I - \frac{\tau}{2}A)^{-1}(I + \frac{\tau}{2}A), \tag{37}$$

i.e., we obtained the operator of the Crank-Nicolson method ("trapezoidal rule"). In this example, the discretization error of the method consists of only the numerical integration part because obviously the splitting error equals zero. This shows that such an approach can lead to an increase of the order: although the applied numerical methods (explicit and implicit Euler methods) are of first order, the combined method has second order accuracy.

4.2 Componentwise Split Crank-Nicolson Method (Second Order Yanenko Method)

Let us consider the ACP (1) and use the sequential splitting and the middle point numerical integration method (CN) with $\tau = \Delta t$, i.e.,

$$\frac{y_1^{n+1} - y_1^n}{\tau} = A(\frac{y_1^{n+1} + y_1^n}{2})$$
$$y_1^n = y_2^{n-1} \tag{38}$$

$$\frac{y_2^{n+1} - y_2^n}{\tau} = B(\frac{y_2^{n+1} + y_2^n}{2})$$
$$y_2^n = y_1^{n+1}, \tag{39}$$

where $y_2^0 = w_0$ and $n = 1, 2, \ldots;\ n\tau \leq T$, which can be written in the form

$$y_1^{n+1} = y_1^n + \tau A(\frac{y_1^n + y_1^{n+1}}{2}),$$
$$y_2^{n+1} = y_1^{n+1} + \tau B(\frac{y_1^{n+1} + y_2^{n+1}}{2}). \tag{40}$$

This method is known as the second-order Yanenko method [12].
An easy computation shows that

$$C(\tau) = (I - \frac{\tau}{2}B)^{-1}(I + \frac{\tau}{2}B)(I - \frac{\tau}{2}A)^{-1}(I + \frac{\tau}{2}A). \quad (41)$$

Obviously, in the above algorithm the inverse operators always exist for sufficiently small values of τ. Using the Neumann series we have

$$(I - \frac{\tau}{2}B)^{-1} = I + \frac{\tau}{2}B + \frac{\tau^2}{4}B^2 + \mathcal{O}(\tau^3) \quad (42)$$

$$(I - \frac{\tau}{2}A)^{-1} = I + \frac{\tau}{2}A + \frac{\tau^2}{4}A^2 + \mathcal{O}(\tau^3) \quad (43)$$

Hence,

$$C(\tau) = I + \tau(A + B) + \frac{\tau^2}{2}(A^2 + B^2 + 2BA) + \mathcal{O}(\tau^3),$$

which shows the consistency (in first order) of this combined method. (If the operators A and B commute then the order is higher.)

The numerical discretization method can be realized as follows:

1. $\xi_1 := (I + \frac{\tau}{2})y^n$
2. $(I - \frac{\tau}{2}A)\xi_2 = \xi_1$
3. $\xi_3 = (I + \frac{\tau}{2}B)\xi_2$
4. $(I - \frac{\tau}{2}B)\xi_4 = \xi_3$
5. $y^{n+1} = \xi_4$,

for $n := 0, 1, \ldots$ and $y^0 = w_0$. This algorithm requires the solution of only two systems of linear equations, with (hopefully) simple linear operators. E.g., when (1) results from a semidiscretization of the heat equation, A and B are the corresponding space discretization matrices, i.e.

$$A = \frac{1}{h_x^2} tridiag(1, -2, 1),\ B := \frac{1}{h_y^2} tridiag(1, -2, 1),$$

then, in practice, this algorithm leads to two one-dimensional problems. Since in this case A and B commute, therefore this method has second order accuracy for this problem.

4.3 Sequential Alternating Marchuk Scheme

Let us denote the Yanenko scheme (40) as $y^{n+1} = \Phi_{AB}(y^n)$. In order to restore the symmetry, we interchange the order of A and B in each step. This leads to the modification

$$y^{n+1} = \Phi_{AB}(y^n);\ y^{n+2} = \Phi_{BA}(y^{n+1}),\ n = 0, 2, 4 \ldots. \quad (44)$$

This method was defined by Marchuk [7] and it corresponds to the case of the Strang splitting with the middle-point numerical integration method and $\tau = \Delta t$.

Finite Difference Approximation of an Elliptic Interface Problem with Variable Coefficients

Boško S. Jovanović[1] and Lubin G. Vulkov[2]

[1] University of Belgrade, Faculty of Mathematics,
Studentski trg 16, 11000 Belgrade, Serbia and Montenegro,
bosko@matf.bg.ac.yu

[2] University of Rousse, Department of Mathematics,
Studentska str. 8, 7017 Rousse, Bulgaria
vulkov@ami.ru.acad.bg

Abstract. General elliptic interface problem with variable coefficients and curvilinear interface is transformed into analogous problem with rectilinear interface. For the numerical solution of transformed problem a finite difference scheme with averaged right–hand side is proposed. Convergence rate estimate in discrete W_2^1 norm, compatible with the smoothness of data, is obtained.

1 Introduction

Interface problems occur in many applications in science and engineering. These kinds of problems can be modelled by partial differential equations with discontinuous or singular coefficients. Various forms of conjugation conditions satisfied by the solution and its derivatives on the interface are known [4], [15]. In [2] a review of results on numerical solution of two-dimensional elliptic and parabolic interface problems in recent years is presented. In [10], [11], [12], [19] convergence of finite difference method for different elliptic, parabolic and hyperbolic problems in which the solution is continuous on the interface curves, while the flow is discontinuous, is studied.

In the present work we investigate a general elliptic interface problem in rectangular domain, crossed by curvilinear interface. By suitable change of variables problem is transformed into analogous one with rectilinear interface. For the numerical solution of transformed problem a finite difference scheme with averaged right–hand side is proposed. Convergence rate estimate in discrete W_2^1 norm, compatible with the smoothness of data, is obtained. Analogous problem for Laplace operator is considered in [8]

The layout of the paper is as follows. In Section 2 the interface boundary value problem (BVP) is defined. Further, it is transformed into BVP with rectilinear interface by suitable change of variables. The main properties of transformed problem are studied in Section 3. In Section 4 finite difference scheme (FDS) approximating considered rectilinear interface problem is constructed. Convergence of FDS is proved in Section 5.

Z. Li et al. (Eds.): NAA 2004, LNCS 3401, pp. 46–55, 2005.

2 Problem with Curvilinear Interface

Let $\Omega = (0,1)^2$, $\Gamma = \partial\Omega$, and let S be a smooth curve intersecting Ω. For clarity, let S be defined by equation $\xi_2 = g(\xi_1)$, where $g \in C^1[0,1]$ and $0 < g_0 \le g(\xi_1) \le g_1 < 1$. In domain Ω we consider Dirichlet boundary value problem

$$\mathcal{M}U + K(\xi)\delta_S(\xi)U = F(\xi) \ \text{ in } \Omega, \qquad U = 0 \ \text{ on } \Gamma, \tag{1}$$

where $\xi = (\xi_1, \xi_2)$,

$$\mathcal{M}U = -\sum_{i,j=1}^{2} A_{ij}(\xi)\,\frac{\partial^2 U}{\partial\xi_i \partial\xi_j} + 2\sum_{i=1}^{2} B_i(\xi)\,\frac{\partial U}{\partial\xi_i} + C(\xi)U\,, \quad A_{ij}(\xi) = A_{ji}(\xi)$$

is elliptic operator and $\delta_S(\xi)$ is Dirac distribution concentrated on S. The equality in (1) is treated in the sense of distributions.

The BVP (1) can be transformed into analogous one with rectilinear interface. It can be easily verified that the change of variables $x = x(\xi)$, where

$$x_1 = \xi_1\,, \qquad x_2 = \frac{[1 - g(\xi_1)]\,\xi_2}{\xi_2 - 2\xi_2 g(\xi_1) + g(\xi_1)} \tag{2}$$

maps Ω onto Ω. The curve S is mapped onto straight line Σ: $x_2 = 1/2$. By change of variables (2) BVP (1) transforms to following one

$$\mathcal{L}u + k(x)\delta_\Sigma(x)u = f(x) \ \text{ in } \Omega, \qquad u = 0 \ \text{ on } \Gamma, \tag{3}$$

where $u(x) = U(\xi)$, $f(x) = F(\xi)$, $\delta_\Sigma(x) = \delta(x_2 - 1/2)$ is Dirac distribution concentrated on Σ, $k(x) = k(x_1) = K(x_1, g(x_1))\sqrt{1 + [g'(x_1)]^2}$, while coefficients of differential operator

$$\mathcal{L}u = -\sum_{i,j=1}^{2} a_{ij}(x)\,\frac{\partial^2 u}{\partial x_i \partial x_j} + 2\sum_{i=1}^{2} \hat{b}_i(x)\,\frac{\partial u}{\partial x_i} + \hat{c}(x)u\,, \quad a_{ij}(x) = a_{ji}(x)$$

can be expressed by coefficients of $\mathcal{M}$ and derivatives $\frac{\partial x_i}{\partial \xi_j}$, for example

$$a_{11}(x) = A_{11}(\xi)\,,$$

$$a_{12}(x) = A_{11}(\xi)\,\frac{\xi_2(1-\xi_2)g'(\xi_1)}{[\xi_2 - 2\xi_2 g(\xi_1) + g(\xi_1)]^2} + A_{12}(\xi)\,\frac{g(\xi_1)[1 - g(\xi_1)]}{[\xi_2 - 2\xi_2 g(\xi_1) + g(\xi_1)]^2}$$

etc. Finally, we represent operator $\mathcal{L}$ in the skew–symmetric form

$$\mathcal{L}u = -\sum_{i,j=1}^{2} \frac{\partial}{\partial x_i}\left(a_{ij}\,\frac{\partial u}{\partial x_j}\right) + \sum_{i=1}^{2}\left[b_i\,\frac{\partial u}{\partial x_i} + \frac{\partial(b_i u)}{\partial x_i}\right] + cu\,, \tag{4}$$

where

$$b_i(x) = \hat{b}_i(x) + \frac{1}{2}\sum_{j=1}^{2}\frac{\partial a_{ij}}{\partial x_j}\,, \qquad c(x) = \hat{c}(x) - \sum_{j=1}^{2}\frac{\partial b_j}{\partial x_j}\,.$$

In such a way, in the sequel we can restrict our investigation to the BVP of the form (3), (4). This problem will be called canonical. Note that differential operator $\mathcal{L}$ is a general convection–diffusion operator [18].

3 Canonical Problem with Rectilinear Interface

Dirac distribution δ_Σ belongs to Sobolev space $W_2^{-\alpha}(\Omega)$, with $\alpha > 1/2$. In such a way, equation (3) must be treated as an equation in this space. For $\alpha = 1$ this means that

$$\langle \mathcal{L}u + k\delta_\Sigma u,\, v\rangle = \langle f, v\rangle, \qquad \forall v \in \overset{\circ}{W}{}_2^1(\Omega), \tag{5}$$

where $\langle f, v\rangle$ denotes duality pairing between spaces $W_2^{-1}(\Omega)$ and $\overset{\circ}{W}{}_2^1(\Omega)$. Using standard rules for differentiation of distributions (see [20]) from (5) we obtain the following weak form of BVP (3), (4): *find* $u \in \overset{\circ}{W}{}_2^1(\Omega)$ *such that*

$$a(u,v) = \langle f, v\rangle\,, \qquad \forall\, v \in \overset{\circ}{W}{}_2^1(\Omega)\,, \tag{6}$$

where

$$a(u,v) = \int_\Omega \left[\sum_{i,j=1}^2 a_{ij}\,\frac{\partial u}{\partial x_j}\,\frac{\partial v}{\partial x_i} + \sum_{i=1}^2 b_i \left(\frac{\partial u}{\partial x_i}\,v - u\,\frac{\partial v}{\partial x_i} \right) + cuv \right] dx \\ + \int_\Sigma kuv\, d\Sigma\,. \tag{7}$$

The following assertion is an immediate consequence of the Lax–Milgram lemma [14] and the imbedding $W_2^1(\Omega) \subset L_2(\Sigma)$.

Lemma 1. *Let the following assumptions*

$$f \in W_2^{-1}(\Omega), \quad a_{ij},\, b_i,\, c \in L_\infty(\Omega), \quad k \in L_\infty(\Sigma), \quad a_{ij} = a_{ji}, \quad c \ge 0, \quad k \ge 0,$$

$$\sum_{i,j=1}^2 a_{ij}(x)\, z_i z_j \ge c_0 \sum_{i=1}^2 z_i^2\,, \quad c_0 = \text{const} > 0, \quad \forall\, z = (z_1, z_2) \in R^2, \quad \text{a. e. in } \Omega$$

hold. Then there exists the unique solution $u \in \overset{\circ}{W}{}_2^1(\Omega)$ *of BVP (6), (7).*

In the case when $f(x)$ does not contains $\delta_\Sigma(x)$ one easily checks that (6), (7) is also the weak form of the following BVP with conjugation conditions on the interface Σ:

$$\mathcal{L}u = f(x) \ \text{ in } \Omega^- \cup \Omega^+, \qquad u = 0 \ \text{ on } \Gamma, \\ [u]_\Sigma = 0\,, \qquad \left[\sum_{j=1}^2 a_{2j}\,\frac{\partial u}{\partial x_j} - b_2 u \right]_\Sigma = ku\Big|_\Sigma \tag{8}$$

where $\Omega^- = (0,1) \times (0,1/2)$, $\Omega^+ = (0,1) \times (1/2,1)$, and $[u]_\Sigma = u(x_1, 1/2+0) - u(x_1, 1/2-0)$. In this sense, BVPs (3), (4) and (8) are equivalent.

Increased smoothness of the solution can be proved under some additional assumptions on the input data.

Lemma 2. *Let the following assumptions hold, besides the assumptions of Lemma 1:* $f \in L_2(\Omega)$, $a_{ij}, b_i \in W_\infty^1(\Omega)$, $k \in W_\infty^1(\Sigma)$. *Then* $\dfrac{\partial^2 u}{\partial x_1^2}, \dfrac{\partial^2 u}{\partial x_1 \partial x_2} \in L_2(\Omega)$

and $\dfrac{\partial^2 u}{\partial x_2^2} \in L_2(\Omega^\pm)$. *If also* $\dfrac{\partial f}{\partial x_1} \in L_2(\Omega)$, $\dfrac{\partial f}{\partial x_2} \in L_2(\Omega^\pm)$, $a_{ij},\, b_i \in W^2_\infty(\Omega) c \in W^1_\infty(\Omega)$, $k \in W^2_\infty(\Sigma)$ *and* $f = a_{12} = b_1 = \dfrac{\partial a_{11}}{\partial x_1} = 0$ *for* $x_1 = 0$ *and* $x_1 = 1$ *then* $\dfrac{\partial^3 u}{\partial x_1^3}, \dfrac{\partial^3 u}{\partial x_1^2 \partial x_2} \in L_2(\Omega)$ *and* $\dfrac{\partial^3 u}{\partial x_1 \partial x_2^2}, \dfrac{\partial^3 u}{\partial x_2^3} \in L_2(\Omega^\pm)$.

For more detailed analysis of elliptic BVPs in domains with corners we refer to [6] and [5].

In the sequel we will assume that the generalized solution of BVP (3), (4) belongs to $\widetilde{W}_2^s(\Omega) = \overset{\circ}{W}{}_2^1(\Omega) \cap W_2^s(\Omega^-) \cap W_2^s(\Omega^+)$, $s > 2$, while the coefficients of equation satisfies the following smoothness conditions: $a_{ij},\, b_i \in W_2^{s-1}(\Omega^-) \cap W_2^{s-1}(\Omega^-) \cap C(\bar\Omega)$, $c \in W_2^{s-2}(\Omega^-) \cap W_2^{s-2}(\Omega^-)$, $k \in W_2^{s-1}(\Sigma)$. We define

$$\|u\|^2_{\widetilde{W}_2^s(\Omega)} = \|u\|^2_{W_2^1(\Omega)} + \|u\|^2_{W_2^s(\Omega^-)} + \|u\|^2_{W_2^s(\Omega^+)} \,.$$

4 Finite Difference Approximation

Let $\bar\omega_h$ be a uniform mesh with step $h = 1/(2n)$ in $\bar\Omega$, $\omega_h = \bar\omega_h \cap \Omega$, $\omega_{1h} = \bar\omega_h \cap \big([0,1) \times (0,1)\big)$, $\omega_{2h} = \bar\omega_h \cap \big((0,1) \times [0,1)\big)$, $\gamma_h = \bar\omega_h \cap \Gamma$, $\sigma_h = \omega_h \cap \Sigma$ and $\sigma_h^- = \sigma_h \cup \{(0,1/2)\}$.

Further, we shall use standard denotations [16]

$$v_{x_i} = \frac{v^{+i} - v}{h}, \quad v_{\bar x_i} = \frac{v - v^{-i}}{h}, \quad v^{\pm i}(x) = v(x \pm h e_i), \quad e_1 = (1,0), \quad e_2 = (0,1)\,.$$

We approximate the BVP (3), (4) on the mesh $\bar\omega_h$ by the following finite difference scheme with averaged right–hand side

$$\mathcal{L}_h\, v + k\delta_{\sigma_h} v = T_1^2 T_2^2 f \ \text{ in } \omega_h, \qquad v = 0 \ \text{ on } \gamma_h, \tag{9}$$

where

$$\mathcal{L}_h v = -\frac{1}{2} \sum_{i,j=1}^{2} \Big[(a_{ij} v_{\bar x_j})_{x_i} + (a_{ij} v_{x_j})_{\bar x_i} \Big]$$
$$+\frac{1}{2} \sum_{i=1}^{2} \Big[b_i v_{\bar x_i} + b_i v_{x_i} + (b_i v)_{\bar x_i} + (b_i v)_{x_i} \Big] + (T_1^2 T_2^2 c) v,$$

$$\delta_{\sigma_h}(x) = \delta_h(x_2 - 1/2) = \begin{cases} 0, & x \in \omega_h \setminus \sigma_h \\ 1/h, & x \in \sigma_h \end{cases}$$

is discrete Dirac delta-function and T_i are Steklov averaging operators [17]:

$$T_1 f(x_1,x_2) = T_1^- f(x_1 + h/2, x_2) = T_1^+ f(x_1 - h/2, x_2) = \frac{1}{h} \int_{x_1 - h/2}^{x_1 + h/2} f(x_1', x_2)\, dx_1',$$

$$T_2 f(x_1,x_2) = T_2^- f(x_1,x_2+h/2) = T_2^+ f(x_1,x_2-h/2) = \frac{1}{h}\int_{x_2-h/2}^{x_2+h/2} f(x_1,x_2')\,dx_2'.$$

Note that these operators are self–commutative and they transform the derivatives to divided differences, for example

$$T_i^- \frac{\partial u}{\partial x_i} = u_{\bar{x}_i}\,, \qquad T_i^+ \frac{\partial u}{\partial x_i} = u_{x_i}\,, \qquad T_i^2 \frac{\partial^2 u}{\partial x_i^2} = u_{x_i \bar{x}_i}\,.$$

We also define

$$T_2^{2-} f(x_1,x_2) = \frac{1}{h}\int_{x_2-h}^{x_2} \Big(1 + \frac{x_2'-x_2}{h}\Big) f(x_1,x_2')\,dx_2',$$

$$T_2^{2+} f(x_1,x_2) = \frac{1}{h}\int_{x_2}^{x_2+h} \Big(1 - \frac{x_2'-x_2}{h}\Big) f(x_1,x_2')\,dx_2'.$$

Analogous FDS for BVP without interface (the case $k = 0$) is investigated in [18].

Let H_h be the space of mesh functions defined on ω_h, equal to zero on γ_h, endowed with the inner product and norm

$$(v,w)_h = h^2 \sum_{x\in\omega_h} v(x)\,w(x)\,, \qquad \|v\|_{L_2(\omega_h)} = (v,v)_h^{1/2}\,.$$

We also define the following discrete inner products, norms and seminorms

$$(v,w)_{ih} = h^2 \sum_{x\in\omega_{ih}} v(x)\,w(x)\,, \qquad \|v\|_{L_2(\omega_{ih})} = (v,v)_{ih}^{1/2}\,,$$

$$\|v\|^2_{W_2^1(\omega_h)} = \|v_{x_1}\|^2_{L_2(\omega_{1h})} + \|v_{x_2}\|^2_{L_2(\omega_{2h})} + \|v\|^2_{L_2(\omega_h)}\,,$$

$$(v,w)_{\sigma_h} = h \sum_{x\in\sigma_h} v(x)\,w(x)\,, \qquad \|v\|_{L_2(\sigma_h)} = (v,v)_{\sigma_h}^{1/2}\,,$$

$$|w|_{W_2^{1/2}(\sigma_h)} = \left(h^2 \sum_{x\in\sigma_h^-} \sum_{x'\in\sigma_h^-,\; x'\neq x} \frac{|w(x)-w(x')|^2}{|x_1-x_1'|^2} \right)^{1/2}.$$

Further we shall make use of the assertion:

Lemma 3. *Let $v \in H_h$ and w be a mesh function defined on σ_h^-. Then*

$$|(w_{\bar{x}_1},\, v)_{\sigma_h}| \le C\, \|v\|_{W_2^1(\omega_h)}\, |w|_{W_2^{1/2}(\sigma_h)}\,.$$

The proof is analogous to the proof of Lemma 2 in [9].

5 Convergence of the Finite Difference Scheme

Let u be the solution of BVP (3), (4) and v – the solution of FDS (9). The error $z = u - v$ satisfies the finite difference scheme

$$\mathcal{L}_h z + k\delta_{\sigma_h} z = \varphi \ \text{ in } \omega_h, \qquad z = 0 \ \text{ on } \gamma_h, \tag{10}$$

where

$$\varphi = \sum_{i,j=1}^{2} \eta_{ij,\bar{x}_i} + \sum_{i=1}^{2} \xi_{i,\bar{x}_i} + \sum_{i=1}^{2} \zeta_i + \chi + \delta_{\sigma_h}\mu\,,$$

$$\eta_{ij} = T_i^+ T_{3-i}^2\Big(a_{ij}\,\frac{\partial u}{\partial x_j}\Big) - \frac{1}{2}\big(a_{ij}u_{x_j} + a_{ij}^{+i}u_{\bar{x}_j}^{+i}\big),$$

$$\xi_i = \frac{1}{2}\big(b_i^{+i}u^{+i} + b_i u\big) - T_i^+ T_{3-i}^2(b_i u),$$

$$\zeta_i = \frac{1}{2}\big(b_i u_{\bar{x}_i} + b_i u_{x_i}\big) - T_1^2 T_2^2\Big(b_i\,\frac{\partial u}{\partial x_i}\Big),$$

$$\chi = (T_1^2 T_2^2 c)u - T_1^2 T_2^2(cu),$$

$$\mu = ku - T_1^2(ku),$$

Let us set

$$\eta_{1j} = \tilde{\eta}_{1j} + \delta_{\sigma_h}\hat{\eta}_{1j}, \quad \zeta_j = \tilde{\zeta}_j + \delta_{\sigma_h}\hat{\zeta}_j, \quad \xi_1 = \tilde{\xi}_1 + \delta_{\sigma_h}\hat{\xi}_1, \quad \chi = \tilde{\chi} + \delta_{\sigma_h}\hat{\chi},$$

where

$$\hat{\eta}_{11} = \frac{h^2}{6}\,T_1^+\bigg(\Big[a_{11}\,\frac{\partial^2 u}{\partial x_1 \partial x_2} + \frac{\partial a_{11}}{\partial x_2}\,\frac{\partial u}{\partial x_1}\Big]_\Sigma\bigg),$$

$$\hat{\eta}_{12} = \frac{h^2}{6}\,T_1^+\bigg(\Big[a_{12}\,\frac{\partial^2 u}{\partial x_2^2} + \frac{\partial a_{12}}{\partial x_2}\,\frac{\partial u}{\partial x_2}\Big]_\Sigma\bigg) - \frac{h^2}{4}\,T_1^+\bigg(\Big[\frac{\partial}{\partial x_1}\Big(a_{12}\,\frac{\partial u}{\partial x_2}\Big)\Big]_\Sigma\bigg),$$

$$\hat{\zeta}_1 = -\frac{h^2}{6}\,T_1^2\bigg(\Big[\frac{\partial}{\partial x_2}\Big(b_1\,\frac{\partial u}{\partial x_1}\Big)\Big]_\Sigma\bigg),$$

$$\hat{\zeta}_2 = \frac{h^2}{4}\bigg[b_2\,T_1^2\Big(\frac{\partial^2 u}{\partial x_2^2}\Big)\bigg]_\Sigma - \frac{h^2}{6}\,T_1^2\bigg(\Big[\frac{\partial}{\partial x_2}\Big(b_2\,\frac{\partial u}{\partial x_2}\Big)\Big]_\Sigma\bigg),$$

$$\hat{\xi}_1 = -\frac{h^2}{6}\,T_1^+\bigg(\Big[\frac{\partial(b_1 u)}{\partial x_2}\Big]_\Sigma\bigg),$$

$$\hat{\chi} = -\frac{h^2}{3}\bigg[(T_1^2 c)\Big(T_1^2\,\frac{\partial u}{\partial x_2}\Big)\bigg]_\Sigma.$$

Using summation by part and Lemma 3, we get the a priori estimate

$$\begin{aligned}\|z\|_{W_2^1(\omega_h)} \le C\bigg[\sum_{j=1}^{2}\Big(&\|\eta_{2j}\|_{L_2(\omega_{2h})} + \|\tilde{\eta}_{1j}\|_{L_2(\omega_{1h})} + |\hat{\eta}_{1j}|_{W_2^{1/2}(\sigma_h)} \\ &+ \|\tilde{\zeta}_j\|_{L_2(\omega_h)} + \|\hat{\zeta}_j\|_{L_2(\sigma_h)}\Big) + \|\xi_2\|_{L_2(\omega_{2h})} + \|\tilde{\xi}_1\|_{L_2(\omega_{1h})} \\ &+ |\hat{\xi}_1|_{W_2^{1/2}(\sigma_h)} + \|\tilde{\chi}\|_{L_2(\omega_h)} + \|\hat{\chi}\|_{L_2(\sigma_h)} + \|\mu\|_{L_2(\sigma_h)}\bigg].\end{aligned} \tag{11}$$

Therefore, in order to estimate the rate of convergence of the difference scheme (10) it is sufficient to estimate the terms at the right–hand side of (10).

Terms η_{2j} are estimated in [7]. After summation over the mesh ω_{2h} we obtain

$$\begin{aligned}\|\eta_{2j}\|_{L_2(\omega_{2h})} \leq Ch^{s-1}\Big(&\|a_{2j}\|_{W_2^{s-1}(\Omega-)}\|u\|_{W_2^s(\Omega-)}\\ &+\|a_{2j}\|_{W_2^{s-1}(\Omega+)}\|u\|_{W_2^s(\Omega+)}\Big), \quad 2 < s \leq 3.\end{aligned} \tag{12}$$

Terms $\tilde{\eta}_{1j}$ for $x \in \omega_{1h} \setminus \sigma_h^-$ can be estimated in the same manner. For $x \in \sigma_h^-$ we set

$$\tilde{\eta}_{11} = \sum_{k=1}^{3}(\eta_{11,k}^- + \eta_{11,k}^+),$$

where

$$\begin{aligned}\eta_{11,1}^{\pm} &= T_1^+T_2^{2\pm}\Big(a_{11}\frac{\partial u}{\partial x_1}\Big) - 2(T_1^+T_2^{2\pm}a_{11})\Big(T_1^+T_2^{2\pm}\frac{\partial u}{\partial x_1}\Big)\\ &\pm\frac{h}{6}\Big(T_1^+\frac{\partial a_{11}}{\partial x_2}\Big)\Big[2\Big(T_1^+T_2^{2\pm}\frac{\partial u}{\partial x_1}\Big) - \Big(T_1^+\frac{\partial u}{\partial x_1}\Big)\Big]\Big|_{x_2=1/2\pm 0}\\ &\pm\frac{h}{6}\Big[\frac{a_{11}+a_{11}^{+1}}{2}\Big(T_1^+\frac{\partial^2 u}{\partial x_1\partial x_2}\Big) - \Big(T_1^+a_{11}\frac{\partial^2 u}{\partial x_1\partial x_2}\Big)\Big]\Big|_{x_2=1/2\pm 0}\\ &\pm\frac{h}{6}\Big[\Big(T_1^+\frac{\partial a_{11}}{\partial x_2}\Big)\Big(T_1^+\frac{\partial u}{\partial x_1}\Big) - \Big(T_1^+\frac{\partial a_{11}}{\partial x_2}\frac{\partial u}{\partial x_1}\Big)\Big]\Big|_{x_2=1/2\pm 0},\\ \eta_{11,2}^{\pm} &= \Big[2(T_1^+T_2^{2\pm}a_{11}) - \frac{a_{11}+a_{11}^{+1}}{2} \mp \frac{h}{3}\Big(T_1^+\frac{\partial a_{11}}{\partial x_2}\Big)\Big]\Big(T_1^+T_2^{2\pm}\frac{\partial u}{\partial x_1}\Big)\Big|_{x_2=1/2\pm 0},\\ \eta_{11,3}^{\pm} &= \frac{a_{11}+a_{11}^{+1}}{4}\Big[2\Big(T_1^+T_2^{2\pm}\frac{\partial u}{\partial x_1}\Big) - u_{x_1} \mp \frac{h}{3}\Big(T_1^+\frac{\partial^2 u}{\partial x_1\partial x_2}\Big)\Big]\Big|_{x_2=1/2\pm 0}.\end{aligned}$$

Terms $\eta_{11,k}^{\pm}$, for $s > 2.5$, also can be estimated as corresponding terms $\eta_{11,k}$ in [7]. In such a way one obtains

$$\begin{aligned}\|\tilde{\eta}_{11}\|_{L_2(\omega_{1h})} \leq Ch^{s-1}\Big(&\|a_{11}\|_{W_2^{s-1}(\Omega-)}\|u\|_{W_2^s(\Omega-)}\\ &+\|a_{11}\|_{W_2^{s-1}(\Omega+)}\|u\|_{W_2^s(\Omega+)}\Big), \quad 2.5 < s \leq 3.\end{aligned} \tag{13}$$

An analogous estimate

$$\begin{aligned}\|\tilde{\eta}_{12}\|_{L_2(\omega_{1h})} \leq Ch^{s-1}\Big(&\|a_{12}\|_{W_2^{s-1}(\Omega-)}\|u\|_{W_2^s(\Omega-)}\\ &+\|a_{12}\|_{W_2^{s-1}(\Omega+)}\|u\|_{W_2^s(\Omega+)}\Big), \quad 2.5 < s \leq 3\end{aligned} \tag{14}$$

holds for $\tilde{\eta}_{12}$.

The value ξ_2 at the node $x \in \omega_{2h}$ is a bounded linear functional of $b_2 u \in W_2^{s-1}(e_2)$, where $e_2 = (x_1 - h, x_1 + h) \times (x_2, x_2 + h)$ and $s > 2$, which vanishes on polynomials of first degree. Using Bramble–Hilbert lemma [1], [3] one immediately obtains

$$|\xi_2(x)| \leq C\,h^{s-2}\,\|b_2 u\|_{W_2^{s-1}(e_2)}, \qquad 2 < s \leq 3,$$

and, after summation over the mesh ω_{2h}

$$\begin{aligned}\|\xi_2\|_{L_2(\omega_{2h})} &\le Ch^{s-1}\Big(\|b_2 u\|_{W_2^{s-1}(\Omega^-)} + \|b_2 u\|_{W_2^{s-1}(\Omega^+)}\Big)\\ &\le Ch^{s-1}\Big(\|b_2\|_{W_2^{s-1}(\Omega^-)}\|u\|_{W_2^s(\Omega^-)}\\ &+\|b_2\|_{W_2^{s-1}(\Omega^+)}\|u\|_{W_2^s(\Omega^+)}\Big), \qquad 2 < s \le 3.\end{aligned} \tag{15}$$

In an analogous way, using Bramble-Hilbert lemma and decomposing some terms if it is necessary, one obtains the following estimates:

$$\begin{aligned}\|\tilde{\xi}_1\|_{L_2(\omega_{1h})} &\le Ch^{s-1}\Big(\|b_1\|_{W_2^{s-1}(\Omega^-)}\|u\|_{W_2^s(\Omega^-)}\\ &+\|b_1\|_{W_2^{s-1}(\Omega^+)}\|u\|_{W_2^s(\Omega^+)}\Big), \quad 2.5 < s \le 3,\end{aligned} \tag{16}$$

$$\begin{aligned}\|\tilde{\zeta}_i\|_{L_2(\omega_h)} &\le Ch^{s-1}\Big(\|b_i\|_{W_2^{s-1}(\Omega^-)}\|u\|_{W_2^s(\Omega^-)}\\ &+\|b_i\|_{W_2^{s-1}(\Omega^+)}\|u\|_{W_2^s(\Omega^+)}\Big), \quad 2.5 < s \le 3,\end{aligned} \tag{17}$$

$$\begin{aligned}\|\tilde{\chi}\|_{L_2(\omega_h)} &\le Ch^{s-1}\Big(\|c\|_{W_2^{s-2}(\Omega^-)}\|u\|_{W_2^s(\Omega^-)}\\ &+\|c\|_{W_2^{s-2}(\Omega^+)}\|u\|_{W_2^s(\Omega^+)}\Big), \quad 2 < s \le 3\end{aligned} \tag{18}$$

and

$$\begin{aligned}\|\mu\|_{L_2(\sigma_h)} &\le Ch^{s-1}\|ku\|_{W_2^{s-1}(\Sigma)}\\ &\le Ch^{s-1}\|k\|_{W_2^{s-1}(\Sigma)}\Big(\|u\|_{W_2^s(\Omega^-)} + \|u\|_{W_2^s(\Omega^+)}\Big), \quad 1.5 < s \le 3.\end{aligned} \tag{19}$$

Terms $\hat{\chi}$ and $\hat{\zeta}_j$ can be estimated directly:

$$\begin{aligned}\|\hat{\chi}\|_{L_2(\sigma_h)} &\le Ch^2\bigg(\|c\|_{L_2(\Sigma+)}\Big\|\frac{\partial u}{\partial x_2}\Big\|_{C(\bar{\Omega}+)} + \|c\|_{L_2(\Sigma-)}\Big\|\frac{\partial u}{\partial x_2}\Big\|_{C(\bar{\Omega}-)}\bigg)\\ &\le Ch^2\Big(\|c\|_{W_2^{s-2}(\Omega^+)}\|u\|_{W_2^s(\Omega^+)} + \|c\|_{W_2^{s-2}(\Omega^-)}\|u\|_{W_2^s(\Omega^-)}\Big), \quad s > 2\,,\end{aligned} \tag{20}$$

$$\begin{aligned}\|\hat{\zeta}_1\|_{L_2(\sigma_h)} &\le Ch^2\bigg(\Big\|\frac{\partial}{\partial x_2}\Big(b_1\,\frac{\partial u}{\partial x_1}\Big)\Big\|_{L_2(\Sigma+)} + \Big\|\frac{\partial}{\partial x_2}\Big(b_1\,\frac{\partial u}{\partial x_1}\Big)\Big\|_{L_2(\Sigma-)}\bigg)\\ &\le Ch^2\Big(\|b_1\|_{W_2^{s-1}(\Omega^+)}\|u\|_{W_2^s(\Omega^+)} + \|b_1\|_{W_2^{s-1}(\Omega^-)}\|u\|_{W_2^s(\Omega^-)}\Big), \quad s > 2.5\,,\end{aligned} \tag{21}$$

$$\begin{aligned}\|\hat{\zeta}_2\|_{L_2(\sigma_h)} &\le Ch^2\bigg(\|b_2\|_{C(\bar{\Omega}+)}\Big\|\frac{\partial^2 u}{\partial x_2^2}\Big\|_{L_2(\Sigma+)} + \|b_2\|_{C(\bar{\Omega}-)}\Big\|\frac{\partial^2 u}{\partial x_2^2}\Big\|_{L_2(\Sigma-)}\\ &+\Big\|\frac{\partial}{\partial x_2}\Big(b_2\,\frac{\partial u}{\partial x_2}\Big)\Big\|_{L_2(\Sigma+)} + \Big\|\frac{\partial}{\partial x_2}\Big(b_2\,\frac{\partial u}{\partial x_2}\Big)\Big\|_{L_2(\Sigma-)}\bigg)\\ &\le Ch^2\Big(\|b_2\|_{W_2^{s-1}(\Omega^+)}\|u\|_{W_2^s(\Omega^+)} + \|b_2\|_{W_2^{s-1}(\Omega^-)}\|u\|_{W_2^s(\Omega^-)}\Big), \quad s > 2.5\,.\end{aligned} \tag{22}$$

Here the following notation is used:

$$\|u\|_{L_2(\Sigma\pm)} = \|u(\,\cdot\,, 1/2 \pm 0)\|_{L_2(0,1)}\,.$$

Seminorm $|T_1^+\phi|_{W_2^{1/2}(\sigma_h)}$, for $\phi \in W_2^\alpha(\Sigma)$, $0 < \alpha \le 1/2$, can be estimated directly:

$$|T_1^+\phi|_{W_2^{1/2}(\sigma_h)} \le 2^{\alpha+1/2} h^{\alpha-1/2} |\phi|_{W_2^\alpha(\Sigma)} \le C h^{\alpha-1/2} \|\phi\|_{W_2^{\alpha+1/2}(\Omega^\pm)},$$

wherefrom follows

$$|\hat{\xi}_1|_{W_2^{1/2}(\sigma_h)} \le C h^{s-1} \left(\left\| \frac{\partial(b_1 u)}{\partial x_2} \right\|_{W_2^{s-2}(\Omega^+)} + \left\| \frac{\partial(b_1 u)}{\partial x_2} \right\|_{W_2^{s-2}(\Omega^-)} \right) \tag{23}$$

$$\le C h^{s-1} \Big(\|b_1\|_{W_2^{s-1}(\Omega^+)} \|u\|_{W_2^s(\Omega^+)} + \|b_1\|_{W_2^{s-1}(\Omega^-)} \|u\|_{W_2^s(\Omega^-)} \Big), \quad 2.5 < s \le 3,$$

$$\begin{aligned} |\hat{\eta}_{11}|_{W_2^{1/2}(\sigma_h)} \le C h^{s-1} \Bigg(& \left\| a_{11} \frac{\partial^2 u}{\partial x_1 \partial x_2} \right\|_{W_2^{s-2}(\Omega^+)} + \left\| a_{11} \frac{\partial^2 u}{\partial x_1 \partial x_2} \right\|_{W_2^{s-2}(\Omega^-)} \\ & + \left\| \frac{\partial a_{11}}{\partial x_2} \frac{\partial u}{\partial x_1} \right\|_{W_2^{s-2}(\Omega^+)} + \left\| \frac{\partial a_{11}}{\partial x_2} \frac{\partial u}{\partial x_1} \right\|_{W_2^{s-2}(\Omega^-)} \Bigg) \end{aligned} \tag{24}$$

$$\le C h^{s-1} \Big(\|a_{11}\|_{W_2^{s-1}(\Omega^+)} \|u\|_{W_2^s(\Omega^+)} + \|a_{11}\|_{W_2^{s-1}(\Omega^-)} \|u\|_{W_2^s(\Omega^-)} \Big), \quad 2.5 < s \le 3$$

and analogously

$$\begin{aligned} |\hat{\eta}_{12}|_{W_2^{1/2}(\sigma_h)} \le C h^{s-1} \Big(& \|a_{12}\|_{W_2^{s-1}(\Omega^+)} \|u\|_{W_2^s(\Omega^+)} \\ & + \|a_{12}\|_{W_2^{s-1}(\Omega^-)} \|u\|_{W_2^s(\Omega^-)} \Big), \quad 2.5 < s \le 3. \end{aligned} \tag{25}$$

Therefore, from (11)–(25) we obtain the main result of the paper.

Theorem. *Let the assumptions from Section 3 are fulfilled. Then FDS (10) converges and the following convergence rate estimate holds*

$$\begin{aligned} \|u - v\|_{W_2^1(\omega_h)} \le C h^{s-1} \Big(& \max_{i,j} \|a_{ij}\|_{W_2^{s-1}(\Omega^+)} + \max_{i,j} \|a_{ij}\|_{W_2^{s-1}(\Omega^-)} \\ & + \max_i \|b_i\|_{W_2^{s-1}(\Omega^+)} + \max_i \|b_i\|_{W_2^{s-1}(\Omega^-)} + \|c\|_{W_2^{s-2}(\Omega^+)} \\ & + \|c\|_{W_2^{s-2}(\Omega^-)} + \|k\|_{W_2^{s-1}(\Sigma)} \Big) \|u\|_{\widetilde{W}_2^s(\Omega)}, \quad 2.5 < s \le 3. \end{aligned}$$

Obtained convergence rate estimate is compatible with the smoothness of data (see [13]).

Acknowledgement

The research of the first author is supported by MSTD of Republic of Serbia under grant 1645, while the research of the second author is supported by the University of Rousse under grant 2002–PF–04.

References

1. Bramble, J.H., Hilbert, S.R.: Bounds for a class of linear functionals with application to Hermite interpolation. Numer. Math. **16** (1971) 362–369
2. Brayanov, I.A., Vulkov, L.G.: Homogeneous difference schemes for the heat equation with concentrated capacity. Zh. vychisl. mat. mat. fiz. **39** (1999) 254–261 (in Russian)
3. Dupont, T. Scott, R.: Polynomial approximation of functions in Sobolev spaces. Math. Comput. **34** (1980) 441–463.
4. Escher, J.: Quasilinear parabolic systems with dynamical boundary conditions. Communs Partial Differential Equations **19** (1993) 1309–1364
5. Dauge, M.: Elliptic boundary value problems on corner domains. Lecture Notes in Mathematics, Springer Verlag, Berlin (1988)
6. Grisvard, P.: Elliptic problems in nonsmooth domains. Pitman, Boston (1985)
7. Jovanović, B.S.: Finite difference method for boundary value problems with weak solutions. Posebna izdanja Mat. Instituta **16**, Belgrade 1993
8. Jovanović, B.S., Kandilarov, J.D., Vulkov, L.G.: Construction and convergence of difference schemes for a model elliptic equation with Dirac delta function coefficient. Lect. Notes Comput. Sci. **1988** (2001) 431–438
9. Jovanović, B.S., Popović, B.Z.: Convergence of a finite difference scheme for the third boundary–value problem for an elliptic equation with variable coefficients. Comp. Methods Appl. Math. **1** No 4 (2001) 356–366
10. Jovanović, B.S., Vulkov, L.G.: Operator's approach to the problems with concentrated factors. Lect. Notes Comput. Sci. **1988** (2001) 439–450
11. Jovanović, B.S., Vulkov, L.G.: On the convergence of finite difference schemes for the heat equation with concentrated capacity. Numer. Math. **89** No 4 (2001) 715–734
12. Jovanović, B.S., Vulkov, L.G.: On the convergence of finite difference schemes for hyperbolic equations with concentrated factors. SIAM J. Numer. Anal. **41** No 2 (2003) 516–538
13. Lazarov, R.D., Makarov, V.L., Samarskiĭ, A.A.: Applications of exact difference schemes for construction and studies of difference schemes on generalized solutions. Math. Sbornik **117** (1982) 469–480 (in Russian)
14. Lax, P.,Milgram, A.N.: Parabolic equations. Annals of Mathematics Studies, No 33, Princeton University Press, Princeton (1954) 167–190
15. Lykov, A.V.: Heat–mass transfer. Energiya, Moscow (1978) (in Russian).
16. Samarskiĭ, A.A.: Theory of Difference Schemes. Nauka, Moscow (1989) (in Russian).
17. Samarskiĭ, A.A., Lazarov, R.D., Makarov, V.L.: Difference schemes for differential equations with generalized solutions. Vysshaya Shkola, Moscow (1987) (in Russian).
18. Vabishchevich, P.N.: Iterative methods for solving convection–diffusion problem. Comp. Methods Appl. Math. **2** No 4 (2002) 410–444
19. Vulkov, L.: Application of Steklov–type eigenvalues problems to convergence of difference schemes for parabolic and hyperbolic equation with dynamical boundary conditions. Lect. Notes Comput. Sci. **1196** (1997) 557–564
20. Wloka, J.: Partial differential equations. Cambridge Univ. Press, Cambridge (1987)

The Finite Element Method for the Navier-Stokes Equations for a Viscous Heat Conducting Gas*

E.D. Karepova, A.V. Malyshev, V.V. Shaidurov, and G.I. Shchepanovskaya

Institute of Computational Modelling SB RAS, Krasnoyarsk, Russia
shidurov@icm.krasn.ru

Abstract. A boundary value problem for the Navier-Stokes equations for a viscous heat conducting gas in a finite computational domain is considered. The space approximation is constructed with the use of the Bubnov-Galerkin method combined with the method of lines.

This paper deals with the numerical solution of a boundary value problem for the Navier-Stokes equations for a viscous heat conducting gas. The space approximation of the two-dimensional Navier-Stokes problem by the finite element method is considered. Notice that a feature of the formulation of the problem used here is that boundary conditions on the boundary of a computational domain relate derivatives of velocity to pressure. These boundary conditions are natural for the variational (integral) formulation, i.e., they do not impose additional conditions on subspaces of trial and test functions as opposed to main boundary conditions (of the Dirichlet type). Moreover, they are "nonreflecting" since they do not distort propagation of local perturbations of velocities outside of the computational domain and have no influence on the values of velocities inside the domain.

To construct the space approximation, the Bubnov-Galerkin method combined with the method of lines is used. For a space of trial and test functions, a space of functions being piecewise bilinear on square meshes is used. For calculation of integrals over an elementary domain the quadrature formulae of the trapezoid method and of its two-dimensional analogue as the Cartesian product are applied.

As a result, we obtain a system of ordinary differential equations in time with respect to four vectors which consist of the values of density, velocities, and energy at the nodes of a square grid and depend on time.

1 The Formulation of the Problem

Let $\Omega = (0,1) \times (0,1)$ be a bounded (computational) domain in R^2 with the boundary Γ. Let also $(0, t_{fin})$ be the time interval. Consider the problem on a

* This work was supported by Russian Foundation of Basic Research (grant N 02-01-00523)

Z. Li et al. (Eds.): NAA 2004, LNCS 3401, pp. 56–65, 2005.

nonstationary flow of a viscous heat conducting gas in the following form. In the cylinder $(0, t_{fin}) \times \Omega$ we write four equations in unknowns ρ, u, v, e which differ from the standard ones by a linear combination on the equations (2) and (3) with (1) in order to simplify the variational formulation to be considered:

$$\frac{\partial \rho}{\partial t} + \frac{\partial}{\partial x}(\rho u) + \frac{\partial}{\partial y}(\rho v) = 0, \tag{1}$$

$$\left(\rho \frac{\partial u}{\partial t} + \frac{u}{2}\frac{\partial \rho}{\partial t}\right) + \left(\rho u \frac{\partial u}{\partial x} + \frac{u}{2}\frac{\partial}{\partial x}(\rho u)\right) + \left(\rho v \frac{\partial u}{\partial y} + \frac{u}{2}\frac{\partial}{\partial y}(\rho v)\right)$$

$$+ \frac{\partial P}{\partial x} - \frac{\partial \tau_{xx}}{\partial x} - \frac{\partial \tau_{xy}}{\partial y} = 0, \tag{2}$$

$$\left(\rho \frac{\partial v}{\partial t} + \frac{v}{2}\frac{\partial \rho}{\partial t}\right) + \left(\rho u \frac{\partial v}{\partial x} + \frac{v}{2}\frac{\partial}{\partial x}(\rho u)\right) + \left(\rho v \frac{\partial v}{\partial y} + \frac{v}{2}\frac{\partial}{\partial y}(\rho v)\right)$$

$$+ \frac{\partial P}{\partial y} - \frac{\partial \tau_{xy}}{\partial x} - \frac{\partial \tau_{yy}}{\partial y} = 0, \tag{3}$$

$$\frac{\partial}{\partial t}(\rho e) + \frac{\partial}{\partial x}(\rho e u) + \frac{\partial}{\partial y}(\rho e v) + P\left(\frac{\partial u}{\partial x} + \frac{\partial v}{\partial y}\right) = Q_t - \frac{\partial q_x}{\partial x} - \frac{\partial q_y}{\partial y} + \Phi. \tag{4}$$

Here we use the following notations: $\rho(t,x,y)$ is density; $e(t,x,y)$ is internal energy of unit mass; $u(t,x,y)$, $v(t,x,y)$ are components of the vector of velocity; $P(t,x,y)$ is pressure ; $\tau_{xx}, \tau_{xy}, \tau_{yy}$ are components of the stress tensor given by the formulae

$$\tau_{xx} = \frac{2}{3}\mu\left(2\frac{\partial u}{\partial x} - \frac{\partial v}{\partial y}\right), \quad \tau_{yy} = \frac{2}{3}\mu\left(2\frac{\partial v}{\partial y} - \frac{\partial u}{\partial x}\right), \quad \tau_{xy} = \mu\left(\frac{\partial u}{\partial y} + \frac{\partial v}{\partial x}\right); \tag{5}$$

$\mu(t,x,y) = \dfrac{1}{Re}\mu^*(t,x,y)$, μ^* is the dynamic coefficient of viscosity:

$$\mu^* = \left((\gamma - 1)\gamma \mathrm{M}_\infty^2\right)^\omega e^\omega, \qquad 0.76 \le \omega \le 0.9; \tag{6}$$

(q_x, q_y) are components of the vector of density of a heat flow given by the formulae:

$$q_x(t,x,y) = -\frac{\gamma}{\mathrm{Pr}}\mu\frac{\partial e}{\partial x}, \qquad q_y(t,x,y) = -\frac{\gamma}{\mathrm{Pr}}\mu\frac{\partial e}{\partial y}; \tag{7}$$

Re is the Reynolds number; Pr is the Prandtl number; M_∞ is the Mach number; γ is the gas constant.

The equation of state has the form:

$$P = (\gamma - 1)\rho e. \tag{8}$$

The function $Q_t(t,x,y)$ in the right-hand side of the equation (4) is given distribution of a source of energy.

The dissipative function Φ will be considered in one of the following forms:

$$\Phi = \mu\left(\frac{2}{3}\left(\frac{\partial u}{\partial x}\right)^2 + \frac{2}{3}\left(\frac{\partial v}{\partial y}\right)^2 + \left(\frac{\partial v}{\partial x} + \frac{\partial u}{\partial y}\right)^2 + \frac{2}{3}\left(\frac{\partial u}{\partial x} - \frac{\partial v}{\partial y}\right)^2\right) \tag{9}$$

$$= \mu\left(\frac{4}{3}\left(\left(\frac{\partial u}{\partial x}\right)^2 - \frac{\partial u}{\partial x}\frac{\partial v}{\partial y} + \left(\frac{\partial v}{\partial y}\right)^2\right) + \left(\frac{\partial v}{\partial x} + \frac{\partial u}{\partial y}\right)^2\right). \tag{10}$$

From (9) it follows that it is nonnegative.

Denote the vector of a unit outer normal to Γ at a point (x, y) by $\mathbf{n}(x, y) = (n_x(x, y), n_y(x, y))$.

To specify boundary conditions for the equation of continuity (1) we dwell on the case where a flow across Γ, defined by the vector $\mathbf{u} = (u, v)$, is directed outward Ω, i.e.,

$$\mathbf{u} \cdot \mathbf{n} \geq 0 \quad \text{on} \quad [0, t_{fin}] \times \Gamma. \tag{11}$$

In this case the characteristics of the equation (1) on the boundary $[0, t_{fin}] \times \Gamma$ are directed outward the domain $(0, t_{fin}) \times \Omega$ and in order for the problem to be well-posed *there is no need for boundary conditions for* ρ.

To close the problem on the boundary Γ of the computational domain Ω we consider the following boundary conditions which are natural in a variational sense:

$$\tau_{xx}\, n_x + \tau_{xy}\, n_y = P\, n_x - P_{ext}\, n_x, \quad \text{on} \quad (0, t_{fin}) \times \Gamma, \tag{12}$$

$$\tau_{yy}\, n_y + \tau_{xy}\, n_x = P\, n_y - P_{ext}\, n_y, \quad \text{on} \quad (0, t_{fin}) \times \Gamma, \tag{13}$$

where $P_{ext}(t, x, y)$ is given external pressure on the boundary of the computational domain. In inward boundary stream P_{ext} is known indeed and equals the pressure in unperturbed medium. In outward boundary stream we take P_{ext} as the value of pressure which is drifted along characteristic of equation (1) from the previous time level.

Boundary conditions of this type are natural for the variational formulation of the problem, i.e., they do not impose additional requirements on spaces of trial and test functions, as opposed to main boundary conditions (for example, when u and v are given on the boundary). Besides, from the computational point of view these boundary conditions are nonreflecting, i.e., they allow perturbations of the functions u and v to pass through the computational boundary Γ leaving their values inside the domain unaffected.

For the energy equation (4) we consider the Neumann boundary conditions:

$$\nabla e \cdot \mathbf{n} = 0 \quad \text{on} \quad \Gamma. \tag{14}$$

Initial conditions are taken in the form

$$\begin{aligned} &\rho(0, x, y) = \rho_0(x, y), \quad && u(0, x, y) = u_0(x, y), \\ &v(0, x, y) = v_0(x, y), \quad && e(0, x, y) = e_0(x, y) \quad \text{on} \quad \Omega. \end{aligned} \tag{15}$$

Notice that in linearization of the equations or their time approximations we will use different approximations of second-order terms like ρu. To distinguish between coefficients and main unknowns, we denote u and v in the coefficients of (2) – (4) by a and b respectively.

2 Space Discretization

In this section we construct the space discretization of the system (1) – (4) using the method of lines.

Before this we substitute unknown nonnegative function ρ by $\sigma = \sqrt{\rho}$ in equation (1) [4]:

$$\frac{\partial \sigma}{\partial t} + \frac{1}{2}\frac{\partial \sigma}{\partial x}u + \frac{1}{2}\frac{\partial (\sigma u)}{\partial x} + \frac{1}{2}\frac{\partial \sigma}{\partial y}v + \frac{1}{2}\frac{\partial (\sigma v)}{\partial y} = 0. \tag{16}$$

In the following it simplifies analysis due to change of trial space $L^1(\Omega)$ for ρ by Hilbert space $L^2(\Omega)$ for σ.

Along with the differential formulation, we will use integral identities which follow from it. To this end, we multiply the equations (16), (2) – (4) by an arbitrary function $w \in W_2^1(\Omega)$ and integrate them over Ω using integration by parts. Taking into account the boundary conditions (12) – (14), we arrive at the following identities:

$$\int_\Omega \left(\frac{\partial \sigma}{\partial t} w + \frac{1}{2} u \frac{\partial \sigma}{\partial x} w - \frac{1}{2} \sigma u \frac{\partial w}{\partial x} + \frac{1}{2} v \frac{\partial \sigma}{\partial y} w - \frac{1}{2} \sigma v \frac{\partial w}{\partial y} \right) d\Omega$$

$$+ \frac{1}{2} \int_\Gamma \sigma w \, \mathbf{u} \cdot \mathbf{n} \, d\Gamma = 0, \tag{17}$$

$$\int_\Omega \left(\rho \frac{\partial u}{\partial t} + \frac{u}{2} \frac{\partial \rho}{\partial t} \right) w \, d\Omega + \frac{1}{2} \int_\Omega \left(-\rho a u \frac{\partial w}{\partial x} + \rho a \frac{\partial u}{\partial x} w \right) d\Omega$$
$$+ \frac{1}{2} \int_\Omega \left(-\rho b u \frac{\partial w}{\partial y} + \rho b \frac{\partial u}{\partial y} w \right) d\Omega + \int_\Omega \left((\tau_{xx}) \frac{\partial w}{\partial x} + (\tau_{xy}) \frac{\partial w}{\partial y} \right) d\Omega$$

$$+ \frac{1}{2} \oint_\Gamma (\rho a u w \, dy - \rho b u w \, dx) = \int_\Omega P \frac{\partial w}{\partial x} \, d\Omega - \oint_\Gamma P_{ext} w \, dy, \tag{18}$$

$$\int_\Omega \left(\rho \frac{\partial v}{\partial t} + \frac{v}{2} \frac{\partial \rho}{\partial t} \right) w \, d\Omega + \frac{1}{2} \int_\Omega \left(-\rho a v \frac{\partial w}{\partial x} + \rho a \frac{\partial v}{\partial x} w \right) d\Omega$$

$$+ \frac{1}{2} \int_\Omega \left(-\rho b v \frac{\partial w}{\partial y} + \rho b \frac{\partial v}{\partial y} w \right) d\Omega + \int_\Omega \left((\tau_{xy}) \frac{\partial w}{\partial x} + (\tau_{yy}) \frac{\partial w}{\partial y} \right) d\Omega$$

$$+ \frac{1}{2} \oint_\Gamma (\rho a v w \, dy - \rho b v w \, dx) = \int_\Omega P \frac{\partial w}{\partial y} \, d\Omega + \oint_\Gamma P_{ext} w \, dx, \tag{19}$$

and

$$(f_v, w^h) = \int_{\Omega} P \frac{\partial w^h}{\partial y} \, d\Omega + \oint_{\Gamma} P_{ext} w^h \, dx, \tag{33}$$

respectively.

Find the function e^h:

$$e^h = \sum_{i,j=0}^{n} \alpha_{i,j}^{(e)}(t) \varphi_{ij}, \tag{34}$$

such that

$$a_e(e^h, w^h) = (f_e, w^h) \quad \forall \, w^h \in H^h \tag{35}$$

with the bilinear form

$$a_e^h(e^h, w^h) = \int_{\Omega} \Big(\frac{\partial}{\partial t}(\rho e) \, w - (e\rho a) \frac{\partial w}{\partial x} - (e\rho b) \frac{\partial w}{\partial y} \tag{36}$$

$$- q_x \frac{\partial w}{\partial x} - q_y \frac{\partial w}{\partial y} \Big) \, d\Omega + \oint_{\Gamma} (e\rho a \, w \, dy - e\rho b \, w \, dx)$$

and the linear form

$$(f_e, w^h) == \int_{\Omega} \left(Q_t - P \left(\frac{\partial u}{\partial x} + \frac{\partial v}{\partial y} \right) + \Phi \right) w \, d\Omega. \tag{37}$$

Since $\{\varphi_{i,j}\}_{i,j=0}^{n}$ is a basis in the space H^h, as the functions w^h in the Bubnov-Galerkin method it is sufficient to consider only the basis functions. Then with the equalities (24) – (25), (27) – (29), (34) – (35) we can associate the systems of equations

$$a_\sigma(\sigma^h, \varphi_{i,j}) = 0; \tag{38}$$
$$a_u(u^h, v^h, \varphi_{i,j}) = (f_u, \varphi_{i,j}), \tag{39}$$
$$a_v(u^h, v^h, \varphi_{i,j}) = (f_v, \varphi_{i,j}); \tag{40}$$
$$a_e(e^h, \varphi_{i,j}) = (f_e, \varphi_{i,j}) \quad i, j = 0, 1, 2, \dots, n. \tag{41}$$

The equalities (38) – (41) involve integrals which can not be calculated exactly in the general case. For their approximation we use the two-dimensional analogue of the trapezoid quadrature formula.

We introduce the vectors $\boldsymbol{\sigma}^h(t)$, $\mathbf{u}^h(t)$, $\mathbf{v}^h(t)$ and $\mathbf{e}^h(t)$ with the components $\sigma_{i,j}^h(t)$, $u_{ij}^h(t)$ $v_{ij}^h(t)$ and $e_{ij}^h(t)$, respectively. We also consider the right-hand side vectors $\mathbf{F}_u^h$, $\mathbf{F}_v^h$ and $\mathbf{F}_e^h$ with the components $(f_u(t), \varphi_{i,j})$, $(f_v(t), \varphi_{i,j})$ and $(f_e(t), \varphi_{i,j})$, respectively.

The approximate replacement of the integrals at each instant t results in linear operators M^h, M_{sqrt}^h (the operator of multiplication by $\sigma_{ij}^h(t)$), $A_\sigma^h(t)$, $A_u^h(t)$, $B_u^h(t)$, $A_v^h(t)$, $B_v^h(t)$, $A_e^h(t, \mathbf{u}^h, \mathbf{v}^h)$. When numbering the nodes of $\bar{\Omega}_h$ from zero to n^2, these operators become isomorphic matrices. We shall use lexicographic ordering. In this case M^h is a diagonal matrix with positive entries and the other matrices are five-diagonal.

Thus, for all $t \in (0, t_{fin})$ we obtain the systems of ordinary differential equations

$$M^h \frac{d\boldsymbol{\sigma}^h}{dt} + A_\sigma^h(t)\boldsymbol{\sigma}^h = 0, \tag{42}$$

$$M_{sqrt}^h \frac{d}{dt}\left(M_{sqrt}^h \mathbf{u}^h\right) + A_u^h(t)\mathbf{u}^h + B_u^h(t)\mathbf{v}^h = \mathbf{F}_u^h(t), \tag{43}$$

$$M_{sqrt}^h \frac{d}{dt}\left(M_{sqrt}^h \mathbf{v}^h\right) + B_v^h(t)\mathbf{u}^h + A_v^h(t)\mathbf{v}^h = \mathbf{F}_v^h(t), \tag{44}$$

$$\frac{d}{dt}\left(M^h \mathbf{e}^h\right) + A_e^h(t, \mathbf{u}^h, \mathbf{v}^h)\mathbf{e}^h = \mathbf{F}_e^h(t). \tag{45}$$

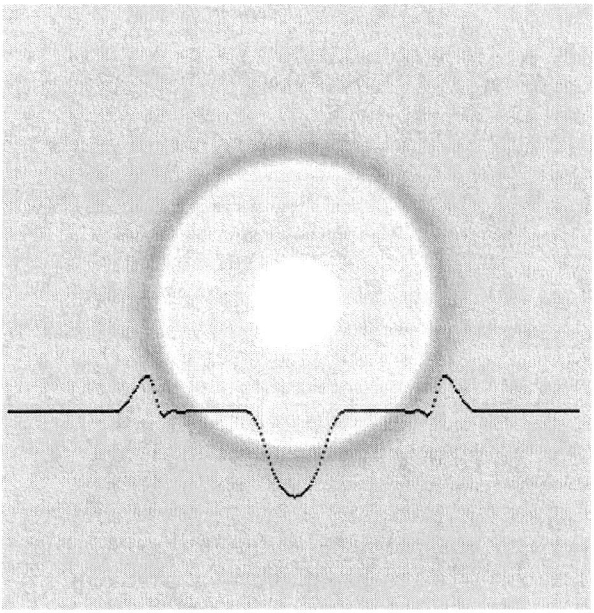

Fig. 1. Density ρ

Notice that the system (42) is written for the derivative and, assuming that $\sigma(t, x, y)$ is known at the stage where u, v, e are determined, we obtain three other systems written for the corresponding derivatives. We shall follow this idea of determination of u, v, e at each time level t once σ is determined in numerical methods as well.

Thus, we obtain a discrete analogue of the Navier-Stokes equations by the finite element method and the systems of ordinary differential equations (42) – (45). Some useful properties being discrete analogues of continuous balance relations, like conservation of mass and total energy, are proved for them. The natural boundary conditions (12) – (13) for the equation of motion and the Neumann condition (14) for the equation of energy are of crucial importance in conservation of balance of total energy.

In addition, some new methods for solving these systems of ordinary differential equations, which conserve basic balance relations, are compared with well-known ones.

Numerical experiments were performed, for example, for the problem (42) – (45) with the initial conditions

$$\rho(x,y,0) = 1, \qquad u(x,y,0) = v(x,y,0) = 0,$$
$$e(x,y,0) = f(0.5,0.5,x,y) + \frac{1}{\gamma\,(\gamma-1)\,M_\infty^2} \quad \text{on} \quad \Omega_h.$$

Here

$$f(A,B,x,y) = \begin{cases} 2\Big(R^2 - (x-A)^2 - (y-B)^2\Big)^8 \left(\frac{1}{R^2}\right)^8, \\ \qquad \text{if} \quad (x-A)^2 + (y-B)^2 \le R^2, \\ 0, \quad \text{otherwise;} \end{cases}$$

$R = 1/30$. Besides, in the numerical experiment we used the following values of the parameters:

$$\gamma = 1.4, \qquad Re = 10^4, \qquad M_\infty^2 = 16, \qquad Pr = 0.7, \qquad \omega = 1.$$

Taking into account the relation between temperature and internal energy:

$$T = e\left(\gamma\,(\gamma-1)\,M_\infty^2\right),$$

initial conditions for the equation of energy were taken so that temperature in the nonperturbed domain is equal to one.

The figure shows the behaviour of density ρ at the instant $t = t^*$ when pressure at the center becomes equal to that in the nonperturbed parts of the computational domain (the deeper is color, the greater is density). In this figure the graph of $\rho(x, 0.5, t^*)$ is shown as well.

References

1. Rannacher R. Finite element method for the incompressible Navier-Stokes equations. – In: Fundamental directions in mathematical fluid mechanics. (Galdi G., Heywood J.G., Rannacher R., eds.), Berlin: Birkhauser Verlag. – 2000.
2. Shaidurov V.V. and Shchepanovskaya G.I. Mathematical and numerical modeling of nonstationary propagation of a pulse of high-power energy in a viscous heat conducting gas. Part I. Mathematical formulation of the problem. – Krasnoyarsk: Institute of Computational Modeling of Russian Academy of Sciences. – 2003. – 50 pp. (Deposited in VINITI 24.10.03, 1860–B2003).

3. Karepova E.D. and Shaidurov V.V. The numerical solution of the Navier-Stokes equations for a viscous heat conducting gas. Part II. Space approximation by the finite element method. – Krasnoyarsk: Institute of Computational Modeling of Russian Academy of Sciences, 2004. – 70 pp. (Deposited in VINITI 13.01.04, 58–B2004).
4. Samarskii A.A., Vabishchevich P.N. Numerical methods for solving problems of convection-diffusion. – Moscow: Publishers of scientific and educational literature. – 2003.

Augmented Strategies for Interface and Irregular Domain Problems

Zhilin Li

Center for Research in Scientific Computation,
Department of Mathematics, North Carolina State University,
Raleigh, NC 27695-8205, USA
zhilin@math.ncsu.edu

Abstract. The augmented strategies for interface problems and problems defined on irregular domains are reviewed in this paper. There are at least two reasons to use augmented strategies. The first one is to get faster algorithms, particularly, to take advantages of existing fast solvers. The second reason is that, for some interface problems, an augmented approach may be the only way to derive an accurate algorithm. Using an augmented approach, one or several quantities of co-dimension one are introduced. The GMRES iterative method is often used to solve the augmented variable(s) that are only defined along the interface or the irregular boundary. Several examples of augmented methods are provided in this paper.

AMS Subject Classification: 65M06, 65M12,76T05

Keywords: interface problem, irregular domain, augmented strategy, discontinuous coefficient, coupled jump conditions, immersed interface method, GMRES method.

1 Introduction

When partial differential equations (PDEs) are used to model interface problems, the coefficients, the source terms, and the solution or/and its derivatives, may be discontinuous, or singular across the interface. It is difficult to solve such PDEs accurately. Various numerical methods have been developed for interface problems. We refer the readers to [14, 16, 20, 18] for an incomplete review of Cartesian grid methods for interface problems.

Among various methods, augmented strategies are relatively new. There are at least two reasons to use augmented strategies. The first one is to get faster algorithms, particularly to take advantages of existing fast solvers. The second reason is that, for some interface problems, an augmented approach may be the only way to derive an accurate algorithm. This will be illustrated through the augmented immersed interface method (IIM) for the incompressible Stokes equations in which the jump conditions for the pressure and the velocity are coupled together. The augmented techniques enable us to decouple the jump

Z. Li et al. (Eds.): NAA 2004, LNCS 3401, pp. 66–79, 2005.

conditions so that the governing equations can be discretized with second order accuracy using the immersed interface method.

For a problem defined on an irregular domain, using the so called fictitious domain method, we can transform the problem to an interface problem on a regular domain. Then a fast solver for the interface problem can be applied with appropriate choice of an augmented variable.

Using augmented strategies, some augmented variable, which can be a vector with several components, of co-dimension one will be introduced. Once we know the augmented variable, it is relatively easy to solve the original PDE assuming that the augmented variable is known. In a finite difference discretization, the approximate solution to the PDE with given augmented variable satisfies a system of linear finite difference equations, but it generally does not satisfy one or several interface relations or the boundary condition, which we call the augmented equation. The two equations, the solution to the original PDE, and the augmented equation, form a large linear system of equations with the solution to the original PDE and the augmented variable as unknowns. If we eliminate the solution to the original PDE from the large linear system of equations, we will get the Schur complement system for the augmented variable that is much smaller than that for the solution to the PDE. Therefore we can use the GMRES iterative method [28] to solve the Schur complement. In implementation, there is no need to explicitly form those coefficient matrices of either the large system or the Schur complement system. The matrix vector multiplication needed for the GMRES iteration includes mainly two steps: (1) solving the original PDE assuming the augmented variable is known; (2) finding the residual of the augmented equation, or the error of the interface conditions or the boundary conditions, given the computed solution and the augmented variables. In this paper, we explain this technique for various problems. While an augmented approach for an interface or irregular domain problem has some similarities with an integral equation, or boundary integral method, to find a source strength, the augmented methods have a few special features: (1) no Green function is needed, and therefore no need to evaluate singular integrals; (2) no need to set up the system of equations for the augmented variable explicitly; (3) applicable to general PDEs with or without source terms; (4) the process for interface problems does not depend on the boundary condition. On the other hand, we may have less knowledge about the condition number of the Schur complement system and how to apply pre-conditioning techniques.

2 The Augmented Approach for Elliptic Interface Problems with Piecewise Constant Coefficient

The first augmented approach may be the fast immersed interface method [17] for elliptic interface problems of the following

$$\nabla \cdot (\beta(x,y)\nabla u) = f(x,y), \quad x \in \Omega - \Gamma,$$
$$[u]\big|_\Gamma = w(s), \qquad [\beta u_n]\big|_\Gamma = v(s), \tag{1}$$

with a specified boundary condition on $\partial\Omega$, where $\Gamma(s)$ is an interface that divides the domain Ω into two sub-domains, Ω^+ and Ω^-, $\Omega = \Omega^+ \cup \Omega^- \cup \Gamma$, and $u_n = \nabla u \cdot \mathbf{n}$ is the normal derivative along the unit normal direction $\mathbf{n}$, s is an arc-length parameterization of Γ. We use $[\,\cdot\,]$ to represent the jump of a quantity across the interface Γ. We assume that $\beta(x,y)$ has a constant value in each sub-domain,

$$\beta(x,y) = \begin{cases} \beta^+ \text{ if } (x,y) \in \Omega^+, \\ \beta^- \text{ if } (x,y) \in \Omega^-. \end{cases} \tag{2}$$

If $\beta^+ = \beta^- = \beta$ is a constant, then we have a Poisson equations $\Delta u = f/\beta$ with the source distributions along the interface that corresponds to the jumps in the solution and the flux. The finite difference method obtained from the immersed interface method [14, 16, 18] yields the standard discrete Laplacian plus some correction terms to the right hand side. Therefore a fast Poisson solver, for example, the Fishpack [1], can be used to solve the discrete system of equations. If $\beta^+ \neq \beta^-$, we can not divide the coefficient β from the flux jump condition. The motivation is to introduce an augmented variable so that we can take advantage of the fast Poisson solver for the interface problem with only singular sources.

2.1 The Augmented Variable and Augmented Equation

There are more than one ways to introduce an augmented variable. Since the PDE (1) can be written as a Poisson equation in each interior domain after we divide β from the original equation excluding the interface Γ, it is natural to introduce $[u_n]$ as the *augmented variable.*

Consider the solution set $u_g(x,y)$ of the following problem as a functional of $g(s)$,

$$\begin{cases} \Delta u = \dfrac{f}{\beta^+}, \text{ if } x \in \Omega^+, \\[2ex] \Delta u = \dfrac{f}{\beta^-}, \text{ if } x \in \Omega^-, \end{cases} \tag{3}$$
$$[u]\Big|_\Gamma = w(s), \qquad [u_n]\Big|_\Gamma = g(s),$$

with the same boundary condition on $\partial\Omega$ as in the original problem. Let the solution of (1) be $u^*(x,y)$, and define

$$g^*(s) = [u_n^*](s) \tag{4}$$

along the interface Γ. Then $u^*(x,y)$ satisfies the PDE and the jump conditions in (3) with $g(s) \equiv g^*$. In other words, $u_{g^*}(x,y) \equiv u^*(x,y)$, and

$$\left[\beta \frac{\partial u_{g^*}}{\partial n}\right] = v(s) \tag{5}$$

is satisfied. The expression (5) is called the *augmented equation.* Therefore, solving the original problem (1) is equivalent to finding the corresponding $g^*(s)$ that satisfies (3) and (5). Note that $g^*(s)$ is only defined along the interface, so it is one dimension lower than that of $u(x,y)$. If we are given $[u_n] = g(s)$, then it is easy to solve (3) using the IIM since only correction terms at irregular grid points are needed to add to the right hand side of the finite difference equations. A fast Poisson solver such as the FFT then can be used. The cost in solving (3) is just a little more than that in solving a regular Poisson equation on the rectangle with a smooth solution.

2.2 The Discrete System of Equations in Matrix Vector Form

Let $\mathbf{G} = [G_1, G_2, \cdots, G_{N_b}]^T$ and $\mathbf{W} = [W_1, W_2, \cdots, W_{N_b}]^T$ be the discrete values of the jump conditions in (3) at a set of points $\mathbf{X}_1$, $\mathbf{X}_2$, $\cdots$, $\mathbf{X}_{N_b}$ on the interface. Given $\mathbf{W}$ and $\mathbf{G}$, using the immersed interface method [14, 16], the discrete form of (3) can be written as

$$L_h U_{ij} = \frac{f_{ij}}{\beta_{ij}} + C_{ij}, \tag{6}$$

where C_{ij} is zero at regular grid points where the interface does not cut through the finite difference stencil, and a correction term at irregular grid points, L_h is the discrete Laplacian operator. If a uniform Cartesian grid and the centered five-point stencil are used, then $L_h U_{ij}$ is

$$L_h U_{ij} = \frac{U_{i+1,j} + U_{i-1,j} + U_{i,j+1} + U_{i,j-1} - 4\,U_{i.j}}{h^2}.$$

The solution U_{ij} depends on G_k, W_k, $k = 1, 2, \ldots, N_b$ linearly. In a matrix and vector form we have

$$A\mathbf{U} + B\mathbf{G} = \mathbf{F} + B_1\mathbf{W} \stackrel{\text{def}}{=} \mathbf{F}_1, \tag{7}$$

where $\mathbf{U}$ and $\mathbf{F}$ are the vectors formed by $\{U_{ij}\}$ and $\{F_{ij}/\beta_{ij}\}$, $A\mathbf{U} = \mathbf{F}$ is the discrete linear system of finite difference equations for the Poisson equation when W_k, G_k, and $[\beta]$, are all zero. We define the residual of the second jump condition as

$$R(\mathbf{G}) = [\beta \mathbf{U}_n](\mathbf{G}) - \mathbf{V} = \beta^+ \mathbf{U}_n^+(\mathbf{G}) - \beta^- \mathbf{U}_n^-(\mathbf{G}) - \mathbf{V}. \tag{8}$$

We are interested in finding $\mathbf{G}^*$ such that $R(\mathbf{G}^*) = \mathbf{0}$, where the components of the vectors $\mathbf{U}_n^+$ and $\mathbf{U}_n^-$ are the discrete approximation of the normal derivative at $\{\mathbf{X}_k\}$, $1 \le k \le N_b$ from the each side of the interface. Once we know the solution $\mathbf{U}$ of the system (7) given $\mathbf{G}$, we can interpolate U_{ij} linearly to get $u_n^\pm$ at $\{\mathbf{X}_k\}$. The interpolation scheme is crucial to the success of the augmented algorithm. Since the interpolation is linear, we can write

$$\frac{\partial \mathbf{U}(\mathbf{G})^\pm}{\partial n} = E^\pm \mathbf{U} + T^\pm \mathbf{G} + P^\pm \mathbf{V}, \tag{9}$$

where E^+, E^-, T^+, T^-, P^+, and P^- are some sparse matrices determined by the interpolation scheme. The matrices are only used for theoretical purposes but not actually constructed in practice. We need to choose such a vector $\mathbf{G}$ that the second interface condition $\beta^+\mathbf{U}_n^+ - \beta^-\mathbf{U}_n^- = \mathbf{V}$ is satisfied along the interface Γ. Therefore we have the second matrix vector equation

$$E\mathbf{U} + T\mathbf{G} - P\mathbf{V} = 0, \tag{10}$$

where

$$E = \beta^+E^+ - \beta^-E^-, \quad T = \beta^+T^+ - \beta^-T^-, \quad P = I + \beta^-P^- - \beta^+P^+.$$

There is no need to generate the matrices E, T, P, explicitly. If we put the two matrix-vector equations (7) and (10) together, then we get

$$\begin{bmatrix} A & B \\ E & T \end{bmatrix} \begin{bmatrix} \mathbf{U} \\ \mathbf{G} \end{bmatrix} = \begin{bmatrix} \mathbf{F_1} \\ P\mathbf{V} \end{bmatrix}. \tag{11}$$

The solution $\mathbf{U}$ and $\mathbf{G}$ are the discrete forms of $u_{g^*}(x,y)$ and g^* , the solution of (3) which satisfies $[u_n] = g^*$ and $[\beta u_n(g^*)] = v$.

One can try to solve (11) directly but it is not very efficient generally. The main reason to use an augmented approach is to take advantage of fast Poisson solvers. Eliminating $\mathbf{U}$ from (11) gives a linear system for $\mathbf{G}$

$$(T - EA^{-1}B)\mathbf{G} = P\mathbf{V} - EA^{-1}\mathbf{F_1} \stackrel{\text{def}}{=} \mathbf{F_2}. \tag{12}$$

This is an $N_b \times N_b$ system for $\mathbf{G}$, a much smaller linear system compared to the one for $\mathbf{U}$. The coefficient matrix is the Schur complement for $\mathbf{G}$ in (11). Since we can not guarantee that the coefficient matrix of the Schur complement is symmetric positive definite, the GMRES iteration [28] method, is preferred. Different ways used to compute (9) will affect the condition number of (12) greatly. The detailed interpolation scheme for the problem (1) can be found in [17] along with the invertibility analysis for the coefficient matrix of (12).

The GMRES method only requires the matrix vector multiplication. We explain below how to evaluate the right hand side $\mathbf{F}_2$ of the Schur complement, and how to evaluate the matrix vector multiplication needed by the GMRES iteration. We can see why we do not need to form the coefficient matrix $T - EA^{-1}B$ explicitly. Note that from (7) we have $A^{-1}B\mathbf{G} = \mathbf{F_1} - \mathbf{U}$, and (12) can be written as

$$T\mathbf{G} - E\left(A^{-1}\mathbf{F_1} - \mathbf{U}\right) = \mathbf{F_2}, \tag{13}$$

whose residual is also defined as

$$R(\mathbf{G}) = T\mathbf{G} + E\mathbf{U} - EA^{-1}\mathbf{F_1} - \mathbf{F_2} = \mathbf{TG} + \mathbf{EU} - \mathbf{EU(0)} - \mathbf{F_2}. \tag{14}$$

Evaluation of the Right Hand Side of the Schur Complement. First we set $\mathbf{G} = 0$ and solve the system (3), or (7) in the discrete form, to get $\mathbf{U}(\mathbf{0})$ which is $A^{-1}\mathbf{F}_1$ from (7). Note that the residual of the Schur complement for $\mathbf{G} = 0$ is

$$\begin{aligned} R(\mathbf{0}) &= (T - EA^{-1}B)\,\mathbf{0} - \mathbf{F}_2 = -\mathbf{F}_2 \\ &= -\left(PV - EA^{-1}\mathbf{F}_1\right) = -\left(PV - E\mathbf{U}(\mathbf{0})\right) \\ &= \beta^+ \frac{\partial \mathbf{U}(\mathbf{0})^+}{\partial n} - \beta^- \frac{\partial \mathbf{U}(\mathbf{0})^-}{\partial n} - \mathbf{V}, \end{aligned} \tag{15}$$

which gives the right hand side of the Schur complement system with an opposite sign.

The Matrix-Vector Multiplication of the Schur Complement. The matrix-vector multiplication of the Schur complement system given $\mathbf{G}$ is obtained from the following two steps:

Step 1: Solve the coupled system (3), or (7) in the discrete form to get $\mathbf{U}(\mathbf{G})$.
Step 2: Interpolate $\mathbf{U}(\mathbf{G})$ to get $R(\mathbf{U}(\mathbf{G}))$ defined in (8). Then the matrix vector multiplication is

$$\begin{aligned} (T - EA^{-1}B)\mathbf{G} &= R(\mathbf{U}(\mathbf{G})) - R(\mathbf{U}(\mathbf{0})) \\ &= \beta^+ \frac{\partial \mathbf{U}(\mathbf{G})^+}{\partial n} - \beta^- \frac{\partial \mathbf{U}(\mathbf{G})^-}{\partial n} \\ &\quad - \left(\beta^+ \frac{\partial \mathbf{U}(\mathbf{0})^+}{\partial n} - \beta^- \frac{\partial \mathbf{U}(\mathbf{0})^-}{\partial n}\right). \end{aligned} \tag{16}$$

This is because

$$\begin{aligned} (T - EA^{-1}B)\mathbf{G} &= T\mathbf{G} - EA^{-1}B\mathbf{G} \\ &= T\mathbf{G} - E\left(A^{-1}\mathbf{F_1} - \mathbf{U}(\mathbf{G})\right) \\ &= T\mathbf{G} + E\mathbf{U}(\mathbf{G}) - \mathbf{F_2} - (\mathbf{T0} + \mathbf{EU}(0) - \mathbf{F_2}) \\ &= \beta^+ \frac{\partial \mathbf{U}(\mathbf{G})^+}{\partial n} - \beta^- \frac{\partial \mathbf{U}(\mathbf{G})^-}{\partial n} - \mathbf{V} \\ &\quad - \left(\beta^+ \frac{\partial \mathbf{U}(\mathbf{0})^+}{\partial n} - \beta^- \frac{\partial \mathbf{U}(\mathbf{0})^-}{\partial n} - \mathbf{V}\right) \end{aligned}$$

from the equalities $A\mathbf{U} + B\mathbf{G} = \mathbf{F}_1$, and $\mathbf{U}(\mathbf{0}) = A^{-1}\mathbf{F}_1$.

Now we can see that a matrix vector multiplication is equivalent to solving the system (3), or (7) in the discrete form, to get $\mathbf{U}$, and using an interpolation scheme to get the residual for the flux condition $[\beta\frac{\partial \mathbf{U}}{\partial n}]$ at $\mathbf{X}_k$, $k = 1, \cdots, N_b$, on the interface.

Once we know the right hand side of the system of linear equations, and the matrix-vector multiplication, the use of the GMRES iterative method is straightforward, see [28]. Numerical examples using the augmented approach to solve the PDE (1) in [17] show that the method is efficient with second order accuracy in the infinity norm. The number of iterations of the GMRES iteration is almost independent of the mesh sizes and the jump in the coefficient β. The three dimensional version of the augmented approach with applications can be found in [5].

3 The Augmented Method for Helmholtz/Poisson Equations on Irregular Domains

The idea of the augmented approach for the elliptic interface problems described in the previous section can be used with a little modifications to solve Helmholtz/Poisson equations

$$\begin{aligned} \Delta u - \lambda u &= f(\mathbf{x}), \quad \mathbf{x} \in \Omega, \\ q(u, u_n) &= 0, \quad \mathbf{x} \in \partial\Omega, \end{aligned} \tag{17}$$

defined on an irregular domain Ω (interior or exterior), where $q(u, u_n)$ is a prescribed linear boundary condition along the boundary $\partial\Omega$. The idea and the detailed algorithm description can be found in [10, 11, 22] with applications.

We embed the domain Ω into a rectangle (a cube in 3D) R and extend the PDE and the source term to the entire rectangle/cube R

$$\begin{aligned} &\Delta u - \lambda u = \begin{cases} f, \text{ if } & \mathbf{x} \in \Omega, \\ 0, \text{ if } & \mathbf{x} \in R - \Omega, \end{cases} \qquad\qquad q(u, u_n)|_{\partial\Omega} = 0, \\ &\begin{cases} [u] \;\; = g, & \text{on} \quad \partial\Omega, \\ [u_n] = 0, & \text{on} \quad \partial\Omega, \end{cases} \quad \text{or} \quad \begin{cases} [u] \;\; = 0, & \text{on} \quad \partial\Omega, \\ [u_n] = g, & \text{on} \quad \partial\Omega. \end{cases} \end{aligned} \tag{18}$$

Again, the solution u is a functional of g. We determine $g(s)$ such that the solution $u(g)$ satisfies the boundary condition $q(u(g), u_n(g)) = 0$ along the boundary $\partial\Omega$ which is treated as an interface in the new settings. Given g, we can use the IIM to solve $u(g)$ with one call to a fast Poisson solver. This corresponds to the first matrix-vector equation in (11), or (7) in the discrete form. The second matrix-vector equation is the boundary condition $q(u(\mathbf{G}), u_n(\mathbf{G})) = 0$ obtained using the least squares interpolation scheme at a set of selected points on the boundary $\partial\Omega$. The interpolation scheme should be third order accurate for the solution u, and second order accurate for the normal derivative u_n, and so on.

For an interior problem, we also need an artificial boundary condition along the fictitious domain ∂R. We usually use a homogeneous Dirichlet boundary condition. The distance between the boundary of the fictitious domain ∂R and the real boundary $\partial\Omega$ is often chosen between 4 to 10 grid sizes.

3.1 A Numerical Example of the Augmented Approach for an Exterior Poisson Equation

We provide one example with different boundary conditions to show the efficiency of the augmented approach. We want to show second order accuracy of the solution, and more importantly also, show that the number of iterations is nearly independent of the mesh size except for a factor of $\log h$.

We construct an exact solution:

$$\begin{aligned} u(x,y) &= -\frac{1}{2}\log r + r^2, \quad r = \sqrt{x^2+y^2}, \\ f(x,y) &= 4, \end{aligned} \tag{19}$$

of an exterior problem. The boundary $\partial\Omega$ is the unit rectangle R: $-1 \le x, y, \le 1$ excluding the ellipse Ω

$$\frac{x^2}{a^2} + \frac{y^2}{b^2} = 1.$$

Case 1: The Dirichlet boundary condition on ∂R and the normal derivative boundary condition on $\partial\Omega$ are given using the exact solution. The first part of Table 1 shows the grid refinement analysis and other information with $a = 0.5$ and $b = 0.4$, while the second part of the table is for $a = 0.5$ and $b = 0.15$, a skinny ellipse with large curvature at two tips.

In Table 1, n is the number of grid lines in the x- and y- directions; e is the error of the computed solution in the maximum norm; n_1 is the total number of irregular grid points in R; n_2 is the number of irregular grid points from $R - \Omega$ side, which is also the dimension of the Schur complement system; k is the number of iterations of the GMRES method which is also the number of the fast Poisson solver called, and r is the ratio of the two consecutive errors.

Table 1. The grid refinement analysis with $a = 0.5$ $b = 0.4$ and $b = 0.15$. Second order convergence is observed. The number of iterations is almost independent of the mesh size

n	a	b	e	r	n_1	n_2	k
40	0.5	0.4	$5.7116\ 10^{-4}$		100	52	16
80	0.5	0.4	$1.4591\ 10^{-4}$	3.9146	204	104	17
160	0.5	0.4	$3.5236\ 10^{-5}$	4.1408	412	208	19
320	0.5	0.4	$8.1638\ 10^{-6}$	4.3161	820	412	21

n	a	b	e	r	n_1	n_2	k
40	0.5	0.15	$4.4820\ 10^{-3}$		68	36	13
80	0.5	0.15	$1.1466\ 10^{-3}$	3.9089	132	68	15
160	0.5	0.15	$2.6159\ 10^{-4}$	4.3832	68	136	17
320	0.5	0.15	$6.7733\ 10^{-5}$	3.8621	68	268	20

Table 2. The grid refinement analysis for pure Neumann type boundary condition

n	a	b	e	r	n_1	n_2	k
40	0.5	0.15	$4.5064\ 10^{-3}$		84	44	14
80	0.5	0.15	$1.2529\ 10^{-3}$	3.5967	164	84	17
160	0.5	0.15	$3.3597\ 10^{-4}$	3.7292	332	168	19
320	0.5	0.15	$7.9409\ 10^{-5}$	4.2309	668	336	21

We observe a second order of convergence in the maximum norm. The number of iterations for the GMRES iteration is almost independent of the mesh size except of a factor $\log h$. Note that the error may not be reduced exactly by a factor of four but rather fluctuate as explained in [17].

Case 2: The normal derivative $\partial u/\partial n$ is prescribed on both ∂R and $\partial \Omega$ using the exact solution. In this case, the solution is not unique and can differ by a constant. To get a unique solution, we specify the solution at one corner using the exact solution. In this way, we can measure the error of the computed solution. Table 2 is the result of the grid refinement analysis. We have similar results as we analyzed above.

The Fortran subroutines for interior/exterior Helmholtz/Poisson equations on irregular domains are available to the public through the anonymous ftp site [19]. There are other fast elliptic solvers for elliptic PDEs defined on irregular domains using embedding or fictitious domain techniques, see [6, 7, 13, 23, 26, 27] for an incomplete list.

4 The Augmented Technique for Biharmonic Equations on Irregular Domains

Biharmonic equations arise in many applications. Classical examples can be found in elasticity, fluid mechanics, and many other areas. The augmented technique for biharmonic equations on irregular domains have been developed in [2, 3].

Consider a biharmonic equation defined on an irregular domain Ω

$$\begin{aligned} \Delta^2 u(x,y) &= f(x,y), & (x,y) &\in \Omega, \\ u(x,y) &= g_1(x,y), & (x,y) &\in \partial\Omega, \\ u_n(x,y) &= g_2(x,y), & (x,y) &\in \partial\Omega, \end{aligned} \tag{20}$$

where

$$\Delta^2 \equiv \nabla^4 = \frac{\partial^4}{\partial x^4} + 2\frac{\partial^4}{\partial x^2 \partial y^2} + \frac{\partial^4}{\partial y^4}. \tag{21}$$

There are limited publications on finite difference methods for biharmonic equations on irregular domains, even fewer with convincing numerical examples.

Among a few finite difference methods for biharmonic equations on irregular domains, the remarkable ones are the fast algorithms based on integral equations and/or the fast multipole method [9, 23, 24]. These methods are most effective for homogeneous source term ($f(x,y) = 0$) and some special boundary conditions. These methods probably still can be applied with some extra efforts for non-homogeneous source terms and the essential boundary condition in (20). The implementations of these methods, especially when they are coupled with the fast multipole method, however, are not trivial.

The augmented technique for Helmholtz/Poisson equations can be easily extended to biharmonic equations on irregular domains based on the following decoupling

$$\begin{cases} \Delta v(x,y) = f(x,y), & (x,y) \in \Omega, \\ v(x,y) = g(x,y), & (x,y) \in \partial\Omega; \end{cases}$$
$$\begin{cases} \Delta u(x,y) = v(x,y), & (x,y) \in \Omega, \\ u(x,y) = g_1(x,y), & (x,y) \in \partial\Omega. \end{cases} \tag{22}$$

Here $g(x,y)$ is the *augmented variable.* The augmented equation is the boundary condition

$$\frac{\partial u(x,y)}{\partial n} = g_2(x,y), \qquad (x,y) \in \partial\Omega. \tag{23}$$

To get the solution $u(x,y)$ given a guess g, we need to solve the two Poisson equations in (22). This can be done using the augmented method for Poisson equations discussed in the previous section. This step corresponds to the first matrix-vector equation in (11) or (7) in the discrete form. Note that the solution $\mathbf{U}$ includes two grid functions $\{U_{ij}\}$ and $\{V_{ij}\}$. The second matrix-vector equation is to set the residual $R(g) = \frac{\partial u(g)}{\partial n} - g_2$ from the computed $u(g)$ to zero. Since the condition number for a bi-harmonic equation is much large than the one for a Poisson equation, it is crucial to choose the interpolation stencil to approximate the boundary condition. The key is to select the interpolation stencil in a cone whose axis is close to the normal direction, see [2, 3] for the details.

5 The Augmented Approach for Stokes Equations with Discontinuous Viscosity and Singular Forces

Solving Stokes equations with discontinuous viscosity is one example that the augmented approach seems to be the only way to derive an efficient method because the jump conditions are coupled together.

Consider the following two-dimensional stationary incompressible Stokes equations

$$\nabla p = \nabla \cdot \mu \left(\nabla \mathbf{u} + (\nabla \mathbf{u})^T\right) + \mathbf{g} + \int_\Gamma \mathbf{f}(s)\, \delta_2(\mathbf{x} - \mathbf{X}(s)) ds, \quad \mathbf{x} \in \Omega, \tag{24}$$

$$\nabla \cdot \mathbf{u} = \mathbf{0}, \quad \mathbf{x} \in \Omega, \tag{25}$$

equation in (11), or (7) in discrete form. The compatibility condition for the two augmented variables are the two *augmented equations* in (34) (which means that the velocity is continuous across the interface). The second matrix-vector equation is to set the residual $R(\mathbf{q}) = ([u(\mathbf{q})/\mu], [v(\mathbf{q})/\mu])$ to zero. The detailed interpolation scheme is explained in [21].

Conclusions and Acknowledgments

In this paper, we give a review of the augmented approaches for interface problems and problems defined on irregular domains. Using an augmented approach, one or several augmented variables are introduced along a co-dimensional interface or boundary. When the augmented variable(s) is known, we can solve the governing PDE efficiently. In the discrete case, this gives a system of equations for the solution with given augmented variable(s). However, the solution that depends the augmented variable(s) usually do not satisfy all the interface relations or the boundary condition. The discrete interface relation or the boundary condition forms the second linear system of equations for the augmented variable whose dimension is much smaller than that of the solution to the PDE. Therefore we can use the GMRES iterative method to solve the Schur complement system for the augmented variable(s). In many instances, the GMRES method converges quickly, although it is still an open question how to preconditioning the system of linear equations without explicitly form the coefficient matrix.

The author was partially supported by USA-NSF grants DMS-0412654 and DMS-0201094; and by a USA-ARO grant 43751-MA.

References

1. J. Adams, P. Swarztrauber, and R. Sweet. Fishpack. http://www.netlib.org/fishpack/.
2. G. Chen. Immersed interface method for biharmonic equations defined on irregular domains and its application to the Stokes flow, Ph.D thesis. North Carolina State University, 2003.
3. G. Chen, Z. Li, and P. Lin. A fast finite difference method for biharmonic equations on irregular domains. CRSC-TR04-09, North Carolina State University, 2004.
4. R. Cortez. The method of regularized Stokeslets. *SIAM J. Sci. Comput.*, 23:1204–1225, 2001.
5. S. Deng, K. Ito, and Z. Li. Three dimensional elliptic solvers for interface problems and applications. *J. Comput. Phys.*, 184:215–243, 2003.
6. M. Dumett and J. Keener. A numerical method for solving anisotropic elliptic boundary value problems on an irregular domain in 2D. *SIAM J. Sci. Comput.*, in press, 2002.
7. A. L. Fogelson and J. P. Keener. Immersed interface methods for Neumann and related problems in two and three dimensions. *SIAM J. Sci. Comput.*, 22:1630–1684, 2000.
8. A. L. Fogelson and C. S. Peskin. Numerical solution of the three dimensional Stokes equations in the presence of suspended particles. In *Proc. SIAM Conf. Multi-phase Flow.* SIAM, June 1986.

9. A. Greenbaum, L. Greengard, and Anita Mayo. On the numerical-solution of the biharmonic equation in the plane. *PHYSICA D*, 60:216–225, 1992.
10. T. Hou, Z. Li, S. Osher, and H. Zhao. A hybrid method for moving interface problems with application to the Hele-Shaw flow. *J. Comput. Phys.*, 134:236–252, 1997.
11. J. Hunter, Z. Li, and H. Zhao. Autophobic spreading of drops,. *J. Comput. Phys.*, 183:335–366, 2002.
12. K. Ito and Z. Li. Interface relations for Stokes equations with discontinuous viscosity and singular sources. preprint, 2004.
13. H. Johansen and P. Colella. A Cartesian grid embedded boundary method for Poisson equations on irregular domains. *J. Comput. Phys.*, 147:60–85, 1998.
14. R. J. LeVeque and Z. Li. The immersed interface method for elliptic equations with discontinuous coefficients and singular sources. *SIAM J. Numer. Anal.*, 31:1019–1044, 1994.
15. R. J. LeVeque and Z. Li. Immersed interface method for Stokes flow with elastic boundaries or surface tension. *SIAM J. Sci. Comput.*, 18:709–735, 1997.
16. Z. Li. *The Immersed Interface Method — A Numerical Approach for Partial Differential Equations with Interfaces.* PhD thesis, University of Washington, 1994.
17. Z. Li. A fast iterative algorithm for elliptic interface problems. *SIAM J. Numer. Anal.*, 35:230–254, 1998.
18. Z. Li. An overview of the immersed interface method and its applications. *Taiwanese J. Mathematics*, 7:1–49, 2003.
19. Z. Li. IIMPACK, a collection of fortran codes for interface problems. Anonymous ftp at ftp.ncsu.edu under the directory: /pub/math/zhilin/Package, last updated: 2001.
20. Z. Li and K. Ito. Maximum principle preserving schemes for interface problems with discontinuous coefficients. *SIAM J. Sci. Comput.*, 23:1225–1242, 2001.
21. Z. Li, K. Ito, and M-C. Lai. An augmented approach for stokes equations with discontinuous viscosity and singular forces. NCSU-CRSC Tech. Report: CRSC-TR04-23, North Carolina State Univeristy, 2004.
22. Z. Li, H. Zhao, and H. Gao. A numerical study of electro-migration voiding by evolving level set functions on a fixed cartesian grid. *J. Comput. Phys.*, 152:281–304, 1999.
23. A. Mayo. The fast solution of Poisson's and the biharmonic equations on irregular regions. *SIAM J. Numer. Anal.*, 21:285–299, 1984.
24. A. Mayo and A. Greenbaum. Fast parallel iterative solution of Poisson's and the biharmonic equations on irregular regions. *SIAM J. Sci. Stat. Comput.*, 13:101–118, 1992.
25. A. Mayo and C. S. Peskin. An implicit numerical method for fluid dynamics problems with immersed elastic boundaries. *Contemp. Math.*, 141:261–277, 1991.
26. A. McKenney, L. Greengard, and Anita Mayo. A fast poisson solver for complex geometries. *J. Comput. Phys.*, 118, 1995.
27. W. Proskurowski and O. Widlund. On the numerical solution of Helmholtz's equation by the capacitance matrix method. *Math. Comp.*, 30:433–468, 1976.
28. Y. Saad. GMRES: A generalized minimal residual algorithm for solving nonsymmetric linear systems. *SIAM J. Sci. Stat. Comput.*, 7:856–869, 1986.
29. C. Tu and C. S. Peskin. Stability and instability in the computation of flows with moving immersed boundaries: a comparison of three methods. *SIAM J. Sci. Stat. Comput.*, 13:1361–1376, 1992.

Accuracy Estimates of Difference Schemes for Quasi-Linear Elliptic Equations with Variable Coefficients Taking into Account Boundary Effect

Volodymyr Makarov[1] and Lyubomyr Demkiv[2]

[1] Institute of Mathematics, National Academy of Sciences of Ukraine,
3 Tereschenkivska St., 01601, Kyiv, Ukraine
makarov@imath.kiev.ua

[2] National University "Lvivska Polytechnica", 12 S.Bandery str.,
79013, Lviv, Ukraine
demkivl@ukrpost.net

Abstract. While solving the elliptic equations in the canonical domain with the Dirichlet boundary conditions by the grid method, it is obviously, that boundary conditions are satisfied precisely. Therefore it is necessary to expect, that close to the domain boundary the accuracy of the corresponding difference scheme should be higher, than in the middle of the domain. The quantitative estimate of this boundary effect first was announced without proves in 1989 in the Reports of the Bulgarian Academy of sciences by the first author. There accuracy of the difference schemes for two-dimensional elliptic equation with variable coefficients in the divergent form has been investigated.

In this paper 'weight' a priori estimates, taking into account boundary effect, for traditional difference schemes, which approximate, with the second order, first boundary problem for quasi-linear elliptic type equation, which main part has a not divergent form, have been obtained.

The paper ends with numerical experiments, which testify to unimprovement, by the order, of the received 'weight' estimates.

1 Introduction

At solving the elliptic equations in canonical domains with Dirichlet boundary conditions by the method of grid, it is obvious, that boundary conditions are satisfied exactly. That is why one should expect that near the boundary of the domain accuracy of the corresponding difference scheme will be higher then in the middle of the domain. The quantitative estimate of this fact first has been announced without proof in 1989 by the first author in Reports of the Bulgarian academy of sciences [1] at investigating the accuracy of the difference schemes for two-dimensional elliptic equation with variable coefficients in divergent form. These ideas for the parabolic equations have received the subsequent development in papers [2], [3]. In this work weight a priori estimates for traditional

Z. Li et al. (Eds.): NAA 2004, LNCS 3401, pp. 80–90, 2005.

difference schemes which approximate the boundary problem

$$\sum_{i,j=1}^{2} a_{ij}(x) \frac{\partial^2 u}{\partial x_i \partial x_j} = f_0(x,u) - f(x), x = (x_1, x_2) \in \Omega \tag{1}$$

$$u(x) = 0, x \in \Gamma$$

with the second order taking into account boundary effect have been obtained, where Ω is a unit square with a boundary Γ , $a_{12}(x) = a_{21}(x)$

$$\nu \sum_{i=1}^{2} \xi_i^2 \leq \sum_{i,j=1}^{2} a_{ij}(x) \xi_i \xi_j \leq \mu \sum_{i=1}^{2} \xi_i^2, (\nu > 0, \mu < \infty), \tag{2}$$

$$\forall x \in \bar{\Omega}, \forall \xi_1, \xi_2 \in R^1,$$

$$f_0(x,0) = 0, |f_0(x,u) - f_0(x,v)| \leq L|u - v|, \forall x \in \bar{\Omega}, u, v \in R^1. \tag{3}$$

here $a_{ij}(x), i,j = \overline{1,2}$, $f_0(x), f(x)$ are sufficiently smooth functions.

The paper ends with numerical experiments, which testify to unimprovement of the obtained weight a priori estimates by the order.

2 Second Main Inequality for Difference Operator of the Second Order. An Estimate of the Green Difference Function

We consider the following linear Dirichlet problem

$$\sum_{i,j=1}^{2} a_{ij}(x) \frac{\partial^2 u}{\partial x_i \partial x_j} = F(x), x \in \Omega, \tag{4}$$

$$u(x) = 0, x \in \Gamma,$$

which corresponds to problem (1) with given function $F(x)$.

This case occupies special place among the multi-dimensional elliptic problems. For it with the help of Bernschtein method (see [5]) it is possible to prove the second main inequality for elliptic operators. It is possible to obtain an corresponding estimate for difference operator which approximate the difference operator of the problem (4). Similar result has been obtained by Djakonov [6], but in different form than that is needed and in another way. On the square Ω, we introduce a grid

$$\omega = \omega_1 \times \omega_2,$$

$$\omega_\alpha = \left\{ x_{i_\alpha} = i_\alpha h_\alpha : i_\alpha = \overline{1, N_\alpha - 1}, h_\alpha = \frac{1}{N_\alpha} \right\}, \alpha = 1, 2$$

a boundary of which we denote as γ . On the grid ω we put the problem (4) in conformity to the following difference scheme

$$Ay \equiv \sum_{i,j=1}^{2} a_{ij}(x)\, y_{\bar{x}_i x_j}(x) = F(x), x \in \omega, \tag{5}$$

$$y(x) = 0, x \in \gamma.$$

Here nonindex notation has been used for difference derivatives (see [7])

$$y_{x_1}(x) = \frac{y(x_1 + h_1, x_2) - y(x_1, x_2)}{h_1},$$

$$y_{\bar{x}_1}(x) = \frac{1}{h_1}\left[y(x_1, x_2) - y(x_1 - h_1, x_2)\right],$$

$$y_{\bar{x}_1 x_1}(x) = \frac{1}{h_1^2}\left[y(x_1 + h_1, x_2) - 2y(x_1, x_2) + y(x_1 - h_1, x_2)\right].$$

Correspondingly, all other difference derivatives

$$y_{\bar{x}_1 x_2}(x), y_{x_1 \bar{x}_2}(x), y_{\bar{x}_2 x_2}(x)$$

have been defined. After simple transformations from an equality (5) it is possible to obtain

$$\frac{a_{11}(x)}{a_{22}(x)} y_{\bar{x}_1 x_1}^2(x) + \frac{a_{12}(x)}{a_{22}(x)} y_{\bar{x}_1 x_1}(x)\left[y_{\bar{x}_1 x_2}(x) + y_{x_1 \bar{x}_2}(x)\right] +$$

$$+\frac{1}{4}\left[y_{\bar{x}_1 x_2}(x) + y_{x_1 \bar{x}_2}(x)\right]^2 = F(x)\frac{y_{\bar{x}_1 x_1}(x)}{a_{22}(x)} + \frac{1}{4}\left[y_{\bar{x}_1 x_2}(x) + y_{x_1 \bar{x}_2}(x)\right]^2 -$$

$$-y_{\bar{x}_1 x_1}(x)\, y_{\bar{x}_2 x_2}(x).$$

Hence, taking into consideration (2), we obtain

$$\frac{\nu}{\mu}\left[y_{\bar{x}_1 x_1}^2(x) + \frac{1}{4}\left(y_{\bar{x}_1 x_1}(x) + y_{x_1 \bar{x}_2}(x)\right)^2\right] \leq$$

$$\leq \frac{1}{2}\left(\sqrt{\frac{\nu}{\mu}} y_{\bar{x}_1 x_1}(x)\right)^2 + \frac{1}{2}\frac{F^2(x)}{a_{22}^2(x)}\frac{\mu}{\nu} +$$

$$+\frac{1}{4}\left[y_{\bar{x}_1 x_2}(x) + y_{x_1 \bar{x}_2}(x)\right]^2 - y_{\bar{x}_1 x_1}(x)\, y_{\bar{x}_2 x_2}(x)$$

or

$$\frac{\nu}{\mu}\left[\frac{1}{2} y_{\bar{x}_1 x_1}^2(x) + \frac{1}{4}\left(y_{\bar{x}_1 x_1}(x) + y_{x_1 \bar{x}_2}(x)\right)^2\right] \leq \tag{6}$$

$$\leq \frac{\mu}{2\nu^3} F^2(x) + \frac{1}{4}\left[y_{\bar{x}_1 x_2}(x) + y_{x_1 \bar{x}_2}(x)\right]^2 - y_{\bar{x}_1 x_1}(x)\, y_{\bar{x}_2 x_2}(x)$$

Correspondingly, we obtain

$$\frac{\nu}{\mu}\left[\frac{1}{4}\left(y_{\bar{x}_1x_1}(x)+y_{x_1\bar{x}_2}(x)\right)^2+\frac{1}{2}y^2_{\bar{x}_2x_2}(x)\right] \tag{7}$$
$$\leq \frac{\mu}{2\nu^3}F^2(x)+\frac{1}{4}\left[y_{\bar{x}_1x_2}(x)+y_{x_1\bar{x}_2}(x)\right]^2-y_{\bar{x}_1x_1}(x)\,y_{\bar{x}_2x_2}(x)$$

From (6), (7), taking into consideration (5), we obtain

$$\|Ay\|^2=\left\|\sum_{i,j=1}^{2}a_{ij}(x)\,y_{\bar{x}_ix_j}(x)\right\|^2\geq$$
$$\frac{\nu^4}{2\mu^2}\left\{\|y_{\bar{x}_1x_1}\|^2+\|y_{\bar{x}_1x_2}+y_{x_1\bar{x}_2}\|^2+\|y_{\bar{x}_2x_2}\|^2\right\} \tag{8}$$
$$\geq\frac{\nu^4}{\mu^2}\left(y_{\bar{x}_1x_1},y_{\bar{x}_2x_2}\right)=\frac{\nu^4}{\mu^2}\|B^*y\|^2_*.$$

Let us explain notations we use in (8). Let us introduce space $\overset{0}{H}_h$ of grid functions, defined on the grid ω , vanishing on γ . Scalar product in this space we define as

$$(y,v)=\sum_{\xi\in\omega}h_1h_2y(\xi)\,v(\xi),$$

and a norm, generated by it, as

$$\|y\|=(y,y)^{1/2}.$$

An operator B^* is a linear grid operator which acts from $\overset{0}{H}_h$ into H_h^* by one of the following formulas

a) $B^*y=-y_{\bar{x}_1\bar{x}_2},\forall y\in\overset{0}{H}_h,$

b) $B^*y=-y_{x_1x_2},\forall y\in\overset{0}{H}_h,$

c) $B^*y=-y_{\bar{x}_1x_2},\forall y\in\overset{0}{H}_h,$

d) $B^*y=-y_{x_1\bar{x}_2},\forall y\in\overset{0}{H}_h.$

Here H_h^* is Hilbert space of grid functions, defined on the grid $\tilde{\omega}$, with scalar product $(y,v)_*=\sum_{x\in\tilde{\omega}}h_1h_2y(x)\,v(x)$ and corresponding generating norm $\|y\|_*^2=(y,y)_*$, where

$$\tilde{\omega}=\begin{cases}\omega\cup\gamma_{+1}\cup\gamma_{+2} & \text{in the case a)},\\ \omega\cup\gamma_{-1}\cup\gamma_{-2} & \text{in the case b)},\\ \omega\cup\gamma_{+1}\cup\gamma_{-2} & \text{in the case c)},\\ \omega\cup\gamma_{-1}\cup\gamma_{+2} & \text{in the case d)},\end{cases}$$

$\gamma_{+1}\,(\gamma_{-1})$ is right (left) vertical boundary of the domain ω , $\gamma_{+2}\,(\gamma_{-2})$ is a top (bottom) horizontal boundary of this domain.

Conjugate operator $B : H_h^* \to \overset{0}{H}_h$ to operator B^* is defined as follows

$$(B^*y, w)_* = (y, Bw), \quad \forall y \in \overset{0}{H}_h, \forall w \in H_h^*, \tag{9}$$

where

$$By = \begin{cases} -y_{x_1x_2} & \text{in the case a),} \\ -y_{\bar{x}_1\bar{x}_2} & \text{in the case b),} \\ -y_{x_1\bar{x}_2} & \text{in the case c),} \\ -y_{\bar{x}_1x_2} & \text{in the case d),} \end{cases} \quad \forall y \in H_h^* \tag{10}$$

Further we will need a statement b) of the main lemma from [4] in a strengthened version.

Lemma 1. *Let A be linear operator, which acts in Hilbert space H , B be linear operator, which acts from $H^* \supseteq H$ into H . Then, if there exists A^{-1} and $\|B^*v\|_* \le \gamma \|Av\|$, $\forall v \in H$, where B^* is an operator, conjugated to B , and $\|\cdot\|_*$ is a norm in the space H^* , which is induced by the scalar product $(\cdot,\cdot)_*$ of the space H^*, then*

$$\|A^{-1}B\| \le \gamma \tag{11}$$

Proof. We proceed from the relation

$$\|u\| = \sup_{\substack{v \ne 0 \\ v \in H}} \frac{(u,v)}{\|v\|}, \quad \forall u \in H. \tag{12}$$

We place

$$u = A^{-1}\varphi, \quad v = Ay,$$

into (12), then we will have

$$\|A^{-1}\varphi\| = \sup_{\substack{y \ne 0 \\ y \in H}} \frac{(\varphi, y)}{\|Ay\|}.$$

Let's put $\varphi = B\psi$ in this equality. Making use of the lemma condition and Cauchy-Bunyakovsky inequality, we obtain

$$\|A^{-1}B\psi\| = \sup_{\substack{y \ne 0 \\ y \in H}} \frac{(B\psi, y)}{\|Ay\|} = \sup_{\substack{y \ne 0 \\ y \in H}} \frac{(\psi, B^*y)_*}{\|Ay\|} \le \sup_{\substack{y \ne 0 \\ y \in H}} \frac{\|\psi\|_* \|B^*y\|_*}{\|Ay\|} \le \gamma \|\psi\|_*,$$

$$\forall \psi \in H^*,$$

which proves (11).

For operators A, B all conditions of the statement b) of the main lemma from [4], p. 54 are satisfied. Consequently

$$\left\|A^{-1}B\right\| \le \frac{\mu}{\nu^2}. \tag{13}$$

For the difference Green function we write down boundary problem, which corresponds to the scheme (5)

$$a_{11}(x)\, G^h_{\bar{x}_1 x_1}(x,\xi) + a_{12}(x)\left[G^h_{\bar{x}_1 x_2}(x,\xi) + G^h_{x_1 \bar{x}_2}(x,\xi)\right] + \tag{14}$$
$$+a_{22}(x)\, G^h_{\bar{x}_2 x_2}(x,\xi) = -\frac{1}{h_1 h_2}\delta(x_1,\xi_1)\,\delta(x_2,\xi_2),\, G^h(x,\xi) = 0, \quad x \in \gamma,$$

where

$$\delta(s,l) = \begin{cases} 1, & s = l, \\ 0, & s \neq l. \end{cases}$$

The Green function $G(x,\xi)$ is symmetric

$$G^h(x,\xi) = G^h(\xi,x)$$

and when x tends to ξ has singularity.

Lemma 2. *The following estimate*

$$\left\|G^h(x,\cdot)\right\| \le \frac{\mu}{\nu^2}\left\{\min\left[x_1 x_2; x_1(1-x_2); (1-x_1)x_2; (1-x_1)(1-x_2)\right]\right\}^{1/2} \equiv \frac{\mu}{\nu^2}\rho^{1/2}(x) \tag{15}$$

is satisfied.

Proof. Making use of the symmetry of the Green function, we write problem (14) in the operator form

$$A_\xi G^h(x,\xi) = B_\xi \mathcal{H}(x-\xi), \tag{16}$$

where

$$\begin{aligned} B_\xi \mathcal{H}(x-\xi) &= -\mathcal{H}(x_1-\xi_1)_{\xi_1}\mathcal{H}(x_2-\xi_2)_{\xi_2} \\ &= -\mathcal{H}(x_1-\xi_1)_{\bar{\xi}_1}\mathcal{H}(x_2-\xi_2)_{\bar{\xi}_2} \\ &= -\mathcal{H}(x_1-\xi_1)_{\xi_1}\mathcal{H}(x_2-\xi_2)_{\bar{\xi}_2} \\ &= -\mathcal{H}(x_1-\xi_1)_{\bar{\xi}_1}\mathcal{H}(x_2-\xi_2)_{\xi_2}. \end{aligned} \tag{17}$$

Here $H(z) = \begin{cases} 1, & z \ge 0, \\ 0, & z < 0 \end{cases}$ is a Heaviside function, operators A_ξ, B_ξ are defined by the formulas (8), (10) by changing x_α to ξ_α, $\alpha = 1, 2$.

Consequently from (13), (16) we obtain

$$\left\|G^h(x,\cdot)\right\| \le \frac{\mu}{\nu^2}\left\|\mathcal{H}(x-\cdot)\right\|,$$

which making use of (17) leads to (15).

Lemma is proved.

3 Difference Scheme for the Problem and Investigation of Its Accuracy in Weight Norm

We approximate the problem (1) with difference scheme of the following form

$$\sum_{i,j=1}^{2} a_{ij}(x)\, y_{\bar{x}_i x_j} - f_0(x,y) = -f(x), \quad x \in \omega, \tag{18}$$

$$y(x) = 0, \quad x \in \gamma.$$

Let us write the equation for the error

$$z(x) = y(x) - u(x).$$

We will have

$$\sum_{i,j=1}^{2} a_{ij}(x)\, z_{\bar{x}_i x_j} = f_0(x,y) - f_0(x,u) - \psi(x), \quad x \notin \omega, \tag{19}$$

$$z(x) = 0, \quad x \in \gamma,$$

$$\psi(x) = \sum_{i,j=1}^{2} \eta_{ij}(x,u), \tag{20}$$

$$\eta_{ij}(x,u) = a_{ij}(x)\left[u_{\bar{x}_i x_j} - \frac{\partial^2 u}{\partial x_i \partial x_j}\right], \quad i,j = 1,2.$$

Making use of the Green difference function $G^h(x,\xi)$ from sec. 2, we submit the solution of the difference scheme (19), (20) implicitly as follows

$$z(x) = \sum_{\xi\in\omega} h_1 h_2 G^h(x,\xi)\left[f_0(\xi,u(\xi)) - f_0(\xi,y(\xi))\right]$$

$$+\sum_{\xi\in\omega} h_1 h_2 G^h(x,\xi)\,\psi(\xi).$$

Therefore, making use of the lemma 2 and Lipschitz condition (3), we obtain

$$|z(x)| \le L\frac{\mu}{\nu^2}\rho^{1/2}(x)\,\|z\| + \frac{\mu}{\nu^2}\rho^{1/2}(x)\,\|\psi\|. \tag{21}$$

We use known inequality

$$(a+b)^2 \le (1+\varepsilon)\,a^2 + \left(1+\frac{1}{\varepsilon}\right)b^2, \quad \varepsilon > 0$$

and the fact that

$$\max_{x\in\bar{\Omega}} \rho(x) = \frac{1}{4}.$$

Consequently, from (21) we obtain the required weight estimate

$$\left\| \frac{z(x)}{\rho^{1/2}(x)} \right\| \leq C \|\psi\|, \tag{22}$$

where

$$C = \frac{\mu}{\nu^2}\left(1+\frac{1}{\varepsilon}\right)^{1/2}\left[1-(1+\varepsilon)\frac{L^2\mu^2}{4\nu^4}\right]^{1/2},$$

which is satisfied under condition

$$\frac{L^2\mu^2}{4\nu^4} < 1. \tag{23}$$

We estimate the norm of the local truncation error $\psi(x)$ by estimating the corresponding items it consists of (see. (20)).

We have

$$\begin{aligned} |\eta_{11}(x,u)| &\leq \|a_{11}\|_{0,\infty,\Omega}\left|u_{\bar{x}_1x_1}(x) - \frac{\partial^2 u(x)}{\partial x_1 \partial x_2}\right| = \|a_{11}\|_{0,\infty,\Omega}|\tilde{\eta}_{11}| \\ &\leq M\frac{|h|^2}{\sqrt{h_1h_2}}|u|_{4,2,\varphi(x)}, \end{aligned} \tag{24}$$

$$\begin{aligned} |\eta_{12}(x)+\eta_{21}(x)| &\leq \|a_{12}\|_{0,\infty,\Omega}\left|u_{\bar{x}_1x_2}(x)+u_{x_1\bar{x}_2}(x) - 2\frac{\partial^2 u(x)}{\partial x_1 \partial x_2}\right| \\ &= \|a_{12}\|_{0,\infty,\Omega}|\tilde{\eta}_{12}| \leq M\frac{|h|^2}{\sqrt{h_1h_2}}|u|_{4,2,\varphi(x)}, \end{aligned} \tag{25}$$

$$\begin{aligned} |\eta_{22}(x,u)| &\leq \|a_{22}\|_{0,\infty,\Omega}\left|u_{\bar{x}_2x_2}(x) - \frac{\partial^2 u(x)}{\partial x_2^2}\right| = \|a_{22}\|_{0,2,\infty}|\tilde{\eta}_{22}| \\ &\leq M\frac{|h|^2}{\sqrt{h_1h_2}}|u|_{4,2,\varphi(x)}, \end{aligned} \tag{26}$$

where M is constant independent of h_1, h_2 and $u(x)$, $|h| = \max\limits_{i=1,2} h_i$, $\varphi(x) = \{(\xi_1,\xi_2) : x_\alpha - h_\alpha \leq \xi_\alpha \leq x_\alpha + h_\alpha, \alpha = 1,2\}$, $|u|_{4,2,\varphi(x)}$ is a semi-norm in the space $W_2^4(\varphi(x))$.

Here at obtaining the inequalities (24)-(26) the Bramble-Hilbert lemma (see [4]) has been used. The fact that each of the linear functionals (concerning $u(x)$) $\tilde{\eta}_{11}, \tilde{\eta}_{12}+\tilde{\eta}_{21}, \tilde{\eta}_{22}$ is bounded in the space $W_2^4(\varphi(x))$ and vanishes on the polynomials of the third order serves as substantiation of the lemma, mentioned above.

Now, according to (20) and (24)-(26), from a priori estimate (22) we come to the statement

Theorem 1. *Let conditions (2), (3), (23) be satisfied, solution of the problem (1) exists, is unique and belongs to the Sobolev space* $W_2^4(\Omega)$ *. Then the difference scheme (18) has a unique solution, accuracy of which is defined by the weight estimate*

$$\left\|\rho^{-1/2}(x)\left[y(x)-u(x)\right]\right\| \leq M\left|h\right|^2\left|u\right|_{4,2,\Omega}. \tag{27}$$

Proof. To prove the theorem it is only enough to additionally ascertain the existence of the unique solution of the problem (18). This fact can be ascertained in the standard way by the fixed point principle, making use of the fact that operator

$$\mathcal{F}(x)y = \sum_{\varsigma\in\omega} h_1h_2G^h(x,\xi)\,f_0(\xi,y(\xi))$$

at using the condition (23) is a contracted operator from $\overset{0}{H}_h$ into $\overset{0}{H}_h$ and it map the sphere

$$S=\left\{y(x)\in\overset{0}{H}_h : \left\|\frac{y(x)}{\rho^{1/2}(x)}\right\|\leq r\right\}$$

into itself.

4 Numerical Experiments

We use difference scheme (18) to find an approximate solution of the following model boundary problem:

Example 1. Solve the problem (1) with $f_0(x,u)\equiv 0$, coefficients

$$a_{11}(x_1,x_2)=(1+x_1)^2,\ a_{12}(x_1,x_2)=a_{21}(x_1,x_2)=\frac{1}{2}(1+x_1)(1+x_2), \tag{28}$$
$$a_{22}(x_1,x_2)=(1+x_2)^2$$

and exact solution

$$u(x_1,x_2)=\sin\pi x_1\sin\pi x_2$$

Let us note, that coefficients (28) satisfy condition (2) ($\nu=\frac{1}{2},\mu=6$) and condition (3), because$f_0(x,u)=0$.

Example 2. Solve the problem (1) with coefficients

$$a_{11}(x_1,x_2)=\sin^2(\pi x_1)+1 \tag{29}$$
$$a_{22}(x_1,x_2)=\sin^2(\pi x_2)+1 \tag{30}$$
$$a_{12}(x_1,x_2)=a_{21}(x_1,x_2)=\frac{1}{2}\sin(\pi x_1)\sin(\pi x_2)+\frac{1}{2}$$

$$f_0 = \frac{1}{8} e^{-|u|}$$

and exact solution

$$u(x_1, x_2) = \ln(\sin \pi x_1 \sin \pi x_2 + 1)$$

The coefficients (29) also satisfy condition (2) ($\nu = \frac{7}{16}, \mu = 3$) and condition (3) $\left(L = \frac{1}{8}\right)$.

For the practical estimation of the rate of convergence we consider the following quantities

$$err = \|z(x)\|_{0,\infty,\omega_h} = \|y(x) - u(x)\|_{0,\infty,\omega_h},$$

$$p = \log_2 \frac{\left\| \frac{z(x)}{\rho^{1/2}(x)} \right\|_{0,\infty,\omega_h}}{\left\| \frac{z(x)}{\rho^{1/2}(x)} \right\|_{0,\infty,\omega_{h/2}}}.$$

The results of calculations are shown in table 1. For computation of the example Fortran PowerStation 4.0 has been used.

Table 1.

	Example 1		Example 2	
#	err	p	err	p
4	$2.792 * 10^{-3}$		$5.782 * 10^{-2}$	
8	$2.405 * 10^{-3}$	1.23	$1.458 * 10^{-2}$	1.78
16	$7.682 * 10^{-4}$	1.77	$3.686 * 10^{-3}$	1.90
32	$1.999 * 10^{-4}$	1.91	$9.241 * 10^{-4}$	1.96
64	$5.093 * 10^{-5}$	1.96	$2.315 * 10^{-4}$	1.98
128	$0.266 * 10^{-5}$	1.98	$8.597 * 10^{-5}$	1.99

The experiments held showed the unimprovement of the theoretical weight a priori estimates (27),by the order.

References

1. Makarov V., On a priori estimates of difference schemes giving an account of the boundary effect, C.R. Acad. Bulgare Sci., 1989, vol.42, No.5. p.41-44.
2. Makarov V.L., Demkiv L.I., Estimates of accuracy of the difference schemes for parabolic type equations taking into account initial-boundary effect. Dopov. Nats. Akad. Nauk. Ukr., No.2, 2003, p.26-32, in Ukrainian
3. Makarov V.L., Demkiv L.I., Improved accuracy estimates of the difference schemes for parabolic equations. // Praci Ukr. matem. conhresu - 2001. - Kyiv:Inst. matem. Nats Acad. Nauk. Ukr., 2002. p.36-47, in Ukrainian.

4. Samarsky A. A., Lazarov R. D., Makarov V. L., Difference schemes for differential equations with generalized solutions. Vysshaya shkola. Moscow, 1987 -296p, in Russian.
5. Ladyzhenskaya O.A., Uraltseva N.N., Linear and Quasi-linear parabolic-type equations, Nauka, Moscow, 1973, in Russian.
6. Djakonov E.G., Difference methods of solving boundary problems. Moscow Government University, Moscow. part , 1971; part , 1972, in Russian.
7. Samarskii A.A., The theory of the difference scheme. Nauka, Moscow, 1977, in Russian.

Nodal Two-Dimensional Solitons in Nonlinear Parametric Resonance

N.V. Alexeeva[1] and E.V. Zemlyanaya[2]

[1] Department of Mathematics, University of Cape Town, South Africa
nora@figaro.mrh.uct.ac.za
[2] Joint Institute for Nuclear Research, Dubna 141980, Russia
elena@jinr.ru

Abstract. The parametrically driven damped nonlinear Schrödinger equation serves as an amplitude equation for a variety of resonantly forced oscillatory systems on the plane. In this note, we consider its nodal soliton solutions. We show that although the nodal solitons are stable against radially-symmetric perturbations for sufficiently large damping coefficients, they are always unstable to azimuthal perturbations. The corresponding break-up scenarios are studied using direct numerical simulations. Typically, the nodal solutions break into symmetric "necklaces" of stable nodeless solitons.

1. Two-dimensional localised oscillating structures, commonly referred to as oscillons, have been detected in experiments on vertically vibrated layers of granular material [1], Newtonian fluids and suspensions [2, 3]. Numerical simulations established the existence of stable oscillons in a variety of pattern-forming systems, including the Swift-Hohenberg and Ginsburg-Landau equations, period-doubling maps with continuous spatial coupling, semicontinuum theories and hydrodynamic models [3, 4]. These simulations provided a great deal of insight into the phenomenology of the oscillons; however, the mechanism by which they acquire or loose their stability remained poorly understood.

In order to elucidate this mechanism, a simple model of a parametrically forced oscillatory medium was proposed recently [5]. The model comprises a two-dimensional lattice of diffusively coupled, vertically vibrated pendula. When driven at the frequency close to their double natural frequency, the pendula execute almost synchronous librations whose slowly varying amplitude satisfies the 2D parametrically driven, damped nonlinear Schrödinger (NLS) equation. The NLS equation was shown to support radially-symmetric, bell-shaped (i.e. nodeless) solitons which turned out to be stable for sufficiently large values of the damping coefficient. These stationary solitons of the amplitude equation correspond to the spatio-temporal envelopes of the oscillons in the original lattice system. By reducing the NLS to a finite-dimensional system in the vicinity of the soliton, its stabilisation mechanism (and hence, the oscillon's stabilisation mechanism) was clarified [5].

In the present note we consider a more general class of radially-symmetric solitons of the parametrically driven, damped NLS on the plane, namely soli-

Z. Li et al. (Eds.): NAA 2004, LNCS 3401, pp. 91–99, 2005.

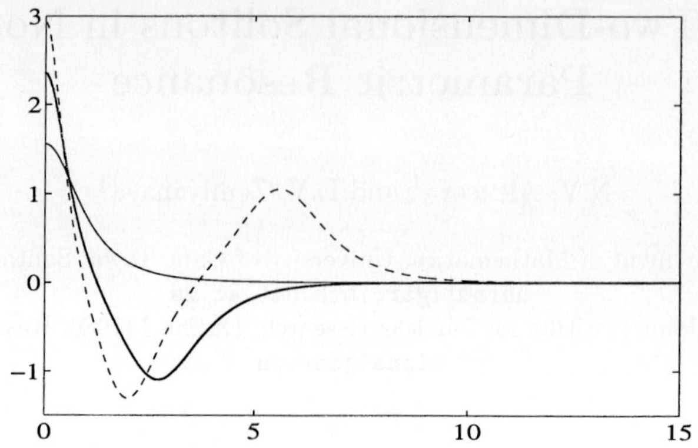

Fig. 1. Solutions of eq.(3): $\mathcal{R}_0(r)$ (thin continuous line), $\mathcal{R}_1(r)$ (thick line), $\mathcal{R}_2(r)$ (dashed)

tons with nodes. We will demonstrate that these solitons are unstable against azimuthal modes, and analyse the evolution of this instability.

2. The parametrically driven, damped NLS equation has the form:

$$i\psi_t + \nabla^2\psi + 2|\psi|^2\psi - \psi = h\psi^* - i\gamma\psi. \tag{1}$$

Here $\nabla^2 = \partial^2/\partial x^2 + \partial^2/\partial y^2$. Eq.(1) serves as an amplitude equation for a wide range of nearly-conservative two-dimensional oscillatory systems under parametric forcing. This equation was also used as a phenomenological model of nonlinear Faraday resonance in water [3]. The coefficient $h > 0$ plays the role of the driver's strength and $\gamma > 0$ is the damping coefficient.

We start with the discussion of its nodeless solitons and their stability. The exact (though not explicit) stationary radially-symmetric solution is given by

$$\psi_0 = \mathcal{A}e^{-i\theta}\mathcal{R}_0(\mathcal{A}r), \tag{2}$$

where $r^2 = x^2 + y^2$,

$$\mathcal{A}^2 = 1 + \sqrt{h^2 - \gamma^2}, \quad \theta = \frac{1}{2}\arcsin\left(\frac{\gamma}{h}\right),$$

and $\mathcal{R}_0(r)$ is the bell-shaped nodeless solution of the equation

$$\mathcal{R}_{rr} + \frac{1}{r}\mathcal{R} - \mathcal{R} + 2\mathcal{R}^3 = 0 \tag{3}$$

with the boundary conditions $\mathcal{R}_r(0) = \mathcal{R}(\infty) = 0$. (Below we simply write $\mathcal{R}$ for $\mathcal{R}_0$.) Solutions of eq.(3) are well documented in literature [6]; see Fig.1.

3. To examine the stability of the solution (2) with nonzero h and γ, we linearise eq.(1) in the small perturbation

$$\delta\psi(\mathbf{x}, t) = e^{(\mu-\Gamma)\tilde{t}-i\theta_{\pm}}[u(\tilde{\mathbf{x}}) + iv(\tilde{\mathbf{x}})],$$

where $\tilde{\mathbf{x}} = \mathcal{A}\mathbf{x}$, $\tilde{t} = \mathcal{A}^2 t$. This yields an eigenvalue problem

$$L_1 u = -(\mu + \Gamma)v, \quad (L_0 - \epsilon)v = (\mu - \Gamma)u, \tag{4}$$

where $\Gamma = \gamma/\mathcal{A}^2$ and the operators

$$L_0 \equiv -\tilde{\nabla}^2 + 1 - 2\mathcal{R}^2(\tilde{r}), \quad L_1 \equiv L_0 - 4\mathcal{R}^2(\tilde{r}), \tag{5}$$

with $\tilde{\nabla}^2 = \partial^2/\partial\tilde{x}^2 + \partial^2/\partial\tilde{y}^2$. (We are dropping the tildas below.) For further convenience, we introduce the positive quantity $\epsilon = 2\sqrt{h^2 - \gamma^2}/\mathcal{A}^2$. Fixing ϵ defines a curve on the (γ, h)-plane:

$$h = \sqrt{\epsilon^2/(2-\epsilon)^2 + \gamma^2}. \tag{6}$$

Introducing

$$\lambda^2 = \mu^2 - \Gamma^2 \tag{7}$$

and performing the transformation [9]

$$v(\mathbf{x}) \to (\mu + \Gamma)\lambda^{-1} v(\mathbf{x}),$$

reduces eq.(4) to a *one*-parameter eigenvalue problem:

$$(L_0 - \epsilon)v = \lambda u, \quad L_1 u = -\lambda v. \tag{8}$$

We first consider the stability with respect to radially symmetric perturbations $u = u(r)$, $v = v(r)$. In this case the operators (5) become

$$L_0 = -\frac{d^2}{dr^2} - \frac{1}{r}\frac{d}{dr} + 1 - 2\mathcal{R}^2(r), \quad L_1 = L_0 - 4\mathcal{R}^2(r). \tag{9}$$

In the absence of the damping and driving, all localised initial conditions in the unperturbed 2D NLS equation are known to either disperse or blow-up in finite time [6, 7, 8]. It turned out, however, that the soliton ψ_0 stabilises as the damping γ is increased above a certain value [5]. The stability condition is $\gamma \geq \gamma_c$, where

$$\gamma_c = \gamma_c(\epsilon) \equiv \frac{2}{2-\epsilon} \cdot \frac{\mathrm{Re}\lambda(\epsilon)\,\mathrm{Im}\lambda(\epsilon)}{\sqrt{(\mathrm{Im}\lambda)^2 - (\mathrm{Re}\lambda)^2}}. \tag{10}$$

We obtained $\lambda(\epsilon)$ by solving the eigenvalue problem (8) directly. Expressing ϵ via γ_c from (10) and feeding into (6), we get the stability boundary on the (γ, h)-plane (Fig.2).

4. To study the stability to asymmetric perturbations we factorise, in (8),

$$u(\mathbf{x}) = \tilde{u}(r)e^{im\varphi}, \quad v(\mathbf{x}) = \tilde{v}(r)e^{im\varphi},$$

where $\tan\varphi = y/x$ and m is an integer. The functions $\tilde{u}(r)$ and $\tilde{v}(r)$ satisfy the eigenproblem (8) where the operators (9) should be replaced by

$$L_0^{(m)} \equiv L_0 + m^2/r^2, \quad L_1^{(m)} \equiv L_1 + m^2/r^2, \tag{11}$$

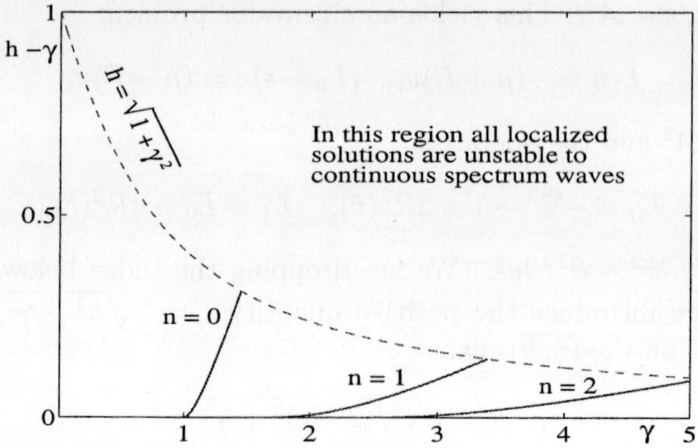

Fig. 2. Stability diagram for two-dimensional solitons. The $(\gamma, h-\gamma)$-plane is used for visual clarity. No localised or periodic attractors exist for $h < \gamma$ (below the horisontal axis). The region of stability of the soliton ψ_0 lies to the right of the solid curve marked "$n = 0$". Also shown are the regions of stability of the solitons ψ_1 and ψ_2 with respect to the radially-symmetric perturbations. (These lie to the right of the corresponding curves in the figure)

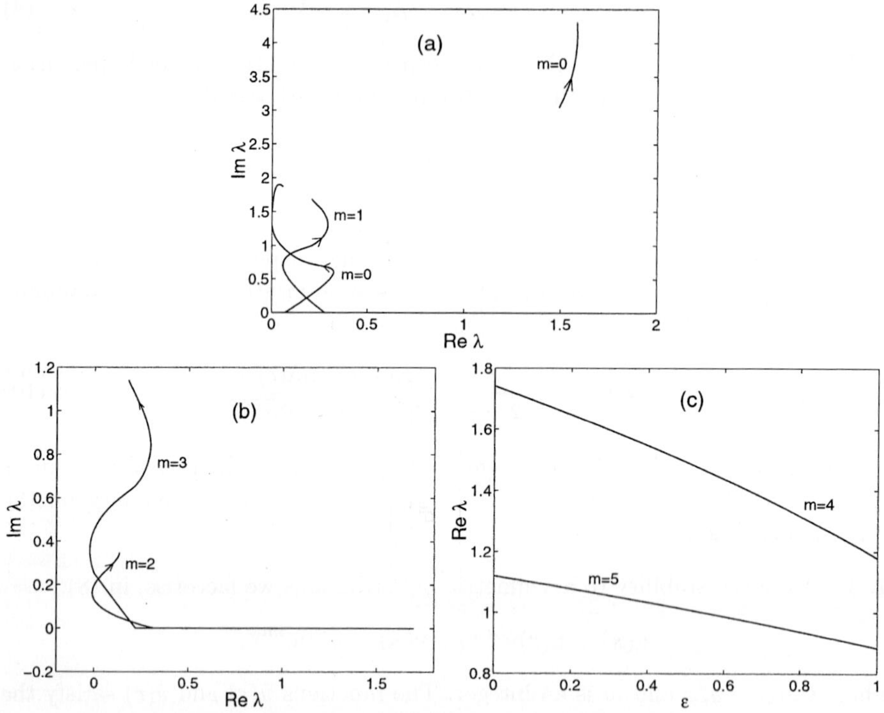

Fig. 3. The discrete eigenvalues of the linearised operator (8) for the one-node soliton, ψ_1. Panels (a) and (b) show the *complex* eigenvalues, $\mathrm{Im}\,\lambda$ vs $\mathrm{Re}\,\lambda$. Arrows indicate the direction of increase of ϵ. Panel (c) shows the *real* eigenvalues, as functions of ϵ

respectively. This modified eigenvalue problem can be analysed in a similar way to eqs.(8). It is not difficult to demonstrate that all discrete eigenvalues of (8) (if any exist) have to be pure imaginary in this case, and hence the azimuthal perturbations do not lead to any instabilities of the solution in question [5].

5. Besides the nodeless solution $\mathcal{R}_0(r)$, the "master" equation (3) has solutions $\mathcal{R}_n(r)$ with n nodes, $n \geq 1$. (See Fig.1). These give rise to a sequence of nodal solutions ψ_n of the damped-driven NLS (1), defined by eq.(2) with $\mathcal{R}_0 \to \mathcal{R}_n$. To examine the stability of the ψ_n, we solved the eigenvalue problem (8) numerically, with operators $L_{0,1}^{(m)}$ as in (11). The *radial* stability properties of the nodal solitons turned out to be similar to those of the nodeless soliton ψ_0. Namely, the ψ_n solutions are stable against radially-symmetric perturbations for sufficiently large γ. The corresponding stability regions for ψ_1 and ψ_2 are depicted in Fig.(2). However, the *azimuthal* stability properties of the nodal solitons have turned out to be quite different.

Both ψ_1 and ψ_2 solutions do have eigenvalues λ with nonzero real parts for orbital numbers $m \geq 1$. (See Fig.3 and 4.) Having found eigenvalues λ for each ϵ, one still has to identify those giving rise to the largest growth rates

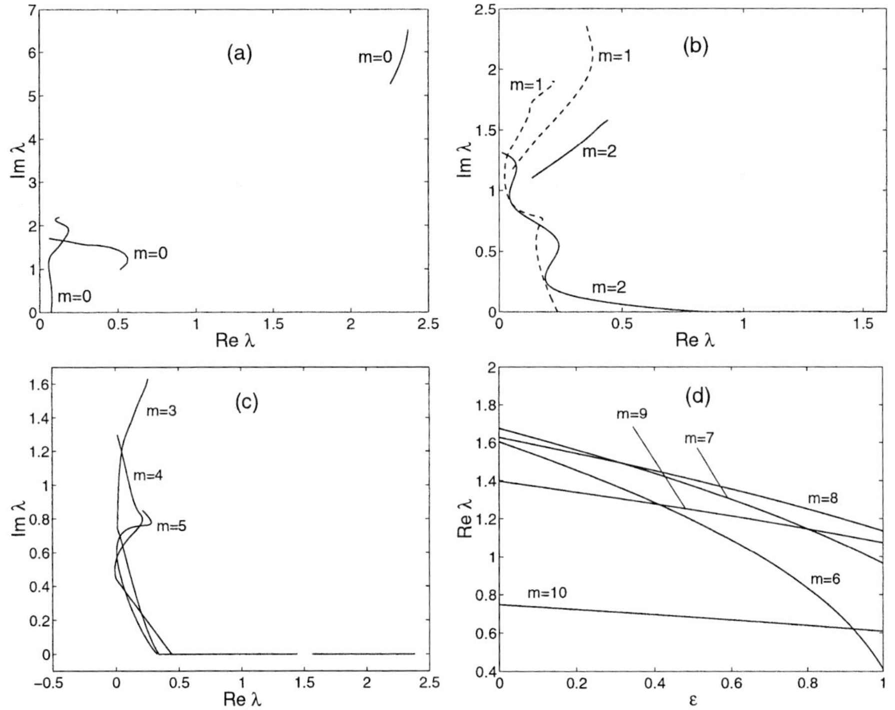

Fig. 4. The discrete eigenvalues of the linearised operator (8) for the two-node soliton, ψ_2. Panels (a),(b), and (c) show the complex eigenvalues, Im λ vs Re λ. Panel (d) shows the real eigenvalues, as functions of ϵ

Table 1. Eigenvalues λ and corresponding growth rates ν for the solitons ψ_1 (left panel) and ψ_2 (right panel). We included only the eigenvalues which can, potentially, give rise to the largest growth rate in each "symmetry class" m. Some other eigenvalues have been filtered out using the above selection rules

m	ν	Reλ	Imλ
0	-0.1620	1.5797	4.2181
0	-1.4827	0.0609	1.8743
1	-0.8255	0.2272	1.6198
1	2.79e-6	0.0033	0.0000
2	-0.3012	0.2213	1.0602
3	0.0872	0.5821	0.0000
4	0.3689	1.2399	0.0000
5	0.2057	0.9076	0.0000

m	ν	Reλ	Imλ
0	-0.3361	2.3531	6.1585
0	-0.5877	0.1819	1.7847
0	-0.4168	0.3093	1.5572
1	-0.8089	0.3818	2.1021
1	-0.4891	0.1111	1.6352
1	1.07e-5	0.0079	0.0000
2	-0.3497	0.3737	1.4597
2	-0.0328	0.2602	0.5128
3	0.5406	1.8686	0.0000
4	0.5286	1.8462	0.0000
5	0.0263	0.3958	0.0000
6	0.1611	0.9898	0.0000
7	0.2490	1.2392	0.0000
8	0.2783	1.3133	0.0000
9	0.2288	1.1861	0.0000
10	0.0720	0.6567	0.0000

$$\nu = \mathrm{Re}\,\mu - \Gamma \tag{12}$$

for each pair (ϵ, γ) [or, equivalently, for each (h, γ)]. In (12), μ is reconstructed using eq.(7). The selection of *real* eigenvalues is straightforward; in this case we have the following two simple rules:

• If, for some ϵ, there are eigenvalues $\lambda_1 > 0$, $\lambda_2 > 0$ such that $\lambda_1 > \lambda_2$, then $\nu_1 > \nu_2$ for this ϵ and all γ. That is, of all real eigenvalues λ one has to consider only the largest one.
• If, for some ϵ, there is a real eigenvalue $\lambda_1 > 0$ and a complex eigenvalue λ_2, with $\mathrm{Re}\,\lambda_2 > 0$ and $\lambda_1 > \mathrm{Re}\,\lambda_2$, then $\nu_1 > \nu_2$ for this ϵ and all γ. That is, one can ignore all complex eigenvalues with real parts smaller than a real eigenvalue — if there is one.

The comparison of two complex eigenvalues is not so straightforward. In particular, the fact that $\mathrm{Re}\,\lambda_1 > \mathrm{Re}\,\lambda_2$ does not necessarily imply that $\nu_1 > \nu_2$. Which of the two growth rates, ν_1 or ν_2, is larger will depend on the imaginary parts of λ, as well as on γ.

In figures 3 and 4, we illustrate the real and imaginary parts of the eigenvalues, arising for different m, for the solitons ψ_1 and ψ_2. The soliton ψ_1 has discrete eigenvalues λ associated with orbital numbers $m = 0, 1, ..., 5$ and the soliton ψ_2 with $m = 0, 1, ..., 10$.

In order to compare the conclusions based on the linearised analysis with direct numerical simulations of the unstable solitons ψ_1 and ψ_2, we fix some h and γ and identify the eigenvalue with the maximum growth rate in each case.

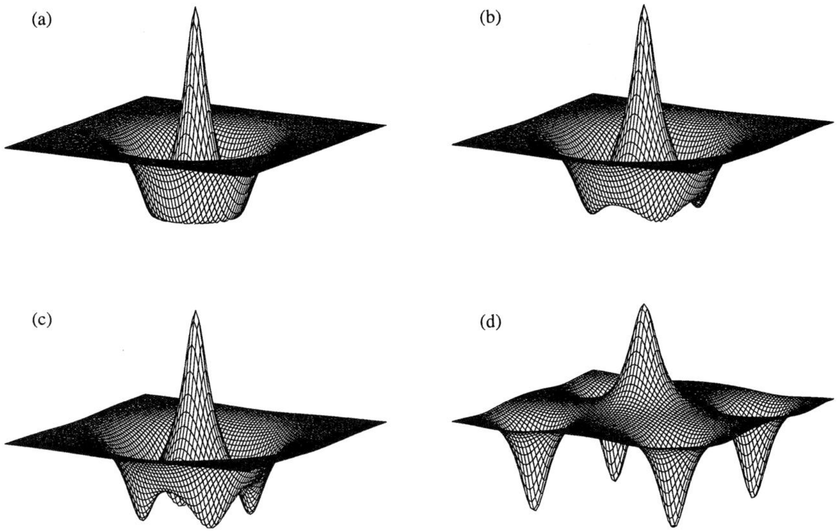

Fig. 5. Evolution of the azimuthal instability of the one-node soliton. (a): the initial condition, soliton ψ_1 ; (b) and (c): dissociation of the ring-like "valley" into 4 nodeless solitons; (d): divergence of the fragments. Here $\gamma = 3.5$ and $h = 3.6$; shown is Re ψ. Note the change of the vertical scale in (d)

In the case of the soliton ψ_1, we choose $\gamma = 3.5$ and $h = 3.6$; the corresponding $\epsilon = 0.9146$. The real and imaginary parts of λ for each m as well as the resulting growth rates ν are given in Table 1 (left panel). The eigenvalue with the largest Reλ is associated with $m = 0$; however, for the given ϵ and γ the resulting $\nu < 0$. (This is because we have chosen a point in the "radially stable" part of the (γ, h)-plane, to the right of the "$n = 1$" curve in Fig.2.) On the contrary, the growth rates corresponding to the real eigenvalues associated with $m = 3, 4, 5$ are positive for all γ. The maximum growth rate is associated with $m = 4$. The corresponding eigenfunctions $u(r)$ and $v(r)$ have a single maximum near the position of the minimum of the function $\mathcal{R}_1(r)$; that is, the perturbation is concentrated near the circular "valley" in the relief of $\psi(x, y)|^2$. This observation suggests that for $\gamma = 3.4$ and $h = 3.5$, the soliton ψ_1 should break into a symmetric pattern of 5 solitons ψ_0: one at the origin and four around it.

Next, in the case of the soliton ψ_2 we fix $\gamma = 4.5$ and $h = 4.53$; this gives $\epsilon = 0.6846$. The corresponding eigenvalues, for each m, are presented in Table 1 (right panel). Again, the eigenvalue with the largest Reλ is the one for $m = 0$ but the resulting ν is negative. The largest growth rates ($\nu_3 = 0.54$ and $\nu_4 = 0.53$, respectively) are those pertaining to $m = 3$ and $m = 4$. The corresponding eigenfunctions have their maxima near the position of the minimum of the function $\mathcal{R}_2(r)$. Therefore, the circular "valley" of the soliton ψ_2 is expected to break into three or four nodeless solitons ψ_0. (Since ν_3 is so close to ν_4, the actual number

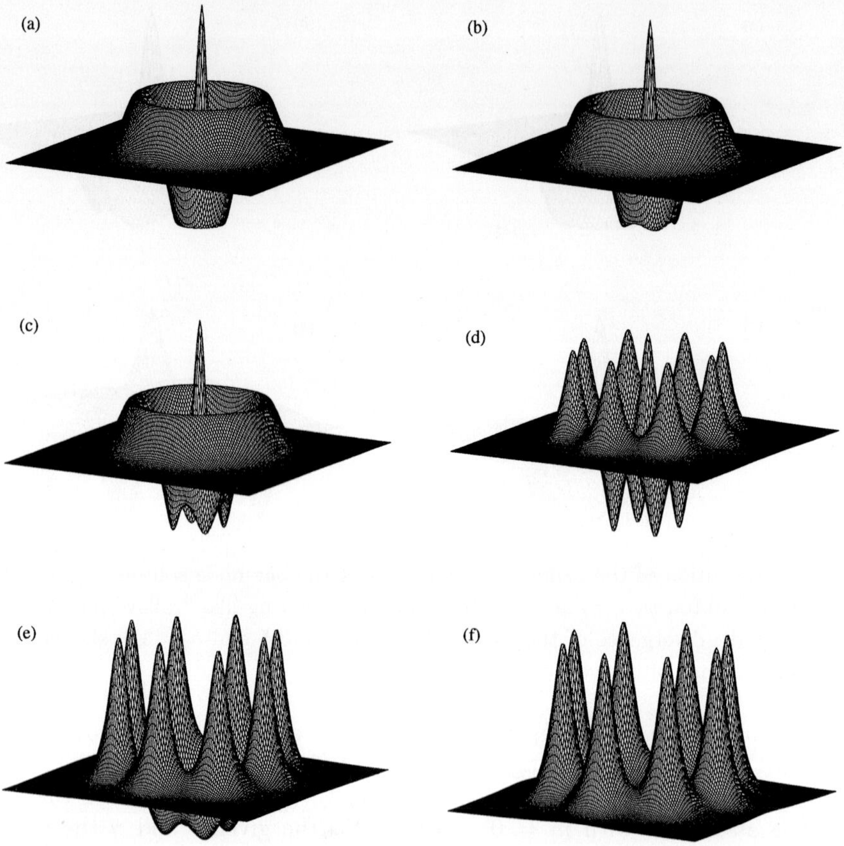

Fig. 6. The evolution of the azimuthal instability of the two-node soliton ψ_2. (a): the initial condition; (b)-(c): the rapid dissociation of the "valley" into 4 nodeless solitons; (c)-(d): a slower decay of the "ridge" into 8 solitons ψ_0; (e)-(f): the annihilation of the internal ring and the central soliton, and the repulsion of the persisting 8 solitons. Here $\gamma = 4.5$ and $h = 4.53$; shown is $\mathrm{Re}\,\psi$. Note the change of the vertical scale in (e)-(f) w.r.t. that in (a)-(d)

of resulting solitons — three or four — will be very sensitive to the choice of the initial perturbation.) Next, eigenfunctions pertaining to $m = 5, 6, ...10$ have their maxima near the second, lateral, maximum of the function $\mathcal{R}_2(r)$. The largest growth rate in this group of eigenvalues arises for $m = 8$. Hence, the circular "ridge" of the soliton ψ_2 should break into 8 nodeless solitons, with this process taking longer than the bunching of the "valley" into the "internal ring" of solitons.

The direct numerical simulations corroborate the above scenarios. The ψ_1 soliton with $\gamma = 3.5$ and $h = 3.6$ splits into a constellation of 5 nodeless solitons:

one at the origin and four solitons of opposite polarity at the vertices of the square centered at the origin. The emerging nodeless solitons are stable but repelling each other, see Fig. 5. Hence, no stationary nonsymmetric configurations are possible; the peripheral solitons escape to infinity. The ψ_2 soliton with $\gamma = 4.5$ and $h = 4.53$ has a more complicated evolution. As predicted by the linear stability analysis, it dissociates into 13 nodeless solitons: one at the origin, four solitons of opposite polarity forming a square around it and eight more solitons (of the same polarity as the one at the origin) forming an outer ring. (The fact that the inner ring consists of four and not three solitons, is due to the square symmetry of our domain of integration which favours perturbations with $m = 4$ over those with $m = 3$.) In the subsequent evolution the central soliton and the nearest four annihilate and only the eight outer solitons persist. They repel each other and eventually escape to infinity, see Fig.6.

In conclusion, our analysis suggests the interpretation of the nodal solutions as degenerate, unstable coaxial complexes of the nodeless solitons ψ_0.

Acknowledgements. This project was supported by the NRF of South Africa under grant 2053723. The work of E.Z. was supported by an RFBR grant 03-01-00657.

References

1. P.B. Umbanhowar, F. Melo, and H.L. Swinney, Nature **382**, 793 (1996)
2. O. Lioubashevski *et al*, Phys. Rev. Lett. **83**, 3190 (1999); H. Arbell and J. Fineberg, *ibid.* **85**, 756 (2000)
3. D. Astruc and S. Fauve, in: Fluid Mechanics and Its Applications, vol.62, p. 39-46 (Kluwer, 2001)
4. L.S. Tsimring and I.S. Aranson, Phys. Rev. Lett. **79**, 213 (1997); E. Cerda, F. Melo, and S. Rica, *ibid.* **79**, 4570 (1997); S.C. Venkataramani and E. Ott, *ibid.* **80**, 3495 (1998); D. Rothman, Phys. Rev. E **57**, 1239 (1998); J. Eggers and H. Riecke, *ibid.* **59**, 4476 (1999); C. Crawford, H. Riecke, Physica D **129**, 83 (1999); H. Sakaguchi, H.R. Brand, Europhys. Lett. **38**, 341 (1997); Physica D **117**, 95 (1998)
5. I.V. Barashenkov, N.V. Alexeeva, and E.V. Zemlyanaya, Phys. Rev. Lett. **89**, 104101 (2002)
6. See e.g. K. Rypdal, J.J. Rasmussen and K. Thomsen, Physica **D 16**, 339 (1985) and references therein.
7. E.A. Kuznetsov and S.K. Turitsyn, Phys. Lett. **112A**, 273 (1985); V.M. Malkin and E.G. Shapiro, Physica D **53**, 25 (1991)
8. For a recent review and references on the blowup in 2D and 3D NLS equations, see e.g. L. Berge, Phys. Rep. **303**, 259 (1998); G. Fibich and G. Papanicolaou, SIAM J. Appl. Math. **60**, 183 (1999)
9. I.V. Barashenkov, M.M. Bogdan and V.I. Korobov, Europhys. Lett. **15**, 113 (1991)

Supercloseness Between the Elliptic Projection and the Approximate Eigenfunction and Its Application to a Postprocessing of Finite Element Eigenvalue Problems

Andrey B. Andreev

Technical University of Gabrovo & IPP - Bulgarian Academy of Sciencies
Andreev@tugab.bg

Abstract. An estimate confirming the supercloseness between the Ritz projection and the corresponding eigenvectors, obtained by finite element method, is hereby proved. This result is true for a large class of self-adjoint $2m-$order elliptic operators. An application of this theorem to the superconvergence postprocessing patch-recovery technique for finite element eigenvalue problems is also presented. Finally, the theoretical investigations are supported by numerical experiments.

2000 Subject Classification: 65N30, 65N25

Keywords: finite elements, eigenvalue problem, superconvergence, post-processing

1 Introduction

Let $(\lambda, u(x))$ be an exact eigenpair of a $2m-$order eigenvalue problem: find $\lambda \in \mathbf{R},\ u(x) \in V$ such that

$$a(u, v) = \lambda(u, v), \quad \forall v \in V, \tag{1}$$

where V is a closed subspace of $H^m(\Omega)$, $H_0^m(\Omega) \subseteq V \subseteq H^m(\Omega)$, $H^m(\Omega)$ and $H_0^m(\Omega)$ being the usual $m-$order Sobolev spaces on the bounded domain $\Omega,\ \Omega \subset \mathbf{R}^{\mathbf{d}},\ d = 1, 2, 3$ with norm $\|.\cdot\|_{m,\Omega}$ (cf. Ciarlet [1]). In order to avoid technical difficulties, we shall always assume that the boundary Γ of Ω is Lipshitz continuous. The notation $(\cdot, \cdot)$ is adopted for the $L_2(\Omega)-$inner product and $a(\cdot, \cdot)$ is the bilinear form on $V \times V$ specified below.

This paper deals with an aspect of superconvergence phenomena in finite element (FE) method for the problem (1). Namely, we prove supercloseness between the approximate eigenfinctions and the corresponding elliptic FE-solutions. Some applications of this result are presented.

We assume that the bilinear form $a(\cdot, \cdot)$ satisfies the following conditions:

(i) The $a-$form is symmetric and coercive on V:

$$a(u, v) = a(v, u), \quad \rho\|v\|_{m,\Omega}^2 \le a(v, v), \quad \forall u, v \in V, \tag{2}$$

where ρ is a positive constant.

Z. Li et al. (Eds.): NAA 2004, LNCS 3401, pp. 100–107, 2005.

(ii) The $a-$form is continuous, i.e.

$$|a(u,v)| \leq C\|u\|_{m,\Omega}\|v\|_{m,\Omega}, \quad \forall u,v \in V.$$

The problem (1) has a countable infinite set of eigenvalues λ_i, all being strictly positive and having finite multiplicity, without a finite accumulation point (see [2, 3]).

The $a-$form could be generated by certain boundary values for the following self-adjoint elliptic operator:

$$\mathcal{A}u(x) = \sum_{|\alpha|\leq m} (-1)^{|\alpha|} D^{\alpha}\left\{p_{\alpha}(x)D^{\alpha}u(x)\right\}, \quad m \in \mathbf{N}.$$

The coefficient functions $p_{\alpha}(x)$ are assumed to be real-valued functions of class $C^{|\alpha|}(\overline{\Omega})$ and $p_{\alpha}(x)$ have all derivatives on Ω that are continuously prolongable on $\overline{\Omega}$.

Problem (1) is caused by the following linear eigenvalue problem:

$$\mathcal{A}u(x) = \lambda u(x), \quad x \in \Omega \tag{3}$$

with the homogeneous boundary conditions

$$\mathcal{B}u(x) = 0, \quad x \in \Gamma, \tag{4}$$

where

$$\mathcal{B} = \{B_j\}_{j=1}^{m}, \quad B_j u(x) = \sum_{|\alpha|\leq 2m-1} s_{j,\alpha}(x)D^{\alpha}u(x), \quad j = 1,\ldots,m$$

are linearly independent conditions.

Problems (3), (4) is a general eigenvalue problem of order $2m$. We assume that the operators $\mathcal{A}$ and $\mathcal{B}$ are such that the symmetry and coerci8vity conditions (2) of the $a-$form are satisfied. The one-dimensional case ($d = 1$ with $\Omega = (a,b)$) allows $2m$ more general homogeneous boundary conditions.

Let us consider a family of regular finite element partitions τ_h of $\overline{\Omega}$ which fulfil standard assumptions [1], and suppose that this family satisfies the inverse assumptions ([1], p.140), i.e. there exists a constant $\nu > 0$ such that $\forall e \in \tau_h$, $\frac{h}{h_e} \leq \nu$, where h_e is the diameter of e, $h = \max_{e\in\tau_h} h_e$.

With a partition τ_h we associate a finite-dimensional subspace V_h of $V \cap C^{m-1}(\overline{\Omega})$ such that the restriction of every function $v \in V_h$ over every finite element $e \in \tau_h$ is a polynomial of degree n at most.

We determine the approximate eigenpairs $(\lambda_h, u_h(x))$ by the finite element method:

$$\lambda_h \in \mathbf{R},\; u_h \in V_h, \quad a(u_h,v) = \lambda_h(u_h,v) \;\; \forall v \in V_h. \tag{5}$$

It is well-known (see [4] ot [2]) that the rate of convergence to the eigenvalues and eigenfunctions provided that $n \geq 2m-1$ is given by

$$|\lambda - \lambda_h| \leq C(\lambda)h^{2(n+1-m)}\|u\|_{n+1,\Omega}, \tag{6}$$

$$\|u - u_h\|_{m,\Omega} \leq C(\Lambda)h^{n+1-m}\|u\|_{n+1,\Omega}. \tag{7}$$

The solutions of (1) and (5) are related to the Rayleigh quotient:

$$\lambda = \frac{a(u,u)}{(u,u)} \quad \text{and} \quad \lambda_h = \frac{a(u_h,u_h)}{(u_h,u_h)}.$$

Let $\mathcal{R}_h : V \to V_h$ be the elliptic or Ritz projection operator defined by:

$$\forall u \in V,\ \forall v \in V_h \ \ a(u - \mathcal{R}_h u, v) = 0.$$

If the function $u(x)$ belongs to $H^{n+1}(\Omega) \cap V$ then [4]

$$\|u - \mathcal{R}_h u\|_{m,\Omega} \leq Ch^{n+1-m}\|u\|_{n+1,\Omega}. \tag{8}$$

2 Main Result

Let (λ_h, u_h) be the FE approximation of any exact eigenpair (λ, u). We shall estimate the difference $u_h - \mathcal{R}_h u$ in H^m–norm of higher order of accuracy as compared to the estimates (7) and 8). Thus, the Ritz projection of any eigenfunction (in the same finite element space) approximates the corresponding FE eigenfunction of better precision. This special feature of both functions is called to satisfy a superclose property, (cf. [5]). Such result is proved by Lin, Yan and Zhou in [6] when eigenpairs of Laplacian operator and its FE-approximations are considered.

Our goal is to prove this superclose property using a substantially different approach. This estimate is valid for a large class of $2m$–order self-adjoint elliptic operators. It is also worthwhile noting that the superclose property is available between the FE eigenfunction u_h and its corresponding Lagrangian FE interpolant u_I of the exact eigenfunction. This result is presented by the author in [7] and it is used to prove that the points of superconvergence of the gradient for the eigenfunctions of second-order elliptic operator are the same as in the case of elliptic boundary value problems.

The following theorem contains the main result of this paper:

Theorem 1. *Let the eigenfunction $u(x)$ belong to $H^{n+1}(\Omega) \cap V$ and let $u_h(x)$ be the corresponding approximation obtained by (5). Then*

$$\|u_h - \mathcal{R}_h u\|_{m,\Omega} \leq Ch^{\alpha(m,n)}\|u\|_{n+1,\Omega}, \tag{9}$$

where $\alpha(m,n) = \min\{2(n+1-m); n+1+s\},\ 0 \leq s \leq m.$

Proof. Using (2) we have

$$\rho\|u_h - \mathcal{R}_h u\|^2_{m,\Omega} \leq a(u_h - \mathcal{R}_h u, u_h - \mathcal{R}_h u).$$

Denote $u_h - \mathcal{R}_h u = z_h \in V_h$. Using the orthogonal property of $\mathcal{R}_h u$, it follows that

$$\begin{aligned}
\rho\|z_h\|_{m,\Omega} &\leq \lambda_h(u_h, z_h) - a(\mathcal{R}_h u, z_h) \\
&= (\lambda_h - \lambda)(u_h, z_h) + \lambda(u_h, z_h) - a(u, z_h) \\
&= (\lambda_h - \lambda)(u_h, z_h) + \lambda(u_h - u, z_h) + \lambda(u, z_h) - a(u, z_h) \\
&= (\lambda_h - \lambda)(u_h, z_h) + \lambda(u_h - u, z_h) \\
&\leq |\lambda_h - \lambda| \|u_h\|_{0,\Omega} \|z_h\|_{m,\Omega} + \lambda \|u_h - u\|_{-s,\Omega} \|z_h\|_{m,\Omega}.
\end{aligned}$$

Note that in the last step we use the duality in negative norms with $0 \leq s \leq m$. The estimate (9) follows from (6) and the inequality [8]

$$\|u - u_h\|_{-s,\Omega} \leq Ch^{n+1+s} \|u\|_{n+1,\Omega}.$$

Remark 1. Obviously we have maximum of $\alpha(m,n)$ when $s = m$. Then $2m - n - 1 \leq -m$, i.e. $3m \leq n+1$. The other extremal case $s = 0$ leads "only" to the superconvergence result $\|u_h - \mathcal{R}_h u\|_{m,\Omega} = \mathcal{O}(h^{n+1})$. This is the case when one uses the lowest possible degree n of the approximation polynomials in the FE method, i.e. $n + 1 = 2m$.

Remark 2. The elliptic projection $\mathcal{R}_h u$ is closer to the approximate eigenfunction u_h than the FE-interpolant u_I (see [7]). This is due to the fact that u_I does not verify the orthogonality with respect to the a−form.

3 Application

In this section a direct application of Theorem 1 to the Zienkiewich-Zhu patch recovery technique will be presented [9]. This application concerns some ultra-convergence properties of the eigenvalues and eigenfunctions, i.e. when the convergence rate is at least two order higher than the optimal global rate. For this aspect of the superconvergence analysis the reader is referred to the book by Ainsworth-Oden ([10], Chapter 4, see also [11]).

To begin with we need the following lemma:

Lemma 1. *Let (λ, u) be the solution of (1) such that $\|u\|_{0,\Omega} = 1$. Then for any $w \in H_0^m(\Omega)$ and $w \neq 0$, there holds*

$$\left| \frac{a(w,w)}{(w,w)} - \lambda \right| \leq C \frac{\|w - u\|_{m,\Omega}^2}{(w,w)}. \tag{10}$$

Proof. Denote $w - u = \varphi \in H_0^m(\Omega)$. Then $w = u + \varphi$.

The relation (10) is equivalent to

$$|a(u+\varphi,u+\varphi)-\lambda(u+\varphi,u+\varphi)|\le C\|\varphi\|^2_{m,\Omega}\quad\forall\varphi\in H_0^m(\Omega).$$

It follows

$$|a(u,u)+2a(u,\varphi)+a(\varphi,\varphi)-\lambda(u,u)-2\lambda(u,\varphi)-\lambda(\varphi,\varphi)|$$
$$=|a(u,u)+a(\varphi,\varphi)-\lambda-\lambda(\varphi,\varphi)|.$$

Having in mind that $a(u,u)=\lambda$, we have

$$\begin{aligned}|a(u+\varphi,u+\varphi)-\lambda(u+\varphi,u+\varphi)| &= |a(\varphi,\varphi)-\lambda(\varphi,\varphi)|\\ &\le C_1\|\varphi\|^2_{m,\Omega}+\lambda\|\varphi\|^2_{0,\Omega}\\ &\le C\|\varphi\|^2_{m,\Omega}.\end{aligned}$$

The last inequality proves the estimate (10).

Now, let us consider a higher order interpolation of the original finite element solution to achieve higher order accuracy. The interpolated FE solution resulting from the postprocessing uses some superclose properties (see [9, 10]). The special interpolation operator is noted by I_{2h} and it is related to the patch recovery technique [6, 11, 12]. So, the interpolation I_{2h} is defined on the mesh of size $2h$ with certain "vertices-edges-element" conditions. Thus the postprocessed FE-solution has increased order of accuracy without increasing the degree of elements.

It is assumed that by the construction, for any $u\in H^{k+1}(\Omega),\ k\ge n+1$,

$$\|I_{2h}\mathcal{R}_hu-u\|_{m,\Omega}\le Ch^k\|u\|_{k+1,\Omega}. \tag{11}$$

Denote

$$\beta=\beta(m,n)=\min\{\alpha(m,n);k\}\quad\text{and}\quad\widetilde{\beta}(m,n)=\max\{\alpha(m,n);k\}.$$

We show that the ultraconvergence (superconvergence) error estimates for eigenvalue approximations for $2m$−order self-adjoint eigenvalue problem are a consequence of two approaches. First, we use the superclose property of the Ritz projection of the eigenfunction $\mathcal{R}_hu$ and the corresponding FE solution u_h (Theorem 1). The second one is the patch recovery postprocessing approach.

Theorem 2. *Let (λ,u) be an eigenpair of the problem (1) and let (λ_h,u_h) be its finite element approximation obtained from (5). Assume that the estimate (11) is fulfilled as a consequence of recovery postprocessing technique and $u\in H^{\widetilde{\beta}+1}(\Omega)$. Then*

$$\|I_{2h}u_h-u\|_{m,\Omega}\le Ch^\beta\|u\|_{\widetilde{\beta}+1,\Omega}, \tag{12}$$

$$\left|\frac{a(I_{2h}u_h,I_{2h}u_h)}{(I_{2h}u_h,I_{2h}u_h)}-\lambda\right|\le Ch^{2\beta}\|u\|^2_{\widetilde{\beta}+1,\Omega}. \tag{13}$$

Proof. For the first estimate (12) we obtain:

$$\|I_{2h}u_h - u\|_{m,\Omega} \le \|I_{2h}u_h - I_{2h}\mathcal{R}_h u\|_{m,\Omega} + \|I_{2h}\mathcal{R}_h u - u\|_{m,\Omega}$$

$$\le \|I_{2h}\| \, \|u_h - \mathcal{R}_h u\|_{m,\Omega} + \|I_{2h}\mathcal{R}_h u - u\|_{m,\Omega}.$$

Consider the interpolation operator $I_{2h} : V_h \to V_h$. This operator has a finite range, i.e. dim range$(I_{2h}) < \infty$. Therefore it is compact. Thus

$$\|I_{2h}\| = \frac{\sup_{v_h \in V_h} \|I_{2h}v_h\|_{m,\Omega}}{\|v_h\|_{m,\Omega}} \le C = const.$$

Using the result of Theorem 1 and (11) we get:

$$\|I_{2h}u_h - u\|_{m,\Omega} \le Ch^{\alpha(m,n)}\|u\|_{n+1,\Omega} + Ch^k\|u\|_{k+1,\Omega}.$$

This inequality proves estimate (12).

For the eigenvalues we apply the result of Lemma 1:

$$\left|\frac{a(I_{2h}u_h, I_{2h}u_h)}{(I_{2h}u_h, I_{2h}u_h)} - \lambda\right| \le C\frac{\|I_{2h}u_h - u\|^2_{m,\Omega}}{\|I_{2h}u_h\|^2_{0,\Omega}} \le Ch^{2\beta}\|u\|^2_{\tilde{\beta}+1,\Omega}.$$

4 Numerical Results

Here follows an illustration of the present theory. Consider the problem

$$-\Delta u = \lambda u, \quad x \in \Omega,$$

$$u_{|\Gamma} = 0,$$

where $\Omega = (0,l) \times (0,l)$ and $\Gamma = \partial\Omega$.

The first exact eigenpairs of this problem are:

$$\lambda_1 = \frac{2\pi^2}{l^2}, \quad u_1(x,y) = C\sin\frac{\pi}{l}x\sin\frac{\pi}{l}y;$$

$$\lambda_2 = \frac{5\pi^2}{l^2}, \quad u_2(x,y) = C\sin\frac{\pi}{l}x\sin\frac{2\pi}{l}y;$$

$$\lambda_3 = \frac{5\pi^2}{l^2}, \quad u_3(x,y) = C\sin\frac{2\pi}{l}x\sin\frac{\pi}{l}y;$$

$$\lambda_4 = \frac{8\pi^2}{l^2}, \quad u_4(x,y) = C\sin\frac{2\pi}{l}x\sin\frac{2\pi}{l}y.$$

The coefficient $C = \frac{2}{l}$ is determined by the normalization

$$\int_\Omega u_j^2(x,y)\,dx\,dy = 1.$$

Table 1 gives the results when $l = 1$ for biquadratic meshes (i.e. $n = 2$ above) of N_e identical 9-nodes rectangular elements ($N_e = 16; 36; 64$). We present the finite element approximation $\lambda_{h,j}$ of the eigenvalues

$$\lambda_1 = 19.7392088022, \lambda_{2,3} = 49.3480220054, \lambda_4 = 78.9568352087,$$

and their improvements $\overline{\lambda}_{h,j}$ by postprocessing patch recoveries. The values of $\overline{\lambda}_{h,j}$ have been obtained by using interpolated finite elements unifying four elements in the patch. It is readily seen that a superclose property arises on the coarse mesh.

Table 1.

N_e	16	36	64
$\lambda_{h,1}$	19.7493180513	19.7412403886	19.7398555788
$\overline{\lambda}_{h,1}$	19.7392377680	19.7392108553	19.7392102567
$\lambda_{h,2}$	49.6506604280	49.4117157727	49.3688522084
$\overline{\lambda}_{h,2}$	49.3494876511	49.3488302530	49.3482370004
$\lambda_{h,3}$	49.7000190559	49.4339097637	49.3921333268
$\overline{\lambda}_{h,3}$	49.3990486874	49.3710485246	49.3584525122
$\lambda_{h,4}$	102.195699803	99.9313207004	98.5022915765
$\overline{\lambda}_{h,4}$	93.8779184523	90.6475349121	88.7411349267

Table 2 illustrates the supercloseness between the eigenfunctions when the same patch recovery strategy is applied.

Table 2.

N_e	16	36	64
$\|u_1 - u_{h,1}\|_{2,\Omega}$	0.14	6.1×10^{-2}	3.2×10^{-2}
$\|u_1 - I_{2h}u_{h,1}\|_{2,\Omega}$	7.2×10^{-3}	1.9×10^{-3}	1.1×10^{-3}
$\|u_2 - u_{h,2}\|_{2,\Omega}$	0.77	0.33	0.19
$\|u_2 - I_{2h}u_{h,2}\|_{2,\Omega}$	5.2×10^{-2}	3×10^{-2}	1.9×10^{-2}
$\|u_3 - u_{h,3}\|_{2,\Omega}$	0.82	0.39	0.26
$\|u_3 - I_{2h}u_{h,3}\|_{2,\Omega}$	0.29	0.19	0.12
$\|u_4 - u_{h,4}\|_{2,\Omega}$	6.79	6.43	6.15
$\|u_4 - I_{2h}u_{h,4}\|_{2,\Omega}$	4.46	4.13	3.42

Acknoledgements. The author would like to thank Prof. Vidar Thomee and Prof. Stig Larson for their assistance during his visit in Chalmers University, Goteborg.

References

[1] Ciarlet, P.G.: The Finite Element Method for Elliptic Problems. North-Holland, Amsterdam 1978.
[2] Babuska I., J. Osborn: Eigenvalue Problems. Handbook of Numer. Anal., Vol. II, North-Holland, Amsterdam 1991.
[3] Pierce J.G., VCarga R.S.: Higher order convergence result for the Rayleigh-Ritz method applied to eigenvalue problems: Improved error bounds for eigenfunctions. Numer. Math. **19**, 1972, pp. 155-169.
[4] Strang G., Fix G.J.: An Analysis of the Finite Element Method. Prentice-Hall, Englewood Cliffs, N.J., 1973.
[5] Wahlbin L.B.: Superconvergence in Galerkin FEM. Springer-Verlag, Berlin, 1995.
[6] Lin Q., Yan N. and Zhou A.: A rectangular test for interpolated finite elements. Proceedings of Systems Science & Systems Engineering, Culture Publish Co., 1991, 217-229.
[7] Andreev A.B.: Superconvergence of the gradient of finite element eigenfunctions. C.R. Acad. Bulg. Sci., **43**, 9-11, 1990.
[8] Thomee V.: Galerkin Finite Element Methods for Parabolic Problems. Springer, 1997.
[9] Zienkiewicz O.C., Zhu J.Z.: The superconvergence patch recovery and a posteriori error estimates. Part 1: The recovery technique. Int. J. Numer. Methods Engrg. **33**, 1331-1364 (1992).
[10] Ainsworth M., Oden J.T.: A Posteriori Error Estimation in Finite Element Analysis. Wiley Interscience, New York, 2000.
[11] Zhang Z., Liu R.: Ultraconcergence of ZZ patch recovery at mesh symmetry points. Numer. Math. (2003) **95**, 781-801.
[12] Andreev A.B., Dimov T.T., Racheva M.R.: One-dimensional patch-recovery finite element method for fourth-order elliptic problems (in this issue).
[13] Racheva M.R., Andreev A.B.: Variational aspects of one-dimensional fourth-order problems with eigenvalue parameter in the boundary conditions. Sib. J.N.M., No4(5), 2002, pp. 373-380.

One-Dimensional Patch-Recovery Finite Element Method for Fourth-Order Elliptic Problems

Andrey B. Andreev[1,2], Ivan, Todor Dimov[2], and Milena R. Racheva[1]

[1]Technical University of Gabrovo
[2]IPP - Bulgarian Academy of Sciencies
{Andreev, Milena}@tugab.bg
TDimov@copern.bas.bg

Abstract. Interpolated one-dimensional finite elements are constructed and applied to the fourth-order self-adjoint elliptic boundary-value problems. A superconvergence postprocessing approach, based on the patch-recovery method, is presented. It is proved that the rate of convergence depends on the different variational forms related to the variety of the corresponding elliptic operators. Finally, numerical results are presented.

2000 Subject Classification: 65N30, 65N25

Keywords: finite elements, eigenvalue problem, superconvergence, post-processing.

1 Introduction

Several types of superconvergence finite element methods have been studied in the last two decades; see for example [1, 2] and the citations therein. This paper dwells upon a superconvergence property of postprocessing type [3]. This procedure increases the order of convergence of the original finite element solution in case of the fourth-order elliptic problem on the single space. Such a postprocessing is no less than a higher order interpolation on a coarser mesh of the original finite element solution.

Our aim is to give explicite postprocessing procedure which to be used in practical calculations. Let us also note that the superclose phenomena for the fourth-order elliptic problem we present is similar to patch-recovery technique, introduced by Zienkiewich-Zhu in 1991 [4].

Let $H^m(\Omega)$ be the usual Sobolev space for the positive integer m [5], provided with the norms and seminorms $\|\cdot\|_{m,\Omega}$ and $|\cdot|_{m,\Omega}$, respectively.

Consider the following model problem for fourth-order self-adjoint elliptic operator ($\Omega \equiv (0,l),\ l>0$):

$$a(u,v) = b(f,v), \quad \forall v \in V, \tag{1}$$

where $V = H^2(0,l)$ and

$$a(u,v) = \int_0^l (u''v'' + pu'v' + quv)\,dx. \tag{2}$$

Z. Li et al. (Eds.): NAA 2004, LNCS 3401, pp. 108–115, 2005.

It is required that the coefficients $p \geq 0$ and $q \geq 0$ are real constants.

Suppose that the bilinear a–form is coercive on V [5]. Obviously it is symmetric. The bilinear b–form contains the first derivatives of the functions at most. It is symmetric and coercive on $H^1(0,l)$.

The one-dimensional fourth-order problems comprise many important mechanical applications. For example, different kind of problems concerning the displacements and stresses of beam-mass systems or cable-mass systems could be mentioned [6]. Variational aspects of the one-dimensional fourth-order problems are presented in [7].

Consider a family of regular finite element partitions τ_h of $[0,l]$ which fulfill standard assumptions (see [5]). The nodes of any partition are s_j, $j = 0,1,\ldots,2k$. Then we define the subintervals $T_j = [s_{j-1}, s_j]$, $j = 1,\ldots,2k$, such that $\tau_h = \bigcup_{j=1}^{2k} T_j$. The following denotations are adopted:

$$h_j = s_j - s_{j-1}, \quad x_j = \frac{s_{j-1} + s_j}{2}, \quad h = \max_{1 \leq j \leq 2k} h_j.$$

With a partition τ_h we associate a finite-dimensional subspace $V_h \subset V \cap C^1[0,l]$ such that the restriction of every function $v \in V_h$ on every interval $T_j \in \tau_h$ is a polynomial of degree n at most.

The patch-recovery technique in considered cases requires $n \geq 4$. For the sake of practical application we provide the investigation of the lowest redundant finite elements, i.e. $n = 4$.

The discrete problem corresponding to (1) is

$$a(u_h, v) = b(f, v), \quad \forall v \in V_h. \tag{3}$$

The proof of superconvergence result consists of two main steps:

(i) Obtaining in the same finite element space an interpolation which approximates of higher order the solution $u(x)$. Thus an interpolant p_h has a special construction and it is called to satisfy a superclose property, cf. [8];

(ii) Construction of higher order interpolation π_{2h} of the original finite element solution to achieve better accuracy. The interpolated finite element solution results from the postprocessing [3].

2 Operator $\mathrm{p_h}$ - Estimates and Properties

In the begining introduce the operator $p_h : C^1(0,l) \to V_h$ under the following conditions (degrees of freedom):

$$p_h v(s_j) = v(s_j), \quad j = 0,\ldots,2k,$$

$$\frac{d\, p_h v}{dx}(s_j) = \frac{dv}{dx}(s_j), \quad j = 0,\ldots,2k,$$

$$\int_{T_j} p_h v(x)\, dx = \int_{T_j} v(x)\, dx, \quad j = 1,\ldots,2k.$$

Obviously, $p_h v(x)$ consists of piecewise polynomials of degree four at most. The basic functions $\{\varphi_i\}_{i=1}^5$ determined the operator $p_h$on the reference element $[0,1]$ are:

$$\begin{pmatrix} \varphi_1(t) \\ \varphi_2(t) \\ \varphi_3(t) \\ \varphi_4(t) \\ \varphi_5(t) \end{pmatrix} = \frac{1}{2} \begin{pmatrix} -30 & 64 & -36 & 0 & 2 \\ -5 & 12 & -9 & 2 & 0 \\ 60 & -120 & 60 & 0 & 0 \\ -30 & 56 & -24 & 0 & 0 \\ 5 & -8 & 3 & 0 & 0 \end{pmatrix} \begin{pmatrix} t^4 \\ t^3 \\ t^2 \\ t \\ 1 \end{pmatrix}, \quad t \in [0,1].$$

We adopt the following denotations:

$$\widehat{v}_{1,j} = v(s_{j-1}), \ \widehat{v}_{2,j} = \frac{dv}{dx}(s_{j-1}),$$

$$\widehat{v}_{3,j} = \int_{T_j} v(x)\,dx, \ \widehat{v}_{4,j} = v(s_j), \ \widehat{v}_{5,j} = \frac{dv}{dx}(s_j).$$

Hence the presentation of any function reads:

$$v(x)|_{T_j} = \widehat{v}_{1,j}.\varphi_1(\frac{x-x_j}{h_j}+\frac{1}{2}) + \widehat{v}_{2,j}.h_j.\varphi_2(\frac{x-x_j}{h_j}+\frac{1}{2})$$

$$+\widehat{v}_{3,j}\frac{1}{h_j}.\varphi_3(\frac{x-x_j}{h_j}+\frac{1}{2}) + \widehat{v}_{4,j}.\varphi_4(\frac{x-x_j}{h_j}+\frac{1}{2}) + \widehat{v}_{5,j}.h_j.\varphi_5(\frac{x-x_j}{h_j}+\frac{1}{2}).$$

Let us introduce the error function for any finite element $T_j \in \tau_h$:

$$E(x) = \frac{1}{2}\left[(x-x_j)^2 - \left(\frac{h_j}{2}\right)^2\right].$$

The following identities are important for proving the superclose property of operator p_h:

$$\begin{aligned} x - x_j &= \tfrac{1}{90}\left[E^3(x)\right]^{(5)}; \\ (x-x_j)^2 &= \tfrac{1}{1260}\left[E^4(x)\right]^{(6)} + \tfrac{h_j^2}{28}; \\ (x-x_j)^3 &= \tfrac{1}{18900}\left[E^5(x)\right]^{(7)} + \tfrac{h_j^2}{5400}\left[E^3(x)\right]^{(5)}; \\ (x-x_j)^4 &= \tfrac{1}{311850}\left[E^6(x)\right]^{(8)} + \tfrac{h_j^2}{9240}\left[E^4(x)\right]^{(6)} + \tfrac{h_j^4}{336}. \end{aligned} \tag{4}$$

Consequently, two other presentations are needed:

$$\begin{aligned} (x-x_j)^3 &= \tfrac{1}{420}\left[E^4(x)\right]^{(5)} + \tfrac{h_j^2}{840}\left[E^3(x)\right]^{(5)}; \\ (x-x_j)^4 &= \tfrac{1}{4725}\left[E^5(x)\right]^{(6)} + \tfrac{h_j^2}{7560}\left[E^4(x)\right]^{(6)} + \tfrac{h_j^4}{336}. \end{aligned} \tag{5}$$

The crucial point in our investigations is the following lemma:

Lemma 1. *Let $u(x)$ be the sufficiently smooth solution of the problem (1). Then for any $v \in V_h$ the following estimates are valid:*

if $p = q = 0$ *then* $a(p_h u - u, v) = 0, \qquad u \in H^2(0,l)$ *else*

if $p = 0$ *then* $a(p_h u - u, v) \le Ch^6 \|u\|_{7,(0,l)} \|v\|_{2,(0,l)}, \quad u \in H^7(0,l)$ *else*

$$a(p_h u - u, v) \le Ch^5 \|u\|_{6,(0,l)} \|v\|_{2,(0,l)}, \quad u \in H^6(0,l).$$

Proof. For $x \in T_j$ and for every $v \in V_h$, there holds

$$v^{(s)}(x) = \sum_{k=s}^{4} \frac{(x - x_j)^{k-s}}{(k-s)!} v^{(k)}(x_j), \quad s = 0, 1, 2. \tag{6}$$

Consider the bilinear form $a(U, v),\ v \in V_h$, where $U = p_h u - u$.

Case 1. $p = q = 0$. From (2), it follows

$$a(U, v) = \int_0^l U'' v''\, dx.$$

We use the properties of the interpolant p_h as well as the equality (6) when $s = 2$. Integrating by parts in the interval T_j reveals that

$$\int_{T_j} U'' v''\, dx = 0,$$

consequently $a(p_h u - u, v) = 0$.

Case 2. $p = 0$. Then from (2)

$$a(U, v) = \int_{(0,l)} [U'' v'' + qUv]\, dx.$$

The first term in the right-hand side is calculated in the previous case. Thus for the second one we use the expansion (6) with $s = 0$:

$$\int_{T_j} U.v\, dx = \sum_{k=0}^{4} \int_{T_j} U(x) \frac{(x - x_j)^k}{k!} v^{(k)}(x_j)\, dx. \tag{7}$$

We transform each term in the right-hand side of (7). The definition of the operator p_h gives

$$\int_{T_j} U(x).v(x_j)\, dx = 0. \tag{8}$$

Using the relations (4) and integrating by parts we obtain successively

$$\int_{T_j} U(x).(x - x_j).v'(x_j)\, dx = -\frac{1}{90} \int_{T_j} U^{(5)}(x).E^3(x).v'(x_j)\, dx; \tag{9}$$

$$\int_{T_j} U(x).\frac{(x-x_j)^2}{2!}.v''(x_j)\,dx = \frac{1}{2520}\int_{T_j} U^{(6)}(x).E^4(x).v''(x_j)\,dx; \qquad (10)$$

$$\int_{T_j} U(x).\frac{(x-x_j)^3}{3!}.v'''(x_j)\,dx = -\frac{1}{113400}\int_{T_j} U^{(7)}(x).E^5(x).v'''(x_j)\,dx$$
$$-\frac{h_j^2}{32400}\int_{T_j} U^{(5)}(x).E^3(x).v'''(x_j)\,dx. \qquad (11)$$

In order to avoid the requirement for the increase of smoothness of the solution $u(x)$, we apply the second equality of (5) for the last term in (8):

$$\int_{T_j} U(x).\frac{(x-x_j)^4}{4!}.v^{IV}(x)\,dx = \frac{1}{113400}\int_{T_j} U^{(6)}(x).E^5(x).v^{IV}(x)\,dx$$
$$+\frac{h_j^2}{181440}\int_{T_j} U^{(6)}(x).E^4(x).v^{IV}(x)\,dx. \qquad (12)$$

Finally, we use the equality

$$v^{(s)}(x_j) = \sum_{k=s}^{4} \frac{(x_j-x)^{k-s}}{(k-s)!} v^{(k)}(x), \quad s = 1;2;3$$

in combination with the inverse inequality. Inserting (8)-(12) into (7) after summing with respect to $T_j \in \tau_h$ leads to

$$a(U,v) \le C.h^6 \|u\|_{7,(0,l)} \|v\|_{2,(0,l)}.$$

Case 3. This is a general situation, namely $a(U,v) = \int_0^l [U''v'' + pU'v" + qUv]dx$. It remains to estimate $\int_0^l pU'v'\,dx$. From (6) for $s = 1$, it follows

$$\int_{T_j} U'v'\,dx = \sum_{k=1}^{4} \int_{T_j} U'(x)\frac{(x-x_j)^{k-1}}{(k-1)!} v^{(k)}(x_j)\,dx, \quad T_j \in \tau_h.$$

Integrating by parts reveals that

$$\int_{T_j} U'v'\,dx = -\sum_{k=2}^{4} \int_{T_j} U(x)\frac{(x-x_j)^{k-2}}{(k-2)!} v^{(k)}(x_j)\,dx. \qquad (13)$$

As a consequence of the properties of the operator p_h we get for each term in the right-hand side of (13):

$$\int_{T_j} U(x)v''(x_j)\,dx = 0; \qquad (14)$$

$$\int_{T_j} U(x)(x-x_j)v'''(x_j)\,dx = \frac{1}{90}\int_{T_j} U^{(5)}(x)E^3(x)v'''(x_j)\,dx; \qquad (15)$$

$$\int_{T_j} U(x)\frac{(x-x_j)^2}{2} v^{IV}(x)\,dx = \frac{1}{2520}\int_{T_j} U^{(6)}(x)E^4(x)v^{IV}(x)\,dx, \qquad (16)$$

where for the last two equalities the relations (4) are used.

It should be noted that $v'''(x_j) = v'''(x) + (x_j - x)v^{IV}(x)$.

¿From the equalities (14)-(16) after summing on $T_j \in \tau_h$ and using the inverse inequality we obtain

$$\int_0^l pU'v'\,dx \le C.h^5\|u\|_{6,(0,l)}\|v\|_{2,(0,l)}. \tag{17}$$

For the third term of $a(U,v)$ we apply the first equality of (5) instead of (11). This is due to the fact that in (17) the requirement of smoothness is $u(x) \in H^6(0,l)$. Thus, it could be written

$$\int_{T_j} U(x).\frac{(x-x_j)^3}{3!}.v'''(x_j)\,dx = -\frac{1}{2520}\int_{T_j} U^{(5)}(x).E^4(x).v'''(x_j)\,dx$$

$$+\frac{h_j^2}{5040}\int_{T_j} U^{(5)}(x).E^3(x).v'''(x_j)\,dx.$$

After summarizing on the elements $T_j \in \tau_h,\ j = 1,\ldots,2k$ for this case we have

$$a(U,v) \le C.h^5\|u\|_{6,(0,l)}\|v\|_{2,(0,l)},$$

which completes the proof.

Remark 1. If $a(u,v) = \int_0^l \alpha u''v''\,dx,\ \alpha = const. > 0$, then p_h is an elliptic projection with respect to this $a-$form. Namely, for any $u \in H^2(0,l)$,

$$a(p_hu - u, v) = 0 \quad \forall v \in V_h.$$

Remark 2. The results of Lemma 1 are valable when some additional terms are added to the $a-$form [6]. For example, consider the problem

$$\widetilde{a}(u,v) = (f,v),$$

where $\widetilde{a}(u,v) = a(u,v) + C_1u'(0)v'(0) + C_0u(0)v(0)$. It is evident that $\widetilde{a}(U,v) = a(U,v)$.

The results of Lemma 1 may be recorded as the following theorem:

Theorem 1. *Let u_h be the finite element solution of (3) and let p_hu be the interpolation of the exact solution $u(x)$ of (1) determined in Lemma 1. Then, the following estimates hold:*

$\|p_hu - u_h\|_{2,(0,l)} = 0$ *if* $p = q = 0$;

$\|p_hu - u_h\|_{2,(0,l)} \le C.h^6\|u\|_{7,(0,l)}$ *if* $p = 0$;

$\|p_hu - u_h\|_{2,(0,l)} \le C.h^5\|u\|_{6,(0,l)}$ *if* $p \ne 0$.

Proof. Having in mind that $p_hu - u_h \in V_h$, from the V_h-ellipticity of $a(\cdot,\cdot)$ it can be written

$$C\|p_hu - u_h\|^2_{2,(0,l)} \le a(p_hu - u_h, p_hu - u_h) = a(p_hu - u, p_hu - u_h).$$

By substituting $v = p_hu - u_h$ in the results of Lemma 1 we complete the proof.

3 Main Result

The superclose properties of the interpolant p_h permit to construct a patch-recovery operator $\pi_{2h} : C^1(0,l) \to \overline{V}_{2h}$, where

$$\overline{V}_{2h} = \{v : \ v|_{T_{2j-1}\cup T_{2j}} \in P_7, \ j = 1,\ldots,k\}.$$

Consider the intervals $T_{2j-1} = [s_{2j-2}, s_{2j-1}]$ and $T_{2j} = [s_{2j-1}, s_{2j}]$, $j = 1,\ldots,k$. Then the following set can be introduced

$$\overline{\tau}_{2h} = \{\overline{T}_j = T_{2j-1} \cup T_{2j}, \ j = 1,\ldots,k\}.$$

On each finite element $\overline{T}_j$ we define the operator $\pi_{2h} : C^1(0,l) \to \overline{V}_{2h}$ such that

$$\begin{aligned} &\pi_{2h}v(s_i) = v(s_i), \quad i = 2j-2, 2j-1, 2j;\\ &\frac{d\pi_{2h}v}{dx}(s_i) = \frac{dv}{dx}(s_i), \quad i = 2j-2, 2j-1, 2j;\\ &\int_{T_j} \pi_{2h}v(x)\,dx = \int_{T_j} v(x)\,dx, \quad i = 2j-1, 2j. \end{aligned} \tag{18}$$

¿From the basic properties (18) it easily follows

$$\pi_{2h} \circ p_h = \pi_{2h}, \tag{19}$$

as well as the approximation result for any $v(x) \in H^7(0,l)$:

$$\|\pi_{2h}v - v\|_{2,(0,l)} \leq C.h^6 \|v\|_{7,(0,l)}. \tag{20}$$

We need also the boundness of the operator π_{2h}, i.e. if $\pi_{2h} : V_h \to \overline{V}_{2h}$, then this operator has a finite range. It follows that $dim\mathcal{R}(\pi_{2h}) < \infty$. Therefore, it is compact. Finally

$$\|\pi_{2h}\| \leq \sup_{v_h \in V_h} \frac{\|\pi_{2h}v_h\|_{2,(0,l)}}{\|v_h\|_{2,(0,l)}} \leq C = const. \tag{21}$$

Our superconvergence result is contained in the following theorem:

Theorem 2. *Let $u(x)$ and $u_h(x)$ be the solutions of (1) and (3), respectively. Supposing that the conditions of Lemma 1 are fulfilled, the following estimates hold:*
$\|\pi_{2h}u_h - u\|_{2,(0,l)} \leq C.h^5 \|u\|_{6,(0,l)}$ *when* $p \neq 0$ *and*

$\|\pi_{2h}u_h - u\|_{2,(0,l)} \leq C.h^6 \|u\|_{7,(0,l)}$ *otherwise.*

Proof. It follows from (19) and (20) that

$$\begin{aligned} \|\pi_{2h}u_h - u\|_{2,(0,l)} &\leq \|\pi_{2h}u_h - \pi_{2h}u\|_{2,(0,l)} + \|\pi_{2h}u - u\|_{2,(0,l)}\\ &= \|\pi_{2h}u_h - \pi_{2h} \circ p_h u\|_{2,(0,l)} + \|\pi_{2h}u - u\|_{2,(0,l)}\\ &\leq \|\pi_{2h}\|.\|u_h - p_h u\|_{2,(0,l)} + \|\pi_{2h}u - u\|_{2,(0,l)}. \end{aligned}$$

In order to complete the proof we apply (21) and the result of Theorem 1.

4 Numerical Example

To illustrate our theoretical results, we consider the following simple problem:

$$y^{IV} = x^3(1-x)^3, \quad x \in (0,1),$$

$$y(x)|_{x=0;1} = y'(x)|_{x=0;1} = 0.$$

The exact solution is the function $u(x) = \frac{1}{7!}(2x^2 - 3x^3 + 6x^7 - 9x^8 + 5x^9 - x^{10})$.

Polynomials of degree four define the interpolation functions of each element. Using the finite element solution u_h we construct a-posteriori interpolated procedure by defining $\pi_{2h}u_h$.

The results in the table below show that the calculations with $h = \frac{1}{16}$ using the operator π_{2h} are comparable with the finite element solution obtained when step h is four time smaller.

h	1/16	1/32	1/64
$\|u - u_h\|_{2,\Omega}$	2.2×10^{-7}	2.8×10^{-8}	3.5×10^{-9}
$\|u - \pi_{2h}u_h\|_{2,\Omega}$	2.8×10^{-9}	4.9×10^{-11}	7.9×10^{-13}

References

[1] Ainsworth M., Oden J. T.: A Posteriori Error Estimation in Finite Element Analysis. Wiley Interscience, New York, 2000.

[2] Brandt J., M. Křížek: History and Future of Superconvergence in Three-Dimensional Finite Element Methods. Preprint N1155, Utrecht University, 2001.

[3] Lin Q., Yan N. and Zhou A.: A rectangular test for interpolated finite elements. Proceedings of Systems Science & Systems Engineering, Culture Publish Co., 1991, 217-229.

[4] Zienkiewicz O. C., Zhu J. Z.: The superconvergence patch recovery and a posteriori error estimates. Part 1: The recovery technique. Int. J. Numer. Methods Engrg. **33**, 1331-1364 (1992).

[5] Ciarlet, P.G.: The Finite Element Method for Elliptic Problems. North-Holland Amsterdam 1978.

[6] Collatz L.: Eigenwertaufgaben mit Technischen Anwendungen. Leipzig, Acad. Verlag, 1963.

[7] Racheva M. R., Andreev A. B.: Variational aspects of one-dimensional fourth-order problems with eigenvalue parameter in the boundary conditions. Sib. JNM, No4(5), 2002, pp. 373-380.

[8] Wahlbin L.B.: Superconvergence in Galerkin FEM. Springer-Verlag, Berlin, 1995.

Modelling of the Elastic Line for Twist Drill with Straight Shank Fixed in Three-Jaw Chuck

Andrey B. Andreev, Jordan T. Maximov, and Milena R. Racheva

Technical University of Gabrovo
{Andreev, Maximov, Milena}@tugab.bg

Abstract. The aim of this study is to present a new approach for investigation of an important problem of mechanical engineering. Namely, the model of twist drill embeded in three-jaw chuck is discussed. This problem could be considered as a variant of a beam on the Winckler's base which is under the influence of a cross-force.

Our principal aim is to deduce the general mathematical model for this type of constructions. In order to determine the dynamic stresses of the drill using any variational numerical methods we present the corresponding variational formulations. These presentations are characterized by mixed formulation. So, the mixed finite element method is convenient for this kind of problems. The possibility for symmetrization of the weak formulation of the model problem is also discussed.

Keywords: twist drill, mathematical model, variational formulation

1 Introduction

The model of continuous beam on elastic base of Winckler's type is well-known in engineering practice (see [6]). Usually, it is assumed that the coefficient of the Winckler's base is a constant at any direction, orthogonal to the axis of the beam. For example, this model is applied to the cantilever straight or taper beams which are fixed in straight or taper holes, respectively [5]. A good illustration to this kind of model is a drill with taper shank fixed at the tailstock of lathe.

Normally in engineering practice it is either the cutting tool with straight shank or the workpiece that is fixed in the three-jaw drill chuck. The free end-point of the drill/workpiece is under impact of the transverse cutting force. In the case of twist drill, for example, such force results in a bending moment at any cross-section. When drilling follows the pattern "rotative tool - stationary workpiece", according to the principle of inversion the transverse force rotates round the axis of the tool. This class of problems presents a model that could be refered to: "Beam on the Winckler's type base with variable rigidity". The authors solved this problem in [2], when bending is in the main and nonchanging plane on inertia (see also [1]). The general case is a beam with continuous and slow changing cross section whereby the main planes of inertia corresponding to the various cross sections are different.

Z. Li et al. (Eds.): NAA 2004, LNCS 3401, pp. 116–124, 2005.

It corresponds to twist drills, when both of the main axes of inertia in the corresponding different cross-sections describe two twisting surfaces.

Accordingly, the principal aim of this paper is to present a new approach for determining of dynamic stresses when constrained frequencies in steady state are considered. The beam (drill) is partially chucked at the elastic base with variable rigidity and its free end is loaded by transverse force which rotates round the axis with constant angular velosity.

Special emphasis is put on mathematical modelling as well as on the variational aspects of the considered problem.

2 General Mathematical Model for the Elastic Line

Consider a straight beam with length L partially located on an elastic foundation. The beam is composed of two parts (see Figure 1). At the free part with length l two groves are cut with a screw pitch of both twisting lines that equals s. A straight beam is represented by the part with length $L-l$. Then the coefficient of the base is not the same in any direction perpendicular to the beam axis.

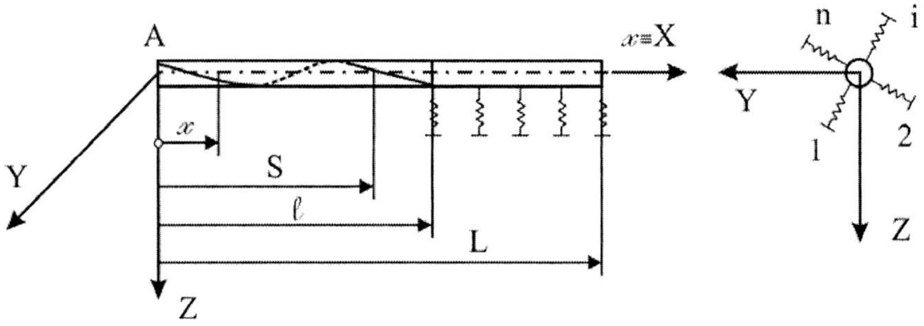

Fig. 1.

The center C_A of the cross-section A is dislocated within a distance $|\overrightarrow{f}|$ from the $x-$axis due to an unknown force P (see Figure 2a). The beam rotates with an angular velocity $\omega = const$. In general, the two vectors $\overrightarrow{f}$ and $\overrightarrow{P}$ are not colinear.

Now we present an *inversion approach*: The considered system rotates with angular velocity ω. Then the beam is fixed and the vector $\overrightarrow{f}$ (or $\overrightarrow{P}$ respectively) rotates with angular velocity ω.

The general coordinates are the displacements v and w of the center of gravity of the variable cross-section in the direction of the main axes of inertia η and ζ (Figure 2b). In the straight part $x \in (l, L)$ the same considerations are adopted for variability of η and ζ inspite of the fact that all central axes for straight cross-section are main axes of inertia. The goal is to ensure a continuity of variation of η and ζ at the point $x = l$.

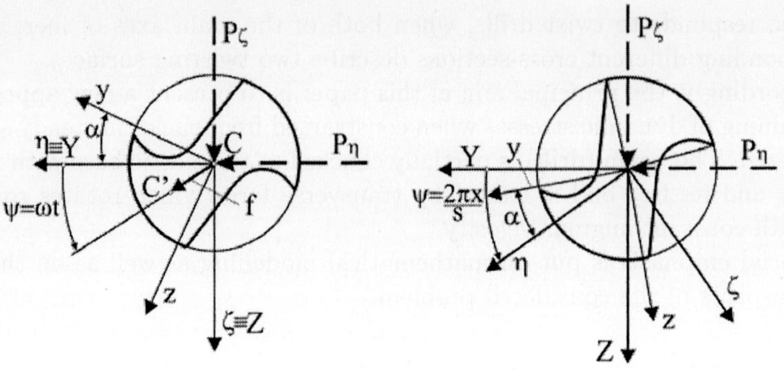

Fig. 2.

Let the straight beam section rest on elastic supports. These supports are spatially located and infinitely close from each other in the direction of the beam axis. It is necessary to determine the stiffness matrix of the elastic foundation in the coordinate system $C\eta\zeta$ (Figure 3). For this purpose, we consider any cross-section uniformly supported by n springs (supports). The i–th spring has the stiffness $C_i, i = 1, \ldots, n$. Springs axes intersect at the center of the gravity of the cross-section and they form equal angles (Figure 3). The stiffness at the node C in the direction $O\eta$ (or $O\zeta$) will present the coefficient of the Winckler's base. This coefficient depends on the angle θ and consequently - on the abscissa x.

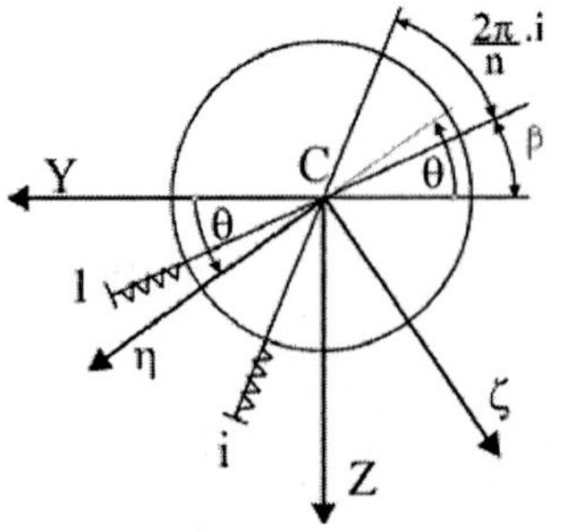

Fig. 3.

Let us consider the i–th spring presented in Figure 4. The projections $R_{\eta,i}$ and $R_{\zeta,i}$ of the reactions at the point C of the simultaneous displacements $v_i = 1, w_i = 1$, are:

$$\begin{aligned} R_{\eta,i} &= C_i \left[\cos\varphi_i \cos\varphi_i + \sin\varphi_i \cos\varphi_i\right], \\ R_{\zeta,i} &= C_i \left[\cos\varphi_i \sin\varphi_i + \sin\varphi_i \sin\varphi_i\right]. \end{aligned} \tag{1}$$

Nodal displacements can be noted by $\{\delta_i\} = [v_i w_i]^T$. Hence, the relation (1) could be written in matrix form: $\{R_i\} = [K_{ii}] . \{\delta_i\}$, where

$$[K_{ii}] = C_i \begin{pmatrix} \cos^2\varphi_i & \sin\varphi_i \cos\varphi_i \\ \sin\varphi_i \cos\varphi_i & \sin^2\varphi_i \end{pmatrix},$$

$\varphi_i = \frac{2\pi}{n} i + \beta - \theta$, and β is the angle between $y-$axes and the first spring, i.e. which corresponds to $i = 1$ (Figure 3). In coupling the matrices of all the n springs, for the global stiffness matrix we obtain:

$$[K] = \sum_{i=1}^{n} C_i \begin{pmatrix} \cos^2 \varphi_i & \sin \varphi_i \cos \varphi_i \\ \sin \varphi_i \cos \varphi_i & \sin^2 \varphi_i \end{pmatrix}.$$

The general mathematical model for the elastic line is deduced under the following conditions:

- The elastic line is a function of the abscissa x and the time-variable t;
- The displacement functions v, w of the centers of gravity for the variable sections are determined in the direction of main axes of inertia η, ζ;
- There is a linear dependence between displacements and deformations. This allows for the curvatures in the main planes of inertia $x\eta$ and $x\zeta$ to be defined only through the second derivative with respect to x:

$$K_{x\eta} = \frac{\partial^2 v}{\partial x^2}, \quad K_{x\zeta} = \frac{\partial^2 w}{\partial x^2};$$

- The bending moments are proportional to the curvatures:

$$M_\eta = -K_{x\eta} E J_\eta, \quad M_\zeta = K_{x\zeta} E J_\zeta,$$

  where E is the modulus of elasticity and $J_\eta(x)$ and $J_\zeta(x)$ are axial moments of inertia.
- There is a relation between the bending moments and distributed loads in the main planes of inertia:

$$\frac{\partial^2 M_\eta}{\partial x^2} = -q_\zeta, \quad \frac{\partial^2 M_\zeta}{\partial x^2} = -q_\eta.$$

- The load of inertia is a result of the transverse oscillations in $x\eta$ and $x\zeta-$plane respectively with intensities:

$$q_\eta = \rho F(x) \frac{\partial^2 v}{\partial t^2}, \quad q_\zeta = \rho F(x) \frac{\partial^2 w}{\partial t^2},$$

  where ρ is the density and $F(x)$ is a cross-section area of the beam.

The distributed load of reaction from the Winckler's type base has an intensity C_η and C_ζ in the directions of η and ζ respectiveely

$$C_\eta = (A + B \cos ax + D \sin ax)\, v + (D \cos ax - B \sin ax)\, w,$$

$$C_\zeta = (A - B \cos ax - D \sin ax)\, w + (D \cos ax - B \sin ax)\, v,$$

where $a = \frac{4\pi}{s}$. Note that the relation between x, θ and the screw pitch s is presented by $\theta = \frac{2\pi}{s} x$.

For any $x \in (0, L)$ (see Figure 1) the bending moments act on the corresponding cross-section

$$M_\eta = P_\eta(x)\sin ax - P_\varsigma(x)\cos ax,$$

$$M_\varsigma = P_\eta(x)\cos ax + P_\varsigma(x)\sin ax.$$

Let us define the continuous functions of displacements $V(x;t)$ and $W(x;t)$ respectively:

$$V(x;t) = \begin{cases} \overline{V}(x;t), \, x \in [0,l] \times [0,T], \\ \widehat{V}(x;t), \, x \in (l,L] \times [0,T], \end{cases} \quad W(x;t) = \begin{cases} \overline{W}(x;t), \, x \in [0,l] \times [0,T], \\ \widehat{W}(x;t), \, x \in (l,L] \times [0,T]. \end{cases}$$

The main moments of inertia J_η, J_ς and the corresponding areas of cross-sections are:

$$J_\eta(x) = \begin{cases} \overline{J}_\eta, \, x \in [0,l], \\ \widehat{J}_\eta, \, x \in (l,L], \end{cases} \quad J_\varsigma(x) = \begin{cases} \overline{J}_\varsigma, \, x \in [0,l], \\ \widehat{J}_\varsigma, \, x \in (l,L], \end{cases} \quad F(x) = \begin{cases} \overline{F}, \, x \in [0,l], \\ \widehat{F}, \, x \in (l,L]. \end{cases}$$

We assume that the distributive damping is negligeable.

The general mathematical model of the elastic line of the beam model is:

$$E\overline{J}_\varsigma \frac{\partial^4 \overline{V}}{\partial x^4} + \rho \overline{F} \frac{\partial^2 \overline{V}}{\partial t^2} - (2P_\varsigma - P_\eta ax)a\cos ax + (2P_\eta - P_\varsigma ax)a\sin ax = 0,$$

$$E\overline{J}_\eta \frac{\partial^4 \overline{W}}{\partial x^4} - \rho \overline{F} \frac{\partial^2 \overline{W}}{\partial t^2} + (2P_\eta + P_\varsigma ax)a\cos ax + (2P_\varsigma - P_\eta ax)a\sin ax = 0,$$

$$0 < x < l; \tag{2}$$

$$E\widehat{J}_\varsigma \frac{\partial^4 \widehat{V}}{\partial x^4} + \rho \overline{F} \frac{\partial^2 \overline{V}}{\partial t^2} - (A + B\cos ax + D\sin ax)\widehat{V} - (D\cos ax - B\sin ax)\widehat{W}$$
$$-(2P_\varsigma - P_\eta ax)a\cos ax + (2P_\eta - P_\varsigma ax)a\sin ax = 0,$$
$$E\widehat{J}_\eta \frac{\partial^4 \widehat{W}}{\partial x^4} - \rho \overline{F} \frac{\partial^2 \overline{W}}{\partial t^2} + (A - B\cos ax - D\sin ax)\widehat{W} + (D\cos ax - B\sin ax)\widehat{V}$$
$$+(2P_\eta + P_\varsigma ax)a\cos ax + (2P_\varsigma - P_\eta ax)a\sin ax = 0, \; l < x < L.$$

We consider the equation (2) with initial conditions

$$\frac{\partial^j \overline{V}}{\partial t^j}(x;0) = \overline{V}^j(x), \; \frac{\partial^j \overline{W}}{\partial t^j}(x;0) = \overline{W}^j(x),$$
$$\frac{\partial^j \widehat{V}}{\partial t^j}(x;0) = \widehat{V}^j(x), \; \frac{\partial^j \widehat{W}}{\partial t^j}(x;0) = \widehat{W}^j(x), \quad j = 0;1. \tag{3}$$

The system (2) formally can be united using the unit Heviside function $\eta(x)$:

$$\frac{\partial^2}{\partial x^2}\left(EJ_\varsigma(x)\frac{\partial^2 V}{\partial x^2}\right) + \rho F(x)\frac{\partial^2 V}{\partial t^2} - (2P_\varsigma - P_\eta ax)a\cos ax + (2P_\eta - P_\varsigma ax)a\sin ax$$
$$-\left[\eta(x-l) - \eta(x)\right]\left[(A + B\cos ax + D\sin ax)V + (D\cos ax - B\sin ax)W\right] = 0, \tag{4}$$

$$\frac{\partial^2}{\partial x^2}\left(EJ_\eta(x)\frac{\partial^2 W}{\partial x^2}\right)-\rho F(x)\frac{\partial^2 W}{\partial t^2}+(2P_\eta+P_\zeta ax)a\cos ax+(2P_\zeta-P_\eta ax)a\sin ax$$

$$+\left[\eta(x-l)-\eta(x)\right]\left[(A-B\cos ax-D\sin ax)W+(D\cos ax-B\sin ax)V\right]=0.$$

Consider the boundary conditions related to the equation (4):

$$\frac{\partial^2 V}{\partial x^2}(0;t)=0,\quad \frac{\partial^2 W}{\partial x^2}(0;t)=0,\quad \frac{\partial^3 V}{\partial x^3}(0;t)=P_\eta,\quad \frac{\partial^3 W}{\partial x^3}(0;t)=P_\zeta,$$

$$\frac{\partial^3 V}{\partial x^3}(l;t)=P_\zeta\sin al+P_\eta\cos al,\quad \frac{\partial^3 W}{\partial x^3}(l;t)=P_\zeta\cos al-P_\eta\sin al, \tag{5}$$

$$\frac{\partial^2 V}{\partial x^2}(L;t)=0,\quad \frac{\partial^2 W}{\partial x^2}(L;t)=0,\quad \frac{\partial^3 V}{\partial x^3}(L;t)=0,\quad \frac{\partial^3 W}{\partial x^3}(L;t)=0.$$

We add also the conditions of smoothness of the functions V, W. The functions $V^{(s)}(x,t), W^{(s)}(x,t),\ s=0,1$ are continuous $\forall x\in[0,L],\ t>0$. In particular

$$\overline{V}^{(s)}(l;t)=\widehat{V}^{(s)}(l;t),\overline{W}^{(s)}(l;t)=\widehat{W}^{(s)}(l;t),\ s=0,1. \tag{6}$$

The last equations are conforming conditions between the unknown functions $\overline{V},\widehat{V}$ and $\overline{W},\widehat{W}$.

Thus the model problem for a beam on the Winckler's type base is expressed by the fourth-order differential equation (4) and with initial and boundary conditions (3) and (5)-(6) respectively.

3 Variational Formulation

Any variational numerical method (for example, finite element method) requires to present the considered problem (4)-(5) with initial conditions (3), in weak form. Let H^k be the usual Sobolev space for positive integeeer k [4].

Our aim here is to present a new approach in order to obtain a weak formulation of the considered problem. That is a possibility of symmetrization of the variational formulation. In order to obtain free vibrations of this problem one sets $P_\eta=P_\zeta=0$. We multiply both equation of (4) by $v(x)$ and $w(x)\in H^2[0,L]$ respectively. Integrating by parts and using (5) reveals that

$$-\int_0^L EJ_\zeta(x)\frac{\partial^2 V}{\partial x^2}\frac{\partial^2 v}{\partial x^2}\,dx+\int_l^L\left[A+B\cos ax+D\sin ax\right]Vv\,dx$$

$$+\int_l^L\left[D\cos ax-B\sin ax\right]Wv\,dx+\left[E\widehat{J}_\zeta\frac{\partial^2 V}{\partial x^2}(l^+)-E\overline{J}_\zeta\frac{\partial^2 V}{\partial x^2}(l^-)\right]\frac{\partial v}{\partial x}(l)$$

$$=\frac{d^2}{dt^2}\int_0^L\rho F(x)Vv\,dx, \tag{7}$$

$$\int_0^L EJ_\eta(x)\frac{\partial^2 W}{\partial x^2}\frac{\partial^2 w}{\partial x^2}\,dx + \int_l^L [A - B\cos ax - D\sin ax]\,Ww\,dx$$

$$+\int_l^L [D\cos ax - B\sin ax]\,Vw\,dx + \left[E\widehat{J}_\eta\frac{\partial^2 W}{\partial x^2}(l^+) - E\overline{J}_\eta\frac{\partial^2 W}{\partial x^2}(l^-)\right]\frac{\partial w}{\partial x}(l)$$

$$= \frac{d^2}{dt^2}\int_0^L \rho F(x)Ww\,dx.$$

Variational formulation (7) is not symmetric in general due to the fact that the functions $J_\eta(x), J_\zeta(x)$, $\frac{\partial^2 V}{\partial x^2}$ and $\frac{\partial^2 W}{\partial x^2}$ are discontinuous at the point $x = l$. Now we propose a smoothness procedure to the coefficient functions as well as to the unknown functions in (4)-(5).

Let ε be a positive number approximating the zero. Introduce the functions $(s = \eta; \zeta)$):

$$\widetilde{J}_s(x) = \begin{cases} \overline{J}_s, & x \in [0, l-\varepsilon], \\ \sum_{i=0}^{3} k_s^i (x-l)^i, & x \in (l-\varepsilon, l+\varepsilon), \\ \widehat{J}_s, & x \in [l+\varepsilon, L], \end{cases}$$

where $k_s^0 = \frac{\overline{J}_s - \widehat{J}_s}{2}, k_s^1 = \frac{3(\widehat{J}_s - \overline{J}_s)}{4\varepsilon}, k_s^2 = 0, k_s^3 = \frac{\overline{J}_s - \widehat{J}_s}{2\varepsilon}$.

Thus the coefficients $k_s^i, i = 0, \ldots, 3$ ensure relations $\frac{\partial^j \widetilde{J}_s}{\partial x^j}(l \pm \varepsilon) = \frac{\partial^j J_s}{\partial x^j}(l \pm \varepsilon)$, $j = 0; 1$. By analogy it could be defined the function $\widetilde{F}(x)$ corresponding to $F(x)$.

Next, for $t \in [0, T]$ let us define the functions:

$$\widetilde{V}(x;t) = \begin{cases} \overline{V}(x;t), & x \in [0, l-\varepsilon], \\ \sum_{i=0}^{7} m_1^i (x-l)^i, & x \in (l-\varepsilon, l+\varepsilon), \\ \widehat{V}(x;t), & x \in [l+\varepsilon, L], \end{cases}$$

$$\widetilde{W}(x;t) = \begin{cases} \overline{W}(x;t), & x \in [0, l-\varepsilon], \\ \sum_{i=0}^{7} m_2^i (x-l)^i, & x \in (l-\varepsilon, l+\varepsilon), \\ \widehat{W}(x;t), & x \in [l+\varepsilon, L], \end{cases}$$

where the coefficients $m_j^i, i = 0, \ldots, 7, j = 1; 2$ are determined in such a way that

$$\frac{\partial^3 \widetilde{V}}{\partial x^3}(l;t) = \frac{\partial^3 V}{\partial x^3}(l;t) = 0, \qquad \frac{\partial^3 \widetilde{W}}{\partial x^3}(l;t) = \frac{\partial^3 W}{\partial x^3}(l;t) = 0.$$

Consequently, the following equality are valid almost everywhere for $x \in [0, L]$ and $\varepsilon \to 0, s = \eta, \zeta$: $\widetilde{J}_s(x) = J_s(x), \widetilde{F}(x) = F(x), \widetilde{V}(x;t) = V(x;t), \widetilde{W}(x;t) = W(x;t)$.

Symmetric variational formulation of considered problem can be obtained by using the functions defined above and integrating by parts:

$$
\begin{gathered}
-\int_0^L E\widetilde{J}_\varsigma(x)\frac{\partial^2\widetilde{V}}{\partial x^2}\frac{\partial^2 v}{\partial x^2}\,dx+\int_l^L[A+B\cos ax+D\sin ax]\,\widetilde{V}v\,dx \\
+\int_l^L[D\cos ax-B\sin ax]\,\widetilde{W}v\,dx=\frac{d^2}{dt^2}\int_0^L\rho\widetilde{F}(x)\widetilde{V}v\,dx, \\
\int_0^L E\widetilde{J}_\eta(x)\frac{\partial^2\widetilde{W}}{\partial x^2}\frac{\partial^2 w}{\partial x^2}\,dx+\int_l^L[A-B\cos ax-D\sin ax]\,\widetilde{W}w\,dx \\
+\int_l^L[D\cos ax-B\sin ax]\,\widetilde{V}w\,dx=\frac{d^2}{dt^2}\int_0^L\rho\widetilde{F}(x)\widetilde{W}w\,dx.
\end{gathered}
\tag{8}
$$

Obviously, this presentation is symmetric. It is appropriate to consider the problem (8) using the mixed formulation (see, for example [3]). Many of the concepts for the mixed methods originated in solid mechanics where it was desirable to have a more accurate approximation of certain derivatives of the displacement.

The mixed method in which there are two approximation spaces for each of both equations (8) and the reaults in this aspect will be the subject of separate investigations.

4 Numerical Example

Consider a twist drill with straight shank. This beam has the following parameters (see Figure 1): $l = 88$ mm; $L = 132$ mm; diameter $d = 10$ mm; screw pitch of the twisting line $s = 54$ mm. The main axial moments of inertia of the cross-section are (see Figure 2): $J_1 = J_\eta = 0.0041d^4$ and $J_2 = J_\varsigma = 0.025d^4$. The following material characteristics are used: Young modulus $E = 2.1 \times 10^{11}$ Pa; $\mu = 0.3$ is the Poisson ratio; density $\rho = 7850$ kg/m^3. The coefficient of Winckler's base is $K = 10^5$ N/mm^2.

The center of gravity variation of cross-section A (see Figure 1) is represented by the function: $J = |\vec{f}|\sin\omega t, z = |\vec{f}|\cos\omega t$, where $\omega = 105$ s$^{-1}, |\vec{f}| = 1$ mm$, t \in [0, 0.06]$ s. The coefficient of material hysteresis is 0.01.

The Winckler's type base when $x \in (0, l)$ has coefficient $K = 10^5$ N/mm^2. This base is reduced by three ($n = 3$) equidistant groups of springs, i.e. the angles are 120°. The number of springs in each group is m. We present the following problem: Find the maximal normal stress σ_x at the critical point $x = l$.

Calculations are made when the number of springs is $m = 30$. The axial rigidity of each spring is

$$K_m = \frac{K(L-l)}{n.m} = 48888.9 \text{ N/mm}.$$

The finite element procedure has been implemented. The beam domain is discretized using finite elements of type BEAM3D. Normal stresses are calculated by the formula

$$\sigma_x(x,\eta,\zeta,t) = E.\zeta\frac{\partial^2 W(x;t)}{\partial x^2} - E.\eta\frac{\partial^2 V(x;t)}{\partial x^2}.$$

The maximal stress is at the point of section with abscissa $x = l$ at the moment $t = t^\star$ when the vector $\overrightarrow{f}$ is orthogonal to the main axis of inertia $C\zeta$. Then $\max\sigma_x = 451.3$ MPa. The critical point of this section is a point of intersection obtained by the second main axis of inertia $C\eta$ and the contour of the section.

References

[1] Andreev A.B., Maximov J.T. and M.R. Racheva: Finite element method for calculations of dynamic stresses in the continuous beam on elastic supports, Sib. JNM, Vol.6, 113-124, 2003.

[2] Andreev A.B., Maximov J.T. and M.R. Racheva: Finite Element Modelling for a Beam on the Winckler's Type Base, SibJNM, 2004 (to appear).

[3] Brenner S.C., L.R. Scott: The Mathematical Theory of Finite Element Methods. Texts in Appl. Math. - Springer-Verlag, 1994.

[4] Ciarlet, P.G.: The Finite Element Method for Elliptic Problems. North-Holland, Amsterdam 1978.

[5] Levina Z.M. and D.N. Reshetov: Contact Stiffness of Machines. Machinery Constraction 1971 (in Russian).

[6] Timoshenko S.P. and J.N. Goodier: Theory of Elasticity. London, McGraw Hill Intern. Editions 1982.

On the Solvability of the Steady-State Rolling Problem

Todor Angelov Angelov

Institute of Mechanics, Bulgarian Academy of Sciences,
"Acad.G.Bonchev" street, block 4, 1113 Sofia, Bulgaria
taa@imbm.bas.bg

Abstract. In this paper a steady-state rolling problem with nonlinear friction, for rigid-plastic, rate sensitive and slightly compressible materials is considered. Its variational formulation is given and existence and uniqueness results, obtained with the help of successive iteration methods are presented. Considering the slight material compressibility as a method of penalisation, it is further shown, that when the compressibility parameter tends to zero the solution of the rolling problem for incompressible materials is approached.

1 Introduction

In spite the significant advances made in the theoretical, computational and experimental study of the rolling processes [1 - 6], they are still far from a complete mechanical and mathematical description. It was recently found, that the rolling problems could be stated analogously to the frictional contact problems in elasticity [7, 8] and thus the rich theory and methods of variational inequalities [9 - 12], to be applied.

In this paper we consider a steady-state rolling problem, assuming a rigid-plastic, rate-sensitive and slightly compressible material model and a nonlinear friction law to hold. For the corresponding variational problem, existence and uniqueness results obtained with the help of successive iteration methods are presented. Considering the slight material compressibility as a method of penalization, it is further shown that when the compressibility parameter tends to zero, the solution of the rolling problem for incompressible materials is obtained.

2 Statement of the Problem

We consider an isotropic, plastically deformable, during rolling, metallic body occupying an open bounded domain $\Omega \subset \mathbb{R}^k (k = 2, 3)$, (Fig.1). The boundary Γ of the domain is supposed sufficiently smooth and consists of four open disjoint subsets, i.e. $\Gamma = \Gamma_1 \cup \Gamma_2 \cup \Gamma_3 \cup \Gamma_4$. The boundaries $\Gamma_1 \cup \Gamma_2$ are free of tractions, Γ_3

Z. Li et al. (Eds.): NAA 2004, LNCS 3401, pp. 125–132, 2005.

is the boundary of symmetry and Γ_4 is the contact boundary. The coordinates of any point of $\bar{\Omega} = \Omega \cup \Gamma$ are $\mathbf{x} = \{x_i\}$. Throughout this work the standard indicial notation and summation convention are employed.

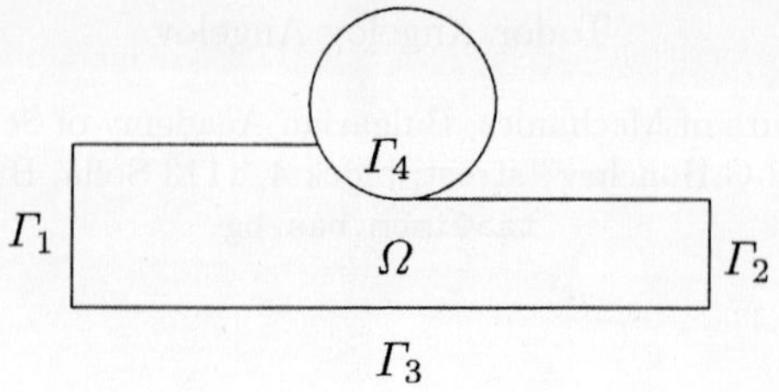

Fig. 1. Illustration of a steady-state rolling process

The material is assumed slightly compressible according [3], as the following yield criterion and associated flow rule hold:

$$F(\sigma_{ij}, \dot{\varepsilon}_{ij}) \equiv \bar{\sigma}^2 - \sigma_p^2(\dot{\bar{\varepsilon}}) = 0, \qquad \dot{\varepsilon}_{ij} = \frac{3}{2}\frac{\dot{\bar{\varepsilon}}}{\bar{\sigma}}\left(s_{ij} + \frac{2}{9} d \sigma_H \delta_{ij}\right), \tag{2.1}$$

where $d > 0$ is the compressibility parameter, assumed small constant, $\sigma_p(\dot{\bar{\varepsilon}})$ is the strain-rate dependent, uniaxial yield limit.

$$\bar{\sigma} = \sqrt{\frac{3}{2} s_{ij} s_{ij} + d\sigma_H^2}, \qquad \dot{\bar{\varepsilon}} = \sqrt{\frac{2}{3}\dot{e}_{ij}\dot{e}_{ij} + \frac{1}{d}\dot{e}_V^2}, \tag{2.2}$$

are the equivalent stress and the equivalent strain-rate expressions, where

$$s_{ij} = \sigma_{ij} - \delta_{ij}\sigma_H, \qquad \dot{e}_{ij} = \dot{\bar{\varepsilon}}_{ij} - \frac{1}{3}\dot{\varepsilon}_V \delta_{ij}, \qquad 1 \leq i, j \leq k, \tag{2.3}$$

are the components of the deviatoric stress and deviatoric strain-rate tensors, σ_{ij} and $\dot{\varepsilon}_{ij}$ are the components of the stress and strain-rate tensors, $\sigma_H = \frac{1}{3}\sigma_{ii}$ and $\dot{\varepsilon}_V = \dot{\varepsilon}_{ii}$ are the hydrostatic stress and the volume dilatation strain-rate. From the flow rule, the following relation between the hydrostatic stress and the volume dilatation strain-rate holds:

$$\sigma_H = \frac{\dot{\varepsilon}_V}{d}\frac{\bar{\sigma}}{\dot{\bar{\varepsilon}}} = \frac{\dot{\varepsilon}_V}{d}\frac{\sqrt{\frac{3}{2}s_{ij}s_{ij}}}{\sqrt{\frac{2}{3}\dot{e}_{ij}\dot{e}_{ij}}} \tag{2.4}$$

Let us consider the following problem:

(P1) Find the velocity $\mathbf{u} = \{u_i\}$ and stress $\boldsymbol{\sigma} = \{\sigma_{ij}\}$, $1 \leq i, j \leq k$, fields satisfying the equations and relations

- equations of equilibrium

$$\sigma_{ij,j} = 0, \text{ in } \Omega \tag{2.5}$$

- constitutive equations

$$\sigma_{ij} = \frac{\sigma_p}{\dot{\bar{\varepsilon}}} \left[\frac{2}{3}\dot{\varepsilon}_{ij} + \left(\frac{1}{d} - \frac{2}{9} \right) \delta_{ij} \dot{\varepsilon}_V \right], \tag{2.6}$$

- Cauchy conditions for strain-rate

$$\dot{\varepsilon}_{ij} = \frac{1}{2}(u_{i,j} + u_{j,i}), \tag{2.7}$$

- boundary conditions

$$\sigma_{ij} n_j = 0, \quad \text{on } \Gamma_1 \cup \Gamma_2 \tag{2.8}$$

$$\boldsymbol{\sigma}_T = \mathbf{0}, \quad u_N = 0, \quad \text{on } \Gamma_3 \tag{2.9}$$

$$u_N = 0, \quad \sigma_N \leq 0 \quad \text{and} \tag{2.10}$$

$$\text{if } |\boldsymbol{\sigma}_T| < \tau_f(\mathbf{u}) \text{ then } \mathbf{u}_T - \mathbf{u}_R = \mathbf{0},$$

$$\text{if } |\boldsymbol{\sigma}_T| = \tau_f(\mathbf{u}) \text{ then } \exists \lambda \geq 0$$

$$\text{such that } \mathbf{u}_T - \mathbf{u}_R = -\lambda \boldsymbol{\sigma}_T(\mathbf{u}) \text{ on } \Gamma_4. \quad \square \tag{2.11}$$

Remark 2.1. The following notations are used above: δ_{ij} is Kronecker's symbol; $\mathbf{n} = \{n_i\}$ is the unit normal vector outward to Γ; $u_N = u_i n_i$, $u_{Ti} = u_i - u_N n_i$ and $\sigma_N = \sigma_{ij} n_i n_j$, $\sigma_{Ti} = \sigma_{ij} n_j - \sigma_N n_i$ are the normal and tangential components of the velocity and the stress vector on Γ;

$$\tau_f(\mathbf{u}) = \min\left\{ \mu_f(\mathbf{x}) \sigma_N^h(\mathbf{u}),\ m_f(\mathbf{x}) \tau_p(\dot{\bar{\varepsilon}}^h(\mathbf{u})) \right\} \tag{2.12}$$

is the nonlinear Coulomb-Siebel friction law, where $\mu_f(\mathbf{x})$ and $m_f(\mathbf{x})$ are the coefficient of friction and friction factor, $\sigma_N^h(\mathbf{u}) \geq 0$ and $\dot{\bar{\varepsilon}}^h(\mathbf{u})$ are the appropriately molified on Γ_4 normal stress and equivalent strain-rate [4-6], $\tau_p(\dot{\bar{\varepsilon}}^h(\mathbf{u}))$ $\sigma_p(\dot{\bar{\varepsilon}}^h(\mathbf{u}))/\sqrt{3}$ is the shear yield limit for the material of Γ_4; $\mathbf{u}_R$ is the rolling velocity. The yield limit $\sigma_p(\dot{\bar{\varepsilon}}) : [0, \infty) \to [0, \infty)$ is assumed monotonously increasing and continuously differentiable function of $\dot{\bar{\varepsilon}}$. Moreover it is supposed that, there exist positive constants Λ_1, Λ_2, such that

$$0 < \Lambda_1 \leq \frac{d\sigma_p}{d\dot{\bar{\varepsilon}}} \leq \frac{\sigma_p}{\dot{\bar{\varepsilon}}} \leq \Lambda_2 < \infty. \quad \square \tag{2.13}$$

3 Variational Statement

Let us denote by $\mathbf{V}$ and $\mathbf{H}$ the following Hilbert spaces

$$\mathbf{V} = \left\{\mathbf{v}: \ \mathbf{v} \in (H^1(\Omega))^k, \ v_N = 0 \ \text{on} \ \Gamma_3 \cup \Gamma_4\right\}, \qquad \mathbf{H} = (H^0(\Omega))^k,$$
$$\mathbf{V} \subset \mathbf{H} \equiv \mathbf{H}' \subset \mathbf{V}',$$

with inner products and norms respectively

$$(\mathbf{u}, \mathbf{v}) = \int_\Omega (u_i v_i + u_{i,j} v_{i,j}) dx, \qquad (\mathbf{u}, \mathbf{v})_0 = \int_\Omega u_i v_i dx,$$

$$\|\mathbf{u}\|_1 = (\mathbf{u}, \mathbf{u})^{1/2}, \qquad \|\mathbf{u}\|_0 = (\mathbf{u}, \mathbf{u})_0^{1/2},$$

where with prime the dual spaces are denoted. We further assume

$$\mu_f(\mathbf{x}), \ m_f(\mathbf{x}) \in L_\infty(\Gamma_4).$$

Multiplying (2.5) by $(\mathbf{v} - \mathbf{u}) \in \mathbf{V}$ in the inner product sense, applying Green's formula and taking into account the boundary conditions, we obtain

$$\int_\Omega \sigma_{ij}(\mathbf{u})(\dot\varepsilon_{ij}(\mathbf{v}) - \dot\varepsilon_{ij}(\mathbf{u})) dx + \int_{\Gamma_4} \tau_f(\mathbf{u}) |\mathbf{v}_T - \mathbf{u}_R| d\Gamma - \int_{\Gamma_4} \tau_f(\mathbf{u}) |\mathbf{u}_T - \mathbf{u}_R| d\Gamma \geq 0. \tag{3.1}$$

Let us denote

$$a(\mathbf{w}; \mathbf{u}, \mathbf{v}) = \int_\Omega \frac{\sigma_p(\mathbf{w})}{\dot{\bar\varepsilon}(\mathbf{w})} \left[\frac{2}{3} \dot\varepsilon_{ij}(\mathbf{u}) \dot\varepsilon_{ij}(\mathbf{v}) + \left(\frac{1}{d} - \frac{2}{9}\right) \dot\varepsilon_V(\mathbf{u}) \dot\varepsilon_V(\mathbf{v})\right] dx \tag{3.2}$$

and

$$j(\mathbf{u}, \mathbf{v}) = \int_{\Gamma_4} \tau_f(\mathbf{u}) |\mathbf{v}_T - \mathbf{u}_R| d\Gamma. \tag{3.3}$$

Then the variational statement of the problem P1 is the following one:

(**P2**) Find $\mathbf{u} \in \mathbf{V}$, satisfying for all $\mathbf{v} \in \mathbf{V}$ the variational inequality

$$a(\mathbf{u}; \mathbf{u}, \mathbf{v} - \mathbf{u}) + j(\mathbf{u}, \mathbf{v}) - j(\mathbf{u}, \mathbf{u}) \geq 0. \tag{3.4}$$

Remark 3.1. Let us remind the properties of the functionals that constitute this inequality [4, 5]. The form $a(\mathbf{w}; \mathbf{u}, \mathbf{v})$ is nonlinear with respect to $\mathbf{w}$, and linear, continuous and symmetric with respect to $\mathbf{u}$ and $\mathbf{v}$. It also follows from (2.13) that there exist constants $\alpha_1 > 0, \ \alpha_2 > 0$, such that for all $\mathbf{u}, \mathbf{v}, \mathbf{w} \in \mathbf{V}$

$$a(\mathbf{w}; \mathbf{u}, \mathbf{u}) \geq \alpha_1 \int_\Omega \dot\varepsilon_{ij}(\mathbf{u}) \dot\varepsilon_{ij}(\mathbf{u}) dx, \quad a(\mathbf{w}; \mathbf{u}, \mathbf{v}) \leq \alpha_2 \left| \int_\Omega \dot\varepsilon_{ij}(\mathbf{u}) \dot\varepsilon_{ij}(\mathbf{v}) dx \right|. \tag{3.5}$$

Using (3.5) and Korn's inequality [7, 8, 12]

$$\int_\Omega \dot\varepsilon_{ij}(\mathbf{u}) \dot\varepsilon_{ij}(\mathbf{u}) dx + \int_\Omega u_i u_i dx \geq c_K \|\mathbf{u}\|_1^2 \quad \forall \mathbf{u} \in (H^1(\Omega))^k, \tag{3.6}$$

for the strain-rates, where $c_K > 0$ is a constant, we have that the form $a(\mathbf{u};\mathbf{u},\mathbf{u})$ is $\mathbf{V}$-elliptic and bounded, i.e. there exists constants $c_1 > 0$ and $c_2 > 0$, such that

$$a(\mathbf{u};\mathbf{u},\mathbf{u}) \geq c_1\|\mathbf{u}\|_1^2, \qquad a(\mathbf{w};\mathbf{u},\mathbf{v}) \leq c_2\|\mathbf{u}\|_1\|\mathbf{v}\|_1 \qquad \forall \mathbf{u},\mathbf{v},\mathbf{w} \in \mathbf{V}. \tag{3.7}$$

Moreover, for all $\mathbf{u},\mathbf{v} \in \mathbf{V}$ there exist constants $m > 0$ and $M > 0$, such that

$$a(\mathbf{v};\mathbf{v},\mathbf{v}-\mathbf{u}) - a(\mathbf{u};\mathbf{u},\mathbf{v}-\mathbf{u}) \geq m\|\mathbf{v}-\mathbf{u}\|_1^2, \tag{3.8a}$$

$$|a(\mathbf{v};\mathbf{v},\mathbf{v}) - a(\mathbf{u},\mathbf{u},\mathbf{v})| \leq M\|\mathbf{v}-\mathbf{u}\|_1\|\mathbf{v}\|_1. \tag{3.8b}$$

For any fixed $\mathbf{u} \in \mathbf{V}$ the functional $\mathbf{V} \ni \mathbf{v} : j(\mathbf{u},\mathbf{v}) \to \mathbb{R}$ is proper, convex and lower semicontinuous and for all $\mathbf{u},\mathbf{v},\mathbf{w} \in \mathbf{V}$ there exist constants $c_f > 0$ and $c > 0$, depending on the friction coefficient, or factor, such that

$$0 \leq j(\mathbf{u},\mathbf{v}) \leq c_f\|\mathbf{u}\|_1\|\mathbf{v}-\mathbf{u}_R\|_1, \tag{3.9}$$

$$|j(\mathbf{u},\mathbf{w}) + j(\mathbf{w},\mathbf{v}) - j(\mathbf{u},\mathbf{v}) - j(\mathbf{w},\mathbf{w})| \leq c\|\mathbf{w}-\mathbf{u}\|_1\|\mathbf{w}-\mathbf{v}\|_1. \quad \square \tag{3.10}$$

4 Existence, Uniqueness and Convergence

Here we shall briefly present existence and uniqueness results for the problem P2, obtained by proving the convergence of the successive linearization methods - the secant-modulus method [5] and the method of contractions [4]. Further we shall show that when compressibility parameter $d \to 0$, the rolling problem for incompressible materials is approached.

Theorem 4.1. Let the conditions (2.13) and the properties given in Remark 3.1 hold. Then for a sufficiently small friction coefficient, or factor, the problem P2 has an unique solution $\mathbf{u} \in \mathbf{V}$.

Proof: *Uniqueness.* Let $\mathbf{u}_1,\mathbf{u}_2 \in \mathbf{V}$ be two solutions of (3.4). Then we have

$$a(\mathbf{u}_1;\mathbf{u}_1,\mathbf{v}-\mathbf{u}_1) + j(\mathbf{u}_1,\mathbf{v}) - j(\mathbf{u}_1,\mathbf{u}_1) \geq 0, \tag{4.1a}$$

$$a(\mathbf{u}_2;\mathbf{u}_2,\mathbf{v}-\mathbf{u}_2) + j(\mathbf{u}_2,\mathbf{v}) - j(\mathbf{u}_2,\mathbf{u}_2) \geq 0. \tag{4.1b}$$

Setting $\mathbf{v} = \mathbf{u}_2$ in (4.1a) and $\mathbf{v} = \mathbf{u}_1$ in (4.1b), after adding and rearranging the inequalities and taking into account Remark 3.1, we obtain that for a sufficiently small coefficient of friction, or friction factor holds

$$\begin{aligned} 0 \geq &- (j(\mathbf{u}_1,\mathbf{u}_2) + j(\mathbf{u}_2,\mathbf{u}_1) - j(\mathbf{u}_1,\mathbf{u}_1) - j(\mathbf{u}_2,\mathbf{u}_2)) + \\ a(\mathbf{u}_1;\mathbf{u}_1,\mathbf{u}_1-\mathbf{u}_2) &- a(\mathbf{u}_2;\mathbf{u}_2,\mathbf{u}_1-\mathbf{u}_2) \geq (m-c)\|\mathbf{u}_1-\mathbf{u}_2\|_1^2 > 0, \end{aligned} \tag{4.2}$$

which yields $\mathbf{u}_1 \equiv \mathbf{u}_2$.

Existence. Let us consider the following problems:

(P3) For an arbitrary initial $\mathbf{u}_0 \in \mathbf{V}$ find $\mathbf{u}_{n+1}$, $n = 0, 1, ...$ satisfying for all $\mathbf{v} \in \mathbf{V}$

$$a(\mathbf{u}_n; \mathbf{u}_{n+1}, \mathbf{v} - \mathbf{u}_{n+1}) + j(\mathbf{u}_n, \mathbf{v}) - j(\mathbf{u}_n, \mathbf{u}_{n+1}) \geq 0; \tag{4.3}$$

(P4) For an arbitrary initial $\mathbf{u}_0 \in \mathbf{V}$ find $\mathbf{u}_{n+1}$, $n = 0, 1, ...$ satisfying for all $\mathbf{v} \in \mathbf{V}$

$$(\mathbf{u}_{n+1}, \mathbf{v} - \mathbf{u}_{n+1}) + \rho j(\mathbf{u}_n, \mathbf{v}) - \rho j(\mathbf{u}_n, \mathbf{u}_{n+1}) \geq$$

$$(\mathbf{u}_n, \mathbf{v} - \mathbf{u}_{n+1}) - \rho a(\mathbf{u}_n; \mathbf{u}_n, \mathbf{v} - \mathbf{u}_{n+1}), \tag{4.4}$$

where $0 < \rho < 2(m - c)/(M^2 - c^2)$, $M > m > c$.

These problems have unique solutions $\mathbf{u}_{n+1}$ for every $n = 0, 1, 2, ...$, [4, 5, 9, 10] and the sequence $\{\mathbf{u}_{n+1}\}$ is such that

$$\|\mathbf{u}_{n+2} - \mathbf{u}_{n+1}\|_1 \leq q\|\mathbf{u}_{n+1} - \mathbf{u}_n\|_1 \leq ... \leq q^{n+1}\|\mathbf{u}_1 - \mathbf{u}_0\|_1. \tag{4.5}$$

But for sufficiently small friction coefficient, or factor and d, we have $0 < q < 1$ and

$$\lim_{n\to\infty} \|\mathbf{u}_{n+2} - \mathbf{u}_{n+1}\|_1 = 0, \tag{4.6}$$

which implies that $\{\mathbf{u}_n\}$ is a fundamental sequence. Therefore there exists an element $\mathbf{u} \in \mathbf{V}$, such that

$$\lim_{n\to\infty} \|\mathbf{u} - \mathbf{u}_n\|_1 = 0. \tag{4.7}$$

Further, taking limit at $n \to \infty$ in (4.3) and (4.4) and having in mind the continuity properties of the functionals, it can be shown that $\mathbf{u}$ satisfies the inequality (3.4), which completes the proof. □

Let us construct, for all sufficiently small compressibility parameters $d > 0$, a sequence $\{\mathbf{u}_d\}$ of solutions of problem P2. Since this sequence is bounded in $\mathbf{V}$, there exists a subsequence, also denoted $\{\mathbf{u}_d\}$, which is weakly convergent at $d \to 0$, to an element $\mathbf{u} \in \mathbf{V}$. We shall further show that $\mathbf{u} \in \mathbf{W} \subset \mathbf{V}$, where

$$\mathbf{W} = \{\mathbf{v} : \ \mathbf{v} \in \mathbf{V}, \ \ v_{i,i} = 0 \ \text{ in } \ \Omega\}$$

and the following result for the problem for incompressible materials holds.

Theorem 4.2. At $d \to 0$, $\mathbf{u} \in \mathbf{W}$ is the unique solution of the problem (3.4) for all $\mathbf{v} \in \mathbf{W}$.

Proof: For $\mathbf{u}_d \in \mathbf{V}$ and all $\mathbf{v} \in \mathbf{V}$, from (3.4) it follows

$$a(\mathbf{u}_d; \mathbf{u}_d, \mathbf{u}_d) + j(\mathbf{u}_d, \mathbf{u}_d) - \frac{9 - 2d}{9d} \int_\Omega \frac{\sigma_p(\mathbf{u}_d)}{\dot{\bar{\varepsilon}}(\mathbf{u}_d)} \dot{\varepsilon}_V(\mathbf{u}_d)\dot{\varepsilon}_V(\mathbf{v})dx \leq$$

$$\int_\Omega \frac{2}{3} \frac{\sigma_p(\mathbf{u}_d)}{\dot{\bar{\varepsilon}}(\mathbf{u}_d)} \dot{\varepsilon}_{ij}(\mathbf{u}_d)\dot{\varepsilon}_{ij}(\mathbf{v})dx + j(\mathbf{u}_d, \mathbf{v}). \tag{4.8}$$

Since the first two terms in the left-hand side of (4.8) are nonnegative, we have

$$-\int_{\Omega} \frac{\sigma_p(\mathbf{u}_d)}{\dot{\bar{\varepsilon}}(\mathbf{u}_d)} \dot{\varepsilon}_V(\mathbf{u}_d)\dot{\varepsilon}_V(\mathbf{v})dx \leq$$

$$\frac{9d}{9-2d}\left[\int_{\Omega} \frac{2}{3}\frac{\sigma_p(\mathbf{u}_d)}{\dot{\bar{\varepsilon}}(\mathbf{u}_d)} \dot{\varepsilon}_{ij}(\mathbf{u}_d)\dot{\varepsilon}_{ij}(\mathbf{v})dx + j(\mathbf{u}_d, \mathbf{v})\right]. \tag{4.9}$$

Let us consider the left-hand side and define the operator $\beta : \mathbf{V} \to \mathbf{V}'$

$$\langle \beta(\mathbf{u}_d), \mathbf{v}\rangle = \int_{\Omega} \frac{\sigma_p(\mathbf{u}_d)}{\dot{\bar{\varepsilon}}(\mathbf{u}_d)} \dot{\varepsilon}_V(\mathbf{u}_d)\dot{\varepsilon}_V(\mathbf{v})dx. \tag{4.10}$$

Since it is bounded, hemicontinuous and monotone, it can be shown that at $d \to 0$ we have $\beta(\mathbf{u}_d) \to \beta(\mathbf{u})$ weakly in $\mathbf{V}'$, i.e.

$$\frac{\sigma_p(\mathbf{u}_d)}{\dot{\bar{\varepsilon}}(\mathbf{u}_d)} \dot{\varepsilon}_V(\mathbf{u}_d) \dot{\to} \frac{\sigma_p(\mathbf{u})}{\dot{\bar{\varepsilon}}(\mathbf{u})} \dot{\varepsilon}_V(\mathbf{u}). \tag{4.11}$$

Further, since the right-hand side of (4.9) is bounded, taking $d \to 0$ in (4.9), we obtain for all $+\mathbf{v}, -\mathbf{v} \in \mathbf{V}$

$$-\int_{\Omega} \frac{\sigma_p(\mathbf{u})}{\dot{\bar{\varepsilon}}(\mathbf{u})} \dot{\varepsilon}_V(\mathbf{u})\dot{\varepsilon}_V(\mathbf{v})dx \leq 0, \qquad \int_{\Omega} \frac{\sigma_p(\mathbf{u})}{\dot{\bar{\varepsilon}}(\mathbf{u})} \dot{\varepsilon}_V(\mathbf{u})\dot{\varepsilon}_V(\mathbf{v})dx \leq 0. \tag{4.12}$$

Therefore we have

$$\int_{\Omega} \frac{\sigma_p(\mathbf{u})}{\dot{\bar{\varepsilon}}(\mathbf{u})} \dot{\varepsilon}_V(\mathbf{u})\dot{\varepsilon}_V(\mathbf{v})dx \equiv 0 \qquad \forall \mathbf{v} \in \mathbf{V} \tag{4.13}$$

and hence $\dot{\varepsilon}_V(\mathbf{u}) \equiv 0$, i.e $\mathbf{u} \in \mathbf{W}$. Let us now show that this $\mathbf{u} \in \mathbf{W}$ is a solution of (3.4) for all $\mathbf{v} \in \mathbf{W}$. For all $\mathbf{w} \in \mathbf{W}$ we have that

$$\begin{aligned}
&a(\mathbf{w}; \mathbf{w}, \mathbf{w} - \mathbf{u}_d) + j(\mathbf{w}, \mathbf{w}) - j(\mathbf{w}, \mathbf{u}_d) = \\
&a(\mathbf{w}; \mathbf{w}, \mathbf{w} - \mathbf{u}_d) + j(\mathbf{w}, \mathbf{w}) - j(\mathbf{w}, \mathbf{u}_d) - \\
&[a(\mathbf{u}_d; \mathbf{u}_d, \mathbf{w} - \mathbf{u}_d) + j(\mathbf{u}_d, \mathbf{w}) - j(\mathbf{u}_d, \mathbf{u}_d)] + \\
&[a(\mathbf{u}_d; \mathbf{u}_d, \mathbf{w} - \mathbf{u}_d) + j(\mathbf{u}_d, \mathbf{w}) - j(\mathbf{u}_d, \mathbf{u}_d)] \geq (m-c)\|\mathbf{w} - \mathbf{u}_d\|_1^2 \geq 0
\end{aligned} \tag{4.14}$$

and taking $d \to 0$ we obtain

$$a(\mathbf{w}; \mathbf{w}, \mathbf{w} - \mathbf{u}) + j(\mathbf{w}, \mathbf{w}) - j(\mathbf{w}, \mathbf{u}) \geq 0, \qquad \forall \mathbf{w} \in \mathbf{W}. \tag{4.15}$$

Setting $\mathbf{w} = \mathbf{u} + t(\mathbf{v} - \mathbf{u}))$, $t \in [0, 1]$, $\forall \mathbf{v} \in \mathbf{W}$ we obtain

$$\begin{aligned}
&0 \leq a(\mathbf{u} + t(\mathbf{v} - \mathbf{u}); \mathbf{u} + t(\mathbf{v} - \mathbf{u}), t(\mathbf{v} - \mathbf{u})) + \\
&j(\mathbf{u} + t(\mathbf{v} - \mathbf{u}), \mathbf{u} + t(\mathbf{v} - \mathbf{u})) - j(\mathbf{u} + t(\mathbf{v} - \mathbf{u}), \mathbf{u}) \leq \\
&ta(\mathbf{u} + t(\mathbf{v} - \mathbf{u}); \mathbf{u} + t(\mathbf{v} - \mathbf{u}), \mathbf{v} - \mathbf{u}) + \\
&(1-t)j(\mathbf{u} + t(\mathbf{v} - \mathbf{u}), \mathbf{u}) + tj(\mathbf{u} + t(\mathbf{v} - \mathbf{u}), \mathbf{v}) - j(\mathbf{u} + t(\mathbf{v} - \mathbf{u}), \mathbf{u}) = \\
&ta(\mathbf{u} + t(\mathbf{v} - \mathbf{u}); \mathbf{u} + t(\mathbf{v} - \mathbf{u}), \mathbf{v} - \mathbf{u}) + \\
&tj(\mathbf{u} + t(\mathbf{v} - \mathbf{u}), \mathbf{v}) - tj(\mathbf{u} + t(\mathbf{v} - \mathbf{u}), \mathbf{u}).
\end{aligned} \tag{4.16}$$

For $t \neq 0$ it follows that

$$a(\mathbf{u} + t(\mathbf{v} - \mathbf{u}); \mathbf{u} + t(\mathbf{v} - \mathbf{u}), \mathbf{v} - \mathbf{u}) +$$

$$j(\mathbf{u} + t(\mathbf{v} - \mathbf{u}), \mathbf{v}) - j(\mathbf{u} + t(\mathbf{v} - \mathbf{u}), \mathbf{u}) \geq 0 \qquad (4.17)$$

and taking $t \to 0$ we finally obtain

$$a(\mathbf{u}; \mathbf{u}, \mathbf{v} - \mathbf{u}) + j(\mathbf{u}, \mathbf{v}) - j(\mathbf{u}, \mathbf{u}) \geq 0, \qquad \forall \mathbf{v} \in \mathbf{W}, \qquad (4.18)$$

which is the variational statement of the rolling problem for incompressible materials. Repeating further the uniqueness part of Theorem 4.1 we obtain that $\mathbf{u}$ is the unique solution of this problem. □

Remark 4.1. The problem P2 can be solved numerically, combining the conventional finite element method and the algorithms, defined by problems P3 and P4. The computational experiments show that a transition from slight material compressibility to material incompressibility is obtained, taking values of d from 10^{-2} to 10^{-6}. More details about the computational treatment of the steady-state rolling problem are presented in [6]. □

References

1. Zienkiewicz, O.C.: Flow formulation for numerical solution of forming processes. In: Pittman, J.F.T., Zienkiewicz, O.C., Wood, R.D., Alexander, J.M. (eds.): Numerical Analysis of Forming Processes. John Willey & Sons (1984) 1 - 44.
2. Kobayashi, S., Oh, S.-I., Altan, T.: Metal Forming and the Finite Element Method. Oxford University Press (1989).
3. Mori, K.-I.: Rigid-plastic finite element solution of forming processes. In: Pietrzyk, M., Kusiak, J., Sadok, L., Engel, Z. (eds.): Huber's Yield Criterion in Plasticity. AGH, Krakow (1994) 73 - 99.
4. Angelov, T., Baltov, A., Nedev, A.: Existence and uniqueness of the solution of a rigid-plastic rolling problem. Int.J.Engng.Sci. **33** (1995) 1251-1261.
5. Angelov, T.: A secant modulus method for a rigid-plastic rolling problem. Int.J.Nonl.Mech. **30** (1995) 169-178.
6. Angelov, T., Liolios, A.: Variational and numerical approach to a steady-state rolling problem. (submitted).
7. Duvaut, G., Lions, J.-L.: Les Inequations en Mechanique en Physique. Dunod, Paris (1972).
8. Kikuchi, N., Oden, J.T.: Contact Problems in Elasticity: A Study of Variational Inequalities and Finite Element Methods. SIAM, Philadelphia (1988).
9. Glowinski, R., Lions, J.-L., Tremolieres, R.: Numerical Analysis of Variational Inequalities. North-Holland,Amsterdam (1981).
10. Glowinski, R.: Numerical Methods for Nonlinear Variational Problems. Springer-Verlag, Berlin (1984).
11. Mikhlin, S.G.: The Numerical Performance of Variational Methods. Walters-Noordhoff, The Netherlands, (1971).
12. Nečas, J., Hlavaček, I.: Mathematical Theory of Elastic and Elasto-Plastic Bodies: An Introduction. Elsevier, Amsterdam, (1981).

A Quasi-Monte Carlo Method for an Elastic Electron Back-Scattering Problem*

Emanouil I. Atanassov and Mariya K. Durchova

Institute for Parallel Processing,
Bulgarian Academy of Sciences,
Acad. G. Bonchev, Bl. 25A, 1113 Sofia, Bulgaria
{emanouil, mabs}@parallel.bas.bg

Abstract. The elastic electron back-scattering is a problem that is important for many theoretical and experimental techniques, especially in the determination of the inelastic mean free paths. This effect arises when a monoenergetic electron beam bombards a solid target and some of the electrons are scattered without energy loss.The description of the flow can be written as an integral equation and may be solved by Monte Carlo methods.

In this paper we investigate the possibility of improving the convergence of the Monte Carlo algorithm by using scrambled low-discrepancy sequences. We demonstrate how by taking advantage of the smoothness of the differential elastic-scattering cross-section a significant decrease of the error is achieved. We show how the contribution of the first few collisions to the result can be evaluated by an appropriate integration method instead of direct simulation, which further increases the accuracy of our computations without increase of the computational time. In order to facilitate these techniques, we use spline approximation of the elastic cross-section, which is more accurate than the widely used tables of Jablonski.

1 Introduction

Elastic electron backscattering problems arise in cases when a monoenergetic electron beam bombards a solid target. Most of the emitted electrons have very low energies but some of them possess the same energy as the incident beam.

The elastic electron backscattering effect plays an important role in many theoretical and experimental techniques. The meaning of elastic electron collisions has been recognized in *Auger Electron Spectroscopy* - AES (see, e.g., [13] for details). Their role in *X-ray photo-electron Spectroscopy* - XPS consists in detecting photo-electrons that are ejected from the material by incident X-rays.

Usually the solid is considered as a semi-infinite medium with properties, depending on the physical characteristics of the material. Often metal targets

* Supported by the Ministry of Education and Science of Bulgaria under contracts NSF I-1201/02 and I-1405/04.

Z. Li et al. (Eds.): NAA 2004, LNCS 3401, pp. 133–140, 2005.

are preferable because of their electrical conductivity. In both XPS and AES, the detected electrons have well known energies which characterize the electron structure of the atoms constructing the target. The escape of the ejected electrons depends on "the average of distance measured along the trajectories, that particles with a given energy travel between inelastic collisions in a substance". This parameter is called *inelastic mean-free-path* - IMFP and was published by Committee E-42 on Surface Analysis of the American Society for Testing and Materials in [16]. The typical energy range for XPS and AES is from 100 eV to a few keV. The IMFP values for various materials can be obtained from the NIST (National Institute of Standards and Technology) database described in [12]. The problem for calculating the elastic backscattering probability for a given experimental setting, considering the IMFP as a parameter, can be solved by Monte Carlo simulation. In practice the people, who make experiments, calculate the yield for various values of the IMFP, obtaining the so-called "calibration curve", and compare them with the measured intensity of the elastic peak. The standard Monte Carlo simulation technique may require several hours of CPU time in order to attain the desired precision.

The model of the interaction between electrons and randomly distributed ionic cores consists in more than one elastic and inelastic collisions. One of the most used techniques is to interpret inelastic collisions as absorption events and realization of such one is presented in [8]. An extensive survey of the various theoretical models and simulation algorithms is presented in [10]. Some variance reduction schemes for solving such problems are discussed in papers [3] and [8]. In this paper we investigate the possibility to obtain even better precision by using low-discrepancy sequences and by using more involved numerical integration schemes.

The problem of the modeling of the electron transport was described in terms of the Boltzmann transport equation for the stationary problem in a homogeneous medium:

$$\bar{\Omega}\bar{\nabla}\Phi\left(\bar{r},\bar{\Omega}\right)+\Sigma_t\Phi(\bar{r},\bar{\Omega})=\int_{4\pi}\Sigma_{el}(\bar{\Omega}'\rightarrow\bar{\Omega})\Phi(\bar{r},\bar{\Omega}')d\bar{\Omega}',\quad z\geq 0 \tag{1}$$

and the boundary condition for vacuum-medium is

$$\Phi(x,y,z=0,\bar{\Omega})=\frac{J_0}{|\mu_0|}\delta(\bar{\Omega}-\bar{\Omega}_0),\quad \bar{\Omega}\bar{l}_z\geq 0. \tag{2}$$

The electron flux $\Phi\left(\bar{r},\bar{\Omega}\right)$ depends on the spatial variable $\bar{r}=(x,y,z)$, (where z is the depth) and on the angular variable $\bar{\Omega}=(\theta,\phi)$. The total cross-section $\Sigma_t=\Sigma_{el}+\Sigma_{in}$, where Σ_{el} is the differential elastic scattering cross-section and $\Sigma_{in}=\dfrac{1}{\lambda_{in}}$. In the boundary condition the incident electron current is denoted with J_0, $\mu=\cos\theta$ and $\mu_0=\alpha$, where α is the incident angle.

From the equations (1) and (2) follows the integral form:

$$\Phi(\bar{r},\bar{\Omega}) = \frac{J_0}{|\mu_0|}\delta(\bar{\Omega}-\bar{\Omega}_0)e^{-\frac{z\Sigma_t}{|\mu_0|}} + \int_0^\infty \{H(\mu)H(z-z')+H(-\mu)H(z'-z)\}$$

$$\times \frac{e^{-\frac{\Sigma_t(z-z')}{|\mu_0|}}}{|\mu|}dz' \int_{4\pi} \Sigma_{el}(\bar{\Omega}'\to\bar{\Omega})\Phi\left(\bar{r}-\frac{\bar{\Omega}(z-z')}{\mu},\bar{\Omega}'\right)d\bar{\Omega}'.$$

The average number of collisions that electrons undergo depends on the material of the target, the electron energy and the experimental setting. Appriory tests show that the biggest yield is obtained from electrons that are detected after only a few collisions. In order to attend to this fact we divide the electrons based on the number of collisions: electrons that escape after having one collision, after having two collisions and after having three or more collisions.

The contribution of electrons, that undergo exactly one collision, can be represented as a two-dimensional integral. For the electrons, that undergo exactly two collisions, the integral becomes 5-dimensional. We estimate the contribution of these two groups of electrons with our quasi-Monte Carlo method for integration of smooth functions. In order to work with sufficiently smooth functions, we develop a spline interpolation of the elastic cross-sections, which is described in Sec. 2. The use of the quasi-Monte Carlo integration method is discussed in Sec. 3.1. The contribution of the collisions after the first two is evaluated with standard quasi-Monte Carlo method. Details are given in Sec. 3.2. Numerical tests are given in Sec. 4 and our conclusions are given in Sec. 5.

2 Spline Approximation of the Elastic Cross Section

In order to be able to apply our integration method [4], we need to work with sufficiently smooth functions. The elastic cross-section, which is used for sampling the scattering angles, is given with an analytical expression, which contains coefficients, obtained from the 3-rd version of the NIST database (see [11]).

The computational procedure is cumbersome and thus can not be used directly in our program. We tabulated interpolations of these functions for various materials and energies, which are sufficiently accurate and easy to compute. We used 3-rd order splines, and we approximated both the density function φ and the function ψ, which is the inverse of $\int_0^x \varphi(t)\,dt$. These approximations were obtained using the Spline Toolbox from MATLAB®. We made sure that the derivative of the spline approximation of ψ is an approximation to $(\varphi(x))^{-1}$. Since computing a spline in a given point is fast, and since we can use an arbitrary number of nodes for the spline, we can achieve any needed precision of the approximations. In our computations we used about 200 nodes.

Table 1. Timing results

Method	Z	En	MC	S	H
A	13		4.68	4.97	4.96
B		300	4.84	5.13	5.08
A	47		4.59	4.85	4.86
B		500	4.72	5.01	4.98
A	29		5.64	5.89	5.88
B		5000	5.76	6.05	6.00
A	79		4.80	5.18	5.08
B		1000	4.96	5.28	5.22

Table 2. Comparison of Monte Carlo and quasi-Monte Carlo (Sobol and Halton) versions

Method	r	$Z = 13$	$Z = 29$	$Z = 47$	$Z = 79$
A	MC	0.00022356	0.00006088	0.00036549	0.00015564
500	S.	0.00083010	0.00029592	0.00074327	0.00044927
	H.	0.00000332	0.00007950	0.00000776	0.00001390
B	MC	0.00176780	0.00409764	0.00802412	0.00209105
500	S.	0.01275007	0.00734996	0.00872082	0.00650686
	H.	0.01271623	0.00829104	0.00529357	0.00842323

and as a computer program and is in orders of magnitude faster than the standard Monte Carlo simulation. The CPU times needed for all calculations for our algorithm A in comparison with the algorithm B can be seen in Table 1. The differences in CPU times are small and actually depend on the time that we are able to spend optimizing the computer code. The same can be said about the comparison between Monte Carlo and quasi-Monte Carlo versions.

We show the difference in accuracy between the algorithms A and B in Table 2 and 3. The approximated relative error of the pseudo-random versions and the quasi-Monte Carlo versions of both algorithms for fixed energy $500eV$ and atomic numbers Z 13, 29, 47, 79 corresponding to the materials aluminum - Al, copper - Cu, silver - Ag, gold - Au is given in Table 2. We observe that the Sobol sequences do not offer any noticeable improvement over the Monte Carlo version, whereas the new algorithm A with the Halton sequences is far superior than any version of the algorithm B. That is why in Table 3 we do not give results with the Sobol sequences.

As an example we have underlined two numbers, so that one can see the improvement that is observed. We do not show here the experiments that we performed trying to integrate the 5-dimensional integral, that represent the contribution of the second collision to the yield, since the results are inferior to

Table 3. Relative error for different atomic numbers and energies

Method	Energ.	r	$Z = 13$	$Z = 29$	$Z = 47$	$Z = 79$
A	100	MC	0.00008113	0.00105891	0.00007185	0.00014770
		H.	0.00003145	0.00168857	0.00001128	0.00010663
B	100	MC	0.00284069	0.00039360	0.00117820	0.00197760
		H.	0.00363741	0.00001094	0.00265598	0.00525238
A	300	MC	0.00016334	0.00022043	0.00022872	0.00008240
		H.	0.00000558	0.00003802	0.00001464	0.00006128
B	300	MC	0.00195670	0.00092463	0.00404158	0.00372215
		H.	0.00535491	0.00807821	0.00340646	0.01149586
A	500	MC	0.00022356	0.00006088	0.00036549	0.00015564
		H.	0.00000332	0.00007950	0.00000776	0.00001390
B	500	MC	0.00176780	0.00409764	0.00802412	0.00209105
		H.	0.01271623	0.00829104	0.00529357	0.00842323
A	1000	MC	0.00029650	0.00010843	0.00014338	0.00011919
		H.	0.00000214	0.00000888	0.00000986	0.00001217
B	1000	MC	0.01533568	0.00166389	0.00092363	0.00824760
		H.	0.01441186	0.01137595	0.01170708	0.01085379
A	5000	MC	0.00006405	0.00014850	0.00009808	0.00013698
		H.	0.00000178	0.00000433	0.00001113	0.00005188
B	5000	MC	0.04280197	0.01696603	0.02436516	0.01659595
		H.	0.08779829	0.06721549	0.03756459	0.04871933

the simple simulation approach. Further investigation is needed to see if these computations can be performed more effectively. In any case this contribution is significantly smaller than the contribution of the first collision.

5 Concluding Remarks

We observe that by applying our quasi-Monte Carlo integration method we obtain accurate estimates of the first term of the Neumann series expansion of the solution, so that the error in their computation is negligible with respect to the total error. This approach is preferable to the analytical or semi-analytical approximation of these terms, discussed in [10], because the error can be decreased arbitrarily.

The application of the method for estimating the contribution of the second collision does not offer increased accuracy due to lack of smoothness in the sub-integral function. We also observe advantage of using the Halton sequence instead of pseudo-random numbers for computing this integral. For computing the contribution of the rest of the collisions we do not observe any noticeable improvement (or worsening) when the Halton or Sobol sequences are used, and the quasi-Monte Carlo method can be preferable in some parallel settings because of the possibility the have reproducible results - we can obtain the same result for any number of processors. It should be noted that both algorithms A and B can be parallelised efficiently, if needed.

References

1. E. Atanassov, A New Efficient Algorithm for Generating the Sobol Sequences with Scrambling, *Lecture Notes in Computer Science, Springer Verlag, Vol. 2542, 2003, pp. 83-90,*
2. E. Atanassov, I. Dimov, A new optimal Monte Carlo method for calculating integrals of smooth functions, *Monte Carlo Methods and Applications,* v 5(2),1999, p. 149-167.
3. E. Atanassov, I. Dimov, A. Dubus, A new weighted Monte Carlo algorithm for elastic electron backscattering from surfaces, *Math. and Comp. in Sim., Vol. 62 (3–6), 2003, pp. 297–305,*
4. E. I. Atanassov, I. T. Dimov, M. K. Durchova, A New Quasi-Monte Carlo Algorithm for Numerical Integration of Smooth Functions *Lecture Notes in Computer Science, Springer Verlag, Vol. 2907, 2004, pp. 128-135,*
5. E. I. Atanassov, M. K. Durchova, Generating and Testing the Modified Halton Sequences, *Lecture Notes in Computer Science, Springer Verlag, Vol. 2542, 2003, pp. 91-98,*
6. N.S. Bachvalov, On the approximate computation of multiple integrals, *Vestnik Moscow State University, Ser. Mat., Mech.,* 4 (1959), p. 3-18.
7. N.S. Bachvalov, Average Estimation of the Remainder Term of Quadrature Formulas, *USSR Comput. Math. and Math. Phys.,* Vol. 1(1) (1961), p. 64-77.
8. I. T. Dimov, E. I. Atanassov, M. K. Durchova, An improved Monte Carlo Algorithm for Elastic Electron Backscattering from Surfaces, *Lecture Notes in Computer Science, Springer Verlag, Vol. 2179, 2001, pp.141–148,*
9. M. Drmota, R. F. Tichy. Sequences, Discrepancies and Applications, *Lecture Notes on Mathematics, Springer, Berlin,* 1997, N 1651,
10. A. Dubus, A. Jablonski, S. Tougaard, Evaluation of theoretical models for elastic electron backscattering from surfaces, *Prog. Surf. Sci., Vol. 63, 2000, pp. 135–175,*
11. A. Jablonski, NIST Electron Elastic Scattering Cross-Section Database, Version 3.0, *http://www.nist.gov/srd,*
12. NIST Electron Inelastic-Mean-Free-Path Database: Version 1.1, *http://www.nist.gov/srd/nist71.htm,*
13. A. Jablonski, Elastic Scattering and Quantification in AES and XPS, *Surf. Interf. Anal. 14, 1989, pp. 659,*
14. L. Kuipers, H. Niederreiter *Uniform distribution of sequences,* John Wiley & sons, New York, 1974.
15. A. B. Owen. Scrambled Net Variance for Integrals of Smooth Functions, *Annals of Statistics,* 25(4), p. 1541-1562.
16. Standard E673, Annual Book of the ASTM Standards, *American Society for Testing and Materials, West Conshohocken, PA, Vol.3.06, 1998,*

Numerical Treatment of Fourth Order Singularly Perturbed Boundary Value Problems

Basem S. Attili

U. A. E. U. - College of Science, P. O. Box 17551,
Department of Mathematics and Computer Science,
Al-Ain - United Arab Emirates
b.attili@uaeu.ac.ae

Abstract. A numerical algorithm is proposed to solve a class of fourth order singularly perturbed two point boundary value problems (BVP). The method starts by transforming the BVP into a system of two second order ordinary differential equations with appropriate boundary conditions. The interval over which the BVP is defined will be subdivided into three disjoint regions. The system will then be solved separately on each subinterval. We combine the obtained solutions to get the solution of the BVP over the entire interval. For the inner regions, the boundary conditions at the end points are obtained through the zero order asymptotic expansion of the solution of the BVP. Examples will be solved to demonstrate the method and its efficiency.

1 Introduction

The problem under consideration is a fourth order and has the form

$$-\epsilon x^{(4)}(t) + a(t)x''(t) - b(t)x(t) = -f(t); \; t \in (0,1) \tag{1}$$

subject to

$$x(0) = \alpha, \;\; x(1) = \beta, \;\; x''(0) = -\gamma, \;\; x''(1) = -\eta, \tag{2}$$

where $\epsilon > 0$ is a small positive perturbation parameter, $a(t)$, $b(t)$ and $f(t)$ are sufficiently smooth functions such that

$$a(t) \geq k_1 > 0; \;\; 0 \geq b(t) \geq -k_2; \;\; k_2 > 0,$$

$$k_1 - 2k_2 \geq k_3 > 0 \;\text{ for some }\; k_3.$$

These conditions are needed to guarantee that the problem is not a turning point problem in addition it is needed together with the maximum principle for stability analysis.

Such singularly perturbed boundary value problems are common in applied sciences. They often occur in different shapes and format in optimal control, fluid dynamics and conviction-diffusion equations, see for example O'Mally[11] and Braianov and Vulkov[2, 3]. The presence of the perturbation parameter leads to

Z. Li et al. (Eds.): NAA 2004, LNCS 3401, pp. 141–148, 2005.

difficulties when classical numerical techniques are used to solve such problems. Convergence in such cases will not be uniform due to the presence of boundary layers, see Kevorkian and Cole[7], Nayfeh[10] and O'Mally[11, 12]. Treatment of problems with importance in fluid dynamics also can be found in Feckan[5], Miller, O'Riordan and Shishkin[8], Natensan and Ramanujam[9], Angelova and Vulkov[1] and O'Mally[12].

Several authors considered the numerical treatment of these problems. For example, Gartland[6] considered graded-mesh difference schemes with the exponentially fitted higher order differences. Finite element method based on a standard C^{m-1} splines and finite element method but for convection-reaction type problems were considered by Sun and Stynes[15]. An excellent collection of works of many authors can be found in Samarskii, Vabishchevich and Vulkov[14].

The system in (1)-(2) is transformed to a second order system of the form

$$\begin{aligned} -x_1''(t) - x_2(t) &= 0 \\ -\epsilon x_2''(t) + a(t)x_2(t) + b(t)x_1(t) &= f(t), \quad t \in (0,1), \end{aligned} \tag{3}$$

subject to

$$x_1(0) = \alpha, \quad x_1(1) = \beta, \quad x_2(0) = \gamma, \quad x_2(1) = \eta.$$

This system is the one we are going to use for computation and analysis from here on. We will consider two types of finite difference schemes; namely, the classical and the fitted finite difference schemes. The region $[0, 1]$ will be divided into three regions, left inner region, outer region and right inner region and the boundary value problem will be treated on each. This means that boundary conditions are needed which will be derived through the use of the zeroth order asymptotic expansion which results from using a standard perturbation technique.

The outline of this paper is will be as follows. In Section 2, we will present standard perturbation techniques in order to obtain appropriate boundary conditions for the different regions mentioned early. The numerical schemes will be presented and analyzed in Section 3. Finally and in Section 4, we will present some numerical examples to show the efficiency of the numerical methods when applied to our problem.

2 Approximation of the Solution

We will employ standard perturbation ideas to obtain an approximation to the solution of (1)-(2). For that reason, let

$$\begin{aligned} x(t,\epsilon) &= x_0 + \epsilon x_1 + O(\epsilon^2) \\ &= (y_0 + z_0 + w_0) + \epsilon(y_1 + z_1 + w_0) + O(\epsilon^2), \end{aligned} \tag{4}$$

with $y_0 = (y_{01}, y_{02})$, $z_0 = (z_{01}, z_{02})$ and $w_0 = (w_{01}, w_{02})$. Substituting back into (1)-(2) and matching the corresponding powers of ϵ, will lead to the zeroth order asymptotic expansion given by $y_{ZE} = (y_0 + z_0 + w_0)$ where y_0 is the solution to

$$-y_{01}''(t) - y_{02}(t) = 0$$

$$a(t)y_{02}(t) + b(t)y_{01}(t) = f(t), \quad t \in (0,1), \tag{5}$$

subject to

$$y_{01}(0) = \alpha, \quad y_{01}(1) = \beta.$$

The left layer correction z_0 is given by

$$z_{01} = 0$$

$$z_{02} = (\alpha - \beta - y_{02}(0) + y_{02}(1))\, e^{-\frac{\sqrt{a(1)}}{\sqrt{\epsilon}}} \cdot \frac{e^{-t\frac{\sqrt{a(0)}}{\sqrt{\epsilon}}}}{\left[1 - e^{\frac{\left(\sqrt{a(0)}+\sqrt{a(1)}\right)}{\sqrt{\epsilon}}}\right]} \tag{6}$$

The right layer correction w_0 is given by

$$w_{01} = 0$$

$$w_{02} = (\eta - \gamma - y_{02}(0) + y_{02}(1))\, e^{-\frac{\sqrt{a(0)}}{\sqrt{\epsilon}}} \cdot \frac{e^{-(1-t)\frac{\sqrt{a(1)}}{\sqrt{\epsilon}}}}{\left[1 - e^{\frac{\left(\sqrt{a(0)}+\sqrt{a(1)}\right)}{\sqrt{\epsilon}}}\right]}. \tag{7}$$

These corrections will be used to obtain boundary values for the three different regions.

3 The Numerical Scheme:

We will consider the exponentially fitted finite difference scheme in addition to the classical finite difference scheme which will start with. To descretize (3), we replace the second derivative term using the central difference formula. For that reason and as usual, we subdivide the interval $[a, b]$ into $N-$equal subintervals of size h where $h = \frac{b-a}{N}$, $a = x_0 < x_1 < x_2 < \ldots.. < x_N = b$ with $x_i = x_0 + ih$. Then (3) can be written in the form

$$\begin{aligned} L_1 x_i \ &:= -\frac{x_{1i+1} - 2x_{1i} + x_{1i-1}}{h^2} - x_{2i} = 0 \\ L_1 x_i \ &:= -\epsilon\frac{x_{2i+1} - 2x_{2i} + x_{2i-1}}{h^2} + a(t_i)x_{2i} + b(t_i)x_{1i} = f(t_i); \\ & i = 0, 1, 2, \ldots.., N-1, \end{aligned} \tag{8}$$

with $x_{10} = \alpha$, $x_{1N} = \beta$, $x_{20} = \gamma$ and $x_{2N} = \eta$.

While the exponentially fitted finite difference scheme will be

$$\begin{aligned} L_1 x_i \ &:= -\frac{x_{1i+1} - 2x_{1i} + x_{1i-1}}{h^2} - x_{2i} = 0 \\ L_1 x_i \ &:= -\epsilon\sigma_i(\rho)\frac{x_{2i+1} - 2x_{2i} + x_{2i-1}}{h^2} + a(t_i)x_{2i} + b(t_i)x_{1i} = f(t_i); \\ & i = 0, 1, 2, \ldots.., N-1, \end{aligned} \tag{9}$$

with $x_{10} = \alpha$, $x_{1N} = \beta$, $x_{20} = \gamma$ and $x_{2N} = \eta$, where

$$\sigma_i(\rho) = \frac{\rho^2 a(t_i)}{4\sin\left[h^2\rho\sqrt{a(t_i)}/2\right]}; \quad \rho = \frac{h}{\sqrt{\epsilon}}. \tag{10}$$

This latter scheme and similar schemes are detailed in Doolan, Miller and Schilders[4].

If y_i is any mesh function satisfying $L_1 y_i \geq 0$ and $L_2 y_i \geq 0$ subject $y_{10} \geq 0$, $y_{1N} \geq 0$, $y_{20} \geq 0$ and $y_{2N} \geq 0$. Then the discrete maximum principle implies that $y_i = (y_{1i}, y_{2i}) \geq 0$ for $i = 1, 2, ..., N$, see Natesan and Ramanujam[9]. This serves as a basis for a needed stability result given by the following.

Lemma 1. *If y_i is any mesh function of (8) or (9), then*

$$\|y_i\| \leq K \max\left\{|y_{10}|, \ |y_{20}|, \ |y_{1N}|, \ |y_{2N}|, \ \max_i |L_1 y_i|, \ \max_i |L_2 y_i|, \right\} \tag{11}$$

where $\|y_i\| = \max\{|y_{1i}|, \ |y_{2i}|\}; \ i = 0, 1, ..., N$ *and* K *constant.*

Proof. Let us define two mesh functions

$$u_i^+ = (u_{1i}^+, \ u_{2i}^+) = \left[k_1(1+d)\left(1 - \frac{t_i^2}{2}\right) + y_{1i}, \ k_1 + y_{2,i}\right] \tag{12}$$

and

$$u_i^- = (u_{1i}^-, \ u_{2i}^-) = \left[k_1(1+d)\left(1 - \frac{t_i^2}{2}\right) - y_{1i}, \ k_1 - y_{2,i}\right], \tag{13}$$

where $0 < d \ll 1$ *and*
$k_1 = K \max\{|y_{10}|, \ |y_{20}|, \ |y_{1N}|, \ |y_{2N}|, \ \max_i |L_1 y_i|, \ \max_i |L_2 y_i|, \}$. *Now using this definition we have* $L_1 u_i^+ \geq 0$, $L_1 u_i^- \geq 0$, $L_2 u_i^+ \geq 0$, $L_2 u_i^- \geq 0$ *and* $u_{10}^{\mp} \geq 0$, $u_{1N}^{\mp} \geq 0$, $u_{20}^{\mp} \geq 0$, $u_{2N}^{\mp} \geq 0$. *These are the conditions of the discrete maximum principle which means* $u_i^+ \geq 0$ *and* $u_i^- \geq 0$. *Now following the discussion before the Lemma with the references there the result follows.*

Following the results obtained in Roos and Stynes[13] on the schemes given by (8) and (9) applied to similar problems, one can show that the schemes converges at the inner grid points with O(h) convergence rate. Note also that (9) converges uniformly over the interval.

4 Numerical Details and Examples

Since there are boundary layers, it is logical to subdivide the interval $[0, 1]$ into three subintervals; namely, $[0, c\epsilon]$ left inner region, $[c\epsilon, 1 - c\epsilon]$ outer region and $[1 - c\epsilon, 1]$ right inner region where $c > 0$ and $c\epsilon \ll 1$. This means three boundary value problems need to be solved in the respective region. To do so there is a need for boundary conditions at the points $c\epsilon$ and $1 - c\epsilon$. This can be done using the zeroth order asymptotic expansion done in Section 2. This leads to the three boundary value problems defined as follows:

1. On $[0, c\epsilon]$:
   Solve (2); that is,

$$
\begin{aligned}
& -x_1''(t) - x_2(t) = 0 \\
-\epsilon x_2'' + a(t)x_2(t) &+ b(t)x_1(t) = f(t); \quad t \in (0, c\epsilon)
\end{aligned}
\tag{14}
$$

   subject to the boundary conditions

$$x_1(0) = \alpha,\ x_2(0) = \gamma,\ x_1(c\epsilon) = \beta_1 \text{ and } \ x_2(c\epsilon) = \eta_1, \tag{15}$$

   where α and γ are as before while

$$
\begin{aligned}
\beta_1 &= y_{01}(c\epsilon) + z_{01}(c\epsilon) + w_{01}(c\epsilon) \\
\eta_1 &= y_{02}(c\epsilon) + z_{02}(c\epsilon) + w_{02}(c\epsilon).
\end{aligned}
\tag{16}
$$

   The exponentially fitted scheme used is the one given by (9) with β replaced by β_1 and η replaced by η_1 given above by (14). Also we use a step size $h_1 = \frac{c\epsilon}{N}$, $t_i = ih_1$.
2. On $[c\epsilon, 1 - c\epsilon]$:
   Using exponentially fitted difference scheme given by (9), solve (14) for $t \in (c\epsilon, 1 - c\epsilon)$ subject to the boundary conditions

$$x_1(c\epsilon) = \alpha_1,\ x_2(c\epsilon) = \gamma_1,\ x_1(1 - c\epsilon) = \beta_1 \text{ and } \ x_2(1 - c\epsilon) = \eta_1, \tag{17}$$

   where β_1 and η_1 are as given by (14) before while

$$
\begin{aligned}
\alpha_1 &= y_{01}(c\epsilon) + z_{01}(c\epsilon) + w_{01}(c\epsilon) \\
\gamma_1 &= y_{02}(c\epsilon) + z_{02}(c\epsilon) + w_{02}(c\epsilon).
\end{aligned}
\tag{18}
$$

   We use a step size $h_2 = \frac{1-2c\epsilon}{N}$, $t_i = c\epsilon + ih_2$.
3. On $[1 - c\epsilon, 1]$:

Again using the scheme given by (9), solve (12) subject to the boundary conditions

$$x_1(1 - c\epsilon) = \alpha_1,\ x_2(1 - c\epsilon) = \gamma_1,\ x_1(1) = \beta \text{ and } \ x_2(1) = \eta, \tag{19}$$

where α_1 and γ_1 are as given by (15). We use $h_2 = \frac{1-(1-c\epsilon)}{N} = \frac{c\epsilon}{N}$ and $t_i = ih_1$ same as before in point one. Note that the resulting systems are linear and are solved usually using self correcting *LU* decomposition. We start with a specific value of c then increase it until the difference between the solutions obtained from the three problems is small enough. Then the solutions will be combined to obtain a solution for the problem.

For numerical testing we consider the following examples:

Example 1. We consider the following examples for numerical testing:

$$-\epsilon y^{(4)}(t) + 4y''(t) = 1 \tag{20}$$

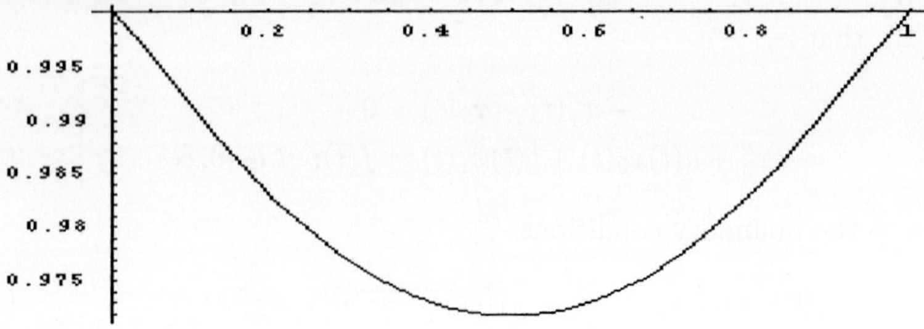

Fig. 1.

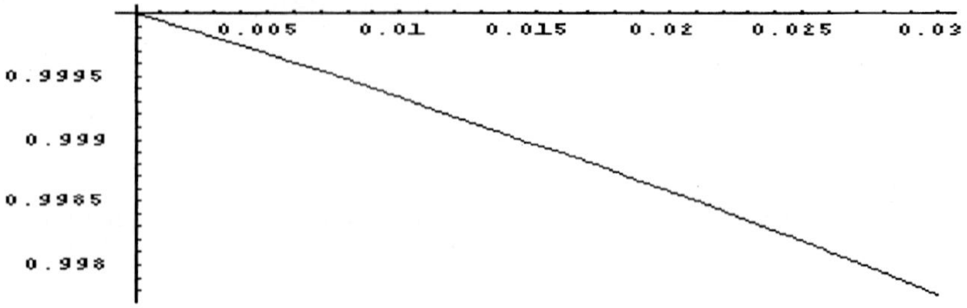

Fig. 2.

subject to

$$y(0) = 1,\ y(1) = 1,\ y''(0) = -1,\ y''(1) = -1. \tag{21}$$

Note that to obtain y_O, z_O, w_O given by (4) and hence y_{ZE}, the system in (5) will be solved first. It is a simple linear second order differential equation with $a(t) = 1$, $b(t) = 0$ and $f(t) = 1$. Once $y_O = (y_{01}, y_{02})$ is computed, one can obtain $z_O = (z_{01}, z_{02})$ and $w_O = (w_{01}, w_{02})$ at $t = c\epsilon$ and $t = 1 - c\epsilon$ from (6) and (7) respectively. This gives the boundary conditions for the three different problems in the three different regions; namely, left, right and inner regions. Doing so, the solutions obtained for the different regions with $\epsilon = 0.01$ and $c = 3$ are given in Figures 1-4.

Example 2. The second example is given as follows:

$$-\epsilon y^{(4)}(t) + 4y''(t) + y(t) = -f(t) \tag{22}$$

subject to

$$y(0) = 1,\ y(1) = 1,\ y''(0) = -1,\ y''(1) = -1 \tag{23}$$

where

$$f(t) = -2 - \frac{t(1-t)}{8} - \frac{5\epsilon}{16} + \frac{5\epsilon}{16}\{e^{\frac{-2t}{\sqrt{\epsilon}}} - e^{\frac{-2(1+t)}{\sqrt{\epsilon}}} + e^{\frac{-2(1-t)}{\sqrt{\epsilon}}} - e^{\frac{-2(2-t)}{\sqrt{\epsilon}}}\}/[1 - e^{\frac{-4}{\sqrt{\epsilon}}}]. \tag{24}$$

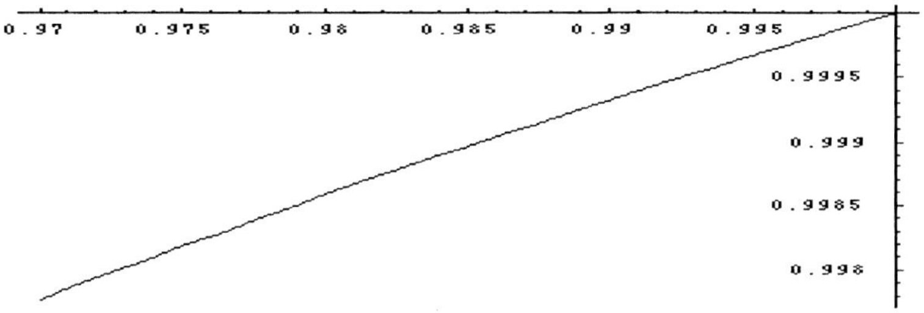

Fig. 3.

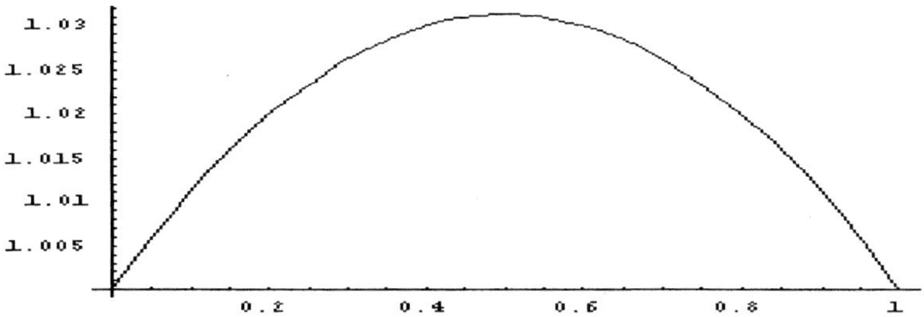

Fig. 4.

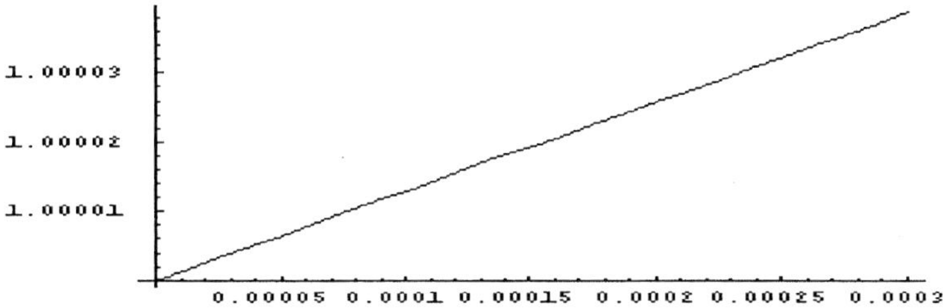

Fig. 5.

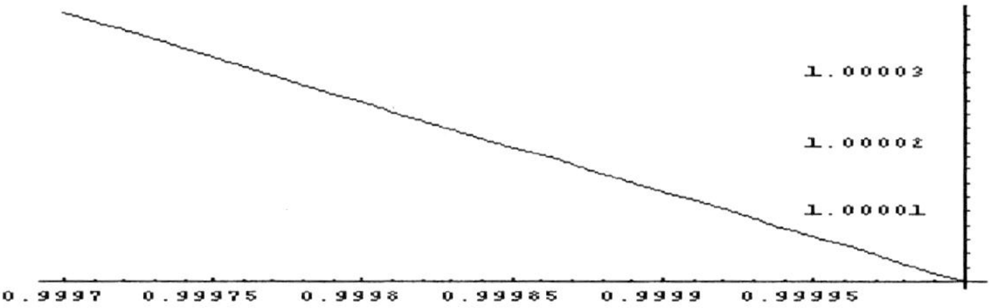

Fig. 6.

Definition 1. *The attainable set* $\mathcal{R}(t, t_0, X_0)$ *at a given time* $t \in I$ *for Problem 1 is defined as*

$$\mathcal{R}(t, t_0, X_0) = \{x(t) \mid x : I \to \mathbb{R}^n \text{ is an absolutely continuous solution of (1)–(2)}\} .$$

The Aumann integral introduced in [1] is an important tool for the following.

Definition 2. *Consider a set-valued function* $F : I \to \mathbb{R}^n$ *with images in* $\mathcal{C}(\mathbb{R}^n)$ *which is measurable and integrably bounded (see [1]).*

Then, Aumann's integral is defined as

$$\int_{t_0}^{T} F(t)dt := \{ \int_{t_0}^{T} f(t)dt \mid f(\cdot) \text{ is an integrable selection of } F(\cdot) \} .$$

It serves as a tool for reducing the approximation of the attainable set at time T to a problem of studying a set-valued quadrature method (see (7)).

Notation 2. *The arithmetic operations of sets*

$$\begin{aligned} \lambda \cdot C &:= \{\lambda \cdot c \mid c \in C\} && \text{(scalar multiple)} , \\ C + D &:= \{c + d \mid c \in C,\ d \in D\} && \text{(Minkowski sum)} , \\ A \cdot C &:= \{Ac \mid c \in C\} && \text{(image under a linear mapping)} \end{aligned}$$

are defined as usual for $C, D \in \mathcal{C}(\mathbb{R}^n)$, $\lambda \in \mathbb{R}$, $A \in \mathbb{R}^{k \times n}$.
We denote with $\mathrm{d}_H(C, D)$ *the Hausdorff-distance of these two sets. The support function for* C *in direction* $l \in \mathbb{R}^n$ *is defined as*

$$\delta^*(l, C) := \max_{c \in C} \langle l, c \rangle .$$

Lemma 1. *Let* $C, D \in \mathcal{C}(\mathbb{R}^n)$, $l \in \mathbb{R}^n$, $\lambda \geq 0$ *and* $A, B \in \mathbb{R}^{m \times n}$. *Then,*

$$\delta^*(l, C + D) = \delta^*(l, C) + \delta^*(l, D), \quad \delta^*(l, \lambda C) = \lambda \delta^*(l, C) , \tag{3}$$

$$\mathrm{d}_H(C, D) = \sup_{\|l\|_2 = 1} |\delta^*(l, C) - \delta^*(l, D)| , \tag{4}$$

$$\mathrm{d}_H(AU, BU) \leq \|A - B\| \cdot \|U\| \text{ with } \|U\| := \sup_{u \in U} \|u\|_2 , \tag{5}$$

$$\mathrm{d}_H((A + B)U, AU + BU) \leq \|A - B\| \cdot \|U\| . \tag{6}$$

In Problem 1, the attainable set at time T

$$\mathcal{R}(T, t_0, X_0) = \Phi(T, t_0)X_0 + \int_{t_0}^{T} \Phi(T, t)B(t)U dt \tag{7}$$

could be rewritten as a sum of the transformed starting set and Aumann's integral of $\Phi(T, \cdot)B(\cdot)U$ (cf. e.g. [6]), where $\Phi(T, t)$ is the corresponding fundamental solution. Scalarization of (7) by support functions and applying the calculus rules in (3) and [4] yields for $l \in S_{n-1}$ (i.e., $l \in \mathbb{R}^n$ with $\|l\|_2 = 1$)

$$\delta^*(l, \mathcal{R}(T, t_0, X_0)) = \delta^*(l, \Phi(T, t_0)X_0) + \int_{t_0}^{T} \delta^*(l, \Phi(T, t)B(t)U)dt \ . \tag{8}$$

2 Quadrature and Combination Methods

Notation 3. *For a given interval $I := [t_0, T]$ and a function $f : I \to \mathbb{R}^n$ consider the point-wise quadrature formula*

$$Q(f; [t_0, T]) := \sum_{\mu=1}^{s} b_\mu f(t_0 + c_\mu(T - t_0))$$

for the approximation of $\int_I f(t)dt$, where $b_\mu \in \mathbb{R}$ are the weights and $c_\mu \in [0, 1]$ determine the nodes ($\mu = 1, \ldots, s$). Introducing the step-size $h = \frac{T-t_0}{N}$ for $N \in \mathbb{N}$ and applying the quadrature formula on each sub-interval $[t_j, t_{j+1}]$ with $t_j = t_0 + jh$, $j = 0, \ldots, N-1$, we arrive at the iterated quadrature formula

$$Q_N(f; [t_0, T]) := h \sum_{j=0}^{N-1} Q(f; [t_j, t_{j+1}]) = h \sum_{j=0}^{N-1} \sum_{\mu=1}^{s} b_\mu f(t_j + c_\mu h) \ .$$

Definition 3. *Consider a point-wise quadrature formula of Notation 3. Using the arithmetic operations of Notation 2, we introduce for a set-valued function $F : I \Rightarrow \mathbb{R}^n$ with images in $\mathcal{C}(\mathbb{R}^n)$ the iterated set-valued quadrature formula*

$$Q_N(F; [t_0, T]) := h \sum_{j=0}^{N-1} \sum_{\mu=1}^{s} b_\mu F(t_j + c_\mu h). \tag{9}$$

These set-valued quadrature methods are studied by several authors, cf. e.g. [12, 6, 4, 9, 2]. Essential for reaching the same order of convergence as in the pointwise case is the smoothness of the function $t \mapsto \delta^*(l, F(t))$ uniformly in $l \in S_{n-1}$ due to the scalarization as in (8). To express the smoothness in a weaker sense, the averaged modulus of smoothness $\tau_k(f; h)$, presented e.g. in [11], is used which is a L_1-norm of the local modulus of smoothness expressed as a certain supremum of the k-th finite difference of the function $f(\cdot)$.

Theorem 4. *Let $F : I \Rightarrow \mathbb{R}^n$ with images in $\mathcal{C}(\mathbb{R}^n)$ be measurable and bounded. Consider a point-wise quadrature formula with precision $p - 1$, $p \in \mathbb{N}$ (cf. [11]) and the set-valued iterated form (9) with step-size $h = \frac{T-t_0}{N}$, $N \in \mathbb{N}$.*

Then, the set-valued quadrature formula fulfills

$$\mathrm{d}_H(\int_I F(t)dt, Q_N(F; I)) \le (1 + \sum_{\mu=1}^{s} \frac{b_\mu}{T - t_0}) \cdot W_p \cdot \sup_{\|l\|_2=1} \tau_p(\delta^*(l, F(\cdot)), \frac{2}{p}h) \ .$$

Proof. For the point-wise result see [2–Satz 1.2.11] which is based on [11–Theorem 3.4]. Apply this result for the function $t \mapsto \delta^*(l, F(t))$ for each $l \in S_{n-1}$ and use the equivalent expression (4) in Lemma 1 for the Hausdorff distance. □

Set-valued quadrature methods could be used to approximate attainable set at the time T, if the values of the fundamental solution are known at the integration nodes $t_j + c_\mu h$, $\mu = 1, \dots, s$, $j = 0, \dots, N-1$. Otherwise, these values of the fundamental solution needs to be approximated carefully so that the order of convergence of the quadrature method is not destroyed. Compare the next proposition with a result in [4] formulated with global disturbances.

Proposition 1. *Let us consider Problem 1, set* $h := \frac{T-t_0}{N}$, $N \in \mathbb{N}$ *and the set-valued quadrature method with precision* $p-1$, $p \in \mathbb{N}$, *studied in Theorem 4 with* $\tau_p(\delta^*(l, \Phi(T, \cdot)B(\cdot)U), h) \le Ch^p$ *uniformly in* $l \in S_{n-1}$. *For* $j = 0, \dots, N-1$ *let the approximations* $\widetilde{\Phi}(t_{j+1}, t_j)$ *of the values of the fundamental solution resp.* $\widetilde{U}_\mu(\cdot)$ *of the images of* $\Phi(t_{j+1}, \cdot)B(\cdot)U$ *for* $t = t_j + c_\mu h$ *fulfill:*

$$\widetilde{\Phi}(t_{j+1}, t_j) = \Phi(t_{j+1}, t_j) + \mathcal{O}(h^{p+1}) \ ,$$

$$\mathrm{d}_H(\widetilde{U}_\mu(t_j + c_\mu h),\ \Phi(t_{j+1}, t_j + c_\mu h)B(t_j + c_\mu h)U) = \mathcal{O}(h^p) \quad (\mu = 1, \dots, s) \ .$$

[$\mathcal{O}(h^q)$ *is understood uniformly in* j *and* μ*.] Then, the combination method*

$$X_{j+1}^N = \widetilde{\Phi}(t_{j+1}, t_j)X_j^N + h \sum_{\mu=1}^{s} b_\mu \widetilde{U}_\mu(t_j + c_\mu h) \quad (j = 0, \dots, N-1) \ , \tag{10}$$

$$X_0^N \in \mathcal{C}(\mathbb{R}^n) \ \text{ with } \ \mathrm{d}_H(X_0, X_0^N) = \mathcal{O}(h^p) \tag{11}$$

defined above satisfies the global estimate

$$\mathrm{d}_H(\mathcal{R}(T, t_0, X_0), X_N^N) = \mathcal{O}(h^p) \ . \tag{12}$$

Especially, if approximations of the values of the fundamental solution

$$\widetilde{\Phi}_\mu(t_{j+1}, t_j + c_\mu h) = \Phi(t_{j+1}, t_j + c_\mu h) + \mathcal{O}(h^p) \quad (\mu = 1, \dots, s) \ ,$$

then the estimation (12) *above also holds with the following setting:*

$$\widetilde{U}_\mu(t_j + c_\mu h) = \widetilde{\Phi}_\mu(t_{j+1}, t_j + c_\mu h)B(t_j + c_\mu h)U \quad (\mu = 1, \dots, s) \ .$$

3 Set-Valued Runge-Kutta Methods

Explicit Runge-Kutta methods could be expressed by the Butcher array (cf. [5])

$$\begin{array}{c|cccccc} c_1 & 0 & 0 & \dots & 0 & 0 & 0 \\ c_2 & a_{21} & 0 & \dots & 0 & 0 & 0 \\ \vdots & \vdots & \vdots & \ddots & \vdots & \vdots & \vdots \\ c_{s-1} & a_{s-1,1} & a_{s-1,2} & \dots & a_{s-1,s-2} & 0 & 0 \\ c_s & a_{s,1} & a_{s,2} & \dots & a_{s,s-2} & a_{s,s-1} & 0 \\ \hline & b_1 & b_2 & \dots & b_{s-2} & b_{s-1} & b_s \end{array} \quad \text{with } c_1 := 0 \ .$$

For a starting value $\eta_0^N \in X_0^N$ and $j = 0, \dots, N-1$, $\nu = 1, \dots, s$ let us define

$$\eta_{j+1}^N = \eta_j^N + h \sum_{\nu=1}^{s} b_\nu \xi_j^{(\nu)} \ , \tag{13}$$

$$\xi_j^{(\nu)} = A(t_j + c_\nu h)(\eta_j^N + h \sum_{\mu=1}^{\nu-1} a_{\nu,\mu} \xi_j^{(\mu)}) + B(t_j + c_\nu h) u_j^{(\nu)} \ , \quad u_j^{(\nu)} \in U \ . \tag{14}$$

X_{j+1}^N and X_0^N consist of all possible iterates η_{j+1}^N in (13)–(14) resp. chosen starting values η_0^N and form the set-valued Runge-Kutta method. Additional restrictions on the selections $u_j^{(\nu)} \in U$ for $\nu = 1, \dots, s$ need to be imposed on each subinterval $I_j = [t_j, t_{j+1}]$ to increase the order of convergence in the set-valued case. These restrictions define different selection strategies.

Modified Euler Method

The modified Euler method resp. the method of Euler-Cauchy/Heun could be described by the Butcher array as

$$\begin{array}{c|cc} 0 & 0 & 0 \\ \frac{1}{2} & \frac{1}{2} & 0 \\ \hline & 0 & 1 \end{array} \quad \text{resp.} \quad \begin{array}{c|cc} 0 & 0 & 0 \\ 1 & 1 & 0 \\ \hline & \frac{1}{2} & \frac{1}{2} \end{array} \ .$$

In [14], the method of Euler-Cauchy is discussed in detail with the result that for this method, one could use either constant selections or two free selections at each subinterval $[t_j, t_{j+1}]$ and reach order of convergence 2 under suitable smoothness conditions. In [13], the proofs are presented for the same method even in the case of strongly convex nonlinear differential inclusions.

Lemma 2. *If we consider Problem 1, then the modified Euler method could be rewritten for the constant selection strategy "$u_j^{(1)} = u_j^{(2)}$" as the combination method* (10) *of Proposition 1 with the iterated midpoint rule and*

$$Q(\Phi(t_{j+1}, \cdot)B(\cdot)U; [t_j, t_{j+1}]) := h\Phi(t_{j+1}, t_j + \frac{h}{2})B(t_j + \frac{h}{2})U \ ,$$

$$\widetilde{\Phi}(t_{j+1}, t_j) := I + hA(t_j + \frac{h}{2}) + \frac{h^2}{2} A(t_j + \frac{h}{2})A(t_j) \ ,$$

$$\widetilde{U}_1(t_j + \frac{h}{2}) := \Big(B(t_j + \frac{h}{2}) + \frac{h}{2} A(t_j + \frac{h}{2})B(t_j)\Big)U \ .$$

For two free selections $u_j^{(1)}, u_j^{(2)} \in U$ we have the iterated trapezoidal rule and

$$Q(\Phi(t_{j+1}, \cdot)B(\cdot)U; I_j) := \frac{h}{2}\big(\Phi(t_{j+1}, t_j)B(t_j)U + \Phi(t_{j+1}, t_{j+1})B(t_{j+1})U\big) \ ,$$

$$\widetilde{\Phi}(t_{j+1}, t_j) := I + hA(t_j + \frac{h}{2}) + \frac{h^2}{2} A(t_j + \frac{h}{2})A(t_j) \ ,$$

$$\widetilde{U}_1(t_j) := B(t_j + \frac{h}{2})U + hA(t_j + \frac{h}{2})B(t_j)U \ , \quad \widetilde{U}_2(t_{j+1}) := B(t_j + \frac{h}{2})U \ .$$

Proposition 2. *Assume that $A'(\cdot)$ and $B(\cdot)$ are Lipschitz in Problem 1 and that $\delta^*(l, \Phi(T, \cdot)B(\cdot)U)$ is absolutely continuous with a L_1-representative of the derivative with bounded variation uniformly in $l \in S_{n-1}$.*

Then, the modified Euler method in Lemma 2 with $p = 2$ in (11) and constant selection converges at least with order 2, whereas the modified Euler method with two independent selections converges at least with order 1.

Proof. Both quadrature methods have precision 1 (cf. [2], [12]), yielding order of convergence 2 in Theorem 4, if the disturbances would be of order $\mathcal{O}(h^2)$. For constant selections the result follows from Proposition 1 together with (5) and careful Taylor expansions in the estimations below:

$$\widetilde{\Phi}(t_{j+1}, t_j) = \Phi(t_{j+1}, t_j) + \mathcal{O}(h^3),$$

$$\mathrm{d_H}(\widetilde{U}_1(t_j + \frac{h}{2}), \Big(I + \frac{h}{2}A(t_j + \frac{h}{2})\Big)B(t_j + \frac{h}{2})U) = \mathcal{O}(h^2) \ ,$$

$$\mathrm{d_H}(\Big(I + \frac{h}{2}A(t_j + \frac{h}{2})\Big)B(t_j + \frac{h}{2})U, \ \Phi(t_{j+1}, t_j + \frac{h}{2})B(t_j + \frac{h}{2})U) = \mathcal{O}(h^2)$$

In the case of two free selections, the reasoning is similar, but only accuracy $\mathcal{O}(h)$ is possible in general (due to (15) and (16)):

$$\mathrm{d_H}(\widetilde{U}_1(t_j), (I + hA(t_j))B(t_j)U) \le \mathrm{d_H}(B(t_j + \frac{h}{2})U, (I + hA(t_j))B(t_j)U)$$

$$+ \mathrm{d_H}(hA(t_j + \frac{h}{2})B(t_j)U, \{0_{\mathbb{R}^n}\}) = \mathcal{O}(h) \ , \tag{15}$$

$$\mathrm{d_H}((I + hA(t_j))B(t_j)U, \Phi(t_{j+1}, t_j)B(t_j)U) = \mathcal{O}(h^2) \ ,$$

$$\mathrm{d_H}(\widetilde{U}_2(t_{j+1}), \Phi(t_{j+1}, t_{j+1})B(t_{j+1})U) = \mathcal{O}(h) \ . \qquad \square \tag{16}$$

The assumptions in Proposition 2 could be weakened by demanding only the bounded variation of $A'(\cdot)$ and $B(\cdot)$. Clearly, for the strategy with two free selections, only $A(\cdot)$ needs to be Lipschitz, $B(\cdot)$ should be bounded and $\delta^*(l, \Phi(T, \cdot) B(\cdot)U)$ be of bounded variation uniformly in $l \in S_{n-1}$.

Since in general (even for the time-independent case, compare also (6)),

$$(B + \frac{h}{2}AB)U \ne (BU + \frac{h}{2}ABU) = \frac{1}{2}(BU + hABU + BU) \ , \tag{17}$$

$$\mathrm{d_H}\left((B + \frac{h}{2}AB)U, \ \frac{1}{2}(BU + (BU + hABU))\right) = \mathcal{O}(h) \ , \tag{18}$$

both selection strategies for modified Euler differ. The proof of (18) uses a similar trick as in (15). This phenomena is also observed in the context of discretization by Runge-Kutta methods of nonlinear optimal control problems in [7]. In this work, additional assumptions on the coercitivity (not fulfilled in Problem 1) and on the smoothness of the optimal control leads to the accuracy up to $\mathcal{O}(h^2)$ for state and control variables using different proof ideas.

Scalarization as in (8) or direct methods for optimal control problems in [3] lead to numerical implementations of both selection strategies. For the scalarization approach, support functions of left-hand and right-hand sides of the equation (10) are calculated. This leads to an iterative method (cf. [2] for more details), if one restricts the computation of the support functions (or points) to a finite number of normed directions $l^{(j)} \in \mathbb{R}^n$, $j = 1, \ldots, M$.

Example 1. (cf. [4]) Let $n = 2$, $m = 1$, $I = [0, 1]$, set $A(t) = \begin{pmatrix} 0 & 1 \\ 0 & 0 \end{pmatrix}$, $B(t) = \begin{pmatrix} 0 \\ 1 \end{pmatrix}$ and $U = [-1, 1]$. Since (17) is fulfilled here, both selection strategies for modified Euler differ (cf. Figure 1). In Figure 1, the reference set (the combination method

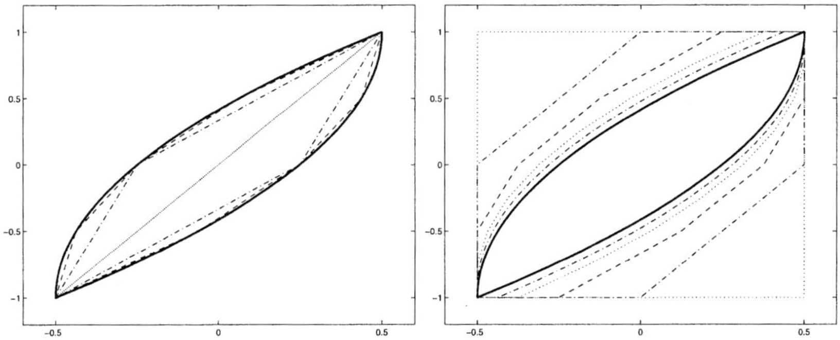

Fig. 1. modified Euler with constant (left picture) resp. 2 free selections (right one) (step sizes $h = 1, 0.5, 0.25, 0.125, 0.0625$)

"iterated trapezoidal rule and Euler/Cauchy" with $N = 10000$ in [2]) is plotted with supporting points in $M = 200$ directions with a thicker solid line, whereas the result for each calculated step size is depicted with dotted ($h = 1, 0.125$), dashed-dotted ($h = 0.5, 0.0625$) and dashed lines ($h = 0.25$). One may recognize the different speed of convergence (2 resp. 1) even by the picture. This is underlined by the computed estimations of the order of convergence in Table 1. Hence, the possible order breakdown to $\mathcal{O}(h)$ in Proposition 2 for modified Euler with two free selections can occur for certain examples.

Table 1. convergence estimation of modified Euler for both selection strategies

N	Hausdorff distance to reference set	estimated order of convergence	Hausdorff distance to reference set	estimated order of convergence
1	0.21434524	——	0.75039466	——
2	0.05730861	1.90311	0.36454336	1.04156
4	0.01517382	1.91717	0.17953522	1.02182
8	0.00384698	1.97979	0.08841414	1.02192
16	0.00096510	1.99498	0.04419417	1.00042
	(constant selections)		(2 free selections)	

4 Conclusions

The presented framework may give a structural outline to proofs for linear differential inclusions with possibly non-optimal order of convergence for a set-valued Runge-Kutta method with a chosen selection strategy. For the modified Euler method, the better selection strategy is formed by the constant selections which fits to the underlying (set-valued) quadrature method (i.e., the midpoint rule).

For the (classical) Runge-Kutta method of order 4 with Butcher array

$$\begin{array}{c|cccc} 0 & 0 & 0 & 0 & 0 \\ \frac{1}{2} & \frac{1}{2} & 0 & 0 & 0 \\ \frac{1}{2} & 0 & \frac{1}{2} & 0 & 0 \\ 1 & 0 & 0 & 1 & 0 \\ \hline & \frac{1}{6} & \frac{1}{3} & \frac{1}{3} & \frac{1}{6} \end{array}$$

first experiments show that the selection strategy with three free selections $u_j^{(\nu)}$, $\nu = 1, 2, 4$ and $u_j^{(2)} = u_j^{(3)}$ lead to a set-valued method of at least order 3 under sufficient smoothness conditions. This fits best to the Simpson's rule as the underlying set-valued quadrature method. All presented selection strategies can be carried over to the case of nonlinear differential inclusions.

References

1. R. J. Aumann. Integrals of Set-Valued Functions. *J. Math. Anal. Appl.*, 12(1):1–12, 1965.
2. R. Baier. Mengenwertige Integration und die diskrete Approximation erreichbarer Mengen. *Bayreuth. Math. Schr.*, 50:xxii + 248 S., 1995.
3. R. Baier, C. Büskens, I. A. Chahma, and M. Gerdts. Approximation of Reachable Sets by Direct Solution Methods of Optimal Control Problems. submitted, 04/2004. 23 pages.
4. R. Baier and F. Lempio. Computing Aumann's integral. In *[10]*, pages 71–92, 1994.
5. J. C. Butcher. *The Numerical Analysis of Ordinary Differential Equations.* John Wiley & Sons, Chichester–New York–Brisbane–Toronto–Singapore, 1987.
6. T. D. Donchev and E. Farkhi. Moduli of smoothness of vector valued functions of a real variable and applications. *Numer. Funct. Anal. Optim.*, 11(5 & 6):497–509, 1990.
7. A. L. Dontchev, W. W. Hager, and V. M. Veliov. Second-Order Runge-Kutta Approximations in Control Constrained Optimal Control. *SIAM J. Numer. Anal.*, 38(1):202–226, 2000.
8. A. L. Dontchev and F. Lempio. Difference methods for differential inclusions: A survey. *SIAM Rev.*, 34(2):263–294, 1992.
9. M. Krastanov and N. Kirov. Dynamic interactive system for analysis of linear differential inclusions. In *[10]*, pages 123–130, 1994.
10. A. B. Kurzhanski and V. M. Veliov, editors. *Modeling Techniques for Uncertain Systems, Proceedings of a Conferences held in Sopron, Hungary, July 6-10, 1992*, volume 18 of *Progress in Systems and Control Theory*, Basel, 1994. Birkhäuser.

11. B. Sendov and V. Popov. *Averaged Moduli of Smoothness.* Applications in Numerical Methods and Approximation. John Wiley and Sons, Chichester–New York–Brisbane–Toronto–Singapore, 1988.
12. V. M. Veliov. Discrete approximations of integrals of multivalued mappings. *C. R. Acad. Bulgare Sci.*, 42(12):51–54, 1989.
13. V. M. Veliov. Second order discrete approximations to strongly convex differential inclusions. *Systems Control Lett.*, 13(3):263–269, 1989.
14. V. M. Veliov. Second Order Discrete Approximation to Linear Differential Inclusions. *SIAM J. Numer. Anal.*, 29(2):439–451, 1992.

Numerical Methods for the Landau-Lifshitz-Gilbert Equation

Ľubomír Baňas

Department of Mathematical Analysis, Ghent University, 9000 Gent, Belgium
lubo@cage.ugent.be
http://cage.ugent.be/~lubo

Abstract. In this paper we give an overview of the numerical methods for the solution of the Landau-Lifshitz-Gilbert equation. We discuss advantages of the presented methods and perform numerical experiments to demonstrate their performance. We also discuss the coupling with Maxwell's equations.

1 Introduction

Numerical simulations based on Landau-Lifshitz-Gilbert (LLG) equation are widely used in the magnetic recording applications. The LLG equation, describing the time evolution of magnetization in a ferromagnetic material, can be written in a normalized form

$$\partial_t \boldsymbol{m} = \gamma M_s \left(\boldsymbol{h}_T \times \boldsymbol{m} + \alpha \boldsymbol{m} \times (\boldsymbol{h}_T \times \boldsymbol{m}) \right) \quad \text{in} \quad \Omega \times (0, T), \tag{1}$$

where $\boldsymbol{h}_T = -\frac{1}{\mu_0 M_s^2} \frac{\partial \boldsymbol{E}}{\partial \boldsymbol{m}}$ is the total field, $\boldsymbol{E}$ is the total free energy in the ferromagnet, M_s is the saturation magnetization, α is the damping constant, γ is the gyromagnetic ratio and μ_0 is the permeability of vacuum. The first term on the right-hand side causes the precession of $\boldsymbol{m}$ around $\boldsymbol{h}_T$ and the second term is the damping term. The magnetization $\boldsymbol{m}$ satisfies an initial condition $\boldsymbol{m}(0) = \boldsymbol{m}_0$ and Neumann boundary condition

$$\frac{\partial \boldsymbol{m}}{\partial \boldsymbol{\nu}} = \boldsymbol{0} \quad \text{on} \quad \partial\Omega, \tag{2}$$

where $\boldsymbol{\nu}$ is the outward unit vector to the boundary.

We take $\boldsymbol{h}_T = \boldsymbol{H}/M_s + \boldsymbol{H}_a + \boldsymbol{H}_{ex}$, where $\boldsymbol{H}$ is the magnetic field usually obtained from the Maxwell's equations. Since we are only concerned with the numerical methods for LLG equation, we will assume $\boldsymbol{H}$ to be a known function. The term $\boldsymbol{H}_a$ is the anisotropy field which, in the case of uniaxial anisotropy in the direction of unit vector $\boldsymbol{p}$, takes the form $\boldsymbol{H}_a = \frac{K}{\mu_0 M_s^2} (\boldsymbol{p} \cdot \boldsymbol{m}) \boldsymbol{p}$; the exchange field $\boldsymbol{H}_{ex} = \frac{A}{\mu_0 M_s^2} \Delta \boldsymbol{m}$ arises due to the exchange interaction between the spins (K, A are the anisotropy and exchange constants).

Z. Li et al. (Eds.): NAA 2004, LNCS 3401, pp. 158–165, 2005.

A scalar multiplication of (1) by $\boldsymbol{m}$ gives $\partial_t \boldsymbol{m} \cdot \boldsymbol{m} = \frac{1}{2}\partial_t |\boldsymbol{m}|^2 = 0$. This directly implies the conservation of magnitude of magnetization $|\boldsymbol{m}(t)| = |\boldsymbol{m}(0)| = 1$, which is a crucial conservation property of the LLG equation. A typical way for solving LLG equations is to first discretize it in space by finite elements or finite differences and than to solve numerically the resulting system of ODE's in time by an appropriate method. It's not difficult to argue that standard time-discretization methods fail to preserve the magnitude of magnetization.

2 Overview of Numerical Methods for LLG

2.1 Projection Methods

The idea of projection methods is simple: first solve LLG by a standard method and then project the solution onto a unit sphere to enforce the constraint $|\boldsymbol{m}| = 1$.

For simplicity we consider LLG in a dimensionless form

$$\boldsymbol{m}_t = -\boldsymbol{m} \times \Delta \boldsymbol{m} - \alpha\, \boldsymbol{m} \times (\boldsymbol{m} \times \Delta \boldsymbol{m}). \tag{3}$$

We took $\boldsymbol{h}_T = \Delta \boldsymbol{m}$, but the extension to the general case is straightforward. From the vector cross product formula $\boldsymbol{a} \times (\boldsymbol{b} \times \boldsymbol{c}) = (\boldsymbol{a} \cdot \boldsymbol{c})\boldsymbol{b} - (\boldsymbol{a} \cdot \boldsymbol{b})\boldsymbol{c}$ and from the fact that $\nabla |\boldsymbol{m}|^2 = 0$ we can rewrite the damping term entering (3),

$$\boldsymbol{m} \times (\boldsymbol{m} \times \Delta \boldsymbol{m}) = -\Delta \boldsymbol{m} - |\nabla \boldsymbol{m}|^2 \boldsymbol{m}.$$

Then (3) can be rewritten in an equivalent form

$$\boldsymbol{m}_t - \alpha \Delta \boldsymbol{m} = \alpha |\nabla \boldsymbol{m}|^2 \boldsymbol{m} - \boldsymbol{m} \times \Delta \boldsymbol{m}. \tag{4}$$

The variational formulation of the equation (4) along with the boundary condition (2) reads as

$$(\boldsymbol{m}_t, \varphi) + \alpha(\nabla \boldsymbol{m}, \nabla \varphi) = \alpha(|\nabla \boldsymbol{m}|^2 \boldsymbol{m}, \varphi) + (\boldsymbol{m} \times \nabla \boldsymbol{m}, \nabla \varphi) \quad \forall \varphi \in V. \tag{5}$$

This problem is nonlinear. However it is possible to avoid solving the nonlinear system by a suitable linearization, while maintaining the accuracy.

Let us denote by $\boldsymbol{m}_j$ the approximation of the solution of (5) at the time t_j. Then, starting from given $\boldsymbol{m}_{j-1}$, $\boldsymbol{m}^*_{j-1}$ we compute $\boldsymbol{m}_j$, $\boldsymbol{m}^*_j$ by the following algorithm [1]:

1. Obtain $\boldsymbol{m}_j$ from backward Euler approximation of (5), viz.

$$(\frac{\boldsymbol{m}_j - \boldsymbol{m}^*_{j-1}}{\tau}, \varphi) + \alpha(\nabla \boldsymbol{m}_j, \nabla \varphi) = \alpha(|\nabla \boldsymbol{m}_{j-1}|^2 \boldsymbol{m}_j, \varphi) + (\boldsymbol{m}^*_{j-1} \times \nabla \boldsymbol{m}_j, \nabla \varphi). \tag{6}$$

2. Project $\boldsymbol{m}_j$ onto a unit sphere to get $\boldsymbol{m}^*_j$ as

$$\boldsymbol{m}^*_j = \frac{\boldsymbol{m}_j}{|\boldsymbol{m}_j|}. \tag{7}$$

The previous semi-implicit scheme is linear and first order accurate.

Another method for the LLG equation, introduced in [2], is based on a splitting procedure. At a time point $t = t_j$, we first obtain the solution of the gyromagnetic part and this is combined with the projection scheme from [3] for the damping part. The gyromagnetic part of (4) reads

$$\boldsymbol{m}_t = -\boldsymbol{m} \times \Delta \boldsymbol{m}, \tag{8}$$

while the damping part is

$$\boldsymbol{m}_t - \alpha \Delta \boldsymbol{m} = \alpha |\nabla \boldsymbol{m}|^2 \boldsymbol{m}.$$

The splitting method consists of two steps:

1. Given the solution $\boldsymbol{m}_{j-1}$ of (4) from the previous time level we discretize (8) by the backward Euler method. The resulting nonlinear system is solved by a Gauss-Seidel based technique (for more details see [2]) in order obtain the approximate solution $\boldsymbol{m}_j^*$ of (8).
2. Having $\boldsymbol{m}_j^*$, we can use the projection method from [3], consisting of

$$\left(\frac{\boldsymbol{m}_j^{**} - \boldsymbol{m}_j^*}{\tau}, \varphi \right) + \alpha (\nabla \boldsymbol{m}_j^{**}, \nabla \varphi) = \mathbf{0},$$

and

$$\boldsymbol{m}_j = \frac{\boldsymbol{m}_j^{**}}{|\boldsymbol{m}_j^{**}|}.$$

The computations in [2] show that the method is stable and faster than a 4-th order Runge-Kutta method.

In [4] the authors propose a backward Euler finite element scheme for the LLG equation, which also uses a projection to conserve $|\boldsymbol{m}|$. The system of nonlinear equations resulting from the implicit discretization of the LLG equations is solved by a GMRES-based method. It is shown in [5] that this method can use larger time steps than an Adams method.

Since the projection type methods don't conserve the norm of magnetization $|\boldsymbol{m}|$ in an implicit way, it can be used as an error indicator during the computations.

2.2 Norm-Conservative Methods

In this section we present another type of methods, where $|\boldsymbol{m}|$ is automatically conserved. These methods are also able to conserve some other physical properties of the micromagnetic systems (cf. [6], [7], [8]).

The LLG equation can be rewritten in the form

$$\boldsymbol{m}_t = \boldsymbol{a}(\boldsymbol{m}) \times \boldsymbol{m}, \tag{9}$$

where $\boldsymbol{a}(\boldsymbol{m}) = \gamma M_s (\boldsymbol{h}_T - \alpha \boldsymbol{h}_T \times \boldsymbol{m})$.

We can discretize the previous equation at $t = t_j$ using the midpoint rule

$$\frac{\boldsymbol{m}_j - \boldsymbol{m}_{j-1}}{\tau} = \boldsymbol{a}_{j-1/2} \times \frac{\boldsymbol{m}_j + \boldsymbol{m}_{j-1}}{2}, \tag{10}$$

where $\boldsymbol{a}_{j-1/2}$ denotes the approximation of vector $\boldsymbol{a}(\boldsymbol{m})$ at the time $t_j - \tau/2$.

After scalar multiplication of (10) by $(\boldsymbol{m}_j + \boldsymbol{m}_{j-1})$ we obtain that

$$\frac{|\boldsymbol{m}_j| - |\boldsymbol{m}_{j-1}|}{\tau} = 0,$$

from which we see that the midpoint rule conserves $|\boldsymbol{m}|$.

A possible choice could be $\boldsymbol{a}_{j-1/2} = \dfrac{\boldsymbol{a}(\boldsymbol{m}_j) + \boldsymbol{a}(\boldsymbol{m}_{j-1})}{2}$. The resulting scheme reads as follows

$$\frac{\boldsymbol{m}_j - \boldsymbol{m}_{j-1}}{\tau} = \frac{\boldsymbol{a}(\boldsymbol{m}_j) + \boldsymbol{a}(\boldsymbol{m}_{j-1})}{2} \times \frac{\boldsymbol{m}_j + \boldsymbol{m}_{j-1}}{2}, \tag{11}$$

and we have to solve a nonlinear system. In [9] a scheme based on the idea of midpoint rule was introduced. The authors constructed an explicit solution for the nonlinear system for materials with uniaxial anisotropy in the absence of exchange field. When the exchange field is included, an explicit solution to the scheme presented in [9] no longer exists and the system has to be solved for instance by Newton's method [6].

In [10] the value of $\boldsymbol{a}_{j-1/2}$ is extrapolated form the values on the previous time levels by the formula $\boldsymbol{a}_{j-1/2} = \frac{3}{2}\boldsymbol{a}(\boldsymbol{m}_{j-1}) - \frac{1}{2}\boldsymbol{a}(\boldsymbol{m}_{j-2}) + O(\tau^2)$. The resulting 2nd order scheme is explicit

$$\frac{\boldsymbol{m}_j - \boldsymbol{m}_{j-1}}{\tau} = \left(\frac{3}{2}\boldsymbol{a}(\boldsymbol{m}_{j-1}) - \frac{1}{2}\boldsymbol{a}(\boldsymbol{m}_{j-2})\right) \times \frac{\boldsymbol{m}_j + \boldsymbol{m}_{j-1}}{2}. \tag{12}$$

We only have to solve a linear system of dimension 3×3 at every spatial mesh point to obtain the values of $\boldsymbol{m}_j$. In [8], the previous method is compared with implicit and explicit Euler methods, and is shown to be more accurate.

In [11] the authors present two explicit first order schemes for LLG equation which conserve $|\boldsymbol{m}|$. They use the fact that for a constant vector $\boldsymbol{a}$ the following linear ODE along with initial data $\boldsymbol{m}(0) = \boldsymbol{m}_0$

$$\boldsymbol{m}_t = \boldsymbol{a} \times \boldsymbol{m},$$

can be solved analytically:

$$\boldsymbol{m} = \boldsymbol{m}_0^{\|} + \boldsymbol{m}_0^{\perp} \cos(|\boldsymbol{a}|t) + \frac{\boldsymbol{a}}{|\boldsymbol{a}|} \times \boldsymbol{m}_0^{\perp} \sin(|\boldsymbol{a}|t), \tag{13}$$

where $\boldsymbol{m}_0 = \boldsymbol{m}_0^{\|} + \boldsymbol{m}_0^{\perp}$, $\boldsymbol{m}_0^{\|}$ is parallel to $\boldsymbol{a}$ and $\boldsymbol{m}_0^{\perp}$ is perpendicular to $\boldsymbol{a}$.

Having the solution $\boldsymbol{m}_{j-1}$ at time level $t = t_{j-1}$ we set $\boldsymbol{a} = \boldsymbol{a}(\boldsymbol{m}_{j-1})$ in (9)

$$\boldsymbol{m}_t = \boldsymbol{a}(\boldsymbol{m}_{j-1}) \times \boldsymbol{m}. \tag{14}$$

We obtain $\boldsymbol{m}_j$ by means of (13) in the time interval (t_{j-1}, t_j), taking $\boldsymbol{m}_{j-1}$ as the initial data.

The second method is based on the analytical solution of the nonlinear ODE (when $\boldsymbol{h}$ is constant): $\boldsymbol{m}_t = \boldsymbol{h} \times \boldsymbol{m} + \alpha \boldsymbol{m} \times (\boldsymbol{h} \times \boldsymbol{m})$. We set $\boldsymbol{h} = \gamma M_s \boldsymbol{h}_T(\boldsymbol{m}_{j-1})$

on the time interval (t_{j-1}, t_j) and proceed analogously as in the first method (for more details see [11]).

In [12], [7] the authors use the Lie Group formalism to develop methods which conserve the modulus of magnetization. Formally, a numerical method of order k for the equation (9) can be written as follow

$$\boldsymbol{m}_j = \mathrm{Exp}(\boldsymbol{A})\boldsymbol{m}_{j-1}, \tag{15}$$

where $\boldsymbol{A}$ is an update determined by $\boldsymbol{a}(\boldsymbol{m})$. By a suitable choice of this update $\boldsymbol{A}$ we can construct explicit or implicit methods of desired order.

The function Exp is an algorithmic exponential of the Lie group $SO(3)$ (for more details and different constructions of the update, see [12], [7]). With the exact matrix exponential we have

$$\exp(\boldsymbol{A})\boldsymbol{m}_{j-1} = \boldsymbol{m}_{j-1} + \frac{\sin(|\boldsymbol{A}|)}{|\boldsymbol{A}|}\boldsymbol{A} \times \boldsymbol{m}_{j-1} + \frac{1-\cos(|\boldsymbol{A}|)}{|\boldsymbol{A}|^2}\boldsymbol{A} \times (\boldsymbol{A} \times \boldsymbol{m}_{j-1}).$$

When we take $\boldsymbol{A} = \tau \boldsymbol{a}(\boldsymbol{m}_{j-1})$ in the previous equation we arrive at a method which is equivalent to method (13). Algorithms of arbitrary order can also be constructed using the Cayley transform, which is a second order approximation of the exact exponential, viz

$$\mathrm{cay}(\boldsymbol{A}) = \left(I - \tfrac{1}{2}\mathrm{skew}[\boldsymbol{A}]\right)^{-1}\left(I + \tfrac{1}{2}\mathrm{skew}[\boldsymbol{A}]\right),$$

where I is the identity matrix and

$$\mathrm{skew}[\boldsymbol{x} = (x_1, x_2, x_3)] = \begin{pmatrix} 0 & -x_3 & x_2 \\ x_3 & 0 & -x_1 \\ -x_2 & x_1 & 0 \end{pmatrix}.$$

When we put $\boldsymbol{A} = \frac{\tau}{2}(\boldsymbol{a}(\boldsymbol{m}_j) + \boldsymbol{a}(\boldsymbol{m}_{j-1}))$, we get a method equivalent to the implicit midpoint rule (11).

With the schemes from this section we can no longer use $|\boldsymbol{m}|$ as an error indicator. A self consistency error control scheme which can be used along with norm-conservative methods was suggested in [13].

3 Numerical Experiments

We will consider a numerical example of a conducting thin film subjected to an in-plane circularly polarized magnetic field, which was suggested in [14]. This problem can be reduced to a 1D problem on the interval $(0, \delta)$, where δ is the thickness of the film. In order to obtain the magnetic field $\boldsymbol{H} = (H_1, H_2, H_3)$, the LLG equation has to be coupled with the eddy current equation. This, in the 1D case takes the form

$$\mu_0 \partial_t H_i - \frac{1}{\sigma}\frac{\partial^2 H_i}{\partial z^2} = -\mu_0 \partial_t M_i \quad i = 1, 2, \tag{16}$$

and $H_3 = -M_3$. We take the vector $\boldsymbol{M} = (M_1, M_2, M_3) = M_s \boldsymbol{m}$.

We solve (16) along with with the boundary condition

$$\boldsymbol{H}(t) = H_s\left(\cos(\omega t), \sin(\omega t), 0\right) \quad z = 0,\, z = \delta. \tag{17}$$

The total field in the LLG equation takes the form $\boldsymbol{h}_T = \dfrac{\boldsymbol{H}}{M_s} + \dfrac{2A}{\mu_0 M_s^2}\dfrac{\partial^2 \boldsymbol{m}}{\partial z^2}$.

The calculations were performed with the following parameters: $\gamma = 2.211 \times 10^5$, $\alpha = 0.01$, $M_s = 8 \times 10^5$, $\sigma = 4 \times 10^6$, $\delta = 1.5 \times 10^{-6}$, $A = 1.05 \times 10^{-11}$, $\omega = 2\pi \times 10^9$, $H_s = 4.5 \times 10^3$, $\mu_0 = 4\pi \times 10^{-7}$. Moreover a uniform initial condition for the LLG equation was used: $\boldsymbol{m}_0 = (1, 0, 0)$. It is expected that the solution of the system (1), (16) with the boundary conditions (2), (17) is periodic in time (Fig. 1).

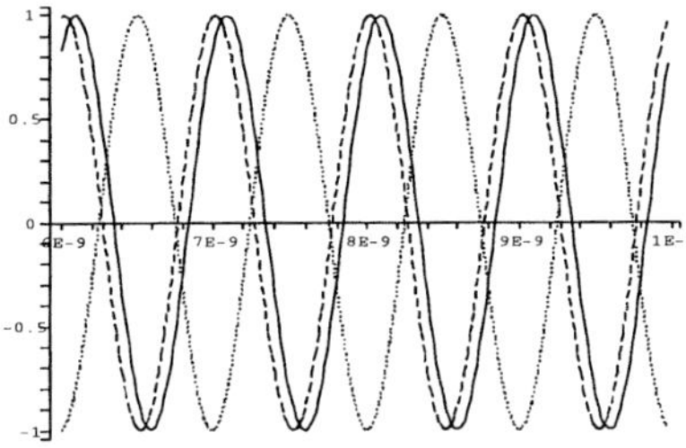

Fig. 1. x-component of $\boldsymbol{H}$ on the boundary (dashed line) and x-component of the magnetization in the points at distance $\delta/6$ (solid line) and $\delta/2$ (dotted line) from the boundary, respectively

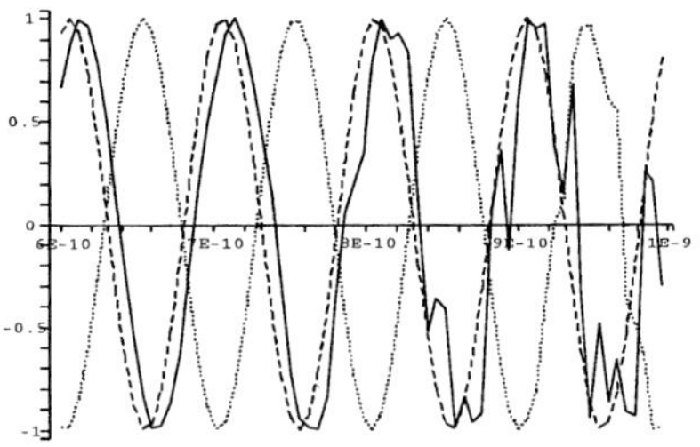

Fig. 2. Unstable solution

The time discretization of the example was performed with the methods described by (6)-(7), (10), (12), (14), and with the classical 4-th order Runge-Kutta method. We will refer to this methods as PR_1, MP_{im}, MP_{ex}, EXP_1, RK_4, respectively. For the time discretization of (16) we used the Crank-Nicholson scheme. which allowed us to use larger time steps for some of the methods. The space discretization was done by standard finite-differences. The nonlinear system in MP_{im} was solved by the Broyden's method.

Although the performance of the methods for the LLG equation is influenced by the coupling with (16), we observed that the errors induced by the discretization of (16) had minor influence on the computation, when compared to the effect of the discretization of the LLG equation. However, some methods were able to use slightly larger time steps when we discretized (16) by the Crank-Nicholson scheme, compared to the situation where we used backward Euler approximation of (16). In practice the magnetic field $\boldsymbol{H}$ is not known and the LLG equation has to be coupled with Maxwell's equations in an appropriate form.

In our experiments we first fixed the mesh parameter $h = \delta/50$ and looked for the largest time step (τ_{max}) for which we could obtain an acceptable numerical solution without oscillations. Then, we decreased the value of h to see if the stability of the method was sensitive to the mesh refinement. An example of an unstable solution computed with MP_{ex} ($h = \delta/60$, $\tau = 6 \times 10^{-12}$) is depicted in Fig. 2. The results of the numerical experiments can be found in Table 1 (by h-sensitive we denote the methods, for which we needed to decrease τ, when we decreased h in order to avoid big changes of the modulus or oscillations). In some cases, the computation of the magnetic field from Maxwell's equation at every time level is a more computationally intensive task than the approximation of the LLG equation. In such a case, the possibility of using larger times steps, gives an obvious advantage. Schemes, which are h-insensitive, can be useful when we want to use adaptive strategies.

The methods MP_{ex} and EXP_1 conserved $|\boldsymbol{m}|$ with an error of order 10^{-15}. The method MP_{im} conserved $|\boldsymbol{m}|$ up to the truncation error of the Broyden's iterations. With the residue of the Broyden's iteration about 10^{-10}, the resulting magnitude drift was of order 10^{-9} and it decreased when we increased Boryden's precision. The method MP_{im} allowed us to use larger time steps than the explicit methods. We expect that more sophisticated nonlinear strategies could speed up the method and give better results.

The projection method PR_1 was the only method for which the choice of the time step was independent of the mesh parameter h. The error in the magnitude, when $\tau = \tau_{max}$, was of order 10^{-3} and decreased with smaller values of τ. Without the projection step (7) the method would blow-up for greater values of τ. From the explicit methods, the RK_4 method could use the largest time-steps, however the magnitude drift was of order 10^{-7}.

Table 1. Performance of the methods

method	τ_{max}	h-sensitive
EXP_1	9×10^{-13}	yes
MP_{ex}	6×10^{-12}	yes
RK_4	1×10^{-11}	yes
MP_{im}	1×10^{-11}	yes
PR_1	1.4×10^{-11}	no

4 Summary

In this paper we have given a comparative overview of various methods for solving the LLG equation of micromagnetics. One of principal goals in micromagnetic computations is to maintain the constraint $|\boldsymbol{m}| = 1$. The projection methods enforce this constraint explicitly at every time level by projecting the solution onto a unit sphere. They seem to be stable with respect to the space discretization and allow us to use large time steps. They might be a good choice when mesh adaptivity is involved. The explicit norm-conservative schemes need to use smaller time steps than the projection methods, but they satisfy the constraint $|m| = 1$ nearly precisely. Because of their fast implementation they have been explored and used in practice. The implicit norm-conservative methods can use larger time steps than explicit methods for the cost of non-linearity of the resulting discrete system. Although the classical RK_4 performed quite well in our numerical example, in more complex problems, we still need to use the projection or small time steps to satisfy the norm constraint. The explicit norm-conservative methods of higher order should be a better choice for their capability of maintaining the physical constraints.

Acknowledgments

The author was supported by the IWT/STWW and IUAP projects of Ghent University. He would like to thank prof. Roger Van Keer for the reading of the manuscript and constructive comments.

References

1. Prohl, A.: Computational micromagnetism. Volume xvi of Advances in numerical mathematics. Teubner, Leipzig (2001)
2. Wang, X.P., García-Cervera, C.J., E, W.: A Gauss-Seidel projection method for micromagnetics simulations. J. Comput. Phys. **171** (2001) 357–372
3. E, W., Wang, X.P.: Numerical methods for the Landau-Lifshitz equation. SIAM J. Numer. Anal. **38** (2000) 1647–1665
4. Yang, B., Fredkin, D.R.: Dynamical micromagnetics by the finite element method. IEEE Trans. Mag. **34** (1998) 3842–3852
5. Suess, D., Tsiantos, V., Schrefl, T., Fidler, J., Scholz, W., Forster, H. Dittrich, R., Miles, J.J.: Time resolved micromagnetics using a preconditioned time integration method. J. Magn. Magn. Mater. **248** (2002) 298–311
6. Monk, P., Vacus, O.: Accurate discretization of a nonlinear micromagnetic problem. Comput. Methods Appl. Mech. Eng. **190** (2001) 5243–5269
7. Lewis, D., Nigam, N.: Geometric integration on spheres and some interesting applications. J. Comput. Appl. Math. **151** (2003) 141–170
8. Spargo, A.W., Ridley, P.H.W., Roberts, G.W.: Geometric integration of the Gilbert equation. J. Appl. Phys. **93** (2003) 6805–6807
9. Joly, P., Vacus, O.: Mathematical and numerical studies of nonlinear ferromagnetic materials. M2AN **33** (1999) 593–626
10. Serpico, C., Mayergoyz, I.D., Bertotti, G.: Numerical technique for integration of the Landau-Lifshitz equation. J. Appl. Phys. **89** (2001) 6991–6993
11. Slodička, M., Baňas, L.: A numerical scheme for a Maxwell-Landau-Lifshitz-Gilbert system. (Appl. Math. Comput.) to appear.
12. Krishnaprasad, P.S., Tan, X.: Cayley transforms in micromagnetics. Physica B **306** (2001) 195–199
13. Albuquerque, G., Miltat, J., Thiaville, A.: Self-consistency based control scheme for magnetization dynamics. J. Appl. Phys. **89** (2001) 6719–6721
14. Mayergoyz, I.D., Serpico, C., Shimizu, Y.: Coupling between eddy currents and Landau Lifshitz dynamics. J. Appl. Phys. **87** (2000) 5529–5531

The Continuous Analog of Newton Method for Nonlinear Data Approximation

N.G. Bankow[1] and M.S. Kaschiev[2]

[1] Space Research Institute,Sofia, Bulgaria
nbankov@argo.bas.bg
[2] Present address: South West University, Blagoevgrad, Bulgaria
Permanent address: Institute of Mathematics and Informatics,
Sofia, Bulgaria
kaschiev@math.bas.bg

Abstract. The experiments, used to measure the cold plasma ionosphere parameters, generate mathematical problems, that usually are "ill posed". There are two basic approaches to solve such problems: deterministic, developed by Tickonov, and Hubber's "robust estimation". In this paper another approach the modified Newton method applied to approximate data by experiment aimed to measure such parameters is discussed. In this case we used the nonlinear Least Square Method. The numerical results show that the last method is stable and has a larger region for convergence than other methods. Some important numerical calculations are given in the paper.

1 Introduction

This paper briefly describes the methods to analyze the experimental data, obtained by cylindrical electrostatic probe on board of "Bulgaria-1300" satellite. The primary purpose of the cylindrical electrostatic probe is to provide accurate measurements of electron temperature Te. A secondary role is to provide measurements of the electron and ion concentrations, Ne and Ni , encountered along the orbits of the satellite.

2 Theoretical Background

The probe is assumed to behave as an idealized cylindrical Langmuir probe and produces as an output so called "volt-ampere characteristics" - measured values of the current of the probe's collector as a function of the applied stepwise voltage. Fig.1a identifies the various regions of the current characteristics. The ion-saturation at the left occurs when the probe is sufficiently negative to cut off all resolvable thermal-electron current. The amplitude of the ion current, Ii, depends of Ni, and its slope is a function of ion mass Mi. As the probe is made less negative, the electron current increases exponentially at a rate determined

Z. Li et al. (Eds.): NAA 2004, LNCS 3401, pp. 166–170, 2005.

entirely by Te. When the probe is swept positive the amplitude of the electron-saturation current depends only upon Ne. It is very important to point out a prior uncertainty of the derived mathematical problem - the point where the volt-ampere curve swept positive is undetectable, as far as the applied voltage to the probe is, in fact, a sum of the generated by the instrument stepwise plus satellite potential, which is unknown.

The probe, launched on board of $B-1300$, provide volt-ampere curves that do not always correspond to the theoretical expectation. We shall not try to diagnose the reasons of the discrepancy of the probe but will just point out two general disagreements with theoretical expectations:

- very short ion-saturation region (more than 80% of the curves)
- the shape of the curve in electron-saturation region do not correspond to the expected behavior of cylindrical probe.

Thus, the regions that could be used to analyze $V-A$ curves, are electron retardation and partially ion-saturation. Some notes about accuracy of the measurements should be made here. In addition to the noise, typical for any experiment, errors generated by the work of other experiments, different condition along the orbit, ionosphere irregularities etc., occurs. Especially roundup errors due to telemetry system should be noted, as far the values of the volt-ampere curves lies into $(-10^{-4}, 7.10^{-5})$, and for each scale, by means of exponent, measured values are encoded by telemetry with 8-bit word, supplying approximately 3 significant digits in each scale. So, requirements for "normally distributed errors" when least-square techniques are to be used, are only partially fulfilled.

3 Numerical Method

Let us denote by X the stepwise voltage, applied to the probe, and by Y measured values of volt-ampere characteristic. Now, denoting $V = X+S$, S - satellite potential, and considering the position where V =0 as known (i.e. S is known), the analysis of the data, traditionally [1], is accomplished by a four-parameter fit to the portion of the curves where $V \leq 0$ using an equation of the form

$$Y(V) = A + BV + C\exp(VD), \tag{1}$$

where $A+BV$ represents a linear approximation of the ion-current component and $C\exp(VD)$ - electron-current on the electron retardation region. The precise formulae for Ii is given by

$$Ii = eLNiW\sqrt{1 + kTi/MiW^2 + 2eV/MiW^2}, \tag{2}$$

where L is the probe area projected normal to W, W is the satellite velocity, e is the electron or ion charge, k is the Boltzmann's constant, Ti is the ion temperature, Mi is the ion mass and V is the probe potential relative to the plasma. As for the electrons, the exact formulae is

$$Ie = eLNe\sqrt{kTe/2Me}\exp(eV/kTe). \tag{3}$$

In these notation full current on the probe will be $I = Ie - Ii$, $Y(V) = I$ correspondingly. The fitting procedure will provide values for Te, A, B and C, which regardless the notes about error distribution, could be acceptable. When S is unknown, it has to be evaluated by the fitting procedure, and serious difficulties had to be considered. First, some of the parameters to be estimated are nonlinearly dependent, so differences between exact values and those of the certain stage of the iteration process, could be accumulated by the others. Second, varying S in order to find proper value, results in rejecting or including new point into the set to be approximated. New point is entering into computation with its own accuracy, and if it is bad enough, cumulative least-square error will be greater than previous, and this value of S will be rejected. Applying sophisticated methods ("robust" estimation [2]) to determine appropriate weight function improves the situation significantly, but the practice show, that the problem still remain unsolved. However, a method that will be proposed here seems to be not so depending from the single erroneous observation. Let us rewrite the problem (1) in the following manner.

$$Y(X - \varphi) = -A + B(X - \varphi) + C\exp(e(X - \varphi)/kTe), \tag{4}$$

where X is such that $Y = 0$ when $X = 0$, and all the parameters to be estimated are positive. In this notation φ is known as plasma potential. Then for $X = 0$ we will have

$$-A - B\varphi + C\exp(-e\varphi/kTe) = 0.$$

This equation could be solved by means of φ, i.e.

$$exp(-e\varphi/kTe) = A/C(1 + B\varphi/A)$$

$$\begin{aligned} -e\varphi/kTe = \ln(A/C) + \ln(1 + B\varphi/A) \le \ln(A/C) \\ +B\varphi/A, \quad (B < A) \end{aligned} \tag{5}$$

$$-\varphi = A/(A + kTe/eB)kTe/e\ln(A/C)$$

Here A and C could be estimated using (2) and (3), and assuming $N = Ni = Ne$ and $T = Ti = Te$, one can get

$$\varphi = -A/(A + BkT/e)kT/2e\ln(2Me/\pi Mi(1 + W^2Mi/kT)),$$

which could be simplified

$$\varphi = -kT/2e\ln(2Me/\pi Mi(1 + W^2Mi/kT)). \tag{6}$$

The last formulae slightly overestimate (6), especially for large values of T, but proves to be convenient in computations. Now, let us note by J the number of the point where $Y = 0$, and note by K the largest number, for which $V_I = h * (I - J) - \varphi$ is les or equal to 0. We obtain the system

$$\begin{aligned} Y_I = -A + BV_I + LNe\sqrt{kT/2\pi Me}\exp(eV_I/kT), \\ I = 1, 2, \ldots K, \end{aligned} \tag{7}$$

where V is undependent variable, A, B, N and T are unknown quntities that have to be determined. Different methods for solving the nonlinear data approximation are considered in the monography [3], page 617. For solving the system (7) we apply the Continuous Analogue of Newton Method (CANM) [4], [5]. Let us now denote left part of (7) by $F(A, B, N, T, V)$. Let us denote $a_1 = A$, $a_2 = B$, $a_3 = N, a_4 = T$. Then the CANM system is described as follows

$$\frac{\partial F(a_1^n, a_2^n, a_3^n, a_4^n, V_i)}{\partial a_1}\hat{a}_1^n + \frac{\partial F(a_1^n, a_2^n, a_3^n, a_4^n, V_i)}{\partial a_2}\hat{a}_2^n +$$

$$\frac{\partial F(a_1^n, a_2^n, a_3^n, a_4^n, V_i)}{\partial a_3}\hat{a}_3^n + \frac{\partial F(a_1^n, a_2^n, a_3^n, a_4^n, V_i)}{\partial a_4}\hat{a}_4^n = \quad (8)$$

$$-(F(a_1^n, a_2^n, a_3^n, a_4^n, V_i) - Y_i), \quad i = 1, 2, \ldots, K.$$

In these formulas $a_1^0, a_2^0, a_3^0, a_4^0$ are initial data and $\hat{a}_1^n, \hat{a}_2^n, \hat{a}_3^n, \hat{a}_4^n$ have to be calculated.

Let us denote by $\mathbf{C}^n$, $K \times 4$ matrix with elements $C_{ij}^n = \dfrac{\partial F(a_1^n, a_2^n, a_3^n, a_4^n, V_i)}{\partial a_j}$, $i = 1, 2, \ldots, K, j = 1, 2, 3, 4$, $(\hat{\mathbf{a}}^n)^T = (\hat{a}_1^n, \hat{a}_2^n, \hat{a}_3^n, \hat{a}_4^n)$, are vectors with 4 components and $\mathbf{z}^n$ is a vector with components $z_i^n = F(a_1^n, a_2^n, a_3^n, a_4^n, V_i) - Y_i$, $i = 1, 2, \ldots, K$. Then the system (8) can be rewrite

$$\mathbf{C}^n \hat{\mathbf{a}}^n = -\mathbf{z}^n.$$

This system is solved using the least square method

$$(\mathbf{C}^n)^T \mathbf{C}^n \hat{\mathbf{a}}^n = -(\mathbf{C}^n)^T \mathbf{z}^n.$$

Now we find the next approximation to the solution $a_j^{n+1} = a_j^n + \tau_n \hat{a}_j^n$, $j = 1, 2, 3, 4$. Here τ_n is a iteration step, $0 < \tau_n \leq 1$ and n is iteration number.

4 Conclusion

On fig. 1b and fig. 1c approximation plot for curves considered as "low noised" and "noised" are provided. For the curve of Fig. 1b the 57 ($K = 57$) experimental points are used, 51 ($K = 51$) - for Fig.1c and initial data for both cases are the same $A = Y_1$, $B = 0$, $N = 10^6$, $T = 3500$. It needed less then 150 iterations to calculate the unknown parameters A, B, N, T with an accuracy 10^{-5} ($\tau = 0.1$) for each case. As result for most important parameters N and T we find the following values $N = 3.66572$, $T = 1.988$ (Fig.1b) and $N = 5.2504$, $T = 3.954$ (Fig.1c).

The described method seems to provide better stability, by means of single or grouped erroneous measurements, then standard optimization methods, which are directed entirely by the level of the sum of squares for error. Another advantage of the method is the possibility to obtain, applying (6), correct values for N from electron retardation region, important when electron saturation

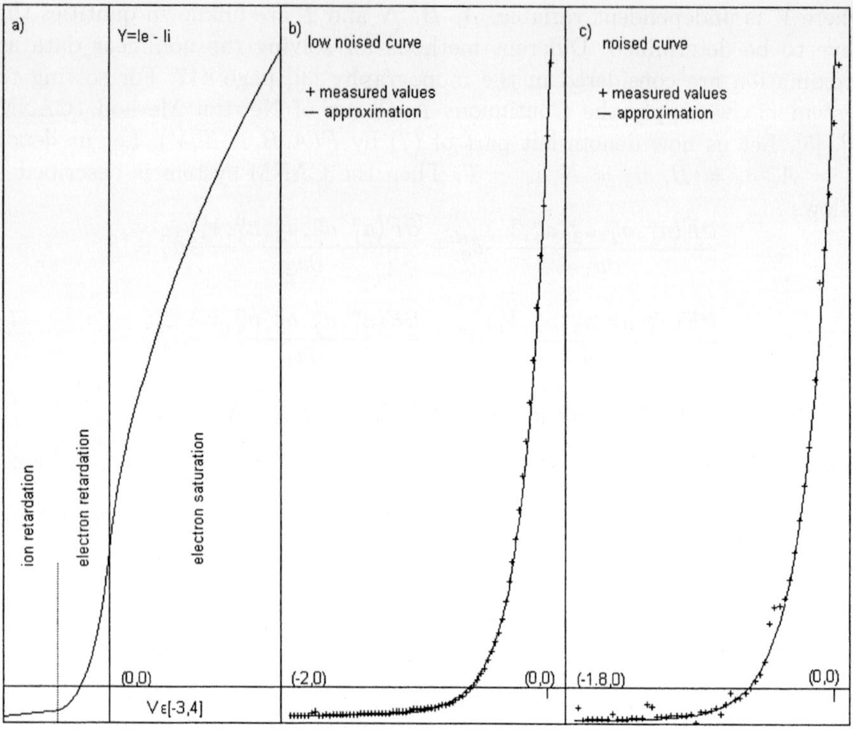

Fig. 1.

part of the curve is corrupt or short. As a disadvantage of the method a bigger computational time should be noted.

Acknowledgement

Authors are grateful for a NFSI grant SE-1309/03.

References

1. Brace L., Theis R., Dalgarno A., The cylindrical electrostatic probes for Atmosphere Explorer-C, Radio Science, Vol. 8, nov. 4, pp 341-348, 1973.
2. Huber,P.J., Robust estimation of a location parameter, Ann. Math. Statist., vol 35, 1964, 73-101.
3. P.G.Ciarlet, J.L.Lions. Handbook of Numerical Analysis, Vol.1, North-Holland, 1990.
4. M.K.Gavurin. Izv. VUZ, Ser. Mathem. v.5(6), pp 18-31, **1958** (*in Russian*).
5. E.P.Zhidkov, G.I.Makarenko and I.V.Puzynin. EPAN, v.4(1),"Atomizdat",M., pp 127-165, **1973** (*in Russian*).

Prestressed Modal Analysis Using Finite Element Package ANSYS

R. Bedri and M.O. Al-Nais

College of Technology at Hail,
P.O. Box. 1690 Hail, Saudi Arabia
Tel: +966 531 7705 ext 240
Fax: +966 531 7704
r_bedri@yahoo.com

Abstract. It is customary to perform modal analysis on mechanical systems without due regards to their stress state. This approach is of course well accepted in general but can prove inadequate when dealing with cases like spinning blade turbines or stretched strings, to name but these two examples.

It is believed that the stress stiffening can change the response frequencies of a system which impacts both modal and transient dynamic responses of the system. This is explained by the fact that the stress state would influence the values of the stiffness matrix.

Some other examples can be inspired directly from our daily life, i.e., nay guitar player or pianist would explain that tuning of his playing instrument is intimately related to the amount of tension put on its cords. It is also expected that the same bridge would have different dynamic responses at night and day in places where daily temperature fluctuations are severe.

These issues are unfortunately no sufficiently well addressed in vibration textbooks when not totally ignored.

In this contribution, it is intended to investigate the effect of prestress on the vibration behavior of simple structures using finite element package ANSYS. This is achieved by first performing a structural analysis on a loaded structure then make us of the resulting stress field to proceed on a modal analysis.

Keywords: Pre-stress, Modal analysis, Vibrations, Finite elements, ANSYS.

1 Scope

In this investigation, we are concerned by the effect of pressure loads on the dynamic response of shell structures.

A modal analysis is first undertaken to ascertain for the eigen-solutions for an unloaded annulus shell using a commercial finite element package ANSYS ([1]).

In the second phase, a structural analysis is performed on the shell. Different pressure loads are applied and the resulting stress and strain fields are determined.

Z. Li et al. (Eds.): NAA 2004, LNCS 3401, pp. 171–178, 2005.

tables 1 and 2. For each loading case, prestress effect is being activated in the analysis option of the program. The resulting stress field is then applied when it comes to performing subsequently modal analysis on the annulus.

3.3 Modal Analysis with Prestress Effect

Once the stress field is being established from the above static analysis, it is applied as prestress to the shell structure through the activation of this option in the subsequent modal analysis. This procedure is reproduced for the twenty different preloading cases.

4 Results

The results of the different analyses i.e., modal analysis of the stress free annulus the static analysis and then the modal analysis of the preloaded structure, are all summarized and displayed in tabular form see tables 1 and 2 in the appendix.

To ascertain the effect of the prestress level on the modes of vibration, some further calculations are done and presented in tables 3 and 4 in the appendix.

Plots of prestress level versus percent increase or decrease in frequencies are plotted respectively in figures 1 and 2.

5 Comments on Results

5.1. Prestress produces no effect on the mode shapes of vibration of the shell structure.
5.2. By examining the results presented in tables 1 and 2, it is evident that the frequencies are impacted by preloading. The effect of such preloading seems to be more apparent on the first modes than on the higher ones. The plotted curves of figures 1 and 2 are here to corroborate these conclusions.
5.3. A closer look at these curves discloses that there seems to be a linear correlation between the prestress level and the percent frequency increase or decrease for each mode of vibration.
5.4. Tensile preloading produces an increase in frequency whereas compressive preloading results in a decrease in frequency.

6 Conclusions

Three pieces of conclusions can be inferred from this study:

6.1. The mode shapes of vibration of the structure are not sensitive to preloading.
6.2. Prestressing seems to impact the dynamic behavior of the structure.
6.3. Tensile prestress acts as a stiffener and enhances the dynamic characteristics of the structure resulting in frequency increase. Whereas compressive prestress has a converse effect on the structure by reducing its frequencies.

References

1. ANSYS, Users manual, revision 5.6, Swanson Analysis Inc., Nov. 1999.
2. Reddy, J.N.: An introduction to the finite element method , McGraw-Hill, New York 1984.
3. Reddy, J.N., energy and variational methods in applied mechanics, John Wiley, New York 1984.
4. Zienkiewicz, O.C., and Taylor, R.L., The finite element method , 4th ed., McGraw-Hill, New York, 1989.
5. Bathe, K.J.. Finite element procedures in engineering analysis, Prentice-Hall, Englewood Cliffs, N.J., 1982.
6. Cheung, Y.K. and Yeo, M. F., a practical introduction to finite element analysis, Pitman, London, 1979.
7. Rao, S.S., The finite element method in engineering, Pergamon Press, Oxford, 1982.
8. Meirovitch, L., Elements of vibration analysis, McGraw-Hill, New York, 1975.
9. Lanczos, C, the variational principles of mechanics, the university of Toronto press, Toronto, 1964.
10. Hutchinson, J.R., Axisymmetric Vibrations of a Solid Elastic Cylinder Encased in a Rigid Container, J. Acoust. Soc. Am., vol. 42, pp. 398-402, 1967.
11. Reddy, J.N., Applied Functional Analysis and Variational Methods in Engineering, McGraw-Hill, New York, 1985.
12. Sandbur, R.B., and Pister, K.S., Variational Principles for Boundary Value and Initial-Boundary Value Problems in Continuum Mechanics, Int. J. Solids Struct., vol. 7, pp. 639-654, 1971.
13. Oden, J.T., and Reddy, J.N., Variational Methods in Theoretical Mechanics, Springer-Verlag, Berlin, 1976.
14. http://caswww.colorado.edu/courses.d/AFEM.d/
15. http://caswww.colorado.edu/courses.d/IFEM.d/
16. Liepins, Atis A., Free vibrations of prestressed toroidal membrane, AIAA journal, vol. 3, No. 10, Oct. 1965, pp. 1924-1933.
17. M. Attaba, M.M. Abdel Wahab, Finite element stress and vibration analyses for a space telescope, ANSYS 2002 Conference, Pittsburg, Pennsylvania, USA, April 22-24, 2002.
18. Alexey I Borovkov, Alexander A Michailov, Finite element 3D structural and modal analysis of a three layered finned conical shell, ANSYS 2002 Conference, Pittsburg, Pennsylvania, USA, April 22-24, 2002.
19. Sun Libin, static and modal analysis of a telescope frame in satellite, ANSYS 2002 Conference, Pittsburg, Pennsylvania, USA, April 22-24, 2002.
20. Erke Wang, Thomas Nelson, structural dynamic capabilities of ANSYS, ANSYS 2002 Conference, Pittsburg, Pennsylvania, USA, April 22-24, 2002.

Appendix

Table 1. The first five modes against the tensile prestress levels

Mode	Tensile Prestress in N/m										
	0	10^3	2.10^3	4.10^3	8.10^3	16.10^3	32.10^3	64.10^3	10^5	2.10^5	3.10^5
	Frequency in kHz										
1	2.602	2.603	2.604	2.606	2.610	2.618	2.633	2.664	2.698	2.789	2.876
2	2.634	2.635	2.636	2.638	2.642	2.649	2.665	2.695	2.729	2.821	2.904
3	2.656	2.657	2.658	2.660	2.664	2.672	2.688	2.718	2.752	2.844	2.932
4	2.790	2.791	2.792	2.794	2.798	2.805	2.821	2.852	2.885	2.977	3.065
5	2.839	2.840	2.841	2.843	2.847	2.855	2.870	2.901	2.935	3.027	3.115

Table 2. The first five modes against the compressive prestress levels

Mode	Pressure Prestress in N/m										
	0	10^3	2.10^3	4.10^3	8.10^3	16.10^3	32.10^3	64.10^3	10^5	2.10^5	3.10^5
	Frequency in kHz										
1	2.602	2.601	2.600	2.598	2.594	2.587	2.571	2.539	2.502	2.396	2.285
2	2.634	2.633	2.632	2.630	2.626	2.618	2.602	2.570	2.533	2.428	2.316
3	2.656	2.655	2.654	2.652	2.648	2.641	2.625	2.593	2.556	2.450	2.338
4	2.790	2.789	2.788	2.786	2.782	2.774	2.758	2.726	2.690	2.585	2.474
5	2.839	2.838	2.837	2.835	2.831	2.823	2.807	2.775	2.739	2.634	2.523

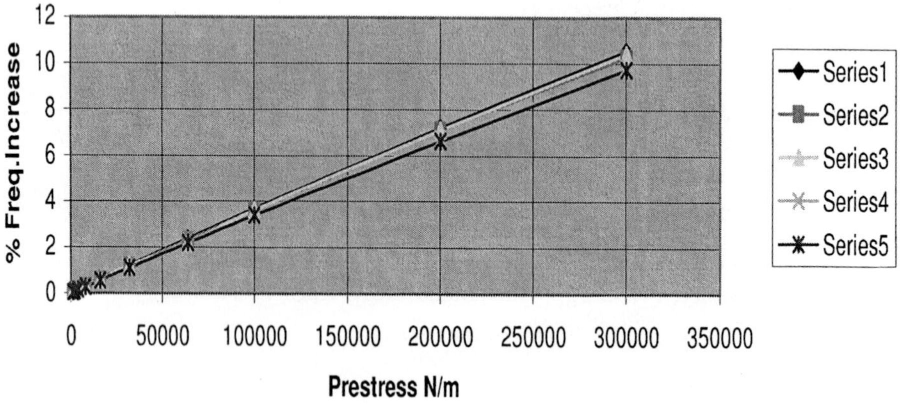

Fig. 1. % Frequency increase versus Prestress

Table 3. The percent frequency increase against prestress levels

Mode	Tensile Prestress in N/m									
	10^3	2.10^3	4.10^3	8.10^3	16.10^3	32.10^3	64.10^3	10^5	2.10^5	3.10^5
	% increase in frequency									
1	0.038	0.077	0.154	0.307	0.615	1.191	2.383	3.689	7.187	10.530
2	0.038	0.076	0.152	0.304	0.569	1.177	2.316	3.607	7.099	10.250
3	0.038	0.075	0.151	0.301	0.602	1.205	2.184	3.614	7.078	10.391
4	0.036	0.072	0.143	0.287	0.538	1.111	2.222	3.405	6.702	9.857
5	0.035	0.070	0.141	0.282	0.563	1.092	2.184	3.381	6.622	9.722

Table 4. The percent frequency decrease against prestress levels

Mode	Pressure Prestress in N/m									
	10^3	2.10^3	4.10^3	8.10^3	16.10^3	32.10^3	64.10^3	10^5	2.10^5	3.10^5
	% decrease in frequency									
1	0.038	0.077	0.154	0.307	0.576	1.191	2.421	3.843	7.917	12.183
2	0.038	0.076	0.152	0.304	0.607	1.215	2.430	3.834	7.821	12.073
3	0.038	0.075	0.151	0.301	0.565	1.167	2.372	3.765	7.756	11.973
4	0.036	0.072	0.143	0.287	0.573	1.147	2.294	3.584	7.348	11.326
5	0.035	0.070	0.141	0.282	0.563	1.127	2.254	3.522	7.221	11.131

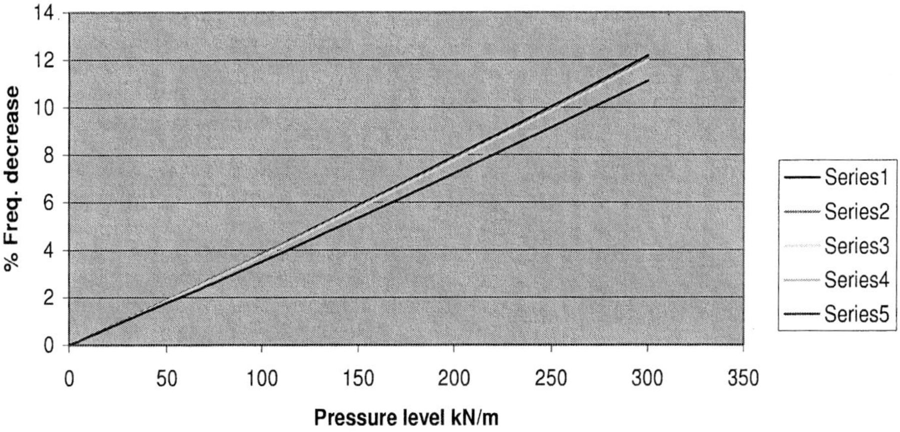

Fig. 2. % Frequency Decrease versus Pressure Level

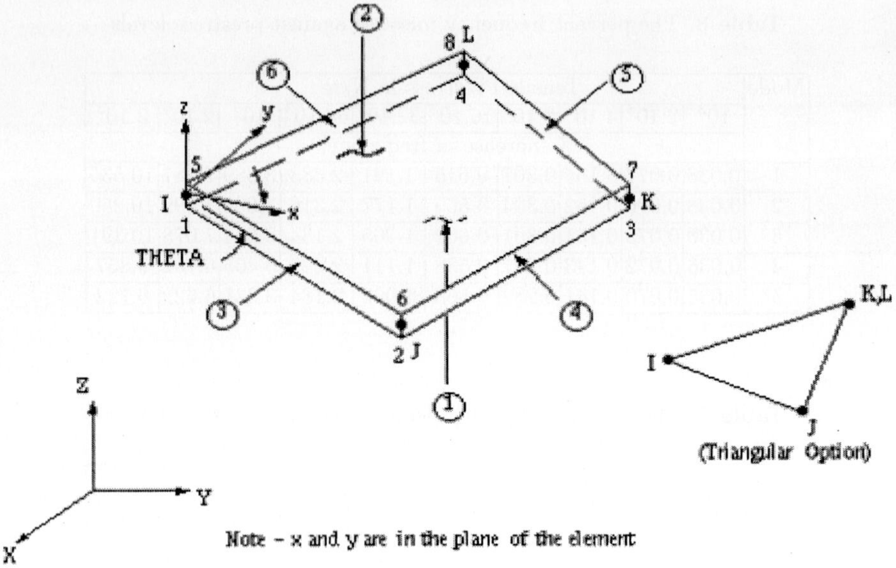
Z
Y
X
z
y
x
I
J
K
L
1
2
3
4
5
6
7
8
THETA
K,L
(Triangular Option)
Note - x and y are in the plane of the element

Fig. 3

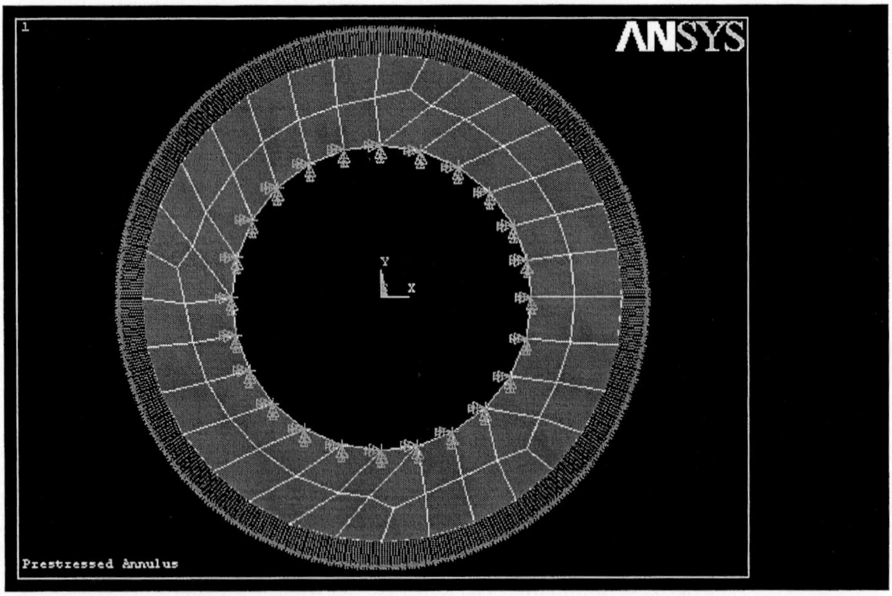
ANSYS
Prestressed Annulus

Fig. 4

Computer Realization of the Operator Method for Solving of Differential Equations

Liepa Bikulčienė[1], Romas Marcinkevičius[2], and Zenonas Navickas[3]

[1] Department of Applied Mathematics, Kaunas University of Technology, Studentu 50-323, Lithuania
liepaite@takas.lt
[2] Department of Software Engineering, Kaunas University of Technology, Studentu 50-415a, Lithuania
romas.marcinkevicius@ktu.lt
[3] Department of Applied Mathematics, Kaunas University of Technology, Studentu 50-325c, Lithuania
zenonas.navickas@ktu.lt

Abstract. The operator method for solving of differential equations and their systems is presented in the paper. Practical applicability of the method – methodology, parallelization of the computational algorithm and the complex solution of concrete differential equations – is described.

Keywords: Operator method, generalized differential operator, differential equations.

1 Introduction

The operator methods allow reduction differential equations to the algebraic ones. Introductory remarks on the operator computational methods can be found in the works of L.Euler (1646-1716), G.W.Leibniz (1707-1783) and others. The modern "algebraic" form was "prescribed" to operator calculus only by O.Viskov, M.Rahula, V.Maslov, Ph. Fensilver, R.Schott and other mathematicians. O. Viskov, using specially selected examples, showed that many classical analysis problems "contain" internally algebraic structures, [1]. Finding of the latter structures simplifies solution of a particular complicated problem. M.Rahula has developed the calculus methodic of Lie-Cartan, i.e. has shown that nonlinear differential equations can be described using ordinary geometrical constructions, [2]. V.Maslov introduced specific operator structures for the description of the solutions of some popular classical differential equations, [3]. Up-to-date modern conception of operator calculus has been presented by Ph. Fensilver and R.Schott in [4]. In this paper, the highly modern structures of operator calculus as well as areas of their practical applicability are described. On the other hand, technological progress makes it possible to develop and employ new powerful computational algorithms. The latter algorithms, in their turn, guarantee more precise solutions of the differential equations.

Often, approximate solutions of differential equations (or their systems) are sought using Runge-Kutta, Adams or some other numerical methods. Those

Z. Li et al. (Eds.): NAA 2004, LNCS 3401, pp. 179–186, 2005.

solutions are presented tabularly and are characterized by a fixed degree of accuracy. Sometimes, it is not sufficient. The operator method appears to be a successful extension of numerical methods. In applying an operator method, the sought-for solutions are represented as operator series, where from polynomials of various degrees – approximate solutions – are obtained. Desirable accuracy is available. To say more, various characteristics of differential equations (their systems) can be found out and analyzed. Finally, application of several independent methods (say, numerical and operator approaches) to solving of a more complicated differential equation facilitates avoidance of errors of various types.

2 The General Part

Operator relationships described in [5] are generalized for numerical realization:

2.1 Solutions of N-th Order Differential Equation

Let a differential equation

$$y_x^{(n)} = P(x, y, y'_x, y''_x, \ldots, y_x^{(n-1)}),$$

$$y(v; s_1, s_2, \ldots, s_{n-1}) = s_1,\ (y(x; s_1, s_2, \ldots, s_{n-1}))'_x|_{x=v} = s_2,$$
$$(y(x; s_1, s_2, \ldots, s_{n-1}))''_x|_{x=v} = s_3,\ \ldots,\ (y(x; s_1, s_2, \ldots, s_{n-1}))_x^{n-1}|_{x=v} = s_{n-1}$$

be given. Then its solution $y(x; s_1, s_2, \ldots, s_{n-1})$ is written:

$$y(x; s_1, s_2, \ldots, s_{n-1}) = \sum_{k=0}^{+\infty} p_k(s_1, s_2, \ldots, s_{n-1}, v)\frac{(x-v)^k}{k!}, \tag{1}$$

where $p_k(s_1, s_2, \ldots, s_{n-1}, v) =$
$(D_v + s_2 D_{s_1} + s_3 D_{s_2} + \ldots + s_{n-1} D_{s_{n-2}} + P(v, s_1, s_2, \ldots, s_{n-1}))^k s_1$. There D_v, D_{s_1}, D_{s_2}, …, $D_{s_{n-2}}$, D are differencial operators and $P(x, y, y'_x, y''_x, \ldots, y_x^{(n-1)})$ is a polynomial or a function.

Example 1. Let a differential equation $y' = y^2$, $y(v) = s$ be given. Using operator expression of solution (1) we get that $y = y(x, s) = \sum_{k=0}^{+\infty} p_k \frac{(x-v)^k}{k!}$, when $p_k = p_k(s, v) = (D_v + s^2 D_s)^k s$.

Then $p_0 = s$, $p_1 = 1 \cdot s^2$, $p_2 = 1 \cdot 2s^3, \ldots, p_n = n! s^{n+1}$, i.e. $y = y(x, s) = s\sum_{k=0}^{+\infty} s^k (x-v)^k$, or $y = \frac{s}{1-s(x-v)}$, when $|s(x-v)| < 1$.

Such analysis of solutions is possible on the occasion, when it is written in the operator form.

2.2 Solutions of a Second Order Differential Equation

Further we shall limit by cases, when all coefficients $|p_k| < M^k$, i.e., the obtained series converge for all $x \in \mathbf{R}$.

Analogically, let the second order differential equation with initial conditions $y''_{xx} = P(x, y, y'_x)$ $y(x, s, t)|_{x=v} = s$; $(y(x, s, t))'_x|_{x=v}$ be given. Then the expression of the solution is

$$y = y(x, s, t) = \sum_{k=0}^{+\infty} p_k(s, t, v)\frac{(x-v)^k}{k!}, v \in \mathbf{R}, \tag{2}$$

where $p_k(s, t, v) = (D_v + tD_s + P(v, s, t)D_t)^k s$ [5].

Besides, using equalities $y(x, s_l, t_l, v_l) = y(x, s_{l+1}, t_{l+1})$ we get

$$y(x, s_l, t_l) = \sum_{k=0}^{+\infty} p_k(s_l, t_l, v_l)\frac{(x-v_l)^k}{k!} = y_l(x, s_l, t_l), l = 1, 2, \ldots,$$

where

$$s_{l+1} = \sum_{k=0}^{+\infty} p_k(s_l, t_l, v_l)\frac{(v_{l+1}-v_l)^k}{k!}, t_{l+1} = \sum_{k=0}^{+\infty} p_{k+1}(s_l, t_l, v_l)\frac{(v_{l+1}-v_l)^k}{k!}, \tag{3}$$

sequence $v_2, v_3, \ldots$ are any variables, but v_1, s_1, t_1 are given.

2.3 Solution of a System of the Second Order Differential Equations

It is possible to generalize the above described methodology for systems of equations. For instance, let a system of differential equations

$$\begin{cases} x''_t = P(t, x, x'_t, y, y'_t, \varphi, \varphi'_t) \\ y''_t = Q(t, x, x'_t, y, y'_t, \varphi, \varphi'_t) \\ \varphi''_t = R(t, x, x'_t, y, y'_t, \varphi, \varphi'_t) \end{cases}$$

with initial conditions $x(v) = s_1$, $x'_t(t)|_{t=v} = t_1$, $y(v) = s_2$, $y'_t(t)|_{t=v} = t_2$, $\varphi(v) = s_3$, $\varphi'_t(t)|_{t=v} = t_3$ be given. Then expressions of the solutions are written:

$$x = \sum_{k=0}^{+\infty} p_k\frac{(t-v)^k}{k!}, y = \sum_{k=0}^{+\infty} q_k\frac{(t-v)^k}{k!}, \varphi = \sum_{k=0}^{+\infty} r_k\frac{(t-v)^k}{k!}, \tag{4}$$

when $p_k = D^k s_1$, $q_k = D^k s_2$, $r_k = D^k s_3$. There $D = D_v + t_1 D_{s_1} + PD_{t_1} + t_2 D_{s_2} + QD_{t_2} t_3 D_{s_3} + RD_{t_3}$ is the generalised differential operator and $P = P(v, s_1, t_1, s_2, t_2, s_3, t_3)$, $Q = Q(v, s_1, t_1, s_2, t_2, s_3, t_3)$, $R = R(v, s_1, t_1, s_2, t_2, s_3, t_3)$.

3 Methodology of Computer Solving Realization

Solving methodology will be presented for differential equation of the second order. For other order differential equations and for their systems of equations this methodology is analogical.

The found solution is series of x with coefficients p_k, which are the functions of initial conditions and center v. A computer can calculate only the finite number of coefficients $p_0, p_1, \ldots, p_N$. So, we get the approximate of solution (2) — polynomial $\hat{y}(x,s,t) = \sum_{k=0}^{N} p_k(s,t,v)\frac{(x-v)^k}{k!}$. Then substituting s, t and v to particular values, we get the approximate of solution — polynomial $\hat{y}(x)$ in neighborhood of a point v. Choosing sequence of the centers and using expression (3) a family of polynomials $\hat{y}_1(x)$, $\hat{y}_2(x)$, … is constructed. Approximations recede from the exact solution, as the variable x recede from the center. Then the approximation $y^*(x)$ of the solution is formed, using the family of approximations $\hat{y}_l(x)$, this way (Fig. 1):

$$y^*(x) = \hat{y}_l(x), \;\; v_l \le x < v_{l+1}, \;\; l = 1, 2, \ldots, n. \qquad (5)$$

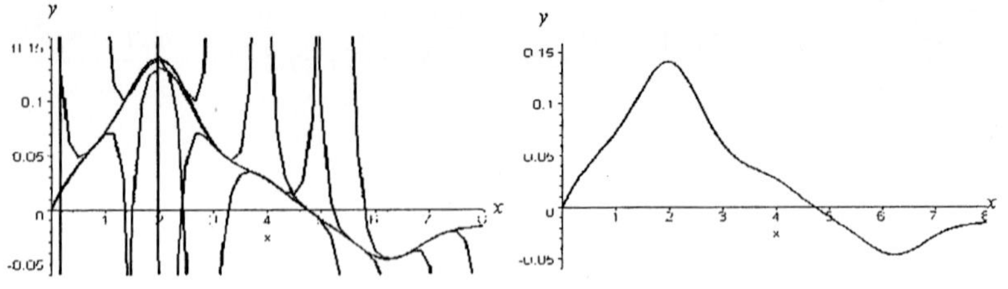

Fig. 1. Approximation of the solution

Other techniques of approximate $y^*(x)$ composition are possible, for instance, average of neighboring polynomials.

Example 2. Let nonlinear Mathieus differential equation $y'' + Hy' + \beta^2(1 + a\cos wx)\sin y = 0$ with initial conditions $y(x)|_{x=v} = s$; $(y(x))'_x|_{x=v}$ be given [6]. There H, β, a, w are numerical real parameters.

Besides, $p_0(s,t,v) = s$;
$p_{k+1}(s,t,v) = (D_v + tD_s - (Ht + \beta^2(1 + a\cos wv)\sin s)D_t)p_k(s,t,v)$, $k = 0, 1, 2, \ldots$, i.e. $p_1(s,t,v) = t$, $p_2(s,t,v) = -(Ht + \beta^2(1 + a\cos wv)\sin s)$, $p_3(s,t,v) = \beta^2 aw \sin s \sin wv - \beta^2 t\cos s(1 + a\cos wv) + H(Ht + \beta^2(1 + a\cos wv)\sin s)\sin s$, etc.

Table 1. The loss-estimates and the time expenditure

N	h	Δ	T,s	N	h	Δ	T,s	N	h	Δ	T,s
8	1	0.034	12	9	1	0.005	42	10	1	0.0011	124
8	0.75	0.028	12	9	0.75	0.003	45	10	0.75	0.0078	137
8	0.5	0.025	14	9	0.5	0.004	54	10	0.5	0.0084	145
8	0.25	0.032	15	9	0.25	0.006	62	10	0.25	0.0014	160

After the calculation and differentiation of the approximation of the solution, $\Delta = \max|\Delta(x)|$ is found, where $\Delta(x) = (y^*(x))'' + h(y^*(x))' + \beta^2(1 + a\cos wx)\sin y^*(x)$.

The loss-estimates Δ, with $y(0) = 0.5$, $y'(0) = 0.2$, $H = 0.7$, $\beta = 0.9$, $a = 2$, $w = 1$, $x = 0..20$ for different values of N and h and the calculation time T are presented in Table 1.

This time expenditure is found using computer Celeron 566 MHz, RAM 128 MB.

4 The Complex Solving Using Independent Methods

Solutions of the Mathieus differential equation have various standing modes. They depend on initial conditions and values of parameters. In practice, finding of attractor zones and their limits is important. For instance, reduced equation $y'' + Hy' + \beta\cos wx \sin y = 0$ with parameter values $H = 0.05$, $\beta = 0.5$, $w = 1$,

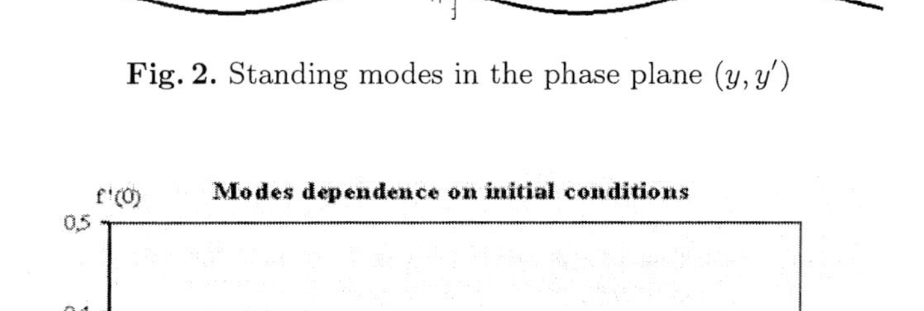

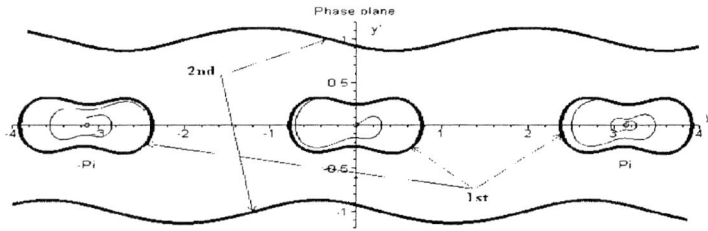

Fig. 2. Standing modes in the phase plane (y, y')

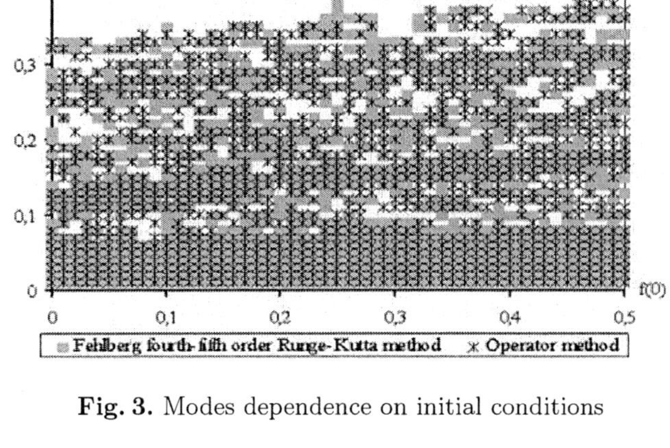

Fig. 3. Modes dependence on initial conditions

have two standing modes, that can be schematically represented in the phase plane (Fig. 2):

It is not possible to estimate attractor zones of the standing modes using ordinary methods, for instance, time back integration. Fig. 3 shows modes dependence on initial conditions. 1-st standing mode is represented by stars or squares, 2-nd is white. It is possible to see transition zone between 1-st and 2-nd modes, Fig. 3. Two qualitative separate methods: Runge-Kutta and operator are used in the precise determination of zone limits. The order of a polynomial of operator method must be increased in order to obtain more accurate modes dependence on initial conditions in the choosed point.

5 Solving of Complicated System, Using Parallelization

Various dynamical systems are described using complicated ordinary differential equations and their systems, comprising a large number of numerical parameters. Solving such differential equations it is necessary to find not only their solutions, but also numerical values of parameters (to provide those solutions with desirable characteristics). It is often required the solutions to be periodical (aperiodical) functions. Also, it is important to know for what values of parameters the dynamical system in question behaves chaotically, etc.

Example 3. Let $x = x(t)$ and $y = y(t)$ describe the vibrating motion of waves or flowing liquid, $\varphi = \varphi(t)$ — the rotational motion, stimulated by this vibration (Fig. 4).

All these functions enter the system of differential equations [7]:

$$x'' + x = f_x, \; y'' + y = f_y, \; \mu^* \varphi'' = f_\varphi,$$

where

$$\begin{aligned} f_x &= a_x - h_x x' + (q_{1x} - q_{3x} x'^2) x' + \mu(\varphi'' \sin\varphi + \varphi'^2 \cos\varphi), \\ f_y &= a_y - h_y y' + (q_{1y} - q_{3y} y'^2) y' + \mu(\varphi'' \cos\varphi - \varphi'^2 \sin\varphi) + g^*, \\ f_\varphi &= \mu_\varphi^* - h_\varphi \varphi' + g^* \cos\varphi + \mu(x'' \sin\varphi - y'' \cos\varphi). \end{aligned}$$

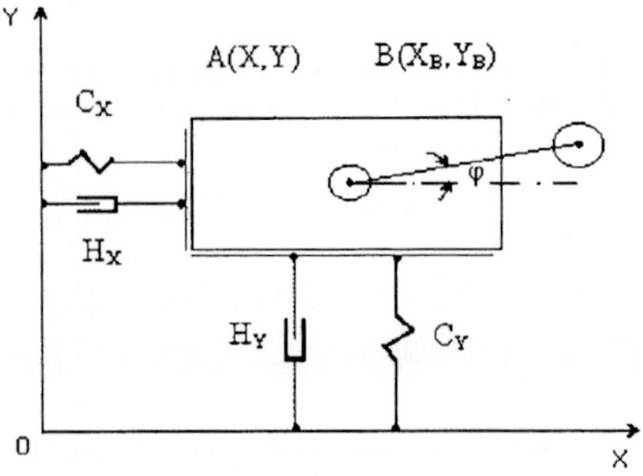

Fig. 4. System model

There a_x, a_y, h_x, h_y, h_φ, q_{1x}, q_{1y}, q_{3x}, q_{3y}, μ, g^*, μ^*_φ, μ are numerical parameters with values, providing both periodicity of the functions $x = x(t)$, $y = y(t)$ and stability of rotational motion $\varphi = \varphi(t)$ to be found.

Solving the system, (for simplicity $\mu^* = 1$, $g^* = 0$) we obtain:

$$\begin{aligned}
x'' &= \tfrac{1}{1-\mu^2}(F_x - \mu^2 \sin\varphi\cos\varphi F_y + \mu\sin\varphi F_\varphi - \mu^2\cos^2\varphi F_x) = P, \\
y'' &= \tfrac{1}{1-\mu^2}(F_y - \mu^2 \sin\varphi\cos\varphi F_x - \mu\cos\varphi F_\varphi - \mu^2\sin^2\varphi F_y) = Q, \\
\varphi'' &= \tfrac{1}{1-\mu^2}(F_\varphi - \mu\sin\varphi F_x - \mu\cos\varphi F_y) = R.
\end{aligned}$$

There

$$\begin{aligned}
F_x &= a_x - x - h_x x' + (q_{1x} - q_{3x}x'^2)x' + \varphi'^2\mu\cos\varphi, \\
F_y &= a_y - y - h_y y' + (q_{1y} - q_{3y}y'^2)y' + \varphi'^2\mu\cos\varphi, \\
F_\varphi &= \mu^*_\varphi - h_\varphi\varphi'.
\end{aligned}$$

Using expressions of solutions (4) and methotodology (5) it is possible to vary the step size of the center variation, the order of the polynomial as well as the number of polynomials, used in forming the approximate solutions. To obtain more accurate approximations, the order of a polynomial must be increased [8]. The accuracy depends on the transition step size between centers, but trials decrease the step size multiply calculation errors as well as calculation time. It is found from experimental investigations that symbolic differentiation of the expressions takes most time of calculation (about 80%). Besides, in relation with Maple peculiarities [9], in solving more complicated differential equations, the necessary amount of random access memory (RAM) becomes critical. Solving ordinary differential equations, we escape problems associated with evaluation of symbolic differential expressions. The calculation time for operator and Runge-Kutta methods realization makes acceptable using (for parallelization) a particular computer network. The symbolic differentiation as well as graphical information is realized using Maple tools, whereas the computer network is used applying MPI [10] tools.

Parallel algorithm of the investigation solution is found from serial algorithm by this:

1. All serial executing actions subject to properties of serial algorithm and are partied to the g groups. There g is odd number between 3 and 15 and depend on parameter values.
   (a) All groups must be executing in series, i.e. all actions from group k are executing before actions from group $(k+1)$ $(k = 1, 2, \ldots, g-1)$ and results of group k are material to the group $(k+1)$.
   (b) Order of action execution inside the group depends to number of group:
      i. Actions from $1, 3, \ldots, g$ group must be executed seriatim;
      ii. Actions from groups $2, 4, \ldots, (g-1)$, can be executed parallel each other inside group.
2. The sequences of serial actions $S_{k,1}$, $S_{k,2}$, $\ldots$, S_{k,n_k} are formed inside each group. Here k is number of group, n_k show number of simultaneously parallel process, besides, $n_k = 1$, if k is odd number and $1 \le n_k \le p$ if k is even number. There p is number of used processors.

3. Sequences of actions are distributed for processors elementary: sequence $S_{k,i}$ is executed using processor i.

One processor is conditional primary: therein calculations initiate and terminate.

If execution time of action sequence $S_{k,i}$ is $t_{k,i}$, then general calculation time using serial algorithm is $T_S = \sum_k \sum_i t_{k,i}$. Parallel algorithm execution time is

$$T_p = t_{1,1} + \max\{t_{2,1}, \ldots, t_{2,n_2}\} + t_{3,1} + \max\{t_{4,1}, \ldots, t_{2,n_4}\} + \ldots + t_{g,1} + T_t;$$

here T_t is total time of transmissions between processors.

Because material must be send just from first processor to reminder and back (not from each to each) losing of time is high and $\max\{t_{k,1}, \ldots, t_{k,n_k}\} < \sum_i t_{k,i}$ for all even k values.

For instance, if values of system parameters are $a_x = a_y = 0.5$, $h_x = h_y = h_\varphi = 0.1$, $q_{1x} = q_{1y} = 0.3$, $q_{3x} = q_{3y} = 1$, $\mu_\varphi^* = 0.5$, $t = 0 \ldots 100$ then distributing actions of algorithm to 3 groups ($g = 3$) and three computers of personal cluster are used. Then execution of actions sequences $S_{1,1}$, $S_{3,1}$ and transmissions between processors time is less than 0.5 s. Calculation time of parallel executed sequences $S_{2,1}$, $S_{2,2}$, $S_{2,3}$, is between 550 s. and 600 s.

References

1. O.V.Viskov, A non commutative approach to classical problem of analysis, Tr. Mat., Inst. Akad. Nauk, SSSR, T 177, (1986), p. 21-32, (in Russian).
2. M.Rahula, Exponential law in the Lie-Cartan calculus, Rend. Sem. Mat. Mesina, Atti del Congresso Internazionale in onore di Pasquale Calapso, (1998), p. 267-291.
3. N.M.Maksimov, Elements of analysis of discrete mappings, In proc. Seminar Discrete Math.Applications, 1, Moscow university Press, Moscow (1989), p. 33-37,(in Russian).
4. Ph. Feinsilver, R.Schott, Algebraic Structures and operator calculus Vol. I: Representation and probability theory, Kluwer Academic Publishers, (1993), p. 225.
5. Z.Navickas, Operator method for solving nonlinear differential equations. Lithuanian Mathematical Journal, Vol. 42, Vilnius, (2002), p.81-92, (in Russian).
6. Ricard H. Rand. Lecture Notes of Nonlinear Vibrations. Cornell University. Version 34, 2000. www.tam.cornell.edu/randocs.
7. L.Bikulčienė, Realization of operator method for the solutions of differential equations. Lithuanian Mathematical Journal, Vol. 42. Vilnius, (2002), p.159-164.
8. K.Ragulskis, Z.Navickas, L.Bikulčienė, Realization of operator method for mechanical system. Journal of Vibroengineering. (2002), No. 1(8). Vilnius, Lithuania, (2002), p.31-34
9. Garvan, Frank, The Maple Book, Chapman & Hill, (2002).
10. Message Passing Interface Forum. (2004). http://www.mpi-forum.org.

The Strong Stability of the Second-Order Operator-Differential Equations

D. Bojović[1], B. Z. Popović[1], and B. S. Jovanović[2]

[1] University of Kragujevac, Faculty of Science,
R. Domanovića 12, 34000 Kragujevac, Serbia and Montenegro
bojovicd@ptt.yu, bpopovic@knez.uis.kg.ac.yu
[2] University of Belgrade, Faculty of Mathematics
Studentski trg 16, 11000 Belgrade, Serbia and Montenegro
bosko@matf.bg.ac.yu

Abstract. In this paper we investigate the stability of the second-order operator-differential equation in Hilbert space, under perturbations of the operators, the initial conditions and right hand side of the equation. The estimates of strong stability in different norms are obtained. As an example, the strong stability of the hyperbolic problem is presented.

1 Introduction

When the correctness of the initial-value problems for the evolution problems of mathematical physics is considered, the attention is mainly paid to the stability of the solution with respect to the initial condition and the right-hand side of the equation. In a more general case it is necessary to require the stability of solution against the perturbation of the operator (or coefficients of the equation). Let us mention, for example, the widely investigated problem of the coefficient stability of the systems of linear equations [2]. This type of stability is usually called strong stability [6]. Analogous problem arise in corresponding numerical methods that approximate differential equations.

This paper deals with the construction of the strong stability estimates for the second-order operator-differential equation in Hilbert spaces in the case of perturbed operators, right-hand side and of the initial conditions. Strong stability and asymptotic stability for the first-order operator-differential equation is considered in [1]. Analogous results hold for the operator-difference schemes [3].

2 Formulation of the Problem

Let H be a Hilbert space, with the inner product $(\cdot,\cdot)$ and norm $\|\cdot\|$. Let A be a selfadjoint positive definite linear operator with domain $D(H)$ dense in H. We use H_A to denote the space with the inner product $(u,v)_A = (Au,v)$ and norm $\|u\|_A = (Au,u)^{1/2}$. Analogously, we define inner product $(u,v)_{A^{-1}} = (A^{-1}u,v)$ and norm $\|u\|_{A^{-1}} = (A^{-1}u,u)^{1/2}$. We have: $H_A \subset H \subset H_{A^{-1}}$. Also, operator

Z. Li et al. (Eds.): NAA 2004, LNCS 3401, pp. 187–194, 2005.

A is bounded in the following cases [7]: $A : H_A \to H_{A^{-1}}$, $A : H_{A^2} \to H$, and $A : H \to H_{A^{-2}}$.

Let us consider the abstract Cauchy problem for the second-order operator-differential equation:

$$Bu'' + Au = f(t)\,,\ 0 < t < T\,, \quad u(0) = u_0\,,\ u'(0) = u_1\,, \tag{1}$$

where A and B are unbounded selfadjoint positive definite linear operators, defined on Hilbert space H, and $u : [0,T] \to H$. We assume that $H_A \subset H_B$ and $A \ge B$, i.e. $(Au, u) \ge (Bu, u)$ for $u \in H_A$. We also assume that the quotient $(Au, u)/(Bu, u)$ is unbounded for $u \in H_A$. Define the next norms:

$$\begin{aligned}
\|u\|^2_{(0)} &= \max_{t\in[0,T]} \|u(t)\|^2_B,\\
\|u\|^2_{(1)} &= \max_{t\in[0,T]} (\|u'(t)\|^2_B + \|u(t)\|^2_A),\\
\|u\|^2_{(2)} &= \max_{t\in[0,T]} (\|u''(t)\|^2_B + \|u'(t)\|^2_A + \|Au(t)\|^2_{B^{-1}}).
\end{aligned}$$

The perturbed problem has the form:

$$\tilde{B}\tilde{u}'' + \tilde{A}\tilde{u} = \tilde{f}(t)\,,\ 0 < t < T\,, \quad \tilde{u}(0) = \tilde{u}_0\,,\ \tilde{u}'(0) = \tilde{u}_1\,, \tag{2}$$

with analogous conditions for operators $\widetilde{A}$ and $\widetilde{B}$.

The problem (1) is strongly stable, by definition [5], if the next estimate holds:

$$\begin{aligned}
\|u - \tilde{u}\|_0 \le C_1\|u_0 - \tilde{u}_0\|_1 + C_2\|u_1 - \tilde{u}_1\|_2 + C_3\|f - \tilde{f}\|_3&\\
+ C_4\|A - \tilde{A}\|_4 + C_5\|B - \tilde{B}\|_5\,,&
\end{aligned} \tag{3}$$

where C_i are some constants and $\|\cdot\|_i$ are some norms.

From (1) and (2) we deduce that the error $z = \tilde{u} - u$ satisfies the next Cauchy problem:

$$Bz'' + Az = F(t)\,,\ t > 0\,,\ \ z(0) = \tilde{u}_0 - u_0\,,\ \ z'(0) = \tilde{u}_1 - u_1\,, \tag{4}$$

where $F(t) = \tilde{f}(t) - f(t) - (\widetilde{A} - A)\tilde{u} - (\widetilde{B} - B)\tilde{u}''$.

The aim of this paper is to prove estimate of the form (3) in the norms $\|\cdot\|_{(0)}$, $\|\cdot\|_{(1)}$, and $\|\cdot\|_{(2)}$. The next a priori estimates for the solution of the problem (1) are proved in [4]:

$$\|u\|^2_{(0)} \le C\Big(\|u_0\|^2_B + \|Bu_1\|^2_{A^{-1}} + \int_0^T \|f(t)\|^2_{A^{-1}} dt\Big), \tag{5}$$

$$\|u\|^2_{(1)} \le C\Big(\|u_0\|^2_A + \|u_1\|^2_B + \int_0^T \|f(t)\|^2_{B^{-1}} dt\Big), \tag{6}$$

$$\|u\|^2_{(2)} \le C\Big(\|Au_0\|^2_{B^{-1}} + \|u_1\|^2_A + \max_{t\in[0,T]} \|f(t)\|^2_{B^{-1}} + \int_0^T \|B^{-1}f(t)\|^2_A dt\Big). \tag{7}$$

3 Strong Stability

In this section the following assertions will be proved to be valid.

Lemma 1. *The problem (1) is strongly stable in the norm* $\|\cdot\|_{(0)}$, *and with conditions* $\tilde{u}_0 - u_0 \in H_B$, $B(\tilde{u}_1 - u_1) \in H_{A^{-1}}$, $\tilde{f} - f \in L_2((0,T); H_{A^{-1}})$, *the next estimate holds:*

$$\|\tilde{u} - u\|_{(0)} \leq C_0^{(0)} \|\tilde{u}_0 - u_0\|_B + C_1^{(0)} \|\tilde{u}_1 - u_1\|_{BA^{-1}B} + C_2^{(0)} \|\tilde{f} - f\|_{L_2((0,T);H_{A^{-1}})}$$
$$+ C_3^{(0)} \|\widetilde{A} - A\|_{H_{\widetilde{A}} \to H_{A^{-1}}} + C_4^{(0)} \|\widetilde{B} - B\|_{H_{\widetilde{B}\widetilde{A}^{-1}\widetilde{B}} \to H_{A^{-1}}} . \tag{8}$$

Proof. Applying estimate of the form (5) to the problem (4) we have:

$$\|z\|_{(0)}^2 \leq C \left(\|\tilde{u}_0 - u_0\|_B^2 + \|B(\tilde{u}_1 - u_1)\|_{A^{-1}}^2 + \int_0^T \|F(t)\|_{A^{-1}}^2 dt \right) . \tag{9}$$

Let us estimate $\int_0^T \|F(t)\|_{A^{-1}}^2 dt$. We have elementary inequality:

$$\|F(t)\|_{A^{-1}}^2 \leq 3 \left(\|\tilde{f}(t) - f(t)\|_{A^{-1}}^2 + \|(\widetilde{A} - A)\tilde{u}\|_{A^{-1}}^2 + \|(\widetilde{B} - B)\tilde{u}''\|_{A^{-1}}^2 \right) .$$

Further:

$$\int_0^T \|(\widetilde{A} - A)\tilde{u}\|_{A^{-1}}^2 dt = \int_0^T \frac{\|(\widetilde{A} - A)\tilde{u}\|_{A^{-1}}^2}{\|\tilde{u}\|_{\widetilde{A}}^2} \|\tilde{u}\|_{\widetilde{A}}^2 dt \leq$$
$$\leq \int_0^T \sup_{\substack{w \neq 0 \\ w \in H_{\widetilde{A}}}} \frac{\|(\widetilde{A} - A)w\|_{A^{-1}}^2}{\|w\|_{\widetilde{A}}^2} \|\tilde{u}\|_{\widetilde{A}}^2 dt = \|\widetilde{A} - A\|_{H_{\widetilde{A}} \to H_{A^{-1}}}^2 \int_0^T \|\tilde{u}\|_{\widetilde{A}}^2 dt$$

Applying estimate of the form (6) to the problem (2) we have:

$$\|\tilde{u}\|_{\widetilde{A}}^2 \leq C \left(\|\tilde{u}_0\|_{\widetilde{A}}^2 + \|\tilde{u}_1\|_{\widetilde{B}}^2 + \int_0^T \|\tilde{f}(t)\|_{\widetilde{B}^{-1}}^2 dt \right) = K_1 .$$

Therefore:

$$\int_0^T \|(\widetilde{A} - A)\tilde{u}\|_{A^{-1}}^2 dt \leq T K_1 \|\widetilde{A} - A\|_{H_{\widetilde{A}} \to H_{A^{-1}}}^2 .$$

Further,

$$\int_0^T \|(\widetilde{B} - B)\tilde{u}''\|_{A^{-1}}^2 dt \leq \|\widetilde{B} - B\|_{H_{\widetilde{B}\widetilde{A}^{-1}\widetilde{B}} \to H_{A^{-1}}} \int_0^T \|\widetilde{B}\tilde{u}''\|_{\widetilde{A}^{-1}}^2 dt =$$
$$= \|\widetilde{B} - B\|_{H_{\widetilde{B}\widetilde{A}^{-1}\widetilde{B}} \to H_{A^{-1}}} \int_0^T \|\tilde{f} - \widetilde{A}\tilde{u}\|_{\widetilde{A}^{-1}}^2 dt \leq$$
$$\leq 2 \|\widetilde{B} - B\|_{H_{\widetilde{B}\widetilde{A}^{-1}\widetilde{B}} \to H_{A^{-1}}} \int_0^T (\|\tilde{f}\|_{\widetilde{A}^{-1}}^2 + \|\tilde{u}\|_{\widetilde{A}}^2) dt \leq K_2 \|\widetilde{B} - B\|_{H_{\widetilde{B}\widetilde{A}^{-1}\widetilde{B}} \to H_{A^{-1}}}$$

where $K_2 = 2 \int_0^T (\|\tilde{f}\|_{\widetilde{A}^{-1}}^2 + K_1) dt$.

Finally, from (9) and previous inequalities, setting $C_0^{(0)} = C_1^{(0)} = \sqrt{C}$, $C_2^{(0)} = \sqrt{3C}$, $C_3^{(0)} = \sqrt{3CTK_1}$, $C_4^{(0)} = \sqrt{3CK_2}$, we obtain estimate (8). □

Lemma 2. *The problem (1) is strongly stable in the norm* $\|\cdot\|_{(1)}$, *and with conditions* $\tilde{u}_0 - u_0 \in H_A$, $\tilde{u}_1 - u_1 \in H_B$, $\tilde{f} - f \in L_2((0,T); H_{B^{-1}})$, *the next estimate holds:*

$$\|\tilde{u} - u\|_{(1)} \le C_0^{(1)} \|\tilde{u}_0 - u_0\|_A + C_1^{(1)} \|\tilde{u}_1 - u_1\|_B + C_2^{(1)} \|\tilde{f} - f\|_{L_2((0,T);H_{B^{-1}})}$$
$$+ C_3^{(1)} \|\widetilde{A} - A\|_{H_{\widetilde{A}\widetilde{B}^{-1}\widetilde{A}} \to H_{B^{-1}}} + C_4^{(1)} \|\widetilde{B} - B\|_{H_{\widetilde{B}} \to H_{B^{-1}}} . \qquad (10)$$

Proof. Applying estimate of the form (6) to the problem (4) we obtain:

$$\|z\|_{(1)}^2 \le C \left(\|\tilde{u}_0 - u_0\|_A^2 + \|\tilde{u}_1 - u_1\|_B^2 + \int_0^T \|F(t)\|_{B^{-1}}^2 dt \right) . \qquad (11)$$

We have:

$$\int_0^T \|(\widetilde{A} - A)\tilde{u}\|_{B^{-1}}^2 dt \le \|\widetilde{A} - A\|_{H_{\widetilde{A}\widetilde{B}^{-1}\widetilde{A}} \to H_{B^{-1}}}^2 \int_0^T \|\widetilde{A}\tilde{u}\|_{\widetilde{B}^{-1}}^2 dt .$$

Applying estimate of the form (7) on the problem (2) we have:

$$\|\widetilde{A}\tilde{u}\|_{\widetilde{B}^{-1}}^2 \le C \left(\|\widetilde{A}\tilde{u}_0\|_{\widetilde{B}^{-1}}^2 + \|\tilde{u}_1\|_{\widetilde{A}}^2 + \max_{t \in [0,T]} \|\tilde{f}(t)\|_{\widetilde{B}^{-1}}^2 + \int_0^T \|\widetilde{B}^{-1}\tilde{f}(t)\|_{\widetilde{A}}^2 dt \right) = K_3 .$$

Therefore,

$$\int_0^T \|(\widetilde{A} - A)\tilde{u}\|_{B^{-1}}^2 dt \le K_3 T \|\widetilde{A} - A\|_{H_{\widetilde{A}\widetilde{B}^{-1}\widetilde{A}} \to H_{B^{-1}}}^2 .$$

Further,

$$\int_0^T \|(\widetilde{B} - B)\tilde{u}''\|_{B^{-1}}^2 dt \le \|\widetilde{B} - B\|_{H_{\widetilde{B}} \to H_{B^{-1}}} \int_0^T \|\tilde{u}''\|_{\widetilde{B}}^2 dt \le K_3 T \|\widetilde{B} - B\|_{H_{\widetilde{B}} \to H_{B^{-1}}} .$$

Finally, from (11) and previous inequalities, setting $C_0^{(1)} = C_1^{(1)} = \sqrt{C}$, $C_2^{(1)} = \sqrt{3C}$, $C_3^{(1)} = C_4^{(1)} = \sqrt{3CTK_3}$, we obtain estimate (10). □

Lemma 3. *The problem (1) is strongly stable in the norm* $\|\cdot\|_{(2)}$, *and with conditions* $A(\tilde{u}_0 - u_0) \in H_{B^{-1}}$, $\tilde{u}_1 - u_1 \in H_A$, $\tilde{f} - f \in C((0,T); H_{B^{-1}}) \cap L_2((0,T); H_{B^{-1}AB^{-1}})$, *the next estimate holds:*

$$\|\tilde{u} - u\|_{(2)} \le C_0^{(2)} \|\tilde{u}_0 - u_0\|_{AB^{-1}A} + C_1^{(2)} \|\tilde{u}_1 - u_1\|_A + C_2^{(2)} \|\tilde{f} - f\|_*$$
$$+ C_3^{(2)} \|\widetilde{A} - A\|_\alpha + C_4^{(2)} \|\widetilde{B} - B\|_\beta , \qquad (12)$$

where

$$\|\tilde{f}-f\|_*=\|\tilde{f}-f\|_{C((0,T);H_{B^{-1}})}+\|\tilde{f}-f\|_{L_2((0,T);H_{B^{-1}AB^{-1}})},$$
$$\|\widetilde{A}-A\|_\alpha=\|\widetilde{A}-A\|_{H_{\widetilde{A}\widetilde{B}^{-1}\widetilde{A}}\to H_{B^{-1}}}+\|\widetilde{A}-A\|_{H_{\widetilde{A}\widetilde{B}^{-1}\widetilde{A}\widetilde{B}^{-1}\widetilde{A}}\to H_{B^{-1}AB^{-1}}},$$
$$\|\widetilde{B}-B\|_\beta=\|\widetilde{B}-B\|_{H_{\widetilde{B}}\to H_{B^{-1}}}+\|\widetilde{B}-B\|_{H_{\widetilde{A}}\to H_{B^{-1}AB^{-1}}}.$$

Proof. Applying estimate of the form (7) to the problem (4) we obtain:

$$\|z\|_{(2)}^2\le C\left(\|Az_0\|_{B^{-1}}^2+\|z_1\|_A^2+\max_{t\in[0,T]}\|F(t)\|_{B^{-1}}^2+\int_0^T\|B^{-1}F(t)\|_A^2dt\right). \quad (13)$$

Let us estimate $\max_{t\in[0,T]}\|F(t)\|_{B^{-1}}^2$. We have:

$$\|(\widetilde{A}-A)\tilde{u}\|_{B^{-1}}^2\le\|\widetilde{A}-A\|_{H_{\widetilde{A}\widetilde{B}^{-1}\widetilde{A}}\to H_{B^{-1}}}^2\|\widetilde{A}\tilde{u}\|_{\widetilde{B}^{-1}}^2\le K_3\|\widetilde{A}-A\|_{H_{\widetilde{A}\widetilde{B}^{-1}\widetilde{A}}\to H_{B^{-1}}}^2.$$

Further,

$$\|(\widetilde{B}-B)\tilde{u}''\|_{B^{-1}}^2\le\|\widetilde{B}-B\|_{H_{\widetilde{B}}\to H_{B^{-1}}}\|\tilde{u}''\|_{\widetilde{B}}^2\le K_3\|\widetilde{B}-B\|_{H_{\widetilde{B}}\to H_{B^{-1}}}.$$

So we obtain:

$$\max_{t\in[0,T]}\|F(t)\|_{B^{-1}}^2\le 3\|\tilde{f}-f\|_{C((0,T);H_{B^{-1}})}$$
$$+3K_3\|\widetilde{A}-A\|_{H_{\widetilde{A}\widetilde{B}^{-1}\widetilde{A}}\to H_{B^{-1}}}^2+3K_3\|\widetilde{B}-B\|_{H_{\widetilde{B}}\to H_{B^{-1}}}. \quad (14)$$

Estimate $\int_0^T\|B^{-1}F(t)\|_A^2dt$. Let us apply operator $(\widetilde{A}\widetilde{B}^{-1})^2$ to the equation (2) and use estimate of the form (6), obtaining:

$$\|B^{-1}(\widetilde{A}-A)\tilde{u}\|_A^2\le\|\widetilde{A}-A\|_{H_{\widetilde{A}\widetilde{B}^{-1}\widetilde{A}\widetilde{B}^{-1}\widetilde{A}}\to H_{B^{-1}AB^{-1}}}^2\|\widetilde{A}\tilde{u}\|_{\widetilde{B}^{-1}\widetilde{A}\widetilde{B}^{-1}}^2$$
$$\le K_4\|\widetilde{A}-A\|_{H_{\widetilde{A}\widetilde{B}^{-1}\widetilde{A}\widetilde{B}^{-1}\widetilde{A}}\to H_{B^{-1}AB^{-1}}}^2$$

where

$$K_4=C\left(\|\tilde{u}_0\|_{\widetilde{A}\widetilde{B}^{-1}\widetilde{A}\widetilde{B}^{-1}\widetilde{A}}^2+\|\tilde{u}_1\|_{\widetilde{A}\widetilde{B}^{-1}\widetilde{A}}^2+\int_0^T\|\tilde{f}(t)\|_{\widetilde{A}^{-1}\widetilde{B}\widetilde{A}^{-1}}^2dt\right).$$

Further, let us apply operator $\widetilde{A}\widetilde{B}^{-1}$ to the equation (2) and use estimate of the form (7), obtaining:

$$\|B^{-1}(\widetilde{B}-B)\tilde{u}''\|_A^2\le\|\widetilde{B}-B\|_{H_{\widetilde{A}}\to H_{B^{-1}AB^{-1}}}^2\|\tilde{u}''\|_{\widetilde{A}}^2\le K_5\|\widetilde{B}-B\|_{H_{\widetilde{A}}\to H_{B^{-1}AB^{-1}}},$$

where

$$K_5=C\left(\|\tilde{u}_0\|_{\widetilde{A}\widetilde{B}^{-1}\widetilde{A}\widetilde{B}^{-1}\widetilde{A}}^2+\|\tilde{u}_1\|_{\widetilde{A}\widetilde{B}^{-1}\widetilde{A}}^2+\max_{t\in[0,T]}\|\tilde{f}\|_{\widetilde{A}\widetilde{B}^{-2}}^2+\int_0^T\|\tilde{f}(t)\|_{\widetilde{B}^{-1}}^2dt\right).$$

Therefore,

$$\int_0^T \|B^{-1}F(t)\|_A^2 dt \le 3\|\tilde f - f\|^2_{L_2((0,T);H_{B^{-1}AB^{-1}})} \tag{15}$$
$$+3K_4T\|\widetilde A - A\|^2_{H_{\tilde A\tilde B^{-1}\tilde A\tilde B^{-1}\tilde A}\to H_{B^{-1}AB^{-1}}} + 3K_5T\|\widetilde B - B\|^2_{H_{\tilde A}\to H_{B^{-1}AB^{-1}}} .$$

Finally, from (13), (14) and (15), setting $C_0 = C_1 = \sqrt{C}$, $C_2 = \sqrt{3C}$, $C_3 = \sqrt{3C\max\{K_3, TK_4\}}$ and $C_4 = \sqrt{3C\max\{K_3, TK_5\}}$, we obtain estimate (12). □

Remark. Estimates (8), (10) and (12) are correct only if $K_i < \infty$. For example, $K_1 < \infty$ for $\tilde u_0 \in H_{\tilde A}$, $\tilde u_1 \in H_{\tilde B}$ and $\tilde f \in L_2((0,t); H_{\tilde B^{-1}})$.

Let us consider full equation:

$$Bu'' + Du' + Au = f(t)\,, \ 0 < t < T\,, \quad u(0) = u_0\,, \ u'(0) = u_1\,, \tag{16}$$

where A, B and D are positive definite selfadjoint linear operators, defined in Hilbert space H.

The perturbed problem has the form:

$$\tilde B\tilde u'' + \widetilde D\tilde u' + \tilde A\tilde u = \tilde f(t)\,, \ 0 < t < T\,, \quad \tilde u(0) = \tilde u_0\,, \ \tilde u'(0) = \tilde u_1\,,$$

with analogous assumptions for operators $\widetilde A$, $\widetilde B$ and $\widetilde D$.

The error $z = \tilde u - u$ satisfies the conditions:

$$Bz'' + Dz + Az = F_1(t)\,, \ t > 0\,, \quad z(0) = \tilde u_0 - u_0\,, \quad z'(0) = \tilde u_1 - u_1\,,$$

where $F_1(t) = \tilde f(t) - f(t) - (\widetilde A - A)\tilde u - (\widetilde D - D)\tilde u' - (\widetilde B - B)\tilde u''$.

By the same technique as in the previous case we can prove the next result.

Lemma 4. *The problem (16) is strongly stable in the norm* $\|\cdot\|_{(1)}$, *and with conditions* $\tilde u_0 - u_0 \in H_A$, $\tilde u_1 - u_1 \in H_B$, $\tilde f - f \in L_2((0,T); H_{B^{-1}})$, *the next estimate holds:*

$$\|\tilde u - u\|_{(1)} \le C_0\|\tilde u_0 - u_0\|_A + C_1\|\tilde u_1 - u_1\|_B + C_2\|\tilde f - f\|_{L_2((0,T);H_{B^{-1}})}$$
$$+C_3\|\widetilde A - A\|_{H_{\tilde A\tilde B^{-1}\tilde A}\to H_{B^{-1}}} + C_4\|\widetilde D - D\|_{H_{\widetilde D\tilde B^{-1}\widetilde D}\to H_{\tilde B^{-1}}} + C_5\|\widetilde B - B\|_{H_{\tilde B}\to H_{B^{-1}}} .$$

4 Hyperbolic Problem

As an example of the previous results, we consider hyperbolic problem:

$$\frac{\partial^2 u}{\partial t^2} = \frac{\partial}{\partial x}\Big(a(x)\,\frac{\partial u}{\partial x}\Big) + f(x,t)\,, \quad x \in (0,1), \ \ t > 0, \tag{17}$$
$$u(0,t) = u(1,t) = 0\,, \quad u(x,0) = u_0(x)\,.$$

Problem (17) can be written in the form (1), for $H = L_2(0,1)$, B – identity operator in $L_2(0,1)$ and $Au = -\dfrac{\partial}{\partial x}\Big(a(x)\dfrac{\partial u}{\partial x}\Big)$. In the case when $a \in C^1[0,1]$, $0 < c_1 \le a(x) \le c_2$ and $|a'(x)| \le c_3$, operator A maps $D(A) = \overset{\circ}{W}{}_2^1(0,1) \cap W_2^2(0,1)$ to $L_2(0,1)$.

The next inequalities hold:

$$c_0 \,\|v\|^2_{W_2^1(0,1)} \le (Av,\, v) = \int_0^1 a(x)\,|v'(x)|^2\,dx \le c_2\,\|v\|^2_{W_2^1(0,1)}\,, \quad v \in \overset{\circ}{W}{}_2^1(0,1)\,,$$

$$c_4\,\|v\|_{W_2^2(0,1)} \le \|Av\|_{L_2(0,1)} \le c_5\,\|v\|_{W_2^2(0,1)}\,, \quad v \in \overset{\circ}{W}{}_2^1(0,1) \cap W_2^2(0,1)\,, \tag{18}$$

where $c_0 = c_1\pi^2/(1+\pi^2)$, and constants c_4 and c_5 depend on c_1, c_2 and c_3. We have: $H_A = \overset{\circ}{W}{}_2^1(0,1)$, $H_{A^{-1}} = W_2^{-1}(0,1)$ and $H_{A^2} = \overset{\circ}{W}{}_2^1(0,1) \cap W_2^2(0,1)$.

The corresponding perturbed problem has the form:

$$\frac{\partial^2 \tilde u}{\partial t^2} = \frac{\partial}{\partial x}\Big(\tilde a(x)\,\frac{\partial \tilde u}{\partial x}\Big) + \tilde f(x,t)\,, \quad x \in (0,1), \ \ t > 0,$$
$$\tilde u(0,t) = \tilde u(1,t) = 0\,, \quad \tilde u(x,0) = \tilde u_0(x)\,.$$

We assume that $\tilde a \in C^1[0,1]$, $0 < \tilde c_1 \le \tilde a(x) \le \tilde c_2$, $|\tilde a'(x)| \le \tilde c_3$, $\tilde B = B =$ identity operator in $L_2(0,1)$ and $\tilde A u = -\dfrac{\partial}{\partial x}\Big(\tilde a(x)\,\dfrac{\partial u}{\partial x}\Big)$.

The next inequalities are valid [1]:

$$\|\tilde A - A\|_{H_A \to H_{A^{-1}}} \le C\|\tilde a - a\|_{C[0,1]}\,, \tag{19}$$
$$\|\tilde A - A\|_{H_{A^2} \to H} \le C\|\tilde a - a\|_{C^1[0,1]}\,, \tag{20}$$
$$\|\tilde A - A\|_{H \to H_{A^{-2}}} \le C\|\tilde a - a\|_{C^1[0,1]}\,, \tag{21}$$

where C is a constant.

For example, we have:

$$\|\tilde A - A\|_{H_A \to H_{A^{-1}}} = \sup_{w \in H_A} \frac{\|(\tilde A - A)w\|_{A^{-1}}}{\|w\|_A} = \sup_{w \in H_A} \frac{|((\tilde A - A)w, w)|}{(Aw, w)} =$$

$$= \sup_{w \in \overset{\circ}{W}{}_2^1} \frac{\left|\int_0^1 \big(\tilde a(x) - a(x)\big)\,|w'(x)|^2\,dx\right|}{\int_0^1 a(x)\,|w'(x)|^2\,dx} \le C\|\tilde a - a\|_{C[0,1]}$$

By the same technique, using (18), we can prove (20) and (21).

On that way, estimates of strong stability (8) and (10), in this case have the following form:

$$\|\tilde{u}-u\|_{(0)} \le C_0^{(0)}\|\tilde{u}_0-u_0\|_{L_2(0,1)} + C_1^{(0)}\|\tilde{u}_1-u_1\|_{W_2^{-1}(0,1)}$$
$$+C_2^{(0)}\|\tilde{f}-f\|_{L_2((0,T);W_2^{-1}(0,1))} + C_3^{(0)}C\|\tilde{a}-a\|_{C[0,1]}\,,$$
$$\|\tilde{u}-u\|_{(1)} \le C_0^{(1)}\|\tilde{u}_0-u_0\|_{\overset{\circ}{W}{}_2^1(0,1)} + C_1^{(1)}\|\tilde{u}_1-u_1\|_{L_2(0,1)}$$
$$+C_2^{(1)}\|\tilde{f}-f\|_{L_2((0,T)\times(0,1))} + C_3^{(1)}C\|\tilde{a}-a\|_{C^1[0,1]}\,.$$

Acknowledgement

This research was supported by MSTD of Republic of Serbia under grant 1840.

References

1. D. Bojović, B. S. Jovanović, P. P. Matus, *Strong stability of the first-order differential-operator equation*, Differents. Uravn., 40, No 5 (2004), pp. 655-661 (in Russian).
2. R. A. Horn, C. R. Johnson, *Matrix analysis,* Cambridge Univ. Press, Cambridge etc., 1986.
3. B. S. Jovanović, P. P. Matus, *On the strong stability of operator-difference schemes in time-integral norms*, Comput. Methods Appl. Math. 1, No 1 (2001), pp. 1-14.
4. B. S. Jovanović, P. P. Matus, *Coefficient stability of the second-order differential-operator equation*, Differents. Uravn. **38**, No 10 (2002), pp. 1371-1377 (in Russian).
5. P. P. Matus, *The maximum principle and some of its applications,* Comput. Methods Appl. Math. **2**, No 1 (2002), pp. 50-91.
6. A. A. Samarskii, *Theory of Difference Schemes,* Nauka, Moscow, 1983 (in Russian).
7. J. Wloka, *Partial differential equations*, Cambridge University Press, Cambridge, 1987.

Semi-Lagrangian Semi-implicit Time Splitting Two Time Level Scheme for Hydrostatic Atmospheric Model

Andrei Bourchtein

Pelotas State University, Department of Mathematics,
Rua Anchieta 4715 bloco K, ap.304, Pelotas 96020-250, Brazil
burstein@terra.com.br

Abstract. A semi-Lagrangian semi-implicit two time level scheme is considered for hydrostatic atmospheric model. Introduced time splitting allows to treat principal physical components of atmospheric waves in semi-implicit manner with second order of accuracy, while insignificant physical modes are approximated by explicit formulas with the first order of accuracy and using a coarser spatial grid. This approach allows to reduce the computational cost with no loss of overall precision of the integrations. Numerical experiments with actual atmospheric fields showed that the developed scheme supplies rather accurate forecasts using time steps up to one hour and it is more efficient than three time level counterparts.

1 Introduction

Semi-Lagrangian semi-implicit (SLSI) approach is an efficient modern method of numerical solution of the hydrothermodynamic equations used in weather prediction and atmospheric modeling. Since the original demonstration of the extended stability and accuracy of the SLSI method in [2,13,16,17], this numerical technique is being used in an increasing range of atmospheric models [9,10,18]. In the last decade, motivated by results of McDonald [14] and Temperton and Staniforth [20] for shallow water equations, numerical modelers have substituted three time level SLSI schemes by two time level ones, which allow to use even larger time steps and achieve the same accuracy with almost doubling of efficiency [9,11,12,15,21].

Although SLSI schemes have been shown to be quite efficient, there are some computationally expensive parts of calculations inherent to any SLSI algorithm, which can be treated in a more optimal manner. Interpolation to the trajectory points, implicit discretization of slow gravitational modes and solution of coupled 3D elliptic problems for linear terms treated implicitly are among these less efficient segments of computation. This research is addressed to the last two issues. For explicit and simple approximation of slow gravitational waves, we split the full SLSI algorithm into two successive steps: in the first step, all terms are treated explicitly and with first overall order of accuracy and the second step

Z. Li et al. (Eds.): NAA 2004, LNCS 3401, pp. 195–202, 2005.

separate semi-Lagrangian method in two steps. In the first step we apply explicit semi-Lagrangian forward-backward discretization (SLFB) to the primitive equations (1):

$$\begin{aligned}
\frac{\hat{u}^{n+1,a}-u^{n,d}}{\tau} &= f_0\frac{\hat{v}^{n+1,a}+v^{n,d}}{2}-\frac{\hat{G}_x^{n+1,a}+G_x^{n,d}}{2}+N_u^{n+1/2},\\
\frac{\hat{v}^{n+1,a}-v^{n,d}}{\tau} &= -f_0\frac{\hat{u}^{n+1,a}+u^{n,d}}{2}-\frac{\hat{G}_y^{n+1,a}+G_y^{n,d}}{2}+N_v^{n+1/2},\\
\frac{\hat{P}^{n+1,a}-P^{n,d}}{\tau} &= D^{n,d}-\dot{\sigma}_\sigma^{n,d},\\
\frac{\hat{T}^{n+1,a}-T^{n,d}}{\tau} &= \frac{RT_0}{c_p}\left(\frac{\hat{P}^{n+1,a}-P^{n,d}}{\tau}+\frac{\dot{\sigma}^{n,d}}{\sigma}\right)+N_T^{n+1/2}.
\end{aligned} \tag{6}$$

This step is computationally much less expensive then a single step by (4) because the formulas (6) are actually explicit. However the scheme (6) has the first order of accuracy and is linearly stable if $\tau \le \sqrt{2}h_g/c_g$, where h_g is a meshsize of spatial grid used for gravitational terms and $c_g \approx 350m/s$ is a maximum velocity of gravitational waves in the system (1). On spatial grid C with principal meshsize $h = 75km$, the minimum gravitational meshsize is $h_g = h/2 = 37.5km$. Then maximum allowable time step is $\tau \approx 2.5\,\text{min}$, which is too small as compared with accuracy requirements.

To improve stability properties of (6) we should introduce more implicit approximation such as in (4). To do this, we consider the differences between SLSI and SLFB schemes:

$$\delta u = \tau/2\cdot(f_0\delta v-\delta G_x),\ \delta v = \tau/2\cdot(-f_0\delta u-\delta G_y), \tag{7}$$

$$\delta P = \tau/2\cdot(-\Delta D-\Delta\dot{\sigma}_\sigma),\ \sigma\delta T = RT_0/c_p\cdot(\sigma\delta P+\tau/2\cdot\Delta\dot{\sigma}), \tag{8}$$

where $\delta\varphi = \varphi^{n+1,a}-\hat{\varphi}^{n+1,a}$, $\Delta\varphi = \varphi^{n+1,a}-\varphi^{n,d}$. Differentiating the last T-equation with respect to σ and using P-equation and hydrostatic equation to eliminate temperature, surface pressure and vertical velocity, we obtain

$$(\sigma\delta C_{\ln\sigma})_\sigma = -\tau RT_0\cdot R/2c_p\cdot\Delta D. \tag{9}$$

Equations (7) and (9) form the closed system for three unknown functions.

3 Vertical and Horizontal Splitting

Discrete analogs of equations (7) and (9) on the vertical K level grid can be written as

$$\delta\mathbf{u} = \tau/2\cdot(f_0\delta\mathbf{v}-\delta\mathbf{G}_x),\ \delta\mathbf{v} = \tau/2\cdot(-f_0\delta\mathbf{u}-\delta\mathbf{G}_x),\ \delta\mathbf{G} = -\tau RT_0/2\cdot\mathbf{A}\cdot\mathbf{\Delta D}, \tag{10}$$

where $\mathbf{u},\mathbf{v},\mathbf{D},\mathbf{G}$ are the vector-columns of the order K and A is matrix $K\times K$ of vertical structure. It can be shown that, under natural restrictions on a choice of vertical approximation, the vertical structure matrix is oscillatory and, therefore, all its eigenvalues are positive and distinct [5].

By using a spectral decomposition $\mathbf{A} = \mathbf{S\Lambda S^{-1}}$, where $\mathbf{\Lambda} = \mathbf{diag}[\lambda_1, \ldots, \lambda_K]$ is eigenvalue matrix and $\mathbf{S}$ is the matrix of eigenvectors (i.e., vertical normal modes) of the $\mathbf{A}$, the system (10) can be rewritten as K decoupled 2D systems

$$u_k = \frac{\tau f_0}{2} v_k - \frac{\tau}{2} G_{k_x},\; v_k = -\frac{\tau f_0}{2} u_k - \frac{\tau}{2} G_{k_y},\; G_k + \frac{\tau c_k^2}{2} D_k = -\frac{\tau c_k^2}{2} D_k^*. \quad (11)$$

Here $k = 1, \ldots, K$ is the index of vertical mode, $c_k^2 = RT_0\lambda_k$ is the square of the gravity-wave phase speed of the k-th vertical mode and

$$\varphi = \mathbf{S}^{-1}(\varphi^{n+1,a} - \hat{\varphi}^{n+1,a}),\; \varphi^* = \mathbf{S}^{-1}(\hat{\varphi}^{n+1,a} - \varphi^{n,d}),\; \varphi = u, v, D, G, \quad (12)$$

that is, φ_k are the coefficients of expansion of physical corrections $\varphi^{n+1,a} - \hat{\varphi}^{n+1,a}$ by the vertical normal modes $\mathbf{s}_k$, which compile the matrix $\mathbf{S}$, and analogously for φ_k^*. Hereafter we suppose that eigenvalues λ_k are numbered in decreasing order.

The second step of two step algorithm consists of solution of some 2D systems (11) with the greatest values of c_k. Each of the systems (11) can be considered as time discretization of the linearized barotropic equations with corresponding equivalent depth $d_k = c_k^2/g$. Since smaller vertical modes have not any significant effect on the accuracy and stability of constructed scheme, we can solve only some first systems (11) in order to improve significally the accuracy and stability properties of the SLFB scheme. Indeed, the analysis of linear stability shows, that by solving the first I systems (11) the CFL criterion for SLFB scheme is substituted by more tolerant condition $\tau \leq \sqrt{2}h_g/c_{I+1}$.

For used vertical discretization on 15 level vertical grid, the fastest six gravity-wave speeds are $\{c_k\}_1^6 = \{342, 188, 101, 62, 42, 30\}$. Therefore, correcting only the first five modes, we increase a scheme stability from 2.5 min to about 30min. Moreover, applying Turkel-Zwas space splitting [22] with gravitational meshsize $h_g = 3h/2 = 112.5\,\text{km}$ for approximation of pressure gradient and divergence terms in the SLFB scheme, one can increase the maximum allowable time step up to 90 min. The last is less restrictive than (5) and large enough for the purposes of numerical weather prediction. Theoretically such scheme has the first order accuracy and use a coarse grid because of low accuracy and large meshsize used in SLFB scheme, but it does not practically result in a loss of accuracy because the slowest internal gravity waves contain only a small fraction of the total available energy.

By eliminating velocity components, solution of each barotropic system (11) can be reduced to 2D Helmholtz equation

$$G_k - \alpha_k \nabla^2 G_k = -\beta_k D_k^*,\; \alpha_k = \tau^2 c_k^2/4 + \tau^2 f_0^2,\; \beta_k = -\tau c_k^2/2. \quad (13)$$

Further horizontal splitting based on modification of (13) to the factored form was applied in [4]:

$$(1 - \alpha_k \partial_{xx})(1 - \alpha_k \partial_{yy}) G_k = -\beta_k D_k^* + \alpha_k^2 G_{k_{xxyy}}^*. \quad (14)$$

The last equation is reduced to a set of 1D elliptic problems solved very efficiently by Gelfand-Thomas algorithm. This is a slight modification for a small values of time step and experiments show that it works well for $\tau \leq 40$ min [4,19]. However, attempts to use this splitting in present model for larger time steps were not successful because of the fast growth of splitting truncation error when time step exceeds 30-40 min. It can be shown that this behavior is the consequence of introduction of additional truncation error in the hydrostatic equation related to the last splitting. Since the principal objective of using the two time level schemes is the increase of time step as compared with three time level ones, this kind of splitting was rejected in current model.

Unfortunately, the same kind of behavior of splitting error was observed applying different splitting techniques [3,6,8]. Therefore, we decide to apply multigrid method for solution of (13), which is well established economic technique for solution of elliptic problems [23]. We choose the BOXMG software [1] because it permits to use the rectangular grids with any point number. Through numerical experiments the optimal version of multigrid algorithm has been determined, which consist of standard V-cyclic method with using two cycles for the first vertical mode ($k = 1$) and one cycle for other corrected modes ($k = 2 : 5$). One four-color Gauss-Seidel point relaxation sweep is performed on any coarse grid both before dropping down to next coarser grid and before interpolation to previous finer grid. Two red-black relaxation sweeps are used on the primitive finest grid. The numerical experiments showed that this multigrid algorithm reduces the computational cost of the solution of the equation (13) about factor of 2 as compared with the traditional SOR method for horizontal grids of 100x100 points.

4 Numerical Experiments

For evaluating efficiency and precision of the constructed two time level scheme (denoted in this section by SLSI-V2TL) we compare its performance with two three time level SLSI schemes developed earlier and with usual leap-frog scheme. Both three time level SLSI schemes were presented in [4] and we keep denotations used there (SLSI-V and SLSI-VH). The first is a three time level version of the studied algorithm and the second uses additional horizontal splitting considered in section 3. The horizontal domain of 7000×7000 km^2 centered at Porto Alegre city ($30^0 S, 52^0 W$) was covered by uniform spatial grid C with meshsize $h = 75$ km. The initial and boundary value conditions were obtained from objective analysis and global forecasts of National Centers for Environmental Prediction (NCEP).

In Table 1 we present the results of comparison of 24-h geopotential height forecasts produced by three different semi-Lagrangian schemes with "exact" forecasts of leap-frog scheme. Each scheme was run with appropriate time step chosen on a base of stability, accuracy and computational cost considerations.

Table 2 shows the results of a series of ten 24-h forecasts of SLSI-V, SLSI-VH and SLSI-V2TL schemes on the same spatial grid and with time steps indicated in Table 1. Two mean objective scores of the geopotential forecasts were cal-

Table 1. Comparative characteristics of the different schemes. τ - time step in minutes used in indicated model; δ_{200}, δ_{500}, δ_{1000} - root-mean-square height differences in meters between the 24-h forecasts produced by chosen scheme and leap-frog scheme at the pressure surfaces 200hPa, 500hPa and 1000hPa, respectively; T_{CPU} - computational time cost of one forecast regarding leap-frog forecast time

scheme	τ	δ_{200}	δ_{500}	δ_{1000}	T_{CPU}
leap-frog	1	0	0	0	1
SLSI-V	60	5.5	5.4	5.1	0.109
SLSI-VH	40	5.6	5.3	5.2	0.112
SLSI-V2TL	60	5.1	4.9	4.7	0.111

Table 2. Mean objective scores of the 24-h geopotential forecasts. $\varepsilon_{200}, \varepsilon_{500}, \varepsilon_{1000}$ - root-mean-square height differences (in meters) between the analysis and the 24-h forecasting fields at the pressure surfaces 200hPa, 500hPa and 1000hPa, respectively; $r_{200}, r_{500}, r_{1000}$ - correlation coefficients (nondimensional) between the analysis and 24-h forecasting fields at the pressure surfaces 200hPa, 500hPa and 1000hPa, respectively

scheme	ε_{200}	r_{200}	ε_{500}	r_{500}	ε_{1000}	r_{1000}
leap-frog	30	0.88	22	0.90	23	0.86
SLSI-V	31	0.89	22	0.91	24	0.86
SLSI-VH	30	0.89	23	0.90	25	0.85
SLSI-V2TL	29	0.90	19	0.93	23	0.87

culated at different vertical levels: the root-mean-square differences in meters between 24-h forecasts and NCEP analysis and the correlation coefficient between observed and forecast changes. Also the leap-frog scheme with 1-min time step was run to compare the relative accuracy of the different schemes. Note that the use of reduced time step of 30 min for SLSI-V and SLSI-VH schemes increases their accuracy up to level of SLSI-V2TL scheme but it requires double computational time. On the other hand, increasing the time step for SLSI-VH scheme up to 60 min. results in drastically decreasing of precision, which is related to additional splitting errors considered in section 3. Obtained evaluations show efficiency and accuracy of the constructed two time level SLSI scheme as compared with three time level counterparts.

Acknowledgements

This research was supported by brazilian science foundation CNPq under grant 302738/2003-7.

References

1. Bandy, V., Sweet, R.: A set of three drivers for BOXMG: a black box multigrid solver. Comm. Appl. Num. Methods, 8 (1992) 563-571

2. Bates, J.R., McDonald A.: Multiply-upstream, semi-Lagrangian advective schemes: Analysis and application to a multilevel primitive equation model. Mon. Wea. Rev., 110 (1982) 1831-1842
3. Bates, J.R.: An efficient semi-Lagrangian and alternating direction implicit method for integrating the shallow water equations. Mon. Wea. Rev. 112 (1984) 2033-2047
4. Bourchtein A., Semi-Lagrangian semi-implicit fully splitted hydrostatic atmospheric model. Lecture Notes in Computer Science, 2657 (2003) 25-34
5. Bourchtein, A., Kadychnikov, V.: Well-posedness of the initial value problem for vertically discretized hydrostatic equations. SIAM J. Numer. Anal. 41 (2003) 195-207
6. Browning, G.L., Kreiss H.-O.: Splitting methods for problems with different timescales. Mon.Wea.Rev. 122 (1994) 2614-2622
7. Burridge, D.M.: A split semi-implicit reformulation of the Bushby-Timpson 10 level model. Quart. J. Roy. Met. Soc., 101 (1975) 777-792
8. Cohn, S.E., Dee, D., Isaacson, E., Marchesin, D., Zwas, G.: A fully implicit scheme for the barotropic primitive equations. Mon. Wea. Rev. 113 (1985) 436-448
9. Côté, J., Gravel, S., Methot, A., Patoine, A., Roch, M., Staniforth, A.: The operational CMC-MRB global environmental multiscale (GEM) model. Part I: Design considerations and formulation. Mon. Wea. Rev. 126 (1998) 1373-1395
10. Durran, D.: Numerical Methods for Wave Equations in Geophysical Fluid Dynamics. Springer, New York (1999)
11. Gospodinov, I.G., Spiridonov, V.G., Geleyn, J.-F.: Second-order accuracy of two-time-level semi-Lagrangian schemes. Quart. J. Roy. Met. Soc., 127 (2001), 1017-1033
12. Hortal, M.: The development and testing of a new two-time-level semi-Lagrangian scheme (SETTLS) in the ECMWF forecast model. Quart. J. Roy. Met. Soc. 128 (2002) 1671-1687
13. McDonald, A.: Accuracy of multiply upstream, semi-Lagrangian advective schemes. Mon. Wea. Rev., 112, (1984) 1267-1275
14. Mcdonald, A.: A semi-Lagrangian and semi-implicit two time level integration scheme. Mon. Wea. Rev. 114 (1986) 824-830
15. McDonald, A., Haugen, J.: A two-time-level, three-dimensional semi-Lagrangian, semi-implicit, limited-area gridpoint model of the primitive equations. Mon. Wea. Rev. 120 (1992) 2603-2621
16. Pudykiewicz, J., Benoit, R., Staniforth, A.: Preliminary results from a partial LRTAP model based on an existing meteorological forecast model. Atmos.-Ocean, 23 (1985) 267-303
17. Robert, A.: A stable numerical integration scheme for the primitive meteorological equations. Atmos.-Ocean, 19 (1981) 35-46
18. Staniforth, A., Côté, J.: Semi-Lagrangian integration schemes for atmospheric models - A review. Mon. Wea. Rev. 119 (1991) 2206-2223.
19. Tanguay, M., Robert, A.: Elimination of the Helmholtz equation associated with the semi-implicit scheme in a grid point model of the shallow water equations. Mon. Wea. Rev. 114 (1986) 2154-2162
20. Temperton, C., Staniforth, A.: An efficient two-time-level semi-Lagrangian semi-implicit integration scheme. Quart. J. Roy. Met. Soc. 113 (1987) 1025-1039
21. Temperton, C., Hortal, M., Simmons, A.J.: A two-time-level semi-Lagrangian global spectral model. Quart. J. Roy. Met. Soc. 127 (2001) 111-126
22. Turkel, E., Zwas, G.: Explicit large-time-step schemes for the shallow-water equations. In: Vichnevetsky R, Stepleman R.S. (eds.): Advances in Computer Methods for Partial Differential Equations, Lehigh University (1979) 65-69
23. Wesseling, P.: An Introduction to Multigrid Methods. Wiley, Chichester (1992)

The Strengthened Cauchy-Bunyakowski-Schwarz Inequality for n-Simplicial Linear Finite Elements

Jan Brandts[1], Sergey Korotov[2], and Michal Křížek[3]

[1] Korteweg-de Vries Institute for Mathematics,
University of Amsterdam, Netherlands
brandts@science.uva.nl
http://staff.science.uva.nl/~brandts
[2] Dept. of Mathematical Information Technology,
University of Jyväskylä, Finland
korotov@mit.jyu.fi
[3] Mathematical Institute, Academy of Sciences, Prague, Czech Republic
krizek@math.cas.cz
http://www.math.cas.cz/~krizek

Abstract. It is known that in one, two, and three spatial dimensions, the optimal constant in the strengthened Cauchy-Bunyakowski-Schwarz (CBS) inequality for the Laplacian for red-refined linear finite element spaces, takes values zero, $\frac{1}{2}\sqrt{2}$ and $\frac{1}{2}\sqrt{3}$, respectively. In this paper we will conjecture an explicit relation between these numbers and the spatial dimension, which will also be valid for dimensions four and up. For each individual value of n, it is easy to verify the conjecture. Apart from giving additional insight into the matter, the result may find applications in four dimensional finite element codes in the context of computational relativity and financial mathematics.

1 Introduction

The topic of this paper can be classified as being on the borderline between computational geometry, finite element theory, and linear algebra. Finite element theory because the result provides bounds for multi-level preconditioning uniform in the size of the system; computational geometry because it concerns uniform partitioning of simplices in spatial dimensions higher than the usual three, and linear algebra because it is the language in which the results and their proofs are formulated. In this section we briefly recall some basics of those three mathematical fields.

1.1 Multi-level Preconditioning

To discuss multi-level preconditioning, it suffices to outline two-level preconditioning and to give the recursive structure that leads to the multi-level variant.

Z. Li et al. (Eds.): NAA 2004, LNCS 3401, pp. 203–210, 2005.

Therefore, we consider here a positive definite and symmetric matrix A that has been block partitioned, and a block diagonal preconditioner K of the form

$$A = \begin{pmatrix} A_{11} & A_{12} \\ A_{21} & A_{22} \end{pmatrix}, \quad \text{and} \quad K = \begin{pmatrix} A_{11} & 0 \\ 0 & A_{22} \end{pmatrix}. \tag{1}$$

It is widely known that if there exists a non-negative number $\gamma < 1$ such that for all vectors v, z of the appropriate dimensions

$$v^* A_{12} z \leq \gamma \sqrt{v^* A_{11} v} \sqrt{z^* A_{22} z}, \tag{2}$$

then the condition number $\kappa(K^{-1}A)$ of the block-diagonally preconditioned matrix satisfies the bound

$$\kappa(K^{-1}A) \leq \frac{1-\gamma}{1+\gamma}. \tag{3}$$

This implies that the block-diagonally preconditioned Richardson iteration (also called block-Jacobi iteration) to approximate the solution of a linear system with system matrix A converges to the exact solution with the right-hand side of (3) as error reduction factor.

Naturally, in each step, two linear systems of the form $A_{11}x = b$ and $A_{22}y = c$ need to be solved. In finite element applications, it is usually the case that one of the two, say A_{22} is well-conditioned without further preconditioning, for instance if the space corresponding to A_{22} is the space of additional details to be added to obtain a red-refined space or to obtain a space of higher polynomial order. The other system, however, typically suffers from the same ill-conditioning as the original matrix A. By recursion, one then proceeds to block-partition A_{11} into a well-conditioned part and an ill-conditioned part. This can result in an optimal complexity solver for the finite element problem at hand. A thorough understanding of the finer subtleties in this process is not needed in the remainder of this paper; we refer to the SIAM Review paper [6] for details and references.

1.2 The Strengthened Cauchy-Bunyakowski-Schwarz Inequality

It is, on the other hand, useful to explain the relation between (2) and the topic of this paper, called the strengthened CBS inequality. For this, consider the usual weak formulation of the Poisson equation with homogeneous Dirichlet boundary conditions, which aims to find $u \in H_0^1(\Omega)$ such that

$$a(u, v) = f(v), \quad \text{for all } v \in H_0^1(\Omega). \tag{4}$$

For ease of explanation, let us assume that the domain Ω is planar and polygonal, and that a triangulation $\mathcal{T}_1$ of Ω is given, as well as the corresponding space V of continuous piecewise linear functions in $H_0^1(\Omega)$ relative to $\mathcal{T}_1$. Moreover, let $\mathcal{T}_2$ be the triangulation of Ω that arises from $\mathcal{T}_1$ by red-refinement: this means that each triangle from $\mathcal{T}_1$ is subdivided into four by connecting each edge midpoint to the other two by a straight line segment. Write W for the space of continuous piecewise linear functions in $H_0^1(\Omega)$ relative to $\mathcal{T}_2$. Clearly $V \subset W$, and we write Z for the complement of V in W,

$$W = V \oplus Z. \tag{5}$$

Discretization of (4) in W results in a stiffness matrix A, coming from the discrete equations

$$a(u_W, w) = f(w), \quad \text{for all } w \in W, \tag{6}$$

together with a choice of a basis for W. If we choose as a basis the nodal basis for the subspace V together with the nodal basis functions in W that correspond to functions in Z, we get a block-partitioning of A as in (1) satisfying (2). In terms of the bilinear form $a(\cdot,\cdot)$ it can be seen that (2) is equivalent to

$$|a(v,z)| \leq \gamma\sqrt{a(v,v)}\sqrt{a(z,z)}, \quad \text{for all } v \in V, z \in Z. \tag{7}$$

This inequality can righteously be called a strengthened CBS inequality; the smallest number γ for which (7) is satisfied, it the angle between the subspaces V and Z in the geometry induced by the energy inner product. For planar domains, it is known that $\gamma = \frac{1}{2}\sqrt{2}$, see [1]. In one space dimension, it is trivial to check that Z is $H_0^1(\Omega)$-orthogonal to V and hence $\gamma = 0$. In three space dimensions, $\gamma = \frac{1}{2}\sqrt{3}$; see for instance [3]. This paper [3] also contains results for elasticity problems and for a more general type of refinement.

1.3 Red-Refinement in Higher Dimensions

Red-refinement in two space dimensions can be defined relatively easily, as we did above. In three dimensions and higher it becomes more troublesome. We will follow the lines from [3] and use the so-called Kuhn's partition [9] of the cube. This elegant partition, and its higher dimensional analogues, is actually due to Freudenthal [7], and can be described as follows.

Proposition 1 (Freudenthal [7]). *Let $K = [0,1]^n$ be the unit (hyper-)cube. Then K can be subdivided into $n!$ simplices S of dimension n. Each of the simplices is the convex hull of a path of n edges of K connecting the origin with the point $(1,\ldots,1)$. Alternatively, these simplices can be characterized as the sets*

$$S_\sigma = \{x \in \mathbb{R}^n \mid 0 \leq x_{\sigma(1)} \leq \ldots \leq x_{\sigma(n)} \leq 1\}, \tag{8}$$

where σ ranges over all $n!$ permutations of the numbers 1 to n.

Definition 1 (Canonical Path Simplex). A simplex with the property of having n mutually orthogonal edges that form a path is called a path simplex [2, 8] or an orthoscheme [5, 10]. The simplex

$$\hat{S} = \{x \in \mathbb{R}^n \mid 0 \leq x_1 \leq \ldots \leq x_n \leq 1\} \tag{9}$$

we will call the *canonical path simplex*.

The unit cube K can be trivially subdivided into 2^n identical subcubes. Each of the subcubes can be partitioned into $n!$ simplices using the above idea in its scaled form, resulting in a total of $n!2^n$ simplices. It can be verified that this partition also constitutes a partition of each of the $n!$ simplices S_σ from (8) in which K could have been subdivided directly; hence we have a way of subdividing the canonical path simplex $\hat{S}$ from (9) into 2^n smaller ones.

Definition 2 (Red-Refinement of an n-Simplex). Given an arbitrary n-simplex S, it can be mapped onto $\hat{S}$ by an affine transformation F. Its image $\hat{S} = F(S)$ can be subdivided into 2^n subsimplices following the above. The pre-images under F of these subsimplices form a partition of S that we will call the red-refinement of S.

1.4 Local Analysis

Our final goal is to compute the strengthened CBS constant for red-refined simplicial partitions in arbitrary space dimensions for the Poisson equation. An important observation [3, 6] in this context is that a local analysis of the CBS constant on the canonical path simplex $\hat{S}$ is sufficient. This is due to the fact that we consider the simple case of the Laplacian. Given $\hat{S}$ and its $n+1$ nodal basis functions ϕ_j, red-refinement introduces $\frac{1}{2}n(n+1)$ additional basis functions, which are the nodal basis functions associated with the midpoints of each edge. This describes two spaces V and Z on $\hat{S}$ of dimensions $n+1$ and $\frac{1}{2}n(n+1)$ respectively. Suppose that

$$|a(v,z)_{\hat{S}}| \leq \hat{\gamma}\sqrt{a(v,v)_{\hat{S}}}\sqrt{a(z,z)_{\hat{S}}}, \quad \text{for all } v \in V \quad \text{and} \quad z \in Z. \tag{10}$$

Then it is well known that the same bound with the same constant $\hat{\gamma}$ will be found on each affine transformation of $\hat{S}$. Also, combination of all local information results in a global result with the same constant. This explains why we will only compute $\hat{\gamma}$ on $\hat{S}$.

2 A Computational Approach

We will now present an approach to find the strengthened CBS constant in higher dimensions. Even though it is a computational approach, it can a posteriori be verified that the computed values are correct.

2.1 The Local Stiffness Matrix for Linear n-Simplicial Elements

In the following, we will use the stiffness matrix $\hat{A}$ for the red-refined canonical path simplex. This patch of 2^n subsimplices counts $\frac{1}{2}(n+1)(n+2)$ degrees of freedom, of which $n+1$ belong to the coarse grid, and $\frac{1}{2}n(n+1)$ to the midpoints of the edges. To compute $\hat{A}$, we need the $(n+1)\times(n+1)$ local stiffness matrix $\hat{B}$ for $\hat{S}$ itself. For this we can use the following lemma.

Lemma 1. *Let P be a nonsingular $n \times n$ matrix, then the convex hull of the origin p_0 and the columns $p_1, \ldots, p_n$ of P is an n-simplex S. Let $\phi_j : S \mapsto \mathbb{R}$ be the nodal linear basis function corresponding to p_j. Then*

$$\begin{pmatrix} (\nabla\phi_1, \nabla\phi_1) \ldots (\nabla\phi_n, \nabla\phi_1) \\ \vdots \qquad\qquad\qquad \vdots \\ (\nabla\phi_1, \nabla\phi_n) \ldots (\nabla\phi_n, \nabla\phi_n) \end{pmatrix} = \mathrm{Vol}(S) Q^* Q, \quad \text{where } Q^* = P^{-1}. \tag{11}$$

Proof. Denote the columns of Q by $q_1, \ldots, q_n$. Since $Q^* P = I$, we see that for $i = 1, \ldots, n$ we have that $q_i^* p_j = \delta_{ij}$ hence $\phi_i(x) \mapsto q_i^* x$, and $\nabla\phi_i = q_i$. Therefore

$$(\nabla\phi_i, \nabla\phi_j) = \int_S q_i^* q_j dx. \tag{12}$$

Since q_i is constant, the statement follows. □

Notice that to get the local stiffness matrix for S, we need to include the inner products of ϕ_0 with all $\phi_j, j = 0, \ldots, n$. For the canonical path simplex $\hat{S}$, this is can be done as follows.

The matrix P corresponding to $\hat{S}$ is the upper triangular matrix for which each upper triangular entry equals one. The transpose Q of the inverse of this particular P can also be written down easily,

$$P = \begin{pmatrix} 1 & \cdots & 1 \\ & \ddots & \vdots \\ & & 1 \end{pmatrix} \quad \text{and} \quad Q = \begin{pmatrix} 1 & & & \\ -1 & 1 & & \\ & \ddots & \ddots & \\ & & -1 & 1 \end{pmatrix}. \tag{13}$$

The gradient q_0 of the nodal basis function ϕ_0 that corresponds to the origin equals minus the first canonical basis vector. Adding this vector the left of Q we find the local stiffness matrix $\hat{B}$ of $\hat{S}$,

$$\hat{B} = \frac{1}{n!}(q_0|Q)^*(q_0|Q) = \frac{1}{n!} \begin{pmatrix} 1 & -1 & & & \\ -1 & 2 & -1 & & \\ & \ddots & \ddots & \ddots & \\ & & -1 & 2 & -1 \\ & & & -1 & 1 \end{pmatrix}. \tag{14}$$

2.2 The Stiffness Matrix for the Red-Refined $\hat{S}$

Using the local stiffness matrix $\hat{B}$, we can now build the stiffness matrix $\hat{A}$ corresponding to all $\frac{1}{2}(n+1)(n+2)$ nodal basis functions ψ_i that are associated with the red-refined canonical path simplex. This can be done in the standard way, by making the usual list of simplices and the numbers of their nodes and doing the global assembly. Without going into too much detail, we mention the following tricks:

- By multiplying everything by $n!$ we can avoid non-integer arithmetic in the process of assembly.
- A suitable numbering of the nodes was used. For example, in 3d, the node with coordinates (x, y, z) got numbered $(2z)(2y)(2x)$, where the latter expression should be interpreted as a base three number, by which we mean that the node with coordinates $(1, 1/2, 1/2)$ got number $1{\cdot}3^2{+}1{\cdot}3^1{+}2{\cdot}3^0 = 14$ assigned to it.

Hence, we are able to compute the stiffness matrix corresponding to the $\frac{1}{2}(n+1)(n+2)$ nodal basis functions ψ_i that correspond to the red-refined canonical simplex $\hat{S}$ exactly, without numerical errors.

2.3 Basis Transformation

The stiffness matrix $\hat{A}$ belonging to the red-refined $\hat{S}$ can be used to get the explicit form of the block partitioned matrix A. For this, we need a basis transformation; after all, apart from the span of the nodal basis functions belonging to edge midpoints, we need the *coarse grid* nodal basis functions corresponding to the $n+1$ vertices of $\hat{S}$ whereas in the computation of $\hat{A}$, the ones belonging to the *red-refined grid* were used.

Remark 1. The gradients of the $n{+}1$ nodal basis functions ϕ_j are linearly dependent. Therefore, we may without loss of generality remove one of them without changing their span. We will remove ϕ_0.

The coarse grid nodal basis function ϕ_i belonging to vertex p_i of $\hat{S}$ can simply be constructed as the sum of the fine grid nodal basis function belonging to p_i, plus the n fine grid nodal basis functions ψ_j belonging to the midpoints of the edges that meet at p_i multiplied by a half. Again, to avoid fractions in the computations, we may multiply everything by two.

All this results in an explicit and exact matrix A of the block form (1) in which

$$A_{11} = \begin{pmatrix} (\nabla\phi_1, \nabla\phi_1) \dots (\nabla\phi_n, \nabla\phi_1) \\ \vdots \qquad\qquad \vdots \\ (\nabla\phi_1, \nabla\phi_n) \dots (\nabla\phi_n, \nabla\phi_n) \end{pmatrix}, \quad A_{22} = \begin{pmatrix} (\nabla\psi_1, \nabla\psi_1) \dots (\nabla\psi_m, \nabla\psi_1) \\ \vdots \qquad\qquad \vdots \\ (\nabla\psi_1, \nabla\psi_m) \dots (\nabla\psi_m, \nabla\psi_m) \end{pmatrix}$$

where $\psi_1, \dots, \psi_m$ with $m = \frac{1}{2}n(n+1)$ are the fine grid nodal basis functions corresponding to the edge midpoints, and

$$A_{12} = \begin{pmatrix} (\nabla\psi_1, \nabla\phi_1) \dots (\nabla\psi_m, \nabla\phi_1) \\ \vdots \qquad\qquad \vdots \\ (\nabla\psi_1, \nabla\phi_n) \dots (\nabla\psi_m, \nabla\phi_n) \end{pmatrix}.$$

Once more, by suitable scaling of the basis functions ϕ_j and ψ_j, non-integer entries were avoided. Notice that both A_{11} and A_{22} are non-singular.

2.4 Computation of $\hat{\gamma}$

Since both A_{11} and A_{22} are non-singular, we have that $\hat{\gamma}$ equals the following supremum,

$$\hat{\gamma} = \sup_{0 \neq v \in \mathbb{R}^n, 0 \neq z \in \mathbb{R}^m} \frac{v^* A_{12} z}{\sqrt{v^* A_{11} v} \sqrt{z^* A_{22} z}}. \tag{15}$$

Now, let the Cholesky decompositions of $A_{11} = U^*U$ and $A_{22} = R^*R$ be computed, then the transformations $v = Ux$ and $z = Ry$ show that

$$\hat{\gamma} = \sup_{\|x\|=1, \|y\|=1} x^* C y, \quad \text{where } C = U^{-*} A_{12} R^{-1}. \tag{16}$$

Elementary linear algebra tells us that his supremum is equal to the largest singular value $\sigma_+(C)$ of C. The square of this singular value is the largest eigenvalue $\lambda_+(CC^*)$ of the $n \times n$ matrix CC^*. Substituting back $C = U^{-*} A_{12} R^{-1}$ gives that we aim for the largest generalized eigenvalue λ of

$$A_{12} A_{22}^{-1} A_{12}^* z = \lambda A_{11} z. \tag{17}$$

We have used MatLab to compute $\lambda_+(C^*C)$ for increasing values of n. The results are tabulated below in 16 decimal places. In the right column, we wrote down the fractions to which the numbers are equal up to machine precision.

n	$\lambda_+(CC^*)$	conjecture
2	0.500000000000000	1/2
3	0.750000000000000	3/4
4	0.875000000000000	7/8
5	0.937500000000000	15/16
6	0.968750000000000	31/32
7	0.984375000000000	63/64

(18)

The numbers in the tabular above are quite convincing, and led us to formulate the following conjecture.

Conjecture 2.3 *The strengthened CBS constant $\hat{\gamma}_n$ for the Laplacian for linear finite elements on red-refined n-simplicial partitions equals, for all $n \geq 1$,*

$$\hat{\gamma}_n = \sqrt{1 - \left(\frac{1}{2}\right)^{n-1}}. \tag{19}$$

Verification. Even though we are not able to prove this formula for all n simultaneously, we are able verify the result for any given value of n *a posteriori*. Since the matrices A_{11}, A_{12} and A_{22} have been computed without numerical error, it is possible to write down the characteristic polynomial

$$p(\lambda) = \det(A_{12} A_{22}^{-1} A_{12}^* - \lambda A_{11}). \tag{20}$$

Once this has been done, it will not be difficult to prove whether $\hat{\gamma}_n$ is its largest root, or whether

$$\hat{\gamma}_n A_{11} - A_{12} A_{22}^{-1} A_{12}^* \tag{21}$$

is singular and positive semi-definite. Even though this may be done relatively straightforwardly in Maple or Mathematica, we did not feel the urge to do so.

3 Conclusions and Remarks

We have computed numerical values for the strengthened Cauchy-Bunyakowski-Schwarz inequality in the context of n-simplicial linear finite elements with red-refinement for the Laplacian. These values led to a conjecture that can easily be verified *a posteriori* for a given value of n, even though we did not succeed to find a way to give a general proof.

The practical scope of the result seems limited; however, in combination with the recently proved superconvergence results in [4] for precisely the red-refinements of a uniform initial n-simplicial partition, it may be of use in developing solvers for elliptic equations in four space dimensions on domains that cannot be easily discretized using (hyper-)block elements.

References

1. O. Axelsson: On multigrid methods of the two-level type. In: Multigrid Methods, Hackbusch, W, Trotenberg U (eds). Lecture Notes in Mathematics, Springer: Berlin, vol **960** (1982) 352–367
2. M. Bern, P. Chew, D. Eppstein, and J. Ruppert: Dihedral Bounds for Mesh Generation in High Dimensions. Proc. 6th Symp. Discrete Algorithms, ACM and SIAM (1995) 189–196
3. R. Blaheta: Nested tetrahedral grids and strengthened CBS inequality. Numer. Linear Algebra Appl. **10** (2003) 619–637
4. J.H. Brandts en M. Křížek : Gradient superconvergence on uniform simplicial partitions of polytopes. IMA J. Num. Anal. **23** (2003) 489-505
5. H.S.M Coxeter: Trisecting an orthoscheme. Computers Math. Applic. **17** (1989) 59–71
6. V.P. Eijkhout and P.S. Vassilevski: The role of the strengthened Cauchy-Buniakowskii-Schwarz inequality in Multilevel Methods. SIAM Review, **33** (1991) 405–419
7. H. Freudenthal: Simplizialzerlegungen von beschraenkter Flachheit. Annals of Math. in Science and Engin. **43** (1942) 580–582
8. S. Korotov en M. Křížek : Local nonobtuse tetrahedral refinements of a cube. Appl. Math. Lett. **16** (2003) 1101-1104
9. H.W Kuhn: Some combinatorial lemmas in topology. IBM J. Res. Develop. **45** (1960) 518–524
10. L. Schläfli. Theorie der vielfachen Kontinuität aus dem Jahre 1852. In: Gesammelte mathematische Abhandlungen, Birkhäuser, Basel, 1952

Uniformly Convergent Difference Scheme for a Singularly Perturbed Problem of Mixed Parabolic-Elliptic Type

Iliya A. Brayanov

Department of Applied Mathematics and Informatics,
University of Rousse, Studentska str.8,
7017 Rousse, Bulgaria,
brayanov@ru.acad.bg

Abstract. One dimensional singularly perturbed problem with discontinuous coefficients is considered. The domain is partitioned into two subdomains and in one of them the we have parabolic reaction-diffusion problem and in the other one elliptic convection-diffusion-reaction equation.The problem is discretized using an inverse-monotone finite volume method on Shishkin meshes. We established almost second-order in space variable global pointwise convergence that is uniform with respect to the perturbation parameter. Numerical experiments support the theoretical results.

1 Introduction

Consider the following problem

$$Lu(x,t) \equiv \frac{\partial u}{\partial t} - \varepsilon \frac{\partial^2 u}{\partial x^2} + q(x,t)u = f(x,t),\ (x,t) \in \Omega^-, \tag{1}$$

$$Lu(x,t) \equiv -\varepsilon \frac{\partial^2 u}{\partial x^2} - r(x,t)\frac{\partial u}{\partial x} + q(x,t)u = f(x,t),\ (x,t) \in \Omega^+, \tag{2}$$

$$[u(x,t)]_\Gamma = u(\xi+0,t) - u(\xi-0,t) = 0, \quad \left[\frac{\partial u(x,t)}{\partial x}\right]_\Gamma = 0, \tag{3}$$

$$u(0,t) = \psi_0(t), \quad u(1,t) = \psi_1(t), \quad u(x,0) = \varphi(x),\ x \in (0,\xi], \tag{4}$$

where $0 < \varepsilon << 1$, $\Omega^- = \{0 < x < \xi, 0 < t < T\}$, $\Omega^+ = \{\xi < x < 1, 0 < t < T\}$, $\Gamma = \{x = \xi, 0 < t < T\}$, and

$$0 < r_0 \le r(x,t) \le r_1, \quad 0 < q_0 \le q(x,t) \le q_1.$$

The functions r, q and f are sufficiently smooth with possible discontinuity on the interface Γ.

Such kind of problems describe for example an electromagnetic field arising in motion of train on air-pillow, see [4]. The asymptotic expansion of this problem is constructed in [8]. Our goal in this paper is to derive a finite difference scheme that is uniformly convergent with respect to the small parameter ε.

Z. Li et al. (Eds.): NAA 2004, LNCS 3401, pp. 211–218, 2005.

2 Decomposition into Regular and Singular Part

For the numerical analysis below we shall need a decomposition of the solution into regular and singular part.

Proposition 1. *The solution u to problem (1)-(4) admits the representation*

$$u(x,t) = v(x,t) + w(x,t),$$

where the regular part $v(x,t)$ satisfies for all k, m, $k+2m = 0,\ldots,4$ the estimates

$$\left\| \frac{\partial^{k+m} v}{\partial x^k \partial t^m} \right\|_{L_\infty(\bar{\Omega})} \le C,$$

and the singular part $w(x,t)$ satisfies for all k, m, $k+2m = 0,\ldots,4$ the estimates

$$\left\| \frac{\partial^{k+m} w}{\partial x^k \partial t^m} \right\|_{L_\infty(\bar{\Omega}^-)} \le C\varepsilon^{-k/2} \left(\exp\left(-\frac{\sqrt{q_0} x}{\sqrt{\varepsilon}} \right) + \exp\left(-\frac{\sqrt{q_0}(\xi - x)}{\sqrt{\varepsilon}} \right) \right),$$

$$\left\| \frac{\partial^{k+m} w}{\partial x^k \partial t^m} \right\|_{L_\infty(\bar{\Omega}^+)} \le C\varepsilon^{1/2-k} \exp\left(-\frac{r_0(\xi - x)}{\varepsilon} \right),$$

for some positive constant C independent of the small parameter ε.

3 Numerical Approximation

3.1 Grid and Grid Functions

It is well known that the singularly perturbed problems are not ε-uniformly convergent on uniform meshes, see [5, 6] for survey. To obtain ε-uniform convergent difference scheme we shall construct a mesh $\omega = \omega_\tau \times \omega_h$, that is uniform with respect to the time variable and partially uniform (Shishkin mesh), condensed near to boundary and around interior layers, with respect to the space variable.

$$w_\tau = \{t_j, t_j = t_{j-1} + \tau, j = 0,\ldots,J, t_0 = 0, t_J = T, \tau = T/J\},$$
$$\bar{\omega}_h = \{x_i, x_i = x_{i-1} + h_i, i = 1,2,\ldots,m+n = N, x_0 = 0, x_m = \xi, x_N = 1\},$$

where

$$h_i = \begin{cases} h^1 = 4\delta_1/m, & i = 1,\ldots,m/4 \cup i = 3m/4+1,\ldots,m, \\ h^2 = 2(\xi - 2\delta_1)/m, & i = m/4+1,\ldots,3m/4 \\ h^3 = 2\delta_2/n, & i = m+1,\ldots,m+n/2, \\ h^4 = 2(1-\xi-\delta_2)/n, & i = m+n/2+1,\ldots,N \end{cases}$$

$$\delta_1 = \min\{\sigma_1\sqrt{\varepsilon}\ln m/\sqrt{q_0}, \xi/4\}, \quad \delta_2 = \min\{\sigma_2\varepsilon\ln n/r_0, (1-\xi)/2\}.$$

Let $v(x_i, t_j)$ be a mesh function of the discrete argument $(x_i, t_j) \in \bar{\omega}$. Let in addition g be a partially continuous function with possible discontinuity at the mesh points (x_m, t_j). Denote by $g_{i,j} = g(x_i, t_j)$ and $g^{\pm}_{m,j} = g(x_m \pm 0, t_j)$. We shall use further the following notations

$$v_{\bar{t},i,j} = \frac{v_{i,j} - v_{i,j-1}}{\tau},\ v_{t,i,j} = v_{\bar{t},i,j+1},\ \hat{h}_i = \frac{h_i + h_{i+1}}{2},$$

$$\bar{g}_{m,j} = \frac{h_{m+1} g^+_{m,j} + h_m g^-_{m,j}}{2\hat{h}_m},\ v_{\bar{x},i,j} = \frac{v_{i,j} - v_{i-1,j}}{h_i},\ v_{\breve{x},i,j} = \frac{v_{i,j} - v_{i-1,j}}{\hat{h}_i}$$

$$v_{x,i,j} = v_{\bar{x},i+1},\ v_{\hat{x},i,j} = \frac{v_{i+1,j} - v_{i,j}}{\hat{h}_i},\ v_{\bar{x}\hat{x},i,j} = \frac{v_{\bar{x},i+1,j} - v_{\bar{x},i,j}}{\hat{h}_i}.$$

Further we shall use the discrete maximum norm

$$\|v\|_{\infty,\bar{\omega}} = \max_{(x_i,t_j)\in\bar{\omega}} |v_{i,j}|.$$

3.2 Finite Difference Scheme

On the left we shall use the standard approximation, derived from the balance equation, see [7] and [2] for discontinuous parabolic problem

$$L^h U_{i,j} = U_{\bar{t},i,j} - \varepsilon U_{\bar{x}\hat{x},i,j} + q_{i,j} U_{i,j} = f_{i,j},\quad i = 1 : m-1,\ j = 1 : J, \tag{5}$$

On the right we use the so called modified Samarskii scheme, see [1, 3]

$$L^h U_{i,j} = -\varepsilon(\kappa^h U_{\bar{x}})_{\hat{x},i,j} - r^h_{i+1,j} U_{\hat{x},i,j} + q^h_{i,j} U_{i,j} = f^h_{i,j},\quad i = m+1 : N-1,\ j = 0 : J, \tag{6}$$

where

$$r^h_{i,j} = r(x_i - h_i/2, t_j),\ R^h_{i,j} = \frac{h_i r^h_{i,j}}{2\varepsilon},\ \kappa^h_{i,j} = (1 + R^h_{i,j})^{-1},$$

$$q^h_{i,j} = q_{i,j} + \frac{1}{2}\left(\frac{hq}{1 + (R^h)^{-1}}\right)_{\hat{x},i,j} + \frac{h^2_{i+1}/\hat{h}_i}{2(1 + (R^h_{i+1,j})^{-1})}\frac{q_{i+1,j} q_{i,j}}{r_{i+1,j}},$$

$$f^h_{i,j} = f_{i,j} + \frac{1}{2}\left(\frac{hf}{1 + (R^h)^{-1}}\right)_{\hat{x},i,j} + \frac{h^2_{i+1}/\hat{h}_i}{2(1 + (R^h_{i+1,j})^{-1})}\frac{q_{i+1,j} f_{i,j}}{r_{i+1,j}}.$$

On the interface we obtain the following difference scheme

$$L^h U_{m,j} = \frac{h_m}{2\hat{h}_m} U_{\bar{t},m,j} - \frac{h^2_m}{6\hat{h}_m} U_{\bar{t}\bar{x},m,j} + \left(\frac{\varepsilon}{\hat{h}_m} - \rho^h_{m,j}\right) U_{\bar{x},m,j}$$
$$- \left(\frac{r^h_{m+1,j} h_{m+1}}{\hat{h}_m} + \frac{\varepsilon\kappa^h_{m+1,j}}{\hat{h}_m}\right) U_{x,m,j} + q^h_{m,j} U_{m,j} = f^h_{m,j}, \tag{7}$$

where

$$\rho^h_{m,j} = \frac{q_{m-0,j} h^2_m}{6\hat{h}_m},\ q^h_{m,j} = \bar{q}_{m,j} + \frac{h_{m+1} q_{m+0,j}}{2\hat{h}_m(1 + (R^h_{m+1,j})^{-1})} - \frac{h^2_m}{6\hat{h}_m} q_{\bar{x},m,j}$$

$$f^h_{m,j} = \bar{f}_{m,j} + \frac{h_{m+1} f_{m+0,j}}{2\hat{h}_m(1 + (R^h_{m+1,j})^{-1})} - \frac{h^2_m}{6\hat{h}_m} f_{\bar{x},m,j}.$$

Since $h_m = O(\sqrt{\varepsilon} m^{-1} \ln m)$ then for sufficiently large m holds

$$\frac{\varepsilon}{\hat{h}_m} - \rho^h_{m,j} = \frac{1}{6\hat{h}_m}\left(6\varepsilon - q_{m-0,j} h_m^2\right) \geq \frac{c_0 \varepsilon}{6\hat{h}_m},$$

where c_0 is independent of m and ε positive constant.

Setting the boundary conditions

$$U_{0,j} = \psi_{0,j}, \quad U_{N,j} = \psi_{1,j}, \quad U_{i,0} = \varphi_i, \; i = 0, \dots, m, \; j = 0, \dots, J, \tag{8}$$

we obtain the discrete problem (P^h): (5)-(8).

Let

$$\tau \geq \frac{h_m^2}{2\varepsilon - h_m^2 q_{m-0,j}} = O(m^{-2} \ln^2 m), \tag{9}$$

then the discrete problem (P^h) satisfies the discrete maximum principle

Lemma 1. *If $U_{0,j} \geq 0, U_{N,j} \geq 0, U_{i,0} \geq 0$, and $L^h U_{i,j} \geq 0$ then $U \geq 0$ on $\bar{\omega}$.*

3.3 A Priori Estimates

Let V^- be a solution to the discrete problem $(P^{h,-})$:

$$L^{h,-} V^-_{i,j} = V^-_{\bar{t},i,j} - \varepsilon V^-_{\bar{x}\hat{x},i,j} + q_{i,j} V^-_{i,j} = f_{i,j}, \quad i = 1 : m-1, \; j = 1 : J,$$

$$L^{h,-} V^-_{m,j} = V^-_{\bar{t},m,j} - \frac{h_m}{3} V^-_{\bar{t}\bar{x},m,j} + \left(\frac{2\varepsilon}{h_m} - \frac{h_m q_{m-0,j}}{3}\right) V^-_{\bar{x},m,j} + q^{h,-}_{m,j} V^-_{m,j} = f^{h,-}_{m,j},$$

$$V^-_{0,j} = 0, \quad V^-_{i,0} = \varphi_i, \quad i = 0 : m, \; j = 0 : J,$$

where

$$q^{h,-}_{m,j} = q_{m-0,j} - \frac{h_m}{3} q_{\bar{x},m,j} \; f^{h,-}_{m,j} = f_{m-0,j} - \frac{h_m}{3} f_{\bar{x},m,j}.$$

Denote $f^{h,-}_{i,j} = f_{i,j}$ and $q^{h,-}_{i,j} = q_{i,j}$ for $i = 1, \dots, m-1$. The problem $(P^{h,-})$ can be written in operator form

$$B^-_j V^-_{\bar{t},j} + A^-_j V^-_j = \Phi^-_j, \; V^-_{0,j+1} = \psi_{0,j+1}, V^-_{i,0} = \varphi_i, \tag{10}$$

where $\Phi^-_j = f^{h,-}_{i,j+1}$,

$$(A^-_j V^-_j)_i = \begin{cases} -\varepsilon V^-_{\bar{x}\hat{x},i,j} + q^{h,-}_{i,j} V^-_{i,j}, & i = 1 : m-1, \\ \left(\frac{2\varepsilon}{h_m} - \frac{h_m q_{m-0,j}}{3}\right) V^-_{\bar{x},m,j} + q^{h,-}_{m,j} V^-_{m,j}, & i = m, \end{cases}$$

and $B_j = D_j + \tau A_j$

$$(D^-_j V^-_j)_i = \begin{cases} V^-_{i,j}, & i = 1 : m-1, \\ V^-_{i,j} - \frac{h_m}{3} V^-_{\bar{x},i,j}, & i = m. \end{cases}$$

For the grid functions defined on the mesh $\bar{\omega}^- = \bar{\omega}^-_h \times \omega_\tau$, $\omega^-_h = \{x_0, \dots, x_m\}$ and vanishing at (x_0, t_j), define the scalar product

$$(y_j, v_j]_{0,\omega_h^-} = \sum_{i=1}^{m-1} \hat{h}_i y_{i,j} v_{i,j} + \tilde{h}_m y_{m,j} v_{m,j}, \; (y_j, v_j]_{*,\omega_h^-} = \sum_{i=1}^{m} h_i y_{i,j} v_{i,j}, \tag{11}$$

where

$$\tilde{h}_m = \frac{3h_m\varepsilon}{6\varepsilon - q_{m-0,j} h_m^2} \leq \frac{3h_m}{c_0}.$$

The operator A_j^- from (10) is selfadjoint and positive definite in the scalar product $[.,.)_{0,\omega_h^-}$ defined in (11). For arbitrary discrete functions y_j, v_j defined on ω_h^- and annulating at $x_0 = 0$ holds

$$(y_j, v_j]_{A_j^-} \equiv (A_j^- y_j, v_j]_{0,\omega_h^-} = \varepsilon(y_{\bar{x},j}, v_{\bar{x},j}]_{*,\omega_h^-} + (q_j^{h,-} y_j, v_j]_{0,\omega_h^-}.$$

Since A_j^- is selfadjoint and positive definite operator, then $(A_j^-)^{-1}$ is also selfadjoint and positive definite operator. So we can define the energy norms

$$\|v_j]|_{A_j^-} = \sqrt{(A_j^- v_j, v_j]_{0,\omega_h^-}}, \quad \|v_j]|_{(A_j^-)^{-1}} = \sqrt{((A_j^-)^{-1} v_j, v_j]_{0,\omega_h^-}}.$$

Lemma 2. *The discrete problem $(P^{h,-})$ is stable with respect to the right hand side and initial condition, and the following estimate holds for the solution V_j^-*

$$\|V_j^-]|_{A_j^-} \leq C\left\{\|\varphi]|_{A_0^-} + \max_{0\leq k\leq j}\left(\|\Phi_k]|_{(A_k^-)^{-1}} + \|\Phi_{\bar{t},k}]|_{(A_k^-)^{-1}}\right)\right\}$$

for some positive constant C independent of the small parameter ε.

Let V_j^+, $j = 0 : J$, be a solution to the discrete problem $(P^{h,+})$:

$$L^{h,+}V_{i,j}^+ = -\varepsilon(\kappa_{i,j}^h V_{\bar{x},i,j}^+)_{\hat{x},i,j} - r_{i+1,j}^h V_{\hat{x},i,j}^+ + q_{i,j}^h V_{i,j}^+ = f_{i,j}^h, \; i = m+1 : N-1,$$

$$V_{N,j}^+ = 0, \quad L^{h,+}V_{m,j}^+ = -\left(\frac{2\varepsilon}{h_m}\kappa_{m,j}^h + 2r_{m+1,j}^h\right)V_{x,m,j}^+ + q_{m,j}^{h,+}V_{m,j}^+ = f_{m,j}^{h,+},$$

where

$$q_{m,j}^{h,+} = q_{m+0,j} + \frac{q_{m+0,j}}{1+(R_{m+1,j}^h)^{-1}}, \; f_{m,j}^{h,+} = f_{m+0,j} + \frac{f_{m+0,j}}{1+(R_{m+1,j}^h)^{-1}}.$$

For the grid functions defined on the mesh $\bar{\omega}^+ = \omega_\tau \times \omega_h^+$, $\bar{\omega}_h^+ = \{x_m, \ldots, x_N\}$, and vanishing at x_N, define the scalar product

$$[y_j, v_j)_{0,\omega_h^+} = \sum_{i=m+1}^{N-1} \hat{h}_i y_{i,j} v_{i,j} + \frac{h_m}{2} y_{m,j} v_{m,j}$$

and the norms

$$|[v_j\|_{1,w_h^+} = [|v_j|, 1)_{0,w_h^+}, \quad \|v_j\|_{\infty,\bar{\omega}_h^+} = \max_{x_i\in\bar{\omega}_h^+} |v_{i,j}|.$$

Now, we consider the Green function $G_j^+(x_i, \eta_k)$ of problem $(P^{h,+})$. As a function of x_i, with η_k held constant, it is defined by the relations

$$L^{h,+}G_j^+(x_i,\eta_k) = \delta^h(x_i,\eta_k),\ i = m : N-1,\ G_j^+(x_N,\eta_k) = 0.$$

where

$$\delta^h(x_i,\eta_k) = \begin{cases} \hat{h}_i^{-1}, & \text{if } x_i = \eta_k,\ i,k = m+1,\dots,N-1, \\ 2/h_m, & \text{if } x_i = \eta_k = x_m, \\ 0, & \text{if } x_i \neq \eta_k. \end{cases}$$

Denote $f_{i,j}^{h,+} = f_{i,j}^h$ for $i = m,\dots,N-1$. It is obvious that the solution to problem $(P^{h,+})$ is expressed in terms of Green function as

$$V^+(x_i,t_j) = [G_j^+(x_i,\eta_k), f_{k,j}^{h,+})_{0,w_h^+}.$$

Lemma 3. *The Green function $G_j^+(x_i,\eta_k)$ is nonnegative and ε - uniformly bounded:*

$$0 \leq G_j^+(x_i,\eta_k) \leq r_0^{-1}.$$

Moreover, the solution to problem $(P^{h,+})$ satisfies the estimate

$$\|V_j^+\|_{\infty,\bar{w}_h^+} \leq r_0^{-1}\|[f_j^{h,+}\|_{1,w_h^+}.$$

3.4 Uniform Convergence

Suppose that $m \approx n \approx N/2$. Let $Z = U - u$ be the error of the discrete problem (P^h). Then Z satisfies

$$L^hZ = (f_{i,j}^h - f_{i,j}) - (L^hu_{i,j} - Lu_{i,j}) \equiv \Psi_{i,j},\ (x_i,t_j) \in \omega,\ Z_{0,j} = Z_{N,j} = Z_{i,0} = 0.$$

The next theorem gives the main result in this paper.

Theorem 1. *Let the conditions in Proposition 1 are fulfilled. If the parameters of the mesh satisfy $\sigma_1,\sigma_2 \geq 2$ and τ satisfies (9), then the solution U of the discrete problem (P^h) is ε-uniformly convergent to the solution u of the continuous problem (1)-(4) in discrete maximum norm and the following estimate hold*

$$\|U - u\|_{\infty,\bar{\omega}} \leq C(\tau + N^{-2}\ln^2 N),$$

for some positive constant C independent of the small parameter ε.

4 Numerical Results

Consider the problem

$$\frac{\partial u}{\partial t} - \varepsilon\frac{\partial^2 u}{\partial x^2} + 4u = 4 + \sin\pi x\left(\frac{\pi}{2}\cos\frac{\pi t}{2} + (4+\varepsilon\pi^2)\sin\frac{\pi t}{2}\right),\ (x,t) \in \Omega^-$$

$$-\varepsilon\frac{\partial^2 u}{\partial x^2} - \frac{\partial u}{\partial x} + u = 1 + \sin\frac{\pi t}{2}\left((1+\varepsilon\pi^2)\sin\pi x - \pi\cos\pi x\right),\ (x,t) \in \Omega^+$$

$$[u(x,t)]_\Gamma = \left[\frac{\partial u(x,t)}{\partial x}\right]_\Gamma = 0,$$

$$u(0,t) = u(1,t) = 1 + \exp(-t/4), u(x,0) = \varphi(x),$$

where $\xi = 0.5$ and $\varphi(x)$ is a solution of the stationary problem with zero right hand side, that is equal to 1 at the boundary points $x = 0$ and $x = 1$. The exact solution is

$$u(x,t) = \begin{cases} 1 + sin(\pi x)\sin(\pi t/2) + exp(-t/4)\varphi(x), & (x,t) \in \bar{\Omega}^-, \\ 1 + cos\,(\pi(x - 1/2))\sin(\pi t/2) + exp(-t/4)\varphi(x), & (x,t) \in \bar{\Omega}^+, \end{cases}$$

For our tests we take $J = N^2/16$. Our goal is to observe the rate of convergence with respect to the space variable. Table 1 displays the results of our numerical experiments. For large N we observe almost second-order ε-uniform convergence. The convergence rate is taken to be

$$\rho_N = \log_2\left(\|E_N\|_{\infty,w}/\|E_{2N}\|_{\infty,w}\right),$$

where $\|E_N\|_{\infty,w}$ is the maximum error norm for the corresponding value of N. Figure 1 shows the approximate solution and the error at different time stages for $\varepsilon = 2^-10$, $N = 32$ and $J = 64$. It illustrates very well the boundary and interior layers behavior of the solution. Thus the numerical results support the theoretical ones and show the effectiveness of special meshes.

Table 1. Error of the solution on Shishkin's meshes

$\varepsilon \backslash N$	$N=8$	$N=16$	$N=32$	$N=64$	$N=128$	$N=256$
$\varepsilon=1$	8.06e-3	1.51e-3	3.03e-4	6.69e-5	1.56e-5	3.77e-6
ρ_N	2.42	2.31	2.18	2.10	2.05	-
$\varepsilon=2^{-2}$	3.20e-1	6.99e-3	1.59e-3	3.77e-4	9.15e-5	2.25e-5
ρ_N	2.20	2.14	2.08	2.04	2.02	-
$\varepsilon=2^{-4}$	6.21e-2	1.56e-2	3.37e-3	7.80e-4	1.89e-4	4.68e-5
ρ_N	2.00	2.21	2.11	2.04	2.02	-
$\varepsilon=2^{-6}$	1.10e-1	3.00e-2	7.59e-3	1.88e-3	5.61e-4	1.78e-4
ρ_N	1.88	1.98	2.01	1.75	1.66	-
$\varepsilon=2^{-8}$	1.18e-1	4.94e-2	1.50e-2	4.16e-3	1.09e-3	2.81e-4
ρ_N	1.25	1.72	1.85	1.93	1.96	-
$\varepsilon=2^{-10}$	9.98e-2	3.74e-2	1.91e-2	8.52e-3	2.99e-3	7.61e-4
ρ_N	1.41	0.97	1.16	1.51	1.98	-
$\varepsilon=2^{-12}$	8.78e-2	2.88e-2	1.91e-2	8.52e-3	3.23e-3	1.11e-3
ρ_N	1.61	0.59	1.16	1.40	1.54	-
$\varepsilon=2^{-14}$	8.61e-2	2.65e-2	1.92e-2	8.53e-3	3.23e-3	1.11e-3
ρ_N	1.70	0.47	1.17	1.40	1.54	-
$\varepsilon=2^{-16}$	8.87e-1	2.64e-2	1.92e-2	8.53e-3	3.23e-3	1.11e-3
ρ_N	1.75	0.46	1.17	1.40	1.54	-
$\varepsilon=2^{-18}$	9.01e-1	2.63e-2	1.92e-2	8.53e-3	3.23e-3	1.11e-3
ρ_N	1.78	0.46	1.17	1.40	1.54	-
$\varepsilon=2^{-20}$	9.07e-1	2.63e-2	1.92e-2	8.53e-3	3.23e-3	1.11e-3
ρ_N	1.79	0.46	1.17	1.40	1.54	-

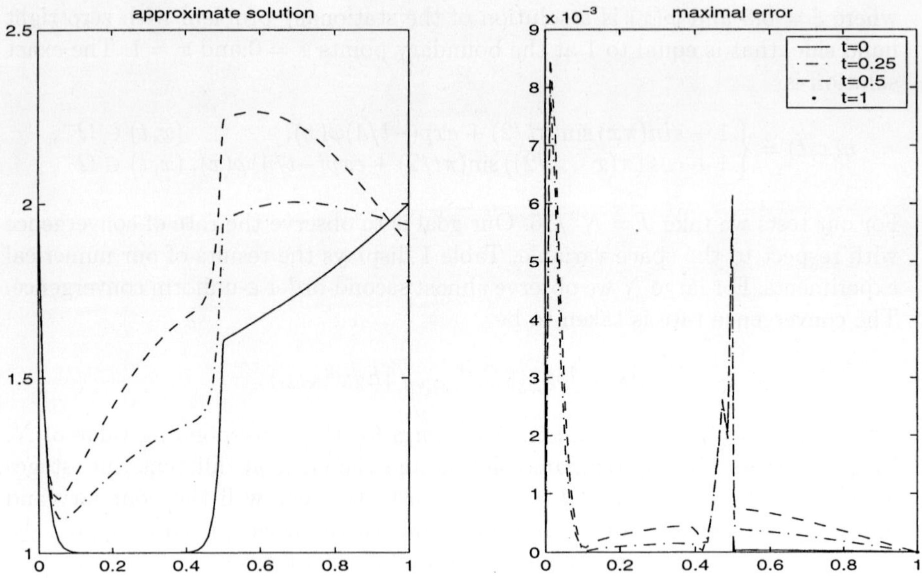

Fig. 1. Approximate solution and error on Shishkin mesh

References

1. Andreev, V.B., Savin, I.A.: On the uniform convergence of the monotone Samarski's scheme and its modification. Zh. Vychisl.Mat. Mat. Fiz. **35** No 5 (1995) 739–752 (in Russian)
2. Braianov, I.A, Vulkov, L.G.: Grid approximation for the solution of the singularly perturbed heat equation with concentrated capacity. J. Math. Anal. and Appl. **237** (1999) 672–697
3. Braianov, I.A., Vulkov, L.G.: Uniform in a small parameter convergence of Samarskii's monotone scheme and its modification for the convection-diffusion equation with a concentrated source. Comput. Math. Math. Phys. **40** No 4 (2000) 534-550
4. Konnor, K.A., Tichy, J.A.: Analysis of an eddy current bearning. Journal of tribology. **110** (1988) 320–326
5. Miller, J.J.H., O'Riordan,E., Shishkin, G.I.: Fitted numerical methods for singular perturbation problems. World scientific, Singapore, (1996)
6. Roos,H.G., Stynes,M., Tobiska, L.: Numerical Methods for singularly perturbed differential equations. Springer-Verlag, Berlin, 1996
7. Samarskii, A.A.: Theory of difference schemes. Nauka, Moscow, 1983 (in Russian)
8. Sushko, V.G.: Asymptotic representation of the solutions of some singularly perturbed problems for mixed type equations, Fundamentalnaya i prikladnaya matematika, **3** No 2 (1997) 579–586

The Numerical Solution for System of Singular Integro-Differential Equations by Faber-Laurent Polynomials

Iurie Caraus

Moldova State University
caraush@usm.md

Abstract. We have elaborated the numerical schemes of reduction method by Faber- Laurent polynomials for the approximate solution of system of singular integro- differential equations. The equations are defined on the arbitrary smooth closed contour. The theoretical foundation has been obtained in Hölder spaces.

Keywords: singular integro-differential equations, reduction method, Hölder spaces

1 The Results of Approximation Functions by Faber-Laurent Polynomials

Let Γ is an arbitrary smooth closed contour limiting the one-spanned area of complex plane D^+, the point $z = 0 \in D^+$ and $D^- = C \setminus \{D^+ \cup \Gamma\}$, C is the full complex plane. The set of these contours we denote by $\Lambda^{(1)}$.

Let $z = \psi(w)$ and $z = \varphi(w)$ are the functions mapping conformably and unambiguously the exterior of unit circle Γ_0 on D^- and on D^+ so that $\psi(\infty) = \infty$, $\psi^{'}(\infty) > 0$ and $\varphi(\infty) = 0$, $\varphi^{'}(\infty) > 0$. We denote by $w = \Phi(z)$ and $w = F(z)$ the reversible functions for $z = \psi(w)$ and $z = \varphi(w)$.

We suppose

$$\lim_{z \to \infty} \frac{1}{z} \Phi(z) = 1 \quad \text{and} \quad \lim_{z \to \infty} zF(z) = 1; \tag{1}$$

(see [1]). We obtain from (1) that the functions $w = \Phi(z)$ and $w = F(z)$ admit in the vicinity of points $z = \infty$ and $z = 0$ the following decompositions accordingly

$$\Phi(z) = z + \sum_{k=0}^{\infty} r_k z^{-k} \quad \text{and} \quad F(z) = \frac{1}{z} + \sum_{k=0}^{\infty} v_k z^k$$

We denote by $\Phi_n(z)$, $n = 0, 1, 2, \ldots$, the set of members by nonnegative degrees z for decomposing $\Phi^n(z)$ and by $F_n(1/z)$, $n = 1, 2, \ldots$ the set of members by negative degrees z for decomposing $F^n(z)$. The polynomials $\Phi_n(z)$, $n = 0, 1, 2, \ldots$ and $F_n(1/z)$, $n = 1, 2, \ldots$, are Faber- Laurent polynomials for Γ (see [1], [2]).

Z. Li et al. (Eds.): NAA 2004, LNCS 3401, pp. 219–223, 2005.

we look for the approximate solution for the problem "(6)- (5)". Most of early approximative methods for SIDE are designed for the case where boundary is a unit circle (see [6,7]). For the contours from $\Lambda^{(1)}$ " the problem(6)-(5)" was solved by quadrature- interpolation and collocation methods. (see [3,8,9]).

We look for the approximate solution of problem "(6)- (5)" in the form

$$\varphi_n(t) = t^q \sum_{k=0}^{n} \alpha_k^{(n)} \Phi_k(t) + \sum_{k=1}^{n} \alpha_{-k}^{(n)} F_k\left(\frac{1}{t}\right), t \in \Gamma, \tag{7}$$

where $\alpha_k = \alpha_k^{(n)}$, $k = \overline{-n, n}$; are unknown vectors of dimension m. Note that function $\varphi_n(t) \in [H_\beta^{(q)}(\Gamma)]_m$. To find the unknown numerical vectors α_k, $k = \overline{-n, n}$ we use the relation

$$S_n[M\varphi_n - f] = 0. \tag{8}$$

We examine the condition (8) as the operator equation

$$S_n M S_n \varphi_n = S_n f \tag{9}$$

for unknown v.f. $\varphi_n(t)$ in subspace R_n, for functions of the form (7), with the same norm as $[H_\beta(\Gamma)]_m$.

We note that the equation (9) is equivalent for the system of linear equations with $2n+1$ dimensional unknown v.f. α_k, $k = \overline{-n, n}$. We do not indicate the obvious form of this system because of difficulty formula.

Theorem 3. *Assume the following conditions are satisfied:*

1. *m. f.* $a_r(t)$, $b_r(t)$ *and* $h_r(t, \tau)$ $r = \overline{0, q}$ *belong to the space* $[H_\alpha(\Gamma)]_{m\times m}$, $0 < \alpha \le 1$;
2. *det* $a_q(t)\cdot$ *det* $b_q(t) \neq 0$, $t \in \Gamma$;
3. *the left partial indexes of m.f.* $a_q(t)$ *are equal zero, the right partial indexes of m.f.* $b_q(t)$ *are equal* q *for any* $t \in \Gamma$;
4. *the operator* $M : [\dot{H}_\beta^{(q)}(\Gamma)]_m \to [H_\beta(\Gamma)]_m$, *where* $0 < \beta < \alpha$ *is linearly reversible.*

Then beginning with numbers n, where the following inequality takes place

$$d_1 \frac{\ln^2 n}{n^{\sigma(\alpha)-\beta}} \le r < 1,$$

(where $\sigma(\alpha) = \alpha$ *for* $\alpha < 1$ *and* $\sigma(1) = 1 - \varepsilon$, $\varepsilon(> 0)$ *is small number) the equation of reduction method (9) has the unique solution.*

The approximate solutions $\varphi_n(t)$, *constructed by formula (7), converge in the norm of space* $[H_\beta^{(q)}(\Gamma)]_m$, *as* $n \to \infty$ *to the exact solution* $\varphi^*(t)$ *of the "problem (6) - (5)" for* $\forall$ *right part* $f(t) \in [H_\alpha(\Gamma)]_m$. *Furthermore, the following error estimate holds:*

$$||\varphi^* - \varphi_n||_{\beta,q} = O\left(\frac{\ln^2 n}{n^{\sigma(\alpha)-\beta}}\right). \tag{10}$$

References

1. Suetin, P.: The series by Faber polynomials.M.: Science, (1984) (in Russian)
2. Kuprin N.: The Faber series and the problem of linear interfacing. Izv. Vuzov Mathematics. **1** (1980) 20–26 (in Russia)
3. Zolotarevskii, V.: Finite methods of solving the singular integral equations on the closed contours of integration . Kishinev: "Stiintsa", (1991) (in Russian)
4. Muskhelishvili, N.: Singular integral equations . M:. Science, (1968) (in Russian)
5. Lifanov, I.: The method of singular integral equtions and numerical experiment . M : TOO "Ianus", (1995) (in Russian)
6. Gabdulahev, B.: The opimal approximation of solutions for lenear problems. Kazani: University of Kazani, (1980) (in Russian)
7. Zolotarevskii, V.: The direct methods for the solution of singular integro- differential equations on the unit circle. Collection "Modern problems of applied mathematics".-Chisinau: Science, (1979) 50–64 (in Russian)
8. Caraus, Iu.: The numerical solution of singular integro- differential equations in Hölder spaces. Conference on Scientific Computation. Geneva, (2002)
9. Caraus Iurie, Zhilin Li. : A Direct methods and convergence analysis for some system of singular integro- differential equations, May 2003, CRSC-TR03-22, North Caroline, USA.
10. Krikunov ,Iu.: The general Reimann problem and linear singular integro- differential equation. The scientific notes of Kazani university. Kazani. **116** (4) (1956) 3–29. (in Russian)
11. Zolotarevskii, V., Tarita, I.: The application of Faber- Laurent polynomials for the approximate solution to singular integral equations.The collection of Harkiv National University **590**, Harkiv, (2003) 124–127. (in Russian)

An Adaptive-Grid Least Squares Finite Element Solution for Flow in Layered Soils

Tsu-Fen Chen[1], Christopher Cox[2], Hasan Merdun[3], and Virgil Quisenberry[4]

[1] Department of Mathematics, National Chung Cheng University, Minghsiung, Chia-Yi, Taiwan
tfchen@math.ccu.edu.tw

[2] Department of Mathematical Sciences, Clemson University, Clemson, South Carolina 29634
clcox@clemson.edu

[3] Faculty of Agriculture, Dept. of Agricultural Structure and Irrigation, Kahramanmaras Sutcu Imam University, Kahramanmaras, Turkey 46060
hmerdun@postaci.com

[4] Department of Crop and Soil Environmental Sciences, 277 Poole Agricultural Center, Clemson University, Clemson, SC 29634
vqsnbrr@clemson.edu

Abstract. Groundwater flow in unsaturated soil is governed by Richards equation, a nonlinear convection-diffusion equation. The process is normally convection-dominated, and steep fronts are common in solution profiles. The problem is further complicated if the medium is heterogeneous, for example when there are two or more different soil layers. In this paper, the least squares finite element method is used to solve for flow through 5 layers with differing hydraulic properties. Solution-dependent coefficients are constructed from smooth fits of experimental data. The least squares finite element approach is developed, along with the method for building an optimized, nonuniform grid. Numerical results are presented for the 1D problem. Generalization to higher dimensions is also discussed.

1 Introduction

1.1 Governing Equations

Flow of water in unsaturated soil is governed by Richards' equation [1], which has several forms including

$$\frac{\partial \theta}{\partial \psi}\frac{\partial \psi}{\partial t} = \frac{\partial}{\partial z}[K(\psi)\frac{\partial \psi}{\partial z}] - \frac{\partial K(\psi)}{\partial z}, \tag{1}$$

where ψ is the pressure head, z is the depth below the surface, K is capillary conductivity, and θ is the water content. Let the flux, q, be defined as

$$q = K(\psi, z)\left(1 - \frac{\partial \psi}{\partial z}\right)$$

Z. Li et al. (Eds.): NAA 2004, LNCS 3401, pp. 224–231, 2005.

Typical boundary and initial conditions for equation (1) are

$$q = F(t), \;\; t \geq 0, \;\; x = 0, \tag{2}$$

$$\lim_{z \to \infty} \psi = \Psi, \;\; t \geq 0, \tag{3}$$

and

$$\psi(z,0) = \psi_0(z), \;\; 0 < z < \infty, \tag{4}$$

where $F(t)$ and $\psi_0(z)$ are continuous functions, and $\Psi = \lim_{z \to \infty} \psi(z)$. For the layered problem, $K(\psi)$ in (1) is replaced by $K(\psi, z)$. In most layered problems θ and $K(\psi, z)$ are discontinuous at the layer boundaries, but flux, q, must be continuous there. Therefore, $\frac{\partial \psi}{\partial z}$ may be discontinuous at layer boundaries as well, though ψ is continuous throughout the domain. In particular, we consider vertical flow through a domain consisting of five soil layers, with data taken from [2]. Coefficient data for the first layer is displayed in Figure 1. Data for the other layers are similar to that for the first layer. Data fits are performed using the splines-under-tension routines described in [3]. Layer interfaces occur at the z values 10, 20, 36, and 53. The rest of this paper is outlined as follows.

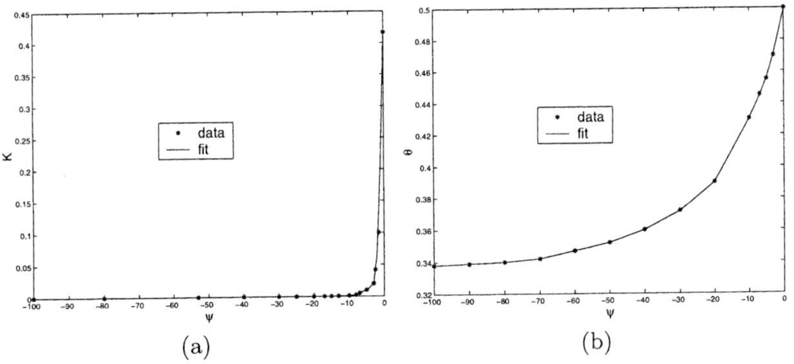

Fig. 1. Coefficient data for soil layer 1: (a) Capillary conductivity (K) (cm/min) vs. pressure head (ψ) (cm) (b) Water content (θ) $\left(\frac{cm^3}{cm^3}\right)$ vs. pressure head (ψ)

This section concludes with a description of a standard finite element solution for the problem represented by equations (1)-(4). The least squares formulation is presented in Section 2. Numerical results, comparing the Galerkin and least squares finite element methods, and conclusions are provided in Section 3.

1.2 Galerkin Finite Element Solution

Multiplying equation (1) by a test function v, integrating both sides and applying integration by parts results in the weak form for the problem corresponding to equations (1)-(4):

$$\int_0^\infty \frac{\partial\theta}{\partial\psi}\frac{\partial\psi}{\partial t} v\, dz = F(t)v(0) + \int_0^\infty K(\psi,z)\left(1-\frac{\partial\psi}{\partial z}\right)\frac{\partial v}{\partial z}\, dz$$
$$+\sum_{i=1}^{b}\left\{\left[K(\psi,z)\left(1-\frac{\partial\psi}{\partial z}\right)v\right]\Bigg|_{\zeta_i^+} - \left[K(\psi,z)\left(1-\frac{\partial\psi}{\partial z}\right)v\right]\Bigg|_{\zeta_i^-}\right\}, \quad (5)$$

where $\zeta_i, i = 1, \ldots, b$ are the values of z at which layer interfaces occur. Specifically, we seek a solution ψ in $S = \{u \mid u \in H^1[0,\infty),\ \lim_{x\to\infty} u = \Psi\}$ which satisfies (5) for all v in $V = \{u \mid u \in H^1[0,\infty),\ \lim_{x\to\infty} u = 0\}$. The discrete solution ψ_h in S_h satisfies (5) for all v_h in V_h, where S_h and V_h and finite dimensional subspaces of S and V, respectively. Computations are performed on a finite domain, so that the upper limit of integration in (5) is replaced by a finite value L. We choose, as the finite element solution, a continuous piecewise-linear function of the form

$$\psi_h(z,t) = \sum_{j=1}^{n-1} c_j(t)\phi_j(z) + \Psi\phi_n(z), \quad (6)$$

where $a_j(t)$ is the magnitude of the approximation at nodal point z_j and $\phi_j(z)$ is the corresponding continuous piecewise linear basis function. We set $L = 120$ for all computations in this manuscript. The grid points used for the discrete solution of (5) are clustered around the layer interfaces. Suppose there are $2m$ subintervals on the interval $[\zeta_i, \zeta_{i+1}]$, and let the interval midpoint be $\zeta_{i+\frac{1}{2}}$. The first m interior points in this interval are computed as

$$z_j^i = \zeta_i + \zeta_{i+\frac{1}{2}}\left(\frac{i}{m}\right)^2,$$

and the remaining points in the interval are clustered towards ζ_{i+1} in a similar manner. Substituting ψ_h defined in (6) for ψ in (5) and letting $v = \phi_j$, $j = 1, \ldots, n-1$ produces the system of nonlinear ordinary differential equations

$$A(\mathbf{c})\frac{d\mathbf{c}}{dt} = B(\mathbf{c})\mathbf{c} + \mathbf{g} \quad (7)$$

where $\mathbf{c} = [c_1, \ldots, c_{n-1}]^T$, and $\mathbf{g}$ is a vector that does not depend on $\mathbf{c}$. The system (7) is solved using the differential-algebraic system solver DASSL [4]. A sampling of solution profiles is displayed in Figure 2. For this calculation, 385 grid points were used. The time step was ramped up from a small starting value at $t = 0$ to $\Delta t = \frac{1}{2}$ at $t = 1$, while the front developed.

2 Least Squares Formulation

A problem which is equivalent to equation (1) is to find the function pair $(\psi,\ q)$ which solves the equations

$$L_1(\psi,\ q) := q + K(\psi,z)\left(\frac{\partial\psi}{\partial z} - 1\right) = 0 \quad (8)$$

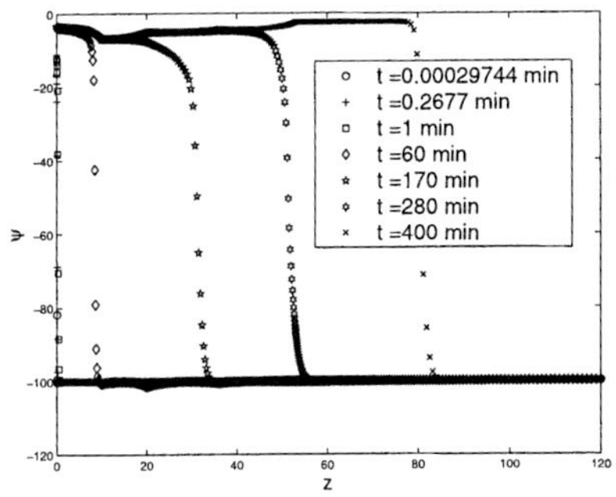

Fig. 2. Finite element solution based on equation (5)

$$L_2(\psi,\ q) := \frac{\partial \theta}{\partial \psi}\frac{\partial \psi}{\partial t} + \frac{\partial q}{\partial z} = 0 \tag{9}$$

along with boundary and initial conditions (2)-(4). This formulation allows for direct solution of two quantities of physical interest. A standard Galerkin finite element solution of this system is subject to strict stability requirements on the solution spaces used for ψ and q, especially in higher dimensions. As pointed out in [5], the least squares finite element approach offers advantages over the standard approach. We apply the approach developed in [5] and [6] to the current problem. The least squares variational principle, in essence, seeks the pair $(\psi,\ q)$ in an appropriate product space which minimizes

$$\|L_1(\gamma,\ s)\|_A^2 + \|L_2(\gamma,\ s)\|_B^2 \tag{10}$$

over all $(\gamma,\ s)$ in a similar product space, with suitably defined norms $\|\cdot\|_A$ and $\|\cdot\|_B$. Specific details will be provided subsequently. To have more control over all aspects of the algorithm during the development stage, we use a simple time discretization along with a quasi-Newton iteration to accommodate the nonlinear terms.

2.1 Time Discretization

Replacing the time derivative in (1) with a difference quotient at $t+\alpha k$, $\alpha \in [0,1]$, we define an approximate solution $\psi^k(z,t)$ for $t = nk = n\Delta t$, (i.e. $k = \Delta t$), $n = 0, 1, 2, \ldots$, satisfying

$$\left.\frac{d\theta}{d\psi}\right|_{\psi^k(z,t+\alpha k)} \left(\frac{\psi^k(z,t+k) - \psi^k(z,t)}{k}\right) + \frac{\partial K(\psi^k(z,t+\alpha k))}{\partial z} - \frac{\partial}{\partial z}\left[K(\psi^k(z,t+\alpha k))\frac{\partial \psi^k(z,t+\alpha k)}{\partial z}\right] = 0. \tag{11}$$

Let $\bar{\psi} = \psi^k(z, t+k)$ and $w = \left.\frac{d\theta}{d\psi}\right|_{\tilde{\psi}} \psi^k(z,t)$, where

$$\tilde{\psi} = \psi^k(z, t+\alpha k) = (1-\alpha)\psi^k(z,t) + \alpha\bar{\psi}, \tag{12}$$

Setting the parameter α to $\frac{1}{2}$ results in the Crank-Nicolson method. The backward difference formula corresponds to $\alpha = 1$. We solve the following equation for $\bar{\psi}$:

$$\left.\frac{d\theta}{d\psi}\right|_{\tilde{\psi}} \bar{\psi} + k\frac{\partial K(\tilde{\psi})}{\partial z} - k\frac{\partial}{\partial z}(K(\tilde{\psi})\frac{\partial\tilde{\psi}}{\partial z}) = w. \tag{13}$$

Given function w and step size $k = \Delta t$, we seek $\bar{\psi}$ satisfying (13) on $z \to \infty$ subject to the boundary conditions

$$K(\tilde{\psi})\left(1 - \frac{\partial\tilde{\psi}}{\partial z}\right) = F(t+\alpha k), \text{ at } z = 0,$$

$$\tilde{\psi} \to \Psi, \text{ as } z \to \infty.$$

An equivalent first order system is:

$$\tilde{q} + K(\tilde{\psi})\left(\frac{\partial\tilde{\psi}}{\partial z} - 1\right) = 0, \qquad 0 < z < \infty, \tag{14}$$

$$\left.\frac{d\theta}{d\psi}\right|_{\tilde{\psi}} \bar{\psi} + k\frac{\partial\tilde{q}}{\partial z} = w, \qquad 0 < z < \infty, \tag{15}$$

$$\tilde{q} = F(t+\alpha k), \qquad \text{at } z = 0, \tag{16}$$

$$\tilde{\psi} \to \Psi, \qquad \text{as } z \to \infty. \tag{17}$$

2.2 Newton Iteration

We linearize equations (14) and (15) before constructing the least-squares approximation. The procedure is described as follows. Let $\tilde{\psi} = \tilde{\psi}_0 + \delta$, where $\tilde{\psi}_0$ is the initial guess and δ is the correction in the Newton iteration. Then equation (14) becomes

$$\tilde{q} + \left(K(\tilde{\psi}_0) + K'(\tilde{\psi}_0)\delta\right)\left[\left(\frac{\partial\tilde{\psi}_0}{\partial z} + \frac{\partial\delta}{\partial z}\right) - 1\right] = 0.$$

Dropping the quadratic terms in δ and rearranging, we have the following approximation of equation (14):

$$\tilde{q} + K(\tilde{\psi}_0)\frac{\partial\delta}{\partial z} + K'(\tilde{\psi}_0)\delta\left[\frac{\partial\tilde{\psi}_0}{\partial z} - 1\right] = K(\tilde{\psi}_0)\left(1 - \frac{\partial\tilde{\psi}_0}{\partial z}\right). \tag{18}$$

Using (12), equation (15) can be written as

$$\frac{1}{\alpha}\left.\frac{d\theta}{d\psi}\right|_{(\tilde{\psi}_0+\delta)}\left[\tilde{\psi}_0+\delta-(1-\alpha)\psi^k(z,t)\right]+k\frac{\partial\tilde{q}}{\partial z}=w.$$

Rewriting this equation and evaluating $\frac{d\theta}{d\psi}$ at $\tilde{\psi}_0$, we have

$$\left.\frac{d\theta}{d\psi}\right|_{\tilde{\psi}_0}\frac{\delta}{\alpha}+k\frac{\partial\tilde{q}}{\partial z}=w-\frac{1}{\alpha}\left(\left.\frac{d\theta}{d\psi}\right|_{\tilde{\psi}_0}\left[\tilde{\psi}_0-(1-\alpha)\psi^k(z,t)\right]\right). \tag{19}$$

Since $\frac{d\theta}{d\psi}$ is evaluated at $\tilde{\psi}_0$ in (19), the linearization corresponding to (18) and (19) is a quasi-Newton approach and the convergence may be slow if the initial guess is not close to the solution. Note that equations (18) and (19) form a linear system in δ and $\tilde{q}$.

2.3 Least Squares Variational Principle

First we define the spaces

$$S=\{\gamma\mid\gamma\in H^1(0,\ L),\ \gamma(L)=\Psi\}\ \text{ and }\ Q=\{s\mid s\in H^1(0,\ L),\ s(0)=F\},$$

where Ψ and F are specified constants. To form the discrete solution, we introduce finite dimensional subspaces

$$S_h\subseteq S,\qquad Q_h\subseteq Q.$$

The (weighted) least squares variational principle, for the approximate problem, is stated as follows. Given $\tilde{\psi}_0\in S_h$, and discrete solution at time t, $\psi_h^k(z,t)\in S_h$, determine the functions $\delta_h\in S_h^0$, $\tilde{q}_h\in Q_h$ which minimize the functional

$$\chi\|s_h+K(\tilde{\psi}_0)\frac{\partial\epsilon_h}{\partial z}+K'(\tilde{\psi}_0)\left[\frac{\partial\tilde{\psi}_0}{\partial z}-1\right]\epsilon_h-\left[K(\tilde{\psi}_0)\left(1-\frac{\partial\tilde{\psi}_0}{\partial z}\right)\right]\|^2$$
$$+\|\left.\frac{d\theta}{d\psi}\right|_{\tilde{\psi}_0}\frac{\epsilon_h}{\alpha}+k\frac{\partial s_h}{\partial z}-\left[w-\frac{1}{\alpha}\left(\left.\frac{d\theta}{d\psi}\right|_{\tilde{\psi}_0}\left[\tilde{\psi}_0-(1-\alpha)\psi_h^k(z,t)\right]\right)\right]\|^2. \tag{20}$$

over all functions $\epsilon_h\in S_h^0$, $s_h\in Q_h$, where χ is a positive weighting function. Note that the discrete pressure at time $t+k$ is then $\bar{\psi}_h\in S_h$, where

$$\bar{\psi}_h=\frac{1}{\alpha}\left[\tilde{\psi}_0+\delta_h-(1-\alpha)\psi_h^k(z,t)\right].$$

The space S_h^0 differs from S_h only in that the (possibly) nonhomogeneous boundary value Ψ is set to zero. The norm in (20) is the L_2 norm.

2.4 Mesh Considerations

As in [6], we use the grading function described in [7]. The problem in the general setting is to construct an appropriate mesh for constructing a continuous piecewise interpolant u_h, of degree k, for a given continuous function u on the interval $[a,\ b]$. We consider the case $k = 1$ in the current work. Suppose that $[a,\ b]$ is to be divided into M subintervals with grid points $\{z_i\}$, $i = 0, 1, \ldots, M$. The strategy is based on determination of a grading function $\xi(z)$ which has the property, after equidistribution, that the inverse of ξ at $\frac{i}{M}$ is z_i. The goal is to minimize the error $e = u - u_h$ in the H^m-seminorm, i.e.

$$|e|_m^2 = \int_a^b (e^{(m)})^2 dx.$$

As shown in [7] and [6], a mesh which is nearly optimal with respect to error minimization and equistribution is found by choosing the grading function as

$$\xi(z) = \frac{\int_a^z (u'')^{2/[2(2-m)+1]} dx}{\int_a^b (u'')^{2/[2(2-m)+1]} dx}.$$

We choose $m = 1$, i.e. we minimize the L_2-norm of the error, so that

$$\xi = \frac{\int_a^z (u'')^{2/5} dx}{\int_a^b (u'')^{2/5} dx}. \tag{21}$$

Since the grading function is nonlinear, the Brent-Dekker scheme is used to locate the grid points and the stopping criterion is measured by the relative error in ξ to be less than a chosen tolerance τ:

$$\frac{\max_i \left| \int_{z_i}^{z_{i+1}} (u'')^{2/5} dx - \frac{1}{M} \int_a^b (u'')^{2/5} dx \right|}{\frac{1}{M} \int_a^b (u'')^{2/5} dx} < \tau.$$

3 Numerical Results and Conclusions

Now we compare the results of the Galerkin finite element formulation and the weighted least squares finite element approach. The weight χ in (20) is determined according to the analysis in [5] and [6] as $\chi = \Delta t \| \left.\frac{d\theta}{d\psi}\right|_{\tilde{\psi}_0} \|_\infty / \ [\alpha \| K(\tilde{\psi}_0) \|_\infty]$. Solution profiles for each method are compared in Figure 3. These results confirm that the profiles developing early in time (up to approximately 1 minute) are nearly identical. The Galerkin finite element solution was computed using the same spatial grid and time-steps as the results given in Section 1. The least squares solution was computed on a grid of up to 121 spatial points, clustered according to the method described in Section 2, using the Crank-Nicolson method with a time step of $\Delta t = 0.01$ seconds. Convergence with respect to time was confirmed for the least squares solution by observing that the solution did not change when calculated with $\Delta t = 0.005$ seconds. The next step in this research effort will be to implement more efficient time-stepping and linearization procedures in the least squares code. The advantages of the least squares scheme over the Galerkin finite element code will be realized more strongly in higher dimensions. The two-dimensional mesh strategy developed in [8] for problems with moving fronts will be generalized to the case with discontinuous coefficients. For that problem, as in the 1D model, solution of a first order system with flux as an unknown will preclude the necessity of jump boundary conditions at layer interfaces. Experimental verification of numerical results, using physical data, is also planned.

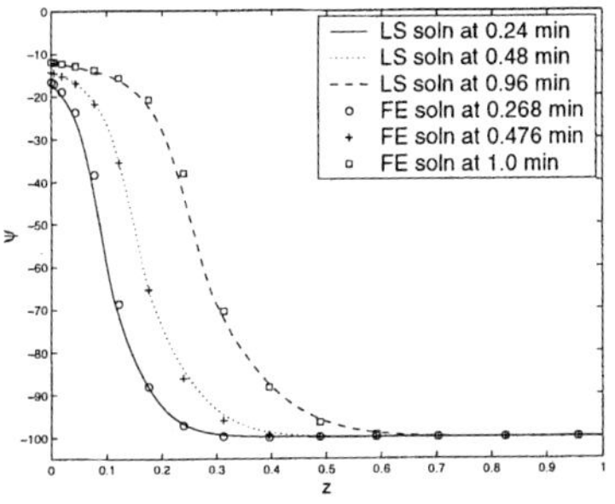

Fig. 3. Comparison of solution profiles at early times

References

1. Richards, L. A., Capillary Conduction of Liquid through Porous Media. Physics **1** 10 (1931) 318-333
2. Romkens, M. J. M., Phillips, R. E., Selim, H. M.,Whisler, F. D., Physical Characteristics of Soils in the Southern Region: Vicksburg, Memphis, and Maury series. South. Coop. Bull. No. 266. MS. Agric. Exp. Stn., Starkville (1985)
3. Cline, A. K., FITPACK - Software Package for Curve and Surface Fitting Employing Splines under Tension, Dept. of Comp. Sci., Univ. of Texas, Austin (1981)
4. Petzold, L. R., A Description of DASSL: A Differential/Algebraic System Solver. In: Stepleman, R. S. and others (eds.): Scientific Computing, Applications of Mathematics and Computing to the Physical Sciences, Volume I. IMACS/North-Holland Publishing Co. (1983)
5. Chen, T. F., Semidiscrete Least Squares Methods for Linear Convection-Diffusion Problems. Computers Math. Applic. **24** 11 (1992) 29-44
6. Chen, T. F., Weighted Least Squares Approximations for Nonlinear Hyperbolic Equations. Computers Math. Applic. To appear.
7. Carey, G. F. and Dinh, H. T., Grading Functions and Mesh Redistribution, Siam J. Numer. Anal. 22 (1985) 1028-1040.
8. Chen, T. F. and Yang, H. D., Numerical Construction of Optimal Grids in Two Spatial Dimensions, Computers and Math. with Applications **39** (2000) 101-120.

New Perturbation Bounds for the Continuous-Time H_∞-Optimization Problem

N. D. Christov[1], M. M. Konstantinov[2], and P.Hr. Petkov[1]

[1] Technical University of Sofia, 1000 Sofia, Bulgaria
{ndchr, php}@tu-sofia.bg
[2] University of Architecture and Civil Engineering, 1421 Sofia, Bulgaria
mmk_fte@uacg.bg

Abstract. A complete perturbation analysis of the H_∞-optimization problem for continuous-time linear systems is presented. Both local and nonlocal perturbation bounds are obtained, which are less conservative than the existing perturbation estimates.

1 Introduction

In this paper we present local and nonlocal perturbation analysis of the H_∞-optimization problem for continuous-time linear multivariable systems. First, nonlinear local perturbation bounds are derived for the matrix equations which determine the problem solution. The new local bounds are tighter than the existing condition number based linear perturbation estimates.

Then, using the nonlocal perturbation analysis techniques developed by the authors, the nonlinear local bounds are incorporated into nonlocal perturbation bounds which are less conservative than the existing nonlocal perturbation estimates for the H_∞-optimization problem. The nonlocal perturbation bounds are valid rigorously in contrast to the local bounds in which higher order terms are neglected.

We use the following notations: $\mathcal{R}^{m\times n}$ – the space of real $m \times n$ matrices; $\mathcal{R}^n = \mathcal{R}^{n\times 1}$; I_n – the unit $n\times n$ matrix; $A^\top$ – the transpose of A; $\|A\|_2 = \sigma_{\max}(A)$ – the spectral norm of A, where $\sigma_{\max}(A)$ is the maximum singular value of A; $\|A\|_{\rm F} = \sqrt{{\rm tr}(A^\top A)}$ – the Frobenius norm of A; $\|.\|$ is any of the above norms; ${\rm vec}(A) \in \mathcal{R}^{mn}$ – the column-wise vector representation of $A \in \mathcal{R}^{m\times n}$; $\Pi \in \mathcal{R}^{n^2\times n^2}$ – the (vec)permutation matrix, i.e. ${\rm vec}(A^\top) = \Pi {\rm vec}(A)$ for $A \in \mathcal{R}^{n\times n}$. The notation ":=" stands for "equal by definition".

2 Statement of the Problem

Consider the linear continuous-time system

$$\begin{aligned}\dot{x}(t) &= Ax(t) + Bu(t) + Ev(t)\\ y(t) &= Cx(t) + w(t)\\ z(t) &= \begin{bmatrix} Dx(t)\\ u(t)\end{bmatrix}\end{aligned} \qquad (1)$$

Z. Li et al. (Eds.): NAA 2004, LNCS 3401, pp. 232–239, 2005.

where $x(t) \in \mathcal{R}^n$, $u(t) \in \mathcal{R}^m$, $y(t) \in \mathcal{R}^r$ and $z(t) \in \mathcal{R}^p$ are the system state, input, output and performance vectors respectively, $v(t) \in \mathcal{R}^l$ and $w(t) \in \mathcal{R}^r$ are disturbances and A, B, C, D, E are constant matrices of compatible dimensions.

The H_∞-optimization problem is stated as follows: Given the system (1) and a constant $\lambda > 0$, find a stabilizing controller

$$\begin{aligned} u(t) &= -K\hat{x}(t) \\ \dot{\hat{x}}(t) &= \hat{A}\hat{x} + L((y(t) - C\hat{x}(t)) \end{aligned}$$

which satisfies

$$\|H\|_\infty := \sup_{Re\ s \geq 0} \|H(s)\|_2 < \lambda$$

where $H(s)$ is the closed-loop transfer matrix from v, w to z.

If such a controller exists, then [1]

$$\begin{aligned} K &= B^T X_0 \\ \hat{A} &= A - Y_0(C^T C - D^T D/\lambda^2) \\ L &= Z_0 Y_0 C^T \end{aligned}$$

where $X_0 \geq 0$ and $Y_0 \geq 0$ are the stabilizing solutions to the Riccati equations

$$A^T X + XA - X(BB^T - EE^T/\lambda^2)X + D^T D = 0 \tag{2}$$

$$AY + YA^T - Y(C^T C - D^T D/\lambda^2)Y + EE^T = 0 \tag{3}$$

and the matrix Z_0 is defined from

$$Z_0 = (I - Y_0 X_0/\lambda^2)^{-1} \tag{4}$$

under the assumption $\|Y_0 X_0\|_2 < \lambda^2$.

In the sequel we shall write equations (2), (3) as

$$A^T X + XA - XSX + Q = 0 \tag{5}$$

$$AY + YA^T - YRY + T = 0 \tag{6}$$

where $Q = D^T D$, $T = EE^T$, $S = BB^T - T/\lambda^2$, $R = C^T C - Q/\lambda^2$.

Suppose that the matrices $A, \ldots, E$ in (1) are subject to perturbations $\Delta A, \ldots, \Delta E$. Then we have the perturbed equations

$$(A + \Delta A)^T X + X(A + \Delta A) - X(S + \Delta S)X + Q + \Delta Q = 0 \tag{7}$$

$$(A + \Delta A)Y + Y(A + \Delta A)^T - Y(R + \Delta R)Y + T + \Delta T = 0 \tag{8}$$

$$Z = (I - YX/\lambda^2)^{-1} \tag{9}$$

where

$$\begin{aligned} \Delta Q &= \Delta D^T D + D^T \Delta D + \Delta D^T \Delta D, \quad \Delta T = \Delta E E^T + E \Delta E^T + \Delta E \Delta E^T \\ \Delta S &= \Delta B B^T + B \Delta B^T + \Delta B \Delta B^T - \Delta T/\lambda^2 \\ \Delta R &= \Delta C^T C + C^T \Delta C + \Delta C^T \Delta C - \Delta Q/\lambda^2. \end{aligned}$$

Denote by $\Delta_M = \|\Delta M\|$ the absolute perturbation of a matrix M. It is natural to use the Frobenius norm $\|.\|_F$ identifying the matrix perturbations with their vector-wise representations.

Since the Fréchet derivatives of the left-hand sides of (5), (6) in X and Y at $X = X_0$ and $Y = Y_0$ are invertible (see the next section) then, according to the implicit function theorem [2], the perturbed equations (7), (8) have unique solutions $X = X_0 + \Delta X$ and $Y = Y_0 + \Delta Y$ in the neighborhoods of X_0 and Y_0 respectively. Assume that $\|YX\| < \lambda^2$ and denote by $Z = Z_0 + \Delta Z$ the corresponding solution of the perturbed equation (9).

The sensitivity analysis of H_∞-optimization problem aims at determining perturbation bounds for the solutions X, Y and Z of equations (5), (6) and (4) as functions of the perturbations in the data A, S, Q, R, T.

In [3, 4] local perturbation bounds for the H_∞-optimization problem have been obtained, based on the condition numbers of equations (5), (6) and (4). However, the local estimates, based on condition numbers, may eventually produce pessimistic results. At the same time it is possible to derive local, first order homogeneous estimates, which are tighter in general [5, 6]. In this paper, using the local perturbation analysis technique developed in [5, 6], we shall derive local first order perturbation bounds which are less conservative than the condition number based bounds in [3, 4].

Local perturbation bounds have a serious drawback: they are valid in a usually small neighborhood of the data $A, \ldots, T$, i.e. for $\Delta = [\Delta_A, \ldots, \Delta_T]^T$ asymptotically small. In practice, however, the perturbations in the data are always finite. Hence the use of local estimates remains (at least theoretically) unjustified unless an additional analysis of the neglected terms is made, which in most cases is a difficult task. In fact, to obtain bounds for the neglected nonlinear terms means to get a nonlocal perturbation bound.

Nonlocal perturbation bounds for the continuous-time H_∞-optimization problem have been obtained in [3, 4] using the Banach fixed point principle. In this paper, applying the method of nonlinear perturbation analysis proposed in [5, 6], we shall derive new nonlocal perturbation bounds for the problem considered, which are less conservative than the nonlocal bounds in [3, 4].

3 Local Perturbation Analysis

Consider first the local sensitivity analysis of the Riccati equation (5). Denote by $F(X, \Sigma) = F(X, A, S, Q)$ the left-hand side of (5), where $\Sigma = (A, S, Q) \in \mathcal{R}^{n.n} \times \mathcal{R}^{n.n} \times \mathcal{R}^{n.n}$. Then $F(X_0, \Sigma) = 0$.

Setting $X = X_0 + \Delta X$, the perturbed equation (7) may be written as

$$\begin{aligned} & F(X_0 + \Delta X, \Sigma + \Delta\Sigma) = \\ & F(X_0, \Sigma) + F_X(\Delta X) + F_A(\Delta A) + F_S(\Delta S) + F_Q(\Delta Q) + G(\Delta X, \Delta\Sigma) = 0 \end{aligned} \tag{10}$$

where $F_X(.)$, $F_A(.)$, $F_S(.)$ and $F_Q(.)$ are the Fréchet derivatives of $F(X, \Sigma)$ in the corresponding matrix arguments, evaluated for $X = X_0$, and $G(\Delta X, \Delta\Sigma)$ contains the second and higher order terms in ΔX, $\Delta\Sigma$.

A straightforward calculation leads to

$$F_X(M) = A_c^T M + M A_c,\ \ F_A(M) = X_0 M + M^T X_0$$
$$F_S(M) = -X_0 M X_0,\ \ F_Q(M) = M$$

where $A_c = A - (BB^T - EE^T/\lambda^2)X_0$. Denote by $M_X \in \mathcal{R}^{n^2.n^2}$, $M_A \in \mathcal{R}^{n^2.n^2}$, $M_S \in \mathcal{R}^{n^2.n^2}$ the matrix representations of the operators $F_X(.)$, $F_A(.)$, $F_S(.)$:

$$M_X = A_c^T \otimes I_n + I_n \otimes A_c^T,\ \ M_A = I_n \otimes X_0 + (X_0 \otimes I_n)\Pi,\ \ M_S = -X_0 \otimes X_0 \quad (11)$$

where $\Pi \in \mathcal{R}^{n^2.n^2}$ is the permutation matrix such that $\mathrm{vec}(M^T) = \Pi \mathrm{vec}(M)$ for each $M \in \mathcal{R}^{n.n}$ and $\mathrm{vec}(M) \in \mathcal{R}^{n^2}$ is the column-wise vector representation of M.

It follows from (10)

$$F_X(\Delta X) = -F_A(\Delta A) - F_S(\Delta S) - \Delta Q - G(\Delta X, \Delta\Sigma). \quad (12)$$

Since A_c is stable, the operator $F_X(.)$ is invertible and (12) yields

$$\Delta X = -F_X^{-1} \circ F_A(\Delta A) - F_X^{-1} \circ F_S(\Delta S) - F_X^{-1}(\Delta Q) - F_X^{-1}(G(\Delta X, \Delta\Sigma)). \quad (13)$$

The operator equation (13) may be written in a vector form as

$$\mathrm{vec}(\Delta X) = N_1 \mathrm{vec}(\Delta A) + N_2 \mathrm{vec}(\Delta S) + N_3 \mathrm{vec}(\Delta Q) - M_X^{-1} \mathrm{vec}(G(\Delta X, \Delta\Sigma)) \quad (14)$$

where$N_1 = -M_X^{-1} M_A,\ N_2 = -M_X^{-1} M_S,\ N_3 = -M_X^{-1}$.

It is easy to show that the well-known condition number based perturbation bound [3, 4] is a corollary of (14). Indeed, it follows from (14)

$$\|\mathrm{vec}(\Delta X)\|_2 \le \|N_1\|_2 \|\mathrm{vec}(\Delta A)\|_2 + \|N_2\|_2 \|\mathrm{vec}(\Delta S)\|_2$$
$$+ \|N_3\|_2 \|\mathrm{vec}(\Delta Q)\|_2 + \mathrm{O}(\|\tilde{\Delta}\|^2)$$

and having in mind that $\|\mathrm{vec}(\Delta M)\|_2 = \|\Delta M\|_F = \Delta_M$ and denoting $K_A^X = \|N_1\|_2$, $K_S^X = \|N_2\|_2$, $K_Q^X = \|N_3\|_2$, we obtain

$$\Delta_X \le K_A^X \Delta_A + K_S^X \Delta_S + K_Q^X \Delta_Q + \mathrm{O}(\|\tilde{\Delta}\|^2) \quad (15)$$

where K_A^X, K_S^X, K_Q^X are the absolute condition numbers of (5) and $\tilde{\Delta} = [\Delta_A, \Delta_S, \Delta_Q]^T$.

Relation (14) also gives

$$\Delta_X \le \|\tilde{N}\|_2 \|\tilde{\Delta}\|_2 + \mathrm{O}(\|\tilde{\Delta}\|^2) \quad (16)$$

where $\tilde{N} = [N_1, N_2, N_3]$.

Note that the bounds in (15) and (16) are alternative, i.e. which one is less depends on the particular value of $\tilde{\Delta}$.

There is also a third bound, which is always less than or equal to the bound in (15). We have

$$\Delta_X \le \sqrt{\tilde{\Delta}^T U(\tilde{N}) \tilde{\Delta}} + \mathrm{O}(\|\tilde{\Delta}\|^2) \quad (17)$$

where $U(\tilde{N})$ is the 3×3 matrix with elements $u_{ij}(\tilde{N}) = \|N_i^T N_j\|_2$. Since $\left\|N_i^T N_j\right\|_2 \leq \|N_i\|_2 \|N_j\|_2$ we get

$$\sqrt{\tilde{\Delta}^T U(\tilde{N})\tilde{\Delta}} \leq \|N_1\|_2 \Delta_A + \|N_2\|_2 \Delta_S + \|N_3\|_2 \Delta_Q.$$

Hence we have the overall estimate

$$\Delta_X \leq f(\tilde{\Delta}) + \mathrm{O}(\|\tilde{\Delta}\|^2),\ \tilde{\Delta} \to 0 \tag{18}$$

where

$$f(\tilde{\Delta}) = \min\{\|\tilde{N}\|_2 \|\tilde{\Delta}\|_2, \sqrt{\tilde{\Delta}^T U(\tilde{N})\tilde{\Delta}}\} \tag{19}$$

is a first order homogeneous and piece-wise real analytic function in $\tilde{\Delta}$.

The local sensitivity of the Riccati equation (6) may be determined using the duality of (5) and (6). For the estimate of Δ_Y we have

$$\Delta_Y \leq g(\hat{\Delta}) + \mathrm{O}(\|\hat{\Delta}\|^2),\ \hat{\Delta} \to 0 \tag{20}$$

where

$$g(\hat{\Delta}) = \min\{\|\hat{N}\|_2 \|\hat{\Delta}\|_2, \sqrt{\hat{\Delta}^T U(\hat{N})\hat{\Delta}}\} \tag{21}$$

$\hat{\Delta} = [\Delta_A, \Delta_R, \Delta_T]^T$ and $\hat{N}$ is determined replacing in (11) A_c and X_0 by $\hat{A}^T$ and Y_0, respectively.

Consider finally the local sensitivity analysis of equation (4). In view of (9) we have

$$\Delta Z = [I_n - (Y_0 + \Delta Y)(X_0 + \Delta X)/\lambda^2]^{-1} - Z_0 = Z_0 W Z_0 + \mathrm{O}(\|W\|^2) \tag{22}$$

where $W = (Y_0 \Delta X + \Delta Y X_0 + \Delta Y \Delta X)/\lambda^2$.

It follows form (22) $\Delta_Z \leq \|Z_0^T \otimes Z_0\|_2 \|W\|_F + \mathrm{O}(\|W\|^2)$ and denoting $\zeta_0 = \|Z_0^T \otimes Z_0\|_2$ we get

$$\begin{aligned}\Delta_Z &\leq \zeta_0(\|Y_0\|_2\, \Delta_X + \|X_0\|_2\, \Delta_Y)/\lambda^2 + \mathrm{O}(\|(\Delta X, \Delta Y)\|^2) \\ &\leq \zeta_0(\|Y_0\|_2\, f(\tilde{\Delta}) + \|X_0\|_2\, g(\hat{\Delta}))/\lambda^2 + \mathrm{O}(\|\Delta\|^2).\end{aligned} \tag{23}$$

The relations (18), (20) and (23) give local first order perturbation bounds for the continuous-time H_∞-optimization problem.

4 Nonlocal Perturbation Analysis

The local perturbation bounds are obtained neglecting terms of order $\mathrm{O}(\|\Delta\|^2$, i.e. they are valid only asymptotically, for $\Delta \to 0$. That is why, their application for possibly small but nevertheless finite perturbations Δ requires additional justification. This disadvantage may be overcome using the methods of nonlinear perturbation analysis. As a result we obtain nonlocal (and in general nonlinear) perturbation bounds which guarantee that the perturbed problem still has a solution and are valid rigorously, unlike the local bounds. However, in some cases the nonlocal bounds may not exist or may be pessimistic.

Consider first the nonlocal perturbation analysis of the Riccati equation (5). The perturbed equation (13) can be rewritten in the form

$$\Delta X = \Psi(\Delta X) \tag{24}$$

where $\Psi\colon \mathcal{R}^{n.n} \to \mathcal{R}^{n.n}$ is determined by the right-hand side of (13). For $\rho > 0$ denote by $\mathcal{B}(\rho) \subset \mathcal{R}^{n.n}$ the set of all matrices $M \in \mathcal{R}^{n.n}$ satisfying $\|M\|_F \le \rho$. For $U, V \in \mathcal{B}(\rho)$ we have

$$\|\Psi(U)\|_F \le a_0(\tilde{\Delta}) + a_1(\tilde{\Delta})\rho + a_2(\tilde{\Delta})\rho^2$$

and

$$\|\Psi(U) - \Psi(V)\|_F \le (a_1(\tilde{\Delta}) + 2a_2(\tilde{\Delta})\rho)\|U - V\|_F$$

where

$$a_0(\tilde{\Delta}) := f(\tilde{\Delta}),\ \ a_2(\tilde{\Delta}) := \|M_X^{-1}\|_2(\|S\|_2 + \Delta_S) \tag{25}$$

$$a_1(\tilde{\Delta}) := 2\|M_X^{-1}\|_2\Delta_A + (\|M_X^{-1}(X_0 \otimes I_n)\|_2 + \|M_X^{-1}(I_n \otimes X_0)\|_2)\Delta_S.$$

Hence, the function $h(\rho, \tilde{\Delta}) = a_0(\tilde{\Delta}) + a_1(\tilde{\Delta})\rho + a_2(\tilde{\Delta})\rho^2$ is a Lyapunov majorant [7] for equation (24) and the majorant equation for determining a nonlocal bound $\rho = \rho(\tilde{\Delta})$ for Δ_X is

$$a_2(\tilde{\Delta})\rho^2 - (1 - a_1(\tilde{\Delta})\rho + a_0(\tilde{\Delta}) = 0. \tag{26}$$

Suppose that $\tilde{\Delta} \in \tilde{\Omega}$, where

$$\tilde{\Omega} = \left\{\tilde{\Delta} \succeq 0 : a_1(\tilde{\Delta}) + 2\sqrt{a_0(\tilde{\Delta})a_2(\tilde{\Delta})} \le 1\right\} \subset \mathcal{R}^3_+. \tag{27}$$

Then equation (26) has nonnegative roots $\rho_1 \le \rho_2$ with

$$\rho_1 = \phi(\tilde{\Delta}) := \frac{2a_0(\tilde{\Delta})}{1 - a_1(\tilde{\Delta}) + \sqrt{(1 - a_1(\tilde{\Delta}))^2 - 4a_0(\tilde{\Delta})a_2(\tilde{\Delta})}}. \tag{28}$$

The operator Ψ maps the closed convex set

$$\mathcal{B}(\tilde{\Delta}) = \left\{M \in \mathcal{R}^{n.n} : \|M\|_F \le \phi(\tilde{\Delta})\right\} \subset \mathcal{R}^{n.n}$$

into itself and according to the Schauder fixed point principle there exists a solution $\Delta X \in \mathcal{B}(\tilde{\Delta})$ of equation (24), for which

$$\Delta_X \le \phi(\tilde{\Delta}),\ \ \tilde{\Delta} \in \tilde{\Omega}. \tag{29}$$

The elements of ΔX are continuous functions of the elements of $\Delta\Sigma$.

If $\tilde{\Delta} \in \tilde{\Omega}_1$, where

$$\tilde{\Omega}_1 = \left\{\tilde{\Delta} \succeq 0 : a_1(\tilde{\Delta}) + 2\sqrt{a_0(\tilde{\Delta})a_2(\tilde{\Delta})} < 1\right\} \subset \tilde{\Omega}$$

then $\rho_1 < \rho_2$ and the operator Ψ is a contraction on $\mathcal{B}(\tilde{\Delta})$. Hence according to the Banach fixed point principle the solution ΔX, for which the estimate (29) holds true, is unique. This means that the perturbed equation has an isolated solution $X = X_0 + \Delta X$. In this case the elements of ΔX are analytical functions of the elements of $\Delta\Sigma$.

In a similar way, replacing A_c with $\hat{A}^T$, S with R, Q with T and X_0 with Y_0, we obtain a nonlocal perturbation bound for ΔY. Suppose that $\hat{\Delta} \in \hat{\Omega}$, where

$$\hat{\Omega} = \left\{ \hat{\Delta} : b_1(\hat{\Delta}) + 2\sqrt{b_0(\hat{\Delta}) b_2(\hat{\Delta})} \leq 1 \right\} \subset \mathcal{R}^3_+$$

and

$$b_0(\hat{\Delta}) = g(\hat{\Delta}), \;\; b_2(\hat{\Delta}) = \|M_Y^{-1}\|_2(\|R\|_2 + \Delta_R)$$

$$b_1(\hat{\Delta}) = 2\|M_Y^{-1}\|_2 \Delta_{\hat{A}} + (\|M_Y^{-1}((Y_0 \otimes I_n))\|_2 + \|M_Y^{-1}((I_n \otimes Y_0))\|_2)\Delta_R.$$

Then

$$\Delta_Y \leq \psi(\hat{\Delta}), \;\; \hat{\Delta} \in \hat{\Omega} \tag{30}$$

where

$$\psi(\hat{\Delta}) = \frac{2b_0(\hat{\Delta})}{1 - b_1(\hat{\Delta}) + \sqrt{(1 - b_1(\hat{\Delta}))^2 - 4b_0(\hat{\Delta}) b_2(\hat{\Delta})}}.$$

Finally, the nonlinear perturbation bound for ΔZ is obtained using (14) and (37), (38). If $1 \notin \text{spect}(WZ_0)$ we have $\Delta Z = Z_0 W Z_0 (I_n - WZ_0)^{-1}$. Hence

$$\Delta_Z \leq \zeta_0 \|W\|_F \|(I_n - WZ_0)^{-1}\|_2.$$

If $\|W\|_2 < 1/\|Z_0\|_2$ we have

$$\Delta_Z \leq \frac{\zeta_0 \|W\|_F}{1 - \|Z_0\|_2 \|W\|_2}.$$

It is realistic to estimate $\|W\|$ when $\Delta X, \Delta Y$ vary independently. In this case one has to assume that

$$\|Y_0\|_2 \phi(\tilde{\Delta}) + \|X_0\|_2 \psi(\hat{\Delta}) + \phi(\tilde{\Delta})\psi(\hat{\Delta}) < \lambda^2 / \|Z_0\|_2$$

and

$$\Delta_Z \leq \frac{\zeta_0 \lambda^2 (\|Y_0\|_2 \phi(\tilde{\Delta}) + \|X_0\|_2 \psi(\hat{\Delta}) + \phi(\tilde{\Delta})\psi(\hat{\Delta}))}{\lambda^2 - \|Z_0\|_2 (\|Y_0\|_2 \phi(\tilde{\Delta}) + \|X_0\|_2 \psi(\hat{\Delta}) + \phi(\tilde{\Delta})\psi(\hat{\Delta}))}. \tag{31}$$

Relations (29), (30) and (31) give nonlocal perturbation bounds for the continuous-time H_∞-optimization problem.

Note finally that one has to ensure the inequality

$$\|YX\|_2 < \lambda^2. \tag{32}$$

Since the unperturbed inequality $\|Y_0 X_0\|_2 < \lambda^2$ holds true, a sufficient condition for (32) to be valid is

$$\|Y_0\|_2 \phi(\tilde{\Delta}) + \|X_0\|_2 \psi(\hat{\Delta}) + \phi(\tilde{\Delta})\psi(\hat{\Delta}) < \lambda^2 - \|Y_0 X_0\|_2.$$

Note that $\tilde{\Delta}, \hat{\Delta}$ depend on λ^2 through Δ_S, Δ_R.

5 Conclusions

The local and nonlocal sensitivity of the continuous-time H_∞-optimization problem have been studied. New local perturbation bounds have been obtained for the matrix equations determining the problem solution. The new local bounds are nonlinear functions of the data perturbations and are tighter than the existing condition number based local bound. Using a nonlinear perturbation analysis technique, nonlocal perturbation bounds for the H_∞-optimization problem are then derived. These bounds have two main advantages: they guarantee that the perturbed problem still has a solution, and are valid rigorously, unlike the local perturbation bounds. The new nonlocal perturbation bounds are less conservative than the existing nonlocal perturbation estimates for the H_∞-optimization problem.

References

1. H. Kwakernaak. Robust control and H_∞-optimization – Tutorial paper. *Automatica*, vol. 29, 1993, pp. 255-273.
2. L.V. Kantorovich and G.P. Akilov. *Functional Analysis in Normed Spaces.* Nauka, Moskow, 1977.
3. N.D. Christov, D.W. Gu, M.M. Konstantinov, P.Hr. Petkov and I. Postlethwaite. *Perturbation Bounds for the Continuous H_∞ Optimisation Problem.* LUED Rpt. 95-6, Dept. of Engineering, Leicester University, UK, 1995.
4. M.M. Konstantinov, P.Hr. Petkov and N.D. Christov. Conditioning of the continuous-time H_∞ optimisation problem. *Proc. 3rd European Control Conf.*, Rome, 1995, vol. 1, pp. 613-618.
5. M.M. Konstantinov, P.Hr. Petkov, N.D. Christov, D.W. Gu, and V. Mehrmann. Sensitivity of Lyapunov equations. In: N.E. Mastorakis (Ed.). *Advances in Intelligent Systems and Computer Science*, World Scientific and Engineering Society Press, 1999, pp. 289-292.
6. M.M. Konstantinov, P.Hr. Petkov and D.W. Gu. Improved perturbation bounds for general quadratic matrix equations. *Numer. Func. Anal. Optimiz.*, vol. 20, 1999, pp. 717-736.
7. E.A. Grebenikov and Yu.A. Ryabov. *Constructive Methods for the Analysis of Nonlinear Systems.* Nauka, Moscow, 1979.

Progressively Refining Discrete Gradient Projection Method for Semilinear Parabolic Optimal Control Problems

Ion Chryssoverghi

Department of Mathematics, National Technical University of Athens,
Zografou Campus, 15780 Athens, Greece
ichriso@math.ntua.gr

Abstract. We consider an optimal control problem defined by semilinear parabolic partial differential equations, with convex control constraints. Since this problem may have no classical solutions, we also formulate it in relaxed form. The classical problem is then discretized by using a finite element method in space and a theta-scheme in time, where the controls are approximated by blockwise constant classical ones. We then propose a discrete, progressively refining, gradient projection method for solving the classical, or the relaxed, problem. We prove that strong accumulation points (if they exist) of sequences generated by this method satisfy the weak optimality conditions for the continuous classical problem, and that relaxed accumulation points (which always exist) satisfy the weak optimality conditions for the continuous relaxed problem. Finally, numerical examples are given.

1 Introduction

We consider an optimal control problem defined by semilinear parabolic partial differential equations, with convex pointwise control constraints. Since this problem may have no classical solutions, we also formulate it in relaxed form. The classical problem is then discretized by using a Galerkin finite element method with continuous piecewise affine basis functions in space and a theta-scheme in time, where the controls are approximated by blockwise constant classical ones. We then propose a discrete, progressively refining, gradient projection method for solving the classical, or the relaxed, problem. We prove that strong accumulation points in L^2 (if they exist) of sequences generated by this method satisfy the weak classical optimality conditions for the continuous classical problem, and that relaxed accumulation points (which always exist) satisfy the weak relaxed optimality conditions for the continuous relaxed problem. For nonconvex problems with convex control constraints, whose solution is not classical, we can apply the above methods to the problem formulated in Gamkrelidze relaxed form, and then approximate the computed Gamkrelidze controls by classical ones. Finally, numerical examples are given. For discretization and optimization methods concerning optimal control problems, see [2], [3] and [4].

Z. Li et al. (Eds.): NAA 2004, LNCS 3401, pp. 240–248, 2005.

2 The Continuous Optimal Control Problems

Let Ω be a bounded domain in $\mathbf{R}^d$ with a Lipschitz boundary Γ, and let $I = (0,T)$, $T < \infty$, be an interval. Consider the semilinear parabolic state equation

$$y_t + A(t)y = f(x,t,y(x,t),w(x,t)) \text{ in } Q = \Omega \times I,$$
$$y(x,t) = 0 \text{ in } \Sigma = \Gamma \times I,\ y(x,0) = y^0(x) \text{ in } \Omega,$$

where $A(t)$ is the second order elliptic differential operator

$$A(t)y = -\sum_{j=1}^{d}\sum_{i=1}^{d}(\partial/\partial x_i)[a_{ij}(x,t)\partial y/\partial x_j].$$

The constraints on the control are $w(x,t) \in U$, in Q, where U is a compact subset of $\mathbf{R}^{d'}$, and the cost functional to be minimized is

$$G(w) = \int_Q g(x,t,y,w)dxdt.$$

Define the set of *classical controls*

$$W = \{w : (x,t) \mapsto w(x,t) \,|\, w \text{ measurable from } Q \text{ to } U\},$$

and the set of *relaxed controls* (Young measures, for the theory, see [7], [8]) by

$$R = \{r : Q \to M_1(U) \,|\, r \text{ weakly measurable}\} \subset L^\infty_w(Q, M(U)) \equiv L^1(Q, C(U))^*,$$

where $M_1(U)$ is the set of probability measures on U. The set W is endowed with the relative strong topology of $L^2(Q)$ and the set R with the relative weak star topology of $L^1(Q,C(U))^*$. The set R is convex, metrizable and compact. If we identify every classical control with its associated Dirac relaxed control $r(\cdot) = \delta_{w(\cdot)}$, then W may be considered as a subset of R, and W is thus dense in R. For a given function $\phi \in L^1(Q,C(U))$ and $r \in R$, we use the notation

$$\phi(x,t,r(x,t)) = \int_U \phi(x,t,u)r(x,t)(du).$$

The relaxed formulation of the above control problem is the following. Setting $V = H_0^1(\Omega)$, the relaxed state equation (in weak form) is

$$< y_t, v > + a(t,y,v) = \int_\Omega f(x,t,y(x,t),r(x,t))v(x)dx, \text{ for } v \in V, \text{ a.e. in } I,$$
$$y(x,t) = 0 \text{ in } \Sigma = \Gamma \times I,\ y(x,0) = y^0(x) \text{ in } \Omega,$$

where $a(t,\cdot,\cdot)$ is the usual bilinear form associated with $A(t)$ and $< \cdot,\cdot >$ the duality bracket between the dual V^* and V. The control constraint is $r \in R$, and the cost functional to be minimized

$$G(r) = \int_Q g(x,t,y(x,t),r(x,t))dxdt.$$

We suppose that the functions f, g are measurable for fixed (y,u) and continuous for fixed (x,t), that f is Lipschitz and sublinear w.r.t. y, g is subquadratic w.r.t y, and $y^0 \in L^2(\Omega)$. For some of the following theoretical results, see [2].

Lemma 1. *The operators $w \mapsto y_w$, from W to $L^2(Q)$, and $r \mapsto y_r$, from R to $L^2(Q)$, and the functionals $w \mapsto G(w)$ on W, and $r \mapsto G(r)$ on R, are continuous.*

Theorem 1. *There exists an optimal relaxed control.*

Note that, even when the control set U is convex, the classical problem may have no classical solutions. We next give some useful results concerning necessary conditions for optimality. We suppose in addition that f'_y, f'_u, g'_y, g'_u are measurable for fixed (y, u) and continuous for fixed (x, t), f'_y is bounded, and g'_y is sublinear w.r.t. y.

Lemma 2. *For $r, r' \in R$, the relaxed directional derivative of G is given by*

$$DG(r, r' - r) = \int_Q H(x, t, y(x, t), z(x, t), r'(x, t) - r(x, t))dxdt,$$

where the Hamiltonian H is defined by

$$H(x, t, y, z, u) = zf(x, t, y, u) + g(x, t, y, u),$$

and the adjoint state $z = z_r$ satisfies the equation

$$-< z_t, v > +a(t, v, z) = (zf'_y(y, r) + g'_y(y, r), v), \text{ for every } v \in V, \text{ a.e. in } I,$$
$$z(x, t) = 0 \text{ in } \Sigma, \quad z(x, T) = 0 \text{ in } \Omega,$$

where $y = y_r$. The mappings $r \mapsto z_r$, from R to $L^2(Q)$, and $(r, r') \mapsto DG(r, r' - r)$, from $R \times R$ to $\mathbf{R}$, are continuous.

Theorem 2. *If $r \in R$ is optimal for either the relaxed or the classical optimal control problem, then r is extremal relaxed, i.e. it satisfies the condition*

$$DG(r, r' - r) \geq 0, \text{ for every } r' \in R,$$

which is equivalent to the strong relaxed pointwise minimum principle

$$H(x, t, y(x, t), z(x, t), r(x, t)) = \min_{u \in U} H(x, t, y(x, t), z(x, t), u), \text{a.e. in } Q.$$

If U is convex, then this minimum principle implies the weak relaxed pointwise minimum principle

$$H'_u(x, t, y(x, t), z(x, t), r(x, t))r(x, t)$$
$$= \min_{\phi} H'_u(x, t, y(x, t), z(x, t), r(x, t))\phi(x, t, r(x, t)), \text{a.e. in } Q,$$

where the minimum is taken over all Caratheodory functions $\phi : Q \times U \to U$, which in turn implies the global weak relaxed condition

$$\int_Q H'_u(x, t, y(x, t), z(x, t), r(x, t))[\phi(x, t, r(x, t)) - r(x, t)]dxdt \geq 0, \text{ for every}$$

Caratheodory function ϕ.
A control r satisfying this condition is called weakly extremal relaxed.

Lemma 3. *If U is convex, then for $w, w' \in W$ the classical directional derivative of G is given by*

$$\mathbf{D}G(w, w' - w) = \int_Q H'_u(x, t, y(x, t), z(x, t))[w'(x, t) - w(x, t)]dxdt,$$

where $z = z_w$ and $y = y_w$. The mappings $w \mapsto z_w$, from W to $L^2(Q)$, and $(w, w') \mapsto \mathbf{D}G(w, w' - w)$, from $W \times W$ to $\mathbf{R}$, are continuous.

Theorem 3. *If U is convex and $w \in W$ is optimal for the classical optimal control problem, then w is weakly extremal classical, i.e. it satisfies the condition*

$$\mathbf{D}G(w, w' - w) \geq 0, \text{ for every } w' \in W,$$

which is equivalent to the weak classical pointwise minimum principle

$$H'_u(x,t,y(x,t),z(x,t))w(x,t) = \min_{u \in U} H'_u(x,t,y(x,t),z(x,t))u, \text{a.e. in } Q.$$

3 The Discrete Optimal Control Problem

We suppose in the sequel that the domain Ω is a polyhedron, $a(t,u,v)$ is independent of t and symmetric, U is convex, the functions f,f'_y,f'_u,g,g'_y,g'_u are continuous on $\bar{Q} \times \mathbf{R} \times U$, and $y^0 \in V$. For each integer $n \geq 0$, let $\{S_i^n\}_{i=1}^{M(n)}$ be an admissible regular quasi-uniform triangulation of $\bar{\Omega}$ into closed d-simplices, with $h^n = \max_i[\mathrm{diam}(S_i^n)] \to 0$ as $n \to \infty$, and $\{I_j^n\}_{j=0}^{N(n)-1}$ a subdivision of the interval $\bar{I}$ into intervals $I_j^n = [t_j^n, t_{j+1}^n]$, of equal length Δt^n, with $\Delta t^n \to 0$ as $n \to \infty$. We suppose that each $S_{i'}^{n+1}$ is a subset of some S_i^n and that each $I_{j'}^{n+1}$ is a subinterval of some I_j^n. We set $Q_{ij}^n = S_i^n \times I_j^n$. Let $V^n \subset V = H_0^1(\Omega)$ be the subspace of functions that are continuous on $\bar{\Omega}$ and affine on each S_i^n. Let $W^n \subset W$ be the set of *discrete controls*

$$W^n = \{w^n \in W \,|\, w^n(x,t) = w_{ij}^n, \text{ on } \overset{o}{Q}{}_{ij}^n\}.$$

For a given discrete control $w^n = (w_0^n, ..., w_{N-1}^n) \in W^n$, with $w_j^n = (w_{0j}^n, ..., w_{Mj}^n)$, and $\theta \in [1/2, 1]$, the corresponding discrete state $y^n = (y_0^n, ..., y_N^n)$ is given by the discrete state equation (implicit θ-scheme)

$$(1/\Delta t^n)(y_{j+1}^n - y_j^n, v) + a(y_{j\theta}^n, v) = (f(t_{j\theta}^n, y_{j\theta}^n, w_j^n), v),$$
$$\text{for every } v \in V^n, \quad j = 0, ..., N-1,$$
$$(y_0^n - y^0, v)_1 = 0, \text{ for every } v \in V^n, \quad y_j^n \in V^n, \quad j = 0, ..., N-1,$$
$$\text{with } y_{j\theta}^n = (1-\theta)y_j^n + \theta y_{j+1}^n, \; t_{j\theta}^n = (1-\theta)t_j^n + \theta t_{j+1}^n.$$

The discrete control constraints are $w^n \in W^n$ and the discrete cost functional

$$G^n(w^n) = \Delta t^n \sum_{j=0}^{N-1} \int_\Omega g(t_{j\theta}^n, y_{j\theta}^n, w_j^n)dx.$$

Lemma 4. *The operators $w^n \mapsto y_j^n$ and the discrete functional $w^n \mapsto G(w_m^n)$ are continuous.*

Lemma 5. *The directional derivative of the functional G^n is given by*

$$\mathbf{D}G^n(w^n, w'^n - w^n) = \Delta t^n \sum_{j=0}^{N-1} (H'_u(t_{j\theta}^n, y_{j\theta}^n, z_{j,1-\theta}^n, w_j^n), w_j'^n - w_j^n),$$

where the discrete adjoint is given by

$$-(1/\Delta t^n)(z_{j+1}^n - z_j^n, v) + a(v, z_{j,1-\theta}^n) = (z_{1-\theta}^n f'_y(t_{j\theta}^n, y_{j\theta}^n, w_j^n) + g'_y(t_{j\theta}^n, y_{j\theta}^n, w_j^n), v),$$

for every $v \in V^n$, $\quad j = N-1, ...0, \quad z_N^n = 0, \quad z_j^n \in V^n, \quad j = 0, ..., N-1.$

Theorem 4. *If $w^n \in W^n$ is optimal for the discrete problem, then it is discrete extremal, i.e. it satisfies the condition*

$$\mathbf{D}G^n(w^n, w'^n - w^n) = \Delta t^n \sum_{j=0}^{N-1} (H'_u(t^n_{j\theta}, y^n_{j\theta}, z^n_{j,1-\theta}, w^n_j), w'^n - w^n) \geq 0,$$

for every $w'^n \in W^n$,
which is equivalent to the discrete blockwise minimum principle

$$w^n_{ij} \int_{S^n_i} H'_u(t^n_{j\theta}, y^n_{j\theta}, z^n_{j,1-\theta}, w^n_{ij})dx = \min_{w'^n_{ij} \in U} w'^n_{ij} \int_{S^n_i} H'_u(t^n_{j\theta}, y^n_{j\theta}, z^n_{j,1-\theta}, w^n_{ij})dx,$$

$$1 \leq i \leq M, \quad 0 \leq j \leq N-1.$$

The following classical control approximation result is proved similarly to the lumped parameter case (see [5]).

Proposition 1. *For every $w \in W$, there exists a sequence of discrete controls $(w^n \in W^n)$ that converges to w in L^2 strongly.*

We suppose in the sequel that if $\theta = 1/2$, there exists a constant C, independent of n, such that $\Delta t^n \leq C(h^n)^2$, for every n. The next lemma is a consistency result.

Lemma 6. *If $w^n \to w \in W$ in L^2 strongly (resp. $w^n \to r$ in R, the w^n considered as relaxed controls), then the corresponding blockwise constant discrete states and adjoints y^n, z^n converge to y_w, z_w (resp. y_r, z_r) in $L^2(Q)$ strongly, as $n \to \infty$, and*

$$\lim_{n\to\infty} G^n(w^n) = G(w) (\textit{resp. } \lim_{n\to\infty} G^n(r^n) = G(r)),$$
$$\lim_{n\to\infty} \mathbf{D}G^n(w^n, w'^n - w^n) = \mathbf{D}G(w, w' - w).$$

4 Discrete Gradient Projection Method

Let $\gamma \geq 0$, $s \in [0,1]$, $b, c \in (0,1)$, and let (β^n) be a positive decreasing sequence converging to zero. The progressively refining discrete gradient projection method is described by the following algorithm.

4.1 Algorithm

Step 1. Set $k = 0$, $n = 0$, and choose an initial control $w^0_0 \in W^0$.
Step 2. Find $v^n_k \in W^n$ such that

$$e_k = \mathbf{D}G^n(w^n_k, v^n_k - w^n_k) + (\gamma/2)\left\|v^n_k - w^n_k\right\|^2$$
$$= \min_{v'^n \in W^n} [\mathbf{D}G^n(w^n_k, v'^n - w^n_k) + (\gamma/2)\left\|v'^n - w^n_k\right\|^2],$$

and set $d_k = \mathbf{D}G^n(w^n_k, v^n_k - w^n_k)$.
Step 3. If $|e_k| \geq \beta^n$, go to Step 4. Else, set

$$w^n = w^n_k, \quad v^n = v^n_k, \quad d^n = d_k, \quad e^n = e_k, \quad n = n+1.$$

Step 4. (Armijo step search) Find the lowest integer value $l \in Z$ such that $\alpha_k = c^l s \in (0,1]$ satisfies the inequality

$$G^n(w_k^n + \alpha_k(v_k^n - w_k^n)) - G^n(w_k^n) \leq \alpha_k b e_k.$$

Step 5. Set

$$w_{k+1}^n = w_k^n + \alpha_k(v_k^n - w_k^n), \quad k = k+1,$$

and go to Step 2.

If $\gamma > 0$, we have a gradient projection method, in which case Step 2 amounts to finding the projection v_{kij}^n of $w_{kij}^n - (1/\gamma)\int_{S_i^n} H_u'(t_{j\theta}^n, y_{j\theta}^n, z_{j,1-\theta}^n, w_{ij}^n)dx$ onto $U = [c_1, c_2]$, for each i, j. If $\gamma = 0$, the above Algorithm is a conditional gradient method and Step 2 reduces to setting, for each i, j

$$v_{kij}^n = c_1, \text{ if } \int_{S_i^n} H_u'(t_{j\theta}^n, y_{j\theta}^n, z_{j,1-\theta}^n, w_{ij}^n)dx \geq 0,$$
$$v_{kij}^n = c_2, \text{ if } \int_{S_i^n} H_u'(t_{j\theta}^n, y_{j\theta}^n, z_{j,1-\theta}^n, w_{ij}^n)dx < 0.$$

Theorem 5. *(i) Let (w^n) be a subsequence, considered as a sequence in R, of the sequence generated by the Algorithm in Step 3 that converges to some r in the compact set R, as $n \to \infty$. Then r is weakly extremal for the continuous relaxed problem.*

(ii) If (w^n) is a subsequence of the sequence generated by the Algorithm in Step 3 that converges to some $w \in W$ in L^2 strongly as $n \to \infty$, then w is weakly extremal for the continuous classical problem.

Proof. We shall first show that $n \to \infty$ in the Algorithm. Suppose, on the contrary, that n remains constant, and so we drop here the index n. Let us show that then $d_k \to 0$. Since W^n is compact, let $(w_k)_{k\in K}$, $(v_k)_{k\in K}$ be subsequences of the generated sequences in Steps 2 and 5 such that $w_k \to \tilde{w}$, $v_k \to \tilde{v}$, in W^n, as $k \to \infty$, $k \in K$. Clearly, by Step 2, $e_k \leq 0$ for every k, hence

$$e = \lim_{k\to\infty,\ k\in K} e_k = \mathbf{D}G(\tilde{w}, \tilde{v} - \tilde{w}) + \tfrac{\gamma}{2}\left\|\tilde{v} - \tilde{w}\right\|^2 \leq 0,$$
$$d = \lim_{k\to\infty,\ k\in K} d_k = \mathbf{D}G(\tilde{w}, \tilde{v} - \tilde{w}) \leq \lim_{k\to\infty,\ k\in K} e_k = e \leq 0.$$

Suppose that $d < 0$. The function $\Phi(\alpha) = G(w + \alpha(v - w))$ is continuous on $[0,1]$. Since the directional derivative $\mathbf{D}G(w, v-w)$ is linear w.r.t. $v - w$, Φ is differentiable on $(0,1)$ and has derivative

$$\Phi'(\alpha) = \mathbf{D}G(w + \alpha(v - w), v - w).$$

Using the Mean Value Theorem, we have, for each $\alpha \in (0,1]$

$$G(w_k + \alpha(v_k - w_k)) - G(w_k) = \alpha \mathbf{D}G(w_k + \alpha'(v_k - w_k), v_k - w_k),$$

for some $\alpha' \in (0, \alpha)$. Therefore, for $\alpha \in [0,1]$, by the continuity of $\mathbf{D}G$ (Lemma 4), we have

$$G(w_k + \alpha(v_k - w_k)) - G(w_k) = \alpha(d + \varepsilon_{k\alpha}),$$

where $\varepsilon_{k\alpha} \to 0$ as $k \to \infty$, $k \in K$, and $\alpha \to 0^+$. Since $d_k = d + \eta_k$, where $\eta_k \to 0$ as $k \to \infty$, $k \in K$, and $b \in (0,1)$, we have

$$d+\varepsilon_{k\alpha} \le b(d+\eta_k) = bd_k,$$

for $\alpha \in [0,\bar{\alpha}]$, for some $\bar{\alpha} > 0$, and $k \ge \bar{k}$, $k \in K$. Hence

$$G(w_k + \alpha(v_k - w_k)) - G(w_k) \le \alpha b d_k,$$

for $\alpha \in [0,\bar{\alpha}]$, $k \ge \bar{k}$, $k \in K$. It follows from the choice of the Armijo step α_k in Step 4 that we must have $\alpha_k \ge c\bar{\alpha}$ for $k \ge \bar{k}$, $k \in K$. Hence

$$G(w_{k+1}) - G(w_k) = G(w_k + \alpha_k(v_k - w_k)) - G(w_k) \le \alpha_k b d_k \le c\bar{\alpha} b d_k \le c\bar{\alpha} b d/2,$$

for $k \ge \bar{k}$, $k \in K$. It follows that $G(w_k) \to -\infty$ as $k \to \infty$, $k \in K$. This contradicts the fact that $G(w_k) \to G(\bar{w})$ as $k \to \infty$, $k \in K$, by continuity of the discrete functional (Lemma 4). Therefore, $d = 0$, hence $e = 0$, and $e_k \to e = 0$ for the whole sequence, since the limit 0 is unique. But by Step 3 we must necessarily have $n \to \infty$, which is a contradiction. Therefore, $n \to \infty$.

(i) Let (w^n) be a subsequence (same notation), considered now as a sequence in R, of the sequence generated in Step 3 that converges to an accumulation point $r \in R$, as $n \to \infty$. By Steps 2 and 3 we have, for every $v'^n \in W^n$

$$\begin{aligned}&\mathbf{D}G^n(w^n, v'^n - w^n) + (\gamma/2)\,\|v'^n - w^n\|_Q^2 \\ &= \int_Q H_u'^n(x, t_\theta^n, y_\theta^n, z_{1-\theta}^n, w^n)(v'^n - w^n)dxdt + (\gamma/2)\int_Q (v'^n - w^n)^2 dxdt \ge e^n,\end{aligned}$$

Choosing any continuous function $\phi : \bar{Q} \times U \to U$ and setting

$$\bar{x}^n(x) = \text{barycenter of } S_i^n, \text{ for } x \in \overset{o}{S_i^n}, \quad i = 1, ..., M,$$

we have

$$\begin{aligned}&\int_Q H_u'^n(x, t_\theta^n, y_\theta^n, z_{1-\theta}^n, w^n)[\phi(\bar{x}^n(x), t_\theta^n(t), w^n(x,t)) - w^n(x,t)]dxdt \\ &+(\gamma/2)\int_Q [\phi(\bar{x}^n(x), t_\theta^n(t), w^n(x,t)) - w^n(x,t)]^2 dxdt \ge e^n, \text{ for every such } \phi.\end{aligned}$$

Using Lemma 6, and Proposition 2.1 in [1], we can pass to the limit in this inequality as $n \to \infty$ and obtain

$$\begin{aligned}&\int_Q H_u'(x,t,y,z,r(x,t))[\phi(x,t,r(x,t)) - r(x,t)]dxdt \\ &+\int_Q [\phi(x,t,r(x,t)) - r(x,t)]^2 dxdt \ge 0, \text{ for every such } \phi.\end{aligned}$$

Replacing ϕ by $u + \lambda(\phi - u)$, with $\lambda \in (0,1]$, dividing by λ, and taking the limit as $\lambda \to 0$, we obtain the weak relaxed condition

$$\int_Q H_u'(x,t,y,z,r(x,t))[\phi(x,t,r(x,t)) - r(x,t)]dxdt \ge 0, \text{ for every such } \phi,$$

which holds also by density for every Caratheodory function ϕ. Therefore, r is weakly extremal relaxed.

(ii) Let now (w^n) be a subsequence (same notation) generated by the Algorithm in Step 3 that converges to some $w \in W$ in L^2 strongly as $n \to \infty$. Let any $w' \in W$ and, by Proposition 1, let $(w'^n \in W^n)$ be a sequence converging to w'. By Step 2, we have

$$\int_Q H_u'^n(x, t_\theta^n, y_\theta^n, z_{1-\theta}^n, w^n)(w'^n - w^n)dxdt + (\gamma/2)\int_Q (w'^n - w^n)^2 dxdt \ge e^n.$$

Using Proposition 2.1 in [1] and Lemma 6, we can pass to the limit and obtain

$\int_Q H'_u(x,t,y,z,w)(w'-w)dxdt + (\gamma/2)\int_Q (w'-w)^2 dxdt \geq 0$, for every $w' \in W$.

It follows then, similarly to (i), that

$\int_Q H'_u(x,t,y,z,w)(w'-w)dxdt \geq 0$, for every $w' \in W$. □

When directly applied to nonconvex optimal control problems whose solutions are non-classical relaxed controls, the classical methods yield often very poor convergence. For this reason, we propose here an alternate approach that uses Gamkrelidze controls in classical form. We suppose that $U = [c_1, c_2]$ is a compact interval. Using the Filippov selection theorem, it can be shown that the relaxed control problem described in §2 is equivalent to the following classical one, with the three-dimensional controlled state equation

$$y_t + A(t)y = \beta(x,t)f(x,t,y,u(x,t)) + [1-\beta(x,t)]f(x,t,y,v(x,t)), \text{ in } Q,$$
$$y = 0 \text{ in } \Sigma, \quad y(x,0) = y^0(x) \text{ in } \Omega,$$

control constraints

$$(\beta(x,t), u(x,y), v(x,t)) \in [0,1] \times U \times U \text{ in } Q,$$

and cost functional

$$\mathbf{G}(\beta,u,v) = \int_Q \{\beta(x,t)g(x,t,y,u)) + [1-\beta(x,t)]g(x,t,y,v)\}dxdt.$$

We can therefore apply the gradient method described above to this classical problem, with the obvious modifications due to the three-dimensional control. The Gamkrelidze relaxed controls computed thus can then be approximated by piecewise constant classical controls using a standard procedure, see [4]. In the general case, i.e. if U is not convex, one can use relaxed methods to solve such strongly nonconvex problems (see [3] and [4]).

5 Numerical Examples

a) Let $\Omega = (0,\pi)$, $I = (0,1)$. Define the functions

$$\bar{y}(x,t) = -e^{-t}\sin x + \tfrac{1}{2}x(\pi - x), \quad \bar{w}(x,t) = \begin{cases} -1, & 0 \leq t \leq 0.3 \\ \frac{t-0.3}{0.35}\sin x, & 0.3 \leq t \leq 1 \end{cases}$$

and consider the following optimal control problem, with state equation

$$y_t - y_{xx} = 1 + \sin y - \sin \bar{y} + w - \bar{w} \text{ in } Q,$$
$$y = 0 \text{ in } \Sigma, \quad y(x,0) = \bar{y}(x,0) \text{ in } \Omega,$$

control constraint set $U = [-1,1]$, and cost functional

$$G(w) = \int_Q 0.5[(y-\bar{y})^2 + (w-\bar{w})^2]dxdt.$$

Clearly, the optimal control and state are $\bar{w}$ and $\bar{y}$. The discrete gradient projection method was applied to this problem, with $\gamma = 0.5$, $\theta = 0.5$ (θ-scheme), and successive discretizations $(M,N) = (30,10), (60,20), (120,40)$. After 12 iterations, we obtained the values $G_0^n(w_k) = 0.1556 \cdot 10^{-7}$, $e_k = -0.1513 \cdot 10^{-12}$,

$\varepsilon_k = 0.3546 \cdot 10^{-4}$, $\zeta_k = 0.1009 \cdot 10^{-3}$, where e_k is defined in Step 2 of the Algorithm, ε_k is the max error for the state at the vertices of the blocks Q_{ij} , and ζ_k the max error for the control at the centers of the blocks.
b) Choosing the set $U = [-0.8, 0.5]$, the control constraints being now active for the method and for the problem, we obtained the values $G_0^n(w_k) = 0.032000304$, $e_k = -0.3124 \cdot 10^{-15}$.
c) Defining the state equation

$$y_t - y_{xx} = 1 + w \text{ in } Q,$$

the convex constraint set $U = [-1, 1]$, and the nonconvex cost functional

$$G(w) = \int_Q [0.5(y - \bar{y})^2 - w^2] dx dt,$$

the unique optimal relaxed control is clearly $\bar{r} = (\delta_{-1} + \delta_1)/2$, where δ denotes the Dirac measure, and the optimal state is $y = \bar{y}$. Note that the optimal relaxed cost $G(\bar{r}) = -\pi$ can be approximated as closely as desired with a classical control, but cannot be attained for such a control. Applying the conditional gradient method (i.e. with $\gamma = 0$) to the problem reformulated in Gamkrelidze form (see end of §4), with initial controls (piecewise constant interpolants of)

$$\beta_0 = 0.5 + 0.25(x/\pi + t), \quad u_0 = -0.5(x/\pi + t), \quad v_0 = -u_0,$$

we obtained after 20 iterations $\beta_k \approx 0.5$ with max error of less than $2 \cdot 10^{-3}$, the controls $u_k = -1$ and $v_k = 1$ exactly, $y_k \approx \bar{y}$ with max error $0.3727 \cdot 10^{-3}$, $G^n(\beta_k, u_k, v_k) = -3.141592636 \approx -\pi$, and $e_k = -0.1521 \cdot 10^{-4}$.

References

1. Chryssoverghi, I.: Nonconvex optimal control problems of nonlinear monotone parabolic systems. Systems Control Lett. **8** (1986) 55–62
2. Chryssoverghi, I., Bacopoulos, A.: Approximation of relaxed nonlinear parabolic optimal control problems. J Optim. Theory Appl. **77**, 1, (1993) 31–50
3. Chryssoverghi, I., Bacopoulos, A., Kokkinis, B., Coletsos, J.: Mixed Frank-Wolfe penalty method with applications to nonconvex optimal control problems. JOTA **94**, 2, (1997) 311–334
4. Chryssoverghi, I., Coletsos, J.,Kokkinis, B.: Discrete relaxed method for semilinear parabolic optimal control problems. Control Cybernet. **28**, 2, (1999), 157–176
5. Polak, E.: Optimization: Algorithms and Consistent Approximations. Springer, Berlin, 1997
6. Roubiček, T.: A convergent computational method for constrained optimal relaxed control problems. Control Cybernet. **69**, (1991), 589–603
7. Roubiček, T.: Relaxation in Optimization Theory and Variational Calculus. Walter de Gruyter, Berlin, 1997
8. Warga, J.: Optimal Control of Differential and Functional Equations. Academic Press, New York, 1972
9. Warga, J.: Steepest descent with relaxed controls. SIAM J. Control **15**, 4, (1977), 674–682

Identification of a Nonlinear Damping Function in a Thermoelastic System

Gabriel Dimitriu

Department of Mathematics and Informatics,
Faculty of Pharmacy, University of Medicine and Pharmacy, Iaşi, Romania
dimitriu@umfiasi.ro

Abstract. In this paper we present an approximation framework and convergence results for the identification of a nonlinear damping function in a thermoelastic system. A functional technique is used to demonstrate that solutions to a sequence of finite dimensional (Galerkin) approximating identification problems in some sense approximate a solution to the original infinite dimensional inverse problem. An example and numerical studies are discussed.

1 Introduction

Based on the approach of Banks, Reich and Rosen in [1], the paper represents a generalization to the nonlinear case of an identification problem for the linear thermoelastic model studied by Rosen and Su in [3].

The paper is organized as follows. In the section 2 we define the abstract thermoelastic system, establish its well-posedness, and state the class of identification problems. The approximation framework and convergence result are developed in the third section, and the example and computational issues are discussed in section 4. The last section contains some concluding remarks.

2 Abstract Thermoelastic System. Identification Problem

Let $\mathcal{Q}$ be a metric space and let Q be a compact subset of $\mathcal{Q}$. The set Q will be known as the admissible parameter set. For $j = 1, 2$ let $\{H_j, \langle\,\cdot\,,\,\cdot\,\rangle_j, |\cdot|_j\}$ be real Hilbert spaces and let $\{V_j, \|\cdot\|_j\}$ be reflexive Banach spaces. We assume that for $j = 1, 2$, V_j is densely and continuously embedded in H_j, with $|\varphi|_j \le \mu_j \|\varphi\|_j$, $\varphi \in V_j$. We let V_j^* denote the continuous dual of V_j. Then with H_j as the pivot space, we have $V_j \hookrightarrow H_j = H_j^* \hookrightarrow V_j^*$ with H_j densely and continuously embedded in V_j^*. We denote the usual operator norm on V_j^* by $\|\cdot\|_{j^*}$, $j = 1, 2$. In the usual manner, $\langle\,\cdot\,,\,\cdot\,\rangle_j$ is understood to denote both the inner product on H_j and the duality pairing on $V_j^* \times V_j$, for $j = 1, 2$.

For each $q \in Q$ we consider the abstract nonlinear thermoelastic system

$$\ddot{u}(t) + C(q)\dot{u}(t) + A_1(q)u(t) + L(q)^*\theta(t) \ni f(t; q), \qquad t > 0, \tag{1}$$

Z. Li et al. (Eds.): NAA 2004, LNCS 3401, pp. 249–256, 2005.

$$\dot{\theta}(t) + A_2(q)\theta(t) - L(q)\dot{u}(t) = g(t;q), \qquad t > 0, \tag{2}$$

$$u(0) = u_0(q), \quad \dot{u}(0) = v_0(q), \quad \theta(0) = \theta_0(q), \tag{3}$$

where for each $t > 0$, $u(t) \in H_1$ and $\theta(t) \in H_2$. Assume that for each $q \in Q$, $A_1(q) \in \mathcal{L}(V_1, V_1^*)$, $A_2(q) \in \mathcal{L}(V_2, V_2^*)$, $L(q) \in \mathcal{L}(V_1, V_2^*)$, $C(q) : \mathrm{Dom}(C(q)) \subset V_1 \to 2^{V_1^*}$, $u_0(q) \in V_1$, $v_0(q) \in \mathrm{Dom}(C(q))$, $\theta_0(q) \in H_2$, and $f(\cdot\,;q) \in L^1(0,T; H_1)$, $g(\cdot\,;q) \in L^1(0,T;H_2)$, for some $T > 0$. $L(q)^* \in \mathcal{L}(V_2, V_1^*)$ is defined by

$$\langle L(q)^*\psi, \varphi\rangle_1 = \langle L(q)\varphi, \psi\rangle_2, \qquad \psi \in V_2, \quad \varphi \in V_1. \tag{4}$$

We shall also require the following further assumptions.

(A1) (Symmetry): For each $q \in Q$ the operator $A_1(q)$ is symmetric in the sense that $\langle A_1(q)\varphi, \psi\rangle_1 \langle A_1(q)\psi, \varphi\rangle_1$, for all $\varphi, \psi \in V_1$.

(A2) (Continuity): For $\varphi \in V_1$ and $\psi \in V_2$, the mappings $q \to A_1(q)\varphi$, $q \to A_2(q)\psi$ are continuous from $Q \subset \mathcal{Q}$ into V_1^* or V_2^* (which ever is appropriate).

(A3) (Uniform Coercivity): For $j = 1, 2$ there exist constants $\alpha_j, \beta_j \in I\!R$, independent of $q \in Q$, $\beta_j > 0$, such that $\langle A_j(q)\varphi, \varphi\rangle_j + \alpha_j|\varphi|_j^2 \geq \beta_j\|\varphi\|_j^2$, $\varphi \in V_j$.

(A4) (Uniform Boundedness) There exist positive constants $\gamma_j, j = 1, 2$, independent of $q \in Q$, for which $\|A_j(q)\varphi\|_{j^*} \leq \gamma_j\|\varphi\|_j$, $\varphi \in V_j, j = 1, 2$.

For each $q \in Q$ the operator $C(q) : \mathrm{Dom}(C(q)) \subset V_1 \to 2^{V_1^*}$ satisfies the following conditions:

(C1) (Domain): $\mathrm{Dom}(C(q)) = \mathrm{Dom}(C)$ is independent of q for $q \in Q$, and $0 \in \mathrm{Dom}(C)$;

(C2) (Continuity): For each $\varphi \in \mathrm{Dom}(C)$, the map $q \to (C(q))\varphi$ is lower semicontinuous from $Q \subset \mathcal{Q}$ into $2^{V_1^*}$;

(C3) (Maximal Monotonicity): For $(\varphi_1, \psi_1), (\varphi_2, \psi_2) \in C_q \equiv \{(\varphi, \psi) \in V_1 \times V_1^* : \varphi \in \mathrm{Dom}(C), \psi \in C(q)\varphi\}$ we have $\langle \psi_1 - \psi_2, \varphi_1 - \varphi_2\rangle_1 \geq 0$ with C_q not properly contained in any other subset of $V_1 \times V_1^*$ for which (C3) holds;

(C4) (Uniform Boundedness): The operators $C(q)$ map V_1-bounded subsets of $\mathrm{Dom}(C)$ into subsets of V_1^* which are uniformly V_1^*-bounded in q for $q \in Q$.

For each $q \in Q$ the operator $L(q) \in \mathcal{L}(V_1, V_2^*)$ and $L(q)^* \in \mathcal{L}(V_2, V_1^*)$ satisfy

(L1) (Continuity): For $\varphi \in V_1$ and $\psi \in V_2$ the mappings $q \in L(q)\varphi$ and $q \to L(q)^*\psi$ are continuous from $Q \subset \mathcal{Q}$ into V_1^* or V_2^* (which ever is appropriate).

(L2) (Uniform Boundedness): There exists positive constant ρ, independent of $q \in Q$, for which $\|L(q)\varphi\|_{2^*} \leq \rho\|\varphi\|_1$, $\varphi \in V_1$.

We require that the mappings $q \to u_0(q), q \to v_0(q)$ and $q \to \theta_0(q)$ are continuous from $Q \subset \mathcal{Q}$ into V_1, H_1, and H_2, respectively, as are the mappings $q \to f(t;q)$ and $q \to g(t;q)$ into H_1 and H_2, respectively, for a.e. $t \in [0,T]$.

(D) (Uniform Domination) There exist $f_0, g_0 \in L^1(0,T)$, independent of $q \in Q$, for which $|f(t;q)|_1 \leq f_0(t)$, $|g(t;q)|_2 \leq g_0(t)$, a.e. $t \in [0,T]$ and every $q \in Q$.

We prove the well-posedness of the system (1)–(3) for each $q \in Q$ by first rewriting it as an equivalent first order system in an appropriate product Hilbert space and then applying results from nonlinear evolution systems theory.

Let X be the Banach space defined by $X = V_1 \times H_1 \times H_2$ with norm $|\cdot|_X$ give by $|(\varphi, \psi, \eta)|_X = (\|\varphi\|_1 + |\psi|_1^2 + |\eta|_2^2)^{1/2}$. For each $q \in Q$, let $X(q)$ denote the Hilbert space which is set equivalent to X and which is endowed with the inner product $\langle \cdot, \cdot \rangle_q$ given by

$$\langle (\varphi_1, \psi_1, \eta_1), (\varphi_2, \psi_2, \eta_2) \rangle_q \langle A_1(q)\varphi_1, \varphi_2 \rangle_1 + \alpha_1 \langle \varphi_1, \varphi_2 \rangle_1 + \langle \psi_1, \psi_2 \rangle_1 + \langle \eta_1, \eta_2 \rangle_2,$$

for $(\varphi_i, \psi_i, \eta_i) \in X$, $i = 1, 2$. We denote the norm on $X(q)$ induced by the inner product $\langle \cdot, \cdot \rangle_q$ by $|\cdot|_q$ and note that it is clear that assumptions (A3) and (A4) imply that the norms $|\cdot|_X$ and $|\cdot|_q$ are equivalent, uniformly in $q \in Q$. That is, there exist positive constants m and M, independent of $q \in Q$, for which

$$m|\cdot|_q \le |\cdot|_X \le M|\cdot|_q \tag{5}$$

For each $q \in Q$ define the operator $A(q) : \mathrm{Dom}(A(q)) \subset X(q) \to X(q)$ by

$$A(q)(\varphi, \psi, \eta) = (-\psi, \{A_1(q)\varphi + C(q)\psi + L(q)^*\eta\} \cap H_1, \{-L(q)\psi + A_2(q)\eta\} \cap H_2),$$

$$\text{with} \quad \begin{aligned} \mathrm{Dom}(A(q)) = \{(\varphi, \psi, \eta) \in V_1 \times V_1 \times V_2 \ :\ & \psi \in \mathrm{Dom}(C), \\ & \{A_1(q)\varphi + C(q)\psi + L(q)^*\eta\} \cap H_1 \neq \emptyset, \\ & \{-L(q)\psi + A_2(q)\eta\} \cap H_2 \neq \emptyset\} \end{aligned} \tag{6}$$

Theorem 1. ([2]) *There exists an $\omega \in \mathbb{R}$, independent of $q \in Q$, for which the operator $A(q) + \omega I$ is m-accretive.*

For each $q \in Q$ define $F \in L^1(0, T; X(q))$ by $F(t; q)(0, f(t; q), g(t; q))$, a.e., $t \in (0, T)$, and set $x_0(q) = (u_0(q), v_0(q), \theta_0(q)) \in X$. Theorem 1 yields that $A(q)$ and $F(\cdot\,; q)$ generate a nonlinear evolution system $\{U(t, s; q) \ :\ 0 \le s \le t \le T\}$ on $\overline{\mathrm{Dom}(A(q))}$. Henceforth we shall assume that $x_0(q) \in \overline{\mathrm{Dom}(A(q))}$ for each $q \in Q$. We refer to the function $x(\cdot\,; q) = (u(\cdot\,; q), v(\cdot\,; q), \theta(\cdot\,; q))$ given by

$$x(t; q) = U(t, 0; q)x_0(q), \qquad t \in [0, T], \tag{7}$$

as the unique mild solution to the abstract thermoelastic system (1)–(3).

We now define an identification problem corresponding to (1)–(3). Let $\mathcal{Z}$ denote an observation space. For $i = 1, 2, \ldots, \nu$ and $z \in \mathcal{Z}$, let $\Phi_i(\cdot, \cdot\,; z) : X \times Q \to \mathbb{R}^+$ denote a continuous map from $X \times Q \subset X \times \mathcal{Q}$ into the nonnegative real numbers. We consider the following parameter estimation problem.

(ID) Given observations $\{z_i\}_{i=1}^{\nu} \subset \times_{i=1}^{\nu} \mathcal{Z}$ at times $\{t_i\}_{i=1}^{\nu} \subset \times_{i=1}^{\nu}[0, T]$ determine parameters $\overline{q} \in Q$ which minimize

$$J(q) = \sum_{i=1}^{\nu} \Phi_i(x(t_i; q), q; z_i), \tag{8}$$

where for each $q \in Q$ and $t_i \in [0, T]$, $x(t_i; q)$ is given by (1)–(3).

3 Approximation Framework and Convergence Result

Using a standard Galerkin technique we construct a sequence of finite dimensional approximations to the abstract thermoelastic system (1)–(3). A sequence of approximating identification problems results. For $j = 1, 2$ and for each $n_j = 1, 2, \ldots$ let $H_j^{n_j}$ be a finite dimensional subspace of H_j with $H_j^{n_j} \subset V_j$, for all n_j. Let $P_j^{n_j} : H_j \to H_j^{n_j}$ denote the orthogonal projection of H_j onto $H_j^{n_j}$ computed with respect to the $\langle \cdot, \cdot \rangle_j$ inner product. Assume that $P_1^{n_1}\mathrm{Dom}(C) \subset \mathrm{Dom}(C)$ and that the following approximation condition holds

(P) (Approximation): For $j = 1, 2 \quad \lim_{n_j \to \infty} \|P_j^{n_j}\varphi - \varphi\|_j = 0, \qquad \varphi \in V_j.$

For each $q \in Q$ we define the operators $A_j^{n_j}(q) \in \mathcal{L}(H_j^{n_j})$, $j = 1, 2$, $C^{n_1}(q) : \mathrm{Dom}(C^{n_1}) \subset H^{n_1} \to 2^{H^{n_1}}$, and $L^n(q) \in \mathcal{L}(H_1^{n_1}, H_2^{n_2})$ using standard Galerkin approximation. More precisely, for $j = 1, 2$, and $\varphi^{n_j} \in H_j^{n_j}$, we set $A_j^{n_j}(q)\varphi^{n_j} = \psi^{n_j} \in H_j^{n_j}$ where ψ^{n_j} is the unique element in $H_j^{n_j}$ guaranteed to exist by the Riesz representation theorem satisfying $\langle A_j(q)\varphi^{n_j}, \chi^{n_j}\rangle_j \langle \psi^{n_j}, \chi^{n_j}\rangle_j$, $\chi^{n_j} \in H_j^{n_j}$. Similarly we set $L^n(q)\varphi^{n_1} = \psi^{n_2}$, where $\chi^{n_2} \in H_2^{n_2}$ satisfies $\langle L(q)\varphi^{n_1}, \chi^{n_2}\rangle_2 \langle \psi^{n_2}, \chi^{n_2}\rangle_2$, $\chi^{n_2} \in H_2^{n_2}$. We define the operator $L^n(q)^* \in \mathcal{L}(H_2^{n_2}, H_1^{n_1})$ to be the Hilbert space adjoint of the operator $L^n(q)$. For $\varphi^{n_1} \in \mathrm{Dom}(C^{n_1}) \equiv \mathrm{Dom}(C) \cap H^{n_1} \neq \emptyset$, let

$$C^{n_1}(q)\varphi^{n_1} = \{\psi^{n_1} : \langle \psi, \chi^{n_1}\rangle = \langle \psi^{n_1}, \chi^{n_1}\rangle, \chi^{n_1} \in H^{n_1}, \text{for some } \psi \in C(q)\varphi^{n_1}\}.$$

We set $u_0^{n_1}(q) = P_1^{n_1}u_0(q)$, $v_0^{n_1}(q) = P_1^{n_1}v_0(q)$ and $\theta_0^{n_2}(q) = P_2^{n_2}\theta_0(q)$, and set $f^{n_1}(t;q) = P_1^{n_1}f(t;q)$ and $g^{n_2}(t;q) = P_2^{n_2}g(t;q)$ for almost every $t \in [0, T]$.

We then consider the finite dimensional system of ordinary differential equations in $H^n = H_1^{n_1} \times H_2^{n_2}$ given by

$$\ddot{u}^n(t) + C^{n_1}(q)\dot{u}^n(t) + A_1^{n_1}(q)u^n(t) + L^n(q)^*\theta^n(t) \ni f^{n_1}(t;q), \qquad t > 0, \quad (9)$$

$$\dot{\theta}^n(t) + A_2^{n_2}(q)\theta^n(t) - L^n(q)\dot{u}^n(t) = g^{n_2}(t;q), \qquad t > 0, \quad (10)$$

$$u^n(0) = u_0^{n_1}(q), \quad \dot{u}^n(0) = v_0^{n_1}(q), \quad \theta^n(0)\theta_0^{n_2}(q) \quad (11)$$

We next rewrite (9)–(11) as an equivalent first order system. For each $n_1, n_2 = 1, 2, \ldots$ and $n = (n_1, n_2)$ let $X^n = H_1^{n_1} \times H_1^{n_1} \times H_2^{n_2}$ be considered as a subspace of the Banach space X, and for each $q \in Q$ let $X^n(q) = X^n$ be considered as a subspace of the Hilbert space $X(q)$. For each $q \in Q$ let $A^n(q) : \mathrm{Dom}(A^n(q)) \subset X^n(q) \to 2^{X^n(q)}$ be given by

$$A^n(q)(\varphi^{n_1}, \psi^{n_1}, \eta^{n_2}) = (-\psi^{n_1}, \{A_1^{n_1}(q)\varphi^{n_1} + C^{n_1}(q)\psi^{n_1} + L^n(q)^*\eta^{n_2}\},$$
$$\{-L^n(q)\psi^{n_1} + A_2^{n_2}(q)\eta^{n_2}\}),$$

$F^n(t;q) = (0, f^{n_1}(t;q), g^{n_2}(t;q))$, a.e. $t \in [0, T]$, $x_0^n(q) = (u_0^{n_1}(q), v_0^{n_1}(q), \theta_0^{n_2}(q))$.

Setting $x^n(t) = (u^n(t), \dot{u}^n(t), \theta^n(t))$, we rewrite (9)–(11) as

$$\dot{x}^n(t) + A^n(q)x^n(t) \ni F^n(t;q), \quad t > 0, \qquad x^n(0) = x_0^n(q). \quad (12)$$

The solution to the initial value problem (12) is given by

$$x^n(t;q) = U^n(t,0;q)x_0^n(q), \tag{13}$$

for $t \in [0,T]$, where $\{U^n(t,s;q) : 0 \leq s \leq t \leq T\}$ is the nonlinear evolution system on X^n (or $X^n(q)$). Note that $F_n(\cdot\,;q) \in L^1(0,T;X^n(q))$ and that the assumptions that $x_0(q) \in \overline{\mathrm{Dom}(A(q))}$ and $P^{n_1}\mathrm{Dom}(C) \subset \mathrm{Dom}(C)$ imply that $x_0^n(q) \in \overline{\mathrm{Dom}(A(q))}$.

It is not difficult to show that $A^n(q) + \omega I$ is m-accretive in $X^n(q)$. It follows that for each n and each $q \in Q$, $A^n(q)$ and $F^n(\cdot\,;q)$ generate a nonlinear evolution system, $\{U^n(t,s;q) : 0 \leq s \leq t \leq T\}$ on $\overline{\mathrm{Dom}(A(q))}$.

Definition 1. By a mild solution $x^n(q) = x^n(\cdot\,;q)$ to the initial value problem (12) we shall mean the $V_1 \times H_1 \times V_2$-continuous function $x^n(\cdot\,;q)(u^n(\cdot\,;q), \dot{u}^n(\cdot\,;q), \theta^n(\cdot\,;q))$ given by $x^n(\cdot\,;q) = U^n(\cdot\,,0;q)x_0^n(q)$. The V_1-continuous first component of $x^n(\cdot\,;q)$ will be taken to be $u^n(\cdot\,;q)$, H_1-continuous second component of $x^n(\cdot\,;q)$ will be taken to be $\dot{u}^n(\cdot\,;q)$ and the V_2-continuous third component of $x^n(\cdot\,;q)$ will be taken to be $\theta^n(\cdot\,;q)$.

We define the finite dimensional approximating identification problems:

(ID^n) Given observations $\{z_i\}_{i=1}^{\nu} \subset \times_{i=1}^{\nu}\mathcal{Z}$ at times $\{t_i\}_{i=1}^{\nu} \subset \times_{i=1}^{\nu}[0,T]$, find parameters $\overline{q}^n \in Q$ which minimize $J^n(q) = \sum_{i=1}^{\nu} \Phi_i(x^n(t_i;q), q; z_i)$, where for each $q \in Q$ and $t_i \in [0,T]$, $x^n(t_i;q)$ is given by (12).

Theorem 2. ([2]) *For each n, (ID^n) admits a solution $\overline{q}_n \in Q$. Moreover, the sequence $\{\overline{q}_n\}_{n=1}^{\infty}$ has a convergent subsequence $\{\overline{q}_{n_k}\}_{k=1}^{\infty}$ with $\lim_{k\to\infty} \overline{q}_{n_k} = \overline{q} \in Q$, where $\overline{q}$ is a solution to problem (ID).*

4 An Example and Numerical Results

We consider the problem of estimating or identifying the nonlinear damping term in the following thermoelastic system:

$$\frac{\partial^2 u}{\partial t^2}(t,x) - a(q_a)\left(\frac{\partial u}{\partial t}(t,x)\right) - q_{A_1}\frac{\partial^2 u}{\partial x^2}(t,x) + q_{L^*}\frac{\partial \theta}{\partial x}(t,x) \ni f_0(t,x), \tag{14}$$

$$\frac{\partial \theta}{\partial t}(t,x) - q_{A_2}\frac{\partial^2 \theta}{\partial x^2}(t,x) - q_L\frac{\partial^3 u}{\partial x^2 \partial t}(t,x) = g_0(t,x), \tag{15}$$

for $0 < x < \ell$, and $t > 0$. With the introduction into the first equation above of the Voigt-Kelvin viscoelastic damping term, equations (14) and (15) describe the longitudinal, or axial, vibrations of a thin visco-thermoelastic rod of length ℓ. Here u denotes the axial displacement, θ the absolute temperature and f_0 and g_0 represent, respectively, an externally applied axial force and thermal input.

We are interested in studying (14), (15) together with initial conditions

$$u(0,x) = u_0(x), \quad \frac{\partial u}{\partial t}(0,x) = v_0(x), \quad \theta(0,x) = \theta_0(x), \tag{16}$$

for $0 < x < \ell$ and the Dirichlet boundary conditions:

$$u(t,0) = u(t,\ell) = 0, \qquad \theta(t,0) = \theta(t,\ell) = 0, \qquad t > 0. \tag{17}$$

We assume that $f_0, g_0 \in L^2((0,T) \times (0,\ell))$, $u_0 \in H_0^1(0,1)$ and $v_0 \in L^2(0,1)$. Let $\mathcal{Q}$ be a metric space and let $Q \subset \mathcal{Q}$ be compact. For each $q \in Q$ we assume that the mapping $a(q_a)(\cdot) : I\!R \to 2^{I\!R} \backslash \emptyset$ satisfies the following conditions:

(a1) We have $0 \in a(q_a)(0)$,
(a2) The mapping $q_a \to a(q_a)(\zeta)$ is lower semi-continuous from $Q_a \subset \mathcal{Q}$ into $2^{I\!R}$ for almost every $\zeta \in I\!R$,
(a3) For each $q_a \in Q_a$ the mapping $a(q_a)(\cdot)$ is nondecreasing and for some $\lambda > 0$ the inclusion $\zeta + \lambda a(q_a)(\zeta) \ni \xi$ has a solution $\zeta \in I\!R$ for each $\xi \in I\!R$ (i.e., $a(q_a)(\cdot)$ is maximal monotone in $I\!R$),
(a4) There exists a polynomial p, independent of $q_a \in Q_a$, for which $|\tilde{\zeta}| \le p(|\zeta|)$ for all $\tilde{\zeta} \in a(q_a)(|\zeta|)$, and for almost every $\zeta \in I\!R$.

We set $H_1 = L^2(0,\ell)$ endowed with the standard inner product (to be denoted by $\langle \cdot\,, \cdot \rangle_1$) and set $H_2(0,\ell)$ with the inner product

$$\langle \varphi, \psi \rangle_2 = \int_0^\ell \varphi\psi.$$

In the case of the boundary conditions (17) we define both V_1 and V_2 to be the Sobolev space $H_0^1(0,\ell)$. For each $q \in Q$ we define the operators $A_j(q) \in \mathcal{L}(V_j, V_j^*)$, $j = 1,2$, and $L(q) \in \mathcal{L}(V_2, V_2^*)$ by

$$(A_j(q)\varphi)(\psi) = \int_0^\ell q_{A_j} D\varphi\, D\psi, \qquad \varphi, \psi \in V_j, \qquad j = 1,2. \tag{18}$$

$$(L(q)\varphi)(\psi) = -\int_0^\ell q_L D\varphi\psi, \qquad \varphi \in V_1, \psi \in V_2. \tag{19}$$

Definition (19) then yields that the operator $L(q)^* \in \mathcal{L}(V_2, V_1^*)$ is given by

$$(L(q)^*\psi)(\varphi) = \int_0^\ell q_{L^*} D\psi\varphi, \qquad \varphi \in V_1, \psi \in V_2. \tag{20}$$

We assume that $u_0 \in H_0^1(0,\ell)$, $v_0, \theta_0 \in L^2(0,\ell)$, $f_0, g_0 \in L^1(0,T; L^2(0,\ell))$ for almost every $(t,x) \in [0,T] \times [0,\ell]$. It is a simple matter to show that the assumptions (A1)–(A4), (L1)–(L2) and (D) stated in Section 2 are satisfied.

The definition of the operator $C(q)$ uses the notion of a subdifferential of a proper convex lower semicontinuous mapping. For each $q_a \in Q_a$, let $a_0(q_a)(\cdot)$ denote the minimal section of the mapping $a(q_a)(\cdot)$ is the single-valued mapping from $I\!R$ into $I\!R$ defined by $a_0(q_a) = \tilde{\zeta}$, where $\tilde{\zeta}$ is the unique element in $a(q_a)(\zeta)$ of minimal absolute value. Since $\mathrm{Dom}(a(q_a)(\cdot)) = I\!R$, the proper, convex, lower semi-continuous function $j(\,\cdot\,; q_a) : I\!R \to I\!R$ can be defined by

$$j(\zeta; q_a) = \int_0^\zeta a_0(q_a)(\xi)\, d\xi.$$

For each $q_a \in Q_a$ define $\gamma(\cdot\,;q_a) : H_0^1(0,1) \to I\!R$ by

$$\gamma(\varphi;q_a) = \int_0^1 j(\varphi(x);q_a)\,dx.$$

Then $\gamma(\cdot\,;q_a)$ is also proper, convex and lower semi-continuous and we define the operators $C(q_a) : V \to 2^{V^*}$ by

$$C(q_a)\varphi = \partial\gamma(\varphi;q_a). \tag{21}$$

It can be shown, that $C(q_a)$ given by (21) satisfies the conditions (a1)–(a4).

Specifically, we are concern with the estimation of the constant parameters $q_a = (\alpha_0, \beta_0, \zeta_0)$ in the saturation function with polynomial growth given by

$$a(q_a)(\zeta) = \begin{cases} \alpha_0|\zeta|^{\beta_0}\mathrm{sgn}(\zeta), & -\zeta_0 \le \zeta \le \zeta_0, \\ \alpha_0|\zeta_0|^{\beta_0}\mathrm{sgn}(\zeta), & |\zeta| > \zeta_0. \end{cases} \tag{22}$$

We take $Q = I\!R^3$, $Q_a = \{(\alpha_0, \beta_0, \zeta_0) \,:\, 0 \le \alpha_0 \le \overline{\alpha}_0,\ 0 \le \beta_0 \le \overline{\beta}_0,\ 0 \le \zeta_0 \le \overline{\zeta}_0\}$, for some $\overline{\alpha}_0, \overline{\beta}_0, \overline{\zeta}_0 \ge 0$ given and fixed. In order to actually test our identification problem numerically, we set

$$q_a^*(\zeta) = \begin{cases} \alpha|\zeta|^{\beta}\mathrm{sgn}(\zeta), & -\zeta_{q_a^*} \le \zeta \le \zeta_{q_a^*}, \\ \alpha|\zeta_{q_a^*}|^{\beta}\mathrm{sgn}(\zeta), & |\zeta| \ge \zeta_{q_a^*}, \end{cases}$$

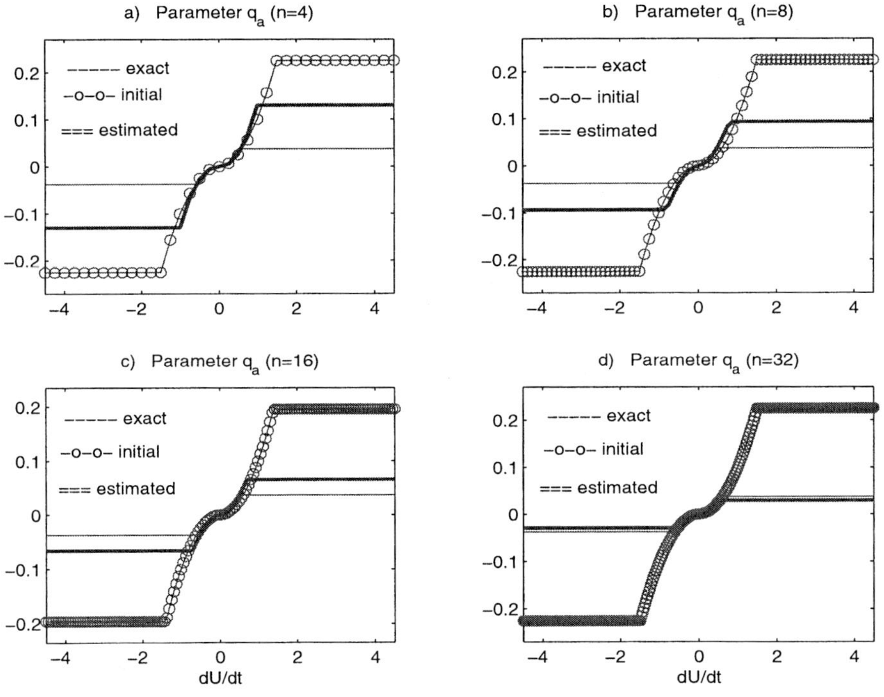

Fig. 1. Initial, true and estimated profiles for the saturation function

with $\alpha = .16$, $\beta = 2$, and $\zeta_{q_a^*} = 2.4$. With $u(t,x) = (\sin t)(\sin \pi x)$ and $\theta(t,x) = e^t(x^3 - x^2)$, for $t > 0$ and $x \in [0,1]$, we set

$$f(t,x) = \frac{\partial^2 u}{\partial t^2}(t,x) + q_a^* \left(\frac{\partial u}{\partial t}(t,x) \right) - q_{A_1}^* \frac{\partial^2 u}{\partial x^2}(t,x) + q_{L^*}^* \frac{\partial \theta}{\partial x},$$

$$g(t,x) = \frac{\partial \theta}{\partial t}(t,x) - q_{A_2}^* \frac{\partial^2 \theta}{\partial x^2}(t,x) - q_L^* \frac{\partial^3 u}{\partial x^2 \partial t}$$

with $q_{A_1}^* = q_{L^*}^* = q_{A_2}^* = q_L^* = 1$, $u^0(x) = u(0,x) = 0$, and $u^1(x) = \frac{\partial u}{\partial t}(0,x) = \sin \pi x$, for $t > 0$ and $x \in [0,1]$. For observations upon which to base our fit, we took $z = \{(z_{i,1}, z_{i,2})\}_{i=1}^{10}$ with $z_{i,1} = u(.5i, .12)$ and $z_{i,2}(x) = \frac{\partial u}{\partial t}(.5, x)$, $x \in [0,1]$, $i = 1, 2, \ldots, 10$. As an initial guess we set

$$q_a^0(\zeta) = \begin{cases} .6\zeta & -1.2 \le \zeta \le 1.5, \\ .6\mathrm{sgn}(\zeta) & |\zeta| \ge 1.5. \end{cases}$$

The results of the estimation for the saturation function $a(q_a)$ given in (22) are illustrated in Fig. 1. We notice that a good estimation is obtained when $n = 32$ (the true and estimated plots become indistinguishable).

5 Concluding Remarks

In this study we considered a parameter identification problem in a thermoelastic system. The approach started from an abstract operator formulation consisting of a coupled second order hyperbolic équation of elasticity and first order parabolic equation for heat conduction. A functional technique is used to demonstrate that solutions to a sequence of finite dimensional (Galerkin) approximating identification problems in some sense approximate a solution to the original infinite dimensional inverse problem. A specific example containing the estimation of a saturation function with polynomial growth is discussed. Future work will be devoted to identification of functional parameters involving spline based numerical schemes, along with theoretical and numerical studies on convergence rates of the estimations.

References

1. H. T. Banks, S. Reich, and I. G. Rosen. Estimation of nonlinear damping in second order distributed parameter systems, *Control Theory and Advanced Technology*, Vol. 6, No. 3, 395–415, 1990.
2. G. Dimitriu. *Parameter Identification for Some Classes of Nonlinear Systems*, PhD Thesis (in Romanian), 1999.
3. I. G. Rosen, C. H. F. Su. *An approximation theory for the identification of linear thermoelastic systems*, Report CAMS # 90–5, January, 1990.

A Monte Carlo Approach for the Cook-Torrance Model*

I.T. Dimov, T.V. Gurov, and A.A. Penzov

Inst. for Par. Proc. - Bulg. Acad. of Sci.,
Acad. G. Bonchev st, bl. 25 A,1113 Sofia, Bulgaria,
ivdimov@bas.bg
{gurov, apenzov}@parallel.bas.bg

Abstract. In this work we consider the rendering equation derived from the illumination model called Cook-Torrance model. A Monte Carlo (MC) estimator for numerical treatment of the this equation, which is the Fredholm integral equation of second kind, is constructed and studied.

1 Introduction

Photorealistic image creation is the main task in the area of computer graphics. The classical Radiosity and Ray Tracing algorithms have been developed to solve the global illumination for diffuse and specular scenes, respectively. However, application of these algorithms to general environments with multiple non-ideal reflection is restricted [8], due to local illumination model usage for image calculation. Monte Carlo algorithms provide with a proper rule for global illumination estimation of a scene, where the light radiated from the light sources is propagated through the scene objects to the human eye.

In order to estimate the global illumination in the scene, it is required to apply a suitable illumination model. The illumination model (see [14] for a survey of illumination models) describes the interaction of the light with a point on the surface in the scene. The simplest illumination model is independent of viewer direction and applicable for perfectly diffuse light reflecting scenes. It considers the light reflection as a sum of two components: ambient and Lambertian diffuse, reflecting light equally in all directions. In 1975 Phong [7] introduces an empirical three-component model with adding to illumination a new specular reflecting component. For the calculation of specular part, the viewer direction becomes more significant.

The first physical based illumination has been developed by Blinn [1] in 1977. And after that in 1982 Cook and Torrance [2] have suggested the complete implementation of the illumination model based on closer look to the physics of a surface. Cook-Torrance illumination model is an isotropic model and considers

* Supported by the EC through SEE-GRID project grant # RI-2002-002356 and by the NSF of Bulgaria through grant # I-1201/02.

Z. Li et al. (Eds.): NAA 2004, LNCS 3401, pp. 257–265, 2005.

the geometry of object surface to be non-ideally smooth but as composed of many microfacets. The microfacet surfaces are V shaped grooves where some microfacets shadow or mask each other. This fact leads up to some attenuation of reflected light. The roughness of the surface is defined by the microfacet distribution. Surface reflectance is described by Fresnel term of a single microfacet and may be obtained theoretically from the Fresnel equations [13].

Many others physical based illumination models develop and/or extend the Cook-Torrance model. An extension presented in [4] splits the diffuse reflection component into directional-diffuse and uniform-diffuse components. Anisotropic model of Ward [9] extends the scope of physical based illumination models. The physical illumination models are also applicable for photorealistic rendering when transparent objects exist in the scene [3].

Further in this paper we consider basically Cook-Torrance illumination model at construction of the Monte Carlo estimator for photorealistic image creation.

2 Rendering with Cook-Torrance Illumination Model

The goal of rendering is to calculate an image from a described 3D scene model. Photorealistic rendering requires realistic description of the scene model with accounting all physical characteristics of the surfaces and environment. The scene model consists of numerical definition of all objects and light sources in the scene, as well the virtual camera from which the scene is observed. Colours in computer graphics are frequently simulated in the RGB colour space, so the radiance $L = (r, g, b)$, where r, g, and b are the intensities for the selected wavelenghts of primary monitor colours (red, green and blue).

2.1 Basic Assumptions

The objects represent real physical objects including solid light sources like lamps with arbitrary defined position and orientation in the scene space. Usually the real solid objects are well modeled by approximation with planar surface primitives like triangles or rectangles. The number M of all surface primitives A_j is very large to ensure good object approximation for realistic rendering. Since the objects are independent scene units the scene domain, $S = \bigcup_{j=1}^{M} A_j$, is union of disjoint two-dimensional surface primitives A_j. The physical properties like reflectivity, roughness, and colour of the surface material are characterized by the bidirectional reflectance distribution function (BRDF), f_r. This function describes the light reflection from a surface point as a ratio of outgoing to incoming light. It depends on the wavelength of the light, incoming, outgoing light directions and location of the reflection point. The BRDF expression receives various initial values for the objects with different material properties. The same values for the BRDF expression are assigned to all surface primitives with equal material characteristics in the scene. Therefore, the number of surfaces with different material properties in the scene is m and the inequality $1 \leq m \leq M$ is hold.

The light sources can be both point light sources or solid light sources. The point light sources are defined by own position in the scene space and radiate light equally to all directions. The solid light sources are treated like ordinary scene objects itself radiating light. The total light intensity of solid light sources is distributed equally in a characteristic set of points generated on its surface and each point is considered like point light source.

The virtual camera frequently is assumed to be a pinhole camera and defines the eye view point position x_{eye} and orientation of scene observation. The image window situated in the projection plane of the camera is divided into matrix of rectangular elements and corresponds to the pixel matrix of the image to be generated. In order to generate an image, we have to calculate the radiance $\overline{L}_P$ propagated in the scene and projected on each pixel area P in the image pixel matrix. This radiance value is radiated through the pixel area P into the eye view point x_{eye}. The radiance $\overline{L}_P$ is mean value integral:

$$\overline{L}_P = \frac{1}{|P|} \int_P L(x_{eye}, x_p) dx_p \tag{1}$$

where $L(x_{eye}, x_p)$ is the radiance incoming from the nearest scene point $x \in S$ seen from the eye view point x_{eye} through the pixel position x_p into direction of the eye view point x_{eye}.

2.2 Rendering Equation by Cook-Torrance BRDF

The light propagation in a scene is described by rendering equation [5], which is a second kind Fredholm integral equation. According to Keller indications in [6], the radiance L, leaving from a point $x \in S$ on the surface of the scene in direction $\omega \in \Omega_x$, where Ω_x is the hemisphere in point x, is the sum of the self radiating light source radiance L^e and all reflected radiance:

$$L(x, \omega) = L^e(x, \omega) + \int_{\Omega_x} L(h(x, \omega'), -\omega') f_r(-\omega', x, \omega) \cos\theta' d\omega', \tag{2}$$

or in an operator form $L = L^e + \mathbb{K}L$. Here $y = h(x, \omega') \in S$ is the first point that is hit when shooting a ray from x into direction ω' and determines the objects visibility in the scene (see Fig. 1). The radiance L^e has non-zero value if the considered point x is a point from solid light source. Therefore, the reflected radiance in direction ω is an integral of the radiance incoming from all points, which can be seen through the hemisphere Ω_x in point x attenuated by the surface BRDF $f_r(-\omega', x, \omega)$ and the projection $\cos\theta'$, which puts the surface $S \times \Omega \to \mathbb{R}^+$ perpendicular to the ray (x, ω'). The angle θ' is the angle between surface normal in x and the direction ω'. The transport operator is physically correct when $\|\mathbb{K}\| < 1$, because a real scene always reflects less light than it receives from the light sources due to light absorption of the objects. The law for energy conservation holds, i.e.: $\alpha(x, \omega) = \int_{\Omega_x} f_r(-\omega', x, \omega) \cos\theta' d\omega' < 1$. That means the incoming photon is reflected with a probability less than 1,

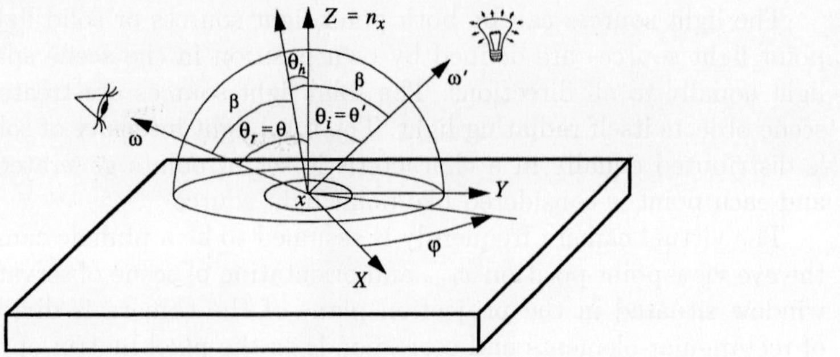

Fig. 1. The geometry for the rendering equation

because the selected energy is less than the total incoming energy. Another important property of the BDRF is the Helmholtz principle: the value of the BRDF will not change if the incident and reflected directions are interchanged, $f_r(-\omega', x, \omega) = f_r(-\omega, x, \omega')$. In the terms of the Cook-Torrance illumination model [2], the BRDF is sum of diffuse $f_{r,d}$ and specular $f_{r,s}$ components:

$$f_r(-\omega', x, \omega) = f_{r,d}(x) + f_{r,s}(-\omega', x, \omega) = \frac{1}{\pi}\left(F(\lambda, \theta' = 0) + \frac{F(\lambda,\theta')D(\theta_h)G}{\cos\theta\cos\theta'}\right),$$

where the angle θ is the angle between surface normal in x and the direction ω. The microfacets distribution function is denoted by $D(\theta_h)$ and $G(\theta_h, \beta, \theta, \theta')$ is geometrical attenuation factor. Fresnel function $F(\lambda, \theta')$ depends on the wavelength λ, incident angle θ' of the light in point x, index of refraction, and absorption coefficient of surface material (see [13]). The diffuse part $f_{r,d}(x)$ of BRDF is the fraction of reflected radiance, independently of incoming and outgoing directions, and may be calculated from the Fresnel equations at angle of incident light $\theta' = 0$, or $f_{r,d}(x) = \frac{F(\lambda,\theta'=0)}{\pi}$. Both values of the microfacets distribution function $D(\theta_h)$ and the geometrical attenuation factor $G(\theta_h, \beta, \theta, \theta')$ are positive and can not exceed a maximum value of 1 (see [1]) in any real scene situation. Therefore, the specular part $f_{r,s}(-\omega', x, \omega)$ of BRDF reaches the maximum value when the Fresnel spectral term has absolute maximum for some light wavelength.

2.3 Analysis of the Neumann Series

Consider the *first-order stationary linear iterative process* for Eq. (2).

$$L_i = \mathbb{K} L_{i-1} + L^e, \quad i = 1, 2, \ldots, \tag{3}$$

where i is the number of the iterations. In fact (3) defines a Neumann series

$$L_i = L^e + \mathbb{K} L^e + \ldots + \mathbb{K}^{i-1} L^e + \mathbb{K}^i L_0, \qquad i > 0,$$

where $\mathbb{K}^i$ means the i-th iteration of the integral operator. If $\mathbb{K}$ is a contraction, then $\lim_{i\to\infty} \mathbb{K}^i L_0 = 0$. Thus $L^* = \sum_{i=0}^{\infty} \mathbb{K}^i L^e$. If $i = k$ and $L_0 = 0$, one can get the value of the truncation error, namely, $L_k - L^* = \sum_{i=k}^{\infty} \mathbb{K}^i L^e$.

It is clear that every iterative algorithm uses a finite number of iterations k. In the presented MC approach below, we evaluate the iterations $L_i,\ 1 \leq i \leq k$ with an additional statistical error. In practice the truncation parameter k is not a priori given. To define it let us denote $\|I\!K\|_{I\!L_1} = \max_{x,\omega} |\alpha(x,\omega)| = q < 1$ and $\|L^e\|_{I\!L_1} = L^e_*$. Further, in order to estimate the error we have

$$\|L_k - L^*\|_{I\!L_1} \leq L^e_* q^k \frac{1}{1-q}$$

Finally, to obtain a desired truncation error ε we have to select $k_\varepsilon = \min\{k \geq c_1|\ln\varepsilon| + c_2\}$, where $c_1 = 1/|\ln(q)|$ and $c_2 = |\ln((1-q)/(L^e_*))|/|\ln(q)|$. In other cases the iteration parameter is obtained from the following condition: the difference between the stochastic approximation of two successive approximations has to be smaller than the given sufficiently small parameter ε.

3 A Monte Carlo Approach

Consider the problem for evaluating the following functional:

$$J_g(L) = (g, L) = \int_S \int_{\Omega_x} g(x,\omega) L(x,\omega) dx d\omega. \tag{4}$$

The radiance $L(.,.) : S \times \Omega \to I\!R^+$ and the arbitrary function $g(.,.) : S \times \Omega \to I\!R^+$ belong to the spaces $I\!L_1$ and $I\!L_\infty$, respectively. The case when $g(x,\omega) = \frac{\chi_P(x)}{|P|}\delta(\omega)$ is of special interest, because we are interested in computing the mean value of the radiance $\overline{L}_P$ over pixel area (1). Here $\chi_P(x) = 1$ when $x \in P$, and $\chi_P(x) = 0$, otherwise. $\delta(\omega)$ is the Dirac delta-function. Since the Neumann series of the integral equation (2) converges, the functional (4) can be evaluated by a MC method. Let us introduce the following notation: $\omega'_0 = \omega_p$, $x_0 = x_p$, $x_1 = h(x_0, -\omega'_0)$, $x_2 = h(x_1, -\omega'_1)$, $x_3 = h(x_2, -\omega'_2) = h(h(x_1, -\omega'_1), -\omega'_2)$, ..., and define the kernels: $K(x_j, \omega'_j) = f_r(\omega'_j, x_j, \omega'_{j-1}) \cos\theta'_j$, $j = 1, 2, \ldots, k_\varepsilon$. Consider a terminated Markov chain $(x_0, -\omega'_0) \to \ldots \to (x_j, -\omega'_j) \to \ldots \to (x_{k_\varepsilon}, -\omega'_{k_\varepsilon})$, such that $(x_j, -\omega'_j) \in S \times \Omega_x$, $j = 1, 2, \ldots, k_\varepsilon$ (see Fig. 2). The initial point

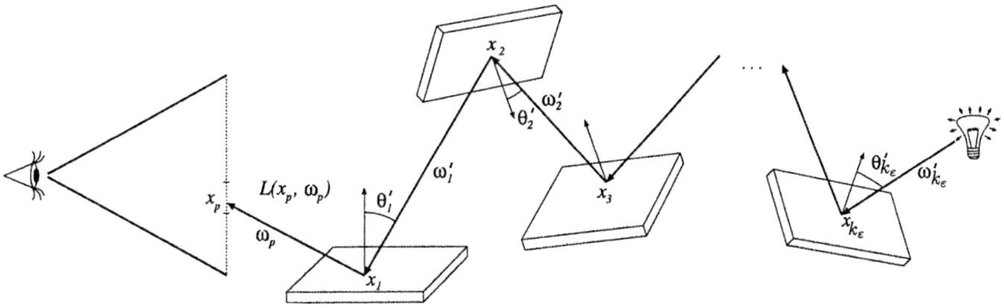

Fig. 2. One simulation of the terminated Markov chain

$(x_0, -\omega'_0)$ is chosen with an initial density $p_0(x, \omega')$. The other points are sampled using an arbitrary transition density function $p(x, \omega')$ which is tolerant (see definition in [10]) to the kernel in equation (2). The biased MC estimator for evaluating the functional (4) has the following form:

$$\xi_{k_\varepsilon}[g] = \frac{g(x_0, \omega'_0)}{p_0(x_0, \omega'_0)} \sum_{j=1}^{k_\varepsilon} W_j L^e(x_j, \omega'_j), \quad j = 1, \ldots, k_\varepsilon\,, \tag{5}$$

where the weights are defined as follows

$$W_0 = 1, \; W_j = W_{j-1} \frac{K(x_j, \omega'_j)}{p(x_j, \omega'_j)}, \; j = 1, \ldots, k_\varepsilon.$$

Theorem 1. *The expected value of the estimator (5) is equal to the functional (4) in case when we replace the radiance L with its iterative solution L_{k_ε}, i.e.*

$$E(\xi_{k_\varepsilon}[g]) = J_g(L_{k_\varepsilon}).$$

Proof. Taking into account the definition for an expected value of a random variable and the Neumann series of the Eq.(3), we obtain:

$$E(\xi_{k_\varepsilon}[g]) = E\left(\frac{g(x_0, \omega'_0)}{p_0(x_0, \omega'_0)} \sum_{j=1}^{k_\varepsilon} W_j L^e(x_j, \omega'_j) \right) =$$

$$= \sum_{j=1}^{k_\varepsilon} E\left(\frac{g(x_0, \omega'_0)}{p_0(x_0, \omega'_0)} W_j L^e(x_j, \omega'_j) \right) = \sum_{j=1}^{k_\varepsilon} (g, \mathbb{K}^j L^e) = (g, L_{k_\varepsilon}) = J_g(L_{k_\varepsilon}).$$

This completes the proof.

It is clear that when $k_\varepsilon \to \infty$ (this is the case of a infinite Markov chain) the MC estimator (5) evaluates the functional (4).

Using N independent samples of the estimator (5) we can compute the mean value:

$$\overline{\xi}_{k_\varepsilon}[g] = \frac{1}{N} \sum_{i=1}^{N} (\xi_{k_\varepsilon}[g])_i \stackrel{Prob}{\longrightarrow} J_g(L_{k_\varepsilon}) \approx J_g(L), \tag{6}$$

where $\stackrel{Prob}{\longrightarrow}$ means stochastic convergence as $N \to \infty$; L_{k_ε} is the iterative solution obtained by the Neumann series of Eq.(2).

The root mean square deviation is defined by the relation (see [11]):

$$E(\xi_{k_\varepsilon}[g] - J_g(L))^2 = Var(\xi_{k_\varepsilon}[g]) + (E(\xi_{k_\varepsilon}[g]) - J_g(L))^2,$$

where $Var(\xi_{k_\varepsilon}[g])$ is the variance of the MC estimator. Hence

$$E(\overline{\xi}_{k_\varepsilon}[g] - J_g(L))^2 = \frac{Var(\overline{\xi}_{k_\varepsilon}[g])}{N} + (J_g(L) - E(\xi_{k_\varepsilon}[g]))^2 \le \frac{d_0}{N} + c_3\varepsilon^2 = \mu^2, \tag{7}$$

where μ is the desired error; d_0 is upper boundary of the variance; ε is a priori given small parameter, and c_3 is the constant. Therefore, if the variance is bounded, the optimal order of the quantities N and ε must be $N = O(\mu^{-2})$ and $\varepsilon = O(\mu)$. In order to estimate the variance let us introduce notation:

$$\theta_j = \frac{g(x_0,\omega_0')}{p_0(x_0,\omega_0')} W_j L^e(x_j,\omega_j'),$$

$$p(x_0,\omega_0',x_1,\omega_1',\ldots,x_j,\omega_j') = p_0(x_0,\omega_0')p(x_1,\omega_1')\ldots p(x_j,\omega_j'),\ j=1,2,\ldots,.$$

Theorem 2. *Let us choose the initial density and transition density in the following way:*

$$p_0(x_0,\omega_0') = \frac{|g(x_0,\omega_0)|}{\int_S\int_{\Omega_{x_0}} g(x_0,\omega_0)dx_0d\omega_0},\quad p(x,\omega') = \frac{K(x,\omega')}{\int_{\Omega_x} K(x,\omega')d\omega'}. \tag{8}$$

Then the variance of the MC estimator (5) is bounded.

Proof. It is enough to prove that $E(\xi^2_{k_\varepsilon}[g])$ is bounded. Taking into account the following inequality (see [10])

$$\left(\sum_{j=1}^{\infty}\theta_j\right)^2 \le \sum_{j=1}^{\infty}\frac{t^{-j}}{1-t}\theta_j^2,\ 0<t<1,$$

we have

$$Var(\xi_{k_\varepsilon}[g]) \le E(\xi^2_{k_\varepsilon}[g]) = E\left(\sum_{j=1}^{k_\varepsilon}\theta_j\right)^2 \le E\left(\sum_{j=1}^{\infty}\theta_j\right)^2 \le$$

$$\sum_{j=1}^{\infty}\frac{t^{-j}}{1-t}E(\theta_j^2) = \sum_{j=1}^{\infty}\frac{t^{-j}}{1-t}\int_{S_{x_0}}\int_{\Omega_{x_0}}\int_{\Omega_{x_1}}\cdots\int_{\Omega_{x_j}}\frac{g^2(x_0,\omega_0')}{p_0^2(x_0,\omega_0')}\times$$

$$W_j^2(L^e)^2(x_j,\omega_j')p(x_0,\omega_0',x_1,\omega_1',\ldots,x_j,\omega_j')dx_0d\omega_0\ldots d\omega_j.$$

Taking in acount the choice of the densities we obtain

$$Var(\xi_{k_\varepsilon}[g]) \le \sum_{j=1}^{\infty}\frac{t^{-j}}{1-t}\left(\int_{S_{x_0}}\int_{\Omega_{x_0}}\int_{\Omega_{x_1}}\cdots\int_{\Omega_{x_j}} g(x_0,\omega_0')\times\right.$$

$$\left.K(x_0,\omega_0')\ldots K(x_j,\omega_j')L^e(x_j,\omega_j')dx_0d\omega_0\ldots d\omega_j\right)^2 \le \sum_{j=1}^{\infty}\frac{t^{-j}}{1-t}g_*^2(L_*^e)^2(q^2)^j,$$

where $g_* = \|g\|_{L_\infty}$. If we choose, $1>t>q$, then we have

$$Var(\xi_{k_\varepsilon}[g]) \le g_*^2(L_*^e)^2\frac{q^2}{(1-t)(t-q^2)}.$$

This completes the proof.

The choice Eq.(8) of the initial and transition densities leads to a reduction of the variance. We note that this choice is used in practice and it is very closely to the importance sampling strategy for variance reduction. The density functions (8) are called "almost optimal" densities [12]. In other cases when is not possible the choice (8), but the densities are chosen to be proportional to the main contribution from the kernel of the rendering equation.

When $g(x,\omega) = \frac{\chi_P(x)}{|P|}\delta(\omega)$ the MC estimator (5) evaluates the mean value of the radiance $\overline{L}_P$ over pixel area (1). In this case, we can take $\varepsilon = 1/(2.2^8)$ because the primary colours in the RGB colour system are presented by 8-bit numbers.

4 Summary and Issues for Future Work

The presented MC estimator evaluates the rendering equation derived from the Cook-Torance illumination model. It is proved that the variance of this estimator is bounded when we use almost optimal initial and transition densities. We obtain condition for balancing of systematic and stochastic errors. The advantages of the studied MC approach lie in the direct estimation of the functional value in fixed phase space points. Also, this approach is easy for parallel realizations over MIMD (multiple instruction - multiple data) architectures and Grid's. Finally, the future research of the MC approach under consideration for Cook-Torance model could be developed in the following directions: 1. Development of computational MC algorithms for creation of photorealistic images. 2. Investigation of the computational complexity of the algorithms for different materials. 3. Creation of parallel MC algorithms for high performance and Grid computing.

References

1. Blinn, James F., Models of Light Reflection for Computer Synthesized Pictures, Computer Graphics, vol. 11, No. 2, pp. 192-198, (1977).
2. Cook, Robert L. and Torrance, Kenneth E., A Reflectance Model for Computer Graphics, ACM Transactions on Graphics, vol. 1, No. 1, pp. 7-24, (1982).
3. Hall, Roy A. and Greenberg, Donald P., A Testbed for Realistic Image Synthesis, IEEE Computer Graphics and Applications, vol. 3, No. 8, pp. 10-20, (1983).
4. He, Xiao D., Torrance, Kenneth E., Sillion, Francois and Greenberg, Donald P., A Comprehensive Physical Model for Light Reflection, Computer Graphics, vol. 25, No. 4, pp. 175-186, (1991).
5. Kajiya, J. T., The Rendering Equation, Computer Graphics, vol. 20, No. 4, pp. 143-150, Proceedings of SIGGRAPH'86, (1986).
6. Keller, Alexander, Quasi-Monte Carlo Methods in Computer Graphics: The Global Illumination Problem, Lectures in Applied Mathematics, vol. 32, pp. 455-469, (1996).
7. Phong, Bui Tuong, Illumination for Computer Generated Pictures, Communication of the ACM, June, vol. 18, No. 6, pp. 311-317, (1975).

8. Szirmay-Kalos, Laszlo, Monte-Carlo Methods in Global Illumination, Script in WS of 1999/2000, http://www.fsz.bme.hu/ szirmay/script.pdf
9. Ward, Gregory, Measuring and modeling anisotropic reflection Computer Graphics, vol. 26 No. 2, pp. 265–272, (1992).
10. Gurov, T.V., Whitlock, P.A.: An efficient backward Monte Carlo estimator for solving of a quantum kinetic equation with memory kernel, *Math. & Comp. in Simul.* **60** (2002) pp. 85–105.
11. Mikhailov, G.A.: New Monte Carlo Methods with Estimating Derivatives. Utrecht, The Netherlands, (1995).
12. I. Dimov, Minimization of the probable error for some Monte Carlo methods. *Proc. of the Summer School on Math. Modeling and Sci. Computations, 23-28.09.1990, Albena, Bulgaria* (Publ. House of the Bulg. Acad. Sci., Sofia, 1991) 159–170.
13. Penzov, A. A., A Data Base Containing Spectral Data of Materials for Creating Realistic Images, Part 1: Theoretical Background, MTA SZTAKI, Report CG-1, Budapest, (1992).
14. Penzov, A. A., Shading and Illumination Models in Computer Graphics - a literature survey, MTA SZTAKI, Research Report CG-4, Budapest, (1992).

Generic Properties of Differential Inclusions and Control Problems

Tzanko Donchev

Department of Mathematics,
University of Architecture and Civil Engineering,
1 "Hr. Smirnenski" str., 1046 Sofia, Bulgaria
tdd51us@yahoo.com

Abstract. We prove that almost all continuous multifunctions are Lipschitz continuous with respect to a Kamke function. We obtain as a corollary that almost every differential inclusion with continuous right-hand side satisfies the relaxation property.

We point out also the possible applications in Bolza problem, given for differential inclusions.

1 Introduction

Consider the following differential inclusion:

$$\dot{x}(t) \in F(t, x(t)), \ x(0) = x_0, \ t \in I = [0,1]. \tag{1}$$

Here F is a map from $I \times \mathbb{R}^n$ into $CC(\mathbb{R}^n)$ (the set of all nonempty convex compact subsets of $\mathbb{R}^n$). If $F(\cdot,\cdot)$ is continuous then $\overline{ext}\ F(\cdot,\cdot)$ is lower semicontinuous (proposition 6.2 of [4]) and hence the system

$$\dot{x}(t) \in \overline{ext}\ F(t, x(t)), \ x(0) = x_0, \ t \in I \tag{2}$$

has a solution. Here $ext\ A := \{a \in A : \ a = \lambda b + (1-\lambda)c, \ b, c \in A, \ \lambda \in (0,1) \Rightarrow a = b \text{ or } a = c\}$.

It is well known the connection between control systems:

$$\dot{x}(t) = f(t, x(t), u(t)), \ t \in I, \ u(t) \in U, \ x(0) = x_0,$$

where U is metric compact and the system (1) (set $F(t,x) = \overline{co}\ f(t,x,U)$).

Central role there plays the so called relaxation property, which means that the solution set of (2) is dense in the solution set of (1) in uniform metric. Other important thing is the numerical approximation of the solution set with respect to $C(I, \mathbb{R}^n)$ and $W^{1,1}(I, \mathbb{R}^n)$.

Let $\mathbb{M}$ be a complete metric space. A set $A \subset \mathbb{M}$ is said to be of the first Baire category if it is a subset of an union of countable many closed subsets of $\mathbb{M}$ every one with an empty interior. A set $B \subset \mathbb{M}$ is said to be residual if $\mathbb{M} \setminus B$ is of the first category. The set C is said to be G_δ if it is intersection of countable

Z. Li et al. (Eds.): NAA 2004, LNCS 3401, pp. 266–271, 2005.

many open and dense subsets of $\mathbb{M}$. We said that some property hold for almost all elements of $\mathbb{M}$ if it holds on a residual subset of $\mathbb{M}$.

Our aim is to prove that almost all (in Baire sense) continuous multifunctions $F(\cdot,\cdot)$ are locally Kamke Lipschitz (CL).

Proving that we will derive as a corollary that for almost all F relaxation property holds and moreover, the usual Euler scheme approximates the solution set of (1) in $C(I,\mathbb{R}^n)$ and $W^{1,1}(I,\mathbb{R}^n)$. Using the later one can show that for almost all optimal control problems the solution is a limit of appropriate discrete optimal solutions. Due to the lenght limit we will not provide all the proofs (they will be given in other paper).

When the right-hand side F is single valued (acting in a Banach space) it is proved in [9] that for almost all F the differential equation (1) admits an unique solution, which depends continuously on the initial condition and on the right-hand side. Using Fort' lemma (cf. [7]) it is shown in [12] that for almost all (multi-functions) F the system (1) has a solution set depending continuously on (F, x_0). We mention also [1], where the existence of solutions for almost all differential inclusions in a Hilbert space is proved. This result is then extended in [5] for functional differential inclusions in Banach spaces. Using Baire categories De Blasi and Pianigiani investigated in a number of papers (among others [2, 3]) existence of solutions and some properties of the solution set of non-convex differential inclusions. However, to the author's knowledge in the literature there are not papers devoted to the genericity of the relaxation property (the solution set of (2) is dense in the solution set of (1)).

We refer to [4, 8] for all the concept used here but not explicitly discussed.

We will denote by $C(\mathbb{R}^n)$ and $CC(\mathbb{R}^n)$ the set of all nonempty compact respectively convex compact subsets of $\mathbb{R}^n$. For a subset A of a Banach space we let $diam(A) = \sup\{|x-y| : x, y \in A\}$. By $co\ A$ $(\bar{A})$ we denote the convex (closed) hull of A. For $A, B \in C(\mathbb{R}^n)$we define $D_H(A,B) = \max\{\max_{a\in A}\min_{b\in B} |a-b|, \max_{b\in B}\min_{a\in A} |a-b|\}$ – the Hausdorff distance between the compact set A and B. A multifunction is called continuous when it is continuous with respect to the Hausdorff metric. We denote by $C(I\times K)$ the set of all continuous (multi)functions from $I\times K$ (K is nonempty compact or $\mathbb{R}^n$) into $CC(\mathbb{R}^n)$.

Definition 1. *A continuous function $w : I\times\mathbb{R}^+ \to \mathbb{R}^+$ is said to be Kamke function if $w(t,0)\equiv 0$ and the unique solution of the differential equation $\dot{s}(t) = w(t,s(t))$, $s(0)=0$ is $s(t)\equiv 0$.*

The multifunction F is called (locally) Kamke Lipschitz (CL) if for every compact K there exists a Kamke function $w(\cdot,\cdot)$ such that $D_H(F(t,x),F(t,y)) \le w(t,|x-y|)$ for every x, y from K.

Multifunction $F(\cdot,\cdot)$ is said to be locally Lipschitz if for every compact $K\subset\mathbb{R}^n$ there exists a constant $L(K,F)$ such that $D_H(F(t,x),F(t,y)) \le L(K,F)|x-y|$ for every $x, y\in K$.

Given a compact set $D\subset\mathbb{R}^n$ we consider the distance:

$$C_D(F,G) := \max_{t\in I}\max_{x\in D} D_H\Big(F(t,x), G(t,x)\Big)$$

To avoid the problem of the continuation of the solutions we will consider in the sequel only multimaps $|F(t,x)| \le \lambda(1+|x|)$, where λ is a positive constant. Under this assumption the solution set of (1) is $C(I,\mathbb{R}^n)$ compact.

The following theorem is well known (cf. [11]).

Theorem 1. *If $F(\cdot,\cdot)$ is locally CL, then the solution set R_2 of (2) is dense in the solution set R_1 of (1). Moreover R_2 (and hence R_1) depends continuously on x_0 and on F.*

2 The Results

In this section we consider the case of jointly continuous right-hand side $F(\cdot,\cdot)$. If $F(\cdot,\cdot)$ is continuous on $I \times \mathbb{R}^n$, then it is uniformly continuous on $I \times K$ for every compact $K \subset \mathbb{R}^n$. Consequently there exists a (jointly) continuous function $w_F : I \times [0, diam(K)] \to \mathbb{R}_+$ such that $w_F(t,0) \equiv 0$ and $D_H(F(t,x),F(t,y)) \le w_F(t,|x-y|)$ for every $x,y \in K$.

We define the following metric, which generates the topology of uniform convergence of bounded subsets ob $\mathbb{R}^n$.

$$\rho_H(F(t,x),G(t,x)) := \sum_{n=1}^{\infty} \frac{C_{n\mathbb{B}}(F,G)}{2^n(1+C_{n\mathbb{B}}(F,G)}), \tag{3}$$

where $\mathbb{B}$ is the closed unit ball in $\mathbb{R}^n$. It is easy to see that endowed with this metric $C(I \times \mathbb{R}^n)$ becomes a complete metric space.

Theorem 2. *For every compact set $K \subset \mathbb{R}^n$ there exists a residual subset $\tilde{C}_K \subset C(I \times K)$ such that every $F \in \tilde{C}_K$ is locally CL.*

Proof. We have only to prove that for almost all continuous $F(\cdot,\cdot)$ the differential equation

$$\dot{s}(t) = w_F(t,s(t)), \;\; s(0) = 0 \tag{CE}$$

has the unique solution $s(t) \equiv 0$. Obviously $s(t) \equiv 0$ is a solution of (CE). Consider the differential inequality:

$$|\dot{s}(t) - w_F(t,s(t))| \le \varepsilon, \;\; s(0) = 0. \tag{IE}$$

Its solution set (denoted by $Sol_F(\varepsilon)$) consists of all absolutely continuous functions $s(\cdot)$ satisfying (IE) for a.a. $t \in I$. If $\lim\limits_{\varepsilon \to 0+} diam(Sol_F(\varepsilon)) = 0$ then the unique solution of (CE) is $s(t) \equiv 0$.

Denote by $(SR)_n := \left\{ F \in C(I \times \mathbb{R}^n) : \lim\limits_{\varepsilon \to 0+} diam(Sol_F(\varepsilon)) < \frac{1}{n} \right\}$.

Claim. For every natural n the set $(SR)_n$ is open and dense.

i) Obviously $(SR)_n$ is dense in $C(I \times \mathbb{R}^n)$, because every locally Lipschitz $F \in (SR)_n$.

ii) If F and G are multimaps with $|w_F - w_G|_C \le \frac{\varepsilon}{2}$ then every $s(\cdot) \in Sol_F(\frac{\varepsilon}{2})$ belongs to $Sol_G(\varepsilon)$ and vice versa. Let $F_i \notin (SR)_n$ and let $F_i \to F$ with respect to $C(I \times \mathbb{R}^n)$. Obviously in this case $w_{F_i}(t,s) \to w_F(t,s)$ uniformly on $I \times [0, diam(K)]$. Consequently $F \notin (SR)_n$. Due to the Claim the set $\tilde{C}_n = \bigcap_{n=1}^{\infty} (SR)_n$ is a residual subset of $C(I \times \mathbb{R}^n)$. □

Corollary 1. *Almost all continuous multifunctions are CL.*

Proof. Denote $K_n = n\bar{B}$. Due to Theorem 2 the set $\tilde{C}_{K_n}$ is a residual subset of $C(I \times \mathbb{R}^n)$ and hence $\bigcap_{n=1}^{\infty} C_{K_n} = \tilde{C}$ is a residual subset of $C(I \times \mathbb{R}^n)$. Obviously every $F \in \tilde{C}$ is CL. □

The following result is a trivial consequence of Theorem 2 and Corollary 1.

Theorem 3. *There exists a residual subset $\tilde{C}$ of $C(I \times \mathbb{R}^n)$, such that for every $F \in \tilde{C}$ relaxation property holds.*

Remark 1. The main result of [12] follows directly from Corollary 1, because the solution set of (1) depend continuously on x_0 and on F on the class of locally CL functions.

To "justify" our results we will prove:

Proposition 1. *The set of all locally Lipschitz multimaps is of first Baire category in $C(I \times \mathbb{R}^n, CC(\mathbb{R}^n))$.*

Proof. Given a compact set $K \subset \mathbb{R}^n$ we denote by $Lip(K,n)$ the set of all Lipschitz on $I \times K$ with a constant n multifunctions. It is well known that the set of all non locally Lipschitz multi-functions is dense in $C(I \times \mathbb{R}^n)$. Let $F_k \in Lip(K,n)$ and let $F_k \to F$. Obviously $F \in Lip(K,n)$. Therefore $NL(n) = C(I \times \mathbb{R}^n) \setminus Lip(K,n)$ is open and dense, i.e. $\tilde{N} = \bigcap_{n=1}^{\infty} NL(n)$ is G_δ set. Obviously every $F \in \tilde{N}$ is not locally Lipschitz. □

Consider now discrete approximations of (1) with respect to $W^{1,1}(I, \mathbb{R}^n)$, i.e. $|x(\cdot) - y(\cdot)|_W := \max_{t \in I} \left\{ |x(t) - y(t)| + \int_0^1 |\dot{x}(t) - \dot{y}(t)|\, dt \right\}$. Consider the uniform grid: $h_k = \frac{1}{k} = t_{j+1} - t_j$. The discrete approximation scheme is:

$$x_{j+1}^k \in x_j^k + h_k F(t_j, x_j^k),\ x(0) = x_0. \tag{4}$$

On $[t_j, t_{j+1}]$ $x(\cdot)$ is defined as a linear function.

A1. There exists a constant M such that $|F(t,x)| \le M$ for all (t,x) and $F(\cdot,\cdot)$ is with nonempty convex compact values.

A2. The multifunction $F(\cdot, x)$ is continuous and $F(t, \cdot)$ is CL.

The following theorem can be proved using the same manner as in [10]:

Theorem 4. *Let $x(\cdot)$ be a trajectory of (1) under* **A1**, **A2**. *Then there exists a sequence $x^k(\cdot)$ converging uniformly to $x(\cdot)$ and $v^k(t) = \frac{x_{j+1} - x_j}{h_k}$ converges to $\dot{x}(\cdot)$ in $L^1(I, \mathbb{R}^n)$.*

Proof. Let Δ^N be a sequence of uniform subdivisions of I and let $v^N(\cdot)$ be a corresponding to Δ^N sequence of step functions converging in $L^1[I, \mathbb{R}^n]$ to $\dot{x}(\cdot)$. Denote $y^N(t) = x_0 + \int_0^t v^N(\tau)\, d\tau$. It is easy to see that

$$\begin{aligned} &dist(v^N(t), F(t, y^N(t)) \le dist(v^N(t), F(t, x(t)) \\ &+ D_H(F(t, x(t)), F(t, y^N(t))) \le |v^N(t) - \dot{x}(t)| + w(t, |x(t) - y^N(t)|). \end{aligned}$$

Using the proximal point algorithm described in [10] (see also [6]) one can complete the proof. □

With the help of the previous result we can approximate the following optimal control problem.

Minimize:

$$J(x(\cdot)) := \varphi(x(1)) + \int_0^1 f(t, x(t), \dot{x}(t))\, dt. \tag{5}$$

Under on the solution set of (1).

Following Mordukhovich [10] we give the following definition:

Definition 2. *The arc $\bar{x}(\cdot)$ is said to be a local minimum if there exist $\varepsilon > 0$ and $\alpha \ge 0$ such that $J[\bar{x}] \le J[x]$ for any other trajectory $x(\cdot)$, satisfying $|\bar{x}(t) - x(t) < \varepsilon$, $\forall t \in I$ and $\alpha \int_0^1 |\dot{x}(t) - \bar{x}'(t)|\, dt < \varepsilon$.*

Given a local minimum $\dot{x}'(\cdot)$ we want to minimize

$$J_k(x^k) = \varphi(x_k^k) + h_k \sum_{j=0}^{k-1} f\left(t_j^k, x_j^k, \frac{x_{j+1}^k - x_j^k}{h_k}\right) + \sum_{j=0}^{k-1} \int_{t_j}^{t_{j+1}} \left|\frac{x_{j+1}^k - x_j^k}{h_k} - \bar{x}'(t)\right| dt, \tag{6}$$

for discrete trajectories $x^k(\cdot)$ such that $|x_j^k - \bar{x}(t_j)| \le \frac{\varepsilon}{2}$, $j = 0, 1, \ldots, k$ and

$$\sum_{j=0}^{k-1} \int_{t_j}^{t_{j+1}} \left|\frac{x_{j+1}^k - x_j^k}{h_k} - \bar{x}'(t)\right| dt \le \frac{\varepsilon}{2}. \tag{7}$$

The functions $f(\cdot, \cdot, \cdot)$ and $\varphi(\cdot)$ are continuous.

Theorem 5. *Let $\bar{x}(\cdot)$ be a local minimum of the optimal control problem (1)–(5). Under the hypotheses above every sequence of solutions of (4)–(6)–(7) converges to $\bar{x}(\cdot)$ in $W^{1,1}(I, \mathbb{R}^n)$ as $k \to \infty$.*

The proof follows closely the proof of theorem 3.3 in [10] and is omitted.

The following result is a consequence of Theorem 2 and Theorem 5.

Theorem 6. *Let $K \subset \mathbb{R}^n$ be a compact set. There exists a residual set $\tilde{C}_K \subset C(I \times K)$ such that if $F \in \tilde{C}_K$ then for every local minimum $\bar{x}(\cdot)$ of (1)–(5) the sequence $x^k(\cdot)$ of solutions of (4–(6)–(7) converges to $\bar{x}(\cdot)$ in $W^{1,1}(I, \mathbb{R}^n)$ as $k \to \infty$.*

Remark 2. From [11] we know that in case of CL right-hand side F the solution set of (2) is a residual subset of the solution set of (1). Thus we have proved also that for almost all (in Baire sense) right-hand sides F the solution set of (2) is residual in the solution set of (1).

It is not difficult also to extend Theorem 2 to the case of Caratheodory (almost continuous) F.

The results of the paper can be easily extended to the case of functional differential inclusions, having the form:

$$\dot{x}(t) \in F(t, x_t), \; x_0 = \varphi. \tag{8}$$

Here $x_t, \varphi \in X = C([-\tau, 0], \mathbb{R}^n)$ is defined with $x_t(s) = x(t+s)$ for $s \in [-\tau, 0]$ ($\tau > 0$).

It will be interesting to extend Theorem 2 and 3 to the case of infinitely dimensional spaces.

Notice that the main results of [5, 9] are proved for differential equations (inclusions) in Banach spaces.

References

1. De Blasi F., Mijak J.: The generic properties of existence of solutions for a class of multivalued differential equations in Hilbert spaces. Funkc. Ekvacioj **21** (1978) 271-278.
2. De Blasi F., Pianigiani G.: On the solution set of nonconvex differential inclusions. J. Differential Equations **128** (1996) 541-555.
3. De Blasi F., Pianigiani G.: On the Baire cathegory method in existence problems for ordinary and partial differential inclusions. SIMAA **4** (2002) 137-148.
4. Deimling K.: Multivalued Differential Equations, De Grujter, Berlin, 1992.
5. Donchev T., Generic properties of functional differential inclusions in Banach spaces. Proc **28** Conf. of the Union Bulgarian Math., Sofia (1999) 80-83.
6. Dontchev A., Farkhi E.: Error estimates for discretized differential inclusions. Computing **41** (1989) 349-358.
7. Fort M.: Points of continuity of semicontinuous functions. Publ. Math. Debrecen **2** (1951) 100-102.
8. Hu S., Papageorgiou N.: Handbook of Multivalued Analysis: vol. I Theory, Kluwer, 1997, vol. II Applications, Kluwer, Dordrecht, 2000.
9. Lasota A., Yorke J.: The generic properties of existence of solutions of differential equations in Banach spaces. J. Differential Equations **13** (1973) 1-12.
10. Mordukhovich B.: Discrete approximations and refined Euler-Lagrange conditions for differential inclusions. SIAM J. Control. Optim. **33** (1995), 882-915.
11. Pianigiani G.: On the fundamental theory of multivalued differential equations. J. Differential Equations **25** (1977) 30-38.
12. Yu J., Yuan X., Isac G.: The stability of solutions for differential inclusions and differential equations in the sense of Baire category theorem. Appl. Math. Lett. **11** (1998) 51-56.

3D Modelling of Diode Laser Active Cavity

N.N. Elkin[1], A.P. Napartovich, A.G. Sukharev, and D.V. Vysotsky

State Science Center Troitsk Institute for Innovation and Fusion Research(TRINITI), 142190, Troitsk Moscow Region, Russia
elkin@triniti.ru

Abstract. A computer program is developed for numerical simulations of diode lasers operating well above threshold. Three-dimensional structures of typical single-mode lasers with the goal of efficient fibre coupling are considered. These devices have buried waveguides with high indices of refraction. The so-called round-trip operator constructed by a beam propagation method is non-linear due to gain saturation and thermal effects. Thus, the problem for numerical modelling of lasing is the eigenvalue problem for the non-linear round-trip operator. Fox-Li iterative procedure is applied for calculation of a lasing mode. A large size 3D numerical mesh is employed to discretize a set of equations describing (a) propagation of two counter-propagating waves using Pade approximation, (b) lateral diffusion of charge carriers within a quantum well, and (c) thermal conductivity. So, many important non-linear effects are properly accounted for: gain saturation, self-focusing, and thermal lens. A serious problem arising for operation far above threshold is the appearance of additional lasing modes that usually cause degradation in optical beam quality. To calculate a critical electric current, at which additional modes appear, the numerical code incorporates a subroutine that calculates a set of competing modes using gain and index variations produced by the oscillating mode. The corresponding linear eigenproblem is solved by the Arnoldi method. The oscillating mode has an eigenvalue equal to 1, while higher-order modes have eigenvalues of amplitude less than 1, which grow with injection current. These eigenvalues are calculated at several values of electric current, and the eigenvalues of the sub-threshold modes are then extrapolated to 1, at which point the device becomes multimode. Results of numerical simulations for typical experimental conditions will be presented.

1 Introduction

The diode laser scheme is the same as in [1]. Fig. 1 illustrates schematic of the structure under consideration in a plane perpendicular to an optical axis not in scale. The layers 3-5 form the waveguide (so-called separate-confinement-heterostructure) with one quantum well (active layer 4). The cross-hatched rectangles in the layer 8 are buried waveguides. The structure has a mirror symmetry relative to plane $y = 0$ and is uniform along optical axis z. Diode laser facets are located at $z = 0$ and $z = L$, where L is length of the laser. The electric current is

Z. Li et al. (Eds.): NAA 2004, LNCS 3401, pp. 272–279, 2005.

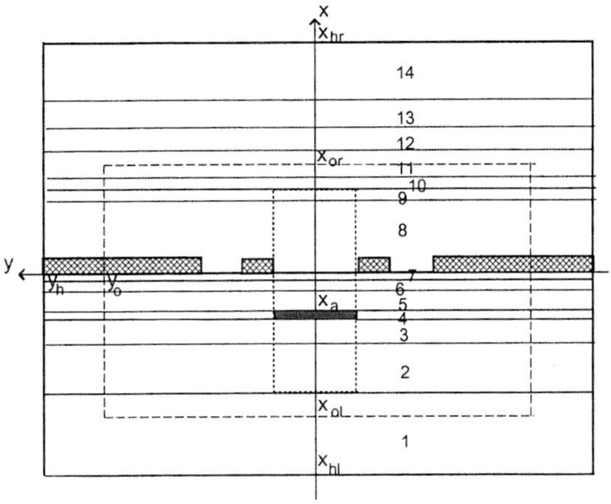

Fig. 1. Scheme of diode laser

Table 1. Allocation of materials

GaAs	AlGaAs	InGaAsP	InGaAs	InGaP	Ti	Au	AuSn	CuW
1,9	2,6,8	3,5	4	7	10	11,13	12	14

assumed to be uniform within the inner rectangle restricted by dot-and-dashed line. Its width is 10 μm. The gain is located within the quantum well. Table 1 assigns the allocation of materials across the layers.

2 Statement of the Problem

We start from Maxwell equations and assume that the polarization effects can be neglected. Eliminating fast oscillatory factor $\exp(-i\omega_0 t)$ and ignoring eventual envelope time-dependence the optical field may be expressed in a scalar functional form

$$E(x, y, z) = (E_+(x, y, z)e^{ik_0 n_0 z} + E_-(x, y, z)e^{-ik_0 n_0 z}),$$

where $k_0 = \omega_0/c$ is the reference wave number, n_0 is the reference index. This expression implies that there exists a preferred direction of light propagation collinear to $z-$ axis. The wave field $E(x, y, z)$ satisfies the Helmholtz equation. Two counter-propagating waves are characterized by slowly varying envelopes $E_\pm(x, y, z)$ where sign " + " corresponds to propagation in a forward direction of z-axis and sign " − " corresponds to propagation in a backward direction. Parabolic equations

$$\pm\frac{\partial E_\pm}{\partial z} - \frac{i}{2k_0 n_0}\Delta_\perp E_\pm - \frac{1}{2}bE_\pm = 0, \tag{1}$$

for waves $E_\pm(x, y, z)$ are used often. Here

$$\Delta_{\perp} = \frac{\partial^2}{\partial x^2} + \frac{\partial^2}{\partial y^2}, \quad b = g + i\frac{k_0}{n_0}(n^2 - n_0^2),$$

g and n are gain and index respectively. For typical diode laser conditions wide-angle diffraction could be important. It is reasonable to use more correct wave equations

$$\pm\left(1 - \frac{i}{4k_0 n_0}b + \frac{1}{4k_0^2 n_0^2}\Delta_{\perp}\right)\frac{\partial E_{\pm}}{\partial z} - \frac{i}{2k_0 n_0}\Delta_{\perp}E_{\pm} - \frac{1}{2}bE_{\pm} = 0, \tag{2}$$

deduced from Helmholtz equation under Pade approximation [2].

We may use boundary conditions $E_-(x,y,L) = E_+(x,y,L)\rho_2$ at the plane $z = L$ and $E_+(x,y,0) = E_-(x,y,0)\rho_1$ at the plane $z = 0$. Here $\rho_{1,2}$ are the reflection coefficients.

Besides, we have to set a boundary conditions for envelopes $E_{\pm}(x,y,z)$ at the lateral boundaries ($y = \pm y_o$) and at the transverse ones ($x = x_{ol}$, $x = x_{or}$). In order to describe leaky waves propagating in transverse direction a non-reflecting boundary conditions must be chosen. The perfectly matched layer boundary condition [3] was set after some trials.

It is necessary to solve the set of 1D non-linear diffusion equations [4]

$$\frac{\partial^2 Y}{\partial y^2} - \frac{Y}{D\tau_{nr}} - \frac{B}{D}N_{tr}Y^2 - \frac{Ig(Y)}{g_{0N}D\tau_{nr}I_s} = -\frac{J}{qDdN_{tr}} \tag{3}$$

for normalized carrier density $Y = N/N_{tr}$ in order to determine gain and index at the active layer ($x = x_a$, $-y_o < y < y_o$, $0 < z < L$). Here N is the carrier density

$$N_{tr} = \frac{1}{2B}\left(-\frac{1}{\tau_{nr}} + \sqrt{\frac{1}{\tau_{nr}^2} + \frac{4BJ_{tr}}{qd}}\right)$$

is the carrier density for the conditions of transparency, D is the diffusion coefficient, τ_{nr} is a recombination time, B is a coefficient of nonlinearity, J_{tr} is the current density for the conditions of transparency, $I = (|E_+|^2 + |E_-|^2)I_s$, $I_s = (\hbar\omega N_{tr})/(g_{0N}\tau_{nr})$ are the light intensity and the intensity of saturation, $\hbar\omega$ is the energy of a light quantum, J is the current density, d is the thickness of the quantum well and q is the elementary charge. Zero boundary conditions for $Y(y)$ are determined at the lateral boundaries of the active layer ($x = x_a$, $y = \pm y_o$, $0 < z < L$). Gain and index at the active layer are approximated by the formulas

$$g(Y) = g_{0N}\ln(\varphi(Y)),\ \varphi(Y) = \begin{cases} \alpha + (1-\alpha)Y^{\frac{1}{1-\alpha}}, & Y < 1 \\ Y, & Y \geq 1 \end{cases}, \quad n = \tilde{n_0} - \frac{R}{2k}g + \nu T, \tag{4}$$

where $\alpha = e^{-1} \cong 0.368$, g_{0N} is a gain parameter, $\tilde{n_0}$ is an index for the conditions of transparency, R is the so-called line enhancement factor, ν is a coefficient of linear temperature dependence of index and T is a temperature relative to a room temperature.

Thermal effects are described by solution of a series of 2D thermal conductivity equations

$$-\frac{\partial}{\partial x}\left(\chi\frac{\partial T}{\partial x}\right)-\frac{\partial}{\partial y}\left(\chi\frac{\partial T}{\partial y}\right)=f \tag{5}$$

in a few transverse planes. Here χ is a heat conductivity, $f = J^2\rho - Ig(Y) + N_{tr}Y\hbar\omega/\tau_{nr} + \hbar\omega B\,(N_{tr}Y)^2$ is a heat source, ρ is ohmic resistance. In concordance with experimental cooling scheme we set the Dirichlet boundary condition $T = 0$ at the top $x = x_{hr}$. Von Neumann conditions $\partial T/\partial n = 0$ were set at the bottom $x = x_{hl}$ and at the left and right edges $y = \pm y_h$ where the thermal fluxes are small.

The temperature field T is interpolated over 3D domain for further recalculations of index by the formula

$$n(T) = n|_{T=0} + \nu T \tag{6}$$

We neglect thermal flow in z direction through given approach.

It is obvious that (2) or (1) are the homogeneous equations having trivial solutions $E_\pm(x,y,z) \equiv 0$. At the above-threshold condition trivial solution is unstable to any noise and optical field will increase until to upper limit determined by gain saturation. Following [5] we construct the so-called round-trip operator $\mathbf{P}$ which transforms the field at a selected cross-section location after a propagation cycle of a laser beam. In detail, if $u(x,y) = E_+(x,y,L)$ then $\mathbf{P}u$ is the composition of {1} reflection of the wave $E_+(x,y,L)$ from the facet $z = L$, {2} propagation of the wave $E_-(x,y,z)$ to the plane $z = 0$ according to equations (2) or (1), (3), (4), (5) and (6), {3} reflection of the wave $E_-(x,y,0)$ from the facet $z = 0$, {4} propagation of the wave $E_+(x,y,z)$ to the plane $z = L$ similar the step {2}. Non-trivial lasing modes are found as the eigen-functions of the operator $\mathbf{P}$ provided the carrier density Y and temperature T are found self-consistently with the wave fields $E_\pm$. Thus, we have an eigen-value problem

$$\mathbf{P}u = \gamma u \tag{7}$$

for a non-linear and non-hermitian operator $\mathbf{P}$. The complex eigenvalue satisfies the condition $|\gamma| = 1$ for steady-state mode, a non-zero phase $\arg(\gamma)$ of the eigenvalue appears due to non-coincidence of a priori unknown exact value of the wave number with the reference value k_0. To analyze stability of the lasing mode we consider also the linear eigen-problem of type (7) when gain and index are established by lasing mode and "frozen". The value $\delta = 1-|\gamma|^2$ has a physical meaning of power loss for one round trip in this case. If all the solutions of a linear eigen-problem have eigenvalues satisfying the condition $|\gamma| \le 1$ ($\delta \ge 0$) then the lasing mode is stable else we have the unstable mode.

3 Numerical Scheme

The basic block of our numerical code is calculation the round trip. It is realized by different ways for the general non-linear problem and for the auxiliary linear eigen-problem. For the second case wave equations (2) or (1) are solved separately using split-operator technique relative to second-order derivatives over lateral and transverse variables and second-order finite-difference scheme. But for the first case wave equations are solved simultaneously with the carrier diffusion equation (3) at each step along a variable z. Resulting non-linear finite-difference equations are solved iteratively at each step. Solving equations (1) and (2) requires nearly the same efforts in programming and computation. An approach based on (2) is more accurate and can be recommended as a favorable one. Thermal equations (5) are solved in uniformly arranged transverse planes after completion of calculations for wave and diffusion equations. The temperature calculated is used in the next iteration of the round-trip operator $\mathbf{P}$. Fast solver for equation (5) was realized using FFT transform over y and sweep method for Fourier coefficients.

The linear eigen-problem with "frozen" gain and index was solved by the Arnoldi algorithm [6] from the family of Krylov's subspace methods. The selective iteration method [7] was applied for a non-linear problem (7). This method expands validity of the traditional Fox-Li [5] iteration technique over the whole range of parameters when single-mode lasing is stable.

4 Simulation Results and Discussion

Some demonstration calculations were performed for length $L = 200\mu$m. We set the following transverse sizes: $x_{hr} - x_{hl} = 339\mu$m, $2y_h = 500\mu$m, $x_{or} - x_{ol} = 4.55\mu$m, $2y_o = 45\mu$m, and reflection coefficients $\rho_1 = 0.99$, $\rho_2 = 0.63$. The optical, electrical and thermal properties of materials were taken from the electronic archive http://www.ioffe.ru/SVA/NSM. Wavelength is $\lambda_0 = 2\pi/k_0 = 0.98\,\mu m$. The parameters of the active layer are taken from the Table 2.

Numerical mesh consisted of $256 \times 256 \times 2000$ cells for wave equations and of $750 \times 2048 \times 10$ cells for heat conductivity equations where 750 nodes were distributed non-uniformly along $x-$ direction. In order to estimate the error of discretization one of the variants was recalculated with the more fine meshes

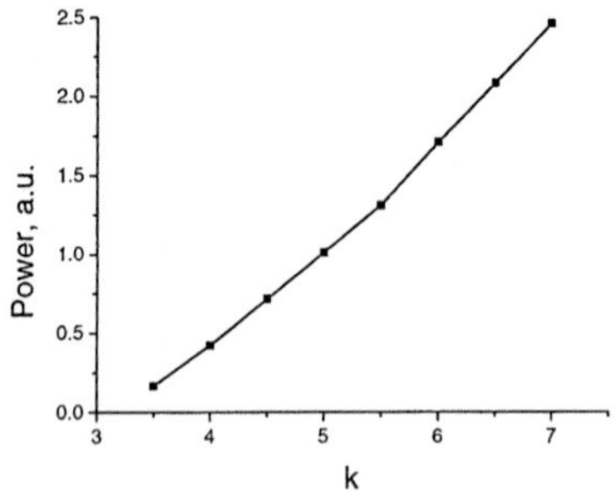

Fig. 2. Output power vs. pump current

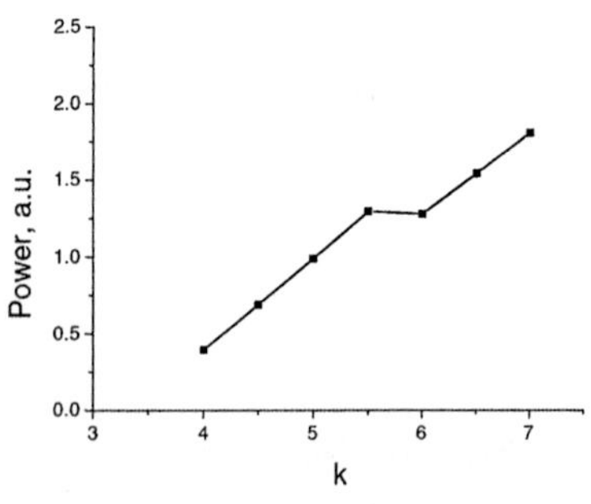

Fig. 3. Output power vs. pump current without thermal effect

Table 2. Parameters of the active layer

J_{tr}	g_{0N}	B	τ_{nr}	D	d	R	$\tilde{n_0}$
$50\,\text{A cm}^{-2}$	$2200\,\text{cm}^{-1}$	$10^{-10}\,\text{cm}^3\,\text{c}^{-1}$	1 ns	$100\,\text{cm}^2\,\text{c}^{-1}$	8.5 nm	2	3.7

$512 \times 512 \times 8000$ and $1500 \times 4096 \times 20$ cells correspondingly. The relative difference in the calculated output power amounted 0.009 what is quite satisfactory. The same variant was recalculated using the parabolic equations (1) for wave propagation. The relative difference in the calculated output power amounted 0.029 which indicates that the small angle (parabolic) approximation (1) is good for rough calculations.

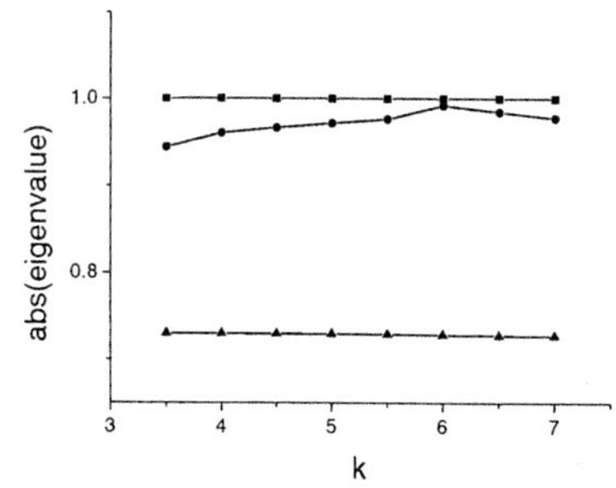

Fig. 4. Absolute value of eigenvalues vs. pump current

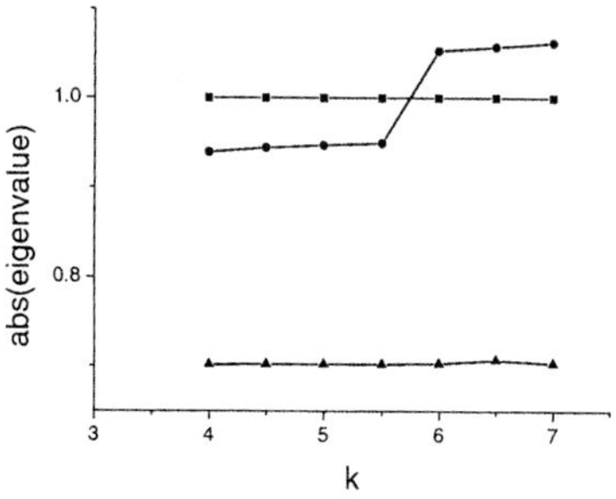

Fig. 5. Absolute value of eigenvalues vs. pump current without thermal effect

From the physical standpoint, dependence of the output power

$$W = \text{const} \times \int |E_+(x, y, L)|^2 \, dxdy$$

on the pump intensity and stability of single-mode lasing are of main interest. The pump intensity is determined by a dimensionless parameter $\kappa = J/J_{tr}$ which means the electric current in units of current for conditions of transparency. The dependence of the output power on κ is shown in Fig. 2. Note that the threshold for lasing value of κ is evaluated by extrapolation this dependence until crossing the $x-$axis. In our example the threshold is 3.2. The results of calculations ignoring thermal effects ($\nu \equiv 0$) are shown in Fig. 3 for comparison. We notice that output power decreases and reveals non-monotonic behavior when thermal effect neglected. The stability of single-mode lasing for both cases was analyzed and results are presented in Figs. 4 and 5. The square-marked curves correspond to operating mode and other curves present the nearest possible competing modes. We see that single-mode operation is stable everywhere if thermal effect is considered. On the contrary, if thermal effects are not taken into consideration single-mode lasing terminates near the value $\kappa \cong 5.7$ when the eigenvalue of the nearest competing mode crosses unity in absolute value.

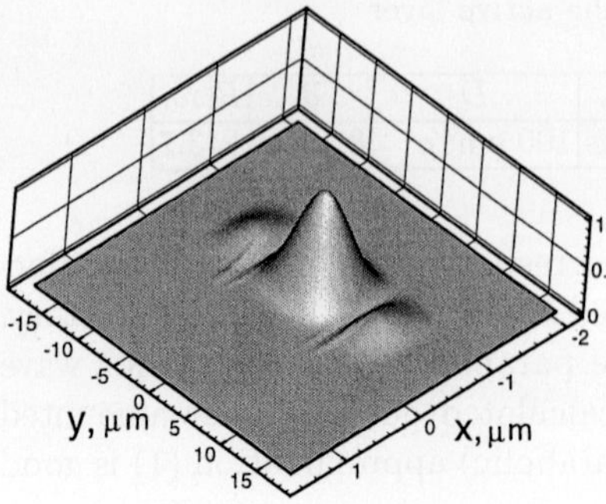

Fig. 6. Near-field intensity distribution of the operating mode. κ=6

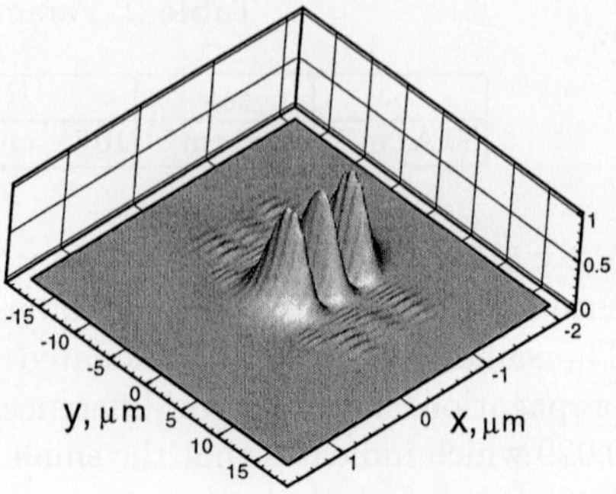

Fig. 7. Near-field intensity distribution of the nearest competing mode. κ=6

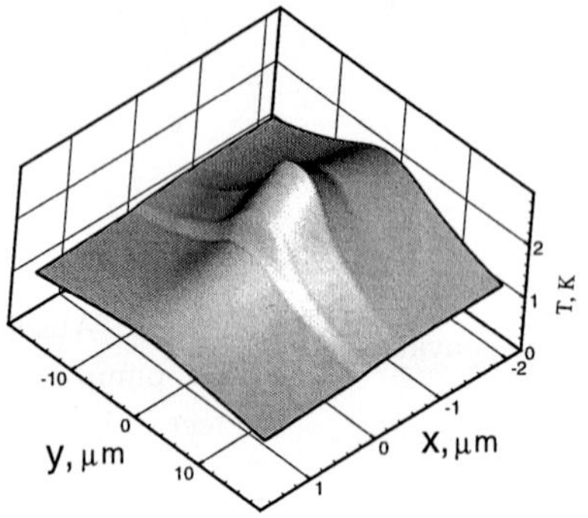

Fig. 8. Temperature distribution. κ=6

Figs. 6 and 7 present the intensity distributions $|E_+(x, y, L)|^2$ of the output beam for the operating mode and the nearest competing mode for $\kappa = 6$. One can see that modes differ in shape strongly. Onset of two-mode lasing causes degradation in optical beam quality. Finally, the Fig. 8 presents the temperature distribution $T(x, y, L)$ in the same area as the intensity distributions above. Note, that it is very small part of the whole area where temperature gradients were calculated. However, in the area in Fig. 8 temperature variations are the strongest.

5 Conclusion

In conclusion, the proposed numerical scheme allows us to gain the physically significant results for such a complicated laser device. As a result, the threshold value of pump current and upper limit of current when single-mode lasing is stable can be found. Calculations provide complete information about spatial distributions of laser intensity and temperature and far-field laser pattern, as well.

Acknowledgments

Work is partially supported by the RFBR project No. 02-02-17101, by the Grant of RF President No. 794.2003.2 and by Alfalight Inc., Madison, WI, USA.

References

1. J.C. Chang, J.J. Lee, A. Al-Muhanna, L.J. Mawst, and D. Botez.: Comprehensive above-threshold analysis of large-aperture (8-10 μm) antiresonant reflecting optical waveguide diode lasers. Applied Physics Letters, **81,** (2002) 1 – 3.
2. G. Ronald Hadley.: Wide-angle beam propagation using Pade approximant operators. Optics Letters, **17,** (1992) 1426 – 1428.
3. W.P. Huang, C.G.Xu, W. Lui, and K. Yokoyama.: The perfectly matched layer boundary condition for modal analysis of optical waveguides: leaky mode calculations. IEEE Photonics Technology Letters, **8,** (1996) 652 – 654.
4. G. R. Hadley.: Modeling of diode laser arrays. Chapter 4 in Diode Laser Arrays, D. Botez and D.R. Scifres, Eds, Cambridge, U.K., Cambridge Univ. Press, (1994) 1-72.
5. A.G.Fox, T.Li.: Effect of gain saturation on the oscillating modes of optical masers. IEEE Journal of Quantum Electronics, **QE-2,** (1966) 774–783.
6. J.M. Demmel.: Applied Numerical Linear Algebra. SIAM, Philadelphia, PA. (1997)
7. Elkin N.N., Napartovich A.P.: Numerical study of the stability of single-mode lasing in a Fabry-Perot resonator with an active medium. Appl. Math. Modelling, **18,** (1994) 513 – 521.

Ant Colony Optimization for Multiple Knapsack Problem and Model Bias

Stefka Fidanova

IPP – BAS, Acad. G. Bonchev str. bl.25A, 1113 Sofia, Bulgaria
stefka@parallel.bas.bg

Abstract. The Ant Colony Optimization (ACO) algorithms are being applied successfully to a wide range of problems. ACO algorithms could be good alternatives to existing algorithms for hard combinatorial optimization problems (COPs). In this paper we investigate the influence of model bias in model-based search as ACO. We present the effect of two different pheromone models for ACO algorithm to tackle the Multiple Knapsack Problem (MKP). The MKP is a subset problem and can be seen as a general model for any kind of binary problems with positive coefficients. The results show the importance of the pheromone model to quality of the solutions.

1 Introduction

There are many NP-hard combinatorial optimization problems for which it is impractical to find an optimal solution. Among them is the MKP. For such problems the reasonable way is to look for algorithms that quickly produce near-optimal solutions. ACO [1, 4, 2] is a meta-heuristic procedure for quickly and efficiently obtaining high quality solutions to complex optimization problems [9]. The ACO algorithms were inspired by the observation of real ant colonies. Ants are social insects, that is, insects that live in colonies and whose behavior is directed more to the survival of the colony as a whole than to that of a single individual component of the colony. An important and interesting aspect of ant colonies is how ants can find the shortest path between food sources and their nest. ACO is the recently developed, population-based approach which has been successfully applied to several NP-hard COPs [5]. One of its main ideas is the indirect communication among the individuals of a colony of agents, called "artificial" ants, based on an analogy with trails of a chemical substance, called pheromones which real ants use for communication. The "artificial" pheromone trails are a kind of distributed numerical information which is modified by the ants to reflect their experience accumulated while solving a particular problem. When constructing a solution, at each step ants compute a set of feasible moves and select the best according to some probabilistic rules. The transition probability is based on the heuristic information and pheromone trail level of the move (how much the movement is used in the past). When we apply ACO algorithm to MKP different pheromone models are possible and the influence on the results is shown.

Z. Li et al. (Eds.): NAA 2004, LNCS 3401, pp. 280–287, 2005.

The rest of the paper is organized as follows: Section 2 describes the general framework for MKP as a COP. Section 3 outlines the implemented ACO algorithm applied to MKP with two different pheromone models. In Section 4 experimental results over test problems are shown. Finally some conclusions are drawn.

2 Formulation of the Problem

The MKP has received wide attention from the operations research community [7,3,8], because it embraces many practical problems. Applications include resource allocation in distributed systems, capital budgeting and cutting stock problems. In addition, MKP can be seen as a general model for any kind of binary problems with positive coefficients [6]. They are optimization problems for which the solutions are sequences of 0 and 1 and the coefficients in objective function are positive and coefficients in constraints are positive or zero. The MKP can be thought as a resource allocation problem, where we have m resources (the knapsacks) and n objects and object j has a profit p_j. Each resource has its own budget c_j (knapsack capacity) and consumption r_{ij} of resource i by object j. We are interested in maximizing the sum of the profits, while working with a limited budget.

The MKP can be formulated as follows:

$$\max \sum_{j=1}^{n} p_j x_j$$

$$\text{subject to} \sum_{j=1}^{n} r_{ij} x_j \le c_i \quad i = 1, \dots, m \tag{1}$$

$$x_j \in \{0, 1\} \quad j = 1, \dots, n$$

x_j is 1 if the object j is chosen and 0 otherwise.

There are m constraints in this problem, so MKP is also called m-dimensional knapsack problem. Let $I = \{1, \dots, m\}$ and $J = \{1, \dots, n\}$, with $c_i \ge 0$ for all $i \in I$. A well-stated MKP assumes that $p_j > 0$ and $r_{ij} \le c_i \le \sum_{j=1}^{n} r_{ij}$ for all $i \in I$ and $j \in J$. Note that the $[r_{ij}]_{m \times n}$ matrix and $[c_i]_m$ vector are both non-negative.

In the MKP we are not interested in solutions giving a particular order. Therefore a partial solution is represented by $S = \{i_1, i_2, \dots, i_j\}$ and the most recent elements incorporated to S, i_j need not be involved in the process for selecting the next element. Moreover, solutions for ordering problems have a fixed length as we search for a permutation of a known number of elements. Solutions for MKP, however, do not have a fixed length. We define the graph of the problem as follows: the nodes correspond to the items, the arcs fully connect nodes. Fully connected graph mens that after the object i we can choose the object j for every i and j if there are enough resources and object j is not chosen yet.

3 ACO Algorithm for MKP

Real ants foraging for food lay down quantities of pheromone (chemical cues) marking the path that they follow. An isolated ant moves essentially at random but an ant encountering a previously laid pheromone will detect it and decide to follow it with high probability and thereby reinforce it with a further quantity of pheromone. The repetition of the above mechanism represents the auto catalytic behavior of real ant colony where the more the ants follow a trail, the more attractive that trail becomes.

The ACO algorithm uses a colony of artificial ants that behave as co-operative agents in a mathematical space were they are allowed to search and reinforce pathways (solutions) in order to find the optimal ones. Solution that satisfies the constraints is feasible. After initialization of the pheromone trails, ants construct feasible solutions, starting from random nodes, then the pheromone trails are updated. At each step ants compute a set of feasible moves and select the best one (according to some probabilistic rules) to carry out the rest of the tour. The transition probability is based on the heuristic information and pheromone trail level of the move. The higher value of the pheromone and the heuristic information, the more profitable it is to select this move and resume the search. In the beginning, the initial pheromone level is set to a small positive constant value τ_0 and then ants update this value after completing the construction stage. ACO algorithms adopt different criteria to update the pheromone level. In our implementation we use Ant Colony System (ACS) [4] approach.

In ACS the pheromone updating stage consists of:

3.1 Local Update Stage

While ants build their solution, at the same time they locally update the pheromone level of the visited paths by applying the local update rule as follows:

$$\tau_{ij} \leftarrow (1-\rho)\tau_{ij} + \rho\tau_0 \qquad (2)$$

Where ρ is a persistence of the trail and the term $(1-\rho)$ can be interpreted as trail evaporation.

The aim of the local updating rule is to make better use of the pheromone information by dynamically changing the desirability of edges. Using this rule, ants will search in wide neighborhood around the best previous solution. As shown in the formula, the pheromone level on the paths is highly related to the value of evaporation parameter ρ. The pheromone level will be reduced and this will reduce the chance that the other ants will select the same solution and consequently the search will be more diversified.

3.2 Global Updating Stage

When all ants have completed their solution, the pheromone level is updated by applying the global updating rule only on the paths that belong to the best solution since the beginning of the trail as follows:

$$\tau_{ij} \leftarrow (1-\rho)\tau_{ij} + \Delta\tau_{ij} \tag{3}$$

where $\Delta\tau_{ij} = \begin{cases} \rho L_{gb} \text{ if } (i,j) \in \text{best solution} \\ 0 \quad \text{otherwise} \end{cases}$,

L_{gb} is the cost of the best solution from the beginning. This global updating rule is intended to provide a greater amount of pheromone on the paths of the best solution, thus intensifying the search around this solution. The transition probability to select the next item is given as:

$$p_{ij}^k(t) = \begin{cases} \frac{\tau_{ij}\eta_{ij}}{\sum_{j \in allowed_k(t)} \tau_{ij}\eta_{ij}} & \text{if } j \in allowed_k(t) \\ 0 & \text{otherwise} \end{cases} \tag{4}$$

where τ_{ij} is a pheromone level to go from i to j, η_{ij} is the heuristic and $allowed_k(t)$ is the set of remaining feasible items. Thus the higher the value of τ_{ij} and η_{ij}, the more profitable it is to include item j in the partial solution.

Let $s_j = \sum_{i=1}^m r_{ij}$. For heuristic information we use:

$$\eta_{ij} = \begin{cases} p_j^{d_1}/s_j^{d_2} \text{ if } s_j \neq 0 \\ p_j^{d_1} \quad \text{if } s_j = 0 \end{cases} \tag{5}$$

Hence the objects with greater profit and less average expenses will be more desirable.

There are two possibilities for placing a pheromone. It can be placed on the arcs of the graph of the problem or on the nodes like in [7]. In the case of arcs τ_{ij}, in the formulas, corresponds to the pheromone quantity on the arc (i,j). In the case of nodes τ_{ij} corresponds to the pheromone quantity on the node j.

The LS method (move-by-move method) [10] perturbs a given solution to generate different neighborhoods using a move generation mechanism. In general, neighborhoods for large-sized problems can be much difficult to search. Therefore, LS attempts to improve an initial solution by a series of local improving changes. A move-generation is a transition from a solution S to another one $S' \in V(S)$ ($V(S)$ is a set of neighbor solutions) in one step. These solutions are selected and accepted according to some pre-defined criteria. The returned solution S' may not be optimal, but it is the best solution in its local neighborhood $V(S)$. A local optimal solution is a solution with the local maximal possible cost value. Knowledge of a solution space is the essential key to more efficient search strategies [11]. These strategies are designed to use this prior knowledge and to overcome the complexity of an exhaustive search by organizing searches across the alternatives offered by a particular representation of the solution. The main purpose of LS implementation is to speed up and improve the solutions constructed by the ACO.

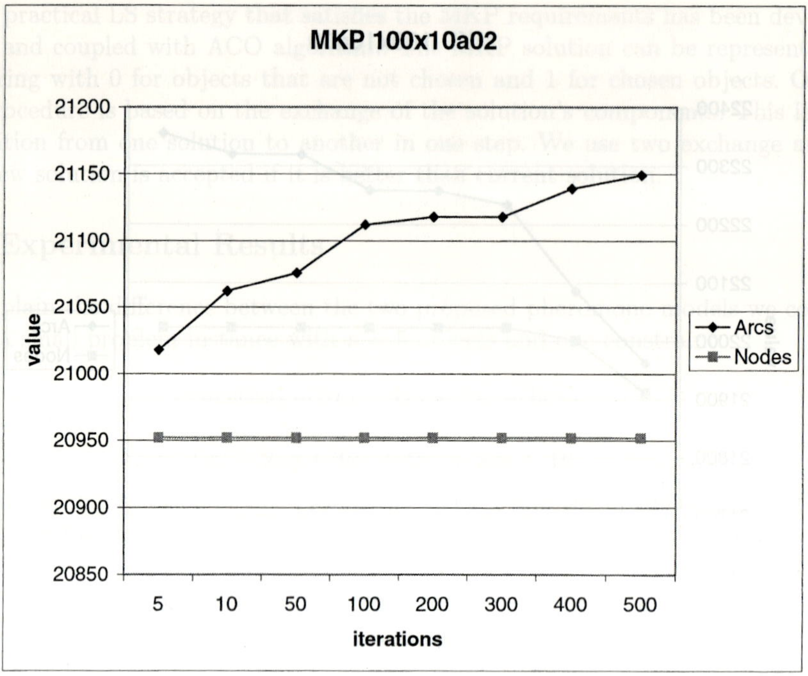

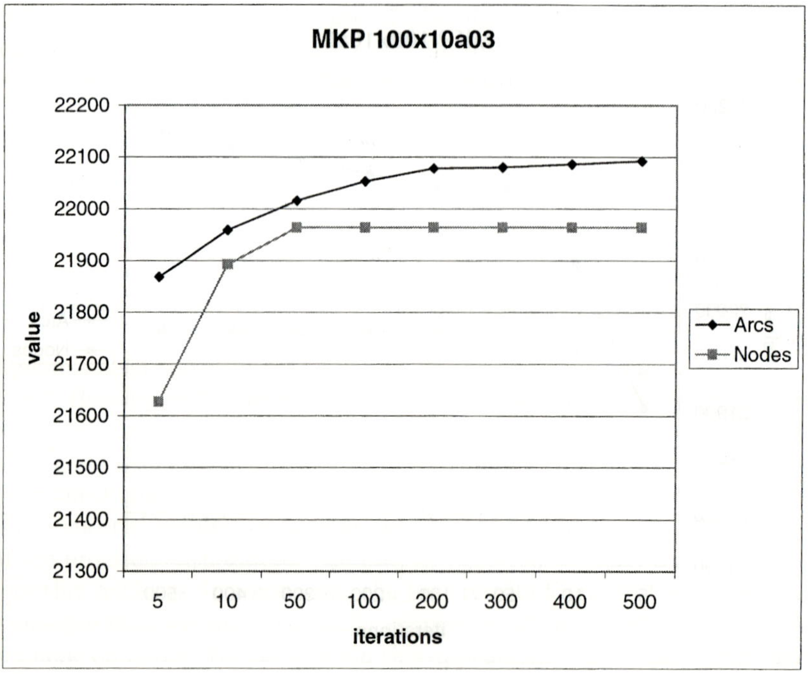

Fig. 1. The figures shows the average solution quality (value of the total cost of the objects in the knapsack) per iteration over 20 runs

5 Conclusion

The design of a meta-heuristic is a difficult task and highly dependent on the structure of the optimized problem. In this paper two models of the pheromone placement have been proposed. The comparison of the performance of the ACS coupled with these pheromone models applied to different MKP problems are reported. The results obtained are encouraging and the ability of the developed models to rapidly generate high-quality solutions for MKP can be seen. For future work another important direction for current research is to try different strategies to explore the search space more effectively and provide good results.

Acknowledgments

Stefka Fidanova was supported by the European Community program "Center of Excellence"BIS 21 under contract number No ICA1-2000-70016 and the European Community program "Structuring the European Research Area" contract No MERG-CT-2005-510714..

References

1. Dorigo, M., Di Caro, G.:*The Ant Colony Optimization metaheuristic.*In: Corne, D., Dorigo, M., Glover, F.(eds.): New Idea in Optimization, McGrow-Hill (1999) 11–32.
2. Dorigo, M., Di Caro, G., Gambardella, L. M.:*Ant Algorithms for Distributed Discrete Optimization.* J of Artificial Life 5 (1999) 137–172.
3. Ferreira C. E., Martin A., Weismantel R.: *Solving Multiple Knapsack Problems by Cutting Planes.* SIAM Journal on Optimization Vol.6(3) (1996) 858–877.
4. Dorigo, M., Gambardella, L.M.:*Ant Colony System: A cooperative Learning Approach to the Traveling Salesman Problem.* IEEE Transaction on Evolutionary Computation 1 (1999) 53–66.
5. Gambardella, M. L., Taillard, E. D., Dorigo, M.: *Ant Colonies for the QAP.* J of Oper. Res. Soc. 50 (1999) 167–176.
6. Kochenberger, G., McCarl, G., Wymann, F.:*A Heuristic for General Integer Programming.* J of Decision Sciences 5 (1974) 34–44.
7. Leguizamon, G., Michalevich, Z.: *A New Version of Ant System for Subset Problems* In: Proceedings of Int. Conf. on Evolutionary Computations, Washington (1999).
8. Marchetti-Spaccamela A., Vercellis C.: *Stochastic on-line Knapsack Problems*J. of Mathematical Programming Vol 68(1) (1995) 73–104.
9. Osman, I. H., Kelley, J. P.:*Metaheuristic: An Overview.* In: Osman, I. H., Kelley, J.P. (eds.), Metaheuristic: Theory and Applications, Kluwer Academic Publishers (1996).
10. Schaffer, A. A., Yannakakis, M.: *Simple Local Search Problems that are Hard to Solve.* Society for Industrial Applied Mathematics Journal on Computing, Vol 20 (1991) 56–87.
11. Wolpert D., Macready W.: *No Free Lunch Theorems for Optimization.* In: IEEE Transactions on Evolutionary Computation,(1997) 67–82.

Discretization Methods with Embedded Analytical Solutions for Convection Dominated Transport in Porous Media

Jürgen Geiser

Weierstrass Institute for Applied Analysis and Stochastics, Mohrenstrasse 39, D-10117 Berlin, Germany
geiser@wias-berlin.de
http://www.wias-berlin.de/~geiser

Abstract. We will present a higher order discretization method for convection dominated transport equations. The discretisation for the convection-diffusion-reaction equation is based on finite volume methods which are vertex centered. The discretisation for the convection-reaction equation is improved with embedded analytical solutions for the mass. The method is based on the Godunovs-method, [10]. The exact solutions are derived for the one-dimensional convection- reaction equation with piecewise linear initial conditions. We introduce a special cases for the analytical solutions with equal reaction-parameters, confer [9]. We use operator-splitting for solving the convection-reaction-term and the diffusion-term. Numerical results are presented and compared the standard- with the modified- method. Finally we propose our further works on this topic.

1 Introduction and Mathematical Model

The motivation for the study came from computational simulations of radioactive contaminant transport in flowing groundwater [4, 5] that are based on the following mathematical equation

$$\partial_t R_i \, c_i \; + \; \nabla \cdot (\mathbf{v} \, c_i - D \nabla c_i) = -R_i \, \lambda_i \, c_i \; + \sum_{k=k(i)} R_k \, \lambda_k \, c_k \; . \tag{1}$$

The unknown concentrations $c_i = c_i(x,t)$ are considered in $\Omega \times (0,T) \subset I\!R^M \times I\!R^+$. The retardation factor is given as $R_i \geq 0$ and is constant. The decay-factors are given as $\lambda_i \leq 0$ and the indices of the predecessors are denotes as $k(i)$. Further D is the Scheidegger diffusion-dispersion tensor and $\mathbf{v}$ is the velocity vector.

The aim of this paper is to derive analytical solutions for the one-dimensional convection-reaction equation and embed this solution in an explicit higher order finite volume discretisation method for the convection-equation.

This will be done by using the Godunovs method. So we have no splitting error and we have an exact discretisation for the one-dimensional problem.

Z. Li et al. (Eds.): NAA 2004, LNCS 3401, pp. 288–295, 2005.

The higher order finite volume method for the convection term is based on the TVD-methods and constructed under the local discrete minimum and maximum-principle, confer [7].

For the new method we will focus us on the convection-reaction-equations with linear decay-chain and our equations are given by

$$\partial_t R_i \, c_i \; + \; \nabla \cdot \mathbf{v} \, c_i = -R_i \, \lambda_i c_i \; + \; R_{i-1} \, \lambda_{i-1} c_{i-1} \; , \tag{2}$$
$$i = 1, \ldots, M \; , \; u_i = (u_1, \ldots, u_M)^T \in I\!R^M \; ,$$

with trivial inflow and outflow boundary conditions where $u = 0$. We use trapezoidal impulses as initial conditions.

The analytical solution is derived for piecewise constant velocities $\mathbf{v} \in I\!R^n$, n is the space-dimension for the equation.

The analytical solution is derived in the next section.

2 Analytical Solutions

For the next section we deal with the following system of one-dimensional convection-reaction-equations without diffusion. The equation is given as

$$\partial_t c_i + v_i \partial_x c_i = -\lambda_i c_i + \lambda_{i-1} c_{i-1} \; , \tag{3}$$

for $i = 1, \ldots, M$. The unknown M is the number of components. The unknown functions $c_i = c_i(x,t)$ denote the contaminant concentrations. They are transported with piecewise constant (and in general different) velocities v_i . They decay with constant reaction rates λ_i. The space-time domain is given by $(0, \infty) \times (0, T)$.

We assume simple (irreversible) form of decay chain, e.g. $\lambda_0 = 0$ and for each contaminant only single source term $\lambda_{i-1} c_{i-1}$ is given. For a simplicity, we assume that $v_i > 0$ for $i = 1, \ldots, M$.
We describe the analytical solutions with piecewise linear initial conditions. But all other piecewise polynom functions could be derived, confer [9].

For boundary conditions we take zero concentrations at inflow boundary $x = 0$ and the initial conditions are defined for $x \in (0,1)$ with

$$c_1(x,0) = \begin{cases} ax + b \, , & x \in (0,1) \\ 0 & \text{otherwise} \end{cases} \; , \tag{4}$$
$$c_i(x,0) = 0 \, , \quad i = 2, \ldots, M \; ,$$

where $a, b \in I\!R^+$ are constants.

We use the Laplace-Transformation for the transformation of the partial differential equation into the ordinary differential equation. We solve the ordinary differential equations, described in [3], and retransformed the solution in the original space of the partial differential equations. We could then use the solution for the one-dimensional convection-reaction-equation, confer [9]. In the next subsection we introduce the special solutions for equal reaction-parameters, i.e. $\lambda_l = \lambda_{a(l)}$.

2.1 Special Solutions

For the special solutions we use the same method as described in the general case. We focus us on the case of 2 components for $\lambda_1 = \lambda_2$, the further special solutions are described in [9].

The solution for $i = 2$ and $(l, a(l)) = (1, 2)$ are given as :

$$c_2 = \lambda_1 \exp(-\lambda_1 t) \left(\frac{1}{v_2 - v_1} \alpha_1 + \frac{1}{v_1 - v_2} \alpha_2 \right) , \tag{5}$$

$$\alpha_l = \begin{cases} 0 & 0 \le x < v_l t \\ a \frac{(x - v_l t)^2}{2} + b(x - v_l t) & v_l t \le x < v_l t + 1 \\ a \frac{1}{2} + b & v_l t + 1 \le x \end{cases} . \tag{6}$$

The solutions are derived in the transformed Laplace-space and the term $\lambda_l - \lambda_{a(l)}$ is skipped because of the singular point. The generalisation of this method is described in [9].

In the next section we derive the analytical solution for the mass.

2.2 Mass Reconstruction

For the embedding of the analytical mass in the discretisation method, we need the mass transfer of the norm-interval $(0, 1)$. We use the construction over the total mass given as

$$m_{i,sum}(t) = m_{i,rest}(t) + m_{i,out}(t) \tag{7}$$

The integrals are computed over the cell $(0, 1)$. We integrate first the mass that retain in the cell i and then we calculate the total mass. The difference between the total mass and the residual mass is the outflowing mass which is used for the discretisation.

The residual mass is described in [9] and is the integration of the analytical solution over the interval $(0, 1)$.

The total mass is calculated by the solution of the ordinary equation and the mass of the initial condition, confer [9]. The outflowing mass is defined for further calculations

$$m_{i,out}(\tau^n) = m_{i,sum}(\tau^n) - m_{i,rest}(\tau^n) , \tag{8}$$

$$m_{i,out}(\tau^n) = m_{i,out}(a, b, \tau^n, v_{1,j}, \ldots, v_{i,j}, R_1, \ldots, R_i, \lambda_1, \ldots, \lambda_i) , \tag{9}$$

whereby τ^n is the time-step, $v_{1,j}, \ldots, v_{i,j}$ are the velocity-, $R_1, \ldots, R_i$ are the retar-dation- , $\lambda_1, \ldots, \lambda_i$ the reaction-parameters and a, b are the initial conditions. In the next section we describe the modified discretisation-method.

3 Discretisation

For the discretisation of the convection-reaction-equation we use finite volume method and reconstruction of higher order with an explicit time-discretisation (e.g. forward Euler). For the diffusion equation we use the finite volume method with central difference method and implicit time-discretisation (e.g. backward Euler). The two equations are coupled together with an Operator-Splitting method, confer [9].

We will focus us on the new discretisation method for the convection-reaction equation and introduce the used notations.

The time steps are defined by $(t^n, t^{n+1}) \subset (0, T)$, for $n = 0, 1, \ldots$. The computational cells are $\Omega_i \subset \Omega$ with $i = 1, \ldots, I$. The unknown I is the number of the nodes on the dual mesh.

To use the finite volumes we have to construct the dual mesh for the triangulation $\mathcal{T}$ of the domain Ω , see [7]. The finite elements for the domain Ω are $T^e, e = 1, \ldots, E$. The polygonal computational cells Ω_i are related with the vertices x_i of the triangulation.

To get the relation between the neighbor cells and to use the volume of each cell we introduce the following notation. Let $V_i = |\Omega_i|$ and the set Θ_j denotes the neighbor-nodes j of the node i. The line segments Γ_{ij}, with $i \neq j$ is given with $\Omega_i \cap \Omega_j$.

The idea of the finite volumes is to construct an algebraic system of equation to express the unknowns $c_i^n \approx c(x_i, t^n)$. The initial values are given with c_i^0. The expression of the interpolation schemes could be given naturally in two ways, the first is given with the primary mesh of the finite elements

$$c^n = \sum_{i=1}^{I} c_i^n \phi_i(x) \,, \tag{10}$$

with ϕ_i are the standard globally finite element basis functions [2]. The second expression is for the finite volumes with

$$\hat{c}^n = \sum_{i=1}^{I} c_i^n \varphi_i(x) \,, \tag{11}$$

where φ_i are piecewise constant discontinuous functions defined by $\varphi_i(x) = 1$ for $x \in \Omega_i$ and $\varphi_i(x) = 0$ otherwise.

3.1 Finite Volume Discretisation of Second Order

For the second order discretisation scheme we use the linear interpolation scheme for the numerical solutions.

The reconstruction is done in the paper [7] and it is here briefly explained for the next steps. We us the following definitions for the element-wise gradient to define the linear construction

$$u^n(x_i) = c_i^n \,, \quad \nabla u^n|_{V_i} = \frac{1}{V_i} \sum_{e=1}^{E} \int_{T^e \cap \Omega_i} \nabla c^n dx \,, \text{ with } \quad i = 1, \ldots, I \,.$$

The piecewise linear function is given by

$$u_{ij}^n = c_i^n + \psi_i \nabla u^n|_{V_i}(x_{ij} - x_i) \ , \text{ with } \quad i = 1, \ldots, I \ ,$$

where $\psi_i \in (0, 1)$ is the limiter which has to fulfill the discrete minimum maximum property, as described in [7].

We also use the limitation of the flux to get no overshooting, when transporting the mass. Therefore the Courant-condition for one cell is $\tau_i = R_i V_i / \nu_i$ with $\nu_i = \sum_{k \in out(i)} v_{ik}$ and the time steps are restricted by $\tau \leq \min\{\tau_i, i = 1, \ldots, I\}$ and we get the restriction for the concentration

$$\tilde{u}_{ij}^n = u_{ij}^n + \frac{\tau_i}{\tau}(c_i^n - u_{ij}^n) \ . \tag{12}$$

Using all the previous schemes the discretisation for the second order is written in the form

$$\phi_i V_i c_i^{n+1} = \phi_i V_i c_i^n - \tau^n \sum_{j \in out(i)} \tilde{u}_{ij}^n v_{ij} + \tau^n \sum_{k \in in(i)} \tilde{u}_{ki}^n v_{ki} \ , \tag{13}$$

whereby $in(j) := \{k \in \Theta_j, v_{jk} < 0\}$ are the inflow-boundaries and $out(j) := \{k \in \Theta_j, v_{jk} \geq 0\}$ are the outflow-boundaries.

3.2 Finite Volume Discretisation Reconstructed with One Dimensional Analytical Solutions

The idea is to apply the Godonovs method, described in [10], for the discretisation. We reduce the equation to a one dimensional problem, solve the equation exactly and transform the one dimensional mass to the multi-dimensional equation.

The equation for the discretisation is given by

$$\partial_t c_i + \nabla \cdot \mathbf{v}_i \ c_i = -\lambda_i c_i + \lambda_{i-1} c_{i-1} \ , \ i = 1, \ldots, M \ . \tag{14}$$

The velocity vector is divided by R_i and M is the number of concentrations. The initial conditions are given by $c_1^0 = c_1(x, 0)$, else $c_i^0 = 0$ for $i = 2, \ldots, M$ and the boundary conditions are trivial $c_i = 0$ for $i = 1, \ldots, M$.

We first calculate the maximal time step for cell j and concentration i with the use of the total outflow fluxes

$$\tau_{i,j} = \frac{V_j R_i}{\nu_j} \ , \quad \nu_j = \sum_{j \in out(i)} v_{ij} \ . \tag{15}$$

We get the restricted time step with the local time steps and the velocity of the discrete equation are given as

$$\tau^n \leq \min_{\substack{i=1,\ldots,I \\ j=1,\ldots,M}} \tau_{i,j} \ , \ v_{i,j} = \frac{1}{\tau_{i,j}} \ . \tag{16}$$

We calculate the analytical solution of the mass with equation (9) with

$$m^n_{i,jk,out} = m_{i,out}(a, b, \tau^n, v_{1,j}, \ldots, v_{i,j}, R_1, \ldots, R_i, \lambda_1, \ldots, \lambda_i) \,, \qquad (17)$$

whereby $a = V_j R_i(c^n_{i,jk} - c^n_{i,jk'})$, $b = V_j R_i c^n_{i,jk'}$ are the parameter for the linear initial-impulse for the finite-volume cell and $c^n_{i,jk'}$ is the concentration on the left, $c^n_{i,jk}$ is the concentration on the right boundary of the cell j.

The discretisation with the embedded analytical mass is calculated by

$$m^{n+1}_{i,j} - m^n_{i,j} = - \sum_{k \in out(j)} \frac{v_{jk}}{\nu_j} m_{i,jk,out} + \sum_{l \in in(j)} \frac{v_{lj}}{\nu_l} m_{i,lj,out} \,, \qquad (18)$$

whereby $\frac{v_{jk}}{\nu_j}$ is the retransformation for the total mass $m_{i,jk,out}$ in the partial mass $m_{i,jk}$. In the next section we apply the modified discretisation method for a benchmark problem.

4 Numerical Experiments

We compare the standard with the modified method using a benchmark problem.

The standard method is based on an operator-splitting method done with the convection- and reaction-term, confer [6].

The modified method is described in the previous section with the embedded analytical solution of the mass.

For the experiment we use an one dimensional problem with triangular initial conditions. The analytical solutions are given, confer [9] and compare the analytical solution with the numerical solutions.

We calculate the solutions on a two dimensional domain, for which the velocity field is constant in the x-direction with the constant value of $\mathbf{v} = (1.0, 0.0)^T$. We use only the convection-reaction equation with 4 components, given as

$$R_i \, \partial_t c_i \; + \; \mathbf{v} \; \cdot \nabla c_i \; = \; -R_i \lambda_i c_i + R_{i-1} \lambda_{i-1} c_{i-1} \,, \quad i = 1, \ldots, 4 \,, \qquad (19)$$

whereby the inflow/outflow boundary condition are $\mathbf{n} \cdot \mathbf{v} \; c_i = 0.0$, with no inflow and outflow. The initial condition for the components are defined as

$$c_1(x, 0) = \begin{cases} x & 0 \le x \le 1 \\ 2 - x & 1 \le x \le 2 \\ 0 & otherwise \end{cases} \,, \quad c_i(x, 0) = 0 \,, \; i = 2, \ldots, 4 \,. \qquad (20)$$

For the problem we could compare the numerical solutions $c^n_{i,j}$ with the analytical solutions $c_i(x_j, y_j, t^n)$ using the L_1-norm.
The error is given on grid-level l as $E^l_{i,L_1} := \sum_{j=1,\ldots,I} V_j |c^n_{i,j} - c_i(x_j, y_j, t^n)|$.

The model domain is presented as a rectangle of 8×1 units. The initial coarse grid is given with 8 quadratic elements. We refine uniformly till the level 7, which has 131072 elements.

The parameters are chosen to get the nearly same maximum values at the end of the calculation to skip the numerical effects with different scalars.

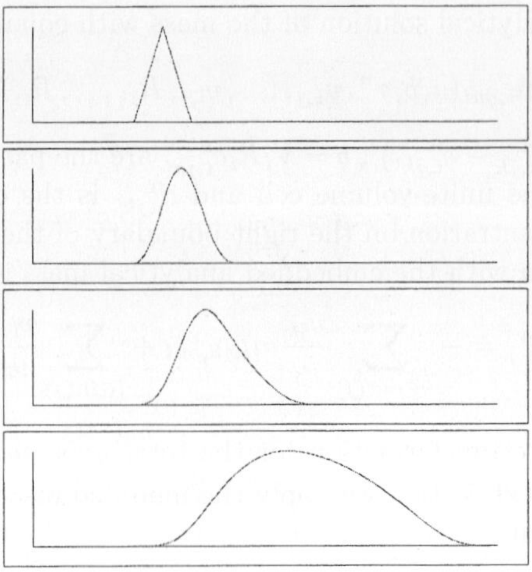

Fig. 1. Concentration for the 4 components with descending retardation factors at time $t = 6$

We use the reaction-parameters as $\lambda_1 = 0.3,\ \lambda_2 = 0.4,\ \lambda_3 = 0.5,\ \lambda_4 = 0.0$ and the retardation-parameters as $R_1 = 16,\ R_2 = 8,\ R_3 = 4,\ R_4 = 2$.

The model time is done from $t = 0, \ldots, 6$. We compared the results at the end-time $t = 6.0$. To do this we compared the L_1-norm and the numerical convergence-rate, which we compute by $\rho_i^l = (\log(E_{i,L_1}^l) - \log(E_{i,L_1}^{l-1}))/\log(0.5)$ for the levels $l = 4, \ldots, 7$.

The results for the standard and modified method are presented by

Table 1. L_1-error and the convergence-rate for the descending retardation-factors for the standard and modified method

	standard method				*modified method*			
l	E_{3,L_1}^l	ρ_{3,L_1}^l	E_{4,L_1}^l	ρ_{4,L_1}^l	E_{3,L_1}^l	ρ_{3,L_1}^l	E_{4,L_1}^l	ρ_{4,L_1}^l
4	$1.069\ 10^{-2}$		$2.502\ 10^{-2}$		$1.43\ 10^{-3}$		$1.255\ 10^{-3}$	
5	$5.16\ 10^{-3}$	1.051	$1.225\ 10^{-2}$	1.02	$3.07\ 10^{-4}$	2.22	$2.82\ 10^{-4}$	2.15
6	$2.52\ 10^{-3}$	1.033	$6.056\ 10^{-3}$	1.01	$7.94\ 10^{-5}$	1.95	$6.81\ 10^{-5}$	2.05
7	$1.24\ 10^{-3}$	1.023	$3.00\ 10^{-3}$	1.01	$2.04\ 10^{-5}$	1.96	$1.68\ 10^{-5}$	2.02

The results confirm our theoretical results. The standard method has the order $O(\tau^n)$ because of the splitting error. The modified method has a higher convergence order for the higher components because of the exact computation of the convection-reaction-term. The convergence order for the modified method is derived from the space-discretisation for the convection term and is $O(h^2)$ (h is the grid-width). The results of the calculation is presented at the end point $t = 6$ in the figure 1.

5 Conclusion

We have presented a modified discretisation method with embedded analytical solutions for the convection-reaction-equation. The idea of the derivation for the general and special analytical solution with Laplace-transformation are described. We confirm the theoretical results for the modified method of higher order discretisation. The analytical solution could also used for further discretisation methods based on flux or mass transfer, e.g. Discontinuous Galerkin methods. The main reason for the derivation of the analytical solutions for the coupled equations are to skip the splitting error. We will work out the application for analytical solutions combined with the diffusion term and kinetic sorption. This methods force the explicit solutions methods for the reaction term and solve the stiffness problem for different reaction-parameters.

References

1. P. Bastian, K. Birken, K. Eckstein, K. Johannsen, S. Lang, N. Neuss, and H. Rentz-Reichert. *UG - a flexible software toolbox for solving partial differential equations.* Computing and Visualization in Science, 1(1):27–40, 1997.
2. P.G. Ciarlet. *The Finite Element Methods for Elliptic Problems.* North Holland Publishing Company, Amsterdam, 1978.
3. G.R.. Eykolt. *Analytical solution for networks of irreversible first-order reactions.* Wat.Res., 33(3):814–826, 1999.
4. E. Fein, T. Kühle, and U. Noseck. *Entwicklung eines Programms zur dreidimensionalen Modellierung des Schadstofftransportes.* Fachliches Feinkonzept, Braunschweig, 2001.
5. P. Frolkovič and J. Geiser. *Numerical Simulation of Radionuclides Transport in Double Porosity Media with Sorption.* Proceedings of Algorithmy 2000, Conference of Scientific Computing : 28–36 , 2000.
6. P. Frolkovič. *Flux-based method of characteristics for contaminant transport in flowing groundwater.* Computing and Visualization in Science, 5 : 73–83, 2002.
7. P. Frolkovič and J. Geiser. *Discretization methods with discrete minimum and maximum property for convection dominated transport in porous media.* Proceeding of NMA 2002, Bulgaria, 2002.
8. J. Geiser. *Numerical Simulation of a Model for Transport and Reaction of Radionuclides.* Proceedings of the Large Scale Scientific Computations of Engineering and Environmental Problems, Sozopol, Bulgaria, 2001.
9. J. Geiser. *Gekoppelte Diskretisierungsverfahren für Systeme von Konvektions-Dispersions-Diffusions-Reaktionsgleichungen.* Doktor-Arbeit, Universität Heidelberg, 2004.
10. R.J. LeVeque. *Finite Volume Methods for Hyperbolic Problems.* Cambridge Texts in Applied Mathematics , Cambridge University Press, 2002.

(i) For all $z \in \mathbb{R}^m$, $y_1^0, y_2^0 \in \mathbb{R}^n$, and solutions $y_1(\cdot)$ of $\dot{y}(t) \in G(z, y(t))$, $y(0) = y_1^0$, there is a solution $y_2(\cdot)$ of $\dot{y}(t) \in G(z, y(t))$, $y(0) = y_2^0$, with

$$\|y_2(t) - y_1(t)\| \leq e^{-\alpha t}\|y_2^0 - y_1^0\|$$

for all $t \geq 0$.
(ii) For all $z \in \mathbb{R}^m$, $y_1^0, y_2^0 \in \mathbb{R}^n$, $S > 0$ the individual finite time averages fulfill the estimate

$$\mathrm{d_H}\left(F_S(z, y_2^0), F_S(z, y_1^0)\right) \leq \frac{L(\|y_2^0 - y_1^0\|)}{\alpha S}.$$

(iii) For all $w, z \in C([0, \infty); \mathbb{R}^m)$, $y^0 \in \mathbb{R}^n$ and all solutions $y_w(\cdot)$ of $\dot{y}(t) \in G(w(t), y(t))$, $y(0) = y^0$, there is a solution $y_z(\cdot)$ of $\dot{y}(t) \in G(z(t), y(t))$, $y(0) = y^0$, with

$$\sup_{t \geq 0} \|y_w(t) - y_z(t)\| \leq \frac{2Le^L}{1 - e^{-\alpha}} \sup_{t \geq 0} \|w(t) - z(t)\|.$$

Proof. This is an easy exercise. (i) follows from the Filippov lemma together with Assumption 1. (ii) follows from (i). (iii) is a consequence of (i) together with the Filippov theorem. □

The exponential stability of the fast subinclusions ensures the existence of a parametrized family of minimal invariant sets, where the slow states serve as parameters. Hence, the equilibria manifold of the Tychonov theory is replaced by more general objects in the multi-valued setting. This is the content of the following lemma.

Lemma 2. *Let Assumption 1 be effective. Then the following statements are valid.*

(i) For any $z \in \mathbb{R}^m$ there is a unique minimal compact forward invariant set $A(z) \subset \mathbb{R}^n$ of the differential inclusion $\dot{y}(t) \in G(z, y(t))$.

(ii) The multi-valued mapping $A : \mathbb{R}^m \to \mathcal{C}(\mathbb{R}^n)$ is Lipschitz continuous with Lipschitz constant $\frac{2Le^L}{1-e^{-\alpha}} \geq 0$.

(iii) For any compact set $M \subset \mathbb{R}^m$ and any compact set $N^0 \subset \mathbb{R}^n$, there is a compact set $N \subset \mathbb{R}^n$ containing N^0, which is forward invariant with respect to the differential inclusion $\dot{y}(t) \in G(z, y(t))$ for any fixed $z \in M$.

Proof. (i) and (iii) follow from Lemma 1 (i). (ii) follows from Lemma 1 (iii). □

Since the forward invariant subsets $A(z)$ in general contain more than one point, it is not sufficient to set $\epsilon = 0$ and to solve the corresponding algebraic equations in order to obtain an appropriate limit system. But the one-sided Lipschitz condition is strong enough to ensure that the finite time averages converge.

Lemma 3. *Let Assumption 1 be effective. Let $M \times N^0 \subset \mathbb{R}^m \times \mathbb{R}^n$ be a compact subset and let $N \subset \mathbb{R}^n$ be the corresponding forward invariant set for the differential inclusions $\dot{y}(t) \in G(z, y(t))$, for $z \in M$. Then there is a Lipschitz continuous multi-valued mapping $F_0 : M \to \mathcal{CC}(\mathbb{R}^m)$ with $\mathrm{d_H}\left(F_S(z, y^0), F_0(z)\right) \to 0$,* *as $S \to \infty$, such that uniformly in $(z, y^0) \in M \times N$ the individual finite time averages satisfy the estimate*

$$\mathrm{dist_H}\left(F_S(z, y^0), F_0(z)\right) = O(S^{-1}), \quad \text{as} \quad S \to \infty.$$

Proof. The existence of the Hausdorff limit follows from Lemma 1 (ii) together with a compactness argument. The Lipschitz continuity follows from immediately from Lemma 1 (iii). A complete proof can be found in [4].

The multi-valued mapping $F_0(\cdot)$ defines an averaged differential inclusion,

$$\dot{z}(t) \in \epsilon F_0(z(t)), \quad z(0) = z^0, \quad t \in \left[0, \frac{1}{\epsilon}\right], \tag{1}$$

which produces all possible limit trajectories of the perturbed system. Let $\mathcal{S}_0$ be the solution mapping of the averaged differential inclusion, and $\mathcal{S}_\epsilon$ be the solution mapping of the perturbed two-scale differential inclusion. The following result is a tightening of a result in [4] for this particular situation, since the order of approximation is computed. It is not known to the author whether the approximation order presented is optimal. For singularly perturbed differential equations, it definitely can be improved.

Theorem 1. *Let Assumption 1 be effective. For any compact sets $M^0 \subset \mathbb{R}^m$, $N^0 \subset \mathbb{R}^n$, we can estimate*

$$\mathrm{dist_H}\left(\Pi_z \mathcal{S}_\epsilon(z^0, y^0), \mathcal{S}_0(z^0)\right) = O(\epsilon^{1/2}), \quad \text{as} \quad \epsilon \to 0,$$

uniformly in $(z^0, y^0) \in M^0 \times N^0$, where Π_z denotes the projection on the z-components of the solution mapping.

Proof. Let $\epsilon \in (0, 1]$. On the time interval $[0, 1/\epsilon]$ the trajectories stay in a compact set $M \subset \mathbb{R}^m$, where the convergence $F_S(z, y^0) \to F_0(z)$, as $S \to \infty$, is uniform. We divide the time interval into subintervals of the form $[t_l, t_{l+1}]$, which all have the same length $S_\epsilon = \epsilon^{-1/2}$, except for the last one, which may be smaller. Accordingly the index l is an element of the index set $I_\epsilon := \{0, .., [1/(\epsilon S_\epsilon)]\}$. We take some initial values $(z^0, y^0) \in M^0 \times N^0$ and a solution $(z_\epsilon(\cdot), y_\epsilon(\cdot)) \in \mathcal{S}_\epsilon(z^0, y^0)$. We have $z_\epsilon(t_{l+1}) = z_\epsilon(t_l) + \int_{t_l}^{t_{l+1}} \dot{z}_\epsilon(s)ds$, where $\dot{z}_\epsilon(\cdot) \in \epsilon F(z_\epsilon(\cdot), y_\epsilon(\cdot))$ is a measurable selection. For $l \in I_\epsilon$ we set $\xi_0 := z^0$ and $\xi_{l+1} := \xi_l + \int_{t_l}^{t_{l+1}} w_l(s)ds$, where $w_l(\cdot) \in \epsilon F(\xi_l, y_{\xi_l}(\cdot, y_\epsilon(t_l)))$ is a measurable selection corresponding to a certain solution $y_{\xi_l}(\cdot, y_\epsilon(t_l))$ to $\dot{y}(t) \in G(\xi_l, y(t))$, $y(t_l) = y_\epsilon(t_l)$. For $t \in [t_l, t_{l+1}]$, we interpolate $\xi_l(t) := \xi_l + \int_{t_l}^{t} w_l(s)ds$. We choose $y_{\xi_l}(\cdot, y_\epsilon(t_l))$ and $w_l(\cdot) \in \epsilon F(\xi_l, y_{\xi_l}(\cdot, y_\epsilon(t_l)))$ in such a way that

$$\Delta_l(t) := \max_{t_l \le s \le t} \|z_\epsilon(s) - \xi_l(s)\| \tag{2}$$

tends to zero as $\epsilon \to 0$. Therefore we also define, for $l \in I_\epsilon$, $D_l(t) := \max_{t_l \le s \le t} \|y_\epsilon(s) - y_{\xi_l}(s, y_\epsilon(t_l))\|$ and $d_l(t) := \max_{t_l \le s \le t} \|z_\epsilon(s) - \xi_l\|$. By Lemma 1 (iii) there exists such a solution $y_{\xi_l}(\cdot, y_\epsilon(t_l))$ with $D_l(t) \leq d_l(t)\frac{2Le^L}{1-e^{-\alpha}}$ for all $t \in [t_l, t_{l+1}]$. In the sequel we abbreviate $C := \frac{2Le^L}{1-e^{-\alpha}}$. According to the Filippov Lemma there is a measurable selection $w_l(\cdot) \in \epsilon F(\xi_l, y_{\xi_l}(\cdot, y_\epsilon(t_l)))$ with $\|\dot{z}_\epsilon(t) - w_l(t)\| \leq \epsilon L d_l(t) + \epsilon L D_l(t)$. We conclude that

$$\begin{aligned}
\Delta_l(t) &\leq \Delta_{l-1}(t_l) + \int_{t_l}^{t} \|\dot{z}_\epsilon(s) - w_l(s)\| ds \\
&\leq \Delta_{l-1}(t_l) + \int_{t_l}^{t} (\epsilon L d_l(s) + \epsilon L D_l(s))\, ds \\
&\leq \Delta_{l-1}(t_l) + \int_{t_l}^{t} (\epsilon L(1+C) d_l(s))\, ds \\
&\leq \Delta_{l-1}(t_l) + \int_{t_l}^{t} (\epsilon L(1+C)(\Delta_l(s) + \epsilon S_\epsilon P))\, ds,
\end{aligned}$$

where $P \geq 0$ is an upper bound for F on $M \times N$. We apply the Gronwall Lemma and obtain $\Delta_l(t_{l+1}) \leq \left(\Delta_{l-1}(t_l) + \epsilon^2 S_\epsilon^2 LP(1+C)\right) e^{\epsilon S_\epsilon L(1+C)}$. We can set $\Delta_{-1}(0) = 0$ and conclude for all $l \in I_\epsilon$ that $\Delta_l(t_{l+1}) \leq \left(\epsilon^2 S_\epsilon^2 LP(1+C)\right) e^{\epsilon S_\epsilon L(1+C)} \sum_{k=1}^{l} e^{(k-1)\epsilon S_\epsilon L(1+C)}$. Since $l \leq [1/\epsilon S_\epsilon]$ we have

$$\Delta_l(t_{l+1}) \leq (\epsilon S_\epsilon LP(1+C))\, e^{\epsilon S_\epsilon L(1+C)} e^{L(1+C)}. \tag{3}$$

We now can choose a $v_l \in \epsilon F_0(\xi_l)$ such that $\left\| v_l - \frac{1}{S_\epsilon} \int_{t_l}^{t_{l+1}} w_l(s) ds \right\| \leq \epsilon O(S_\epsilon^{-1})$. For all $l \in I_\epsilon$ we define $\eta_0 := z^0$ and $\eta_{l+1} := \eta_l + S_\epsilon v_l$. We interpolate piecewise linearly and set $\eta_l(t) := \eta_l + (t - t_l)v_l$. for $t \in [t_l, t_{l+1}]$. Obviously, for all $l \in I_\epsilon$ and all $t \in [t_l, t_{l+1}]$, the estimate

$$\|\eta_l(t) - \xi_l(t)\| \leq O(S_\epsilon^{-1}) \tag{4}$$

is valid. On the other hand, this piecewise linear curve is almost a solution of the averaged multi-valued differential equation, since we can estimate $\mathrm{dist}(\eta_l(t), \epsilon F_0(\eta_l(t))) \leq \mathrm{d_H}(\epsilon F_0(\xi_l), \epsilon F_0(\eta_l)) + \mathrm{d_H}(\epsilon F_0(\eta_l), \epsilon F_0(\eta_l(t))) \leq \epsilon C O(S_\epsilon^{-1}) + \epsilon^2 C S_\epsilon P$. According to the Filippov Theorem there is a solution $z_0(\cdot) \in S_0(z^0)$ of the averaged multi-valued differential equation with

$$\|z_0(t) - \eta_l(t)\| \leq (CO(S_\epsilon^{-1}) + C\epsilon S_\epsilon P)e^C \tag{5}$$

for $t \in [t_l, t_{l+1}]$. For $t \in [0, 1/\epsilon]$, we can estimate $\|z_\epsilon(t) - z_0(t)\| \leq \|z_\epsilon(t) - \xi_l(t)\| + \|\xi_l(t) - \eta_l(t)\| + \|\eta_l(t) - z_0(t)\|$, where $l \in I_\epsilon$ is chosen such that $t \in [t_l, t_{l+1}]$ and considering (3), (4), (5), the proof is finished. □

Actually one can even prove that the convergence stated in the theorem above is in the Hausdorff sense, see e.g. [4]. However, the rate of approximation w.r.t. the Hausdorff metric is reduced. Moreover, the distance estimation above is more useful than its converse, since it can be utilized for verifying exponential stability of the perturbed system by investigating the exponential stability of the averaged system.

3 Order Reduction of Three-Scale Differential Inclusions

We re-iterate the averaging method presented in the previous section in order to construct a limiting system for the slowest motion. Consider the three-scale differential inclusion

$$\begin{aligned} \dot{z}(t) &\in \epsilon F(z(t), y(t), x(t)), \quad z(0) = z^0, \\ \dot{y}(t) &\in G(z(t), y(t), x(t)), \quad y(0) = y^0, \\ \delta \dot{x}(t) &\in H(z(t), y(t), x(t)), \quad x(0) = x^0, \quad t \in \left[0, \frac{1}{\epsilon}\right], \end{aligned}$$

in $\mathbb{R}^m \times \mathbb{R}^n \times \mathbb{R}^p$, where $\delta, \epsilon > 0$ are small real parameters.

Aassumption 2 (Decoupled One-Sided Lipschitz Continuity). The multi-valued mappings $F : \mathbb{R}^m \times \mathbb{R}^n \times \mathbb{R}^p \to \mathcal{CC}(\mathbb{R}^m)$, $G : \mathbb{R}^m \times \mathbb{R}^n \times \mathbb{R}^p \to \mathcal{CC}(\mathbb{R}^n)$ and $H : \mathbb{R}^m \times \mathbb{R}^n \times \mathbb{R}^p \to \mathcal{CC}(\mathbb{R}^p)$ are Lipschitz continuous with Lipschitz constant $L \geq 0$. Furthermore, there is a constant $\alpha > 0$ such that the multi-valued mappings G, H possess the following properties. For all $(z, y_1, x) \in \mathbb{R}^m \times \mathbb{R}^n \times \mathbb{R}^p$, $v_1 \in G(z, y_1, x)$, $y_2 \in \mathbb{R}^n$ there is a $v_2 \in G(z, y_2, x)$ with

$$\langle v_2 - v_1, y_2 - y_1 \rangle \leq -\alpha \|y_2 - y_1\|^2.$$

For all $(z, y, x_1) \in \mathbb{R}^m \times \mathbb{R}^n \times \mathbb{R}^p$, $w_1 \in H(z, y, x_1)$, $x_2 \in \mathbb{R}^p$ there is a $w_2 \in H(z, y, x_2)$ with

$$\langle w_2 - w_1, x_2 - x_1 \rangle \leq -\alpha \|x_2 - x_1\|^2.$$

According to Lemma 3 we can set

$$F_{(\epsilon,0)}(z, y) := \lim_{S \to \infty} \bigcup_{x(\cdot)} \frac{1}{S} \int_0^S \epsilon F(z, y, x(t))dt,$$

$$G_{(\epsilon,0)}(z, y) := \lim_{S \to \infty} \bigcup_{x(\cdot)} \frac{1}{S} \int_0^S G(z, y, x(t))dt,$$

where the union is taken of all solutions $x(\cdot)$ of $\dot{x}(t) \in H(z, y, x(t))$, $x(0) = x^0$, $t \in [0, \infty)$. Let $\mathcal{S}_{(\epsilon,0)}$ be the solution mapping of the averaged differential inclusion

$$\begin{aligned} \dot{z}(t) &\in F_{(\epsilon,0)}(z(t), y(t)), \quad z(0) = z^0, \\ \dot{y}(t) &\in G_{(\epsilon,0)}(z(t), y(t)), \quad y(0) = y^0, \quad t \in [0, 1], \end{aligned}$$

and $\mathcal{S}_{(\epsilon,\delta)}$ be the solution mapping of the perturbed three-scale differential inclusion.

Lemma 4. *Let Assumption 2 be effective. For any compact subset* $M^0 \times N_y^0 \times N_z^0 \subset \mathbb{R}^m \times \mathbb{R}^n \times \mathbb{R}^p$ *we can estimate*

$$\frac{1}{\epsilon} \mathrm{dist_H} \left(\Pi_z \mathcal{S}_{(\epsilon,\delta)}(z^0, y^0, x^0), \Pi_z \mathcal{S}_{(\epsilon,0)}(z^0, y^0) \right) = O(\delta^{1/2}),$$

$$\mathrm{dist_H}\left(\Pi_y S_{(\epsilon,\delta)}(z^0,y^0,x^0), \Pi_y S_{(\epsilon,0)}(z^0,y^0)\right) = O(\delta^{1/2}),$$

as $\delta \to 0$, *uniformly in* $\epsilon \in (0,1]$, $(z^0,y^0,x^0) \in M^0 \times N_y^0 \times N_z^0$, *where* Π_z, *respectively* Π_y, *are the projections on the* z*-components, respectively on the* y*-components, of the corresponding solution mappings.*

Proof. The second estimate is an immediate consequence of Theorem 1 applied to $\epsilon = 1$ for the case that $\delta \to 0$. As for the first estimate, we can apply Theorem 1 to the rescaled variable $Z(t) = z(t)/\epsilon$. Note that the bound $P \geq 0$ for the sets F remains unchanged and that the Lipschitz constant $L \geq 0$ of F can be used for $Z(t)$ as well. □

Note that Lemma 4 only gives an estimation on the time interval $[0,1]$, and not on the whole time interval $[0,1/\epsilon]$. For this reason we have to employ Assumption 2 in order to improve Lemma 4.

Lemma 5. *Let Assumption 2 be effective. For all* $(z,y_1) \in \mathbb{R}^m \times \mathbb{R}^n$, $v_1 \in G_{(\epsilon,0)}(z,y_1)$, $y_2 \in \mathbb{R}^n$ *there is a* $v_2 \in G_{(\epsilon,0)}(z,y_2)$ *with*

$$\langle v_2 - v_1, y_2 - y_1 \rangle \leq -\alpha \|y_2 - y_1\|^2.$$

Proof. Follows in a straight-forward way from Assumption 2. □

Lemma 6. *Let Assumption 2 be effective. For any compact subset* $M^0 \times N_y^0 \times N_x^0 \subset \mathbb{R}^m \times \mathbb{R}^n \times \mathbb{R}^p$, *the solution set* $S_{(\epsilon,0)}(z^0,y^0)$, *obtained on the time interval* $t \in [0,1/\epsilon]$, *fulfils the estimate*

$$\mathrm{dist_H}\left(\Pi_z S_{(\epsilon,\delta)}(z^0,y^0,x^0), \Pi_z S_{(\epsilon,0)}(z^0,y^0)\right) = O(\delta^{1/2}), \quad \text{as} \quad \delta \to 0,$$

uniformly in $\epsilon \in (0,1]$, $(z^0,y^0,x^0) \in M^0 \times N_y^0 \times N_x^0$, *where* Π_z *denote the projections on the* z*-component of the corresponding solution mappings.*

Proof. We divide the time interval $[0,1/\epsilon]$ into subintervals of the form $[k,k+1]$, $k \in \{1,2,\ldots,[1/\epsilon]\}$. In each subinterval, Lemma 4 is applicable. However, we still do not obtain a limit trajectory on the whole interval $[0,1/\epsilon]$, since there are jumps at $t_k = k$, $k \in \{1,2,\ldots,[1/\epsilon]\}$. To overcome these jumps we make use of a refined version of the Filippov theorem, as presented in [10]. By the one-sided Lipschitz continuity of $G_{(\epsilon,0)}(z,\cdot)$, the $\epsilon O(\delta^{1/2})$ approximation of the y-components given by Lemma 4 on the subintervals $[k,k+1]$ even takes place on the whole time interval $[0,1/\epsilon]$, whereas the $\epsilon O(\delta^{1/2})$ approximation of the z-components on the subintervals $[k,k+1]$ is reduced to an $O(\delta^{1/2})$ approximation on the whole interval $[0,1/\epsilon]$. □

According to Lemma 3 and Lemma 5 we can set

$$F_{(0,0)}(z) := \lim_{S\to\infty} \left(\bigcup_{y(\cdot)} \frac{1}{S} \int_0^S F_{(\epsilon,0)}(z,y(t))dt \right),$$

where the union is taken of all solutions $y(\cdot)$ of $\dot{y}(t) \in G_{(\epsilon,0)}(z, y(t))$, $y(0) = y^0$, $t \in [0, \infty)$. Let $\mathcal{S}_{(0,0)}$ be the solution mapping of the (twice) averaged differential inclusion

$$\dot{z}(t) \in F_{(0,0)}(z(t)), \quad z(0) = z^0, \quad t \in [0, 1/\epsilon].$$

Theorem 2. *Let Assumption 2 be effective. For any compact subset $M^0 \times N_y^0 \times N_x^0 \subset \mathbb{R}^m \times \mathbb{R}^n \times \mathbb{R}^p$, the solution set $\mathcal{S}_{(0,0)}(z^0)$ fulfils the estimate*

$$\mathrm{dist}_{\mathrm{H}}\left(\Pi_z \mathcal{S}_{(\epsilon,\delta)}(z^0, y^0, x^0), \Pi_z \mathcal{S}_{(0,0)}(z^0)\right) = O(\epsilon^{1/2} + \delta^{1/2}), \quad \text{as} \quad (\epsilon, \delta) \to 0,$$

uniformly in $(z^0, y^0, x^0) \in M^0 \times N_y^0 \times N_x^0$, where Π_z denote the projections on the z-component of the corresponding solution mappings.

Proof. Follows immediately from Lemma 6 and from Theorem 1. □

References

1. Dontchev, A.L. Time Scale Decomposition of the Reachable Set of Constrained Linear Systems, Mathematics of Control Signals Systems, **5** (1992), 327-340
2. Gaitsgory, V, M.-T. Nguyen, M.-T., Averaging of Three Time Scale Singularly Perturbed Control Systems, Systems & Control Letters, **42** (2001), 395–403
3. Gaitsgory, V., Nguyen, M.-T.: Multicsale Singularly Perturbed Control Systems: Limit Occupational Measure Sets and Averaging, SIAM Journal on Control and Optimization, **41** (2002), 354–974
4. Grammel, G.: Singularly Perturbed Differential Inclusions: An Averaging Approach, Set-Valued Analysis, **4** (1996), 361–374
5. Grammel, G.: Order Reduction for Nonlinear Systems by Re-iterated Averaging, Proceedings IEEE Singapore International Symposium on Control Theory and Applications, 29-30 July 1997, Singapore, 37–41
6. Grammel, G.: On Nonlinear Control Systems with Multiple Time Scales, Journal of Dynamical and Control Systems, **10** (2004), 11–28
7. Hoppensteadt, F: On Systems of Ordinary Differential Equations with Several Parameters Multiplying the Derivatives, Journal of Differential Equations, **5** (1969), 106–116
8. Ladde, G.S., Siljak, D.D.: Multiparameter Singular Perturbations of Linear Systems with Multiple Time Scales, Automatica, **19** (1983), 385–394
9. Tychonov, A.N.: Systems of Differential Equations Containing Small Parameters in the Derivatives, Matematicheskiu i Sbornik, **31** (1952), 575–586
10. Veliov, V.M.: Differential Inclusions with Stable Subinclusions, Nonlinear Analysis TMA, **23** (1994), 1027–1038

Computing Eigenvalues of the Discretized Navier-Stokes Model by the Generalized Jacobi-Davidson Method

G. Hechmé and M. Sadkane

Laboratoire de Mathématiques,
6, Av. Le Gorgeu, CS 93837
29238 BREST Cedex 3. FRANCE

Abstract. In this work the stability analysis of a $3D$ Navier-Stokes model for incompressible fluid flow is considered. Investigating the stability at a steady state leads to a special generalized eigenvalue problem whose main part of the spectrum is computed by the Jacobi-Davidson-QZ algorithm.

1 Introduction

We consider a three dimensional incompressible fluid flow Navier-Stokes problem, described by the following system of equations:

$$\begin{cases} \frac{\partial \tilde{v}}{\partial t} + < \tilde{v}, \nabla \tilde{v} >= -\nabla p + \triangle_{\mu,\nu} \tilde{v}, \\ \\ \text{div } \tilde{v} = 0. \end{cases} \tag{1}$$

The operator $\triangle_{\mu,\nu}$ denotes $\mu \left(\frac{\partial^2}{\partial x^2} + \frac{\partial^2}{\partial y^2} \right) + \nu \frac{\partial^2}{\partial z^2}$, where x and y are horizontal Cartesian coordinates, z is the vertical Cartesian downward coordinate, and μ, ν are the coefficients of horizontal and vertical turbulent diffusivities; p is the effective pressure and $\tilde{v} = (u, v, w)$ the fluid velocity; u, v and w are the velocity components in the directions of x, y and z respectively.

The study is carried out in the rectangular domain defined below:

$$\{(x, y, z) : \ 0 < x < x_{max}, \ 0 < y < y_{max}, \ 0 < z < z_{max}\},$$

with boundary conditions set as follows:

. $u = \frac{\partial v}{\partial x} = \frac{\partial w}{\partial x} = 0$ at lateral surfaces $x = 0$ or $x = x_{max}$,

. $\frac{\partial u}{\partial y} = v = \frac{\partial w}{\partial y} = 0$ at lateral surfaces $y = 0$ or $y = y_{max}$,

. $\frac{\partial u}{\partial z} = \frac{\partial v}{\partial z} = w = 0$, at the bottom surface $z = z_{max}$,

Z. Li et al. (Eds.): NAA 2004, LNCS 3401, pp. 304–311, 2005.

. $\nu\frac{\partial u}{\partial z} = \tau \sin\frac{\pi x}{x_{max}} \cos\frac{\pi y}{y_{max}}, \quad \nu\frac{\partial v}{\partial z} = -\tau \cos\frac{\pi x}{x_{max}} \sin\frac{\pi y}{y_{max}}, \quad w = 0,$

at the top surface $z = 0$.

After discretization, the Navier-Stokes equations are linearized at a steady state. Stability analysis around this steady state requires the solution of a non-symmetric generalized eigenvalue problem with the characteristic of being singular. We discuss how to reduce the eigenvalue problem to a regular one, more stable for computations, along with the techniques adapted to investigate its spectral characteristics.

The outline of the paper is as follows in section 2 we give the adapted approximation and discretization schemes. Section 3 presents how stability analysis at a steady state leads to the abovementioned generalized eigenvalue problem. In section 4 the generalized Jacobi-Davidson method to compute the main part of the spectrum is briefly discussed. Section 5 lists numerical experiments.

2 Discretization

Applying the marker and cell grid discretization [4, 7] to the rectangular domain, with $I,\ J,\ K \in \mathbf{N}^+$ discretization parameters for the three axis respectively, we get a $3D$ grid with IJK cells. A vertex $(x_\alpha, y_\beta, z_\gamma)$ is defined by

$$x_\alpha = \frac{x_{max}}{I}\alpha, \quad y_\beta, = \frac{y_{max}}{J}\beta, \quad z_\gamma = \frac{z_{max}}{K}\gamma,$$

where α, β and γ are integer or half integer numbers.

The velocity and pressure components are discretized on the grid as follows: u, v and w are approximated at vertices of type $(i, j-1/2, k-1/2)$, $(i-1/2, j, k-1/2)$ and $(i-1/2, j-1/2, k)$ respectively, where i, j and k are integers, while the pressure p components are approximated at vertices of type $(i-1/2, j-1/2, k-1/2)$. Thus the velocity is assigned at $n = (I-1)JK + I(J-1)K + IJ(K-1)$ total number of the inner grid vertices and the pressure at $m = IJK$ total inner grid vertices.

Centered finite difference approximations applied to (1) in the rectangular domain discretized as described above leads to the following ordinary differential and algebraic system of equations

$$\begin{cases} \frac{dv_d}{dt} = F(v_d) - Gp_d, \\ \\ G^T v_d = 0, \end{cases} \tag{2}$$

where $v_d \in \mathbf{R}^n$ denotes the discrete fluid velocity components, $p_d \in \mathbf{R}^m$ the discrete values of the pressure. G is a real n-by-m matrix having a one-dimensional null-space. G results from the discretization of the gradient of the pressure and has the following block form:

$$G = \begin{pmatrix} \text{-}G_1 \\ \text{-}G_2 \\ \text{-}G_3 \end{pmatrix}.$$

The one dimensional null-space property of G is due to the fact that the pressure is known up to an additive constant. F is a quadratic function $\mathbf{R}^n \to \mathbf{R}^n$ of the form:

$$F = \begin{pmatrix} -A_1(v_d) + D_1 & 0 & 0 \\ 0 & -A_2(v_d) + D_2 & 0 \\ 0 & 0 & -A_3(v_d) + D_3 \end{pmatrix}.$$

The $\{D_l\}_{l=1}^{3}$ blocks of F result from the discretization of $\triangle_{\mu,\nu}$ applied to the velocity components and the $\{-A_l(v_d)\}_{l=1}^{3}$ blocks come from the discretization of the $< \tilde{v}, \nabla\tilde{v} >$ term in (1). It is clear that the quadratic terms are due to the presence of the $-A_l(v_d)$ blocks in F.

3 Eigenvalue Problem at the Steady State

A steady state (v_s, p_s) of the Navier-Stokes model is a solution of (2) such that v_s is constant. To compute (v_s, p_s) we use the Crank-Nicolson scheme [4, 5]. At each time stepping the solution of an algebraic nonlinear system is required, we achieve this by using the Newton-GMRES method [1].

To examine the behavior of the Navier-Stokes model around this steady state, we consider small deviations v' and p' such that $v_d = v' + v_s$ and $p_d = p' + p_s$. By substituting v_d, p_d in (2) and taking into account that $F(v_d)$ is quadratic in v_d, we get

$$\begin{cases} \frac{dv'}{dt} = Jv' - Gp' + \mathcal{O}(\|v'\|^2), \\ \\ G^T v' = 0, \end{cases} \tag{3}$$

where $J \in \mathbf{R}^{n\times n}$ is the Jacobian of F at the steady state. We assume now that $v' = v_0 e^{\lambda_0 t}$ and $p' = p_0 e^{\lambda_0 t}$, where v_0 and p_0 are some fixed vectors. Keeping only linear terms in (3), we derive the following system of approximate equations for λ_0, v_0 and p_0:

$$\begin{cases} \lambda_0 v_0 = Jv_0 - Gp_0, \\ G^T v_0 = 0. \end{cases} \tag{4}$$

The latter is equivalent to the generalized eigenvalue problem :

$$\lambda_0 \begin{pmatrix} I & 0 \\ 0 & 0 \end{pmatrix} \begin{pmatrix} v_0 \\ p_0 \end{pmatrix} = \begin{pmatrix} J & -G \\ G^T & 0 \end{pmatrix} \begin{pmatrix} v_0 \\ p_0 \end{pmatrix}, \tag{5}$$

where I is the identity matrix of order n. Hence a normal mode (v_0, p_0) of (3) is an eigenvector of (5) and λ_0 the corresponding eigenvalue. Thus to determine stability and study the behavior of system (1) around the steady state, we need to compute the set of eigenvalues with largest real parts and the corresponding "leading modes" or eigenvectors of (5).

The matrix pencil

$$\lambda \begin{pmatrix} I & 0 \\ 0 & 0 \end{pmatrix} - \begin{pmatrix} J & -G \\ G^T & 0 \end{pmatrix} \tag{6}$$

is singular since vectors of the form $\begin{pmatrix} 0 \\ q \end{pmatrix}$ with $q \in \text{null}(G)$ belong to null spaces of both $\begin{pmatrix} I & 0 \\ 0 & 0 \end{pmatrix}$ and $\begin{pmatrix} J & -G \\ G^T & 0 \end{pmatrix}$.

To overcome this difficulty, let q be a non-zero vector in null(G) and consider the matrix L

$$L = \begin{pmatrix} I_{m-1} & \vdots \; q \\ 0 & \end{pmatrix}.$$

Multiplying (6) on the right with diag(I, L) and on the left by its transpose, leads to a strictly equivalent matrix pencil [3, 6] with a regular leading part of order $n + m - 1$:

$$\begin{pmatrix} J - \lambda I & -\bar{G} \\ \bar{G}^T & 0 \end{pmatrix},$$

and zero rest. $\bar{G}$ is the n-by-$(m-1)$ matrix formed by the first $m-1$ columns of G. Hence deleting the last zero row and the last zero column of (6) gives us a regular matrix pencil $\lambda B - A$ of order $n + m - 1$ with :

$$A = \begin{pmatrix} J & -\bar{G} \\ \bar{G}^T & 0 \end{pmatrix}, \quad B = \begin{pmatrix} I & 0 \\ 0 & 0 \end{pmatrix}. \tag{7}$$

4 Generalized Jacobi-Davidson Method

When the size of the pencil (7) is not large, the computation of the whole spectrum can then be done using the QZ method applied to the pencil (7). For large sizes we are confronted with the high computational cost and storage requirements. Fortunately, in practice we need only to find a main part of the spectrum, i.e. a few finite eigenvalues of largest real parts and the corresponding eigenvectors. To this end, we have chosen to use the Jacobi-Davidson method for generalized eigenvalue problems [2]. This method is a subspace iteration variant of the QZ algorithm. It computes an approximation of the partial generalized Schur decomposition of the pencil (7) in the form

$$(A - \lambda B)Q = Z(S_A - \lambda S_B),$$

where Q and Z are rectangular matrices with orthonormal columns and $S_A - \lambda S_B$ is the corresponding upper triangular pencil.

The method works as follows: suppose we are interested in the rightmost eigenvalue and the corresponding eigenvector of (7). The algorithm constructs two orthonormal bases of the same dimension which increase at each step. Denote by X and $\widetilde{X}$ two matrices whose columns form such orthonormal bases, then in each step, the method finds a scalar λ and a vector q that satisfy the Petrov-Galerkin condition:

$$q \in \text{range}X \quad \text{and} \quad r = (A - \lambda B)q \perp \text{range}\widetilde{X}.$$

This leads to a generalized eigenvalue problem with the small pencil $H - \lambda L$, $H = \widetilde{X}^T AX$ and $L = \widetilde{X}^T BX$.

The generalized Schur decomposition $W(S_H - \lambda S_L)U^T$ of the pencil $H - \lambda L$ can be used to approximate the desired rightmost eigenvalue. More precisely, the decomposition can be reordered such that the rightmost eigenvalue of the pencil $S_H - \lambda S_L$ appears in the top left corner. The matrices U and W are sorted accordingly, hence the right and left approximate Schur vectors corresponding to the rightmost eigenvalue are $q = XU(:,1)$ and $z = \widetilde{X}W(:,1)$. When an approximate eigenpair (λ, q) is not accurate enough the correction equation

$$t \perp q \quad \text{and} \quad [(I - zz^*)(A - \lambda B)(I - qq^*)]\, t = -r$$

is solved for t and used to expand the bases X and $\widetilde{X}$. If several rightmost eigenvalues are also of interest, the method proceeds by deflation once an eigenvalue has been computed. It is important to note that the methods restarts to limit storage requirements and keep orthogonality of both bases. In our simulations we use the *JDQZ* implementation coded by G.W. Sleijpen et al. [2] with a choice of bases suitable for our matrix pencil.

5 Numerical Experiments

For numerical experiments we set $\mu = 10^7 m^2/s$, $\nu = 10^{-4} m^2/s$ and $\tau = 10^{-3} m^2/s^2$ for the hydrodynamics equations, $x_{max} = y_{max} = 6 \times 10^6 m$ and $z_{max} = 4 \times 10^3 m$ for the rectangular domain. We apply discretization and approximation schemes discussed in section 2 to get system (2). After computation of the steady state, we form the regular pencil of order $n + m - 1$ and compute "the stability determining" rightmost eigenvalues and the corresponding eigenvectors using *JDQZ*. Because of space limitations, this paper concentrates only on the latter point. For the three simulations given below the correction equation is solved with the method GMRES. The parameter l_{max} denotes the maximum

Table 1. Rightmost eigenvalues computed by *JDQZ* with $tol = 10^{-10}$ in the case n+m-1=1375

Eigenvalues	Residuals	# iteration	# MVP
$-2.7065\ 10^{-6}\ -6.4518\ 10^{-9}i$	$1.2856\ 10^{-11}$	34	6849
$-2.7065\ 10^{-6}\ +6.4524\ 10^{-9}i$	$2.3594\ 10^{-11}$	34	6849
$-2.7067\ 10^{-6}\ -1.3548\ 10^{-13}i$	$7.3112\ 10^{-12}$	53	10668
$-2.7069\ 10^{-6}\ -1.0372\ 10^{-13}i$	$1.9581\ 10^{-11}$	54	10869
$-2.7071\ 10^{-6}\ -2.9205\ 10^{-13}i$	$2.6093\ 10^{-11}$	56	11271
$-2.7080\ 10^{-6}\ -3.2344\ 10^{-8}i$	$7.2179\ 10^{-12}$	59	11874
$-2.7080\ 10^{-6}\ +3.2344\ 10^{-8}i$	$1.6681\ 10^{-11}$	61	12276
$-2.7067\ 10^{-6}\ -4.1508\ 10^{-15}i$	$2.3064\ 10^{-11}$	80	16095
$-2.7069\ 10^{-6}\ +5.2544\ 10^{-14}i$	$1.7039\ 10^{-11}$	83	16698
$-2.7071\ 10^{-6}\ +2.9451\ 10^{-15}i$	$1.5182\ 10^{-11}$	88	17703

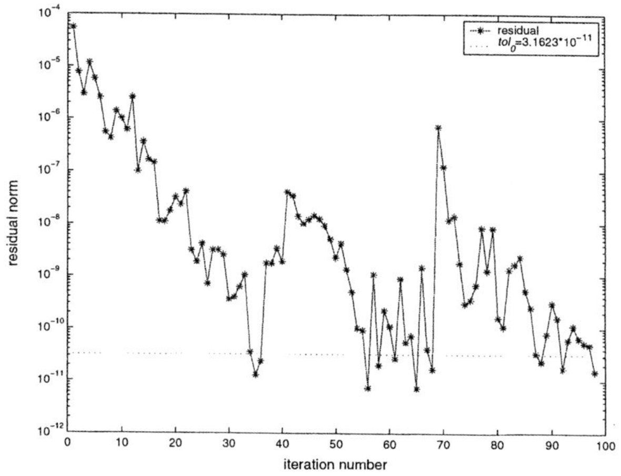

Fig. 1. Residual norm history for n+m-1=1375

dimension for the bases and l_{min} the number of columns of U and W kept at every restart. We set $l_{min} = 15$, $l_{max} = 50$. For the convergence test the program uses $tol_0 = tol/\sqrt{p}$, with p the number of required eigenvalues.

5.1 Matrix Pencil of Order n+m-1=1375 ($I = J = 8$, $K = 6$)

For the given values of the discretization parameters we get a matrix pencil with small dimensions, therefore computing the entire spectrum by the QZ algorithm is not very expensive. We obtained 766 infinite eigenvalues and 609 finite ones. We computed 10 rightmost eigenpairs with $JDQZ$ with the threshold tol equal to 10^{-10}. Table 1 shows the computed eigenvalues and the corresponding residual norms, the iteration number at convergence along with the total number of matrix-vector products. Figure 1 plots the residual norm history versus the iteration number. Comparison between the ten exact eigenvalues with largest real parts given by QZ and the ones given by $JDQZ$ shows that the latter computed the same eigenvalues as QZ with minor differences in the imaginary parts.

5.2 Matrix Pencil of Order n+m-1=11647 ($I = J = 16$, $K = 12$)

For this size of the matrix pencil the computation of all the finite eigenvalues is very expensive. We ran $JDQZ$ with $tol = 10^{-8}$ to compute 8 eigenvalues with largest real parts, the results are presented in Table 2 and Figure 2. We observe that although the first two eigenvalues, which form a complex conjugate pair, both converged at iteration 50, the third one required 60 more iterations. This behavior is due to the fact that the matrix pencil we deal with has very close clustered eigenvalues, which stalls the method when locating several eigenvalues in the same cluster, having very close real parts.

Table 2. Rightmost eigenvalues computed by *JDQZ* with $tol = 10^{-8}$ in the case n+m-1=11647

Eigenvalues	Residuals	# iteration	# MVP
$-2.7328\ 10^{-6}\ -1.5635\ 10^{-8}i$	$2.8767\ 10^{-9}$	50	10056
$-2.7328\ 10^{-6}\ +1.5641\ 10^{-8}i$	$2.8419\ 10^{-9}$	50	10056
$-2.7328\ 10^{-6}\ -4.5108\ 10^{-11}i$	$3.3209\ 10^{-9}$	110	22125
$-2.7341\ 10^{-6}\ +1.4949\ 10^{-9}i$	$2.9352\ 10^{-9}$	111	22326
$-2.7342\ 10^{-6}\ -1.5796\ 10^{-9}i$	$2.5173\ 10^{-9}$	111	22326
$-2.7353\ 10^{-6}\ +7.8288\ 10^{-10}i$	$2.4670\ 10^{-9}$	111	22326
$-2.7354\ 10^{-6}\ -6.5655\ 10^{-10}i$	$2.0403\ 10^{-9}$	111	22326
$-2.7398\ 10^{-6}\ -6.8790\ 10^{-8}i$	$2.4546\ 10^{-9}$	112	22527

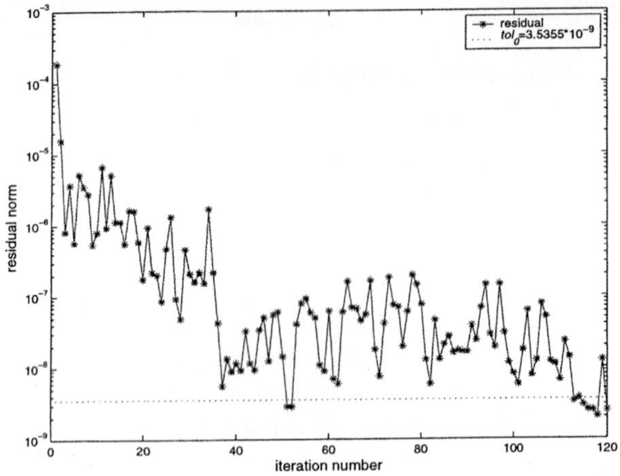

Fig. 2. Residual norm history for n+m-1=11647

5.3 Matrix Pencil of Order n+m-1=95743 ($I = J = 32$, $K = 24$)

The last simulation deals with a matrix pencil of very large dimensions, thus we compute 4 eigenvalues with $tol = 10^{-7}$. The resulted eigenvalues as well as information about the convergence are given in Table 3, residual history is plotted in Figure 3. For this case, the method required more iterations to locate the first eigenvalue, this is natural since the large size of the matrix pencil makes

Table 3. Rightmost eigenvalues computed by *JDQZ* with $tol = 10^{-7}$ in the case n+m-1=95743

Eigenvalues	Residuals	# iteration	# MVP
$-2.7402\ 10^{-6}\ +3.4028\ 10^{-8}i$	$4.9120\ 10^{-8}$	83	16698
$-2.7400\ 10^{-6}\ -3.4195\ 10^{-8}i$	$4.7387\ 10^{-8}$	84	16899
$-2.7695\ 10^{-6}\ -1.4179\ 10^{-7}i$	$3.0006\ 10^{-8}$	84	16899
$-2.7695\ 10^{-6}\ +1.4184\ 10^{-7}i$	$4.7360\ 10^{-8}$	84	16899

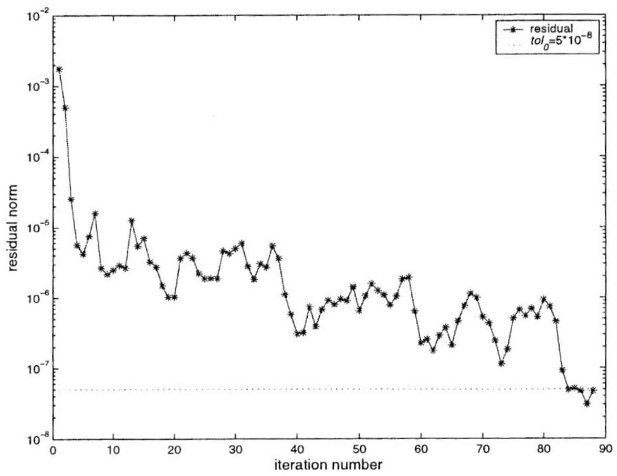

Fig. 3. Residual norm history for n+m-1=95743

the solution of the correction equation more difficult. We note that the rightmost eigenvalues are all located in the left side of the complex plane which confirms stability. This is true for the previous two simulations as well.

References

1. Brown N.P., Saad Y.: Hybrid Krylov methods for nonlinear systems of equations. SIAM J. Sci. Stat. Comput. 1986. v.11. n.3. pp.450-481.
2. Fokkema D.R., Sleijpen G.L., Van Der Vorst H.A.: Jacobi-Davidson style QZ and QR algorithms for the reduction of matrix pencils. SIAM J. Sci. Comput. 1999. v.20. n.1. pp. 94-125.
3. Gantmacher F.R.: Theorie des matrices. Dunod. -Paris, 1966.
4. Peyret R., Taylor T.D.: Computational methods for fluid flow. Springer Series in Computational Physics. -New-York: Springer-Verlag, 1983.
5. Roache P.J.: Computational fluid dynamics. Hermosa publishers. -New Mexico,1972.
6. Stewart G.W., Sun J.G.: Matrix perturbation theory. Academic Press, 1990.
7. Wesseling P.: An introduction to multigrid methods. Pure and Applied Mathematics. -Chichester: John Wiley & Sons, 1992.

where $\mathcal{N}_T$ is the set of noise candidates, φ_α is an edge-preserving potential function, β is a regularization parameter, $\mathcal{V}_{ij}$ denotes the four closest neighbors of (i, j), not including (i, j), and $\mathcal{V}_{\mathcal{N}_T} = \left(\bigcup_{(i,j)\in\mathcal{N}_T} \mathcal{V}_{ij}\right) \setminus \mathcal{N}_T$. Possible choices for φ_α are:

$$\varphi_\alpha(t) = |t|^\alpha, \quad 1 < \alpha \leq 2,$$
$$\varphi_\alpha(t) = 1 + \frac{|t|}{\alpha} - \log(1 + \frac{|t|}{\alpha}), \quad \alpha > 0,$$
$$\varphi_\alpha(t) = \log\left(\cosh(\frac{t}{\alpha})\right), \quad \alpha > 0,$$
$$\varphi_\alpha(t) = \sqrt{\alpha + t^2}, \quad \alpha > 0,$$

see [15], [16], [17], [18], [19].

The minimization algorithm is given in [12]. It is a Jacobi-type relaxation algorithm and works on the residual $\mathbf{z} = \mathbf{x} - \mathbf{y}$. The minimization scheme is restated as follows.

Minimization Scheme

1. Initialize $z_{ij}^{(0)} = 0$ for each (i, j) in the noise candidate set $\mathcal{N}_T$.
2. At each iteration k, do the following for each $(i, j) \in \mathcal{N}_T$:
   (a) Calculate
   $$\xi_{ij}^{(k)} = \beta \sum_{(m,n)\in\mathcal{V}_{ij}} \varphi'_\alpha(y_{ij} - z_{mn} - y_{mn}),$$
   where z_{mn}, for $(m, n) \in \mathcal{V}_{ij}$, are the latest updates and φ'_α is the derivative of φ_α.
   (b) If $|\xi_{ij}^{(k)}| \leq 1$, set $z_{ij}^{(k)} = 0$. Otherwise, find $z_{ij}^{(k)}$ by solving the nonlinear equation
   $$\beta \sum_{(m,n)\in\mathcal{V}_{ij}} \varphi'_\alpha(z_{ij}^{(k)} + y_{ij} - z_{mn} - y_{mn}) = \operatorname{sgn}(\xi_{ij}^{(k)}). \quad (2)$$
3. Stop the iteration when
   $$\max_{i,j}\{|z_{ij}^{(k+1)} - z_{ij}^{(k)}|\} \leq \tau_A \text{ and } \frac{F_{\mathbf{y}}(\mathbf{y} + \mathbf{z}^{(k)}) - F_{\mathbf{y}}(\mathbf{y} + \mathbf{z}^{(k+1)})}{F_{\mathbf{y}}(\mathbf{y} + \mathbf{z}^{(k)})} \leq \tau_A,$$
   where τ_A is some given tolerance.

It was shown in [12] that the solution $z_{ij}^{(k)}$ of (2) satisfies

$$\operatorname{sgn}(z_{ij}^{(k)}) = -\operatorname{sgn}(\xi_{ij}^{(k)}),$$

and that $\mathbf{z}^{(k)}$ converges to $\hat{\mathbf{z}} = \hat{\mathbf{x}} - \mathbf{y}$ where $\hat{\mathbf{x}}$ is the minimizer for (1).

The second phase is equivalent to solving the one-dimensional nonlinear equation (2) for each noise candidate. Typically, Newton's method is preferred to solve the (2), such as [1], [12], [13]. Since the convergence domain of Newton's method can be very narrow, care must be exercised in choosing the initial guess. See [14].

This two-phase scheme has successfully suppressed the noise while preserving most of the details and the edges in both cases, even when the noise level is high.

3 Secant-Like Method

Newton's method is one way to solve the minimization problem and from [14] we know how to find the initial guess such that Newton's method is guaranteed to converge. However, if we use $\varphi_\alpha(t) = |t|^\alpha$, which is applied widely for its simplicity, as our edge-preserving function, the complexity of the algorithm is not very good [1]. Fox example, for 30% noise, it will take 30 times more CPU time than ACWMF. To improve the timing, we describe a simple method which is called secant-like method to solve (2). This method converges in fewer number of iterations than Newton's method with both methods requiring the same number of function and derivative evaluations per iteration.

According to Step 2(b) of Minimization Scheme, we only need to solve (2) if $|\xi_{ij}^{(k)}| > 1$. We first consider the case where $\xi_{ij}^{(k)} > 1$. When solving (2), $z_{mn} + y_{mn} - y_{ij}$, for $(m,n) \in \mathcal{V}_{ij}$, are known values. Let these values be denoted by d_j, for $1 \le j \le 4$, and be arranged in an increasing order: $d_j \le d_{j+1}$. Then (2) can be rewritten as

$$H(z) \equiv -1 + \alpha\beta \sum_{j=1}^{4} \operatorname{sgn}(z - d_j)|z - d_j|^{\alpha-1} = 0, \tag{3}$$

hence (2) has a unique solution $z^* > d_1$, see [14]. By evaluating $\{H(d_j)\}_{j=2}^4$, we can check that if any one of the d_j, $2 \le j \le 4$, is the root z^*. If not, then z^* lies in one of the following intervals:

$$(d_1, d_2), (d_2, d_3), (d_3, d_4), \text{ or } (d_4, \infty).$$

We first consider the case where z^* is in one of the finite intervals (d_j, d_{j+1}). For simplicity, we give the details only for the case where $z^* \in (d_2, d_3)$. The other cases can be analyzed similarly. Since H is a monotone function in (d_2, d_3), see [14], it has an inverse G with $G(H) = z$. The goal is to compute $G(0)$. Suppose the following data are known:

$$G(H_1) = z_1, \; G(H_2) = z_2, \; G'(H_1) = p_1, \; G'(H_2) = p_2. \tag{4}$$

Note that

$$G'(H_i) = \frac{1}{\frac{dH(z_i)}{dz}}, \qquad i = 1, 2.$$

We approximate G by a cubic polynomial P which satisfies the conditions (4). Let

$$z = P(H) = z_1 + a(H - H_1) + b(H - H_1)(H - H_2) + cH(H - H_1)(H - H_2)$$

for some constants a, b, c. From the conditions $z_i = P(H_i)$, $p_i = P'(H_i)$, $i = 1, 2$, we obtain

$$a = \frac{z_2 - z_1}{H_2 - H_1}, \qquad b = \frac{H_2(a - p_1) + H_1(a - p_2)}{(H_1 - H_2)^2}.$$

Thus the new approximate zero is given by

$$P(0) = z_1 - aH_1 + bH_1H_2.$$

In summary, the iteration is given by

$$z_{n+1} = z_n - a_n H(z_n) + b_n H(z_n) H(z_{n-1}), \qquad n \geq 1$$

where

$$a_n = \frac{z_n - z_{n-1}}{H(z_n) - H(z_{n-1})}, \qquad b_n = \frac{H(z_n)(a_n - \frac{1}{H'(z_{n-1})}) + H(z_{n-1})(a_n - \frac{1}{H'(z_n)})}{(H(z_n) - H(z_{n-1}))^2}.$$

Given z_0, the iterate z_1 is taken as the Newton iterate. Then apply the secant-like method to obtain the solution up to a given tolerance τ_B, that is, $|z_n - z_{n-1}| \leq \tau_B$.

Finally, we turn to the case where $\xi^{(k)} < -1$. The nonlinear equation (3) becomes:

$$1 + \alpha\beta \sum_{j=1}^{4} \operatorname{sgn}(z - d_j)|z - d_j|^{\alpha-1} = 0,$$

we can use almost the same method to solve this equation.

The secant-like method is equivalent to scheme 3 on p. 233 of [20], where it is stated that the order of convergence is $1 + \sqrt{3}$. Hence the method converges faster than Newton's method. In the next section, we see the secant-like method always takes fewer number of iterations than Newton's method. Experimentally, the secant-like method is as robust as Newton's method although we do not have any theoretical result in this direction.

4 Numerical Experiments

In this section, we simulate the restoration of the 256-by-256 gray scale image *Lena* corrupted by 30% and 50% random-valued impulse noise with dynamic range $[0, 255]$, see Figure 1.

In the simulations, for each noise pixel, we use ACWMF in the first phase, and detail-preserving regularization in the second phase. Also, we use the secant-like method to solve (2) with the potential function $\varphi(t) = |t|^{1.2}$. We choose $\beta = 2$ for all settings. The restored images are shown in Figure 2. We see that the noise are successfully suppressed while the edges and details are well preserved.

We also compare the number of iterations of secant-like method with Newton's method with different magnitudes of α. The tolerances is chosen to be $\tau_A = (n_{\max} - n_{\min}) \times 10^{-4}$ and $\tau_B = 5 \times 10^{-4}$. In Figure 3, we give, for different values of α, the total number of iterations.

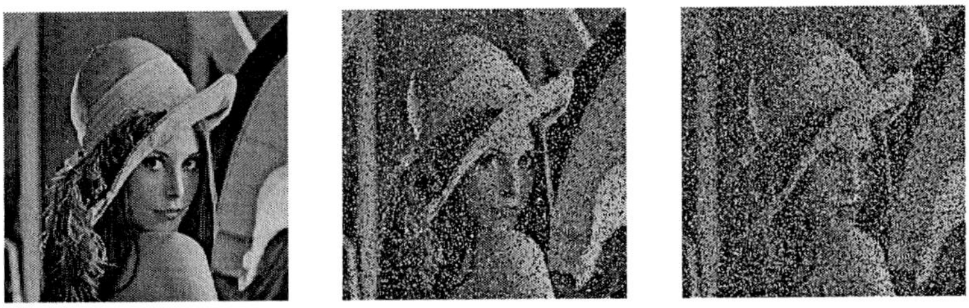

(a) (b) (c)

Fig. 1. (a) Original Image, (b) noisy image corrupted with 30% random-valued impulse noise, (c) noisy image corrupted with 50% random-valued impulse noise

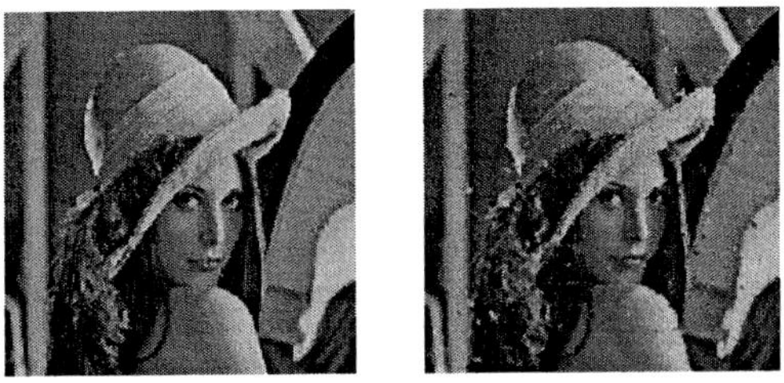

(a) (b)

Fig. 2. Restored images by secant-like solver using $\alpha = 1.2$ and $\beta = 2$. (a) 30% noise, (b) 50% noise

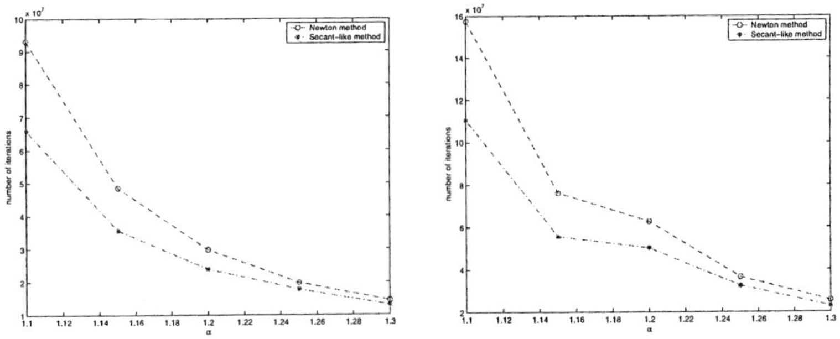

(a) (b)

Fig. 3. Compare the number of iterations of secant-like method with Newton's method (a) restore 30% noisy image (b) restore 50% noisy image

From the figures, we see that the secant-like method converges faster than Newton's method. The gain is greater as α approaches 1.

5 Conclusion

In this report, we give an overview of a 2-phase scheme for cleaning random-valued impulse noise and show that the second phase is equivalent to solving a nonlinear equation. We describe a simple secant-like method to solve this equation. This method converges faster than Newton's method. Simulation results show that the images are restored satisfactorily even at very high noise levels.

Acknowledgment

We thank Professor Raymond Chan for his guidance in this project.

References

1. Chan, R.H., Hu, C., Nikolova,M.: An iterative procedure for removing random-valued impulse noise. IEEE Singal Processing Letters (to appear)
2. Arce, G.R., Foster, R.E.: Detail-preserving ranked-order based filters for image processing. IEEE Transactions on Acoustics, Speech, and Signal Processing **37** (1989) 83–98
3. Han, W.-Y., Lin, J.-C.: Minimum-maximum exclusive mean (MMEM) filter to remove impulse noise from highly corrupted images.Electronics Letters **33** (1997) 124–125
4. Lee, Y.H., Kassam, S.A.: Generalized median filtering and related nonlinear filtering techniques. IEEE Transactions on Acoustics, Speech, and Signal Processing **33** (1985) 672–683
5. Chen, T.,Wu, H.R.: Space variant median filters for restoration of impulse noise corrupted images. IEEE Transactions on Circuits and Systems II. **48** (2001) 784–789
6. Chen, T.,Wu, H.R.: Adaptive impulse detection using center-weighted median filters. IEEE Singal Processing Letters **8** (2001) 1–3
7. T. S. Huang, T.S., Yang, G.J., Tang, G.Y.: Fast two-dimensional median filtering algorithm.IEEE Transcactions on Acoustics, Speech, and Signal Processing. **1** (1979) 13–18
8. H. Hwang, H. Haddad, R.A.: Adaptive meian filters: new algorithms and results. IEEE Transactions on Image Processing **4** (1995) 499–502
9. Ko, S.-J., Lee, Y.H.: Center weighted median filters and their applications to image enhancement.IEEE Transactions on Circuits and Systems **38** (1991) 984–993
10. Pok, G., Liu, J.-C., Nair, A.S.: Selective removal of impulse noise based on homogeneity level information. IEEE Transactions on Image Processing **12** (2003) 85–92
11. Sun, T., Neuvo,Y.: Detail-preserving median based filters in image processing. Pattern Recognition Letters.**15** (1994) 341–347
12. Nikolova, M.: A variational approach to remove outliers and impulse noise. Journal of Mathematical Imaging and Vision.**20** (2004)99–120
13. Chan, R.H., Ho, C.-W., Nikolova, M.: Impulse noise removal by median-type noise detectors and edge-preserving regularization. Report, Department of Mathematics, The Chinese University of Hong Kong, 2003-28 (302)

14. Chan, R.H., Ho, C.-W., Nikolova, M.: Convergence of Newton's method for a minimization problem in impulse noise removal. Journal of Computational Mathematics.**22** (2004) 168–177
15. Black, M., Rangarajan, A.:On the unification of line processes, outlier rejection, and robust statistics with applications to early vision. International Journal of Computer Vision **19** (1996) 57–91
16. Bouman, C., Sauer, K.: A generalized Gaussian image model for edge-preserving MAP estimation. IEEE Transactions on Image Processing **2**(1993) 296–310
17. Bouman, C., Sauer, K.: On discontinuity-adaptive smoothness priors in computer vision. IEEE Transactions on Pattern Analysis and Machine Intelligence.**17** (1995) 576–586
18. Charbonnier, P., Blanc-Féraud,L., Aubert, G., Barlaud, M.: Deterministic edge-preserving regularization in computed imaging. IEEE Transactions on Image Processing **6** (1997) 298–311
19. Greeen, P.J.: Bayesian reconstructions from emission tomography data using a modified EM algorithm. IEEE Transactions on Medical Imaging, MI-9 (1990) 84–93
20. Traub, J.F.: Iterative methods for the solution of equations. Englewood Cliffs, N. J.: Prentice Hall, (1964)

Adaptive Filters Viewed as Iterative Linear Equation Solvers

John Håkon Husøy

University of Stavanger, Department of Electrical and Computer Engineering,
PO Box 8002, 4068 Stavanger, Norway

Abstract. Adaptive filtering is an important subfield of digital signal processing having been actively researched for more than four decades and having important applications such as noise cancellation, system identification, and telecommunications channel equalization. In this paper we provide a novel framework for adaptive filtering based on the theory stationary iterative linear equation solvers. We show that a large number of established, and some quite novel, adaptive filtering algorithms can be interpreted as special cases of a generic update equation forming the cornerstone of our framework. The novelty of our contribution is the casting of the adaptive filtering problem as a problem in numerical linear algebra facilitating a clearer understanding and a unified analysis of the various algorithms.

1 Introduction

Adaptive filtering is an important subfield of digital signal processing having been actively researched for more than four decades and having important applications such as noise cancellation, system identification, telecommunications channel equalization, and telephony acoustic and network echo cancellation. The various adaptive filtering algorithms that have been developed have traditionally been presented without a unifying theoretical framework: Typically, each adaptive filter algorithm is developed from a particular objective function whose iterative or direct minimization gives rise to the various algorithms. This approach obscures the relationships, commonalities and differences, between the numerous adaptive algorithms available today. The objective of the present paper is to provide a novel unifying framework for adaptive filtering based on stationary methods for the iterative solution of sets of linear equations. We show how many known adaptive filtering algorithms can be seen as simple special cases of a *generic update equation* within this framework which is based solely on numerical linear algebra. While this is important in its own right, it also facilitates a clearer and complementary understanding paving the way for possible future contributions from scientists with other backgrounds than electrical engineering who, to the present time, have dominated the field.

We have organized our paper as follows: The next section gives a short review of optimum filtering, the Wiener filter, and adaptive filters. Following this, we develop our *generic filter update equation* which can be viewed as an iteration

Z. Li et al. (Eds.): NAA 2004, LNCS 3401, pp. 320–327, 2005.

step in the solution of an *estimated and weighted Wiener-Hopf equation.* In the following two sections we first show how established algorithms [1, 2] such as the Least Mean Squares (LMS) algorithm, its normalized version (NLMS), the Affine Projection Algorithm (APA), and the Recursive Least Squares (RLS) algortithm all can be seen as special cases of our unifying theory. Following this, we also demonstrate that two recently introduced adaptive filter algorithms, the Fast Euclidean Direction Search (FEDS) algorithm [3] and the Recursive Adaptive Matching Pursuit (RAMP) algorithm [4], fit nicely into our unified framework.

2 Optimum and Adaptive Filtering

The optimum filtering problem in the discrete time domain is briefly reviewed with reference to Figure 1. The problem can, in an application independent

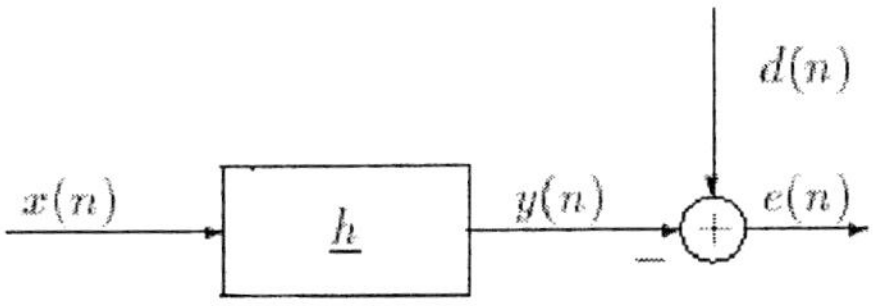

Fig. 1. The optimum filtering problem: $\min_{\underline{h}}\{E(e^2(n))\}$

setting, be stated as: Given a time series, or signal, $d(n)$, often referred to as the *desired signal* and some measurements represented by the signal $x(n)$: *How do we filter* $x(n)$ *such that the filtered signal is an optimum estimate of* $d(n)$ *in a mean square sense?* Obviously, we have to use a *filter vector*

$$\underline{h} = [h_0, h_1, \ldots, h_{M-1}]^T, \tag{1}$$

where T denotes vector transpose, found as the solution to the minimization problem

$$\min_{\underline{h}} E\{[d(n) - \underline{h}^T \underline{x}(n)]^2\}, \tag{2}$$

where $E(.)$ denotes the expectation operator,

$$\underline{x}(n) = [x(n), x(n-1), \ldots, x(n-M+1)]^T, \tag{3}$$

is an M-vector of signal samples to be filtered, and $\underline{h}^T\underline{x}(n)$ is the filter output denoted by $y(n)$ if Figure 1. The optimum filter under *statistically stationary conditions*, $\underline{h}$, is found as the solution to the Wiener-Hopf equation

$$\mathbf{R}\underline{h} = \underline{r}, \tag{4}$$

where

$$\mathbf{R} = E\{\underline{x}(n)\underline{x}^T(n)\}, \tag{5}$$

and

$$\underline{r} = E\{\underline{x}(n)d(n)\} \tag{6}$$

are the autocorrelation matrix and cross correlation vector, respectively.

In many important digital signal processing applications, we cannot assume statistical stationarity, or, if we can, we cannot assume prior knowledge of the exact nature of the stochastic properties of the signals involved as manifest in $\mathbf{R}$ and $\underline{r}$. In both cases there is a demand for a an algorithm that computes the filter vector in real time based on the the signals $d(n)$ and $x(n)$. Computing such a filter vector for each sample time instant, n, we get a time varying filter vector denoted $\underline{h}(n)$. In the latter case (unknown statistically stationary conditions), we would require the resulting filter $\underline{h}(n)$ to converge as rapidly as possible to some value close to the solution of the underlying Wiener-Hopf equation, Equation 4. In the former case (non stationary situation) which *in most cases* can be viewed as a sequence of shorter or longer time intervals of locally stationary situations, we require both fast convergence to the Wiener-Hopf solution appropriate for each time interval as well as an ability to track slow and fast transitions from one locally stationary situation to the other. Algorithms employed for these purposes are called *adaptive filters*. Given the high sampling rates of real life signals to which adaptive filters are applied (e.g. digital telephony with 8000 samples per second), the art of devising good adaptive filters involves a tradeoff between performance and computational complexity.

3 The Estimated Wiener-Hopf Equation

A natural, but neglected, approach to the adaptive filtering problem stated above is as follows: Estimate the quantities $\mathbf{R}$ and $\underline{r}$ of the Wiener-Hopf equation, Equation 4, at each time instant n. This gives us an *estimated Wiener-Hopf equation* for each n. At each sample time n we propose the application of some technique for the iterative solution of this estimated equation. In the present context we interleave the updates for each n of the estimated quantities and the computations required by the iterative solution procedure. In doing this we have several choices:

- The number of signal samples to use in the estimates of $\mathbf{R}$ and $\underline{r}$. A large number of samples should give reliable estimates, whereas a smaller number of samples used will give less reliable estimates.
- The selection of a method for the iterative linear equation solver.
- At each sample time, n, we can perform one iteration of the selected method, we can perform several iterations, or, we can perform parts of an iteration[1]

It is intuitively plausible that all three choices directly impact both the performance and the computational complexity of the resulting adaptive filter algorithm.

[1] Example: If the Gauss Seidel method has been selected, one iteration corresponds to one update of all elements of $\underline{h}(n)$, whereas it is natural to view each single element update of $\underline{h}(n)$ as a partial iteration.

Natural estimates of $\mathbf{R}$ and $\underline{r}$ are

$$\hat{\mathbf{R}}(n) = \frac{1}{L}\begin{pmatrix} x(n) & x(n-1) & \dots & x(n-L+1) \\ x(n-1) & x(n-2) & & x(n-L) \\ \vdots & & \ddots & \\ \vdots & & & \\ x(n-M+1) & x(n-M) & \dots & x(n-M-L+2) \end{pmatrix}$$
$$\cdot \begin{pmatrix} x(n) & x(n-1) & \dots & x(n-M+1) \\ x(n-1) & x(n-2) & & x(n-M) \\ \vdots & & \ddots & \\ \vdots & & & \\ x(n-L+1) & x(n-L) & \dots & x(n-M-L+2) \end{pmatrix} \tag{7}$$

and

$$\hat{\underline{r}}(n) = \frac{1}{L}\begin{pmatrix} x(n) & x(n-1) & \dots & x(n-L+1) \\ x(n-1) & x(n-2) & & x(n-L) \\ \vdots & & \ddots & \\ \vdots & & & \\ x(n-M+1) & x(n-M) & \dots & x(n-M-L+2) \end{pmatrix} \cdot \begin{pmatrix} d(n) \\ d(n-1) \\ \vdots \\ \\ d(n-L+1) \end{pmatrix}. \tag{8}$$

With obvious definition for $\mathbf{X}(n)$ and $\underline{d}(n)$, the estimates can be expressed compactly as $\hat{\mathbf{R}}(n) = \mathbf{X}(n)\mathbf{X}^T(n)$ and $\hat{\underline{r}}(n) = \mathbf{X}(n)\underline{d}(n)$. Furthermore, for future use, we introduce the notation $\underline{x}(n-i)$ for column no. i of $\mathbf{X}(n)$ and $\tilde{\underline{x}}(n-j)$ for column no. j of $\mathbf{X}^T(n)$. From the above, we can state our estimated Wiener-Hopf equation as

$$\mathbf{X}(n)\mathbf{X}^T(n)\underline{h}(n) = \mathbf{X}(n)\underline{d}(n). \tag{9}$$

In many situations involving estimates it is convenient to use *weighted* estimates. Various weighted estimates of $\mathbf{R}$ and $\underline{r}$ can be expressed through an $L \times L$ weighting matrix $\mathbf{W}(n)$ giving $\hat{\mathbf{R}}(n) = \mathbf{X}(n)\mathbf{W}(n)\mathbf{X}^T(n)$ and $\hat{\underline{r}}(n) = \mathbf{W}(n)\mathbf{X}(n)\underline{d}(n)$ in which case our *weighted*, estimated Wiener-Hopf equation can be stated as

$$\mathbf{X}(n)\mathbf{W}(n)\mathbf{X}^T(n)\underline{h}(n) = \mathbf{W}(n)\mathbf{X}(n)\underline{d}(n). \tag{10}$$

4 Solving the Estimated Wiener-Hopf Equation

A stationary iterative solution procedure applied to a system of linear equations, $\mathbf{A}\underline{x} = \underline{b}$ is expressed as [5]

$$\underline{x}^{(k+1)} = [\mathbf{I} - \mathbf{M}^{-1}\mathbf{A}]\underline{x}^{(k)} + \mathbf{M}^{-1}\underline{b}, \tag{11}$$

where $\mathbf{I}$ is the identity matrix, and $\mathbf{M}$ is an invertible matrix given by the *splitting* of $\mathbf{A}$

$$\mathbf{A} = \mathbf{M} - \mathbf{N} = \mathbf{M} - (\mathbf{M} - \mathbf{A}). \tag{12}$$

We point out that the iteration of Equation 11 is meaningful also in the case of equation sets with singular $\mathbf{A}$ matrices [6]. This observation is implicitly invoked in the following.

Splitting the coefficient matrix of Equation 10 we have

$$\mathbf{X}(n)\mathbf{W}(n)\mathbf{X}^T(n) = \mathbf{C}^{-1}(n) - [\mathbf{C}^{-1}(n) - \mathbf{X}(n)\mathbf{W}(n)\mathbf{X}^T(n)], \tag{13}$$

where $\mathbf{C}^{-1}(n)$ is some full rank $M \times M$ matrix[2]. If one iteration according to Equation 11 is performed for Equation 10 at each time instant, n, we can use common indexing for the signal quantities and the iteration index, i.e. we have

$$\begin{aligned}\underline{h}(n+1) &= [\mathbf{I} - \mathbf{C}(n)\mathbf{X}(n)\mathbf{W}(n)\mathbf{X}^T(n)]\underline{h}(n) \\ &\quad + \mathbf{C}(n)\mathbf{X}(n)\mathbf{W}(n)\underline{d}(n) \\ &= \underline{h}(n) + \mathbf{C}(n)\mathbf{X}(n)\mathbf{W}(n)\underline{e}(n),\end{aligned} \tag{14}$$

where $\underline{e}(n)$, the *error vector*, is defined by $\underline{e}(n) = \underline{d}(n) - \mathbf{X}^T(n)\underline{h}(n)$. This clearly has the flavor of an adaptive filter. We now identify several special cases of Equation 14 giving rise to well established apaptive filtering algorithms.

5 Classical Adaptive Filters

5.1 Special Case No. 1: LMS/NLMS

It is trivial to observe that if we set $L = 1$, – that is $\mathbf{X}(n) = \underline{x}(n)$, $\mathbf{W}(n) = 1$, and $\mathbf{C}^{-1}(n) = \mu^{-1}\mathbf{I}$ in Equation 14, the classical LMS algorithm

$$\begin{aligned}\underline{h}(n+1) &= [\mathbf{I} - \mu\underline{x}(n)\underline{x}^T(n)]\underline{h}(n) + \mu\underline{x}(n)d(n) \\ &= \underline{h}(n) + \mu\underline{x}(n)e(n),\end{aligned} \tag{15}$$

results. Note that the splitting employed corresponds to the *Richardson splitting* [5]. Keeping the choices above, but with $\mathbf{W}(n) = \|\underline{x}(n)\|^{-2} = [\underline{x}^T(n)\underline{x}(n)]^{-1}$ in Equation 14, the standard version of the NLMS algorithm

$$\underline{h}(n+1) = \underline{h}(n) + \frac{\mu}{\|\underline{x}(n)\|^2}\underline{x}(n)e(n) \tag{16}$$

results.

5.2 Special Case No. 2: The Affine Projection Algorithm (APA)

Selecting $1 < L < M$, $\mathbf{C}^{-1}(n)$ as above, and $\mathbf{W}(n) = [\mathbf{X}^T(n)\mathbf{X}(n)]^{-1}$, i.e. we use what is commonly referred to as data normalized estimates of the quantities of the Wiener-Hopf equation, the iteration of Equation 14 takes the form

$$\underline{h}(n+1) = \underline{h}(n) + \mu\mathbf{X}(n)\{\mathbf{X}^T(n)\mathbf{X}(n)\}^{-1}\underline{e}(n), \tag{17}$$

[2] We prefer to denote the splitting matrix $\mathbf{C}^{-1}(n)$ rather than $\mathbf{C}(n)$ which would be more in line with Equation 12.

which we immediately identify as the standard version of the APA adaptive filter [1]. There are many versions of the APA, see for example [7] for a tabular classification of the most common ones. It is important to observe that all versions of the APA are captured by Equation 17 if appropriate adjustments to the definitions of $\mathbf{W}(n)$ and $\mathbf{X}(n)$ are made.

5.3 Special Case No. 3: RLS

With $L = 1$, – again implying that $\mathbf{X}(n) = \underline{x}(n)$, $\mathbf{W}(n) = 1$, and $\mathbf{C}^{-1}(n) = \tilde{\mathbf{X}}(n)\tilde{\mathbf{X}}^T(n)$, where $\tilde{\mathbf{X}}(n)$ is an $M \times K, K > M$ data matrix of the same form as $\mathbf{X}(n)$, but with K taking the role of L., the recursion of Equation 14 becomes

$$\underline{h}(n+1) = \underline{h}(n) + \{\tilde{\mathbf{X}}(n)\tilde{\mathbf{X}}^T(n)\}^{-1}\underline{x}(n)e(n). \tag{18}$$

Realizing that the Kalman gain vector in common presentations of the RLS algorithm is given by [1]

$$\underline{k}(n) = \{\tilde{\mathbf{X}}(n)\tilde{\mathbf{X}}^T(n)\}^{-1}\underline{x}(n), \tag{19}$$

we see that the iteration of Equation 18 is equivalent to a sliding window RLS algorithm. It is well known that the RLS algorithm gives an exact solution to $\tilde{\mathbf{X}}(n)\tilde{\mathbf{X}}^T(n)\underline{h}(n) = \tilde{\mathbf{X}}(n)\underline{d}(n)$ for each n. A consequence of the above is the interesting alternative interpretation of RLS as a time sequential iterative scheme applying the matrix splitting

$$\begin{aligned}\underline{x}(n)\underline{x}^T(n) = {} & \tilde{\mathbf{X}}(n)\tilde{\mathbf{X}}^T(n) \\ & - [\tilde{\mathbf{X}}(n)\tilde{\mathbf{X}}^T(n) - \underline{x}(n)\underline{x}^T(n)]\end{aligned} \tag{20}$$

to Equation 10 with $L = 1$, the same equation that formed the basis for the derivation of the LMS algorithm above. Thus, the only difference in the derivation of the LMS and RLS algorithms within the presented framework is the selection of the rank one coefficient matrix splitting: In Equation 13, the matrix $\mathbf{C}^{-1}(n)$ of the splitting for the RLS is selected as $\tilde{\mathbf{X}}(n)\tilde{\mathbf{X}}^T(n)$, which for a sufficiently high $K > M$ is a good estimate of the the autocorrelation matrix of the adaptive filter input signal. Accepting the premise that the use of a good estimate of the autocorrelation matrix as the $\mathbf{C}^{-1}(n)$ matrix in the splitting of Equation 13 gives rise to well performing adaptive algorithms, the use of $\mathbf{C}(n) = \mu^{-1}\mathbf{I}$, the crudest possible estimate/assumption on an autocorrelation matrix[3], naturally positions the LMS and RLS algorithms at opposite ends of the performance scale for adaptive filtering algorithms.

As a final interpretation of the RLS algorithm, we point out that the application of a Richardson splitting to the *preconditioned* equation set

$$\{\tilde{\mathbf{X}}(n)\tilde{\mathbf{X}}^T(n)\}^{-1}\underline{x}(n)\underline{x}^T(n)\underline{h}(n) = \{\tilde{\mathbf{X}}(n)\tilde{\mathbf{X}}^T(n)\}^{-1}\underline{x}(n)d(n) \tag{21}$$

also gives the RLS algorithm of Equation 18. Preconditioning is an extensively used technique for speeding up the convergence of an iterative linear equation

[3] That is, we assume a white input signal whether this is the case or not.

solver [5]. Within the present framework we can thus interpret the RLS algorithm as *a particular preconditioned* Richardson iteration applied to the rank one ($L = 1$) matrix equation, Equation 10, forming the basis of the LMS algorithm.

Novel Adaptive Filters

The recently introduced FEDS and RAMP adaptive filtering algorithms [4, 3] were originally derived from very different perspectives [8]. Nevertheless, as shown in [8], both algorithms can be interpreted as a Gauss Seidel single element update at time n applied to the *normal equations*, Equation 9, with $L > M$. The update, when updating only element no. $j(n)$, can be expressed as

$$\begin{aligned} \underline{h}(n+1) &= \underline{h}(n) + \frac{1}{\|\underline{\tilde{x}}(n-j(n))\|^2} \mathbf{i}_{j(n)} \mathbf{X}(n)\{\underline{d}(n) - \mathbf{X}^T(n)\underline{h}(n)\} \\ &= \underline{h}(n) + \frac{1}{\|\underline{\tilde{x}}(n-j(n))\|^2} \mathbf{i}_{j(n)} \mathbf{X}(n)\underline{e}(n), \end{aligned} \quad (22)$$

where $\mathbf{i}_{j(n)}$ is the $M \times M$ matrix with a 1 in position $(j(n), j(n))$ and zeros in all other positions, i.e. a matrix leaving intact row no. $j(n)$ of the matrix with which it is pre multiplied and zeroing out all the other rows. The FEDS and RAMP algorithms differ only in the sequence in which the element updates are performed: FEDS updates each element of $\underline{h}(n)$ in a sequential fashion in exactly the same way as a standard Gauss Seidel procedure. The RAMP algorithm, on the other hand, will, for each element update, select the element of $\underline{h}(n)$ corresponding to the maximum (in absolute value) element of the residual vector associated with Equation 9, $\mathbf{X}(n)\underline{e}(n)$ and perform the update on this element. This variation of the Gauss Seidel scheme is sometimes referred to as Southwell's method. Now, associating $\mathbf{C}(n)$ with $\frac{1}{\|\underline{\tilde{x}}(n-j(n))\|^2}\mathbf{i}_{j(n)}$, we see that also these two adaptive filter algorithms can be interpreted as special cases of the generic iteration of Equation 14. Generalizing the above to the case where more than one update according to Equation 22 is performed for each sample time index n is trivial.

6 Summary and Conclusions

Our objective in this paper has been the reformulation of the adaptive filtering problem in such a way that most, possibly all, adaptive filters can be viewed as simple special cases of a unified theory. This is in sharp contrast to established approaches in which each adaptive filter algorithm is derived and analyzed more or less in without reference to its position in some wider context. Our results are summarized in Table 1, in which we tabulate selections for the various quantities of Equation 14, identify the stationary iterative methods employed, and relate them to the various adaptive filter algorithms. We believe that the introduced framework facilitates a clearer and complementary understanding of adaptive filters that will pave the way for future contributions from scientists with other

Table 1. Correspondence between special cases of Equation 14 and various adaptive filtering algorithms

L	$\mathbf{W}(n)$	$\mathbf{C}(n)$	Algorithm
1	1	$\mu\mathbf{I}$ (Richardson iter.)	LMS
1	$\|\underline{x}(n)\|^{-2}$	$\mu\mathbf{I}$ (Richardson iter.)	NLMS
$1 < L < M$	$[\mathbf{X}^T(n)\mathbf{X}(n)]^{-1}$	$\mu\mathbf{I}$ (Richardson iter.)	APA
1	1	$[\tilde{\mathbf{X}}^T(n)\tilde{\mathbf{X}}(n)]^{-1}$ (Precond. Richardson)	RLS
$L > M$	$\mathbf{I}$	$\mathbf{i}_{j(n)}\|\tilde{\underline{x}}(n-j(n))\|^{-2}$ ($j(n) = n \oplus M$, Partial Gauss Seidel)	FEDS
$L > M$	$\mathbf{I}$	$\mathbf{i}_{j(n)}\|\tilde{\underline{x}}(n-j(n))\|^{-2}$ ($j(n) = \arg\max_i \mid \underline{e}^T(n)\tilde{\underline{x}}(n-i)/\|\tilde{\underline{x}}(n-i)\| \mid .$) (Partial Gauss Seidel)	RAMP

backgrounds than electrical engineering who, to the present time, have dominated the field. Its usefulness in the analysis of the performance of some specific adaptive filtering algorithms has already been demonstrated in [9].

References

1. S. Haykin, *Adaptive Filter Theory*, Fourth ed. Upper Saddle River, NJ, USA: Prentice Hall, 2002.
2. J. R. Treichler, C. R. Johnson, and M. G. Larimore, *Theory and design of Adaptive filters.* Upper Saddle River, NJ: Prentice Hall, 2001.
3. T. Bose and G. F. Xu, "The Euclidian direction search algorithm in adaptive filtering," *IEICE Trans. Fundamentals*, vol. E85-A, no. 3, pp. 532–539, Mar. 2002.
4. J. H. Husøy, "RAMP: An adaptive filter with links to matching pursuits and iterative linear equation solvers," in *Proc. ISCAS*, Bangkok, Thailand, May 2003, pp. IV–381–384.
5. Y. Saad, *Iterative Methods for Sparse Linear Systems.* PWS Publishing, 1996.
6. H. B. Keller, "On the solution of singular and semidefinite linear systems by iteration," *SIAM J. Numer. Anal.*, vol. 2, no. 2, pp. 281–290, 1965.
7. H.-C. Shin and A. H. Sayed, "Transient behavior of affine projection algorithms," in *Proc. Int. Conf. Acoust. Speech, Signal Proc.*, Hong Kong, Apr. 2003, pp. VI–353–356.
8. J. H. Husøy and M. S. E. Abadi, "Interpretation and convergence speed of two recently introduced adaptive filters (FEDS/RAMP)," 2004, Submitted.
9. ——, "A common framework for transient analysis of adaptive filters," in *Proc. Melecon*, Dubrovnik, Croatia, May 2004, pp. 265–268.

A Rothe-Immersed Interface Method for a Class of Parabolic Interface Problems

Juri D. Kandilarov

Center of Applied Mathematics and Informatics
University of Rousse, 8 Studentska str.,
Rousse 7017, Bulgaria
juri@ami.ru.acad.bg

Abstract. A technique combining the Rothe method with the immersed interface method (IIM) of R. Leveque and Z. Li, [8] for numerical solution of parabolic interface problems in which the jump of the flux is proportional to a given function of the solution is developed. The equations are discretized in time by Rothe's method. The space discretization on each time level is performed by the IIM. Numerical experiments are presented.

AMS Subject Classification: 65N06, 35K55, 35K57.

Keywords: immersed interface method, nonlinear parabolic problems, delta-Dirac function, finite difference schemes, Cartesian grids.

1 Introduction and Statement of the Differential Problem

Let Ω be a bounded domain in $\mathbf{R}^2$ with piecewise smooth boundary $S = \partial\Omega$. Consider the following problem:

$$u_t - \Delta u = Ff(x, y, t, u), \quad (x, y, t) \in Q_T = \Omega \times (0, T), \tag{1}$$

$$u(x, y, t) = 0, \quad (x, y, t) \in S_T = S \times [0, T], \tag{2}$$

$$u(x, y, 0) = u_0(x, y), \quad (x, y) \in \Omega, \tag{3}$$

where F is a functional form $C(\overline{\Omega}) \to R^1$, generated by the Dirac-delta function, $Ff = \delta_\Gamma(x, y)f$, defined only on a smooth curve Γ:

$$\int_\Omega \delta(x - X(s)\delta(y - Y(s)f(x, y)dxdy = \int_\Gamma f(x, y)d\sigma \ , \ f \in C(\overline{\Omega}).$$

The problem (1)-(3) describes some physical phenomena in which the reactions in a dynamical system take place only on some curves in Ω. This causes the chemical concentration to be continuous, but the gradient of the concentration to have a jump on these curves. The magnitude of the jump typically depends

Z. Li et al. (Eds.): NAA 2004, LNCS 3401, pp. 328–336, 2005.

on the concentration. Similar processes are also observed in biological systems, on chemically active membranes, [1], [2], [10] and processes in which the ignition of a combustible medium is accomplished through the use of either a heated wire or a pair of small electrodes to supply a large amount of energy to a very confined area, [3].

The equivalent formulation of the problem (1)-(3) is described as follows:

$$u_t - \Delta u = 0 \quad in \quad Q_T \backslash \Gamma_T, \tag{4}$$

$$u(x,y,t) = 0, \;\; (x,y,t) \in S_T = S \times [0,\; T], \tag{5}$$

$$u(x,y,0) = u_0(x,y), \;\; (x,y) \in \Omega, \tag{6}$$

and on the interface $\Gamma_T = \Gamma \times [0,T]$

$$[u] := u^+(x,y,t) - u^-(x,y,t) = 0, \tag{7}$$

$$\left[\frac{\partial u}{\partial \mathbf{n}(x,y)}\right] = -f(u(x,y,t)), \tag{8}$$

where $\mathbf{n}(x,y)$ is the normal direction at $(x,y) \in \Gamma$, pointing form Ω^- into Ω^+ ($\Omega = \Omega^- \cup \Omega^+ \cup \Gamma$, $\Omega^- \cap \Omega^+ = \emptyset$) and

$$\left[\frac{\partial u(x,y,t)}{\partial \mathbf{n}(x,y)}\right] = \left[\frac{\partial u(x,y,t)}{\partial x}\right] \cos(\mathbf{n},x) + \left[\frac{\partial u(x,y,t)}{\partial y}\right] \cos(\mathbf{n},y).$$

The global solvability and blow-up of the solution in finite time are studied in [2].

Various approaches have been used to solve differential equations with discontinuous coefficients and concentrated factors numerically. Most of standard methods are not second order accurate or require the grid points to lie along the interface. The IIM, [8], [9], [12] does not have such a requirement and is still a second order accurate even though for arbitrary curve Γ.

The conjugation conditions (7)-(8) are specific, which leads to some new difficulties, [5], [6]. We must to incorporate the conjugation conditions and to approximate the unknown solution from the interface to the grid points, which follows some times to the extension of the standard stencil, [7], [11].

In this paper we develop a Rothe method, [4], combined with the IIM, which provides first order accuracy in time and second order in space dimension on a Cartesian grids.

The organization of the paper is as follows: in Section 2 the semidiscretization in time, using Rothe method is done; in Section 3 the full discretization for 2D is discussed and in Section 4 numerical experiments are presented.

2 Time Discretization

Consider the problem (4)-(8) and let for simplicity the local reaction term be linear, i.e. $f(x,y,t,u) = -Ku(x,y,t)$, $K = const.$

The method of Rothe we used consists in:

- Divide the interval $[0,T]$ into N subintervals $[t_{n-1},t_n]$, $n=1,...,N$, $t_n=n\tau$, $\tau=T/N$;
- Discretize the time derivative $\partial u/\partial t$, introducing level set functions $u^n(x,y)$ at time level t_n;
- Solve by recurrence the received system of elliptic equations for u^n, $n=1,...,N$;
- Construct the Rothe function

$$u_R(x,y,t)=u^{n-1}(x,y)+\frac{(t-t_{n-1})}{\tau}(u^n(x,y)-u^{n-1}(x,y)), \qquad t\in[t_{n-1},t_n].$$

To overcome some additional difficulties the function f is evaluated on the previous time level and the semidiscretization looks as follows:

$$\begin{aligned}
\frac{u^{n+1}(x,y)-u^n(x,y)}{\tau}-\Delta u^{n+1}(x,y)&=0, && (x,y)\in\Omega\backslash\Gamma,\\
\left[u^{n+1}(x,y)\right]&=0, && (x,y)\in\Gamma, \qquad (9)\\
\left[\frac{\partial u^{n+1}(x,y)}{\partial \mathrm{n}}\right]&=Ku^n(x,y), && (x,y)\in\Gamma,\\
u^n(x,y)=0,\ (x,y)\in\partial\Omega,\quad u^0(x,y)&=u_0(x,y),\ (x,y)\in\Omega.
\end{aligned}$$

Proposition 1. *Let $u_0(x)\in C^\alpha(\overline{\Omega})\cap C^{2+\alpha}(\Omega\backslash\Gamma)$ with $u_0(x)=0$ on $\partial\Omega$. Then the problem (4)-(8) admits a unique solution $u(x,y,t)\in C^\alpha(\overline{Q}_T)\cap C^{2+\alpha,1+\alpha/2}(Q_T\backslash\Gamma)$. If more the solution $u(x,y,t)\in C^\alpha(\overline{Q}_T)\cap C^{2+\alpha,2+\alpha/2}(Q_T\setminus\Gamma)$ and $K\geq 0$ then for sufficiently small τ the solution of the elliptic system (9) convergence to the solution of the parabolic problem*

$$|u(x,y,t)-u_R(x,y,t)|\leq C\tau,$$

where the constant C does not depend on τ.

3 Total Discretization Using the IIM

We wish to approximate the solution $u^n(x,y)$ on a uniform mesh grid in $\Omega=[0,1]\times[0,1]$ with mesh sizes $h_1=1/M_1$, $h_2=1/M_2$, and let $x_i=ih_1$ for $i=0,1,...,M_1$, $y_j=jh_2$ for $j=0,1,...,M_2$. Then the difference scheme with IIM can be written as:

$$\begin{aligned}
\frac{u_{ij}^{n+1}-u_{ij}^n}{\tau}&=\frac{u_{i+1,j}^{n+1}-2u_{ij}^{n+1}+u_{i-1,j}^{n+1}}{h_1^2}+D_{x,ij}^n\\
&+\frac{u_{i,j+1}^{n+1}-2u_{ij}^{n+1}+u_{i,j-1}^{n+1}}{h_2^2}+D_{y,ij}^n, \qquad (10)
\end{aligned}$$

where u_{ij}^n is an approximation to $u^n(x_i,y_j)$ and $D_x^n=D_{xl}^n+D_{xr}^n$, $D_y^n=D_{yt}^n+D_{yb}^n$ are additional terms, choosing later to decrease the local truncation error (LTE).

With l, r, t, b, we denote, that the interface crosses the left, right, top or bottom arm of the standard five point stencil respectively at (x_i, y_j).

Let us introduce the **level set function** $\phi(x,y)$, so that $\phi(x,y) = 0$ when $(x,y) \in \Gamma$, $\phi(x,y) < 0$ in Ω^- and $\phi(x,y) > 0$ in Ω^+. The unit normal is $\mathbf{n}(n^1, n^2)$, points from Ω^- into Ω^+. We call the grid point (x_i, y_j) **regular**, if there is no grid crossing with the arms of the stencil and **irregular** in other case.

At regular grid points the correction terms $D_x^n = D_y^n = 0$. Let (x_i, y_j) be irregular one. We assume without loss of generality that the interface Γ crosses the right and top arm of the stencil at the points (x_{ij}^*, y_j) and (x_i, y_{ij}^*), see Fig.1. Then $D_{xl}^n = D_{yb}^n = 0$, but D_{xr}^n and D_{yt}^n need more attention.

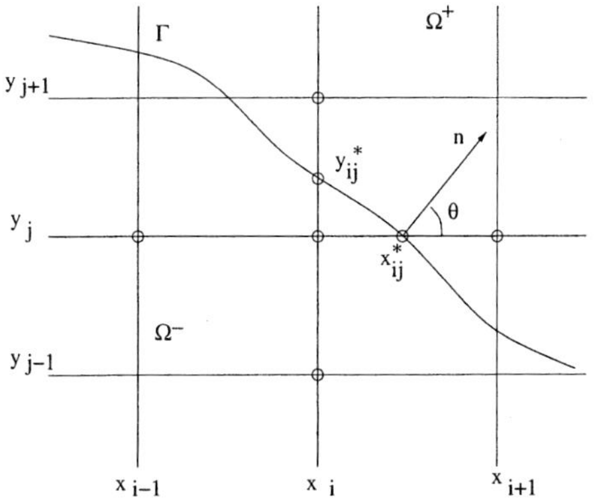

Fig. 1. The geometry at the irregular grid point (x_i, y_j)

Using Taylor expansion near (x_i, y_j) for the derivatives u_{xx} and u_{yy} we have:

$$\left.\frac{\partial^2 u^{n+1}}{\partial x^2}\right|_{(x_i,y_j)} = \frac{u_{i+1,j}^{n+1} - 2u_{ij}^{n+1} + u_{i-1,j}^{n+1}}{h_1^2} + D_{x,ij}^n + O(h_1),$$

$$\left.\frac{\partial^2 u^{n+1}}{\partial y^2}\right|_{(x_i,y_j)} = \frac{u_{i,j+1}^{n+1} - 2u_{ij}^{n+1} + u_{i,j-1}^{n+1}}{h_2^2} + D_{y,ij}^n + O(h_2),$$

where

$$D_{x,ij}^n = \begin{cases} -\frac{\rho_{i+1,j}}{h_1}[u_x^n]_{(x_{ij}^*,y_j)} - \frac{\rho_{i+1,j}^2}{2}[u_{xx}^n]_{(x_{ij}^*,y_j)} & \text{if } \phi_{i,j} \le 0, \text{ and } \phi_{i+1,j} > 0, \\ \frac{\rho_{i+1,j}}{h_1}[u_x^n]_{(x_{ij}^*,y_j)} + \frac{\rho_{i+1,j}^2}{2}[u_{xx}^n]_{(x_{ij}^*,y_j)} & \text{if } \phi_{i,j} > 0, \text{ and } \phi_{i+1,j} \le 0, \end{cases}$$

$$D_{y,ij}^n = \begin{cases} -\frac{\rho_{i,j+1}}{h_2}[u_y^n]_{(x_i,y_{ij}^*)} - \frac{\rho_{i,j+1}^2}{2}[u_{yy}^n]_{(x_i,y_{ij}^*)} & \text{if } \phi_{i,j} \le 0 \text{ and } \phi_{i,j+1} > 0, \\ \frac{\rho_{i,j+1}}{h_2}[u_y^n]_{(x_i,y_{ij}^*)} + \frac{\rho_{i,j+1}^2}{2}[u_{yy}^n]_{(x_i,y_{ij}^*)} & \text{if } \phi_{i,j} \le 0 \text{ and } \phi_{i,j+1} \le 0. \end{cases}$$

Here $\rho_{i+1,j} = (x_{i+1} - x_{ij}^*)/h_1$, $\rho_{i,j+1} = (y_{i+1} - y_{ij}^*)/h_2$ and $\phi_{i,j} = \phi(x_i, y_j)$.

We need to find the jumps $[u_x]$, $[u_y]$, $[u_{xx}]$ and $[u_{yy}]$ in terms of the known information for $[u]$ and $[u_n]$ (for simplicity we omit the superscript n). Using the idea of Li, [9], we introduce a local coordinates transformation at (x^*_{ij}, y_j) (for the jumps in x-direction and (x_i, y^*_{ij}) for the jumps in y-direction):

$$\xi = (x - x^*_{ij}) \cos\theta + (y - y_j) \sin\theta,$$
$$\eta = -(x - x^*_{ij}) \sin\theta + (y - y_j) \cos\theta,$$

where θ is the angle between x-axis and the normal direction $\mathbf{n}$ at the interface point. In a neighborhood of this point the interface lies roughly in tangential direction η and Γ can be locally parameterized by $\xi = \chi(\eta)$, η. So we have:

$$\begin{aligned} u_\xi &= u_x \cos\theta + u_y \sin\theta = u_n, \\ u_\eta &= -u_x \sin\theta + u_y \cos\theta, \end{aligned} \tag{11}$$

The jumps in the new coordinate system are:

$$\begin{aligned} &[u_\xi] = [u_n] = Ku, \qquad && [u_{\xi\xi}] = \chi''[u_\xi], \\ &[u_\eta] = [u]_\eta = 0, && [u_{\eta\eta}] = -\chi''[u_\xi], \\ & && [u_{\eta\eta}] = \chi''[u_\eta] + Ku_\eta. \end{aligned} \tag{12}$$

We come back to the jumps in the x and y direction by the formulas:

$$\begin{aligned} [u_x] &= [u_\xi] \cos\theta - [u_\eta] \sin\theta, \\ [u_y] &= [u_\xi] \sin\theta + [u_\eta] \cos\theta, \\ [u_{xx}] &= [u_{\xi\xi}] \cos^2\theta - 2\,[u_{\xi\eta}] \cos\theta \sin\theta\, [u_{\eta\eta}] \sin^2\theta, \\ [u_{yy}] &= [u_{\xi\xi}] \sin^2\theta + 2\,[u_{\xi\eta}] \cos\theta \sin\theta\, [u_{\eta\eta}] \cos^2\theta. \end{aligned} \tag{13}$$

Plugging (12) in (13) at the point (x^*_{ij}, y_j) we have:

$$\begin{aligned} [u_x] &= Ku \cos\theta, \\ [u_y] &= Ku \sin\theta, \\ [u_{xx}] &= -2Ku_\eta \cos\theta \sin\theta + \chi'' Ku(\cos^2\theta - \sin^2\theta), \\ [u_{yy}] &= 2Ku_\eta \cos\theta \sin\theta - \chi'' Ku(\cos^2\theta - \sin^2\theta). \end{aligned}$$

The next difficulty for the full discretization is to approximate $u_\eta(x^*_{ij}, y_j)$ and $u(x^*_{ij}, y_j)$ using the nodes of the stencil. We use bi-variable Lagrange interpolation by linear polynomials in two variables on three nodes $(\tilde{x}_1, \tilde{y}_1)$, $(\tilde{x}_2, \tilde{y}_2)$, and $(\tilde{x}_3, \tilde{y}_3)$:

$$u(x^*_{ij}, y_j) \approx L_1 u(\tilde{x}_1, \tilde{y}_1) + L_2 u(\tilde{x}_2, \tilde{y}_2) + L_3 u(\tilde{x}_3, \tilde{y}_3), \tag{14}$$

where

$$\begin{bmatrix} L_1 \\ L_2 \\ L_3 \end{bmatrix} = \frac{1}{S} \begin{bmatrix} \tilde{x}_2\tilde{y}_3 - \tilde{x}_3\tilde{y}_2 & \tilde{y}_2 - \tilde{y}_3 & \tilde{x}_3 - \tilde{x}_2 \\ \tilde{x}_3\tilde{y}_1 - \tilde{x}_1\tilde{y}_3 & \tilde{y}_3 - \tilde{y}_1 & \tilde{x}_1 - \tilde{x}_3 \\ \tilde{x}_1\tilde{y}_2 - \tilde{x}_2\tilde{y}_1 & \tilde{y}_1 - \tilde{y}_2 & \tilde{x}_2 - \tilde{x}_1 \end{bmatrix}, \qquad S = det \begin{pmatrix} 1 & \tilde{x}_1 & \tilde{y}_1 \\ 1 & \tilde{x}_2 & \tilde{y}_2 \\ 1 & \tilde{x}_3 & \tilde{y}_3 \end{pmatrix}. \tag{15}$$

Lemma 1. *If the points* $(\tilde{x}_1, \tilde{y}_1)$, $(\tilde{x}_2, \tilde{y}_2)$, *and* $(\tilde{x}_3, \tilde{y}_3)$ *lie not on a line, then* $S \neq 0$ *and the approximation (14) is correct.*

Lemma 2. *(***Curve Condition***) Let the interface curve* $\Gamma \in C^2$. *Then for sufficiently small* $h = \max(h_1, h_2)$ *there exist three nodes, which all lie in* Ω^- *or* Ω^+ *and the condition of Lemma 1 is fulfilled.*

To approximate the tangential derivative we use:

$$\begin{aligned} \frac{\partial u}{\partial x} &\approx \frac{1}{S}\left((\tilde{y}_2 - \tilde{y}_3)u(\tilde{x}_1, \tilde{y}_1) + (\tilde{y}_3 - \tilde{y}_1)u(\tilde{x}_2, \tilde{y}_2) + (\tilde{y}_1 - \tilde{y}_2)u(\tilde{x}_3, \tilde{y}_3)\right), \\ \frac{\partial u}{\partial y} &\approx \frac{1}{S}\left((\tilde{x}_3 - \tilde{x}_2)u(\tilde{x}_1, \tilde{y}_1) + (\tilde{x}_1 - \tilde{x}_3)u(\tilde{x}_2, \tilde{y}_2) + (\tilde{x}_2 - \tilde{x}_1)u(\tilde{x}_3, \tilde{y}_3)\right). \end{aligned} \tag{16}$$

Then we put this expressions in (11). From (15) it follows, that $0 \leq |L_i| \leq 2$, $i = 1, 2, 3$ and the coefficients in (16) are of order h^{-1}.

Proposition 2. *Let the solution of the differential problem (4)-(8)* $u(x, y, t) \in C^{\alpha}(\overline{Q}_T) \cap C^{4+\alpha, 2+\alpha/2}(Q_T \setminus \Gamma)$, $f(u) = -Ku(x, y, t)$, $K \geq 0$ *and the curve* $\Gamma \in C^2$. *Then for sufficiently small* τ/h *the solution obtained by the difference scheme (10) convergence to the solution of the parabolic problem and*

$$|u(x_i, y_j, t_n) - u_{ij}^n| \leq C(\tau + h^2),$$

where the constant C *does not depend on* τ *and* h.

4 Numerical Results

Example 1. 1D diffusion problem

Consider the problem

$$\begin{aligned} u_t &= Du_{xx}, \quad x \in (0, 1), x \neq \xi, \quad 0 < t \leq T, \\ [u]_\xi &= 0, \quad D[u_x]_\xi = -Ku(\xi, t), \quad 0 < t \leq T, \\ u(0, t) &= u(1, t) = 0, \quad 0 < t \leq T, \end{aligned}$$

where for $D = 1$, $K = 1$, the exact solution is

$$u_0(x) = \begin{cases} \sin \lambda x / \sin \lambda \xi, & 0 \leq x \leq \xi, \\ \sin \lambda(1 - x) / \sin \lambda(1 - \xi), & \xi \leq x \leq 1, \end{cases}$$

and λ is a root of the equation

$$\lambda(\cot(\lambda x) + \cot(\lambda(1 - x))) = 1.$$

In Table 1 we present the results of the *Example 1* for $\xi = 0.5$ and $\lambda = 2.7865$. We control the absolute error $||E||_\infty$ and relative error $||E_r||_\infty$ in maximum norm:

$$\|E\|_\infty = \max_{i,j,n} |u(x_i, y_j, t_n) - u_{ij}^n|, \quad \|E_r\|_\infty = \max_{i,j,n} \|E\|_\infty / |u(x_i, y_j, t_n)|.$$

Table 1. Mesh refinement analysis for *Example 1* with time step $\tau = .000005$

	$T = 0.2$, $n = 20000$				$T = 1$, $n = 200000$			
M	$\|E\|_\infty$	$ratio$	$\|E_r\|_\infty$	$ratio$	$\|E\|_\infty$	$ratio$	$\|E_r\|_\infty$	$ratio$
21	1.2106e-03	-	5,8047e-02	-	1.1791e-05	-	2.8173e-02	-
41	3.0275e-04	4.00	1.4397e-03	4.03	2.9089e-06	4.05	6.8960e-03	4.09
81	8.2494e-05	3.67	3.9098e-04	4.40	7.9016-07	3.68	1.8670e-03	3.69
161	2.8311e-05	2.91	1.3397e-04	2.92	2.9192e-07	2.71	6.4148e-04	2.91

Table 2. Mesh refinement analysis for *Example 2* with time step $\tau = 0.0001$

		$T = 0.0001$, $n = 1$				$T = 1$, $n = 10000$			
M_1	M_2	$\|E\|_\infty$	$ratio$	$\|Tr\|_\infty$	$ratio$	$\|E\|_\infty$	$ratio$	$\|Tr\|_\infty$	$ratio$
10	10	1.8801e-05	-	0.1896	-	3.7936e-02	-	3.7936e+02	-
20	20	4.2898e-05	0.44	0.4476	0.42	5.5056e-03	6.89	5.5293e+01	6.75
40	40	2.1489e-05	2.00	0.2495	1.79	1.2764e-03	4.31	1.2897e+01	4.28
80	80	9.1365e-06	2.35	0.1421	1.76	2.4941e-04	4.34	2.5887e+00	4.98

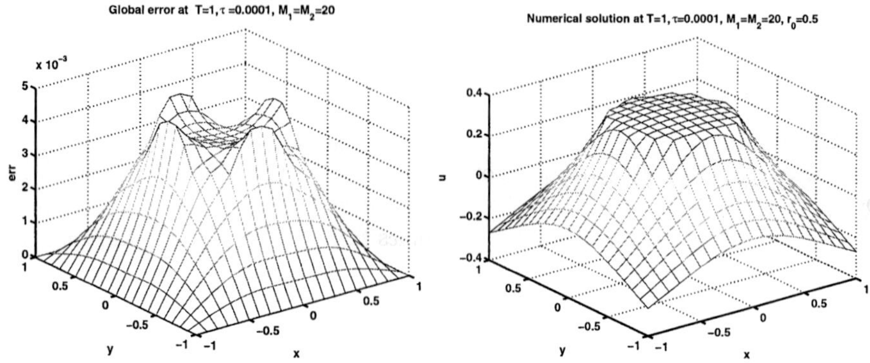

Fig. 2. Global error and numerical solution for *Example 2* at $T = 1$, $\tau = 0.0001$, $M_1 = M_2 = 20$ and $r_0 = 0.5$

The ratio of the successive errors near four confirm second order with respect to the mesh parameter h for small M. With growing of M we observe the decreasing of the accuracy. The reason is in $O(\tau/h)$-term in the LTE.

Example 2. 2D diffusion problem

Let consider the following problem:

$$u_t = u_{xx} + u_{yy} + \int_\Gamma C(t)u\delta(x - X(s))\delta(y - Y(s))ds$$

on a square $[-1, 1] \times [-1, 1]$ and $\Gamma : x^2 + y^2 = r_0^2$. With

$$C(t) = \left(\frac{Y_0'(r_0)J_0(r_0)}{Y_0'(r_0)} - J_0'(r_0) \right)$$

we take the exact solution to be

$$u(x,y,t) = \begin{cases} \exp(-t)J_0(r), & r \le r_0, \\ \exp(-t)J_0(r_0)Y_0(r)/Y_0(r_0), & r > r_0. \end{cases} \tag{17}$$

The Dirichlet boundary condition and initial condition are taken from (17).

In Table 2 the maximum error $\|E^n_{M_1,M_2}\|_\infty$ between the exact and numerical solution at the n-th time layer and the maximum LTE $\|Tr^n\|_\infty$ are presented:

$$\|E^n_{M_1,M_2}\|_\infty = \max_{i,j}\{|u(x_i,y_j,t_n) - u^n_{ij}|\}.$$

We also control the ratios of the successive errors:

$$\text{ratio} = \|E^n_{M_1,M_2}\|_\infty / \|E^n_{2M_1,2M_2}\|_\infty.$$

The results confirm first order of the LTE on the first time layer and second order for the solution at the last time $T = 1$.

5 Conclusions

Numerical procedures for parabolic problems with interface jump conditions, in which the solution is continuous, but the jump of the flux is proportional to the solution, are developed. The Rothe method for discretization in time and the IIM for discretization in space on a Cartesian grid are used. The numerical experiments confirm $O(\tau+h^2)$ rate of accuracy for arbitrary curvelinear interface. Our next work is to take in (9) the singular term $f(x,y,t) = Ku(x,y,t)$ on the current time layer. Then the scheme becomes fully implicit, which improves the stability of the method.

References

1. Bimpong-Bota, K., Nizan, A., Ortoleva, P., Ross, J.: Cooperative Phenomena in Analysis of Catalytic Sites. J. Chem. Phys. 66 (1970) 3650–3678
2. Chadam, J.M., Yin, H.M.: A Diffusion Equation With Localized Chemical Reactions. Proc. of Edinburgh Math. Soc. 37 (1993) 101–118
3. Chan, C.Y., Kong, P.C.: Channel Flow of a Viscous Fluid in the Boundary Layer. Quart. Appl. Math. 55 (1997) 51-56
4. Kačur, J.: Method of Rothe in Evolution Equations. Leipzig, BSB Teubner Verlagsges (1985)
5. Kandilarov, J.D.: The Iimmersed Interface Method for a Reaction Diffusion Equation with a Moving Own Concentrated Source. In: Dimov, I., Lirkov, I., Margenov, S., Zlatev Z. (eds.): Numerical Methods and Applications. Lect. Notes in Comp. Science, Vol. 2542. Springer-Verlag, Berlin Heidelberg New York (2003) 506-513
6. Kandilarov, J.D.: Immersed-Boundary Level Set Approach for Numerical Solution of Elliptic Interface Problems. Lect. Not. in Comp. Sci., Vol. 2907. (2004) 456-464
7. Kandilarov, J., Vulkov, L.: The Immersed Interface Method for a Nonlinear Chemical Diffusion Equation with Local Sites of Reactions. Numer. Alg. (to appear)

8. Leveque, R.J., Li, Z.: The Immersed Interface Method for Elliptic Equations with Discontinuous Coefficients and Singular Sources. SIAM J. Numer. Anal. 31 (1994) 1019–1044
9. Li, Z.: The Immersed Interface Method - a Numerical Approach to Partial Differential Equations with Interfaces. Ph. D. thesis, University of Washington, Seatle (1994)
10. Pierce, A.P., Rabitz, H.: An Analysis of the Effect of Defect Structures on Catalytic Surfaces by the Boundary Element Technique. Surface Science 202 (1988) 1–31
11. Vulkov, L.G., Kandilarov, J.D.: Construction and Implementation of Finite-Difference Schemes for Systems of Diffusion Equations with Localized Nonlinear Chemical Reactions. Comp. Math. Math. Phys. 40 (2000) 705–717
12. Wiegmann, A., Bube, K.P.: The Eexplicit Jump Immersed Interface Method: Finite Difference Methods for PDE with Piecewise Smoth Solutions. SIAM J. Numer. Anal. 37 (2000) 827-862

Volterra Series and Numerical Approximations of ODEs

Nikolay Kirov and Mikhail Krastanov

Institute of Mathematics and Informatics,
Bulgarian Academy of Sciences,
Acad. G. Bonchev str. bl.8, 1113 Sofia
{nkirov, krast}@math.bas.bg

Abstract. A numerical approach for solving systems of nonautonomous ordinary differential equations (ODEs) is proposed under suitable assumptions. This approach is based on expansion of the solutions of ODEs by Volterra series and allows to estimate the distance between the obtained approximation and the true trajectory.

1 Introduction

We propose a numerical approach for solving systems of nonautonomous ordinary differential equations (ODEs). This approach is based on Volterra expansion of the solutions and allows to estimate the accuracy of the approximation. For simplicity, we assume that the right-hand side depends on the state-variables in an analytic way.

The approximation of trajectories of nonlinear control systems is reduced to the problem of approximation the solutions of nonautonomous systems of ODEs, whose right-hand sides depend on the time in a discontinuous way. But applying the traditional numerical schemes of higher order (such as Runge-Kutta schemes) is a nontrivial task (cf. for example [4], [9], etc.). An approach for solving this problem for affinelly controlled systems is proposed in [5] and [6]. It is based on the well known (in the control theory) expansion of the solution of systems of ODEs by Volterra series (cf. for example [7]). Combining this approach with the ideas developed in [8], we obtain a method for approximation the trajectories of analytic control systems with guaranteed accuracy.

The organization of this paper is as follows: We start with some general results concerning expansion of the solutions of ODEs systems by Volterra series. Next, we prove a priori criterion for existence of a solution of an ODEs system. The relation between the local and global approximation errors is also investigated. At the end, a computational procedure is proposed. Some numerical results are presented too.

[1] This research was partially supported by the Ministry of Science and Higher Education – National Fund for Science Research under the contract MM-1104/01.

Z. Li et al. (Eds.): NAA 2004, LNCS 3401, pp. 337–344, 2005.

2 Systems of ODEs and Volterra Series

First, we introduce briefly some notations and notions: For every point $y = (y^1, \ldots, y^n)^T$ from R^n we set $\|y\| := \sum_{i=1}^{n} |y^i|$ and let B be the open unit ball in R^n (according to this norm) centered at the origin. Let $x_0 \in R^n$ and Ω be a convex compact neighbourhood of the point x_0. C denotes the set of the complex numbers. If $z \in C$, then $|z|$, Re z and Im z denote the norm, the real and the imaginary part of z, respectively. For some $\sigma > 0$ we set

$$\Omega^\sigma := \{z = (z_1, \ldots, z_n) \in C^n : \ (\text{Re } z_1, \ldots, \text{Re } z_n) \in \Omega + \sigma B;$$

$$|\text{Im } z_i| < \sigma, \ i = 1, \ldots, n\}.$$

By $\mathcal{F}_\Omega^\sigma$ we denote the set of all real analytic functions ϕ defined on Ω with bounded analytic extensions $\bar{\phi}$ on Ω^σ. We define a norm in the set $\mathcal{F}_\Omega^\sigma$ as follows:

$$\|\phi\|_\Omega^\sigma = \sup \left\{ |\bar{\phi}(z)| \ : z \in \Omega^\sigma \right\}.$$

If $h(x) = (h_1(x), h_2(x), \ldots, h_n(x)), x \in \Omega$ is a vector field defined on Ω, we identify h with the corresponding differential operator

$$\sum_{i=1}^{n} h_i(x) \frac{\partial}{\partial x_i}, \quad x \in \Omega.$$

Let $\mathcal{V}_\Omega^\sigma$ be the set of all real analytic vector fields h defined on Ω such that every $h_i, i = 1, \ldots, n$ belongs to $\mathcal{F}_\Omega^\sigma$. We define the following norm in $\mathcal{V}_\Omega^\sigma$:

$$\|h\|_\Omega^\sigma = \max\{\|h_i\|_\Omega^\sigma, \ i = 1, \ldots, n\}.$$

An integrable analytic vector field $X_t(x) = (X_1(t,x), X_2(t,x), \ldots, X_n(t,x))$, $x \in \Omega$, $t \in R$ (parameterized by t) is a map $t \to X_t \in \mathcal{V}_\Omega^\sigma$ such that:

i) for every $x \in \Omega$ the functions $X_1(.,x), X_2(\cdot,x), \ldots, X_n(\cdot,x)$ are measurable;
ii) for every $t \in R$ and $x \in \Omega$, $|X_i(t,x)| \leq m(t)$, $i = 1, 2, \ldots, n$, where m is an integrable (on every compact interval) function.

Let t_0 be a real number, M be a compact set contained in the interior of Ω and $x_0 \in M$, and let X_t be an integrable analytic vector field defined on Ω. Then there exists $T(M, X_t) > t_0$ such that for every point x of M the solution $y(.,x)$ of the differential equation

$$\dot{y}(t,x) = X_t(y(t,x)), \quad y(t_0,x) = x \tag{1}$$

is defined on the interval $[t_0, T(M, X_t)]$ and $y(T,x) \in \Omega$ for every T from $[t_0, T(M, X_t)]$. In this case we denote by $\exp \int_{t_0}^{T} X_t \, dt : \ M \to \Omega$ the diffeomorphism defined by

$$\exp \int_{t_0}^{T} X_t \, dt \, (x) := y(T,x).$$

According to Proposition 2.1 from [1], the positive number $T(M, X_t)$ can be chosen in such a way that $T(M, X_t) > t_0$ and for every T from the open interval $(t_0, T(M, X_t))$, for every point x from M and for every function ϕ from $\mathcal{F}_\Omega^\sigma$, the following expansion of $\phi\left(\exp\int_{t_0}^T X_t\, dt\ (x)\right)$ in Volterra series holds true:

$$\phi\left(\exp\int_{t_0}^T X_t\, dt\ (x)\right) = \tag{2}$$

$$\phi(x) + \sum_{N=1}^{\infty}\int_{t_0}^{T}\int_{t_0}^{\tau_1}\int_{t_0}^{\tau_2}\cdots\int_{t_0}^{\tau_{N-1}} X_{\tau_N}X_{\tau_{N-1}}\ldots X_{\tau_2}X_{\tau_1}\phi(x)d\tau_N d\tau_{N-1}\ldots d\tau_1$$

and the series is absolutely convergent. The proof is based on the following estimate:

$$\max\left\{|X_{\tau_N}\ldots X_{\tau_2}X_{\tau_1}\phi(x)|, x\in M\right\}\leq$$

$$\leq N!\left(\frac{2n}{\sigma}\right)^N \|X_{\tau_N}\|_M^\sigma\cdots\|X_{\tau_2}\|_M^\sigma\cdot\|X_{\tau_1}\|_M^\sigma\cdot\|\phi\|_M^\sigma, \tag{3}$$

for every positive real number $\sigma > 0$, for every point x of M, for every function ϕ from $\mathcal{F}_\Omega^\sigma$ and for every points $\tau_N, \tau_{N-1}, \ldots, \tau_2, \tau_1$ from $[t_0, T]$. This estimate implies the following technical lemma.

Lemma 1. *Let M be a convex compact subset of Ω, $\psi\in\mathcal{F}_\Omega^\sigma$, $X_t\in\mathcal{V}_\Omega^\sigma$ and $t_0\leq\tau_1\leq\tau_2\leq\ldots\leq\tau_\mu\leq T$. We set*

$$\Psi_{\tau_1,\tau_2,\ldots,\tau_\mu} := X_{\tau_\mu}\ldots X_{\tau_2}X_{\tau_1}\psi.$$

Then for every $\sigma > 0$ and for every two points y_1, y_2 of M the following inequality holds true:

$$|\Psi_{\tau_1,\tau_2,\ldots,\tau_\mu}(y_2) - \Psi_{\tau_1,\tau_2,\ldots,\tau_\mu}(y_1)|\leq$$

$$\leq(\mu+1)!\left(\frac{2n}{\sigma}\right)^{\mu+1}\|X_{\tau_\mu}\|_M^\sigma\cdots\|X_{\tau_2}\|_M^\sigma\cdot\|X_{\tau_1}\|_M^\sigma\cdot\|\psi\|_M^\sigma\cdot\|y_2-y_1\|.$$

Let the functions $E_i:\Omega\to R$, $i = 1,\ldots,n$, be defined as follows: $E_i(y) = y_i$. We set $E := (E_1,\ldots,E_n)^T$ and $X_t\ E := (X_t\ E_1,\ldots,X_tE_n)^T$. Applying Lemma 1 we obtain that for every two points y_1 and y_2 from M the following estimate holds true:

$$\|X_{\tau_\mu}\ldots X_{\tau_2}X_{\tau_1}E(y_2) - X_{\tau_\mu}\ldots X_{\tau_2}X_{\tau_1}E(y_1)\|\leq$$

$$\leq n\mu!\left(\frac{2n}{\sigma}\right)^{\mu}\|X_{\tau_\mu}\|_M^\sigma\cdots\|X_{\tau_2}\|_M^\sigma\cdot\|X_{\tau_1}\|_M^\sigma\cdot\|y_2-y_1\|. \tag{4}$$

We use this estimate to prove a priori existence criterion for $\exp\int_{t_0}^T X_t\, dt\ (x)$, where x belongs to a given compact set.

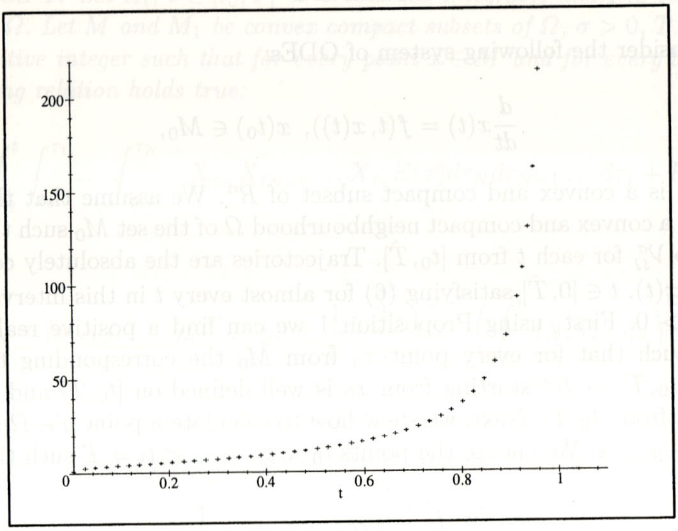

Fig. 1. The true solution

Example 1. Let us consider the following ODE:

$$\dot{x} = \frac{\gamma}{t^{1-\gamma}} x^2, \quad x(0) = x_0 > 0,$$

where $\gamma \in (0, 1]$. For $T \geq (1/x_0)^{1/\gamma}$ it can be directly checked that there does not exist a compact set M_1 satisfying the assumptions of Proposition 1. Hence, we can not conclude that the solution of this ODE is well defined on $[0, T]$ (actually, such a solution does not exist). Now, let us choose an arbitrary T, $0 \leq T < (1/x_0)^{1/\gamma}$. By setting $f(t, x) := \gamma x^2 / t^{1-\gamma}$, it can be directly calculated that

$$\int_0^t \int_0^{\tau_1} \cdots \int_0^{\tau_{k-1}} f(\tau_k, \cdot) \ldots f(\tau_1, \cdot) E(x_0) d\tau_k \ldots d\tau_1 = x_0^{k+1} t^{k\gamma}.$$

Hence, the series (2) for this ODE is:

$$x_0 + x_0^2 t^{\gamma} + x_0^2 t^{3\gamma} + \cdots x_0^{k+1} t^{k\gamma} + \ldots,$$

which is convergent for our choice of t and tends to its analytic solution

$$x(t) = \frac{x_0}{1 - x_0 t^{\gamma}}, \quad t \in \left[0, (1/x_0)^{1/\gamma}\right),$$

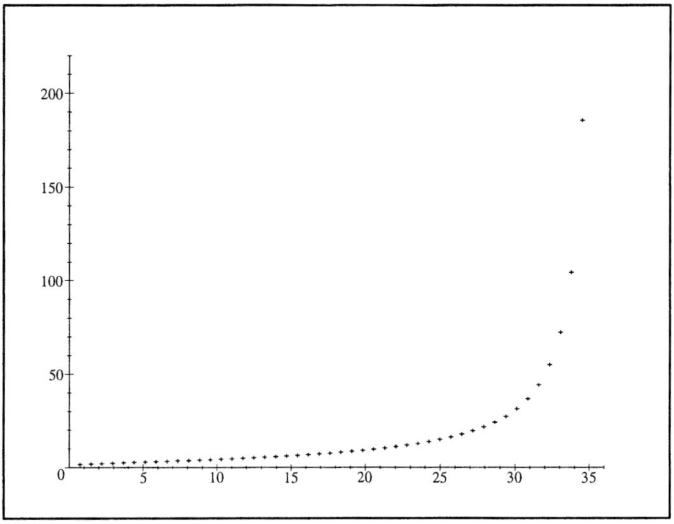

Fig. 2. A Runge-Kuta approximation

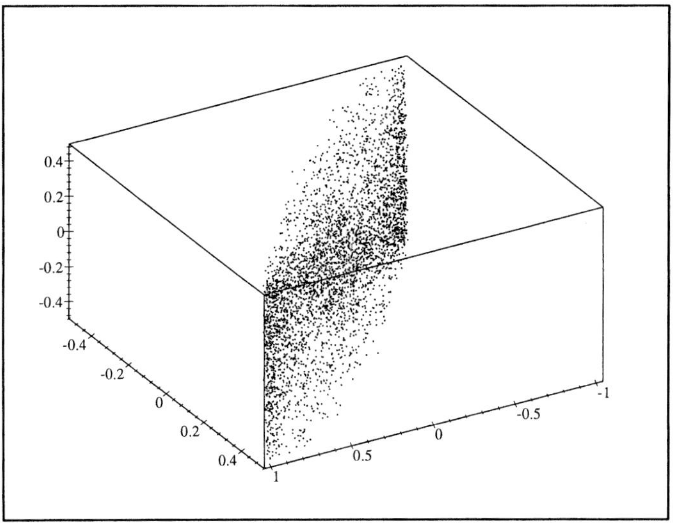

Fig. 3. Projection of the reachable set with two jump controls

as k tends to infinity. We make some numerical experiments with MAPLE for $\gamma = 0.125$ and $x_0 = 1$. The corresponding results are shown on Fig. 1 and Fig. 2.

Example 2. Let us consider the following control system:

$$\left|\begin{array}{l}\dot{x}_1 = u \\ \dot{x}_2 = x_1 \\ \dot{x}_3 = v \\ \dot{x}_4 = x_3 \\ \dot{x}_5 = x_1 v\end{array}\right. \qquad \begin{array}{l}|u| \leq 1 \\ x(0) = (0,0,0,0,0).\end{array}$$

The Volterra series of the system is finite. For every control we can calculate the exact trajectory. The reachable set $R(1)$ consists of all end-trajectory points at the moment 1. Figure 3 presents a 3-dimensional projection of the set $R(1)$ in the space of variables x_1, x_2 and x_5 using piecewise controls with two jumps (see also `http://www.math.bas.bg/~nkirov/2003/csys/consys.html`).

References

1. A. Agrachev, R. Gamkrelidze, The exponential representation of flows and the chronological calculus, Math. USSR Sbornik **107** (1978), 467-532.
2. U. Boscain, B. Piccoli A Short Introduction to Optimal Control, Lecture Notes for the CIMPA School, Tlemcen, Algeria, 2003.
3. F. Clarke, Yu. Ledyaev, R. Stern, P. Wolenski, Nonsmooth analysis and control theory, Graduate text in Mathematics **178**, Springer, 1998.
4. A. Dontchev, F. Lempio, Difference methods for differential inclusions, SIAM J. Review **34** (1992), 263-294.
5. L. Grüne, P. Kloeden, Higher order numerical schemes for affinely controlled nonlinear systems, Numer. Math. **89** (2001), 669-690.
6. L. Grüne, P. E. Kloeden, Numerical schemes of higher order for a class of nonlinear control systems, In: "Numerical Methods and Applications", I. Dimov, I. Lirkov, S. Margenov, Z. Zlatev, (Eds.), Springer Lecture Notes in Computer Science 2542, 213-220, 2003.
7. A. Isidori, Nonlinear control systems. An introduction, Springer, 1995.
8. M. Krastanov, N. Kirov, Dynamic Interactive System for Analysis of Linear Differential Inclusions, In: "Modelling techniques for uncertain systems", A. Kurzhanski, V. Veliov (Eds.), Birkhauser, 123-130, 1994.
9. V. Veliov, On the time discretization of control systems, SIAM J. Control Optim. **35** (1997), 1470-1486.

A Property of Farey Tree

Ljubiša Kocić[1] and Liljana Stefanovska[2]

[1] University of Niš, 18 000 Niš, Serbia and Montenegro
kocic@elfak.ni.ac.yu
[2] SS Cyril and Methodius University, Skopje, R. of Macedonia
liljana@ian.tmf.ukim.edu.mk

Abstract. The Farey tree is a binary tree containing all rational numbers from $[0,1]$ in ordered way. It is constructed hierarchically, level by level, using the Farey mediant sum. Some extreme properties of the Farey sum tree, i.e., the set of points $\{p+q\}$ associated with Farey tree in the way that rationals $\{p/q\}$ belong to k-th level of the Farey tree are investigated.

1 Introduction

Farey tree is the collection of sets (called levels) $\mathrm{FT} = \{\mathrm{T}_{-1}, \mathrm{T}_0, \mathrm{T}_1, ...\}$, where the n-th level $T_n = \{r_{2^n}, ..., r_{2^{n+1}-1}\}$, $(n = 0, 1, ...)$ is the decreasing sequence of rationals $r_j \in \mathbb{Q}(0,1)$ and $T_{-1} = \{r_{-1} = 0/1,\ r_0 = 1/1\}$ is called *seed* of the tree. The 0-th level, $T_0 = \{r_1 = 1/2\}$ is the *root* of FT. Further levels are $T_1 = \{r_2 = 2/3,\ r_3 = 1/3\}$, $T_2 = \{r_4 = 3/4,\ r_5 = 3/5,\ r_6 = 2/5,\ r_7 = 1/4\}$, etc (see Figure 1). Hierarchic construction of Farey tree is related to the commutative semi-group binary operation, called *Farey sum*, defined on $\mathbb{Q}[0,1]$ by $(p/q) \oplus (r/s) = (p+r)/(q+s)$, called *mediant* of p/q and r/s, due to double inequality $p/q < p \oplus q < r/s$. It is customary to identify FT with the infinite binary graph whose vertices form the set of rationals isomorphic to $\mathbb{Q}[0,1]$.

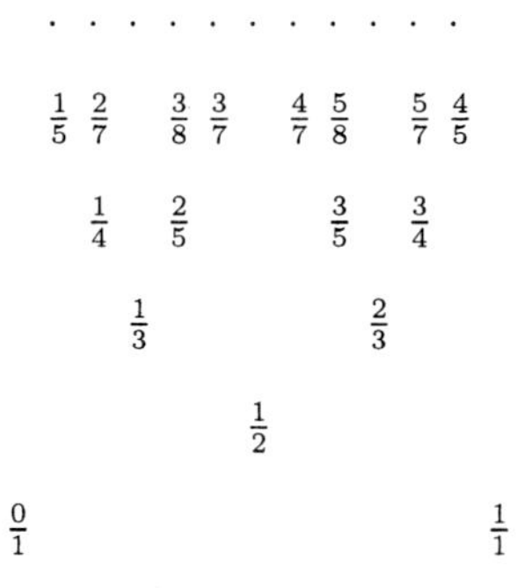

Fig. 1. Farey tree (first five levels)

Farey tree introduces order in the set $[0,1]$, i.e. embodies the mapping $\mathbb{N} \to \mathbb{Q}$, which is precisely given in the following lemma:

Z. Li et al. (Eds.): NAA 2004, LNCS 3401, pp. 345–351, 2005.

Lemma 1 ([3]). *Every rational $\rho \in (0,1)$ can be expressed uniquely as a mediant of two distinctive rationals $\rho_1, \rho_2 \in (0,1)$, i.e., $\rho = \rho_1 \oplus \rho_2$. Every such rational occurs uniquely as a vertex of the Farey tree.*

There are several algorithms for Farey tree construction, see [1], [2], [4], [6]. Further, two rationals p/q and r/s are called adjacents (or Farey neighbors) if $|ps - qr| = 1$. Note that the binary relation of adjacency is non-reflexive and non-transitive, but symmetric.

Lemma 2. *If ρ is the mediant of adjacent rationals $\rho_1, \rho_2 \in \mathbb{Q}(0,1)$, then ρ is adjacent to ρ_1 and ρ_2.*

Proof. Let $\rho = p/q$, $\rho_1 = p_1/q_1$ and $\rho_2 = p_2/q_2$. Since p_1/q_1 and p_2/q_2 are adjacents, it is valid $|p_1q_2 - p_2q_1| = 1$. By definition, $p = p_1 + p_2$ and $q = q_1 + q_2$, and therefore $|pq_1 - p_1q| = |(p_1 + p_2)q_1 - p_1(q_1 + q_2)| = |p_2q_1 - p_1q_2| = 1$. Similarly, $|pq_2 - p_2q| = |(p_1 + p_2)q_2 - p_2(q_1 + q_2)| = |p_1q_2 - p_2q_1| = 1$.

Let $R = \{\rho_1, \rho_2, ..., \rho_n\}$ be an increasing sequence of rationals from $(0,1)$, such that consecutive pairs (ρ_i, ρ_{i+1}) are adjacent. Consider the following "Farey power" operation:

$$R^{\oplus} = \{\rho_i \oplus \rho_{i+1} \mid \rho_i, \rho_{i+1} \in R,\ i = 1, \ldots, n-1\} \cup R, \tag{1}$$

where the result is taken in the increasingly ordered form. By virtue of Lemma 2, any consecutive pair from $R^{\oplus}$ is a pair of adjacents too. This means that Farey power can be iterated

$$R^{(k+1)\oplus} = (R^{k\oplus})^{\oplus},\ k = 1, 2, \ldots\ ,$$

assuming the convention $R^{0\oplus} = R$, $R^{1\oplus}R^{\oplus}$.

The Farey power (1) alowes the following generic definition of Farey tree.

Definition 1. *Farey tree is given by* $\mathrm{FT} = \{\mathrm{T}_{-1}, \mathrm{T}_0, \mathrm{T}_1, ...\}$, *where*

$$T_{-1} = \{0/1, 1/1\},\ T_k = T_{-1}^{(k+1)\oplus} \setminus T_{-1}^{k\oplus}, k0, 1, 2, \ldots,$$

($\setminus$ denotes the set-difference operation).

Definition 2. *The sequence $R_{n-1}T_{-1}^{n\oplus} = \cup_{j=-1}^{n-1} T_j$, will be called the n-th shadow of Farey tree ($n = 0, 1, 2, \ldots$).*

The first few shadows are $R_{-1} = \{0/1,\ 1/1\}$, $R_0 = \{0/1,\ 1/2,\ 1/1\}$, $R_1 = \{0/1,\ 1/3,\ 1/2,\ 2/3,\ 1/1\}$, ..., $R_n = \{r_1^n,\ r_2^n, \ldots, r_{2^{n+1}+1}^n\}$.

2 Sum-Tree

Let the simple mapping $\sigma : p/q \mapsto p+q$, be applied on Farey tree and let call the result *Farey sum tree* or FST$= \{U_{-1}, U_0, U_1, \ldots\}$ in the sense that $\sigma(T_n) = U_n$. The first five levels of the FST (Figure 2) are $U_{-1} = \{1, 2\}$, $U_0 = \{3\}$, $U_1 = \{4, 5\}$, $U_2 = \{5, 7, 8, 7\}$, $U_3 = \{6, 9, 11, 10, 11, 13, 12, 9\}$, etc.

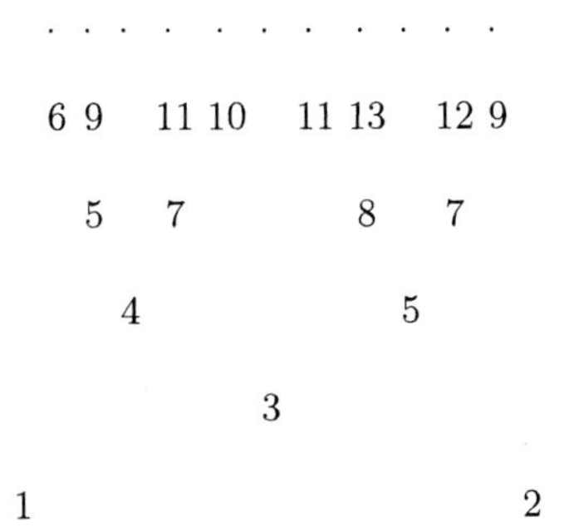

Fig. 2. Farey sum tree (first five levels)

Definition 3. *The set* $S_n = \{s_1^n, s_2^n, \ldots, s_{2^{n+1}+1}^n\}$ *denotes the* n*-th shadow of the Farey sum tree provided* $S_n = \sigma(R_n)$, $n = -1, 0, 1, \ldots$.

Obviously, FST and FT are isomorphic structures, as well as their shadows.

It can be easily verified that the shadow of FST is defined by the following recurrence (being implied by Definitions 2 and 3):

$$s_{2j-3}^{n+1} = s_{j-1}^n,\ s_{2j-1}^{n+1} = s_j^n,\ s_{2j+1}^{n+1} = s_{j+1}^n,\ s_{2j-2}^{n+1} = s_{j-1}^n + s_j^n,$$
$$s_{2j}^{n+1} = s_j^n + s_{j+1}^n, j = 2, 3, \ldots, 2^{n+1}, n = 0, 1, 2, \ldots; \tag{2}$$
$$s_1^0 = 1,\ s_2^0 = 3,\ s_3^0 = 2;$$

For instance, $S_0 = \{1, 3, 2\}$, $S_1 = \{1, 4, 3, 5, 2\}$, $S_2 = \{1, 5, 4, 7, 3, 8, 5, 7, 2\}$, etc.

Note that S_n can be split into 2^n triples "with offset 1", i.e, triples of the form

$$\ldots,\ (s_{j-1}^n,\ s_j^n,\ s_{j+1}^n),\ (s_{j+1}^n,\ s_{j+2}^n,\ s_{j+3}^n),\ \ldots,$$

where $j = 2, 3, \ldots, 2^{n+1} - 2$, and $n = 0, 1, 2, \ldots$, with the following property:

Lemma 3. *Any triple* $t_j^n = (s_{j-1}^n,\ s_j^n,\ s_{j+1}^n) \in S_n$, $j = 2, \ldots, 2^{n+1}$, $n = 0, 1, 2, \ldots$, *satisfies*

$$s_j^n = s_{j-1}^n + s_{j+1}^n \tag{3}$$

$$s_{j+1}^n > s_{j-1}^n \geq 1 \quad \text{or} \quad s_{j-1}^n > s_{j+1}^n \geq 1 \tag{4}$$

Proof. By induction. The "root" triple $t_2^0 = (1, 3, 2)$ satisfies (3) and (4). Suppose that $t_j^n = (s_{j-1}^n,\ s_j^n,\ s_{j+1}^n)$ satisfies (3-4), or, using abbreviations $a = s_{j-1}^n$ and $b = s_{j+1}^n$, the conditions become $s_j^n = a + b$ and $a > b \geq 1$ or $b > a \geq 1$. Now, by the generic recurrence (2), t_j^n produces two "children", the "left" one t_{2j-2}^{n+1} and the "right" one t_{2j}^{n+1} with the values $t_{2j-2}^{n+1} = (a, 2a + b, a + b)$ and $t_{2j}^{n+1} = (a + b, a + 2b, b)$. So, it is evident that both descendant triples satisfy the condition (3).

Further, if t_j^n satisfies $a > b \geq 1$, then, $a + b > a \geq 1$, which is $s_{2j-1}^{n+1} > s_{2j-3}^{n+1} \geq 1$ and t_{2j-2}^{n+1} satisfies (4) as well. Similarly, for t_{2j}^{n+1}, where $s_{2j-1}^{n+1} > s_{2j+1}^{n+1} \geq 1$.

It follows from Lemma 3 that the middle term of any triple is maximal. The triple $t_k^n = (s_{k-1}^n, s_k^n, s_{k+1}^n) \in S_n$ will be called *dominant* of the level S_n if $s_k^n > s_j^n$ for all $1 \leq j \leq 2^{n+1}+1$, $j \neq k$.

Lemma 4. *There is one and only one dominant triple in S_n, $n = 0, 1, 2, \ldots$.*

Proof. Also by induction. The "root" level S_0 has one dominant triple. Suppose that $t_k^n = (s_{k-1}^n, s_k^n, s_{k+1}^n) = (a, a+b, b)$ is a dominant triple in S_n. Then, $a+b$ is strictly bigger than any other middle term of other triples of the level. So, the only candidates for the S_{n+1} level dominant are "children" of t_k^n, i.e., $t_{2j-2}^{n+1} = (a, 2a+b, a+b)$ and $t_{2j}^{n+1} = (a+b, a+2b, b)$. By Lemma 3, only two possibilities exist: $a > b \geq 1$ or $b > a \geq 1$. In the first case, $a > b$ implies $2a+b \geq a+2b$, so the triple t_{2j-2}^{n+1} is dominant. In the second case, $a < b$, and therefore $a+2b > 2a+b$ which makes t_{2j}^{n+1} the dominant triple.

Let $L(t)$ and $R(t) \in S_{n+1}$ be the left and right "child" of a triple $t \in S_n$. The left and right "child" of $L(t)$ will be $LL(t) = L(L(t))$ and $LR(t) = R(L(t))$. Similarly, $RL(t)$ and $RR(t)$ will be "children" of $R(t)$. So, t will have four descendants in S_{n+2}, the $(n+2)$-level of the FST shadow. Continuing this process produces a sub-tree in triples of FST (and also in the corresponding FST and FT).

Lemma 5. *If the triple $t = (a, a+b, b) \in S_n$ satisfies $a > b$, then the sequence of descendants $L(t)$, $LR(t)$, $LRL(t)$,... is the sequence of dominants in the sub-tree of the FST shadow triples rooted in t. If $b > a$, then the dominant sequence is $R(t)$, $RL(t)$, $RLR(t)$,...*

Proof. It follows by proof of Lemma 4 that if $a > b$ then the left "child" $L(t) = (a, 2a+b, a+b) = (a_1^L, a_1^L + b_1^L, b_1^L)$ dominates its right "child" $R(t) = (a+b, a+2b, b)$. At the same time, $b_1^L > a_1^L$ and now the right "child" $LR(t) = (a_2, a_2+b_2, b_2)$ dominates $LL(t)$. But, since $a_2^L = a_1^L + b_1^L$ and $b_2^L = b_1^L$, now $a_2^L > b_2^L$, so the left "child" is dominant, etc. Of course, the reversal inequality, $a < b$ yields dual sequence as dominant within the subtree.

Lemma 6. *If $a > b$ in $t = (a, a+b, b) \in S_n$, then the sequence of descendants $R(t)$, $RR(t)$, $RRR(t)$,... is the sequence of triples with minimal middle element within the sub-tree rooted at t. If $b > a$, then the minimal middle element sequence is $L(t)$, $LL(t)$, $LLL(t)$,...*

Proof. With notations used in the proof of the previous Lemma, $a > b$ implies that the right "child" $(a_1^R, a_1^R + b_1^R, b_1^R)$ has smaller middle element than the left one, and moreover $a_1^R > b_1^R$, so that it again produces smaller right "child" etc. Under the inverse condition, $a < b$, the sequence of left descendants is smaller.

Let $\phi = (1+\sqrt{5})/2$ be the *golden ratio.* Then, $f_n = (\phi^n - (-\phi)^{-n})/\sqrt{5}$, $n = 0, 1, 2, \ldots$ denotes the n-th Fibonacci number. The following theorem unifies results of the preceding lemmas.

Theorem 1. *Minimal and maximal elements of the n-th shadow of the Farey sum tree are $s_{\min}^n = n+3$ and $s_{\max}^n = f_{n+4}$, $n = 0, 1, 2, \ldots$.*

Proof. Applying Lemma 6 to the root triple $t_2^0 = (1,3,2)$ $(a < b)$, gives $L(t_2^0) = (1,4,3) \in S_1$, $LL(t_2^0) = (1,5,4) \in S_2$, $LLL(t_2^0) = (1,6,5) \in S_3$,...,$L^n(t_2^0) = (1, n+3, n+2) \in S_n$.

Also, by Lemma 5, the sequence $R(t_2^0)$, $RL(t_2^0)$, $RLR(t_2^0)$,..., gives $(3,5,2)$, $(3,8,5)$, $(8,13,5)$,..., the sequence of dominants in S_1, S_2, S_3, etc. Note that $t_2^0 = (1,3,2) = (f_2, f_4, f_3)$, where f_i is i-th Fibonacci number. By Lemma 3, $R(t_2^0) = (f_4, f_5, f_3)$, $RL(t_2^0) = (f_4, f_6, f_5)$, $LRL(t_2^0) = (f_6, f_7, f_5)$, etc.

The minimal sequence $\{s^n_{\text{min}}\}$ of the n-th shadow of the Farey sum tree corresponds to the harmonic sequence $\{1/n\}_{i=2}^{+\infty}$ of the Farey tree, while the maximal one $s^n_{\text{max}} = f_{n+4}$ corresponds to the *golden sequence* $\{f_{i+1}/f_{i+2}\}_{i=1}^{+\infty}$ converging to $1/\phi = (\sqrt{5}-1)/2 \approx 0.6180339887498948482$. This golden sequence is important in Chaos theory, providing the quickest quasi-periodic route to chaos ([1], [4], [6]).

3 Applications and Examples

Farey tree is an important construction in both theoretical (Number theory, Topology, Graph theory) and applied topics (Dynamics theory, Chaos, Fractal analysis). The common issue that connects all these areas is the *coupled oscillators phenomena* first noticed by Huygens in 17-th century. If a coupled system oscillates in two different frequencies ω_1 and ω_2 $(\omega_1 < \omega_2)$, then their ratio $\Omega = \omega_1/\omega_2$ has tendency to be a rational number, i.e. the node of Farey tree and to belong to the lowest level possible. The nature of oscillation regime is indicated by the sequence $\{\theta_n\}_{n=0}^{+\infty}$ generated by the recurrence $\theta_{n+1} = \phi(\theta_n)$, where $\phi(\theta) = \theta + \Omega - (K/2\pi)\sin 2\pi\theta$, $K \in \mathbb{R}$, is so called *circle map.* Even if Ω is irrational, the "long term" behavior of the system, embodied in the quantity $\omega = \lim_{n\to\infty}(\theta_n - \theta_0)/n$ (known as *winding number*) "locks" into ratios with small denominators. This is known as *mode locking* regime of coupled oscillators. Parameter K that is usually a real number from [0, 1] represents the coupling strength in the system. It is known that the graph $\omega = \omega(\Omega)$ is fractal function having "devil staircase" structure with plateaus of finite length. This length is bigger if the ratio Ω is closer to the simple rationals. So, the widest plateau corresponds to $\Omega = 1/2$, then there are two equally wide for $\Omega = 1/3$ and $2/3$ (T_1 level). Then, there are four plateaus for the ratios from level T_2, and so on. The diagram representing width of these plateaux as K increases from 0 to 1 is known as *Arnold tongues* diagram. Knowing the extremal property of Farey tree, given in Theorem 1, enables understanding scaling properties of the set of rationals from the specific level of the FT or its shadow.

Another application, concerning density of plane covering by Lissajous parametric curves $t \mapsto (\sin t,\ \sin((p/q)t))$, $0 \le t \le 2\pi$, $p/q \in$ FT is given in [5]. There, it is conjectured that the Lissajous figures provide the densest cover of the unit square provided p/q has maximal sum $p+q$.

EXAMPLE 1. In the following table, we compare the CPU time needed for calculating certain levels (7 to 20) of Farey tree (an average Pentium PC is used). The left column presents CPU times consumed by the "classic" algorithm ([6], [2]) based on binary conversion to continuous fraction. The right column gives times needed for the algorithm based on Farey power operation (1) (see Example 3). The last ones are considerably shorter. For levels below seven CPU time is too small to be measured.

Table 1. CPU time comparison

level	Classic alg.	Farey power
7	0.078125	0.
8	0.23437	0.03125
9	0.25	0.046875
10	0.53125	0.07812
11	1.171875	0.1875
12	2.390625	0.359375
13	5.03125	0.71875
14	10.625	1.609375
15	22.265625	3.203125
16	46.9375	6.171875
17	97.828125	11.578125
18	206.125	22.765625
19	424.734375	47.453125
20	879.890625	93.5375

EXAMPLE 2. Another example gives accumulation points for the $(LR)^\infty$ or $(RL)^\infty$ chains of the Golden sequence $\{f_i/f_{i+1}\}_{i=2}^{+\infty}$, different from $1/\phi$. Clearly, for even i, $(RL)^\infty(f_i/f_{i+1}) = 1/\phi$ and for odd i, $(LR)^\infty(f_i/f_{i+1}) = 1/\phi$. But, what are limits in the opposite cases? Computation gives:

$$l_1 = (LR)^\infty(\frac{1}{2}) = \frac{3-\sqrt{5}}{2}, \qquad l_2 = (RL)^\infty(\frac{2}{3}) = \frac{5+\sqrt{5}}{10},$$

$$l_3 = (LR)^\infty(\frac{3}{5}) = \frac{15-\sqrt{5}}{22}, \qquad l_4 = (RL)^\infty(\frac{5}{8}) = \frac{37+\sqrt{5}}{62},$$

$$l_5 = (LR)^\infty(\frac{8}{13}) = \frac{99-\sqrt{5}}{158}, \qquad l_6 = (RL)^\infty(\frac{13}{21}) = \frac{257+\sqrt{5}}{418},$$

$$l_7 = (LR)^\infty(\frac{21}{34}) = \frac{675-\sqrt{5}}{1090}, \qquad l_8 = (RL)^\infty(\frac{34}{55}) = \frac{1765+\sqrt{5}}{2858} \ldots$$

The sequence of limits $\{l_k\}$ converges to $1/\phi$. For our example, the difference $1/\phi - l_k$, for $k = 1, \ldots, 8$ yields 0.236068, −0.105573, 0.0378553, −0.0148058, 0.00560404, −0.00214799, 0.000819372 and −0.000313131.

EXAMPLE 3. Finally, here is the code in *MATHEMATICA* programming language for Farey tree construction by using Farey power operation (1). The command **fareyTree**[n], $n \in I\!N$ causes displaying the tree, rooted at $\{1/2\}$ up to the n-th level.

```
fareySum=Numerator[First[#]]+Numerator[Last[#]])
        /(Denominator[First[#]]+ Denominator[Last[#]])&;
ins=Insert[#,fareySum[#],2]&;
ffs=Union@@ins/@Partition[#,2,1]&;
fareyPow=Nest[ffs,{0, 1},#]&;
ftLevel=Complement[fareyPow[#+1],fareyPow[#]]&;
fareyTree=ColumnForm[ftLevel/@ Range[0,#],Center]&;
```

This code almost literally follows Definition 1. The CPU times elapsed in calculating certain levels are diplayed in Table 1.

4 Conclusion

This note gives some properties of the Farey sum tree (Fig. 2) by using the concept of Farey power and Farey tree shadow. It is shown that the sum $p + q$ of Farey tree elements p/q reaches its maximum along the "golden sequence" $\frac{1}{2}$, $\frac{2}{3}$, $\frac{3}{5}$, $\frac{5}{8}$, $\frac{8}{13}$, $\frac{13}{21}$, ... that paves the famous "golden route to chaos".

References

1. Cvitanovic, P., *Circle maps: irrationally winding*, Number Theory and Physics (C. Itzykson, P. Moussa and M. Waldschmidt, eds.), Les Houches 1989 Spring School, Springer, New York 1992.
2. Kappraff, J and Adamson, G. W., *A Fresh Look at Number*, Vis. Math. **2**, (2003) no. 3 (electronic: http://members.tripod.com/vismath4/ kappraff1/index.html).
3. Lagarias, J. C., *Number Theory and Dynamical Systems*, The Unreasonable Effectiveness of Number Theory, (S. A. Burr, ed.), Proceedings of Symposia in Applied Mathematics 46, (AMS, 1992), pp. 35-72.
4. Kocic, Lj. M., Stefanovska, R. L., *Golden Route to Chaos*, Facta Universitatis, Ser. Mechanics, Automatic Control and Robotics, **4** (2004), no. 16, 69-75.
5. Kocic, Lj. M., Stefanovska, R. L., *Lissajous Fractals*, The 10th International Symposium of Mathematics and its Applications (N. Boja, ed.), "Politehnica" University of Timisoara, Timisoara 2003, pp. 416-423.
6. Schroeder, M. R., *Fractals, Chaos, Power Laws: Minutes from an Infinite Paradise*, Freeman and Co., New York 1991.

Comparison of a Rothe-Two Grig Method and Other Numerical Schemes for Solving Semilinear Parabolic Equations

Miglena N. Koleva

Center of Applied Mathematics and Informatics
University of Rousse, 8 Studentska str., Rousse 7017, Bulgaria
mkoleva@ru.acad.bg

Abstract. A technique combined the Rothe method with two-grid (coarse and fine) algorithm of Xu [18] for computation of numerical solutions of nonlinear parabolic problems with various boundary conditions is presented. For blow-up solutions we use a decreasing variable step in time, according to the growth of the solution. We give theoretical results, concerning convergence of the numerical solutions to the analytical ones. Numerical experiments for comparison the accuracy of the algorithm with other known numerical schemes are discussed.

1 Introduction and Statement of the Differential Problems

A flexible approach in the numerical solution of parabolic differential equations (PDEs) is obtained by combining the Rothe method with the two-grid method of Xu [18]. The numerical solution process may be considered as to consist of two parts, viz. time semi-discretization and space integration. In the time discretization the PDEs are converted on each time level into an elliptic boundary semilinear problem. Then the technique of Xu [18], based on finite element spaces defined on two grids of different sizes (coarse and fine), is applied.

In this paper we consider one-dimensional problems:

$$u_t - u_{xx} = f_0(x,t,u) \ \text{ in } Q_T = D \times I,\ D=(0,1),\ I=(0,T),\ T<\infty, \quad (1)$$
$$c_0 \ u_t(0,t) - d_0 u_x(0,t) = f_1(u(0,t)),\ \ 0<t<T, \quad (2)$$
$$c_1 \ u_t(1,t) + d_1 u_x(1,t) = f_2(u(1,t)),\ \ 0<t<T, \quad (3)$$
$$u(x,0) = u_0(x),\ x \in \bar{D} = [0,1], \quad (4)$$

where c_0, c_1, d_0 and d_1 are equal to 0 or 1. Each combination of these parameters leads to the corresponding boundary value problem (b.v.p.). For example

- Homogeneous Dirichlet b.v.p.: $c_0 = c_1 = d_0 = d_1 = 0$, $f_1(s) = f_2(s) = s$;
- Neumann b.v.p.: $c_0 = c_1 = 0$, $d_0 = d_1 = 1$, $f_1 = f_2 = 0$;
- Robin b.v.p.: $c_0 = c_1 = 0$;
- Dynamical boundary conditions: $c_0 = c_1 = 1$.

Z. Li et al. (Eds.): NAA 2004, LNCS 3401, pp. 352–359, 2005.

Semilinear parabolic equations with classical boundary conditions(Dirichlet's, Neumann's and Robin's) are well studied for global existence and blow-up of the solutions, cf. the monographs [12, 15] and review papers [1, 7, 16].

Problems of type (1)-(4), $c_0 + c_1 \neq 0$ arise in the theory of heat-diffusion conduction, cf. [3, 4], in chemical reactor theory [13], in colloid chemistry [17], in modelling of semiconductor devices [8], as well as in heat transfer in a solid in contact with a fluid, see the reference of the paper [7, 10] for more information. In [11] finite difference blow-up solutions was studied.

Under appropriate assumption on f_0, f_1, f_2 the solution blows up in finite time for some initial data or exist globally, see [7].

Definition 1. *A function u is called a weak solution to the problem (1)-(4) if :*

i) $u \in L_2(I, \mathcal{H}^1(D)) \bigcap L_\infty(I \times D)$;
ii) u satisfies the integral identity

$$\int_0^T\int_D u_x(t,x)v_x(t,x)dxdt - \int_0^T\int_D u(t,x)v_t(t,x)dxdt = \int_D u_0(x)v(0,x)dx +$$
$$\frac{c_0}{d_0}\int_0^T u(0,t)v_t(0,t)dt + \frac{c_1}{d1}\int_0^T u(1,t)v_t(1,t)dt + \int_0^T f_1(u(0,t))v(0,t)dt +$$
$$\int_0^T f_2(u(1,t))v(1,t)dt + \frac{c_0}{d_0}u(0,0)v(0,0) + \frac{c_1}{d_1}u(1,0)v(1,0)$$

for all $v \in L_2(I \times H^1(D))$ such that $v_t \in L_\infty(I \times D)$ and $v(.,T) = 0$.

It worth to note that the problem has unique local weak solution (see [5]). In [9] for the proof of the solution existence a Rothe method is used.

The rest of the paper is organized as follows. In Section 2 we present time discretization, using Rothe method for problem (1)-(4). Section 3 is devoted to the method of Xu, applied for generated semilinear elliptic problem. This section also contain convergence results. Section 4 gives numerical examples.

In principle, the proposed method works for two and three dimensional problems. However, in order to simplify the presentation, we have discussed only the one-dimensional case.

2 Time Discretization

We divide the interval $I = [0,T]$ into n subintervals I_j, $j = 1, 2, ..., n$ of the length τ_j, with points of division t_j. For every $t = t_j$ we approximate the unknown function $u(x,t_j)$ by function $z_j(x)$ and the derivative $\frac{\partial u}{\partial t}$ by the difference quotient

$$Z_j(x) = \frac{z_j(x) - z_{j-1}(x)}{\tau_j},$$

the function z_{j-1} is the solution, obtained in the previous time level.
Starting by the function $z_0(x) = u_0(x)$, the functions $z_1(x), z_2(x), ..., z_n(x)$ are determined subsequently as solutions of ordinary boundary value problems :

$$\frac{z_1 - z_0}{\tau_1} - z_1'' = f_0(x, t_1, z_1), \quad c_0\frac{z_1 - z_0}{\tau_1}(0) - d_0 z_1'(0) = f_1(z_1(0)),$$

$$c_1\frac{z_1 - z_0}{\tau_1}(1) + d_1 z_1'(1) = f_2(z_1(1)),$$

$$\frac{z_2 - z_1}{\tau_2} - z_2'' = f_0(x, t_2, z_2), \quad c_0\frac{z_2 - z_1}{\tau_2}(0) - d_0 z_2'(0) = f_1(z_2(0)),$$

$$c_1\frac{z_2 - z_1}{\tau_2}(1) + d_1 z_2'(1) = f_2(z_2(1)),$$

$$\cdots$$

Having obtained the function $z_1(x), z_2(x), ..., z_n(x)$ the so-called Rothe function $u_n(t)$ is defined in the whole region $(0,1) \times (0,T)$ by

$$u_n(t) = z_{j-1} + \frac{z_j + z_{j-1}}{\tau_j}(t - t_j), \quad \text{for } t_{j-1} < t < t_j, \ j = 1, ..., n,$$

assuming the values z_j to every $t = t_j$. By refining the original division (τ_k, $k = 1, 2, ..., \tau_k \to 0, k \to \infty$), we obtain the sequence $u_{n_k}(t)$ for corresponding Rothe functions, which can be expected to converge (in an appropriate space) to the solution u (in an appropriate sense) of the given problem.

Theorem 1. *Let $u_0 \in \mathcal{H}^1(D)$ and f_i, $i = 0, 1, 2$ are smooth functions. Then there exists exactly one weak solution of the problem (1)-(4) and $u_n \to u$ ($n \to \infty$) in $L_2(I, \mathcal{H}^1(D))$ with rate of convergence $O(\tau)$.*

3 Space Discretization

For each time level $t = t_j$, $j = 1, ..., n$, the following semilinear elliptic problem is generated by using Rothe method.

$$-z_{xx} + g_0^j(x, z(x)) = 0, \quad x \in D, \quad g_0^j(x, z) = -f_0^j(x, z) + Z_j(x), \qquad (5)$$

$$-d_0 z_x(0) + g_1^j(z(0)) = 0, \qquad g_1^j(z(0)) = -f_1^j(z(0)) + c_0 Z_j(0), \qquad (6)$$

$$d_1 z_x(1) + g_2^j(z(1)) = 0, \qquad g_2^j(z(1)) = -f_2^j(z(1)) + c_1 Z_j(1), \qquad (7)$$

and $f_0^j(x, z) = f_0(x, t_j, z_j)$, $f_i^j(z) = f_i(z_j)$, $i = 1, 2$.

Now we use the algorithm proposed in [18]. We assume that the above problem has solution $z \in \mathcal{H}^2(D)$. Let V_h is a piecewise linear finite element space, defined on an uniform mesh with size h in $\bar{D}$. The standard finite element discretization of (5)-(7) for $d_0 = d_1 = 1$, is to find $z_h \in V_h$ so that

$$\chi(0)g_1(z_h(0)) + \chi(1)g_2(z_h(1)) + (\nabla z_h, \nabla\chi) + (g_0(z_h), \chi) = 0, \quad \forall\, \chi \in V_h,$$

where $(u, v) = \int\limits_D uvdx$. We have dropped the dependence of variable x in $g_0(x, z)$.

We introduce another finite element space V_H ($V_H \subset V_h$) defined on a coarser uniform mesh (with mesh size $H > h$) of $\bar{D}$. The three-step algorithm of Xu, based on two (coarse and fine) grids for problem (5)-(7) is:

Find $z_h^\star = z_H + e_h + e_H$ such that

1. $z_H \in V_H$,
   $\varphi(0)g_1(z_H(0)) + \varphi(1)g_2(z_H(1)) + (\nabla z_H, \nabla\varphi) + (g_0(z_H), \varphi) = 0, \ \forall\varphi \in V_H;$
2. $e_h \in V_h$,
   $\chi(0)g_1'(z_H(0))e_h(0) + \chi(1)g_2'(z_H(1))e_h(1) + (\nabla e_h, \nabla\chi) + (g_0'(z_H)e_h, \chi) =$
   $-(\nabla z_H, \nabla\chi) - (g_0(z_H), \chi) - \chi(0)g_1(z_H(0)) - \chi(1)g_2(z_H(1)), \qquad \forall\chi \in V_h;$
3. $e_H \in V_H$,
   $(\nabla e_H, \nabla\varphi) + (g_0'(z_H)e_H, \varphi) + \varphi(0)g_1'(z_H(0))e_H(0) + \varphi(1)g_2'(z_H(1))e_H(1) =$
   $-\frac{1}{2}(g_0''(z_H)e_h^2, \varphi) - \frac{1}{2}g_1''(z_H(0))e_h^2(0) - \frac{1}{2}g_2''(z_H(1))e_h^2(1), \qquad \forall\varphi \in V_H.$

Theorem 2. *If $u \in \mathcal{H}^2(D)$ then*

i) $\|z - (z_H + e_h)\|_\infty \le C(\tau + h^2 + H^4)$,
ii) $\|z - z_h^\star\|_\infty \le C(\tau + h + H^5)$, $\qquad \|z - z_h^\star\|_{L_2} \le C(\tau + h^2 + H^6)$,

where C is independent of τ, h, H.

4 Numerical Experiments

In this section we present some numerical results. We show the efficiency of the proposed algorithm for bounded and unbounded solution of the semilinear parabolic problem with dynamical boundary conditions.

4.1 Bounded Solutions

The test problem is (1)-(4), where $c_0 = c_1 = d_0 = d_1 = 1$ and

$$f_0(x, t, u) = u^2 + f(x, t), \quad f_1(u) = f_2(u) = \lambda u, \quad u_0(x) = \cos(\pi x),$$

$f(x, t)$ is chosen such that $u(x, t) = e^{\lambda t}\cos(\pi x)$ is the exact solution.

Example 1. To avoid the influence of time step τ and two different mesh sizes over the error results, we chose $h = H^3$, $\tau \le h^2$, $V_H = 5$, $V_h = 65$, $\tau = 0.0002$, $\lambda = -1$. We compare the results, computed by applying Rothe-two-grid method

Table 1. Absolute error of bounded solution, $\tau = 0.0002$

T	Rothe-two-grid method		Semi-implicit scheme	
	max *norm*	L_2 *norm*	max *norm*	L_2 *norm*
0.001	6.024656e-6	2.387063e-6	2.114412e-3	1.039873e-3
0.01	3.803958e-5	1.720149e-5	1.330760e-2	6.647705e-3
0.1	1.266726e-4	8.034069e-5	4.567016e-2	2.715512e-2
1	2.245534e-4	1.775041e-4	6.141114e-2	5.026921e-2
1.5	1.782726e-4	1.468143e-4	4.551890e-2	3.730650e-2

Table 2. Absolute error and convergence rate on second step of algorithm

	$V_H = 5$ ($V_h = 17$)	$V_H = 9$ ($V_h = 65$)	$V_H = 17$ ($V_h = 256$)	$V_H = 33$ ($V_h = 1025$)
Max norm	2.52109025e-4	1.49044011e-5	9.11314558e-7	5.70021684e-8
ratio in V_H		4.0798	4.0316	3.9989
ratio in V_h		2.0199	2.0079	1.9997
L_2 *norm*	1.64754271e-4	9.36265703e-6	2.47604815e-7	1.51749375e-8
ratio in V_H		4.1369	5.2479	4.0283
ratio in V_h		2.0339	2.2908	2.0071

and Semi-implicit scheme (65 grid points, $u^\alpha \simeq (u^{\alpha-1})^n u^{n+1}$). In Table 1 we give absolute error in corresponding discrete norms.
The bounded solution of problem (1)-(4), computed with Rothe method together with two-grid technique of Xu, is more precisely.

Example 2. Now we check the convergence rate of the second step of algorithm. We chose $h = H^2$ and fix the ratio $\frac{\tau}{h^2} = \frac{\tau}{H^4}$, $(\tau < h^2)$. In Table 2 we give the results after second step on one and the same time level. Decreasing the coarse step size twice (fine step size decrease four times) yields to decreasing the error approximately $2^4 = 4^2$ times. The convergence rate is $O(H^4) = O(h^2)$.

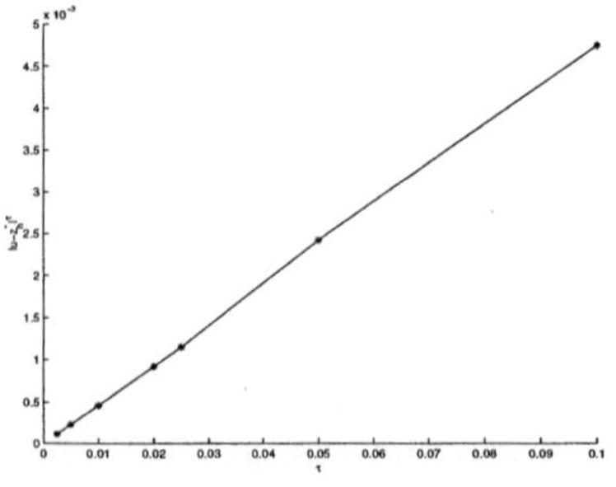

Fig. 1. Semi-discrete error

Table 3. Semi-discrete error and convergence rate

τ	$\|u - z_h^\star\|^\tau$	*ratio*
0.10000	0.00475	
0.05000	0.00242	0.9769
0.02500	0.00115	1.0733
0.02000	9.19507e-4	
0.01000	4.57323e-4	1.0094
0.00500	2.27857e-4	1.0051
0.00250	1.13553e-4	1.0048

Example 3. For comparing the exact $u(x,t)$ and numerical solution $z_h^\star$ we use the semi-discrete norm $|u - z_h^\star|^\tau = \frac{1}{N_\tau}\sum_{i=0}^{N_\tau}(\int_0^1(u(t_i) - z^\star(t_i))^2)^{\frac{1}{2}}$, (see [2]), where N_τ is the number of time steps in the interval $[0,T]$. The evolution of this error is depicted in Figure 1 and Table 3 .
$V_H = 11$, $V_h = 1001$.

4.2 Blow-Up Solutions

We consider the problem (1)-(4), where $c_0 = c_1 = d_0 = d_1 = 1$ and

$$f_0(x,t,u) = |u|\,u + f(x,t), \quad f_1(u) = f_2(u) = -\,|u|\,u, \quad u_0(x) = -\frac{1}{T_b}\cos(\pi x),$$

$f(x,t)$ is chosen that $u(x,t) = \frac{1}{t-T_b}\cos(\pi x)$ is the exact solution. T_b is the blow-up time.
For successive computation of the blow-up solution in each example below, we use a decreasing variable time step $\tau_j = \tau_0 \times \min(1, \frac{1}{\|u\|_{\infty\ or\ L_2}})$, $j = 1, ..., n$ as in [14], τ_0 is the initial time step. Depending on numerical method the solution grows with different speed and the time step τ became different. In order to compare the results in one and the same time level, we use a linear interpolation in time.

Example 4. In Table 4 we show the relative error of the solution, in corresponding discrete norms, computed for $h = H^2$, $\tau \leq h^2$, $V_H = 9$, $V_h = 65$, $\tau_0 = 0.0006$, $T_b = 0.3$, with Rothe two-grid method, Semi-implicit scheme (65 grid points, i.e. $V_H = 65$) and Fully implicit scheme($V_H = 65$, $V_H = 9$), using Rothe discretization in time and then Newton method for solving the generated nonlinear system.

Semi-implicit scheme and Rothe two-grid method have similar efficiency for computation of the blow-up solution of problem (1)-(4). It seems that the Rothe two-grid method maintains an optimal approximation of the solution even for large H.

4.3 Two-Dimensional Example

Now we apply Rothe two-grid method in two-dimensional case. It's worth to note that Theorems 1 and 2 are still hold. We show the efficiency of the algorithm for the following problem

$$\begin{aligned}
&u_t - \Delta u = u^2 + f(x,y,t), \quad (x,y) \in (0,1) \times (0,1), \quad 0 < t < T,\\
&u(0,y,t) = u(x,0,t) = u(1,y,t) = u(x,1,t) = 0, \quad 0 < t < T,\\
&u(x,y,0) = u_0(x,y), \quad (x,y) \in [0,1], \times [0,1].
\end{aligned}$$

We chose $f(x,y,t)$, such that $u(x,y,t) = e^{\lambda t}\sin(\pi x)\cos(\pi x)$ is the exact solution.

Example 5. Let V_H (V_h) is a finite element space defined on a coarse (fine) uniform triangulation (with mesh step size $H > h$) of $(0,1) \times (0,1)$. We chose

Table 4. Relative error of unbounded solution, $\tau_0 = 0.0006$

method	norm	T=0.1002	T=0.2497	T=0.2797	T=0.2971
Rothe two-grid method $V_H = 9$ $V_h = 65$	*max norm*	4.676127e-5	2.114390e-4	4.267488e-4	2.210406e-3
	L_2 *norm*	2.942311e-5	9.616564e-5	1.739034e-4	8.312089e-4
Semi-implicit scheme $V_H = 65$	*max norm*	1.519233e-4	5.145033e-4	1.082693e-3	5.747927e-3
	L_2 *norm*	9.157006e-5	2.502561e-4	4.183389e-4	1.171946e-3
Fully implicit scheme $V_H = 65$	*max norm*	3.681883e-5	1.848127e-4	4.382562e-4	2.938198e-3
	L_2 *norm*	1.428532e-5	6.929959e-5	1.614487e-4	9.581974e-4
Fully implicit scheme $V_H = 9$	*max norm*	1.408391e-2	5.253866e-2	1.179672e-1	4.715143e-1
	L_2 *norm*	1.024675e-2	3.522179e-2	7.066371e-2	2.5999441e-1

Table 5. Absolute error in max norm

	T=0.1	T=0.5	T=0.8	T=1	T=1.5
k=1/(4H)	2.836656e-3	3.050820e-3	2.231112e-3	1.812540e-3	1.084325e-3
k=1/(2H)	1.531089e-3	1.422669e-3	1.038991e-3	8.434673e-4	5.039424e-4
k=1/H	1.096637e-3	1.016676e-3	7.416862e-4	6.017683e-4	3.591656e-4
k=1/(4H^2)	5.354021e-4	4.902133e-4	1.305285e-4	8.773251e-5	5.122130e-5

$V_H = 9 \times 9$, $H = kh$, $\tau = 0.05$, $\lambda = -1$. In Table 5 we give the absolute error in maximal norm of the numerical solution for different k and T.

We can attain the necessary accuracy of the numerical solution even for large τ, by solving only a few fixed number nonlinear equations and increasing the linear equations, which we solve (i. e. decreasing the fine mesh step size).

5 Conclusions

• For bounded solutions of problem (1)-(4) the Rothe two-grid method gives more precisely results than Semi-implicit scheme.
• For blow-up solutions of problem (1)-(4) the Rothe two-grid method and Semi-implicit scheme have similar efficiency. The computational efforts of Semi-implicit scheme are less.
• Nevertheless the Newton method is very effective for solving the nonlinear systems (Fully implicit scheme), the approach of Xu is more attractive, because the procedure involves a nonlinear solve (based on Newton method) on the coarse

space and a linear solve on the fine space. The coarse mesh can be quite coarse and still maintains an optimal approximation [18].

References

1. Andreuci D., Gianni R. (1996), Global existence and blow up in a parabolic problem with nonlocal dynamical boundary conditions, *Adv. Diff. Equations 1*, **5**, pp. 729-752
2. Cimrák I. (2002), Numerical solution of degenerate convection-diffusion problem using broyden scheme, *Proceedings of ALGORITMY*, Conf. on Sci. Comput., pp.14-22
3. Courant R., Hilbert D. (1965), *Methoden der Mathematishen Physika, Springer, Berlin.*
4. Crank J. (1973),*The Mathematics of Diffusion. Clarendon Press, Oxford.*
5. Escher J.(1995), On the quasilinear behaviour of some semilinear parabolic problems, *Diff. Integral Eq.*, **8**, pp. 247-267.
6. Fila M. and Quittner P.(1997), Global solutions of the Laplace equation with a nonlinear dynamical boundary condition, *Math. Appl. Sci.*, **20**, pp. 1325-1333.
7. Fila M. and Quittner P.(1999), Large time behavior of solutions of semilinear parabolic equations with nonlinear dynamical boundary conditions, in: Topics in Nonlinear Analysis, *Progr.. Nonl. Diff. Eq. Appl.*, **35**, pp. 251-272
8. Gröger K.(1987), Initial boundary value problems from semiconductor device theory, *ZAMM*, **67**, pp. 345-355.
9. Kačur J. (1990), Nonlinear parabolic equations with mixed nonlinear and nonstationary boundary conditions, *Math Slovaca*, **30**, pp. 213-237
10. Kirane M.(1992), Blow-up for some equations with semilinear dynamical boundary conditions of parabolic and hyperbolic type, *Hokkaido Math J.* , **21** , pp. 221-229.
11. Koleva M. Vulkov L., On the blow-up of finite difference solutions to the heat-diffusion equation with semilinear dynamical boundary conditions, *Appl. Math. & Comp.*, in press.
12. Ladyzhenskaya A., Solnnikov A., Uraltseva N.(1968), *Linear and Quasi-Linear Equations of Parabolic Type. Translations of Mathematical Monographs , American Mathematical Society, Providence ,RI*, **23**.
13. Lapidus L., Amundson R. N.(1977), *Chemical Reactor Theory. Prentice - Hall, Englewood Cliffs.*
14. Nakagava T.(1976), Blowing up of a finite difference solution to $u_t = u_{xx} + u^2$, *Appl. Math. & Optimization*, **2**, pp. 337-350.
15. Samarskii A., Galakttionov V., Kurdyumov, Mikhailov A. (1995), *Blow-up in problems for quasilinear parabolic equations (Walter de Gruyter Trans., Expositionin Mathematics)* Nauka, Moskow, **19**, 1987 (Original work published in 1987), Berlin, 1995 (in Russian) .
16. Velazquez L. (1993), Blow-up for semilinear parabolic equations, *Recent Adv. in PDEs, Eds. M. A. Herrero, E. Zuazua, John Wiley and ons*, pp. 131-145.
17. Vold R. D., Vold M. J.(1983), *Colloid and Interface Chemistry. Addision - Wesley, Reading - Mass.*
18. Xu J. (1994), A novel two-grid method for semilinear elliptic equations, *SIAM J. Sci. Comput.*, **15, N 1**, pp.231-237 .

An Economic Method for Evaluation of Volume Integrals

Natalia T. Kolkovska
Institute of Mathematics and Informatics, Bulgarian Academy of Sciences
natali@math.bas.bg

Abstract. Approximations to the volume integrals with logarithmic kernels are obtained as solutions to finite difference schemes on the regular grid in a rectangular domain, which contains the domain of integration. Two types of right-hand sides in the finite difference equations are constructed - the first one includes some double and line integrals and the second one is fully discrete. The error estimates (optimal with respect to the smoothness of the volume integral) for both types of right-hand sides are obtained for volume integrals in appropriate Besov spaces.

Keywords: volume integrals, integral equations, particular solution, Poisson equation.

1 Description of the Method

Let D be a bounded simply connected domain in R^2 and let Γ denotes its boundary. We shall consider the Poisson equation

$$\triangle u(r) = f(r), \quad r \in D$$

with Dirichlet or Neumann boundary conditions on Γ.

Integral equation method, applied for solving this problem, requires a particular solution of the non-homogeneous equation $\triangle u(r) = f(r), r \in D$. It is well known, that the volume integral

$$u(r) = -\frac{1}{2\pi} \int_D f(s) \ln |r - s| \, ds, \quad r \in R^2 \tag{1}$$

is such a particular solution. Once this particular solution has been evaluated, the initial problem is reduced to the homogeneous equation in D with Dirichlet or Neumann boundary conditions on Γ. Then standard boundary element methods can be applied to solve the homogeneous elliptic problem.

As shown in [4] and [10], the direct evaluation of the volume integral with good accuracy can be very expensive and time-consuming procedure. The problem arises from the unboundness of the kernel near the point of evaluation r, whenever $r \in D$. Thus one needs special quadrature formulae for evaluation

Z. Li et al. (Eds.): NAA 2004, LNCS 3401, pp. 360–367, 2005.

of singular integrals. In [2] high order quadratures based on Chebishev polynomial approximation to f and fast multipole method are applied to evaluation of volume integral.

Alternative method for evaluation of volume and surface potentials in R^2 and R^3 is proposed independently by Lazarov and Mokin in [7] and by Mayo, in a series of papers [4], [8], [9], [10].

For the method, the domain D is imbedded into a rectangle Ω with sides parallel to the axes of a Cartesian coordinate system so that $dist\,(\Gamma, \partial\Omega) > 0$. Then the volume integral is considered as a C^1 solution to the differential equation

$$\begin{aligned} \triangle\, u(r) &= f(r), \quad r \in D, \\ \triangle u(r) &= 0, \quad r \in \Omega \backslash D. \end{aligned} \tag{2}$$

If one only needs some particular solution to the Poisson equation, then the boundary condition

$$u(r) = 0, \qquad r \in \partial\Omega \tag{3}$$

is applied. If the volume integral (1) has to be evaluated, then its values on $\partial\Omega$ are considered as a boundary condition instead of (3). These two cases can be treated in similar ways. Therefore we shall consider only (3) as a boundary condition.

Problem (2), (3) is discretized using an uniform mesh Ω_h in Ω with mesh sizes $h = (h_1, h_2)$ and the standard five-point discrete Laplacian Λ_h:

$$\begin{aligned} \Lambda_h u_h\,(r) &= \varphi\,(r)\,, \quad r \in \Omega_h, \\ u_h\,(r) &= 0, \quad r \in \partial\Omega_h. \end{aligned} \tag{4}$$

If the right-hand side φ is defined, then the solution u_h to (4) can be effectively found at $N_1 N_2$ points on the rectangular grid Ω_h ($N_i = O(h_i^{-1}), i = 1, 2$), applying fast solvers with $O(N_1 N_2 \ln(N_1 + N_2))$ operations.

Therefore, the direct computation on a regular grid of the volume integral u given by (1) is replaced with finding the solution u_h to a standard finite difference scheme (4).

The main difficulty is the proper construction (in order to ensure good approximation of u_h to u) of a discrete function φ at grid points near the boundary Γ. For example, a combination of Steklov averaging operators with the direct evaluation of u is applied in [7]. On the other hand, in [8] and [9] Mayo includes explicitly known jumps of different derivatives of u across Γ in the right-hand side of (4).

In this paper we follow the method proposed by Mayo. We suppose that the boundary Γ is sufficiently smooth and that the restrictions on D and on $\Omega \backslash D$ of the volume integral (1) belong to some Besov spaces $B^{\theta+2}_{p,p}$, $1/p < \theta < 1 + 1/p$, $1 < p < \infty$. (Some sufficient conditions are given in Lemma 1.) We utilize new right-hand sides φ in (4). The first right-hand side includes one double integral and two line integrals over Γ. If the parameter of smoothness θ is greater than

$2/p$, then the integrands are continuous functions on D and on Γ resp. and, using simple quadratures, we define the second, fully discrete right-hand side φ_1. The right-hand side, proposed by Mayo in [8], [9], can be obtained as a special case of the function φ_1.

In Theorem 1 and Theorem 2 we prove that the solution u_h to (4) with any of the above right-hand sides has $O(|h|^\theta)$ rate of convergence to the function u in the discrete Sobolev norm W_p^2, provided $u|_D \in B_{p,p}^{\theta+2}(D)$, $u|_{\Omega\backslash D} \in B_{p,p}^{\theta+2}(\Omega\backslash D)$ and $1/p < \theta < 1 + 1/p$. This rate of convergence is the best one can generally expect under condition imposed on the smoothness of u in D.

An important consequence of the the investigations in L_p norm is that one can easily get error estimates (optimal up to logarithmic factor) in the max-norm – see remarks at the end of paper.

The paper is organized as follows. In Section 2 we give preliminary notations and assertions. The jumps in the second and third derivatives of the volume integral across Γ, which are essential part of the right-hand side, are explicitly given in Lemma 1. In Section 3 we construct a right-hand side φ and investigate the error of the method. In Section 4 new, fully discrete right-hand sides are constructed. It is proved, that the error of the method is unaffected. In Section 5 the case of higher smoothness of the volume integral is briefly studied.

2 Preliminaries

We define function γ by $\gamma(r) = 1$, if $r \in D$, and $\gamma(r) = -1$, if $r \notin D$. We use $(x)_+^n$ for the truncated power notation: $(x)_+^n = x^n$, if $x > 0$ and $(x)_+^n = 0$ otherwise. At a point $\overline{r} \in \Gamma$ we denote the jump of u by $[u(\overline{r})] = u_e(\overline{r}) - u_i(\overline{r})$, where $u_e(\overline{r})$ and $u_i(\overline{r})$ are the limit values of u from the exterior $\Omega\backslash D$ and from the interior D resp.

Let $R(\lambda) = (X(\lambda), Y(\lambda))$, $\lambda \in [0, 2\pi)$ be a parametric representation of the curve Γ.

For $1 \le p \le \infty$ we use the Sobolev spaces W_p^k (k – integer) and the Besov spaces $B_{p,p}^\theta(D)$, $\theta > 0$ of functions, where p' is the conjugate index of p, i.e. $1/p + 1/p' = 1$. For general results on Sobolev and Besov spaces see e.g. [1].

For mesh functions we follow notations from [11]. For $r = (x, y)$ we define the second finite difference Λ_1 in x direction by

$$\Lambda_1 u(r) = \left(u(x + h_1, y) - 2u(x, y) + u(x - h_1, y)\right) / h_1^2$$

and by analogy - the second finite difference Λ_2 in y direction. Then $\Lambda_h = \Lambda_1 + \Lambda_2$ is the five-point discrete Laplacian. We define the discrete Sobolev space $W_p^2(\Omega_h)$ of mesh functions u, which vanish on $\partial\Omega_h$, with the norm

$$\|u\|_{W_p^2(\Omega_h)} = \|\Lambda_1 u\|_{L_p(\Omega_h)} + \|\Lambda_2 u\|_{L_p(\Omega_h)} + \|u\|_{L_p(\Omega_h)}.$$

Let E be the basic rectangle $\{(s,t) : |s| \le h_1, |t| \le h_2\}$. For $r = (x, y) \in \Omega_h$ define E_r to be the union of the four cells with the common vertex r: $E_r = \{(s,t) : |s - x| \le h_1, |t - y| \le h_2\}$. Set $\Gamma_r = \Gamma \cap E_r$.

We define the set Ω_h^1 of "regular" grid points r as those grid points, for which all points from E_r are on the same side of Γ: $\Omega_h^1 = \{r \in \Omega_h : E_r \cap \Gamma = \oslash\}$. We denote the set of the remaining grid points in Ω_h by Ω_h^2 and call the points in Ω_h^2 "irregular". All points from Ω_h^2 lie in a strip of width $O(|h|)$ around Γ.

If $\Gamma_r \cap \{(x, y+t) : |t| \leq h_2\} \neq \oslash$, then denote this intersection point by $(x, \overline{y})$, thus $(x, \overline{y}) = \Gamma_r \cap \{(x, y+t) : |t| \leq h_2\}$. In a similar way, if $\Gamma_r \cap \{(x+s, y) : |s| \leq h_1\} \neq \oslash$, then we denote this point by $(\overline{\overline{x}}, y)$, thus $(\overline{\overline{x}}, y) = \Gamma_r \cap \{(x+s, y) : |s| \leq h_1\}$. If some of the points $\overline{\overline{x}}$ or $\overline{y}$ does not exist, in the next formulae we take $\overline{\overline{x}} = x + 2h_1$ or $\overline{y} = y + 2h_2$ resp.

In the sequel by B we denote arbitrary function $B \in W^1_{\infty,\infty}(R^2)$, supported on the set E, which is non-negative, even (with respect to s and t) and such that

$$\int_{-\infty}^{\infty} B(s,t)dt = h_2\Big(1 - \frac{|s|}{h_1}\Big)_+, s \in R^1; \quad \int_{-\infty}^{\infty} B(s,t)ds = h_1\Big(1 - \frac{|t|}{h_2}\Big)_+, t \in R^1.$$

A simple example of such function is $B(s,t) = \left(1 - \frac{|s|}{h_1}\right)_+ \left(1 - \frac{|t|}{h_2}\right)_+$.

Throughout this paper C will denote positive constants independent of the functions like u, f, φ, of the mesh size h and of the norm index p. The values of C are not necessarily the same at different occurrences.

In the next lemma we summarize some properties of the volume integrals:

Lemma 1. *Assume* $f \in B^{\theta}_{p,p}(D)$, $\Gamma \in B^{\theta+1/p'}_{p,p}[0, 2\pi)$, $\theta > 2/p$. *Then:*

(a) *(see [12], [13]) The volume integral u given by (1) and its first derivatives in x and y directions are continuous functions in R^2. The restiction of u on D is in $B^{\theta+2}_{p,p}(D)$ and the restriction of u on $\Omega \backslash D$ is in $B^{\theta+2}_{p,p}(\Omega \backslash D)$.*

(b) *Using the density f and the parametrization $r = R(\lambda) \in \Gamma$ of Γ, we have the following representations for the jumps of the second and third derivatives of u:*

$$\left[\frac{\partial^2 u}{\partial x^2}\right] = -\frac{Y'^2}{X'^2 + Y'^2} f, \quad \left[\frac{\partial^2 u}{\partial y^2}\right] = -\frac{X'^2}{X'^2 + Y'^2} f.$$

If $\theta > 1 + 2/p$, then the jumps in the third derivatives are:

$$\left[\frac{\partial^3 u}{\partial x^3}\right] = kf\frac{Y'(Y'^2 - 3X'^2)}{(X'^2 + Y'^2)^{3/2}} - \frac{\partial f}{\partial y}\frac{2X'Y'^3}{(X'^2 + Y'^2)^2} - \frac{\partial f}{\partial x}\frac{Y'^2(Y'^2 + 3X'^2)}{(X'^2 + Y'^2)^2},$$

$$\left[\frac{\partial^3 u}{\partial y^3}\right] = kf\frac{X'(X'^2 - 3Y'^2)}{(X'^2 + Y'^2)^{3/2}} - \frac{\partial f}{\partial y}\frac{X'^2(X'^2 + 3Y'^2)}{(X'^2 + Y'^2)^2} - \frac{\partial f}{\partial x}\frac{2Y'X'^3}{(X'^2 + Y'^2)^2},$$

where k is the curvature of Γ.

The formulae for the second derivatives jumps are obtained in [3], [9]. The proof of the formulae for the third derivatives jumps are lengthily but straightforward.

3 Finite Difference Method

We use exact integral representations of the discrete Laplacian, valid for any function u with restrictions in $B^{\theta+2}_{p,p}(D)$ and in $B^{\theta+2}_{p,p}(\Omega\backslash D)$, $1/p < \theta < 1+1/p$. At regular grid points $r \in \Omega^1_h$ the representation is:

$$h_1 h_2 \Lambda_h u(r) = \iint\limits_E \Delta u(x+s,y+t)B(s,t)dsdt$$

$$+ \iint\limits_E B(s,t)\left(\frac{\partial^2 u}{\partial x^2}(x+s,y) + \frac{\partial^2 u}{\partial y^2}(x,y+t) - \Delta u(x+s,y+t)\right)dsdt.$$

At irregular points $r \in \Omega^2_h$ the Laplacian is spread over E_r as a sum of double integrals over the domains $E_r \cap D$ and $E_r \cap (\Omega\backslash D)$ and some line integrals over Γ_r – see [5].

Including the leading terms of these representations into the right-hand side φ, we define function φ at the point $r=(x,y)$ by:

$$\left|\begin{array}{ll}\varphi(r) = 0, \quad r \in \Omega^1_h \cap (\Omega\backslash D), \\ h_1 h_2 \varphi(r) = \iint\limits_E f(x+s,y+t)B(s,t)dsdt, \quad r \in \Omega^1_h \cap D, \\ h_1 h_2 \varphi(r) = \iint\limits_{E_r \cap D} f(\alpha,\beta)B(\alpha-x,\beta-y)d\alpha d\beta \\ -\int_{\Gamma_r} G_1(\lambda)\left[\frac{\partial^2 u}{\partial x^2}(R(\lambda))\right]d\lambda - \int_{\Gamma_r} G_2(\lambda)\left[\frac{\partial^2 u}{\partial y^2}(R(\lambda))\right]d\lambda, r \in \Omega^2_h,\end{array}\right. \tag{5}$$

where

$$G_1(\lambda) = \gamma(X(\lambda),y)X'(\lambda)\int_{|Y(\lambda)-y|}^{h_2} B(X(\lambda)-x,t)\,dt,$$

$$G_2(\lambda) = \gamma(x,Y(\lambda))Y'(\lambda)\int_{|X(\lambda)-x|}^{h_1} B(s,Y(\lambda)-y)\,ds.$$

The accuracy of the finite difference method (4) with the above defined right-hand side φ is given by

Theorem 1. *Let u be the solution to (2) and (3). Suppose $u|_D \in B^{\theta+2}_{p,p}(D)$ and $u|_{\Omega\backslash D} \in B^{\theta+2}_{p,p}(\Omega\backslash D)$, $1/p < \theta < 1+1/p$. If $\theta = 1$ and $p > 2$, suppose additionally $u \in W^3_p(D)$ and $u \in W^3_p(\Omega\backslash D)$.*

Let u_h be the solution to (4) with the right-hand side φ given by (5). Then the following error estimate holds

$$\|u-u_h\|_{W^2_p(\Omega_h)} \le C\frac{p^8}{(p-1)^4}|h|^\theta \left\{\|u\|_{B^{\theta+2}_{p,p}(D)} + \|u\|_{B^{\theta+2}_{p,p}(\Omega\backslash D)}\right\}.$$

For the proof of this theorem, the error $z = u - u_h$ is considered as the solution to the problem

$$\Lambda_h z(r) = \psi(r), \quad r \in \Omega_h; \qquad z(r) = 0, \quad r \in \partial\Omega_h,$$

where $\psi = \Lambda_h u - \varphi$. We apply the a-priori inequality

$$\|z\|_{W_p^2(\Omega_h)} \leq C \frac{p^8}{(p-1)^4} \|\psi\|_{L_p(\Omega_h)},$$

and evaluate the discrete L_p norm of ψ separately in Ω_h^1 and in Ω_h^2.

4 Fully Discrete Right-Hand Side

Assume $f \in B_{p,p}^{\theta}(D)$, $2/p < \theta < 1 + 1/p$. Then $f \in C(D)$ and the second derivatives jumps of u in (5) are also continuous functions. Therefore one can apply simple one-point quadrature formulae to all integrals in the definition of φ and to obtain a fully discrete right-hand side φ_1:

$$\left|\begin{array}{ll} \varphi_1(r) = 0, & r \in \Omega_h^1 \cap (\Omega \backslash D), \\ \varphi_1(r) = f(r), & r \in \Omega_h^1 \cap D, \\ h_1 h_2 \varphi_1(r) = f(r^*) \displaystyle\iint_{E_r \cap D} B(\alpha - x, \beta - y) d\alpha d\beta & \\ -\left[\dfrac{\partial^2 u}{\partial x^2}\left(R(\overline{\overline{\lambda}})\right)\right] \displaystyle\int_{\Gamma_r} G_1(\lambda) d\lambda - \left[\dfrac{\partial^2 u}{\partial y^2}\left(R(\overline{\lambda})\right)\right] \displaystyle\int_{\Gamma_r} G_2(\lambda) d\lambda, & r \in \Omega_h^2. \end{array}\right. \tag{6}$$

The only restrictions on the parameters r^*, $\overline{\lambda}$ and $\overline{\overline{\lambda}}$ in (6) are $r^* \in E_r \cap D$, $R(\overline{\lambda}) \in \Gamma_r$ and $R(\overline{\overline{\lambda}}) \in \Gamma_r$. The integrals in (6) are independent of f and u. They depend on the relative position of r and Γ and can be simply computed with high order of accuracy.

The error estimate from Theorem 1 remains true:

Theorem 2. *Let u be the solution to (2) and (3). Assume $u|_D \in B_{p,p}^{\theta+2}(D)$, $u|_{\Omega\backslash D} \in B_{p,p}^{\theta+2}(\Omega\backslash D)$ and $2/p < \theta < 1 + 1/p$. If $\theta = 1$ and $p > 2$, assume additionally $u \in W_p^3(D)$ and $u \in W_p^3(\Omega\backslash D)$.*

Let u_h^1 be the solution to (4) with right-hand side φ_1 given by (6). Then the following error estimate holds

$$\left\|u - u_h^1\right\|_{W_p^2(\Omega_h)} \leq C \frac{p^8}{(p-1)^4} |h|^{\theta} \left\{\|u\|_{B_{p,p}^{\theta+2}(D)} + \|u\|_{B_{p,p}^{\theta+2}(\Omega\backslash D)}\right\}.$$

Note that the discretization of the integrals in (5) by the simplest one-point quadrature provides sufficient accuracy for our purposes.

If we make some special choices of parameters r^*, $\overline{\lambda}$ and $\overline{\overline{\lambda}}$ in the third equation in (6), we can obtain the well-known right-hand side of A. Mayo [9]:

$$\begin{aligned} \varphi_1^M(r) = f(x,y) + \gamma(r) \left[\frac{\partial^2 u}{\partial x^2}\left(\overline{\overline{x}}, y\right)\right] \frac{\left(h_1 - \left|\overline{\overline{x}} - x\right|\right)_+^2}{2h_1^2} \\ + \gamma(r) \left[\frac{\partial^2 u}{\partial y^2}(x, \overline{y})\right] \frac{\left(h_2 - |\overline{y} - y|\right)_+^2}{2h_2^2}, \quad r \in \Omega_h^2. \end{aligned} \tag{7}$$

5 Higher Smoothness Case

The same scheme is applied when we treat the higher smoothness case $f \in B^{\theta}_{p,p}(D)$, $\Gamma \in B^{\theta+1/p'}_{p,p}[0,2\pi)$, $1+2/p<\theta\leq 2$: first, we use integral representations of the discrete Laplacian from [6] valid for this parameters of smoothness. Then we include the leading terms into the right hand side of (4). At the end we simplify the right hand side by using one-point quadratures. In each case we prove $O(|h|^{\theta})$ rate of convergence of the approximate function u_h to the volume integral u in $W^2_p(\Omega_h)$ norm. The full treatment of this case will be given in another paper. Here we consider the special, but important for the practice case, where the right-hand side φ_2 is defined by:

$$\left|\begin{array}{l} \varphi_2(r)=0, \qquad r\in\Omega^1_h\cap(\Omega\backslash D),\\ \varphi_2(r)=f(x,y), \qquad r\in\Omega^1_h\cap D,\\ \varphi_2(r)=\varphi^M_1(r)+\gamma(r)\mathrm{sign}(\overline{\overline{x}}-x)\left[\frac{\partial^3 u}{\partial x^3}(\overline{\overline{x}},y)\right]\frac{(h_1-|\overline{\overline{x}}-x|)^3_+}{6h_1^2}\\ \qquad +\gamma(r)\mathrm{sign}(\overline{y}-y)\left[\frac{\partial^3 u}{\partial y^3}(x,\overline{y})\right]\frac{(h_2-|\overline{y}-y|)^3_+}{6h_2^2}, \qquad r\in\Omega^2_h. \end{array}\right. \tag{8}$$

The above right-hand side was proposed by Mayo in [9]. For the accuracy of this finite difference scheme we prove the following estimate.

Theorem 3. *Let u be the solution to (2) and (3). Suppose $u|_D\in B^{\theta+2}_{p,p}(D)$, $u|_{\Omega\backslash D}\in B^{\theta+2}_{p,p}(\Omega\backslash D)$ and $1+2/p<\theta\leq 2$. If $\theta=2$ and $p>2$, suppose additionally $u\in W^4_p(D)$ and $u\in W^4_p(\Omega\backslash D)$.*

Let u^2_h be the solution to (4) with the right hand side φ_2 given by (8). Then the following error estimate holds

$$\left\|u-u^2_h\right\|_{W^2_p(\Omega_h)}\leq C\frac{p^8}{(p-1)^4}|h|^{\theta}\left\{\|u\|_{B^{\theta+2}_{p,p}(D)}+\|u\|_{B^{\theta+2}_{p,p}(\Omega\backslash D)}\right\}.$$

6 Remarks

1. An attractive feature of the error estimates proved in Theorems 1, 2 and 3 is their rate of convergence in the $W^2_p(\Omega_h)$ mesh norm. The rate is optimal with respect to the smoothness in D of the exact solution u although the second derivatives of u have jumps across the curve Γ.

2. The presence of the factor $\frac{p^8}{(p-1)^4}$ in the inequalities in Theorems 1, 2 and 3 allows the proof of max-norm estimate with additional logarithmic factor. For example, under the conditions of Theorem 1, the following error estimate in the max-norm holds:

$$\max_{r\in\Omega_h}\left(|\Lambda_1(u-u_h)(r)|+|\Lambda_2(u-u_h)(r)|\right)\leq$$
$$C|h|^{\theta}\left(\ln\frac{1}{h}\right)^4\left\{\|u\|_{B^{\theta+2}_{\infty,\infty}(D)}+\|u\|_{B^{\theta+2}_{\infty,\infty}(\Omega\backslash D)}\right\}.$$

3. The extra condition in Theorems 1, 2 and 3 in the case of integer parameter of smoothness θ in the Besov space and $p>2$ is needed because $B^3_{p,p}(D)\not\hookrightarrow$

$W_p^3(D)$, i.e. in this case functions from the Besov space $B_{p,p}^3(D)$ may not have third derivatives in $L_p(D)$ (see [1], Theorem 18.9).

4. A variety of right-hand sides φ are discussed in this article. From practical point of view the appropriate scheme is chosen as follows. Fix p. Given Γ and f, we determine their smoothness parameter θ in therms of Besov spaces. Then Lemma 1 implies that $u|_D \in B_{p,p}^{\theta+2}(D)$ and $u|_{\Omega\backslash D} \in B_{p,p}^{\theta+2}(\Omega\backslash D)$. The inequality $2/p < \theta \leq 1 + 1/p$ gives the right-hand sides (6) and (7) and $1 + 2/p < \theta \leq 2$ – the right-hand side (8). In the case $1/p < \theta \leq 2/p$ the right-hand side is (5), but there are functions u, for which the line integrals in (5) exist, but they cannot be approximated by quadratures, because the integrands are essentially unbounded.

References

1. Besov, O., Il'in, V., Nikolskii, S.: Integral representations of functions and imbedding theorems. New York, John Wiley & Sons (1979)
2. Ethridge, F., Greengard, L.: A new fast-multipole accelerated Poisson solver in two dimensions. Siam J. Sci. Comp. **23** (2001) 741–760
3. Friedman, A.: Mathematics in industrial problems. part 5, Springer (1992)
4. Greenbaum, A., Mayo, A.: Rapid parallel evaluation of integrals in potential theory on general three-dimensional regions. J Comput. Physics **145** (1998) 731–742
5. Kolkovska, N.: Numerical methods for computation of the double layer logarithmic potential. Lecture Notes in Comp. Science **1196** (1997) 243–249
6. Kolkovska, N.: A finite difference scheme for computation of the logarithmic potential. in: Finite difference methods, Theory and Applications, Nova Science Publishers (1999) 139–144
7. Lazarov, R., Mokin, Yu.: On the computation of the logarithmic potential. Soviet Math. Dokl. **272** (1983) 320–323
8. Mayo, A.: The fast solution of Poisson's and the biharmonic equations on irregular regions. SIAM J. Numer. Anal. **21** (1984) 285–299
9. Mayo, A.: The rapid evaluation of volume integrals of potential theory on general regions. J Comput. Physics **100** (1992) 236–245
10. McKenney, A., Greengard, L., Mayo, A.: A fast Poisson solver for complex geometries. J Comp Physics **118** (1995) 348–355
11. Samarskii, A., Lazarov, R., Makarov, V.: Difference schemes for differential equations having generalized solutions. Moscow, Vysshaya Shkola Publishers, (1987) (in Russian)
12. Triebel, H.: Theory of function spaces. Birkhauser (1983)
13. Vladimirov, V.: Equations in mathematical physics. Moscow, Nauka (1976) (in Russian)

Sensitivity Analysis of Generalized Lyapunov Equations

M.M. Konstantinov[1], P.Hr. Petkov[2], and N.D. Christov[2]

[1] University of Architecture and Civil Engineering,
1421 Sofia, Bulgaria
mmk_fte@uacg.bg
[2] Technical University of Sofia,
1000 Sofia, Bulgaria
{php, ndchr}@tu-sofia.bg

Abstract. The sensitivity of generalized matrix Lyapunov equations relative to perturbations in the coefficient matrices is studied. New local and non-local perturbation bounds are obtained.

1 Introduction

In this paper we study the sensitivity of the generalized Lyapunov equations (GLE) arising in the theory of linear descriptor systems. First a new, non-linear local bound is obtained for the perturbation in the solution of GLE, which is less conservative than the existing condition number based local perturbation bounds [1].

The local perturbation bounds, however, are valid only asymptotically and their application for possibly small but nevertheless finite perturbations in the data requires additional justification. The disadvantage of the local bounds is overcome using the techniques of non-local perturbation analysis [2] which is aimed at two things simultaneously: first to show that a solution of the perturbed equation exists, and second to find a non-local perturbation bound for GLE.

The following notations are used later on: $\mathcal{R}^{m\times n}$ – the space of real $m \times n$ matrices; $\mathcal{R}^n = \mathcal{R}^{n\times 1}$; I_n – the unit $n \times n$ matrix; $A^\top$ – the transpose of A; $\|A\|_2 = \sigma_{\max}(A)$ – the spectral norm of A, where $\sigma_{\max}(A)$ is the maximum singular value of A; $\|A\|_{\mathrm{F}} = \sqrt{\mathrm{tr}(A^\top A)}$ – the Frobenius norm of A; $\|.\|$ is any of the above norms; $\mathrm{vec}(A) \in \mathcal{R}^{mn}$ – the column-wise vector representation of $A \in \mathcal{R}^{m\times n}$; $\Pi \in \mathcal{R}^{n^2\times n^2}$ – the (vec)permutation matrix, i.e. $\mathrm{vec}(A^\top) = \Pi\mathrm{vec}(A)$ for $A \in \mathcal{R}^{n\times n}$. The notation ":=" stands for "equal by definition".

2 Statement of the Problem

Consider the GLE

$$\mathcal{L}(X, S, Q) := A^\top XE + E^\top XA + Q = 0 \tag{1}$$

Z. Li et al. (Eds.): NAA 2004, LNCS 3401, pp. 368–374, 2005.

where $A, E, Q \in \mathcal{R}^{n.n}$ are given matrix coefficients such that rank$E = n$, $Q = Q^\top \geq 0$, and $X \in \mathcal{R}^{n.n}$ is the unknown matrix. We assume that $E^{-1}A$ is a stable matrix, which guarantees the existence of an unique nonnegative solution $X = X_0$ of (1). Further on we denote the pair (E, A) by S. Let Σ be the set of all S which satisfy the above conditions.

Let ΔE, ΔA and $\Delta Q = \Delta Q^\top$ be perturbations in E, A and Q, respectively. Denote $\Delta S = (\Delta E, \Delta A)$ and suppose that $S + \Delta S \in \Sigma$. Then the perturbed GLE

$$\mathcal{L}(X, S + \Delta S, Q + \Delta Q) = 0 \tag{2}$$

has an unique solution $X = X_0 + \Delta X$.

The above restrictions on the perturbation ΔS are too weak and in particular it is possible that for perturbations ΔS_1 with smaller norms, the perturbed pair $S + \Delta S_1$ will not belong to Σ. For this reason we restrict the class of admissible perturbations in the following way.

Denote by $\Gamma \subset \mathcal{R}^2_+$ the set of all pairs (α, ε) such that $\Delta_E < \varepsilon$, $\Delta_A < \alpha$ implies invertibility of $E + \Delta E$ and stability of $(E + \Delta E)^{-1}(A + \Delta A)$. Further on we suppose that $(\Delta_E, \Delta_A) \in \Gamma$.

The aim of the sensitivity analysis of GLE is to find local and non-local perturbation bounds for $\Delta_X := \|\Delta X\|_{\mathrm{F}}$ as functions of $\Delta := [\Delta_E, \Delta_A, \Delta_Q]^\top \in \mathcal{R}^3_+$, where $\Delta_M := \|\Delta M\|_{\mathrm{F}}$.

Note that it is natural to work with the Frobenius norms of the perturbations in the data and the solution. Indeed, often it is not clear how to determine the spectral norm in the perturbation of a coefficient matrix which is due to uncertainties in the data and/or rounding errors during the computational process. At the same time when we have uncertainties in the data it is reasonable to identify a matrix A with its vector representation vec(A) and to work with the Euclidean norm $\|\mathrm{vec}(\Delta A)\|_2$ of the vector perturbation vec(ΔA), which is equal to the Frobenius norm $\|\Delta A\|_{\mathrm{F}}$ of the matrix perturbation ΔA.

The perturbation problem is to find a bound

$$\Delta_X \leq f(\Delta),\ \Delta \in \Omega \subset \mathcal{R}^3_+ \tag{3}$$

where Ω is a given set and f is a continuous function, non-decreasing in each of its arguments and satisfying $f(0) = 0$.

A first order local bound

$$\Delta_X \leq f_1(\Delta) + O(\|\Delta\|^2),\ \Delta \to 0,$$

shall be first derived, which will then be incorporated in the non-local bound (3). The inclusion $\Delta \in \Omega$ also guarantees that the perturbed equation (2) has an unique solution $X = X_0 + \Delta X$.

3 Local Perturbation Analysis

Consider first the conditioning of the GLE (1). Since $\mathcal{L}(X, S, Q) = 0$, the perturbed equation (2) may be written as

$$\mathcal{L}_X(\Delta X) = -\mathcal{L}_A(\Delta A) - \mathcal{L}_E(\Delta E) - \Delta Q - G(\Delta X, \Delta S) \tag{4}$$

where

$$\mathcal{L}_X(Z) = A^\top ZE + E^\top ZA$$

$$\mathcal{L}_A(Z) = E^\top X_0 Z + Z^\top X_0 E$$

$$\mathcal{L}_E(Z) = A^\top X_0 Z + Z^\top X_0 A$$

are the Fréchet derivatives of $\mathcal{L}(X, S, Q)$ in X, A and E, respectively, and $G(\Delta X, \Delta S)$ contains second and higher order terms in ΔX, ΔS.

Under the assumptions made, the linear operator $\mathcal{L}_X(.)$ is invertible and (4) yields

$$\Delta X = \Psi(\Delta X, \Delta S, \Delta Q) \tag{5}$$

where

$$\Psi(Z, \Delta S, \Delta Q) := -\mathcal{L}_X^{-1} \circ \mathcal{L}_A(\Delta A) - \mathcal{L}_X^{-1} \circ \mathcal{L}_E(\Delta E) - \mathcal{L}_X^{-1}(\Delta Q)$$
$$- \mathcal{L}_X^{-1}(G(\Delta X, \Delta S)).$$

The relation (5) gives

$$\Delta_X \leq K_A \Delta_A + K_E \Delta_E + K_Q \Delta_Q + \mathrm{O}(\|\Delta\|^2), \ \Delta \to 0 \tag{6}$$

where

$$K_A = \| \mathcal{L}_X^{-1} \circ \mathcal{L}_A \|, \ \ K_E = \| \mathcal{L}_X^{-1} \circ \mathcal{L}_E \|, \ \ K_Q = \| \mathcal{L}_X^{-1} \|$$

are the absolute condition numbers of the GLE (1). Here $\|\mathcal{L}\|$ is the norm of the operator $\mathcal{L}$, induced by the F-norm.

The calculation of the condition numbers K_A, K_E, K_Q is straightforward. Denote by L_X, L_A, L_E the matrix representations of the operators $\mathcal{L}_X(\cdot)$, $\mathcal{L}_A(\cdot)$, $\mathcal{L}_E(\cdot)$:

$$L_X = E^\top \otimes A^\top + A^\top \otimes E^\top$$

$$L_A = (E^\top X_0 \otimes I_n)\Pi + I_n \otimes E^\top X_0$$

$$L_E = I_n \otimes A^\top X_0 + (A^\top X_0 \otimes I_n)\Pi.$$

Then

$$K_A = \| L_X^{-1} L_A \|_2, \ \ K_E = \| L_X^{-1} L_E \|_2, \ \ K_Q = \| L_X^{-1} \|_2.$$

Local estimates, based on the condition numbers, may eventually produce pessimistic results. At the same time it is possible to derive local, first order homogeneous estimates, which are better in general.

The operator equation (5) may be written in a vector form as

$$\mathrm{vec}(\Delta X) = N_A \mathrm{vec}(\Delta A) + N_E \mathrm{vec}(\Delta E) + N_Q \mathrm{vec}(\Delta Q) - L_X^{-1} \mathrm{vec}(G(\Delta X, \Delta S)) \tag{7}$$

where

$$N_A := -L_X^{-1} L_A, \ \ N_E := -L_X^{-1} L_E, \ \ N_Q := -L_X^{-1}.$$

It is easy to show that the condition number based estimate (6) is a corollary of (7). Indeed, since $\|\mathrm{vec}(\Delta Z)\|_2 \leq \Delta_Z$, one has

$$\begin{aligned}\Delta_X = \|\mathrm{vec}(\Delta X)\|_2 &\leq \mathrm{est}_1(\Delta, N) + \mathrm{O}(\|\Delta\|^2)\\ &:= \|N_A\|_2\Delta_A + \|N_E\|_2\Delta_E + \|N_Q\|_2\Delta_Q + \mathrm{O}(\|\Delta\|^2)\\ &= K_A\Delta_A + K_E\Delta_E + K_Q\Delta_Q + \mathrm{O}(\|\Delta\|^2),\ \Delta \to 0\end{aligned}$$

where $N := [N_1, N_2, N_3] := [N_A, N_E, N_Q]$.

Relation (7) also gives

$$\Delta_X \leq \mathrm{est}_2(\Delta, N) + \mathrm{O}(\|\Delta\|^2) := \|N\|_2\|\Delta\|_2 + \mathrm{O}(\|\Delta\|^2),\ \Delta \to 0.$$

The bounds $\mathrm{est}_1(\Delta, N)$ and $\mathrm{est}_2(\Delta, N)$ are alternative, i.e. which one is less depends on the particular value of Δ.

There is also a third bound, which is always less than or equal to $\mathrm{est}_1(\Delta, N)$. We have

$$\Delta_X \leq \mathrm{est}_3(\Delta, N) := \sqrt{\Delta^\top S(N)\Delta} + \mathrm{O}(\|\Delta\|^2),\ \Delta \to 0$$

where $S(N)$ is the 3×3 matrix with elements $s_{ij}(N) = \|N_i^\top N_j\|_2$. Since

$$\|N_i^\top N_j\|_2 \leq \|N_i\|_2\|N_j\|_2$$

we get $\mathrm{est}_3(\Delta, N) \leq \mathrm{est}_1(\Delta, N)$.

Hence we have the overall estimate

$$\Delta_X \leq \mathrm{est}(\Delta, N) + \mathrm{O}(\|\Delta\|^2),\ \Delta \to 0 \tag{8}$$

where

$$\mathrm{est}(\Delta, N) := \min\{\mathrm{est}_2(\Delta, N), \mathrm{est}_3(\Delta, N)\}. \tag{9}$$

The local bound $\mathrm{est}(\Delta, N)$ in (8), (9) is a non-linear, first order homogeneous and piece-wise real analytic function in Δ.

4 Non-local Perturbation Analysis

The local perturbation bounds are usually used neglecting the terms of order $\mathrm{O}(\|\Delta\|^2)$. Unfortunately, it is usually impossible to say, having a small but a finite perturbation Δ, whether the neglected terms are indeed negligible. We can not even claim that the magnitude of the neglected terms is less, or of the order of magnitude of the local bound. Even for simple linear equations the local bound may always underestimate the actual perturbation in the solution for a class perturbations in the data. Moreover, for some critical values of the perturbations in the coefficient matrices the solution may not exist (or may go to infinity when these critical values are approached). Nevertheless, even in such cases the local estimates will still produce a 'bound' for a very large or even for a non-existing solution, which is a serious drawback.

The disadvantages of the local estimates may be overcome using the techniques of non-linear perturbation analysis. As a result we get a domain $\Omega \subset \mathcal{R}_+^3$ and a non-linear function $f : \Omega \to \mathcal{R}_+$ such that $\Delta_X \leq f(\Delta)$ for all $\Delta \in \Omega$. In connection with this we would like to emphasis two important issues. First, the inclusion $\Delta \in \Omega$ guarantees that the perturbed equation has a solution (this is an independent and essential point). And second, the estimate $\Delta_X \leq f(\Delta)$ is rigorous, i.e. it is true for perturbations with $\Delta \in \Omega$, unlike the local bounds. However, in some cases the non-local bounds may not exist or may be pessimistic.

The operator equation (5) may be rewritten as

$$\Delta X = \Psi_0(\Delta S, \Delta Q) + \Psi_1(\Delta X, \Delta S) \tag{10}$$

where

$$\Psi_0(\Delta S, \Delta Q) := -\mathcal{L}_X^{-1}(G_0(\Delta S, \Delta Q)), \quad \Psi_1(\Delta X, \Delta S) := -\mathcal{L}_X^{-1}(G_1(\Delta X, \Delta S))$$

and

$$G_0(\Delta S, \Delta Q) = \Delta Q + R_1(X, \Delta S) + R_2(X, \Delta S)$$

$$G_1(\Delta X, \Delta S) = R_1(\Delta X, \Delta S) + R_2(\Delta X, \Delta S).$$

Here $R_k(\cdot, \Delta S)$ are linear operators of asymptotic order k relative to ΔS, $\Delta S \to 0$, determined from

$$R_1(Z, \Delta S) := \Delta A^\top Z E^\top + A^\top Z \Delta E^\top + \Delta E^\top Z A + E^\top Z \Delta A$$

and

$$R_2(Z, \Delta S) := \Delta A^\top Z \Delta E + \Delta E^\top Z \Delta A.$$

Suppose that $\|Z\|_{\mathrm{F}} \leq \rho$. Then we have

$$\|\Psi_0(\Delta S, \Delta Q)\|_{\mathrm{F}} \leq a_0(\Delta), \quad \|\Psi_1(Z, \delta S)\|_{\mathrm{F}} \leq a_1(\Delta)\rho$$

where

$$a_0(\Delta) := a_{01}(\Delta) + a_{02}(\Delta), \quad a_1(\Delta) := a_{11}(\Delta) + a_{12}(\Delta). \tag{11}$$

The quantities $a_{ik}(\Delta)$ are of asymptotic order $\mathrm{O}(\|\Delta\|^k)$ for $\Delta \to 0$ and are determined as

$$a_{01}(\Delta) := \mathrm{est}(\Delta, N), \quad a_{02}(\Delta) := 2K_Q\|X_0\|_2\Delta_A\Delta_E \tag{12}$$

$$a_{11}(\Delta) := \|L_X^{-1}(I_{n^2} + \Pi_{n^2})(E^\top \otimes I_n)\|_2\Delta_A + \|L_X^{-1}(I_{n^2} + \Pi_{n^2})(A^\top \otimes I_n)\|_2\Delta_E$$

$$a_{12}(\Delta) := 2K_Q\Delta_A\Delta_E.$$

Let $\|Z\|_{\mathrm{F}}, \|\widetilde{Z}\|_{\mathrm{F}} \leq \rho$. The Lyapunov majorant for equation (10) is a function $(\rho, \Delta) \mapsto h(\rho, \Delta)$, defined on a subset of $\mathcal{R}_+ \times \mathcal{R}_+^3$ and satisfying the conditions

$$\|\Psi(Z, \Delta S, \Delta Q)\|_{\mathrm{F}} \leq h(\rho, \Delta)$$

and

$$\|\Psi(Z, \Delta S, \Delta Q) - \Psi(\widetilde{Z}, \Delta S, \Delta Q)\|_{\mathrm{F}} \leq h'_\rho(\rho, \Delta)\|Z - \widetilde{Z}\|_{\mathrm{F}}.$$

According to the above considerations, the Lyapunov majorant is linear and is determined from

$$h(\rho, \Delta) = a_0(\Delta) + a_1(\Delta)\rho.$$

In this case the fundamental equation $h(\rho, \Delta) = \rho$ for determining the non-local bound $\rho = \rho(\Delta)$ for Δ_X gives

$$\Delta_X \leq f(\Delta) := \frac{a_0(\Delta)}{1 - a_1(\Delta)}, \quad \Delta \in \Omega \tag{13}$$

where

$$\Omega := \{\Delta \succeq 0 : a_1(\Delta) < 1\} \subset \mathcal{R}^3_+. \tag{14}$$

Inequality (13) in view of (11), (12) and (14) gives the nonlinear nonlocal perturbation bound for the solution of the generalized matrix Lyapunov equation (1).

5 Numerical Example

Consider the GLE with matrices

$$A = \begin{bmatrix} 5 & 8 & -8 & 18 \\ -3 & -6 & 6 & -7 \\ 2 & 2 & -5 & 8 \\ -1 & -2 & -1 & -4 \end{bmatrix}, \quad E = \begin{bmatrix} -5 & 1 & -1 & -3 \\ 3 & 0 & 0 & 1 \\ -2 & 1 & 0 & -1 \\ 1 & 0 & 1 & 1 \end{bmatrix}, \quad Q = I_4.$$

The solution of this equation is (to four digits)

$$X = \begin{bmatrix} 0.5000 & -0.3333 & 0.3333 & -0.1667 \\ -0.3333 & 0.5833 & -0.3833 & 0.2417 \\ 0.3333 & -0.3833 & 0.5167 & -0.1655 \\ -0.1667 & 0.2417 & -0.1655 & 0.2461 \end{bmatrix}.$$

The perturbations in the matrices A, E, Q are taken in the form

$$\Delta A = \Delta A_0 \times 10^{-14+k}, \quad \Delta E = \Delta E_0 \times 10^{-14+k}, \quad \Delta Q = \Delta Q_0 \times 10^{-14+k}$$

where

$$\Delta A_0 = \begin{bmatrix} 4.56 & -1.75 & 0.74 & -0.35 \\ 2.54 & -0.96 & 0.40 & -0.19 \\ -5.40 & 2.08 & -0.88 & 0.42 \\ 5.32 & -2.05 & 0.87 & -0.42 \end{bmatrix},$$

$$\Delta E_0 = \begin{bmatrix} 2.58 & 1.77 & -3.23 & 2.04 \\ 1.35 & 0.97 & -1.73 & 1.09 \\ -3.10 & -2.11 & 3.87 & -2.43 \\ 3.04 & 2.07 & -3.80 & 2.39 \end{bmatrix},$$

$$\Delta Q_0 = \begin{bmatrix} 9.02 & -2.67 & 0.92 & -0.42 \\ -2.67 & 1.21 & -0.56 & 0.27 \\ 0.92 & -0.56 & 0.33 & -0.16 \\ -0.42 & 0.27 & -0.16 & 0.08 \end{bmatrix}$$

and k is an integer.

The actual changes in the solution along with the local and non-local estimates (8) and (13) for different values of k are shown in Table 1. The case when the non-local estimate is not valid is shown by asterisk.

Table 1.

k	$\Delta_A, \Delta_E, \Delta_Q$	Δ_X	Est. (8)	Est. (13)
1	1.0×10^{-12}	1.1×10^{-11}	3.0×10^{-11}	5.8×10^{-8}
2	1.0×10^{-11}	1.1×10^{-10}	3.0×10^{-10}	5.8×10^{-7}
3	1.0×10^{-10}	1.1×10^{-9}	3.0×10^{-9}	5.8×10^{-6}
4	1.0×10^{-9}	1.1×10^{-8}	3.0×10^{-8}	5.8×10^{-5}
5	1.0×10^{-8}	1.1×10^{-7}	3.0×10^{-7}	5.8×10^{-4}
6	1.0×10^{-7}	1.1×10^{-6}	3.0×10^{-6}	5.8×10^{-3}
7	1.0×10^{-6}	1.1×10^{-5}	3.0×10^{-5}	5.8×10^{-2}
8	1.0×10^{-5}	1.1×10^{-4}	3.0×10^{-4}	6.0×10^{-1}
9	1.0×10^{-4}	1.1×10^{-3}	3.0×10^{-3}	8.7×10^{0}
10	1.0×10^{-3}	1.1×10^{-2}	3.0×10^{-2}	*

References

1. M.M. Konstantinov, P.Hr. Petkov, N.D. Christov, V. Mehrmann, A. Barraud and S. Lesecq. Conditioning of the generalized Lyapunov and Riccati equations. *Proc.1999 American Control Conf.*, San Diego, 1999, pp. 2253-2254.
2. M.M. Konstantinov, P.Hr. Petkov, N.D. Christov, D.W. Gu, and V. Mehrmann. Sensitivity of Lyapunov equations. In: N.E. Mastorakis (Ed.). *Advances in Intelligent Systems and Computer Science*, World Scientific and Engineering Society Press, 1999, pp. 289-292.

An Algorithm to Find Values of Minors of Skew Hadamard and Conference Matrices

C. Kravvaritis[1], E. Lappas[2], and M. Mitrouli[1]

[1] Department of Mathematics, University of Athens,
Panepistimiopolis 15784, Athens, Greece
ckrawaritis@yahoo.com
mmitroul@math.uoa.gr

[2] Department of Mathematics, National Technical University of Athens,
Zografou 15773, Athens, Greece
elappas@central.ntua.gr

Abstract. We give an algorithm to obtain formulae and values for minors of skew Hadamard and conference matrices. One step in our algorithm allows the $(n-j) \times (n-j)$ minors of skew Hadamard and conference matrices to be given in terms of the minors of a $2^{j-1} \times 2^{j-1}$ matrix. In particular we illustrate our algorithm by finding explicitly all the $(n-3) \times (n-3)$ minors of such matrices.

1 Introduction

An Hadamard matrix H of order n is an $n \times n$ matrix with elements ± 1 and $HH^T = nI$. For more details and construction methods of Hadamard matrices we refer the interested reader to the books [1] and [4]. If an Hadamard matrix, H, of order n can be written as $H = I + S$ where $S^T = -S$ then H is called *skew–Hadamard.*

A $(0, 1, -1)$ matrix $W = W(n, k)$ of order n satisfying $WW^T = kI_n$ is called a *weighing matrix of order* n *and weight* k or simply a *weighing matrix.* A $W(n, n)$, $n \equiv 0 \,(\mathrm{mod}\, 4)$, is a Hadamard matrix of order n. A $W = W(n, k)$ for which $W^T = -W$ is called a *skew–weighing matrix.* A $W = W(n, n-1)$ satisfying $W^T = W$, $n \equiv 2 \,(\mathrm{mod}\, 4)$, is called a *symmetric conference matrix.* Conference matrices cannot exist unless $n-1$ is the sum of two squares: thus they cannot exist for orders $22, 34, 58, 70, 78, 94$. For more details and construction of weighing matrices the reader can consult the book of Geramita and Seberry [1].

For the conference matrix $W(n, n-1)$ since $WW^T = (n-1)I$ we have that $det(W) = (n-1)^{\frac{n}{2}}$. It also holds that the maximum $(n-1) \times (n-1)$ minors denoted by $W(n-1)$ are: $W(n-1) = (n-1)^{\frac{n}{2}-1}$ and the maximum $(n-2) \times (n-2)$ minors are $W(n-2) = 2(n-1)^{\frac{n}{2}-2}$. [3]. In [3] it was also proved that for a matrix A which is skew-Hadamard or conference matrix of order n, the $(n-3) \times (n-3)$ minors are $W(n-3) = 0,\ 2(n-1)^{\frac{n}{2}-3}$, or $4(n-1)^{\frac{n}{2}-3}$ for $n \equiv 0(mod\ 4)$ and $2(n-1)^{\frac{n}{2}-3}$, or $4(n-1)^{\frac{n}{2}-3}$ for $n \equiv 2(mod\ 4)$.

Z. Li et al. (Eds.): NAA 2004, LNCS 3401, pp. 375–382, 2005.

In the present paper we give a useful method for finding the $(n-3)\times(n-3)$ minors of skew Hadamard and conference matrices. This method points the way to finding other minors such as the $(n-j)\times(n-j)$ minors.

Notation 1. We use $-$ for -1 in matrices in this paper. Also, when we say the determinants of a matrix we mean the absolute values of the determinants.

Notation 2. We write

$$J_{b_1,b_2,\cdots,b_z}$$

for the all ones matrix with diagonal blocks of sizes $b_1\times b_1,\ b_2\times b_2\ \cdots\ b_z\times b_z$. Write

$$a_{ij}J_{b_1,b_2,\cdots,b_z}$$

for the matrix for which the elements of the block with corners $(i+b_1+b_2+\cdots+b_{j-1}, i+b_1+b_2+\cdots+b_{i-1})$, $(i+b_1+b_2+\cdots+b_{j-1}, b_1+b_2+\cdots+b_i)$, $(b_1+b_2+\cdots+b_j, i+b_1+b_2+\cdots+b_{i-1})$, $(b_1+b_2+\cdots+b_j, b_1+b_2+\cdots+b_i)$ are a_{ij} an integer.

Write

$$(k-a_{ii})I_{b_1,b_2,\cdots,b_z}$$

for the direct sum $(k-a_{11})I_{b_1}+(k-a_{22})I_{b_2}+\cdots+(k-a_{zz})I_{b_z}$.

2 Preliminary Results

We first note a very useful lemma as it allows us to obtain bounds on the column structure of submatrices of a weighing matrix.

Lemma 1. (The Distribution Lemma for $W(n,n-1)$). *Let W be any $W(n,n-1)$ of order $n>2$. Then, writing $\varepsilon=(-1)^{\frac{n+2}{2}}$ and with $a,b,c\in\{1,-1\}$ for every triple of rows containing*

$$\begin{matrix} 0 & a & b \\ \varepsilon a & 0 & c \\ \varepsilon b & \varepsilon c & 0 \end{matrix}$$

the number of columns which are

(a) $(1,1,1)^T$ *or* $(-,-,-)^T$ *is* $\frac{1}{4}(n-3-bc-\varepsilon ac-ab)$
(b) $(1,1,-)^T$ *or* $(-,-,1)^T$ *is* $\frac{1}{4}(n-3-bc+\varepsilon ac+ab)$
(c) $(1,-,1)^T$ *or* $(-,1,-)^T$ *is* $\frac{1}{4}(n-3+bc-\varepsilon ac+ab)$
(d) $(1,-,-)^T$ *or* $(-,1,1)^T$ *is* $\frac{1}{4}(n-3+bc+\varepsilon ac-ab)$

Proof. Let the following rows represent three rows of an weighing matrix $W=W(n,n-1)$ of order n.

$$\begin{matrix} & & & u_1 & u_2 & u_3 & u_4 & u_5 & u_6 & u_7 & u_8 \\ 0 & a & b & 1\ldots1 & 1\ldots1 & 1\ldots1 & 1\ldots1 & -\ldots- & -\ldots- & -\ldots- & -\ldots- \\ \varepsilon a & 0 & c & 1\ldots1 & 1\ldots1 & -\ldots- & -\ldots- & 1\ldots1 & 1\ldots1 & -\ldots- & -\ldots- \\ \varepsilon b & \varepsilon c & 0 & 1\ldots1 & -\ldots- & 1\ldots1 & -\ldots- & 1\ldots1 & -\ldots- & 1\ldots1 & -\ldots- \end{matrix}$$

where $u_1, u_2, \ldots, u_8$ are the numbers of columns of each type. Then from the order and the inner product of rows we have

$$\begin{aligned} u_1+u_2+u_3+u_4+u_5+u_6+u_7+u_8 &= n-3 \\ u_1+u_2-u_3-u_4-u_5-u_6+u_7+u_8 &= -bc \\ u_1-u_2+u_3-u_4-u_5+u_6-u_7+u_8 &= -\varepsilon ac \\ u_1-u_2-u_3+u_4+u_5-u_6-u_7+u_8 &= -ab \end{aligned} \tag{1}$$

Solving we have

$$\begin{aligned} u_1+u_8 &= \frac{1}{4}(n-3-bc-\varepsilon ac-ab) \\ u_2+u_7 &= \frac{1}{4}(n-3-bc+\varepsilon ac+ab) \\ u_3+u_6 &= \frac{1}{4}(n-3+bc-\varepsilon ac+ab) \\ u_4+u_5 &= \frac{1}{4}(n-3+bc+\varepsilon ac-ab) \end{aligned}$$

□

Theorem 1. (Determinant Simplification Theorem) *Let*

$$CC^T = (k-a_{ii})I_{b_1,b_2,\cdots,b_z} + a_{ij}J_{b_1,b_2,\cdots,b_z}$$

then

$$det\ CC^T = \Pi_{i=1}^{z}(k-a_{ii})^{b_i-1} det\ D \tag{2}$$

where

$$D = \begin{bmatrix} k+(b_1-1)a_{11} & b_2a_{12} & b_3a_{13} \cdots & b_za_{1z} \\ b_1a_{21} & k+(b_2-1)a_{22} & b_3a_{23} \cdots & b_za_{2z} \\ \vdots & \vdots & \vdots & \vdots \\ b_1a_{z1} & b_2a_{z2} & b_3a_{z2} \cdots & k+(b_z-1)a_{zz} \end{bmatrix}$$

Proof. We note the matrix CC^T has k down the diagonal and elsewhere the elements are defined by the block of elements a_{ij}.

We start with the first row and subtract it from the 2nd to the b_1th row. Then take the first row of the 2nd block (the b_1+1st row) and subtract it from the b_1+2th to b_1+b_2th rows. We continue this way with each new block.

Now the first column has: $a_{11}-k$ b_1-1 times, then a_{21} followed by b_2-1 rows of zero, then a_{31} followed by b_3-1 rows of zero, and so on until we have a_{z1} followed by b_z-1 rows of zero. We now add columns 2 to b_1 to the first column: each of the columns b_1+2nd to $b_1+\ldots+b_2$th to the b_1+1st column; and so on until finally add each of the columns $b_1+\cdots+b_{z-1}+2$nd to $b_1+\cdots+b_{z-1}$th to the $b_1+\cdots+b_{z-1}+1$st column.

The rows which contain zero in the first column will have $k - a_{ii}$ on the diagonal and all other elements zero. They can then be used to zero every element in their respective columns.

We now expand the determinant, taking into the coefficient those rows and columns which contain $k - a_{ii}$ as required. The remaining matrix to be evaluated is D given in the enunciation.

□

3 The Algorithm

3.1 The Matrix U_j

Let $\underline{x}_{\beta+1}^T$ the vectors containing the binary representation of each integer $\beta + 2^{j-1}$ for $\beta = 0, \ldots, 2^{j-1} - 1$. Replace all zero entries of $\underline{x}_{\beta+1}^T$ by -1 and define the $j \times 1$ vectors

$$\underline{u}_k = \underline{x}_{2^{j-1}-k+1}, \quad k = 1, \ldots, 2^{j-1}$$

We write U_j for all the $j \times (n-j)$ matrix in which $\underline{u}_k$ occurs u_k times. So

$$U_j = \begin{array}{ccccc} \overbrace{1...1}^{u_1} & \overbrace{1...1}^{u_2} & \ldots & \overbrace{1...1}^{u_{2^{j-1}-1}} & \overbrace{1...1}^{u_{2^{j-1}}} \\ 1...1 & 1...1 & \ldots & -...- & -...- \\ \cdot & \cdot & \ldots & \cdot & \cdot \\ \cdot & \cdot & \ldots & \cdot & \cdot \\ 1...1 & 1...1 & \ldots & 1...1 & -...- \\ 1...1 & -...- & \ldots & 1...1 & -...- \end{array} = \begin{array}{ccccc} u_1 & u_2 & \ldots & u_{2^{j-1}-1} & u_{2^{j-1}} \\ 1 & 1 & \ldots & 1 & 1 \\ 1 & 1 & \ldots & - & - \\ \vdots & \vdots & & \vdots & \vdots \\ 1 & 1 & \ldots & - & - \\ 1 & - & \ldots & 1 & - \end{array} \quad (3)$$

Example 1.

$$U_3 = \begin{array}{cccc} u_1 & u_2 & u_3 & u_4 \\ 1 & 1 & 1 & 1 \\ 1 & 1 & - & - \\ 1 & - & 1 & - \end{array}$$

Example 2.

$$U_4 = \begin{array}{cccccccc} u_1 & u_2 & u_3 & u_4 & u_5 & u_6 & u_7 & u_8 \\ 1 & 1 & 1 & 1 & 1 & 1 & 1 & 1 \\ 1 & 1 & 1 & 1 & - & - & - & - \\ 1 & 1 & - & - & 1 & 1 & - & - \\ 1 & - & 1 & - & 1 & - & 1 & - \end{array}$$

3.2 The Matrix D

Any $W = W(n, n-1)$ matrix can be written:

$$W = \begin{bmatrix} M & U_j \\ \varepsilon U_j^T & C \end{bmatrix}, \quad (4)$$

where M,C are $j \times j$ and $(n-j) \times (n-j)$ matrices respectively, with diagonal entries all 0, such that $M = \varepsilon M^T$ and $C = \varepsilon C^T$. The elements in the $(n-j) \times (n-j)$ matrix CC^T obtained by removing the first j rows and columns of the weighing matrix W can be permuted to appear in the form

$$CC^T = (n-1)I_{u_1,u_2,\cdots,u_{2^{j-1}}} + a_{ik}J_{u_1,u_2,\cdots,u_{2^{j-1}}}$$

where $(a_{ik}) = (-\underline{u}_i \cdot \underline{u}_k)$, with $\cdot$ the inner product. By the determinant simplification theorem

$$det\ CC^T = (n-1)^{n-2^{j-1}-j} det\ D$$

where D, of order 2^{j-1} is given by

$$D = \begin{bmatrix} n-1-ju_1 & u_2a_{12} & u_3a_{13} \cdots & u_za_{1z} \\ u_1a_{21} & n-1-ju_2 & u_3a_{23} \cdots & u_za_{2z} \\ \vdots & \vdots & \vdots & \vdots \\ u_1a_{z1} & u_2a_{z2} & u_3a_{z2} \cdots & n-1-ju_z \end{bmatrix}$$

The $(n-j) \times (n-j)$ minor of the $W(n, n-1)$ is the determinant of C for which we have

$$detC = ((n-1)^{n-2^{j-1}-j} det\ D)^{1/2}$$

Remark 1. *The algorithm described above works on weighing matrices $W(n, n-1)$ only when all the zeros are on the diagonal, or equivalently the matrix M has j zeros and the C matrix has (n-j) zeros. Consequently it might loose some expected values for some minors.*

4 Application of the Algorithm on the (n-3)×(n-3) Minors of Weighing Matrices

The matrix $(a_{ik}) = (-\underline{u}_i \cdot \underline{u}_k)$ is

$$(a_{ik}) = \begin{bmatrix} -3 & -1 & -1 & 1 \\ -1 & -3 & 1 & -1 \\ -1 & 1 & -3 & -1 \\ 1 & -1 & -1 & -3 \end{bmatrix}$$

4.1 Case 1, ε=1

Without loss of generality we can assume that the matrix M is

$$M = \begin{bmatrix} 0 & 1 & 1 \\ 1 & 0 & 1 \\ 1 & 1 & 0 \end{bmatrix}$$

then by the Distribution Lemma $u_1 = \frac{n-6}{4}$, $u_2 = \frac{n-2}{4}$, $u_3 = \frac{n-2}{4}$, $u_4 = \frac{n-2}{4}$. The matrix D is:

$$D = \frac{1}{4^4}\begin{bmatrix} n+14 & -n+2 & -n+2 & n-2 \\ -n+6 & n+2 & n-2 & -n+2 \\ -n+6 & n-2 & n+2 & -n+2 \\ n-6 & -n+2 & -n+2 & n+2 \end{bmatrix}$$

and its determinant is $detD = 1024n - 1024$. Finally

$$detCC^T = (n-1)^{n-7}\frac{1024n - 1024}{256}$$

4.2 Case 2, ε=-1

Without loss of generality we can assume that the matrix M is

$$M = \begin{bmatrix} 0 & 1 & 1 \\ -1 & 0 & 1 \\ -1 & -1 & 0 \end{bmatrix}$$

then by the Distribution Lemma $u_1 = \frac{n-4}{4}$, $u_2 = \frac{n-4}{4}$, $u_3 = \frac{n}{4}$, $u_4 = \frac{n-4}{4}$. The matrix D is:

$$D = \frac{1}{4^4}\begin{bmatrix} n+8 & -n+4 & -n & n-4 \\ -n+4 & n+8 & n & -n+4 \\ -n+4 & n-4 & n-4 & -n+4 \\ n-4 & -n+4 & -n & n+8 \end{bmatrix}$$

and its determinant is $detD = 0$. Finally

$$detCC^T = 0$$

5 Application to Numerical Analysis

Let $A = [a_{ij}] \in \mathcal{R}^{n\times n}$. We reduce A to upper triangular form by using Gaussian Elimination (GE) operations. Let $A^{(k)} = [a_{ij}^{(k)}]$ denote the matrix obtained after the first k pivoting operations, so $A^{(n-1)}$ is the final upper triangular matrix. A diagonal entry of that final matrix will be called a pivot. Matrices with the property that no exchanges are actually needed during GE with complete pivoting are called completely pivoted (CP) or feasible. Let $g(n, A) = \max_{i,j,k} |a_{ij}^{(k)}|/|a_{11}^{(0)}|$ denote the growth associated with GE on a CP A and $g(n) = \sup\{\, g(n, A)/A \in \mathcal{R}^{n\times n}\,\}$. The problem of determining $g(n)$ for various values of n is called the growth problem.

The determination of $g(n)$ remains a mystery. Wilkinson in [6] proved that

$$g(n) \leq [n\, 2\, 3^{1/2} \ldots n^{1/n-1}]^{1/2} = f(n)$$

Wilkinson's initial conjecture seems to be connected with Hadamard matrices. Interesting results in the size of pivots appears when GE is applied to CP skew-Hadamard and conference matrices of order n. In these matrices, the growth is also large, and experimentally, we have been led to believe it equals $n-1$ and special structure appears for the first few and last few pivots. These results give rise to new conjectures that can be posed for this category of matrices.

The Growth Conjecture for Skew Hadamard and Conference Matrices Let W be a CP skew-Hadamrad or conference matrix. Reduce W by GE. Then

(i) $g(n, W) = n-1$.
(ii) The two last pivots are equal to $\frac{n-1}{2}, n-1$.
(iii) Every pivot before the last has magnitude at most $n-1$.
(iv) The first four pivots are equal to 1, 2, 2, 3 or 4, for $n > 14$.

The magnitude of the pivots appearing after GE operations on a CP matrix W is

$$p_j = \frac{W(j)}{W(j-1)}, \quad j = 1, \ldots, n, \quad W(0) = 1 \tag{5}$$

This relationship gives the connection between pivots and minors, so it is obvious that the calculation of minors is very important. Theoretically it can be proved [3] the following results:

- The first 4 pivots for every $W(n, n-1)$ are

$$1,\ 2,\ 2,\ 3 \textit{ or } 4$$

- The 3 last pivots for every $W(n, n-1)$ are

$$n-1 \textit{ or } \frac{n-1}{2}, \frac{n-1}{2}, n-1$$

An experimental study of all the possible pivot patterns of $W(8,7)$ and $W(10,9)$ gave the following results:

- Pivot patterns of $W(8,7)$:

$$\{1,\ 2,\ 2,\ 4,\ 7/4,\ 7/2,\ 7/2,\ 7\}$$

or

$$\{1,\ 2,\ 2,\ 3,\ 7/3,\ 7/2,\ 7/2,\ 7\}.$$

- Pivot patterns of $W(10,9)$:

$$\{1,2,2,3,3,4,9/4,9/2,9/2,9\}$$

or

$$\{1,2,2,4,3,3,9/4,9/2,9/2,9\}$$

or

$$\{1,2,2,3,10/4,18/5,9/3,9/2,9/2,9\}.$$

An interesting numerical analysis problem concerns the theoretical specification of the pivot patterns of various $W(n, n-1)$ and the study of their growth factor. We hope that the algorithm developed in this paper, computing the $(n-j) \times (n-j)$ minors of skew-Hadamard and conference matrices, can be applied for the study of this problem. In our future research we will also study the extension of the algorithm for larger values of j. This will allow us to specify more values of pivots.

References

1. A.V.Geramita, and J.Seberry, *Orthogonal Designs: Quadratic Forms and Hadamard Matrices*, Marcel Dekker, New York-Basel, 1979.
2. C. Koukouvinos, E. Lappas, M. Mitrouli, and J. Seberry, An algorithm to find formulae and values of minors for Hadamard matrices: II, *Linear Algebra and its Applications*, 371 (2003), 111-124.
3. C. Koukouvinos, M. Mitrouli and J. Seberry, Growth in Gaussian elimination for weighing matrices, $W(n, n-1)$, *Linear Algebra and its Appl.*, 306 (2000), 189-202.
4. J. Seberry Wallis, Hadamard matrices, Part IV, *Combinatorics: Room Squares, Sum-Free Sets and Hadamard Matrices*, Lecture Notes in Mathematics, Vol. 292, eds. W. D. Wallis, Anne Penfold Street and Jennifer Seberry Wallis, Springer-Verlag, Berlin-Heidelberg-New York, 1972.
5. F. R. Sharpe, The maximum value of a determinant, *Bull. Amer. Math. Soc.*, 14 (1907), 121-123.
6. J. H. Wilkinson, Error analysis of direct methods of matrix inversion, *J. Assoc. Comput. Mach.*, 8 (1961), 281-330.

Parallel Performance of a 3D Elliptic Solver

Ivan Lirkov

Central Laboratory for Parallel Processing, Bulgarian Academy of Sciences,
Acad.G.Bonchev, Bl. 25A,
1113 Sofia, Bulgaria
ivan@parallel.bas.bg

Abstract. It was recently shown that block-circulant preconditioners applied to a conjugate gradient method used to solve structured sparse linear systems arising from 2D or 3D elliptic problems have good numerical properties and a potential for high parallel efficiency. The asymptotic estimate for their convergence rate is as for the incomplete factorization methods but the efficiency of the parallel algorithms based on circulant preconditioners are asymptotically optimal. In this paper parallel performance of a circulant block-factorization based preconditioner applied to a 3D model problem is investigated. The aim of this presentation is to analyze the performance and to report on the experimental results obtained on shared and distributed memory parallel architectures. A portable parallel code is developed based on Message Passing Interface (MPI) and OpenMP (Open Multi Processing) standards. The performed numerical tests on a wide range of parallel computer systems clearly demonstrate the high level of parallel efficiency of the developed parallel code.

1 Introduction

In this article we are concerned with the numerical solution of 3D linear boundary value problems of elliptic type. After discretization, such problems lead to find the solution of linear systems of the form $A\mathbf{x} = \mathbf{b}$. We shall only consider the case where A is symmetric and positive definite. In practice, large problems of this class are often solved by iterative methods, such as the conjugate gradient method. At each step of these iterative methods only the product of A with a given vector $\mathbf{v}$ is needed. Such methods are therefore ideally suited to exploit the sparsity of A.

Typically, the rate of convergence of these methods depends on the condition number $\kappa(A)$ of the coefficient matrix A: the smaller $\kappa(A)$ leads to the faster convergence. Unfortunately, for elliptic problems of second order, usually $\kappa(A) = \mathcal{O}(n^2)$, where n is the number of mesh points in each coordinate direction, and hence grows rapidly with n. To somewhat facilitate this problem, these methods are almost always used with a preconditioner M. The preconditioner is chosen with two criteria in mind: to minimize $\kappa(M^{-1}A)$ and to allow efficient computation of the product $M^{-1}\mathbf{v}$ for a given vector $\mathbf{v}$. These two goals are often conflicting ones and much research has been done into devising preconditioners

Z. Li et al. (Eds.): NAA 2004, LNCS 3401, pp. 383–390, 2005.

Table 1. Parallel time, speed-up and parallel efficiency for the CBF preconditioner on Athlon and Macintosh clusters, Cray T3E, and NEC server Azusa Express5800/1160Xa

		Athlon			Macintosh			Cray T3E			NEC		
n	p	T_p	S_p	E_p	T_p	S_p	E_p	T_p	S_p	E_p	T_p	S_p	E_p
32	1	0.111			0.109			0.133			0.077		
	2	0.082	1.35	0.673	0.061	1.78	0.891	0.068	1.96	0.982	0.040	1.95	0.975
	4	0.057	1.95	0.487	0.050	2.18	0.544	0.034	3.88	0.969	0.020	3.86	0.965
	8	0.031	3.61	0.452	0.036	3.05	0.382	0.020	6.64	0.830	0.011	7.23	0.904
	16							0.014	9.61	0.601			
	32							0.011	12.58	0.393			
48	1	0.758			0.761			0.898			0.470		
	2	0.469	1.62	0.809	0.405	1.88	0.939	0.452	1.99	0.993	0.234	2.01	1.004
	3	0.340	2.23	0.744	0.318	2.39	0.798	0.304	2.95	0.984	0.156	3.01	1.004
	4	0.261	2.91	0.726	0.265	2.87	0.718	0.227	3.95	0.988	0.117	4.01	1.002
	6	0.189	4.01	0.669	0.200	3.81	0.634	0.154	5.82	0.971	0.079	5.94	0.989
	8	0.203	3.74	0.468	0.164	4.63	0.579	0.116	7.71	0.964	0.060	7.78	0.973
	12							0.080	11.29	0.941	0.042	11.30	0.942
	16							0.062	14.58	0.911			
	24							0.044	20.27	0.845			
	48							0.027	33.84	0.705			
64	1	1.095			1.191						0.744		
	2	0.753	1.45	0.727	0.705	1.69	0.845	0.583			0.364	2.05	1.023
	4	0.429	2.55	0.638	0.495	2.41	0.602	0.297		0.981	0.177	4.20	1.051
	8	0.268	4.08	0.511	0.327	3.64	0.456	0.153		0.953	0.092	8.06	1.007
	16							0.079		0.922			
	32							0.045		0.810			
	64							0.028		0.651			
96	1	6.787			6.701						4.332		
	2	4.107	1.65	0.826	3.748	1.79	0.894				2.185	1.98	0.991
	3	2.975	2.28	0.760	2.871	2.33	0.778				1.467	2.95	0.984
	4	2.202	3.08	0.770	2.398	2.79	0.699				1.083	4.00	1.000
	6	1.530	4.44	0.739	1.743	3.84	0.641	1.256			0.733	5.91	0.985
	8	1.160	5.85	0.731	1.374	4.88	0.610	0.944		0.998	0.561	7.73	0.966
	12							0.639		0.983	0.397	10.91	0.909
	16							0.484		0.973			
	24							0.322		0.975			
	32							0.243		0.969			
	48							0.170		0.924			
	96							0.095		0.826			
128	1										7.240		
	2										3.576	2.02	1.012
	4										1.897	3.82	0.954
	8										1.045	6.93	0.866
192	1										37.901		
	2										18.096	2.09	1.047
	3										12.280	3.09	1.029
	4										9.345	4.06	1.014
	6										6.627	5.72	0.953
	8										5.348	7.09	0.886
	12										4.082	9.28	0.774
256	1										64.039		
	2										30.626	2.09	1.045
	4										15.636	4.10	1.024
	8										9.437	6.79	0.848

Table 2. Parallel time, speed-up and parallel efficiency for the CBF preconditioner on SUNfire 6800 and on Cray Opteron cluster using MPI

	SUNfire								Cray Opteron							
p	n	T_p	S_p	E_p	n	T_p	S_p	E_p	n	T_p	S_p	E_p	n	T_p	S_p	E_p
1	32	0.050			128	8.445			32	0.034			128	2.768		
2		0.034	1.48	0.741		4.386	1.93	0.963		0.017	2.06	1.030		1.428	1.94	0.969
4		0.015	3.27	0.818		2.105	4.01	1.003		0.009	3.94	0.985		0.750	3.69	0.923
8		0.009	5.84	0.729		0.898	9.40	1.175		0.005	7.16	0.895		0.402	6.88	0.861
16		0.004	11.28	0.705		0.386	21.88	1.367		0.003	12.07	0.754		0.222	12.47	0.779
32										0.003	13.17	0.412		0.127	21.80	0.681
64														0.068	40.67	0.635
128														0.045	61.22	0.478
1	48	0.404			192	41.063			48	0.232			192	16.576		
2		0.242	1.67	0.834		20.448	2.01	1.004		0.116	2.00	0.998		8.591	1.93	0.965
3		0.146	2.77	0.924		14.141	2.90	0.968		0.079	2.95	0.984		5.715	2.90	0.967
4		0.111	3.65	0.912		10.279	3.99	0.999		0.059	3.96	0.989		4.391	3.78	0.944
6		0.068	5.97	0.995		7.004	5.86	0.977		0.040	5.74	0.957		3.023	5.48	0.914
8		0.053	7.62	0.953		5.155	7.96	0.996		0.031	7.44	0.930		2.258	7.34	0.918
12										0.021	11.26	0.938		1.533	10.81	0.901
16		0.024	16.73	1.045		2.346	17.50	1.094		0.016	14.67	0.917		1.188	13.95	0.872
24		0.019	21.65	0.902		1.737	23.63	0.985		0.011	20.66	0.861		0.816	20.31	0.846
32														0.633	26.17	0.818
48										0.007	31.80	0.662		0.445	37.27	0.776
64														0.344	48.14	0.752
96														0.246	67.38	0.702
192														0.139	118.99	0.620
1	64	0.672			256				64	0.307			256	24.442		
2		0.388	1.73	0.866		38.417				0.162	1.90	0.950		13.104	1.87	0.933
4		0.160	4.20	1.050		18.642		1.030		0.084	3.66	0.914		6.896	3.54	0.886
8		0.075	8.97	1.122		9.826		0.977		0.044	7.00	0.875		3.510	6.96	0.870
16		0.031	22.01	1.375		4.624		1.039		0.026	11.91	0.744		1.915	12.76	0.798
32										0.013	23.68	0.740		1.078	22.68	0.709
64										0.009	35.57	0.556		0.587	41.62	0.650
128														0.356	68.65	0.536
1	96	4.695							96	1.973						
2		2.302	2.04	1.020						1.010	1.95	0.977				
3		1.499	3.13	1.044						0.688	2.87	0.956				
4		1.091	4.30	1.076						0.516	3.82	0.956				
6		0.682	6.88	1.147						0.361	5.47	0.912				
8		0.494	9.49	1.187						0.272	7.26	0.907				
12										0.184	10.73	0.894				
16		0.242	19.43	1.214						0.145	13.63	0.852				
24		0.191	24.63	1.026						0.100	19.79	0.825				
32										0.076	25.87	0.808				
48										0.053	36.97	0.770				
96										0.031	64.58	0.673				

Table 3. Parallel time, speed-up and parallel efficiency for the CBF preconditioner on SUNfire 6800 using OpenMP

p	n	T_p	S_p	E_p	n	T_p	S_p	E_p
1	32	0.056			128	8.313		
2		0.028	1.97	0.983		4.233	1.96	0.982
3		0.020	2.75	0.918		2.757	3.01	1.005
4		0.015	3.69	0.922		2.029	4.10	1.024
5		0.014	4.10	0.820		1.633	5.09	1.018
6		0.012	4.65	0.775		1.387	5.99	0.999
7		0.011	5.11	0.730		1.196	6.95	0.993
8		0.009	6.43	0.803		1.039	8.00	1.000
16		0.006	9.71	0.607		0.538	15.44	0.965
24		0.006	9.70	0.404		0.398	20.89	0.870
1	48	0.386			192	40.471		
2		0.196	1.97	0.984		20.181	2.01	1.003
3		0.130	2.96	0.988		13.437	3.01	1.004
4		0.099	3.91	0.977		10.048	4.03	1.007
5		0.082	4.70	0.941		8.044	5.03	1.006
6		0.067	5.79	0.964		6.751	6.00	0.999
7		0.060	6.48	0.926		5.808	6.97	0.996
8		0.052	7.40	0.925		5.086	7.96	0.995
16		0.029	13.42	0.838		2.514	16.10	1.006
24		0.020	19.38	0.807		1.885	21.47	0.894
1	64	0.650			256			
2		0.339	1.92	0.958		51.349		
3		0.230	2.82	0.940		33.793		1.013
4		0.170	3.82	0.954		25.664		1.000
5		0.140	4.65	0.930		20.521		1.001
6		0.119	5.46	0.910		17.093		1.001
7		0.104	6.27	0.896		14.707		0.998
8		0.087	7.46	0.933		13.242		0.969
16		0.049	13.23	0.827		6.918		0.928
24		0.040	16.40	0.683		5.116		0.836
1	96	3.973						
2		1.880	2.11	1.057				
3		1.224	3.25	1.082				
4		0.930	4.27	1.068				
5		0.759	5.24	1.048				
6		0.622	6.39	1.065				
7		0.542	7.33	1.047				
8		0.471	8.44	1.055				
16		0.245	16.23	1.015				
24		0.168	23.67	0.986				

The memory on one processor of Cray computer is sufficient only for the discretization with coarse grid. For larger problems we report the parallel efficiency related to the results on 2 and 6 processors respectively.

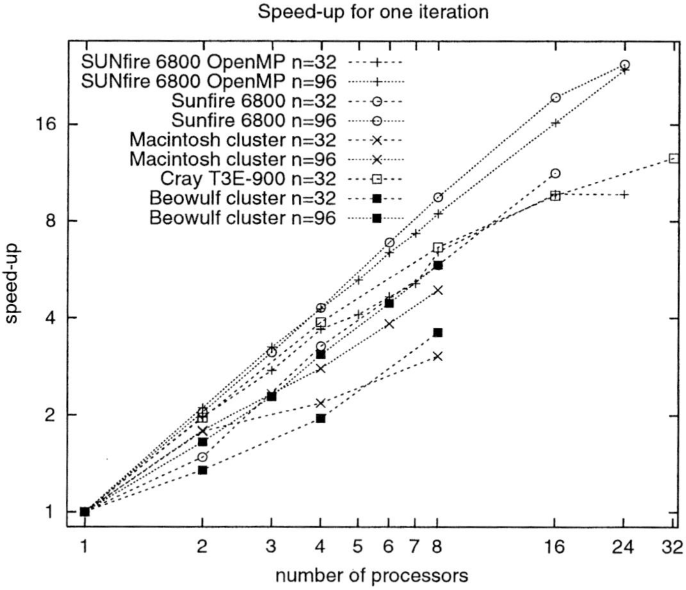

Fig. 1. Speed-up for one iteration on three parallel computer systems

Tables 2 and 3 shows results obtained on SUNfire 6800 and on Cray Opteron cluster. As expected, the parallel efficiency increases with the size of the discrete problems. The parallel efficiency for relatively large problems is above 80% which confirms our general expectations. There exist at least two reasons for the reported high efficiency: (a) the network parameters *start-up time* and *time for transferring of single word* are relatively small for the multiprocessor machines; (b) there is also some overlapping between the computations and the communications in the algorithm. Moreover, the super-linear speed-up can be seen in some of the runs. This effect has a relatively simple explanation. When the number of processors increases, the size of data per processor decreases. Thus the stronger *memory locality* increases the role of the cache memories. (The level 2 cache on the SUNfire 6800 is 8 MB.) The obtained speed-up on 16 processors on SUNfire 6800 in some cases is close to 22. In these cases the whole program is fitted in the cache memory and there is no communication between processors and main memory but only between processors.

Finally, we compare results on Cray, SUN, NEC, and Linux clusters. Fig. 1 shows parallel speed-up for execution of one iteration on different parallel systems.

4 Summary

We are concerned with the numerical solution of 3D elliptic problems. After discretization, such problems reduce to the solution of linear systems. We use a preconditioner based on a block-circulant approximation of the blocks of the stiffness matrix. We exploit the fast inversion of block-circulant matrices. The computation and the inversion of these circulant block-factorization preconditioners are highly parallelizable on a wide variety of architectures. The developed code provide new effective tool for solving of large-scale problems in realistic time on a coarse-grain parallel computer systems.

Acknowledgments

This research was supported by the European Commission through grants number HPRI-CT-1999-00026 (the TRACS Programme at EPCC) and number 506079 (HPC-Europa). The experimental results on Cray T3E and SunFire 6800 were obtained during the period of extended access to the EPCC supercomputing facilities after a TRACS research visit in EPCC and the results on NEC and Cray Opteron cluster were obtained during the HPC-Europa research visit at the University of Stuttgart. It was supported also by grant I-1402/2004 from the Bulgarian NSF and by SEE-GRID grant FP6-2002-Infrastructures-2 #002356.

References

1. R.H. Chan and T.F. Chan, Circulant preconditioners for elliptic problems, *J. Num. Lin. Alg. Appl.* **1** (1992) 77–101.
2. I. Lirkov and S. Margenov, Conjugate gradient method with circulant block-factorization preconditioners for 3D elliptic problems, in: *Proceedings of the Workshop # 3 Copernicus 94-0820 "HIPERGEOS" project meeting, Rožnov pod Radhoštěm* (1996).
3. I. Lirkov and S. Margenov, Parallel complexity of conjugate gradient method with circulant block-factorization preconditioners for 3D elliptic problems, in: O.P. Iliev, M.S. Kaschiev, Bl. Sendov, and P.V. Vassilevski, eds., *Recent Advances in Numerical Methods and Applications* (World Scientific, Singapore, 1999) 455–463.
4. I. Lirkov, S. Margenov, and P.S. Vassilevski, Circulant block-factorization preconditioners for elliptic problems, *Computing* **53** (1) (1994) 59–74.
5. I. Lirkov, S. Margenov, and L. Zikatanov, Circulant block–factorization preconditioning of anisotropic elliptic problems, *Computing* **58** (3) (1997) 245–258.
6. C. Van Loan, *Computational frameworks for the fast Fourier transform* (SIAM, Philadelphia, 1992).
7. Y. Saad and M.H. Schultz, Data communication in parallel architectures, *Parallel Comput.* **11** (1989) 131–150.
8. M. Snir, St. Otto, St. Huss-Lederman, D. Walker, and J. Dongara, *MPI: The Complete Reference.* Scientific and engineering computation series (The MIT Press, Cambridge, Massachusetts, 1997) Second printing.
9. D. Walker and J. Dongara, MPI: a standard Message Passing Interface, *Supercomputer* **63** (1996) 56–68.
10. Web-site for Open MP tools, http://www.openmp.org

Generalized Rayleigh Quotient Shift Strategy in QR Algorithm for Eigenvalue Problems

Yifan Liu and Zheng Su

Stanford University, Stanford CA 94305, USA
liuyifan@stanford.edu

Abstract. QR algorithm for eigenproblems is often applied with single or double shift strategies. To save computation effort, double implicit shift technique is employed. In Watkins and Elsner[1], a higher-order shift strategy, generalized Rayleigh quotient shift strategy, is introduced. In this paper, we give a generalization of the double implicit shift technique for the higher-order strategy, and compare the computation cost for different orders with numerical experiments.

1 Introduction to the Generic GR Algorithm and Generalized Rayleigh Quotient Strategy

Let A be an $n \times n$ matrix whose eigenvalues we would like to know. The GR algorithm generates a sequence $\{A_i\}$ of similar matrices as follows:

- A_0 is taken to be A or some convenient matrix similar to A, say $A_0 = G_0^{-1}AG_0$.
- Given A_{i-1}, let p_i be some polynomial,
  1. take the GR decomposition, $p_i(A_{i-1}) = G_iR_i$
  2. $A_i = G_i^{-1}A_{i-1}G_i$

Here GR decomposition stands for any well-defined rule for certain class of matrices. In this paper, we concentrate on QR decomposition, while there're some other cases, such as LR, SR, HR etc..

As to p_i, we must make a clever choice if we need rapid convergence. The generalized Rayleigh quotient shift strategy chooses p_i as the characteristic polynomial of A_{22}^{i-1}, the trailing $m \times m$ submatrix of A_{i-1}.

If $m = 1$, then we recover the Rayleigh quotient strategy $p_i(A_{i-1}) = A_{i-1} - \sigma_i I$, where σ_i is the (n, n) entry of A_{i-1}. If $m = 2$, $p_i(A_{i-1}) = (A_{i-1} - \sigma_i I)(A_{i-1} - \tau_i I)$, where σ_i, τ_i are the eigenvalues of the lower-right 2×2 submatrix of A_{i-1}. This is often used for the unsymmetric case to avoid complex computation.

In this paper, we discuss the choice of m for faster convergence.

2 Upper-Hessenberg Structure

It is well-known that, before applying the QR algorithm, we need to reduce the matrix to upper-Hessenberg, so that the computation of QR in each iteration can be reduced from $O(n^3)$ to $O(n^2)$.

Z. Li et al. (Eds.): NAA 2004, LNCS 3401, pp. 391–398, 2005.

In this section, we show that the generalized Rayleigh quotient shift strategy in QR algorithm preserves the upper-Hessenberg structure, just as the single or double shift strategy does.

Theorem 1. *Suppose A is an upper-Hessenberg matrix, p is an m-order polynomial, then there exists a QR factorization of $p(A)$, $p(A) = QR$, such that $Q^{-1}AQ$ is still an upper-Hessenberg matrix.*

Proof. Suppose the m roots of p are $a_1, ..., a_m$. Without loss of generality, we assume the coefficient of the highest-order term is 1, i.e.,

$$p(A) = (A - a_1 I)(A - a_2 I) \cdots (A - a_m I).$$

Take the following steps:
$H_0 = A$
for $i = 1 : m$
  $H_{i-1} - a_i I = U_i R_i$ (QR factorization by Givens rotations)
  $H_i = R_i U_i + a_i I$
end
let $\tilde{A} = H_m$

It is easy to see that $\tilde{A} = (U_1, ..., U_m)^H A (U_1, ..., U_m)$, since H_0 is upper-Hessenberg, so is $H_0 - a_1 I$, we can use $(n-1)$ Givens rotations to get $U_1 R_1$, and obviously U_1 is upper-Hessenberg, thus $H_1 = R_1 U_1 + a_1 I$ is also upper-Hessenberg. For the same reason, $H_1, H_2, ..., H_m = \tilde{A}$ are all upper-Hessenberg. Thus, the only thing left is to show that $(U_1 U_2, ..., U_m)^H p(A)$ is an upper triangular R, then we take $Q = U_1, ..., U_m$, and the theorem can be proved.

Actually

$$p(A) = (A - a_1 I)(A - a_2 I)...(A - a_m I) = U_1 ... U_m R_m ... R_1.$$

We show this by induction, $A - a_1 I = U_1 R_1$ is obvious according to the procedure above. Suppose we have $(A - a_1 I)...(A - a_{i-1} I) = U_1 ... U_{i-1} R_{i-1} ... R_1$, then

$$\begin{aligned}
&(A - a_1 I)...(A - a_{i-1} I)(A - a_i I) \\
&= U_1 ... U_{i-1} R_{i-1} ... R_1 (A - a_i I) \\
&= U_1 ... U_{i-1} R_{i-1} ... R_1 (A - a_1 I) + (a_1 - a_i) U_1 ... U_{i-1} R_{i-1} ... R_1 \\
&= U_1 ... U_{i-1} R_{i-1} ... R_1 (U_1 R_1) + (a_1 - a_i) U_1 ... U_{i-1} R_{i-1} ... R_1 \\
&= U_1 ... U_{i-1} R_{i-1} ... R_2 (H_1 - a_1 I) R_1 + (a_1 - a_i) U_1 ... U_{i-1} R_{i-1} ... R_1 \\
&= U_1 ... U_{i-1} R_{i-1} ... R_2 (H_1 - a_2 I) R_1 + (a_2 - a_i) U_1 ... U_{i-1} R_{i-1} ... R_1 \\
&= U_1 ... U_{i-1} R_{i-1} ... R_2 U_2 R_2 R_1 + (a_2 - a_i) U_1 ... U_{i-1} R_{i-1} ... R_1 \\
&= ... \\
&= U_1 ... U_{i-1} R_{i-1} U_{i-1} R_{i-1} ... R_1 + (a_{i-1} - a_i) U_1 ... U_{i-1} R_{i-1} ... R_1 \\
&= U_1 ... U_{i-1} (H_{i-1} - a_{i-1} I) R_{i-1} ... R_1 + (a_{i-1} - a_i) U_1 ... U_{i-1} R_{i-1} ... R_1 \\
&= U_1 ... U_{i-1} (H_{i-1} - a_i I) R_{i-1} ... R_1 \\
&= U_1 ... U_i R_i ... R_1.
\end{aligned}$$

3 Implicit Shifts

Since the cost for $p(A)$ is $O(mn^3)$, it is not a practical approach to explicitly compute $p(A)$. Thus we follow the implicit skills as follows, which is a generalization of the double implicit shift method in Golub and van Loan [2].

For simplicity, we assume $p(A)$ is non-singular. In this case, the QR factorization of $p(A)$ is unique up to a diagonal and orthogonal matrix. That is,

Lemma 1. *Suppose matrix A is $n \times n$ non-singular, and it has two QR factorizations*

$$A = Q_1 R_1 = Q_2 R_2,$$

then there exists a dianognal matrix $D = diag(d_1, ..., d_n)$, which satisfies that $|d_i| = 1$ for $i = 1, ..., n$, and $Q_2 = Q_1 D$

Proof. Use induction to prove that $Q_2(:, j) = d_j Q_1(:, j)$.

The implicit shift algorithm is based on the following theorem.

Theorem 2. *Suppose A is an upper-Hessenberg matrix, p is an m-order polynomial, $p(A)$ is non-singular. Q is an orthogonal matrix, such that $Q^H A Q$ is unreduced upper-Hessenberg matrix, and the first column of Q equals that of another orthogonal matrix P, which satisfies $p(A) = PR$ is QR factorization. Then $Q^H p(A)$ is upper-triangular.*

Proof. According to Theorem 1, there exists an orthogonal $\tilde{Q}$, such that, $p(A) = \tilde{Q}\tilde{R}$ is a QR factorization of $p(A)$, and $\tilde{Q}^H A \tilde{Q}$ is upper-Hessenberg. Now that $p(A) = PR$ is another QR factorization, then, from lemma 2, there exists a diagonal and orthogonal D, s.t. $P = \tilde{Q}D$. Thus

$$P^H A P = D^H \tilde{Q}^H A \tilde{Q} D$$

is still upper-Hessenberg. Since the first columns of P and Q are the same, and they both reduce A to an upper-Hessenberg matrix, and $Q^H A Q$ is unreduced, by Implicit Q Theorem, there exists another diagonal and orthogonal $\tilde{D}$, s.t. $Q = P\tilde{D}$, thus $Q^H p(A)$ is also upper-triangular.

According to Theorem 3, given the upper-Hessenberg A_j, we can proceed one step QR algorithm in the following way. First, compute the first column of $p(A_j)$. Second, use Householder or Givens transformation to find orthogonal P_1, s.t. $P_1^H p(A_j) e_1$ is a multiple of e_1. According to the procedure of Householder or Givens, there exists another orthogonal P_2, which we are not interested, s.t. $P_2^H P_1^H p(A_j)$ is upper-triangular, and $P_2 e_1 = e_1$, i.e., if we let $P = P_1 P_2$, then $P e_1 = P_1 e_1$. Third, compute the matrix $P_1^H A_j P_1$. Fourth, use Givens rotation to find Q_1 s.t. $Q_1 e_1 = e_1$ and $Q^H A_j Q$ is upper-Hessenberg, where $Q = P_1 Q_1$. Thus, Q satisfies $p(A_j) = QR$, and $Q^H A_j Q$ is just the A_{j+1} we need. This gives the following algorithm:

Algorithm 1. one-step QR with implicit generalized Rayleigh quotient shift (Compute the first column of $p(A) = \sum_{i=0}^{m} a_i A^i$, suppose A is upper-Hessenberg)

```
p(A)e_1 = a_0 e_1, v = e_1
for i = 1 : m
    v(1 : i+1) = H(1 : i+1, 1 : i)v(1 : i)
    p(A)e_1(1 : i+1) = p(A)e_1(1 : i+1) + a_i v(1 : i+1)
end
(compute P_1^H A P_1 via Givens rotations)
for i = 2 : m+1
    compute Givens matrix G(1, i) to eliminate p(A)e_1(i)
    A = G(1, i)AG(1, i)^H
end
(Restore A to upper-Hessenberg)
for j = 1 : n-2
    for i = j+2 : min(j+m+1, n)
        compute Givens matrix G(i-1, i) to eliminate A(i, j)
        A = G(i-1, i)AG(i-1, i)^H
    end
end
```

The complexity of Algorithm 1 is $O(m^2)$ for computing the first column of $p(A)$, $O(mn)$ for computing $P_1^H AP_1$, and $O(mn^2)$ for restoring A to upper-Hessenberg. Thus, the total cost is $O(mn^2)$.

4 Choice of m

The overall process of QR method is a deflation process. That is, After some iterations, we hope a certain subdiagonal entry $A(i+1, i)$ will converge to zero, and then the large matrix A can be deflated into two smaller submatrices $A(1 : i, 1 : i)$ and $A(i+1 : n, i+1 : n)$. Then we repeat the iterations recursively for the submatrices, until each submatrix is small enough for us to get the eigenvalues easily.

According to the complexity of Algorithm 1, the cost per iteration is proportional to the order of the polynomial m, i.e., the larger m is, the more computation it takes per iteration. However, there are two factors which can offset the disadvantage of larger m in the generalized Rayleigh quotient shift strategy.

The first is the speed of convergence. Our experiments show that, sometimes, it may take much less iterations with a larger m than a smaller one to reach the convergence.

The second is the speed of deflation. Our experiments show that, if m is not too large, say $m = 3, 4, 5, 6...$, then in most cases, though not always, m equals the size deflated. That is, if we apply an m-order shift on an $n \times n$ matrix, chances are that it will deflate into an $(n-m) \times (n-m)$ matrix and an $m \times m$ one. Thus, the speed of deflation is almost proportional to m.

From these two factors, we see that the second one has already offset the disadvantage of larger m in a lot of cases. So what counts more is the speed of convergence.

The following theorem in [1] gives the heuristic idea of choosing m.

Theorem 3. *Let $A_0 \in C^{n\times n}$, and let p be a polynomial. Let $\lambda_1, ..., \lambda_n$ denote the eigenvalues of A_0, ordered so that $|p(\lambda_1)| \geq |p(\lambda_2)| \geq \cdots \geq |p(\lambda_n)|$. Suppose k is a positive integer less than n such that $|p(\lambda_k)| > |p(\lambda_{k+1})|$, let $\rho = |p(\lambda_{k+1})|/|p(\lambda_k)|$, and let (p_i) be a sequence of polynomials such that $p_i \longrightarrow p$ and $p_i(\lambda_j) \neq 0$ for $j = 1, ..., k$ and all i. Let $\mathcal{T}$ and $\mathcal{U}$be the invariant subspaces of A_0 associated with $\lambda_1, ..., \lambda_k$ and $\lambda_{k+1}, ..., \lambda_n$, respectively, and suppose $< e_1, ..., e_k > \cap \mathcal{U} = \{0\}$. Let (A_i) be the sequence of iterates of the GR algorithm using these p_i, starting from A_0. If there exists a constant $\hat{\kappa}$ such that the cumulative transformation matrices $\hat{G}_i$ all satisfy $\kappa_2(\hat{G}_i) \leq \hat{\kappa}$, then (A_i) tends to block triangular form, in the following sense. Write*

$$A_i = \begin{bmatrix} A_{11}^{(i)} & A_{12}^{(i)} \\ A_{21}^{(i)} & A_{22}^{(i)} \end{bmatrix},$$

where $A_{11}^{(i)} \in k \times k$. Then for every $\hat{rho}$ satisfying $\rho < \hat{\rho} < 1$ there exists a constant C such that $\| A_{21}^{(i)} \| \leq C\hat{\rho}^i$ for all i.

Proof. See Watkins and Elsner [1]

From this theorem, we know that if the m roots of p were eigenvalues of A_i, then $p(\lambda_{n-m+1}) = p(\lambda_{n-m+2}) = \cdots = p(\lambda_n) = 0$, so the subdianognal entry $A_i(m+1, m)$ would become zero after one iteration with this shift.

However, the eigenvalues are unknown, so what we need to do is to find a polynomial whose roots are close to eigenvalues. If

$$A_i = \begin{bmatrix} A_{11}^{(i)} & A_{12}^{(i)} \\ A_{21}^{(i)} & A_{22}^{(i)} \end{bmatrix},$$

where the upper-right entry of $A_{21}^{(i)}$ (which is the only non-zero entry, because of upper-Hessenberg) is small, then the characteristic polynomial of $A_{21}^{(i)}$ would have the roots close to the eigenvalues, and $p(\lambda_{n-m+1})/p(\lambda_{n-m})$ would be near zero, which makes $A(n-m+1, n-m)$ converge to zero rapidly.

Thus, we may search for the subdiagonal entry, whose absolute value is the smallest among those near the lower-right corner of the matrix, say, $A(k+1, k)$, then probably $m = n - k$ is a good choice.

Two other issues need to be concerned. First, m cannot be too large. Anyway, the disadvantage of large m in each iteration is inevitable, while the advantage is not always guaranteed. So the larger the m is, the higher risk we're taking slowing down the algorithm. That is why we only search near the lower-right corner. For example, we may require $m < M$, where M may be chosen as 4, 5, 6,etc. The other way is to set a threshold constant c, and find the m, s.t. $|A(n-m+1, n-m)|/c^m$ is the smallest. Second, when an m is chosen in this way, we are not yet ensured a fast convergence, so we need to set a limit for the iteration times. If unfortunately, after a certain number of iterations, it hasn't converged, then we'd better re-choose m by testing $m = 1, 2, 3...$ one by one.

Based on these ideas, we may use the following algorithm:

Algorithm 2. Overall process of QR algorithm with implicit generalized Rayleigh quotient shift

If we skip the initial choice, then,

modified m	*iterations*	*matrix order after deflation*
2	1	99
2	5	97
2	3	95
1	1	94
1	2	93
1	2	92
1	4	91
1	3	90
1	4	89
1	3	88
1	5	86
1	3	85
1	7	84
1	4	83
1	4	82
1	3	81

Again, algorithm 2 has some advantage.

There seems to be some similar results as to the situation when a subdiagonal in a leading (rather than trailing) submatrix has small modulus, and we'll have a better choice of m in these cases.

References

1. Watkins, D., Elsner, L.: Convergence of Algorithms of Decomposition Type for the Eigenvalue Problem. Linear Algebra Appl. **78** (1982) 315–333
2. Golub, G., Van Loan, C.: Matrix Computations. 3rd Edition, The Johns Hopkins University Press, Baltimore and London, 1996
3. Dubrulle, A., Golub, G.: A Multishift QR Iteration without Computation of the Shifts. SCCM technical report, NA-93-04

Parameter Estimation of Si Diffusion in Fe Substrates After Hot Dipping and Diffusion Annealing

B. Malengier

Research Group for Numerical Functional Analysis and Mathematical Modelling (*NfaM*^{2}),
Department of Mathematical Analysis,
Ghent University, Galglaan 2, B-9000 Gent, Belgium
bm@cage.UGent.be
http://cage.UGent.be/nfam2

Abstract. In this paper a general model is developed for the simulation of one dimensional diffusion annealing. Our main interest is the determination of the diffusion coefficient from measured values at discrete space-time points within the sample. The method is based on a suitable reduction of the PDE to a system of ODEs by a second order finite difference space discretization. The inverse problem is solved by implementation of the Levenberg-Marquardt method. This allows the estimation of the parameters and the determination of Cramér-Rao lower bounds.

1 Introduction

Iron silicon alloys are extensively used as soft magnetic materials in electrical devices. A higher silicon content increases the electrical resistance and improves the magnetic properties. Several methods have been developed to obtain high Si steel, see [3] and references therein.

In [3] the authors propose a hot dipping process followed by diffusion annealing. During the hot dipping a normal steel substrate (low Si) is dipped into a hypereutectic Al-Si-bath. This produces a thin sheet with Al and Si at the surface. To obtain the required Si amount over the entire thickness of the steel sheet, the sample is placed into a furnace to allow diffusion annealing in order to enrich the Fe-matrix with Si. At the micro level this process is difficult to model: the thin sheet consists of different compounds (eg. $Fe_3Al_3Si_2$), holes can develop, etc. On the macro level, the picture is much clearer as Fick's law of diffusion can be used. Therefore, we use the following phenomenological equation for the Si concentration C, expressed in atomic percentage (At%),

$$\partial_t C - \nabla \cdot (D(C)\ \nabla C) = 0,$$

where the diffusion coefficient D depends on the Si concentration. This diffusion coefficient must be determined experimentally for each type of Fe-matrix.

Z. Li et al. (Eds.): NAA 2004, LNCS 3401, pp. 399–407, 2005.

An outline of the paper is in order. In Section 2 we present the mathematical model for diffusion annealing of a thin steel sheet. In Section 3 the numerical approximation in terms of a system of ODEs is developed, and in Section 4 some numerical examples are given.

2 Mathematical Model

2.1 Setup

We describe diffusion annealing in a thin sheet by a 1D-model. Let the sheet be a rectangular beam with thickness $2L$, width W and heigth H. Typically, we have $2L = 800\mu$m, $W = 5$cm and $H = 10$cm. During hot dipping a surface layer is formed over the entire sample. We need only consider the diffusion annealing over the thickness of the sample, away from the edges. By symmetry reasons this diffusion is uniform over the beam, thus we get a 1D model. By symmetry reasons we may reduce the diffusion phenomenon to half of the thickness of the sheet, i.e. over the interval $(0, L)$, when imposing a homogeneous Neumann BC at the middle of the sheet ($x = L$). Thus, we arrive at the following 1D-model, consisting of the diffusion equation

$$\partial_t u - \partial_x \left(D(u)\, \partial_x u\right) = 0 \quad \text{in } (0, L) \times (0, T), \tag{1}$$

with $D(u) \in L^\infty$, $D(u) \geq 0$, along with boundary conditions

$$u(0, t) = c_0 \quad \text{or} \quad - D(u(0, t))\partial_x u = 0, \tag{2}$$

$$-D(u(L, t))\partial_x u = 0, \quad 0 < t < T, \tag{3}$$

and an initial condition

$$u(x, 0) = u^0(x), \quad x \in (0, L). \tag{4}$$

Here, u is the concentration of silicon (Si) in steel (Fe). Here, $(0, T)$ is a given time interval, u^0 represents the given initial concentration profile and c_0 is a constant.

2.2 The Substrate Problem

The initial concentration profile $u^0(x)$ entering (4) results from the hot dipping process. We call this setup the *substrate problem.*

Definition 1. *The substrate problem is defined by Equations (1)-(4) where the initial condition $u^0(x)$ is given by*

$$u^0(x) > \delta \quad x \in [0, s_0[, \quad u^0(x) = \delta \quad x \in [s_0, L], \tag{5}$$

with $0 < s_0 < L$, $\delta > 0$. We assume further that $u^0(x)$ is smooth up to s_0.

It follows immediately that $u^0(x)$ is not analytic in s_0. The value δ is the amount of Si in Fe before the hot dipping. It is typically between 0 and 3 At%. As a special case we have the *zero substrate problem* when $\delta = 0$. A characteristic property of the substrate problem is the movement of the *contact point* $x = s(t)$.

Definition 2. *The contact point is the point $x = s(t)$ where $u(s(t), t) = \delta$, with $s(0) = s_0$.*

The speed of the contact point is denoted as $\dot{s}(t)$. We have the following property

Proposition 1. *Assume that the speed of the contact point is finite in the substrate problem. Let (1) be satisfied in $x = s(t)$ in the limit sense. Then the following holds*

$$\dot{s}(t) = -\lim_{x \to s(t)^-} \left(D'(u)\partial_x u + D(u) \frac{\partial_x^2 u}{\partial_x u} \right) \tag{6}$$

$$= -\lim_{x \to s(t)^-} \left(\partial_x \int_0^u \frac{D(z)}{z - \delta} \, \mathrm{d}z \right) \tag{7}$$

Proof. From the definition of the contact point we have that $u(s(t), t)$ is independent on t. Therefore, if $\dot{s}(t)$ is finite,

$$0 = \frac{\mathrm{d}}{\mathrm{d}t} u(s(t), t) = \partial_t u + \dot{s}(t) \partial_x u.$$

Therefore,

$$\begin{aligned} \dot{s}(t) &= -\lim_{x \to s(t)-} \frac{\partial_t u}{\partial_x u} = -\lim_{x \to s(t)^-} \left(D'(u)\partial_x u + D(u) \frac{\partial_x^2 u}{\partial_x u} \right) \\ &= -\lim_{x \to s(t)^-} \frac{D(u)\partial_x u}{u - \delta} = -\lim_{x \to s(t)^-} \left(\partial_x \int_0^u \frac{D(z)}{z - \delta} \, \mathrm{d}z \right). \end{aligned}$$

Here, for the second line we noted that the flux $D(u)\partial_x u$ and $u - \delta$ both tend to zero at the interface allowing the use of de l'Hospital's rule in reverse (this implies $D(\delta) = 0$) which is indeed a requirement for finite propagation speed, practically we will have this only for $\delta = 0$). □

To ensure that (1) holds in $x = s(t)$ in the limit sense, we have that $u^0(x)$ is not C^∞ apart from being not analytic in s_0.

Remark 1. From (7) it follows that the function $\frac{D(u)}{u - \delta}$ must be integrable.

Remark 2. From (6) some deductions can be made on the form of the concentration profile $u(x, t)$ in the contact point in order that for a given diffusion coefficient the speed $\dot{s}(t)$ is indeed finite. If for example $D(u) = u^p$, $p > 0$, then we have $\lim_{x \to s(t)-} D(u) = 0$ in the zero substrate problem. The propation speed (6) is finite in following cases.

– If $\lim_{x \to s(t)-} \frac{\partial_x^2 u}{\partial_x u}$ is finite or behaves as $\frac{1}{u^{p-k}}$, where $0 \le k < p$, the second term of (6) cancels.

- If $p = 1$, then $D'(u) = c$, a constant. Therefore $\lim_{x \to s(t)-} \partial_x u = f(t)$, with $f(t) > 0$.
- If $p > 1$, then $\lim_{x \to s(t)-} D'(u) = 0$. In order to have a finite speed, it is necessary that $\partial_x u \sim u^{1-p}$. Like this, $\lim_{x \to s(t)^-} D'(u)\partial_x u$ is still a finite function of t. This means however that $\lim_{x \to s(t)^-} \partial_x u = \infty$.
- If $p < 1$, the same deduction can be made, now $\lim_{x \to s(t)-} \partial_x u = 0$.

Remark 3. From Remark 2 it follows that $\partial_x u$ should be avoided in numerical computations. Therefore the form (7) should be used. If the speed of the contact point is finite, the function $F(x)$ defined by

$$F(x) \equiv \tilde{F}(u(x)) = \int_0^{u(x)} \frac{D(z)}{z},$$

will have a finite derivative in $x = s(t)$.

In general, the initial concentration profile will only be a given set of data-points. Therefore, $u^0(x)$ will be constructed by suitable polynomial interpolation, so that $\partial_x u$ and $\partial_x^2 u$ are two constants, not zero. From (6) we obtain that $\dot{s}(t) = 0$ for $p > 1$, and $\dot{s}(t) = \infty$ for $0 < p < 1$. This does not allow us to recover $s(t)$, but indicates what will happen numerically: the initial function will transform to a profile with the desired derivative in the contact point.

2.3 The Inverse Problem Setup

Assume that we have the measured concentrations

$$C^*(x_i, t_i) \quad \text{for } x_i \in (0, L),\ t_i \in (0, T), \quad i = 1, \ldots, n. \tag{8}$$

From (8) the function $D(C)$, $C \in (0, 1)$, has to be restored. Our aim is to determine the unknown function $D(s)$, $s \in (0, 1)$, so that the measured values (8) are well-approximated by the numerical results from the corresponding direct problem.

To this end we look for a function D in a class of explicit functions parametrized by a vector $\mathbf{p} = (p_1, \ldots, p_m)$. More explicitely, we take $D = D(s, p)$, where $s \in (0, 1)$ and $\mathbf{p} \in U_{\mathrm{ad}} \subset \mathbf{R}^m$, with U_{ad} an admissible compact subset of $\mathbf{R}^m$. Possible choices for $D(s, \mathbf{p})$ are

$$D(s, \mathbf{p}) = p_1 s^{p_2}(1 + p_3 s + p_4 s^2), \tag{9}$$

with $p_2 > 0$, or

$$D(s, \mathbf{p}) = p_j + \frac{p_{j+1} - p_j}{h}(s - jh) \quad \text{for } s \in (jh, (j+1)h), \quad j = 0, \ldots, m-1, \tag{10}$$

with $h = \frac{1}{m}$. Note that for (9) and (10) we indeed have that $\frac{D(u)}{u-\delta}$ is integrable, as required from Remark 1.

We look for an optimal vector-parameter $\widehat{\mathbf{p}} = (\hat{p}_1, \ldots, \hat{p}_m)$ such that the cost functional

$$\mathcal{F}(\mathbf{p}) \equiv \mathcal{F}(C, \mathbf{p}) \sum_i [C(x_i, t_i, \mathbf{p}) - C^*(x_i, t_i)]^2 \, dx \, dt, \tag{11}$$

attains its minimum on U_{ad} at $\mathbf{p} = \widehat{\mathbf{p}}$. Here, $C(x, t, \mathbf{p})$ is the solution of (1)-(4). The vector $\widehat{\mathbf{p}}$ is obtained as the limit of a sequence $\{\mathbf{p}_k\}_{k=1}^{\infty}$ such that $\mathcal{F}(\mathbf{p}_{k+1}, C_{k+1}) < \mathcal{F}(\mathbf{p}_k, C_k)$ and $\mathcal{F}(\mathbf{p}_k, C_k) \to \mathcal{F}(\widehat{\mathbf{p}}, \widehat{C})$, where $C_k = C(x, t, \mathbf{p}_k))$. We will construct this sequence by a Levenberg-Marquardt method.

Remark 4. In the special case $D(C, p) = (p+1)C^p$, $p > 0$, a closed-form solution of (1)-(4) exists in case the initial profile is $C(x, 0) = E\delta(x)$ (the Dirac measure). It is the Barenblatt-Pattle solution, [5] p.31, taking the form

$$C(x,t) \begin{cases} t^{-1/(p+2)}(1 - (x/s(t))^2)^{1/p}, \text{for } |x| < s(t); \\ 0, \text{for } |x| \geq s(t), \end{cases} \tag{12}$$

with the interface given by $x = s(t)$, $s(t) = \sqrt{\frac{2(p+1)(p+2)}{p}} t^{1/(p+2)}$. The solution (12) has a singularity at $x = s(t)$. We see that this solution shows the properties mentioned above. Thus, we have that at $x = s(t)$ $\partial_x C$ is 0 or ∞ depending on p, and that $\partial_x C \sim C^{1-p}$.

3 Numerical Realization for a 1D Problem

3.1 The Direct Problem

Let $\mathbf{p}$ be given. We construct the solution $C(x, t, \mathbf{p})$ of (1)-(4) in an approximative way as follows. The problem (1)-(4) is reduced to an initial value problem for a nonlinear system of ODEs by means of a nonequidistant finite difference discretization with respect to the space variable. Next, a stiff ODE solver is used to solve the system of ODEs.

We consider two possibilities. In the first we assume there is a contact point $x = s(t)$ of which the speed is determined by (7). The second possibility is that there is no contact point. In that case we set $s(t) = L$ and $\dot{s}(t) = 0$.

Our procedure is as follows. We split $(0, L)$ into two domains $\Omega_1 \equiv (0, s(t))$ and $\Omega_2 \equiv (s(t), L)$. We transform the PDE (1) on $\Omega_1(t)$ to the fixed domain $(0, 1)$ using Landau's transformation $y = \frac{x}{s(t)}$. Denoting the corresponding solution by $\overline{C^1}(y, t)$, we have that

$$\partial_t C^1 = \partial_t \overline{C^1} - y \frac{\dot{s}(t)}{s(t)} \partial_y \overline{C^1}, \quad \partial_x C^1 = \frac{1}{s(t)} \partial_y \overline{C^1}, \quad \partial_x^2 C^1 = \frac{1}{s^2(t)} \partial_y^2 \overline{C^1}. \tag{13}$$

Consequently the PDE we need to solve becomes

$$\partial_t \overline{C^1} - \frac{1}{s^2(t)} \partial_y \left(D(C^1) \partial_y \overline{C^1} \right) - y \frac{\dot{s}(t)}{s(t)} \partial_y \overline{C^1} = 0. \tag{14}$$

The interval $(0,1)$ is partitioned by the set of grid points $\{y_i\}_{i=0}^N$, with $y_i = \sum_{l=0}^i \alpha_l$, $i = 0,\dots,N$, where $\alpha_0 = 0$ and $\sum_{l=0}^N \alpha_l = 1$. We can choose these gridpoints so as to obtain a more dense discritization around the point $y = 1$. We denote $C_i^1(t) \approx \overline{C^1}(y_i,t)$ and let $l_2(y,i)$ stand for the Lagrange polynomial of the second order interpolating the points (y_{i-1}, C_{i-1}^1), (y_i, C_i^1) and (y_{i+1}, C_{i+1}^1). Then, we approximate $\partial_y \overline{C^1}$ by $dl_2(y_i,i)/dy \equiv (dl_2(y,i)/dy)_{y=y_i}$ and $\partial_y^2 \overline{C^1}$ by $d^2l_2(y_i,i)/dy^2 \equiv (d^2l_2(y,i)/dy^2)_{y=y_i}$. In the case of a Dirichlet BC, the nodal point y_0 need not be considered. In the case of a Neumann BC, we extend the governing PDE to the boundary point and discretize it similarly as for the inner points by the introduction of a fictive point y_{-1}. Equation (14) leads to the system of ODE's

$$\frac{d}{dt}C_i^1(t) - \frac{1}{s^2(t)}D(C_i^1)\frac{d^2}{dy^2}l_2(y_i,i) - \frac{1}{s^2(t)}D'(C_i^1)\left[\frac{d}{dy}l_2(y_i,i)\right]^2 - y_i\frac{\dot{s}(t)}{s(t)}\frac{d}{dy}l_2(y_i,i) = 0, \quad (15)$$

for $i = 1,\dots,N-1$, and, in the case of a Neumann BC, also for $i = 0$. In the second domain $\Omega_2(t)$ the concentration remains constant, $C(x,t) = \delta$.

In the interface point $x = s(t)$ the following ODE needs to be satisfied

$$\dot{s}(t) = -\lim_{y\to 1^-} \frac{\frac{d}{dy}F_l}{s(t)} \quad (16)$$

where F_l is the second degree Lagrange polynomial interpolating the points $(y_{N-2}, \tilde{F}(C_{N-2}^1))$, $(y_{N-1}, \tilde{F}(C_{N-1}^1))$ and $(1, \tilde{F}(s(t)) \equiv 0)$. This equation only applies as long as $s(t) < L$. When $s(t) \geq L$, Equation (16) is replaced by a homogeneous Neumann condition in $y = 1$, so that in (15) $i = 1,\dots,N$. Next, we solve (15)-(16) by means of a standard package for stiff ODEs, e.g. LSODA based on a backward finite difference formula.

Remark 5. In the case of a power type diffusion coefficient (9) an additional transformation is performed, see also [2]. This transformation is suggested by Remark 2 where we noted that $\partial_x u \sim u^{1-p}$. Under the transformation $u = v^{1/p}$, this leads to the expression $p \sim \partial_x v$. Therefore, all power degeneracies are removed, making $v(x)$ more suitable to apply Lagrange interpolation. For example, (12) is transformed into an expression for v that is quadratic in x. Using Lagrange polynomials of the second order for the space discretization, the space interpolation will be exact. Hence, the differences between the exact and the numerical solution of the Barenblatt-Pattle problem will be solely due to time integration errors. Generally, under the transformation $C = v^{1/p}$, instead of (15) we obtain the ODE system

$$\frac{d}{dt}v_i^1(t) - \frac{1}{s^2(t)}D((v_i^1)^{1/p})\frac{d^2}{dy^2}\tilde{l}_2(y_i,i) - \frac{\left[\frac{d}{dy}\tilde{l}_2(y_i,i)\right]^2}{s^2(t)pv_i^1} \Big(D'((v_i^1)^{1/p})(v_i^1)^{1/p} + (1-p)D((v_i^1)^{1/p})\Big) - y_i\frac{\dot{s}(t)}{s(t)}\frac{d}{dy}\tilde{l}_2(y_i,i) = 0, \quad (17)$$

and likewise for (16), where $\tilde{l}_2(y_i, i)$ is now the Lagrange polynomial of the second order interpolating the points (y_{i-1}, v^1_{i-1}), (y_i, v^1_i) and (y_{i+1}, v^1_{i+1}).

3.2 The Inverse Problem

We use a Levenberg-Marquardt method for the inverse problem. This method only requires the implementation of the direct problem. The gradient $\nabla_p \mathcal{F}(\mathbf{p})$ is approximated numerically. The algorithm reads as follows. Starting from an initial parameter set $\mathbf{p}^k$ we obtain a new set by

$$\mathbf{p}^{k+1} = \mathbf{p}^k + \left[\frac{1}{J_k^T J_k + \lambda I}(J_k^T D_k)\right] \tag{18}$$

with $(J_k)_{ij} = \partial_{\mathbf{p}_j} C(x_i, t_i, \mathbf{p}^k)$, the Jacobian which is determined numerically, and $(D_k)_i = C(x_i, t_i, \mathbf{p}) - C^*(x_i, t_i)$. If the new parameter set has a smaller cost functional than the old one, the set is retained, and λ is divided by an ever increasing integer so as to get quadratic convergence. If not, the parameter set is discarded, and another set is sought with a bigger λ value. The algorithm stops when λ becomes larger than a preset value, or when sufficient precision is reached. We denote the obtained set of parameters as $\widehat{\mathbf{p}}$.

The advantage of the Levenberg-Marquardt method is the possibility to obtain the Cramér-Rao lower error bound (CRB) on $\widehat{\mathbf{p}}$. Indeed we have, [1],

$$\mathrm{Cov}(\widehat{\mathbf{p}}) \geq \mathrm{CRB} = \left[J(\widehat{\mathbf{p}})^T J(\widehat{\mathbf{p}})\right]^{-1}, \tag{19}$$

where $J(\widehat{\mathbf{p}})$ is the Jacobian at the minimum $\widehat{\mathbf{p}}$. The determination of the CRB in inverse problems is important, as it allows to determine the deviation and the correlation between the different parameters, as well as confidence regions. The correlation ρ is given by

$$\rho_{p_i p_j} = \frac{CRB_{ij}}{\sqrt{CRB_{ii} CRB_{jj}}}. \tag{20}$$

Here $\rho = 0$ indicates the situation without correlation; a value close to 1 reflects a strong correlation.

4 Numerical Experiments

Direct Problem Example. In this example, we consider the data from an experiment of hot dipping and diffusion annealing in a steel sample with no Si present at the beginning. In Fig. 1 the Si concentration profile after hot dipping is given. This is the initial condition for the zero substrate problem.

The other smooth curves are the concentration profiles obtained for $D(C) = 0.4C^{1.59}$ ($\mathbf{p} = (0.4, 1.59, 0, 0)$ in (9)) after 1, 7 and 30 minutes. An experimental Si profile measured after 30 minutes of diffusion annealing is also plotted. This curve matches very well the modeled curve.

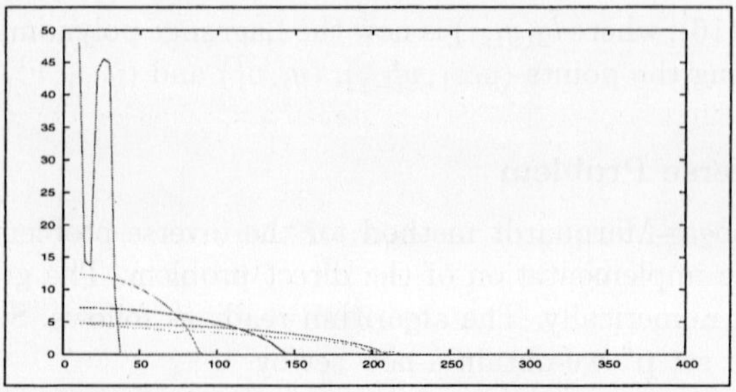

Fig. 1. Diffusion profiles of Si (At%) into steel (μm). A typical Si profile after hot dipping is the initial condition. Other curves are simulations after 1, 7 and 30 minutes, and an experimental Si profile after 30 min

Inverse Problem Example. For the inverse problem we construct the following setup. The direct problem is solved for $\mathbf{p} = (0.5, 1.5, 0.2, 0.2)$ in (9), and data is stored after 1 and 2 minutes. With these measurements, the inverse problem is run starting from $\mathbf{p}^0 = (1, 1, 0, 0)$. We stop the iterations of the algorithm when the root-mean square error (RMS) is smaller than 0.015. After 143 iteration steps we obtained $\widehat{\mathbf{p}} = (0.547, 1.549, 0.0445, 0.189)$, with CRB given by

$$\text{CRB} = \begin{bmatrix} 0.14 & -0.0074 & -0.19 & -0.022 \\ -0.0074 & 0.0034 & 0.0075 & 0.0009 \\ -0.19 & 0.0075 & 0.27 & 0.031 \\ -0.022 & 0.0009 & 0.031 & 0.0039 \end{bmatrix} .$$

This yields $\sigma_{\widehat{\mathbf{p}}} = (0.37, 0.06, 0.5, 0.06)$ for the deviation . This is in correspondance with the real value of $\mathbf{p}$, and, in particular, indicates that p_3 is not well determined yet. Further, note that there is a big correlation between p_1 and p_3 ($\rho_{p_1 p_3} = 0.97$), between p_1 and p_4 ($\rho_{p_1 p_4} = -0.94$), and between p_3 and p_4 ($\rho_{p_3 p_4} = 0.96$). It follows that the power type of diffusion coefficient, (9), is perhaps not the best candidate. A diffusion coefficient dependent on parameters that are less correlated would be favorable. The high correlation is also reflected in a RMS in parameter space that exhibits a broad valley. Therefore, all inverse methods based on minimisation of (11) will loose efficiency. In this example, we had for $\mathbf{p}^0$ an RMS=4.6, and $\mathbf{p}^{20} = (0.19, 1.9, 0.32, 0.30)$ with RMS=0.17, $\mathbf{p}^{50} = (0.61, 1.83, -0.28, 0.16)$ with RMS=0.08. Only from here on, p_3 starts to converge slowly to its correct value.

Acknowledgment

The author thanks Prof. Y. Houbaert of the laboratory LabMet of Ghent University for suggesting the problem, Dr. J. Barros of the same laboratory for providing experimental data, and Dr. D. Constales of the research group *NfaM*^{2} of Ghent University for providing parts of the code. Also many thanks to Prof. R. Van Keer, Prof. J. Kačur, and Dr. D. Constales for the many useful discussions.

References

1. R.C. Aster, B. Borchers, C. Thurber, *Parameter Estimation and Inverse Problems.* Academic Press, 2004 (ISBN 0120656043)
2. D. Constales and J. Kačur. Determination of soil parameters via the solution of inverse problems in infiltration. *Computtional Geosciences*, **5**, 25–46, 2001.
3. T. Ros-Yañez and Y. Houbaert. High-silicon steel produced by hot dipping and diffusion annealing. *Journal of Applied Physics*, **91**, 10, 7857–7859, 2002.
4. E. Polak, *Optimization: algorithms and consistent approximations.* Springer, 1997.
5. A.D. Polyanin and V.F. Zaitsev, *Handbook of Nonlinear partial differential equations.* Chapman&Hall/CRC, 2004. (ISBN 1-58488-355-3)

Numerical Design of Optimal Active Control for Seismically-Excited Building Structures

Daniela Marinova[1] and Vasil Marinov[2]

[1] Technical University-Sofia, Sofia 1756, Bulgaria
dmarinova@dir.bg
[2] Technical University-Sofia, Sofia 1756, Bulgaria

Abstract. This paper presents a dynamical model of an active control system for seismic-resistant building structures. Three optimal performance indexes: LQR, discrete time-dependent non-integral and generalized LQR, based on linear quadratic optimization are considered. The maximum structural response and maximum active control force versus control design parameters are studied. The influence of the time increment used in response analysis on the algorithm is investigated. Numerical examples illustrate the effectiveness of the proposed algorithms in reducing the structural response.

1 Introduction

The vibrations of flexible structures, like tall buildings, during earthquakes or gust winds may influence the comfort of the users or even lead to damages and destruction of the structural system. Hence, the reducing of the vibrations of civil engineering structures during impact force has considerable attention. The dynamical response of a linear elastic structure depends on the mass and elasticity properties. Passive design is based on appropriate choice of the materials and on the dimensions of the structure. It produces energy dissipating mechanisms activated by the motion of the structure itself. A new trend is applying of a control system which can modify the internal stresses, the mass or the damping of the dynamical system and can keep the response of the resulting "intelligent" structure within required limits. The active control system requires external energy for its operation [1].

Several control techniques have been recently developed as a possible way of reducing the vibrations of civil engineering structures during seismic excitations or strong wind gusts. Based on system control theory active control systems have been promoted. One effective approach is the optimal control theory. Each structural control system should have appropriate optimal control algorithm suitable to the system's characteristics and its external loads. The optimal control law can be chosen using different ways. Linear quadratic control refers to the body of the powerful machinery of optimal control and is widely used in structural control [2]. The linearity of this controller and phase margin theoretically provided by this technique are the main reasons for its choice. Other control

Z. Li et al. (Eds.): NAA 2004, LNCS 3401, pp. 408–415, 2005.

algorithms as robust H_2 and H_{inf} based on the minimizing the H_2 and H_{inf} norms of the corresponding transfer matrices have been considered for building structures [3-5].

This paper presents two dimensional dynamical model of active controlled seismic-resistant building structures. A linear quadratic optimization problem is considered. Three optimal performance indexes are considered. The first criterion is connected with Linear Quadratic Regulator (LQR) problem that uses an integral performance index representing the balance between structural response and control energy and leads to control forces proportional to the structural response. The second criterion is a discrete time-dependent non-integral performance index in which the optimality is achieved at each instant time and leads to optimal control forces that are proportional to the time step and the structural response. The third criterion represents an integral generalized LQR performance index that generalizes the ideas of the first and second proposed criteria. For the proposed algorithms the maximum structural response and maximum active control force versus control design parameters are studied. The influence of the time increment on the feedback control gain matrix produced by every of the three algorithms is investigated. Numerical examples illustrate the effectiveness of the proposed algorithms in reducing structural response for an active control building under instantaneous external excitation.

2 Equation of Motion

We consider a linear building structure with n degree of freedom subjected to instantaneous external load. The equation of motion of the seismic resistant building structure with active controller can be expressed as [6]

$$\ddot{X}(t) + C\dot{X}(t) + KX(t) = Hu(t) + F(t) \tag{1}$$

where $X(t)$ of (n, l) is the state vector of the displacements, $u(t)$ of (m, l) is the control vector, M of (n, n) is the mass matrix, K of (n, n) is the stiffness matrix. The damping matrix C of (n, n) is supposed to be proportional to a linear combination of mass and stiffness matrices. H of (n, m) is a matrix defining the locations of the control inputs. $F(t)$ is the external load. For control design the equation (1) can be transformed in state-space form of a set of first order differential equations

$$\dot{x} = Ax + Bu + DF \tag{2}$$

where $x(t) = \begin{bmatrix} X(t) \; \dot{X}(t) \end{bmatrix}^T$ of $(2n, 1)$ is the state vector, $A = \begin{bmatrix} 0 & I \\ -M^{-1}K & -M^{-1}C \end{bmatrix}$ of $(2n, 2n)$ is square matrix determining the system dynamics, $B = \begin{bmatrix} 0 \; M^{-1}H \end{bmatrix}^T$ of $(2n, m)$ is the control input matrix, $D = \begin{bmatrix} 0 \; M^{-1} \end{bmatrix}^T$ is a matrix determining the external load distribution. The corresponding initial conditions for the equation (2) are

$$x(0) = 0, \quad u(0) = 0, \quad F(0) = 0 \tag{3}$$

The equation (2) cannot be solved directly because there are only $2n$ equations but $(2n+m)$ unknown variables, i.e. $x(t)$ of $(2n,1)$ and $u(t)$ of $(m,1)$. In order to solve the active control problem shown in equation (2), m more equations are needed.

Consider the solution of the equation (2) assuming that the optimal control force $u^\star(t)$ is obtained and the external excitation $F(t)$ has been measured. The structural response $x(t)$ can be found analytically using the modal approach. Let the state vector be expressed as

$$x(t) = Tz(t) \tag{4}$$

where T of $(2n,2n)$ is a matrix constructed from the eigenvectors of the matrix A. The corresponding modal state equation is

$$\dot{z}(t) = \pi z(t) + \psi(t) \tag{5}$$

$$\pi = T^{-1}AT, \quad \psi(t) = T^{-1}Bu^\star(t) + T^{-1}DF(t) \tag{6}$$

The modal plant matrix is a decoupled matrix. The initial conditions are

$$z(0) = 0, \quad u(t) = 0, \quad F(0) = 0 \tag{7}$$

The solution of equation (5) can be obtained by the integral

$$z(t) = \int_0^t exp(\pi(t-\tau))\psi(\tau)d\tau \tag{8}$$

Using the initial conditions (7), the integration of the equation (8) can be solved numerically. In order to evaluate $z(t)$ at time t, only one measurement of the force $F(t)$ is needed. We shall determine the vector of control forces $u^\star(t)$ such that it will satisfy the dynamic equations (5) and it will be subjected to optimal performance criterion. The control force is regulated by the feedback state vector $x(t)$. Therefore, we need to measure only the response of the system at time instant t that can be obtained by placing displacement and velocity sensors at each floor of the building structure. Three different algorithms of linear quadratic optimal control law will be considered.

3 LQR Control Law

The Linear Quadratic Regulator algorithm gives a control law proportional to the structural states. The optimal control $u^\star(t)$ is obtained by the minimizing the standard quadratic performance index

$$J = \int_{t_0}^{t_f} [x^T(t)Qx(t) + u^T(t)Rx(t)]dt \tag{9}$$

and satisfying the state equation (2). Here t_0 is the initial and t_f is the final time instants. In equation (9) both of the two boundary values of the integrand are specified by the following boundary conditions

$$x(t_0) = 0, \quad x(t_f) = 0, \quad u(t_0) = 0, \quad u(t_f) = 0 \tag{10}$$

The matrix Q of $(2n, 2n)$ is positive semi definite matrix and R of (m, m) is positive definite matrix. Performance index J represents a balance between structural response and control energy. The matrixes Q and R are the main design parameters in this algorithm.

Let now assume that the system (2) is controllable and observable. Than the solution to the optimization problem (2),(9) can be obtained by using the variational calculus approach that satisfy optimality and transversality conditions. We consider the case, when the time interval $[t_0, t_f]$ is enough large. The optimal control force can be expressed by the equation

$$u^\star = -R^{-1}B^T P x(t) \tag{11}$$

where P can be derived from the following Algebraic Riccati Equation (ARE)

$$PA + A^T P - PBR^{-1}B^T P + Q = 0 \tag{12}$$

which in this case is time invariant. This formulation of the problem is known as Linear Quadratic Regulator (LQR) optimal control algorithm. The control gain

$$K = R^{-1}B^T P \tag{13}$$

in this formulation is time invariant and is a constant that depends only on the system matrices A and B and on the choice of the performance index weight matrices Q and R. The equation (11) shows that the optimal control force vector is proportional to the structural response. Note, that an actual closed loop control LQR system requires measurements of the full state vector, i.e. $2n$ sensors are needed.

4 Generalized LQR Law

Let now suppose that the time interval $[t_0, t_f]$ is divided in several smaller intervals $[t_{i-1}, t_i]$ with size $(i = l, 2, ..., n)$ and instead of minimizing the integral performance (9) for the whole interval, an optimal control problem is solved consequently for every subinterval [8-10]. It leads to the design of an optimal controller per interval that means that optimality was achieved at each instant of time.

The cost functional can be still chosen as quadratic function of the states and the control forces, but not in an integral form. The optimal control force $u^\star(t)$ can be derived by minimizing the following instantaneous time-dependent non-integral performance index $J_p(t)$ defined as

$$J_p(t) = x^T(t)Qx(t) + u^T(t)Ru(t) \tag{14}$$

and satisfies the state equation (2). Thus, the performance index $J_p(t)$ is minimized at every time instant t_i of the interval $[t_0, t_f]$. The time increment δt

to reducing the maximum of the displacement's picks and with respect to the time of vibration suppression compared with the LQR approach (dot line).

Let consider the relationships between the maximum structural response and maximum active control force versus control design parameters Q and R. It can be observed that of the maximum structural response and maximum active control force vary with the change of control design parameters. Figure 2 illustrates that the maximum horizontal displacement for the top floor of the building (solid dot line) decreases with increasing the ratio of the weight parameters Q and R. In the same time the maximum control force (solid star line) increases with the increase of the corresponding weight parameters. The shapes of the curves are similar for both LQR and generalized LQR algorithms. The numerical calculations show that the integral performance criterion in equation (9) for LQR approach keeps growing when the ratio Q/R increases while the integral performance criterion in equation (17) for generalized LQR approach first decreases reaching a minimum value and after that increases. The amount of the ratio Q/R where this minimum is achieved can be accepted as the best for the considering building structure.

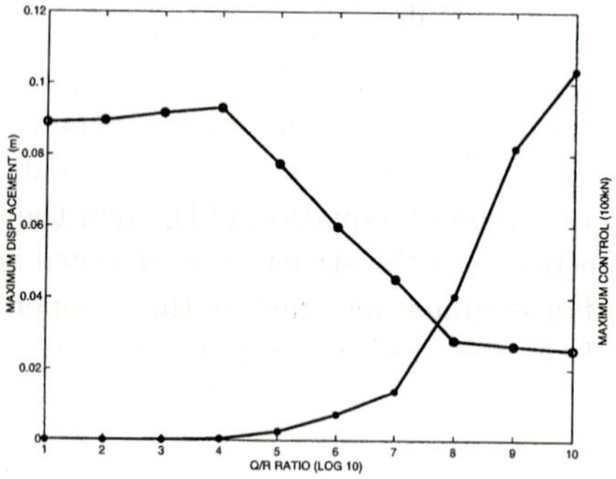

Fig. 2. Relationships among maximum displacement and maximum control force

6 Conclusions

Optimal control design of smart building structures is considered in this work. The aims are to determine the vector of active control forces subjected to some performance criteria of optimality and satisfying the system's dynamic equations such that to reduce in an optimal way the earthquake and wind excitations. Finite element model for eight storey two-dimensional building structures is utilized for numerical experiments.

This paper is dedicated to LQR, instantaneous linear quadratic non-integral and generalized LQR control algorithms for active structural control systems. The three approaches lead to linear control laws that are easy to analyze and implement. Numerical results demonstrate that the considered algorithms suppress the adverse vibrations of the building good but the generalized LQR control algorithms shows better results in the simulation. In the latter discrete time non-integral algorithm the control gain matrix is sensitive to the time increment used in the response analysis. The LQR and the generalized LQR control algorithms do not depend on the time step. For the LQR approach an optimum amount for the weight parameters ratio is not found. The generalized LQR approach overcomes this shortcoming of the LQR algorithm and an optimal ratio for the weight control parameters can be obtained. The generalized LQR approach suggests an additional flexibility including the weight matrix S, that can be chosen depending on the design requirements.

References

1. Arvanitis, K., Zaharenakis, E.,Stavroulakis, G.: New trends in optimal structural control. In: *Selected topics in structronic and mechatronic systems*, Ed. Belyaev A. and Guran A. Singapore, World Academic Publishers (2003)
2. Marinova, D., Stavroulakis, G., Zaharenakis, E.: Numerical experiments in optimal structural control. Proc. 4 GRACM Congress on computational mechanics Patra (2002)
3. Marinova, D., Stavroulakis, G.: Optimal control for smart structures. Proc. 28-th Int. Summer school AMEE, Bulgaria (2002) 216–224
4. Schmitendorf, N., Jabbari F., Yang J.: Robust control techniques for buildings under earthquake excitation. J. Earthquake Engineering and Structural Dynamics **23** (1994)
5. Marinova, D., Stavroulakis, G.: Numerical investigation of optimal control design for two-dimensional building structures subjected earthquake and wind loadings. Isv. RAN: Teoria i sistemy upravleniya **3** (2004)
6. Yang, J., Li, Z., Vongchavalitkuls, S.: Generalization of optimal control theory: linear and nonlinear theory. J. Engineering Mechanics **120** (1994) 266–282
7. Shahian, B., Hassul, M.: Control system design using MATLAB. Prentice Hall. Englewood Cliffs NJ (1994)
8. Yang, J., Akbarpour, A., Ghaemmaghami, P.: Instantaneous optimal control laws for tall buildings under seismic excitations. Technical Report, NCEER-87-0007, National Center for Earthquake Engineering, SUNY-Buffalo (1987)
9. Cheng, F., Suthiwong, S., Tian, P.: Generalized active control with embedded and half-space soil structure interaction. Proc. ASCE 11-th Analysis and Computation Conference, Atlanta (1994) 337-346
10. Cheng, F., Pantelides, C.: Algorithm development for using optimal control in structural optimization subjected to seismic and wind forces. NSF Report, U.S. Department of Commerce, National Technical Information Service, NTIS No.PB90-1333471 (1998)
11. Cheng, F.: Generalized optimal active control algorithm for seismic-resistant structures with active and hybrid control. Int.J. of Computer Applications in Technology bfseries 13 Special Issue (2000) 42-51
12. Saleh, A., Adeli, H.: Optimal control of adaptive/smart ulti-storey building structures. J. Computer-Aided Civil and Infrastructure Engineering. bfseries 13 (1998)

Computing Transitive Closure Problem on Linear Systolic Array

I. Ž. Milovanović, E. I. Milovanović, and B. M. Randjelović

Faculty of Electronic Engineering,
Beogradska 14a, P.O. Box 73, 18000 Niš, Serbia

Abstract. We consider the problem of computing transitive closure of a given directed graph on the regular bidirectional systolic array. The designed array has n PEs, where n is a number of nodes in the graph. This is an optimal number for a given problem size.

1 Introduction

A directed graph G is a doublet $G = (V, E)$, where V is a set of vertices and E is the set of direct edges in the graph. The graph which has the same vertex set V, but has a directed edge from vertex v to vertex w if there is a directed path (of length zero or more) from vertex v to vertex w in G, is called the (reflexive and) transitive closure of G. A graph $G = (V, E)$ can be represented as an adjacency matrix A, whose elements $a_{ij} = 1$ if there is an edge directed from vertex i to vertex j, or $i = j$; otherwise $a_{ij} = 0$. The transitive closure problem is to compute the adjacency matrix A^* for $G^* = (V, E^*)$ from A. The problem of computing transitive closure (TC) of a digraph was first considered in 1959 by B. Roy [1]. Since then, the TC has been studied extensively in the literature because it is an abstraction of many practical problems that are used in wide variety of applications in mathematics, computer science, engineering and business (see, for example, [2]). The computational complexity of TC problem is $0(n^3)$, where n corresponds to number of vertices in the graph . To speed-up the computation of TC, various parallel architectures can be used. In this paper we consider the computation of TC on linear bidirectional systolic array (BLSA) with two-dimensional links.

The rest of the paper is organized as follows. In Section 2 describe some important properties of the systolic arrays. In Section 3 we define a systolic algorithm for computing TC which is suitable for the implementation on BLSA. The BLSA synthesis and the example of TC computation on BLSA are described in Section 4. Conclusion is given in Section 5.

2 Systolic Arrays

In the last decade, the rapid development of VLSI computing techniques has had a significant impact on the development of novel computer architectures.

Z. Li et al. (Eds.): NAA 2004, LNCS 3401, pp. 416–423, 2005.

One class of architectures, the so-called systolic arrays, first introduced by Kung [3], has gained popularity because of its ability to exploit massive parallelism and pipelining to achieve high performance. Informally, a systolic system can be envisaged as an array of synchronized processors (or processing elements, abbreviated as PEs) which process data in parallel by passing them from PE to PE in regular rhythmic fashion. Systolic arrays have balanced, uniform, grid-like architectures of special PEs that process data like n-dimensional pipeline. The fundamental concept behind a systolic architecture is that von Neumann bottleneck is greatly alleviated by repeated use of fetched data item in a physically distributed array of PEs. The regularity of these arrays leads to inexpensive and dense VLSI implementations, which imply high performance at low cost.

Three factors have contributed to the evolution of systolic arrays into leading approach for computationally intensive applications [4]:

- Technology Advances: The growth of VLSI/WSI technology. Smaller and faster gates allow a higher rate of on-chip communication as data have to travel a shorter distance. Higher data densities permit more complex PEs with higher performance and granularity. Progress in design tools and simulation techniques ensure that a systolic PE can be fully simulated before fabrication, reducing the chances it will fail to work as designed. The regular modular arrays also decrease time to design and test , as fully tested unique cells can be copied quickly and arranged into systolic array.
- Parallel processing: Efforts to add concurrency to conventional von Neumann computers have yielded a variety of techniques such as coprocessors, multiple functional units, data pipelining (and parallelism), and multiple homogenous processors. Systolic arrays combine features from all of these architectures in a massively parallel architecture that can be integrated into existing platforms without a complete redesign.
- Demanding scientific applications: The technology growth in the past three decades has produced the computers that make it feasible to attack scientific applications on a larger scale. New applications requiring increased computational performance have been developed that were not possible earlier. Examples of these application include interactive language (or speech) recognition, text recognition, virtual reality, database operations, and real-time image and signal processing. These applications require massive, repetitive parallel processing, and hence, systolic computing.

The most important aspect in the design of a systolic computer is the mapping of the algorithm to the processor array. In the systolic paradigm, every algorithm requires a specialized systolic design in which communication data streams, PE definitions, and input-output patterns are customized. Only highly regular algorithms with the structure of nested loops are suitable for systolic implementation.

This paper deals with systolic implementation of TC problem on bidirectional linear systolic array (BLSA) since it has many desirable properties, like maximal data-flow, possibilities for fault-tolerant calculations and applicability in many scientific and technical problems.

3 Systolic Algorithm

Let $G = (V, E)$ be a given weighted directed or undirected graph, where $V = \{1, 2, \ldots, n\}$. Graph $G = (V, E)$ can be described by the adjacency matrix $A = (a_{ij})$ of order $n \times n$, where

$$a_{ij} = \begin{cases} 1, & \text{if } (i,j) \in E \\ 0, & \text{if } (i,j) \notin E \end{cases},$$

for each $i = 1, 2, \ldots, n$ and $j = 1, 2, \ldots, n$. Similarly, the transitive closure, $G^* = (V, E^*)$ of G, is uniquely defined by its adjacency matrix $A^* = (a^*_{ij})$ of order $n \times n$. Therefore, transitive closure problem is to compute the A^* from A.

A well-known method for finding TC is Wars hall's algorithm. This algorithm computes $(n-1)$ intermediate matrices $A^{(k)} = (a_{ij}^{(k)})$, $1 \leq k \leq n-1$, starting with $A^{(0)} = A$ and the last matrix $A^{(n)} = (a_{ij}^{(n)})$ computed is A^*. At the k-th iteration, matrix $A^{(k-1)}$ is used to compute $A^{(k)}$. A Pascal-like description of the algorithm is given below

Algorithm(*Warshall*)
 for $k := 1$ **to** n **do**
  for $j := 1$ **to** n **do**
   for $i := 1$ **to** n **do**
   $a_{i,j}^{(k)} := a_{i,j}^{(k-1)} \vee (a_{i,k}^{(k-1)} \wedge a_{k,j}^{(k-1)})$;

where $\vee$ and $\wedge$ are Boolean operators. Each algorithm can be represented by the corresponding data dependency graph. The data dependency graph for TC is not completely regular and therefore it is difficult to map on regularly connected planar SAs. Accordingly, the obtained SAs are irregular in the sense that interconnection pattern is not the same in all parts of the array (for 2D arrays)(see [6]) or delay elements are inserted between the processing elements (1D arrays) (see [7]).

Our goal is to design regular BLSA with two-dimensional links for computing the TC problem. Since the TC is a three-dimensional problem and its data dependency graph is not completely regular, Warshall algorithm is not suitable for direct synthesis of BLSA. To overcome this problem we partition the computations in Warshall algorithm into appropriate number of two-dimensional entities. These entities are equal with respect to operational complexity. Each entity must be suitable for the implementation on BLSA. The final result is obtained by repeating the computation of entities on the designed BLSA. We require that designed BLSA be space-optimal with respect to a problem size and the execution time should be as minimal as possible for a given size of BLSA.

In order to obtain two-dimensional entities suitable for the synthesis of BLSA with desired properties, we set index variable k on some constant value $1 \leq k \leq n$. Then we design BLSA that computes $A^{(k)}$, for some constant k. Starting from $A^{(0)} = A$, the final result is obtained by repeating the computation on the BLSA n times, such that the result from the $(k-1)$-st step is used as input in the k-th step, for $k = 2, 3, \ldots, n$.

4 BLSA Synthesis

To ease the presentation we use the following denotation

$$c(i,j,k) \equiv a_{ij}^{(k)}, \quad d(i,0,k) \equiv a_{ik}^{(k-1)} \quad b(0,j,k) \equiv a_{kj}^{(k-1)} \tag{1}$$

for each $i = 1,2,\ldots,n$ and $j = 1,2,\ldots,n$. Then the corresponding Warshall algorithm for some fixed k can be written as

Algorithm_1
  for $i := 1$ **to** n **do**
    for $j := 1$ **to** n **do**
    $c(i,j,k) := c(i,j,k-1) \vee (d(i,0,k) \wedge b(0,j,k));$

Yet, this is not a systolic algorithm that can be used for BLSA synthesis because it has a global data dependencies as illustrated in Fig. 1a) for the case $n = 3$. The corresponding systolic algorithm with localized dependencies has the following form

Algorithm_2
  for $i := 1$ **to** n **do**
    for $j := 1$ **to** n **do**
    $d(i,j,k) := d(i,j-1,k);$
    $b(i,j,k) := b(i-1,j,k);$
    $c(i,j,k) := c(i,j,k-1) \vee (d(i,j,k) \wedge b(i,j,k));$

The corresponding data dependency graph with localized dependencies is illustrated in Fig.1 b).

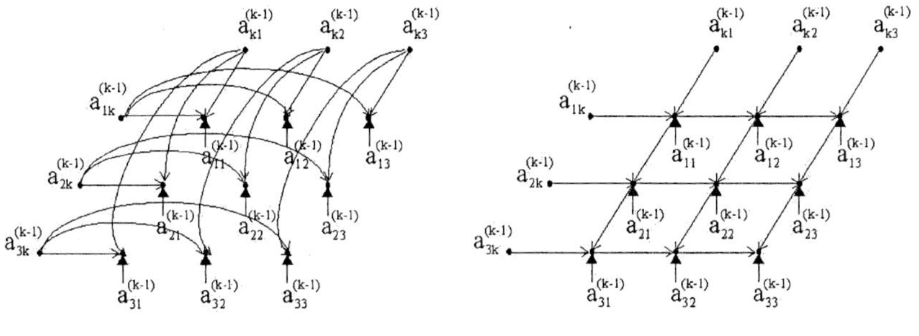

a) b)

Fig. 1. Data dependency graph for $n = 3$ a) Algorithm_1 b) Algorithm_2

The inner computation space of Algorithm_2, denoted as $\bar{P}_{int}$, is

$$\bar{P}_{int} = \{(i,j,k) | 1 \leq i \leq n, 1 \leq j \leq n\} \tag{2}$$

and the corresponding data dependency matrix is

$$D = \left[e_b^3 \; e_d^3 \; e_c^3 \right] = \begin{bmatrix} 1\,0\,0 \\ 0\,1\,0 \\ 0\,0\,1 \end{bmatrix} . \tag{3}$$

The systolic algorithm Algorithm_2 is completely defined by a pair $(\bar{P}_{int}, D)$, i.e. by directed graph $\bar{\Gamma} = (\bar{P}_{int}, D)$, where $\bar{P}_{int}$ represents a set of nodes, while matrix D, i.e. its column vectors, represent directions of arcs between adjacent nodes. The systolic array is obtained by mapping $(\bar{P}_{int}, D)$ along some possible projection direction vector into the projection plane. The mapping is defined by transformation matrix T that corresponds to a particular projection direction vector. The possible projection direction vectors that can be used to obtain 1D SA are $\boldsymbol{\mu} = [1\,0\,0]^T$, $\boldsymbol{\mu} = [0\,1\,0]^T$, $\boldsymbol{\mu} = [1\,-1\,0]^T$ and $\boldsymbol{\mu} = [1\,1\,0]^T$. Since $\boldsymbol{\mu} = [1\,0\,0]^T$ and $\boldsymbol{\mu} = [0\,1\,0]^T$ give orthogonal projections the corresponding SAs are static, which means that one of the variables is resident in the array. By the direction $\boldsymbol{\mu} = [1\,-1\,0]$ we would obtain unidirectional 1D SA which requires delay elements to be inserted between the PEs since data elements have to propagate with different speed through the array. Therefore, this array is irregular. Thus, only direction $\boldsymbol{\mu} = [1\,1\,0]^T$ can be used to obtain BLSA.

If we had tried to design BLSA directly from Algorithm_2 and direction $\boldsymbol{\mu} = [1\,1\,0]^T$ the obtained BLSA would not be optimal with respect to a number of PEs. Namely, the obtained BLSA would have $2n - 1$ PEs, instead of n which is optimal number for a given problem size. The data dependency graph of Algorithm_2 and its projection (i.e. the corresponding SA) are illustrated in Fig.2 for $n = 3$.

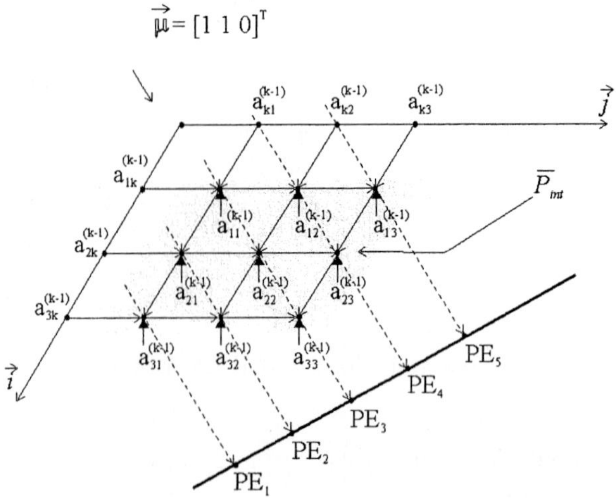

Fig. 2. Data dependency graph of Algorithm_2 and its projection for the case $n = 3$

In order to obtain BLSA with optimal number of PEs , i.e. n, we use the fact that computations in Algorithm_2 can be performed over arbitrary permutation $\{i_1, i_2, \ldots, i_n\}$ of index variable i, where $\{1, 2, \ldots, n\}$ is the basic permutation. The same is valid for index variable j. This means that Algorithm_2 belongs to a class of adaptable algorithms and that its index space $\bar{P}_{int}$ can be adjusted to a given projection direction vector (see [8]). An algorithm derived from Algorithm_2 by adjusting $\bar{P}_{int}$ to the direction $\boldsymbol{\mu} = [1\,1\,0]^T$ has the following form

Algorithm_3
for $i := 1$ **to** n **do**
for $j := 1$ **to** n **do**
$d(i, i+j-1, k) := d(i, i+j-2, k);$
$b(i, i+j-1, k) := b(i-1, i+j-1, k);$
$c(i, i+j-1, k) := c(i, i+j-1, k-1) \vee (d(i, i+j-1, k) \wedge b(i, i+j-1, k));$

where $d(i, j+n, k) \equiv d(i, j, k) \equiv d(i, 0, k)$, $b(i, j+n, k) \equiv b(i, j, k) \equiv b(0, j, k)$, $c(i, j+n, k) \equiv c(i, j, k)$, for $i = 1, 2, \ldots, n$ and $j = 1, 2, \ldots, n$.

The inner computation space of Algorithm_3 is given by

$$P_{int} = \{(i, i+j-1, k) | 1 \le i \le n, 1 \le j \le n\} \tag{4}$$

while data dependency matrix is given by (3). The data dependency graph of Algorithm_3 and its projection are illustrated in Fig.3 for $n = 3$. As one can see the number of points (i.e. number of PEs) in the projection plane is 3, which is optimal number for a given problem size n.

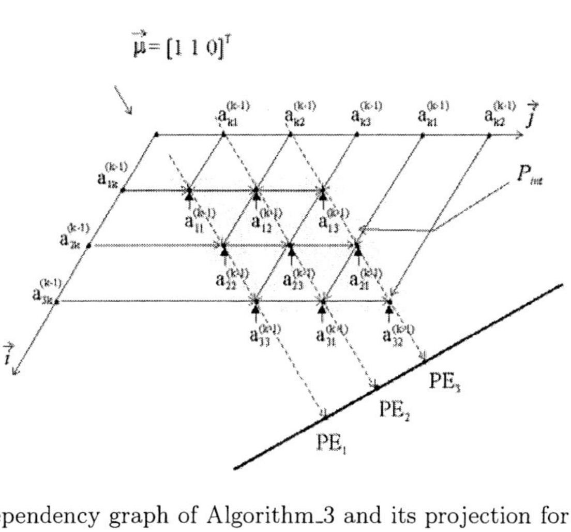

Fig. 3. Data dependency graph of Algorithm_3 and its projection for the case $n = 3$

To obtain BLSA with optimal number of PEs, from a set of valid transformations T that correspond to the direction $\boldsymbol{\mu} = [1\,1\,0]^T$ (see, for example [8]), we chose

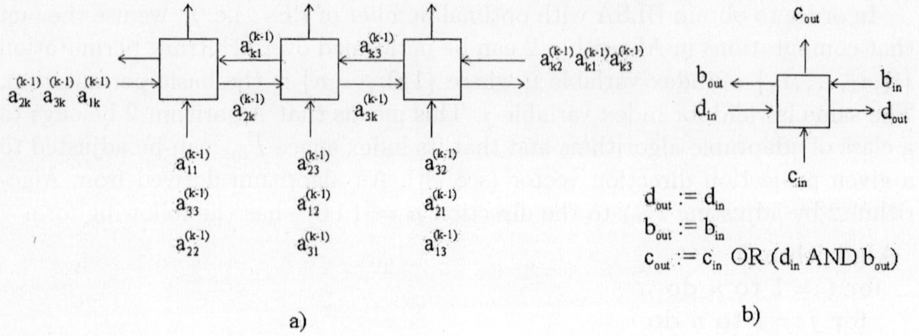

Fig. 4. a) Data flow in the BLSA for the case $n = 3$; b) Functional property of the PE

Table 1. Step by step diagram for the computation of TC in the BLSA for the case $n=3$

	PE_1				PE_2				PE_3			
clk	d_{in}	b_{in}	c_{in}	c_{out}	d_{in}	b_{in}	c_{in}	c_{out}	d_{in}	b_{in}	c_{in}	c_{out}
0	-	-	-	-	-	-	-	-	-	-	-	-
1	$a_{31}^{(0)}$	-	-	-	-	-	-	-	-	$a_{11}^{(0)}$	-	-
2	$a_{21}^{(0)}$				$a_{31}^{(0)}$	$a_{11}^{(0)}$				$a_{13}^{(0)}$		
3	$a_{11}^{(0)}$	$a_{11}^{(0)}$	$a_{11}^{(0)}$		$a_{21}^{(0)}$	$a_{13}^{(0)}$	$a_{23}^{(0)}$		$a_{31}^{(0)}$	$a_{12}^{(0)}$	$a_{32}^{(0)}$	
4	$a_{31}^{(0)}$	$a_{13}^{(0)}$	$a_{33}^{(0)}$	$a_{11}^{(1)}$	$a_{11}^{(0)}$	$a_{12}^{(0)}$	$a_{12}^{(0)}$	$a_{23}^{(1)}$	$a_{21}^{(0)}$	$a_{11}^{(0)}$	$a_{21}^{(0)}$	$a_{32}^{(1)}$
5	$a_{21}^{(0)}$	$a_{12}^{(0)}$	$a_{22}^{(0)}$	$a_{33}^{(1)}$	$a_{31}^{(0)}$	$a_{11}^{(0)}$	$a_{31}^{(0)}$	$a_{12}^{(1)}$	$a_{11}^{(0)}$	$a_{13}^{(0)}$	$a_{13}^{(0)}$	$a_{21}^{(1)}$
6	$a_{32}^{(1)}$	$a_{11}^{(0)}$		$a_{22}^{(1)}$	$a_{21}^{(0)}$	$a_{13}^{(0)}$		$a_{31}^{(1)}$	$a_{31}^{(0)}$	$a_{21}^{(1)}$		$a_{13}^{(1)}$
7	$a_{22}^{(1)}$	$a_{13}^{(0)}$			$a_{32}^{(1)}$	$a_{21}^{(1)}$			$a_{21}^{(0)}$	$a_{23}^{(1)}$		
8	$a_{12}^{(1)}$	$a_{21}^{(1)}$	$a_{11}^{(1)}$		$a_{22}^{(1)}$	$a_{23}^{(1)}$	$a_{23}^{(1)}$		$a_{32}^{(1)}$	$a_{22}^{(1)}$	$a_{32}^{(1)}$	
9	$a_{32}^{(1)}$	$a_{23}^{(1)}$	$a_{33}^{(1)}$	$a_{11}^{(2)}$	$a_{12}^{(1)}$	$a_{22}^{(1)}$	$a_{12}^{(1)}$	$a_{23}^{(2)}$	$a_{22}^{(1)}$	$a_{21}^{(1)}$	$a_{21}^{(1)}$	$a_{32}^{(2)}$
10	$a_{22}^{(1)}$	$a_{22}^{(1)}$	$a_{22}^{(1)}$	$a_{33}^{(2)}$	$a_{32}^{(1)}$	$a_{21}^{(1)}$	$a_{31}^{(1)}$	$a_{12}^{(2)}$	$a_{12}^{(1)}$	$a_{23}^{(1)}$	$a_{13}^{(1)}$	$a_{21}^{(2)}$
11	$a_{33}^{(2)}$	$a_{21}^{(1)}$		$a_{22}^{(2)}$	$a_{22}^{(1)}$	$a_{23}^{(1)}$		$a_{31}^{(2)}$	$a_{32}^{(1)}$	$a_{31}^{(2)}$		$a_{13}^{(2)}$
12	$a_{23}^{(2)}$	$a_{23}^{(1)}$			$a_{33}^{(2)}$	$a_{31}^{(2)}$			$a_{22}^{(1)}$	$a_{33}^{(2)}$		
13	$a_{13}^{(2)}$	$a_{31}^{(2)}$	$a_{11}^{(2)}$		$a_{23}^{(2)}$	$a_{33}^{(2)}$	$a_{23}^{(2)}$		$a_{33}^{(2)}$	$a_{32}^{(2)}$	$a_{32}^{(2)}$	
14	$a_{33}^{(2)}$	$a_{33}^{(2)}$	$a_{33}^{(2)}$	$a_{11}^{(3)}$	$a_{13}^{(2)}$	$a_{32}^{(2)}$	$a_{12}^{(2)}$	$a_{23}^{(3)}$	$a_{23}^{(2)}$	$a_{31}^{(2)}$	$a_{21}^{(2)}$	$a_{32}^{(3)}$
15	$a_{23}^{(2)}$	$a_{32}^{(2)}$	$a_{22}^{(2)}$	$a_{33}^{(3)}$	$a_{33}^{(2)}$	$a_{31}^{(2)}$	$a_{31}^{(2)}$	$a_{12}^{(3)}$	$a_{13}^{(2)}$	$a_{33}^{(2)}$	$a_{13}^{(2)}$	$a_{21}^{(3)}$
16		$a_{31}^{(2)}$		$a_{22}^{(3)}$	$a_{23}^{(2)}$	$a_{33}^{(2)}$		$a_{31}^{(3)}$	$a_{33}^{(2)}$			$a_{13}^{(3)}$
17		$a_{33}^{(2)}$							$a_{23}^{(2)}$			
18												

$$T = \begin{bmatrix} \boldsymbol{\Pi} \\ S \end{bmatrix} = \begin{bmatrix} 1\,1\,1 \\ -1\,1\,0 \\ 0\,0\,1 \end{bmatrix}.$$

The BLSA is obtained by mapping $T: \quad (P_{int}, D) \mapsto (\hat{P}_{int}, \Delta)$. The (x, y) coordinates of the PEs in the projection plane are defined by the set $\hat{P}_{int} = \{[x\, y]^T\}$, where

$$\begin{bmatrix} x \\ y \end{bmatrix} = S \begin{bmatrix} i \\ i+j-1 \\ k \end{bmatrix} = \begin{bmatrix} -1\,1\,0 \\ 0\,0\,1 \end{bmatrix} \cdot \begin{bmatrix} i \\ i+j-1 \\ k \end{bmatrix} = \begin{bmatrix} j-1 \\ k \end{bmatrix}, \tag{5}$$

where k is some constant value and $j = 1, 2, \ldots, n$.

The interconnections between the PEs in the BLSA are implemented along the propagation vectors $\Delta = \left[e_b^2 \; e_d^2 \; e_c^2 \right] = S \cdot D$

According to (5) it is not difficult to conclude that the number of PEs in the obtained BLSA is $\Omega = n$, which is optimal number for a given problem size. The execution time of the BLSA can be minimized by the procedure described in [9], so we will omit the details here.

Data schedule in the BLSA at the beginning of the computation of Algorithm_3, for the case $n = 3$, is depicted in Fig.4. Table 1. shows the complete, step by step, diagram for computing transitive closure problem on the BLSA for the case $n = 3$

5 Conclusion

We have discussed the problem of computing transitive closure of a given directed graph on the regular bidirectional systolic array. Our array has n PEs, where n is a number of nodes in the graph. This is an optimal number for a given problem size.

References

1. B. Roy, *Transitivite et connexite*, C. R. Acad. Sci. Paris, 249 (1959), 216-218.
2. A. V. Aho, J. E. Hopcroft, J. D. Ullman, *Design and analysis of computer algorithms*, Reading, MA: Addison-wesley, 1975.
3. H. T. Kung, *Why systolic architectures*, Computer, Vol. 15, 1 (1982), 37-46.
4. K. T. Johnson, A. R. Hurson, B. Shirazi, *General purpose systolic arrays*, Computer, Vol. 26, 11(1993), 20-31.
5. S. Warshall, *A Theorem on Boolean matrices*, J. ACM, 9(1962), 11-12.
6. S. Y. Kung, S. C. Lo, P. S. Lewis, *Optimal systolic design for the transitive closure problem*, IEEE Trans. Comput., Vol. C-36, 5 (1987), 603-614.
7. W. Shang, J. A. B. Fortes, *On mapping of uniform dependence algorithms into lower dimensional processor arrays*, IEEE Trans. Parall. Distr. Systems, Vol. 3, 3(1992), 350-363.
8. M. Bekakos, E. Milovanović, N. Stojanović, T. Tokić, I. Milovanović, I. Milentijević, *Transformation matrices for systolic array synthesis*, J. Electrothn. Math., 7 (2002), 9-15.
9. M. P. Bekakos, I.Ž. Milovanović, E. I. Milovanović, T. I. Tokić, M. K. Stojčev, *Hexagonal systolic arrays for matrix multiplications*, (M. P. Bekakos, ed.), Series: advances in High Performance Computing, Vol. 5, WITpress, Southampton-Boston, UK, 2001, 175-209.

$$x \pm \alpha = (x_0 \pm \alpha) + \sum_{i=1}^{n} x_i \varepsilon_i \tag{7}$$

$$\alpha x = (\alpha x_0) + \sum_{i=1}^{n} (\alpha x_i) \quad . \tag{8}$$

Let f be a nonlinear unary operation. Generally, $f(x)$ for an affine form (3) is not able to be represented directly as an affine form. Therefore, we consider linear approximation of f and representation of the approximation error by introducing a new noise symbol ε_{n+1}. First, X, range of x (domain of f), is calculated by the formula (5). Next, $ax + b$, linear approximation of $f(x)$ in X is calculated (see the Fig. 1).

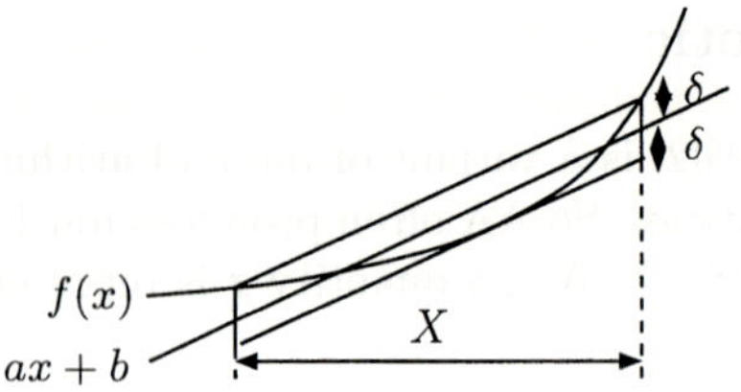

Fig. 1. Linear approximation of nonlinear function

And maximum approximation error δ is calculated as follows:

$$\delta = \max_{x \in X} |f(x) - (ax + b)| \quad . \tag{9}$$

By introducing a new noise symbol ε_{n+1}, the result of nonlinear unary operations is represented as follows:

$$f(x) = ax + b + \delta\varepsilon_{n+1} = a(x_0 + x_1\varepsilon_1 + \cdots + x_n\varepsilon_n) + b + \delta\varepsilon_{n+1} \quad . \tag{10}$$

Nonlinear binomial operations are calculated similarly.

3 New Method

In this section, we propose the new method which finds the maxima and minima of a multivariable function (1) applying AA. Here, without loss of generality, we consider finding maxima of a two-dimensional function $y = f(x_1, x_2)$ in the box $X^{(0)} = (X_1^{(0)}, X_2^{(0)}) = ([\underline{X_1^{(0)}}, \overline{X_1^{(0)}}], [\underline{X_2^{(0)}}, \overline{X_2^{(0)}}])$. For a interval J, let the center and the width of J be $c(J)$ and $w(J)$. And for a box X, let the range boundary of f in X applying AA be $F_A(X)$ and let the upper bound of $I(F_A(X))$ be $\overline{F_A(X)}$. The concrete step of the new method is as follows:

step1. Initialize the list $\mathcal{S}$ and $\mathcal{T}$, which memorize boxes and range boundaries of f in the each box applying AA, as $\mathcal{S} = \mathcal{T} = \phi$. And set the stopping criteria ϵ_r and ϵ_b.

step2. If $w(X_1^{(0)}) < w(X_2^{(0)})$, divide $X^{(0)}$ into $X^{(1)} = ([\underline{X_1^{(0)}}, \overline{X_1^{(0)}}], [\underline{X_2^{(0)}}, c(X_2^{(0)})])$ and $X^{(2)} = ([\underline{X_1^{(0)}}, \overline{X_1^{(0)}}], [c(X_2^{(0)}), \overline{X_2^{(0)}}])$. Otherwise divide $X^{(0)}$ into $X^{(1)} = ([\underline{X_1^{(0)}}, c(X_1^{(0)})], [\underline{X_2^{(0)}}, \overline{X_2^{(0)}}])$ and $X^{(2)} = ([c(X_1^{(0)}), \overline{X_1^{(0)}}], [\underline{X_2^{(0)}}, \overline{X_2^{(0)}}])$.

step3. Calculate $F_{\mathrm{A}}(X^{(1)})$ and $F_{\mathrm{A}}(X^{(2)})$. And calculate $\underline{f_{\max}^{(1)}}$, the candidate of lower bound of maxima by utilizing $X^{(1)}$ and $F_{\mathrm{A}}(X^{(1)})$ and by applying algorithm 1 (details are described in the section 3.1). Calculate $\underline{f_{\max}^{(2)}}$ similarly. Let the lower bound of maxima $\underline{f_{\max}}$ be $\underline{f_{\max}} = \max(\underline{f_{\max}^{(1)}}, \underline{f_{\max}^{(2)}})$.

step4. If $\overline{F_{\mathrm{A}}(X^{(1)})} < \underline{f_{\max}}$, insert $X^{(2)}$ and $F_{\mathrm{A}}(X^{(2)})$ into $\mathcal{S}$ (here, $X^{(1)}$ is discarded). If $\overline{F_{\mathrm{A}}(X^{(2)})} < \underline{f_{\max}}$, insert $X^{(1)}$ and $F_{\mathrm{A}}(X^{(1)})$ into $\mathcal{S}$ (here, $X^{(2)}$ is discarded). Otherwise insert $X^{(1)}, F_{\mathrm{A}}(X^{(1)}), X^{(2)}$ and $F_{\mathrm{A}}(X^{(2)})$ into $\mathcal{S}$.

step5. Repeat the following steps:

step5-1. If $\mathcal{S} = \phi$, go to the **step6**. Otherwise, find the box $X^{(i)}$ in $\mathcal{S}$ for which $\overline{F_{\mathrm{A}}(X^{(i)})}$ is largest. Select $X^{(i)}$ and $F_{\mathrm{A}}(X^{(i)})$ as the box and the range boundary to be processed and discard $X^{(i)}$ and $F_{\mathrm{A}}(X^{(i)})$ from $\mathcal{S}$.

step5-2. Calculate $\underline{f_{\max}^{(i)}}$, the candidates of $\underline{f_{\max}}$, by utilizing $X^{(i)}$and $F_{\mathrm{A}}(X^{(i)})$ and by applying algorithm 1. If $\underline{f_{\max}} < \underline{f_{\max}^{(i)}}$, renew $\underline{f_{\max}}$ as $\underline{f_{\max}} = \underline{f_{\max}^{(i)}}$.

step5-3. Discard any box X and range boundary $F_{\mathrm{A}}(X)$ from $\mathcal{S}$ and $\mathcal{T}$ for which $\overline{F_{\mathrm{A}}(X)} < \underline{f_{\max}}$.

step5-4. Narrow $X^{(i)}$ down by utilizing $X^{(i)}$, $F_{\mathrm{A}}(X^{(i)})$ and $\underline{f_{\max}}$ and by applying the algorithm 2 (details are described in the section 3.2).

step5-5. If $w(X_1^{(i)}) < w(X_2^{(i)})$, divide $X^{(i)}$ into $X^{(j)} = ([\underline{X_1^{(i)}}, \overline{X_1^{(i)}}], [\underline{X_2^{(i)}}, c(X_2^{(i)})])$ and $X^{(k)} = ([\underline{X_1^{(i)}}, \overline{X_1^{(i)}}], [c(X_2^{(i)}), \overline{X_2^{(i)}}])$. Otherwise divide $X^{(i)}$ into $X^{(j)} = ([\underline{X_1^{(i)}}, c(X_1^{(i)})], [\underline{X_2^{(i)}}, \overline{X_2^{(i)}}])$ and $X^{(k)} = ([c(X_1^{(i)}), \overline{X_1^{(i)}}], [\underline{X_2^{(i)}}, \overline{X_2^{(i)}}])$.

step5-6. Calculate $F_{\mathrm{A}}(X^{(j)})$ and $F_{\mathrm{A}}(X^{(k)})$. If $\max\limits_{1 \le h \le m} w(X_h^{(j)}) < \epsilon_r \wedge w(I(F_{\mathrm{A}}(X^{(j)}))) < \epsilon_b$, insert $X^{(j)}$ and $F_{\mathrm{A}}(X^{(j)})$ into $\mathcal{T}$. Otherwise, insert $X^{(j)}$ and $F_{\mathrm{A}}(X^{(j)})$ into $\mathcal{S}$. For $X^{(k)}$ and $F_{\mathrm{A}}(X^{(k)})$, implement the similar procedure.

step5-7. Go back to the **step5-1**.

step6. In the boxes remained in $\mathcal{T}$, group the boxes which posses the common point each other. Let the boxes which belong to a group be $Y^{(1)}, \cdots, Y^{(l)}$ where l is the number of the box belonging in the group. The maximum value in the group is calculated as $\bigcup\limits_{h=1}^{l} I(F_{\mathrm{A}}(Y^{(h)}))$. And the point which the maximum value occurs in the group is calculated as $\bigcup_{h=1}^{l} Y^{(h)}$. For the other groups, implement the similar procedure.

step7. Terminate.

By this step, we are able to calculate the all maxima and the all points which the maxima occur with guaranteed accuracy. If $2 < m$, the step is similar. And if minima are calculated, let f be $-f$.

3.1 Algorithm 1

In this section, we proposed the algorithm 1 introduced in the **step3** and **step5-2**. In this algorithm, when we find maxima, larger candidate of $\underline{f_{\max}}$ than that of the Fujii's method is calculated by utilizing a box X and the range boundary $F_A(X)$. To simplify the explanation, we consider finding maxima of f where $m = 1$. In this case, for example, the shape of $F_A(X)$ becomes as the slant line part in the Fig. 2 because in AA, the image of function is bounded in the form of "linear approximation + error term"(without loss of generality, suppose that the slope of the linear approximation is positive). In Fujii's method, candidates

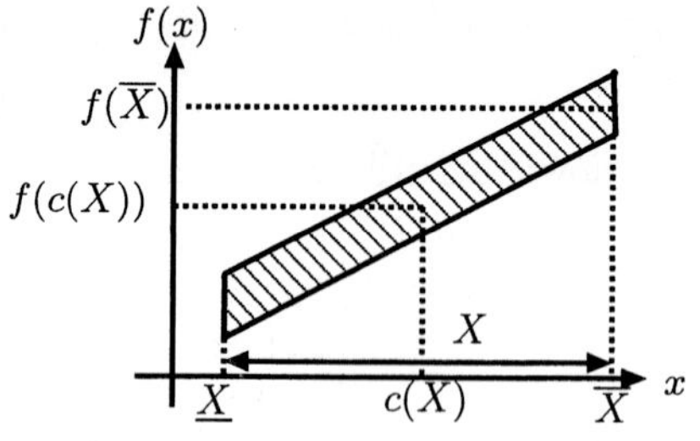

Fig. 2. An example of the algorithm 1

of $\underline{f_{\max}}$ is always calculated as $f(c(X))$. As opposed to this, in this algorithm, candidates of $\underline{f_{\max}}$ are calculated as $f(\overline{X})$ in the case of Fig.2. The reason is that the possibility of $f(c(X)) < f(\overline{X})$ is high because in the Fig. 2, approximation of f's slope in X is positive. If the slope of the linear approximation is negative, candidates of $\underline{f_{\max}}$ are calculated as $f(\underline{X})$.

Being based on the above description, we describe the concrete step of this algorithm when we find maxima of f where $1 \le m$.

step1. Suppose that $F_A(X)$ is calculated as follows:

$$F_A(X) = a_0 + a_1\varepsilon_1 + \cdots + a_m + a_{m+1} + \cdots + a_n\varepsilon_n \quad (m < n) \quad . \quad (11)$$

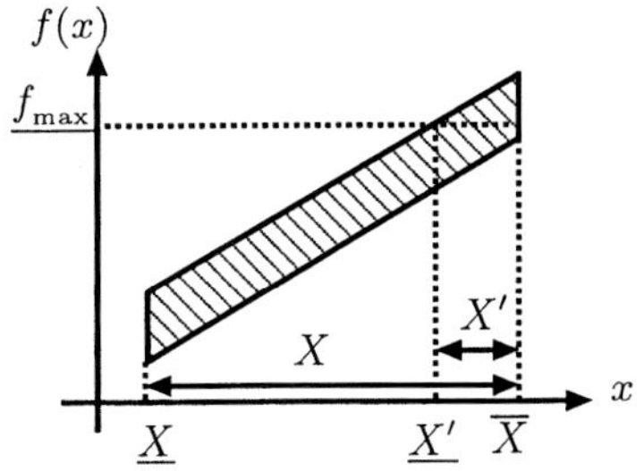

Fig. 3. An example of the algorithm 2

step2. Let the point (vector) $y = (y_1, \cdots, y_m)$ be as follows:

$$y_i = \begin{cases} \overline{X_i} & (0 < a_i) \\ \underline{X_i} & (a_i < 0) \\ c(X_i) & (\text{otherwise}) \end{cases} \qquad (i = 1, \cdots, m) \quad . \tag{12}$$

step3. The candidate of $\underline{f_{\max}}$ is calculated as $f(y)$.

step4. Terminate.

By this algorithm, we are able to calculate the candidate of $\underline{f_{\max}}$ larger than that of the Fujii's method. Therefore, we are able to discard more subregions (boxes) in initial stage.

3.2 Algorithm 2

In this section, we proposed the algorithm 2 introduced in the **step5-4**. In this algorithm, a box X is narrowed down by utilizing X and the range boundary $F_A(X)$. Namely, the parts of X is able to be discarded while it is not able to be discarded in the Fujii's method. Similar to the section3.1, to simplify the explanation, we consider finding maxima of f where $m = 1$. In this case, for example, the shape of $F_A(X)$ becomes as the slant line part in the Fig. 3 (without loss of generality, suppose that the slope of the linear approximation is positive). In the Fig.3, maxima of $f(x)$ in X exist obviously in X'. Therefore, we are able to narrow X down to X' i.e. we are able to discard $[\underline{X}, \underline{X'})$. Being based on the above description, we describe the concrete step of this algorithm when we find maxima of f where $1 \leq m$.

step1. Suppose that $F_A(X)$ is calculated as the formula (11).

step2. Calculate α as $\alpha = \sum_{i=m+1}^{n} |a_i|$.

step3. Implement the following procedure for all $i = 1, \cdots, m$: If $a_i \neq 0$, calculate the interval $\varepsilon_i^* = [\underline{\varepsilon_i^*}, \overline{\varepsilon_i^*}]$ applying IA as follows:

$$\varepsilon_i^* = \frac{1}{a_i}(\underline{f_{\max}} - a_0 - \alpha - \sum_{j=1, j \neq i}^{m} (a_j \times [-1, 1])) \quad . \tag{13}$$

Otherwise let the interval ε_i^* be as $[-1,1]$. If $\varepsilon_i^* \subset [-1,1]$, narrow X_i down as follows:

$$\begin{cases} [\underline{X_i} + r(X_i)(\underline{\varepsilon_i^*}+1), \overline{X_i}] \ (0 < a_i) \\ [\underline{X_i}, \overline{X_i} - r(X_i)(1-\overline{\varepsilon_i^*})] \ (a_i < 0) \end{cases} \text{ where } r(X_i) = \frac{\overline{X_i} - \underline{X_i}}{2} . \tag{14}$$

If $\underline{\varepsilon_i^*} \le -1 \wedge \overline{\varepsilon_i^*} \in [-1,1) \wedge a_i < 0$, narrow X_i down as $[\underline{X_i}, \overline{X_i} - r(X_i)(1-\overline{\varepsilon_i^*})]$.
If $\underline{\varepsilon_i^*} \in (-1,1] \wedge 1 \le \overline{\varepsilon_i^*} \wedge 0 < a_i$, narrow X_i down as $[\underline{X_i} + r(X_i)(\underline{\varepsilon_i^*}+1), \overline{X_i}]$.
Otherwise, we are not able to narrow X_i down.

step4. Terminate.

4 Numerical Examples

In this section, to show the efficiency of the new method, some numerical examples are implemented. Here, our computer environment is CPU: PentiumIV 3.2GHz, memory: 2.0GB, OS: Free BSD 4.9 and compiler: g++ 2.7.2.1. We set ϵ_r and ϵ_b as $\epsilon_r = \epsilon_b = 10^{-7}$. In the Table 1 and the Table 2, the notation – means that memory over occurred. At first, we found the maximum value of the following Rosenbrock function:

$$f(x_1, \cdots, x_m) = \sum_{i=1}^{m-1} (100(x_{i+1} - x_i^2)^2 + (1 - x_i)^2) \tag{15}$$

in the box $X_i = [-1.4, 1.5] \quad (i = 1, \cdots, m)$. The Table 1 shows the dividing time and the calculating cost when we apply the Fujii's method and the new method for various m.

Next, we found the minimum value of the following Ridge function:

$$f(x_1, \cdots, x_m) = \sum_{i=1}^{m} (\sum_{j=1}^{i} x_j)^2 \tag{16}$$

Table 1. Comparison of the efficiency for the function(15)

m	dividing time		calculating cost (s)	
	Fujii's method	New method	Fujii's method	New method
10	13546	20	0.4687500	0.0078125
12	90458	31	5.5781250	0.0234375
14	473666	57	201.5078125	0.0468750
16	–	101	–	0.1328125
18	–	168	–	0.2812500
20	–	291	–	0.5937500
30	–	4864	–	27.7109375
40	–	82294	–	3376.8515625

Table 2. Comparison of the efficiency for the function(16)

m	dividing time		calculating cost (s)	
	Fujii's method	New method	Fujii's method	New method
2	2479	62	0.0078125	0.0069531
3	269913	93	118.4296875	0.0234375
4	–	124	–	0.1093750
5	–	155	–	0.3515625
6	–	186	–	0.9140625
7	–	217	–	2.2578125
8	–	248	–	5.0859375
9	–	279	–	10.6015625
10	–	310	–	20.4609375
20	–	620	–	1961.2265625

in the box $X_i = [-64, 64] \quad (i = 1, \cdots, m)$. The Table 2 shows the dividing time and the calculating cost when we apply these methods for various m.

By the Table1 and the Table2, we are able to confirm that the new method is much faster than the Fujii's method. And in the both example, even if m increase, the new method was able to find the maximum and minimum value although the Fujii's method was not able to find them because of momery over. This fact also shows the efficiency of the new method.

5 Conclusion

In this paper, a new method was proposed for finding all maxima and minima of a multivariable function in a box applying AA. And to show the efficiency of the new method, some numerical examples were implemented.

References

1. Marcus Vinícius A. Andrade, João L. D. Comba and Jorge Stolfi:"Affine Arithmetic", INTERVAL'94, St. petersburg (Russia), March 5-10, (1994).
2. Luiz H. de Figueiredo and Jorge Stolfi:"Self-Validated Numerical Methods and Applications", Brazilian Mathematics Colloquium monograph, IMPA, Rio de Janeiro, Brazil; July (1997).
3. A. Neumaier: "Interval methods for systems of equations", Cambrige University Press (1990).
4. Yasuo Fujii, Kozo Ichida and Takeshi Kiyono:"A Method for Finding the Greatest Value of a Multivariable Function Using Interval Arithmetic", J. IPS Japan, Vol. 18, No. 11, pp. 1095-1101 (1977) (In Japanese).

i.e.

$$|\varphi(z) - \alpha| \le \frac{M\rho^p}{r^{p-1}(r-\rho)} . \tag{6}$$

Next we impose to ρ the stronger restriction

$$\rho < \frac{r}{2} . \tag{7}$$

Then, from (6) and (7) we obtain

$$|\varphi(z) - \alpha| \le \gamma\rho^p , \tag{8}$$

where

$$\gamma = \frac{2M}{r^p},\ \rho = |z-\alpha|,\ z \in K\left(\alpha, \frac{r}{2}\right) .$$

Let the first approximation z_0 of (4) is in

$$K(\alpha, \tilde{\rho}) = \{z : |z-\alpha| < \tilde{\rho}\} , \quad \tilde{\rho} = \min\left(\frac{r}{2}, \gamma^{\frac{1}{1-p}}\right)$$

and $\rho_0 = |z_0 - \alpha| > 0$. The case $\rho_0 = 0$ is trivial. Then we consider the case $\rho_0 > 0$. It is easy to prove

$$0 < \rho_k = |z_k - \alpha| = |\varphi(z_{k-1}) - \alpha| \le \gamma\rho_{k-1}^p < \rho_{k-1} , \tag{9}$$

for every $k = 1, 2, \ldots$.

Hence,

$$0 \le \rho_{k-1} < \gamma^{\frac{1}{1-p}} . \tag{10}$$

From (9) and (10) it follows that $\rho_k \to 0$. But we are to prove something more, namely

$$\rho_k \le cq^{p^k} , \quad \text{where} \quad c = \gamma^{\frac{1}{1-p}} , \ q = \gamma^{\frac{1}{p-1}}\rho_0 \in (0,1) . \tag{11}$$

Indeed,

$$\rho_k \le \gamma\rho_{k-1}^p = \gamma\left(\gamma\rho_{k-2}^p\right)^p = \gamma^{1+p}\rho_{k-2}^{p^2} \le \cdots \le \gamma^{1+p+\ldots p^{k-1}}\rho_0^{p^k}$$

$$= \gamma^{\frac{p^k-1}{p-1}}\rho_0^{p^k} = \gamma^{\frac{1}{1-p}}\left(\gamma^{\frac{1}{p-1}}\rho_0\right)^{p^k} = cq^{p^k} ,$$

where $q \in (0,1)$, according to (10).

The estimate (11) means, that (4) has an order of convergence p.

Example 1. From Theorem 2 it follows, that if $f(z)$ is analytic in K and $f'(z) \neq 0$ for every $z \in K(\alpha, r)$ and if $f(z)$ has a simple root α in $K(\alpha, r)$, then the Newton's iterative process

$$z_{k+1} = z_k - \frac{f(z_k)}{f'(z_k)} , \quad k = 0, 1, 2, \ldots \tag{12}$$

converges to α with an order of convergence $p = 2$ for $z_0 \in K(\alpha, r)$, provided z_0 is sufficiently close to α.

Example 2. From Theorem 2 it follows, that if $f(z)$ is analytic in K and $f'(z) \neq 0$ for every $z \in K(\alpha, r)$ and if $f(z)$ has a simple root α in $K(\alpha, r)$, then the modified Newton's method

$$z_{k+1} = z_k - u(z_k) - \frac{f(z_k - u(z_k))}{f'(z_k)}, \quad k = 0, 1, 2, \dots \tag{13}$$

where $u(z_k) = \frac{f(z_k)}{f'(z_k)}$ converges to α for $z_0 \in K(\alpha, r)$ with an order of convergence $p = 3$, when z_0 is sufficiently close to α.

These two examples illustrate the two of the most popular iterative methods for solving nonlinear equations(see [4, 5, 7]).

3 Iterative Function of Several Complex Variables

Iterative functions of several complex variables are used in iterative methods for solving systems of equations. Before considering this problem, we remind some definitions, related to it.

Let C^n be the set of all vectors $z = (z_1, z_2, \dots, z_n)$ with n complex components. Then a function of n complex variables has the form

$$u = f(z) \quad \text{or} \quad u = f(z_1, z_2, \dots, z_n), \tag{14}$$

where $u \in U \subset C^1$, $z \in Z \subset C^n$.

If $\alpha = (\alpha_1, \alpha_2, \dots, \alpha_n) \in C^n$ and $\tilde{r} = (r_1, r_2, \dots, r_n)$ is a vector with positive components, then

$$\bar{P}(\alpha, \tilde{r}) = \{z : z = (z_k) \in C^n, \ |z_k - \alpha_k| \leq r_k, k = 1, 2, \dots, n\} \tag{15}$$

is called closed polydisc, and

$$P(\alpha, \tilde{r}) = \{z : z = \{z_k\} \in C^n, \ |z_k - \alpha_k| < r_k, k = 1, 2, \dots, n\} \tag{16}$$

open polydisc.

Further we assume, that $r_1 = r_2 = \dots = r_n = r > 0$ and we denote the corresponding polydiscs by $\bar{P}(\alpha, r)$ and $P(\alpha, r)$.

Here we use the vectorial norm

$$\| z \| = \max_k |z_k| . \tag{17}$$

Let us consider the following nonlinear system

$$z_1 = \varphi_1(z), z_2 = \varphi_2(z), \dots, z_n = \varphi_n(z) \quad \text{or} \quad z = \varphi(z), \tag{18}$$

where $z = (z_1, z_2, \dots, z_n)$. (We do not consider linear systems here. There are many special iterative methods for them.) Let $\alpha = (\alpha_1, \alpha_2, \dots, \alpha_n)$ be a solution of (18), or a fixed point of φ, localized in the polydisc $P(\alpha, r) \subset C^n$. Moreover, let $z^0 = \{z_i^0\}$ be an initial approximation of α such that the iterative process

$$z^{k+1} = \varphi(z^k), \quad k = 0, 1, 2, \dots \tag{19}$$

converges to α, i.e. $\| z^k - \alpha \| \to 0$ for $k \to \infty$. We say, that the order of convergence is $p > 1$, if

$$\| z^k - \alpha \| \leq cq^{p^k}\ , \ c > 0\,,\ q \in (0,1)\,,\ k = 0,1,2,\dots \tag{20}$$

Theorem 3. *Let $\varphi(z)$ from (18) has a unique fixed point α in $K(\alpha, r)$ and let every $\varphi_i(z)$ be analytic on $\bar{K}(\alpha, r)$. If for an integer $p > 1$ we have*

$$\frac{\partial^s \varphi_i(\alpha)}{\partial z_1^{s_1} \partial z_2^{s_2} \dots \partial z_n^{s_n}} = 0\,, \tag{21}$$

for $s = 1, 2, \dots, p-1$; $s_1 + s_2 + \dots + s_n = s$; $i = 1, 2, \dots, n$ and for at least one combination $(i; p_1, p_2, \dots, p_n)$

$$\frac{\partial^p \varphi_i(\alpha)}{\partial z_1^{p_1} \partial z_2^{p_2} \dots \partial z_n^{p_n}} \neq 0\,,\quad p_1 + p_2 + \dots + p_n = p \tag{22}$$

is fulfilled, i.e. $\varphi(z)$ has an order p, then for z_0 sufficiently close to α, the iterative process

$$z^{k+1} = \varphi(z^k)\,,\quad k = 0, 1, 2, \dots \tag{23}$$

converges to α with an order of convergence p.

Proof. For every $\varphi_i(z)\,,\ z \in K(\alpha, r)$ we have

$$\varphi_i(z) = \varphi_i(\alpha) + \sum_{k_1+k_2+\dots+k_n \geq p} \alpha_{k_1,k_2,\dots,k_n} (z_1 - \alpha_1)^{k_1} (z_2 - \alpha_2)^{k_2} \dots (z_n - \alpha_n)^{k_n} \tag{24}$$

where $z = (z_1, z_2, \dots, z_n)$, and

$$|\alpha_{k_1,k_2,\dots,k_n}| \leq \frac{M_i}{r^{k_1+k_2+\dots+k_n}}\,,$$

where

$$M_i = \max_{z \in \partial \bar{K}} |\varphi_i(z)|$$

and $\partial \bar{K}$ is the boundary of $\bar{K}$.

From (24) under the stronger restriction $\rho = \| z - \alpha \| \leq \frac{r}{2n}$ we get

$$\begin{aligned}
|\varphi_i(z) - \alpha_i| &\leq M_i \sum_{k_1+k_2+\dots+k_n \geq p} \left(\frac{\rho}{r}\right)^{k_1+k_2+\dots+k_n} = M_i \sum_{s \geq p} \binom{n+s-1}{s} \left(\frac{\rho}{r}\right)^s \\
&\leq M_i \sum_{s=p}^{\infty} \left(\frac{n\rho}{r}\right)^s = M_i \left(\frac{n\rho}{r}\right)^p \frac{1}{1 - \frac{n\rho}{r}} \\
&\leq 2M_i \left(\frac{n\rho}{r}\right)^p
\end{aligned}$$

or

$$|\varphi_i(z) - \alpha_i| \leq 2M_i \left(\frac{n\rho}{r}\right)^p .$$

Further,

$$|\varphi_i(z) - \alpha_i| \le \gamma\rho^p, \quad \gamma = 2\left(\frac{n}{r}\right)^p \max_i M_i, \; i = 1, 2, \ldots, n. \tag{25}$$

From (25), it follows the only one inequality

$$\| \varphi(z) - \alpha \| \le \gamma\rho^p. \tag{26}$$

Let $\tilde{\rho} = \min\left(\frac{r}{2n}, \gamma^{\frac{1}{1-p}}\right)$ and the initial approximation of the iteration (23) $z^0 \in K(\alpha, \tilde{\rho})$, $z^0 \neq \alpha$, and $\rho_k = \|z^k - \alpha\|$. Then from (26) we get

$$\rho_k \le \gamma\rho_{k-1}^p \le \rho_{k-1}, \quad k = 1, 2, 3, \ldots \tag{27}$$

Further, as in Section 2, we receive

$$\|z_k - \alpha\| \le c q^{p^k}, \tag{28}$$

where $c = \gamma^{\frac{1}{1-p}}$ and $q \in (0, 1)$. The condition (28) means, that the iteration process (23) has an order of convergence p.

Example 3. Let $f(z)$ be a polynomial of degree n with zeros $\alpha_1, \alpha_2, \ldots, \alpha_n$ ($\alpha_i \neq \alpha_j$ for $i \neq j$). Then the well known Weierstrass-Dochev iterative method for simultaneous calculation of all zeros α_i is

$$z_i^{k+1} = z_i^k - \frac{f(z_i^k)}{P_i(z^k)}, \quad i = 1, \ldots, n, \quad k = 0, 1, 2, \ldots, \tag{29}$$

where $z^k = (z_1^k, \ldots, z_n^k)$ and $P_i(z^k) = \prod_{j \neq i}^n (z_i^k - z_j^k)$.

But the iterative process (29) could be considered as an iterative process for solving the system of equations

$$z_i = \varphi_i(z), \quad i = 1, \ldots, n, \tag{30}$$

where $\varphi_i(z) = z_i - \frac{f(z_i)}{P_i(z)}$.

It is easy to see, that for all $z \in K(\alpha, \rho)$, $\alpha = (\alpha_1, \alpha_2, \ldots, \alpha_n)$ and ρ sufficiently small the iterative functions $\varphi_i(z)$ are analytic in $C(\alpha, \rho)$, all $\frac{\partial \varphi_i(\alpha)}{\partial z_s} = 0$ and at least one $\frac{\partial^2 \varphi_i(\alpha)}{\partial z_s \partial z_j} \neq 0$. Thus from Theorem 3 it follows, that the iterative process (29) converges to α with order of convergence $p = 2$ for initial approximation z^0 sufficiently close to α.

Example 4. One possible modification of the iterative process (29) for solving of the same problem is

$$z_i^{k+1} = \psi_i(z^k), \quad i = 1, \ldots, n, \tag{31}$$

where

$$\psi_i(z) = \varphi_i(z) - \frac{f(\varphi_i(z))}{P_i(z)}, \quad i = 1, \ldots, n \tag{32}$$

and $\varphi_i(z)$ is the iterative function (30).

It is easy to verify, that in sufficiently small polydisc surrounding α the iterative functions $\psi_i(z)$ are analytic and all derivatives of order 1 and 2 of these functions in α are zero, and at least one derivative of order 3 is non zero. From these properties and from Theorem 3 we get again, that for z^0 sufficiently close to α the iterative process (31) converges to α with order of convergence 3.

The iterative methods from examples 3 and 4 are studied in [5, 7, 8, 9], too.

Example 5. It is easy to apply the same approach to the Newton's method

$$z^{k+1} = z^k - f'(z_k)^{-1} f(z^k) \equiv \lambda(z^k) \tag{33}$$

and to the following modification of the Newton's method

$$z^{k+1} = \lambda(z^k) - f'(z_k)^{-1} f(\lambda(z^k)) \tag{34}$$

for solving of the nonlinear system

$$f(z) = (f_1(z), f_2(z), \ldots, f_n(z)) = 0\,.$$

The described approach for investigation of iterative processes sometimes is more natural and shorter(see [7] for the Dochev's method). In the same time this approach gives alternative options for location of the solution α and the choice of the initial approximation z^0.

4 Numerical Experiments

In this section we compare the behavior of the *single-root* algorithms: Newton's method (12) and the modified Newton's method obtained by (13); and the *simultaneous* methods: Weierstrass-Dochev's method (29) and modified Weierstrass-Dochev's method obtained by (32). The software implementation of the algorithms has been made in the MATLAB programming language.

Example 6. Let consider the equation $f(z) = z^6 - 1 = 0$ with a root $\alpha_1 = 1$. From (12) we obtain $\varphi(z) = z - \frac{z^6-1}{6z^5}$. Obviously the function φ is analytic in the disc $K(1, 1/2)$. It is easy to prove that $M = \max_{|z-1|=\frac{1}{2}} |\varphi(z)| < 7$ and $\gamma = 56$. From Theorem 2 it follows that the Newton's method is convergent to the root $\alpha_1 = 1$ for initial approximations $z_0 \in K(1, \tilde{\rho})$, where $\tilde{\rho} = \min(r/2, \gamma^{-1}) = \frac{1}{56}$.

Table 1. Experimental results of Example 6

Initialapproximation x^0	*iter* (12)	(13)
$x^0 = 1.006 + 0.01\,i$	5	4
$x^0 = 1.1 - 0.1\,i$	6	5
$x^0 = 1.3 + 0.2\,i$	8	6
$x^0 = 1.5 + 0.4\,i$	10	7

Table 2. Experimental results of Example 7

Initialapproximation x^0	*iter* (29)	(32)
$x^0 = (0.1 - 1.1\,i,\ 0.2 + 1.9\,i,\ 3.8 + 0.1\,i)$	5	4
$x^0 = (0.2 - 0.8\,i,\ 0.3 + 2.2\,i,\ 4.2 + 0.2\,i)$	5	4
$x^0 = (0.4 - 1.3\,i,\ 0.3 + 2.5\,i,\ 4.4 + 0.3\,i)$	6	4
$x^0 = (0.5 - 0.5\,i,\ 0.4 + 1.5\,i,\ 3.5 - 0.3\,i)$	7	5

Example 7. Let consider the equation $f(z) = (x + i)(x - 2\,i)(x - 4) = 0$ with solution $\alpha = (-i\,, 2\,i\,, 4)$.

We have denoted: x^0- initial approximation; *iter*- the number of iterations for which the exact root α is reached.

References

1. Cartan, H.: Theorie elementaire des functions analytiques d'une ou plusieurs variables complexes, Herman, Paris, (1961), (in Russian).
2. Chakalov, L.: Introduction in theory of analytic functions, Science and art, Sofia, (1975) (in Bulgarian).
3. Bizadze, A.: Bases of the Theory of the Analytical Functions of the complex variables, Science, Main edition of Mathematical-Physical Literatures, Moscow, (1969).
4. Traub, J.: Iterative Methods for the Solution of Equations, Prentice Hall, Englewood Cliffs, New Jersey (1964).
5. Petkov, M., Kjurkchiev, N.: Numerical Methods for Solving Nonlinear equations, University of Sofia, Sofia, (2000) (in Bulgarian).
6. Schröder, E.: Über unendlich viele Algorithmen zur Auflösung der Gleichungen, Math. Ann., **2**, (1870), 317–365.
7. Sendov, B., Popov, V.: Numerical Methods, Part I, Nauka i Izkustvo, Sofia, (1976).
8. Nedzhibov, G., Petkov, M.: A Family of Iterative Methods for Simultaneous Computing of All Zeros of Algebraic Equation, XXVIII Summer School "Application of Mathematics in Engineering and Economics'28", Sozopol'2003, Sofia, (2004).
9. Nedzhibov, G., Petkov, M.: On a Family of Iterative Methods for Simultaneous Extraction of All Roots of Algebraic Polynomial, Applied Mathematics and Computation, (2004) (in press).

Parallel Implementation and One Year Experiments with the Danish Eulerian Model

Tzvetan Ostromsky[1], Ivan T. Dimov[1], and Zahari Zlatev[2]

[1] Institute for Parallel Processing, Bulgarian Academy of Sciences, Acad. G. Bonchev str., bl. 25-A, 1113 Sofia, Bulgaria
ceco@parallel.bas.bg
ivdimov@bas.bg http://www.bas.bg/clpp/

[2] National Environmental Research Institute, Department of Atmospheric Environment, Frederiksborgvej 399 P. O. Box 358, DK-4000 Roskilde, Denmark
zz@dmu.dk
http://www.dmu.dk/AtmosphericEnvironment

Abstract. Large scale air pollution models are powerful tools, designed to meet the increasing demand in different environmental studies. The atmosphere is the most dynamic component of the environment, where the pollutants can be moved quickly on far distnce. Therefore the air pollution modeling must be done in a large computational domain. Moreover, all relevant physical, chemical and photochemical processes must be taken into account. In such complex models operator splitting is very often applied in order to achieve sufficient accuracy as well as efficiency of the numerical solution.

The Danish Eulerian Model (DEM) is one of the most advanced such models. Its space domain (4800×4800 km) covers Europe, most of the Mediterian and neighboring parts of Asia and the Atlantic Ocean. Efficient parallelization is crucial for the performance and practical capabilities of this huge computational model. Different splitting schemes, based on the main processes mentioned above, have been implemented and tested with respect to accuracy and performance in the new version of DEM. Some numerical results of these experiments are presented in this paper.

1 Introduction

The problem for air pollution modelling has been studied for years [8, 15]. An air pollution model is generally described by a system of partial differential equations for calculating the concentrations of a number of chemical species (pollutants and other components of the air that interact with the pollutants) in a large 3-D domain (part of the atmosphere above the studied geographical region). The main physical and chemical processes (horizontal and vertical wind, diffusion, chemical reactions, emissions and deposition) should be adequately represented in the system.

The Danish Eulerian Model (DEM) [1, 10, 14, 15, 16] is mathematically represented by the following system of partial differential equations:

Z. Li et al. (Eds.): NAA 2004, LNCS 3401, pp. 440–447, 2005.

$$\frac{\partial c_s}{\partial t} = -\frac{\partial(uc_s)}{\partial x} - \frac{\partial(vc_s)}{\partial y} - \frac{\partial(wc_s)}{\partial z} + \\ + \frac{\partial}{\partial x}\left(K_x \frac{\partial c_s}{\partial x}\right) + \frac{\partial}{\partial y}\left(K_y \frac{\partial c_s}{\partial y}\right) + \frac{\partial}{\partial z}\left(K_z \frac{\partial c_s}{\partial z}\right) + \\ + E_s + Q_s(c_1, c_2, \dots c_q) - (k_{1s} + k_{2s})c_s, \quad s = 1, 2, \dots q \ . \tag{1}$$

where

- c_s – the concentrations of the chemical species;
- u, v, w – the wind components along the coordinate axes;
- K_x, K_y, K_z – diffusion coefficients;
- E_s – the emissions;
- k_{1s}, k_{2s} – dry / wet deposition coefficients;
- $Q_s(c_1, c_2, \dots c_q)$ – non-linear functions describing the chemical reactions between species under consideration (Gery et al. (1989)).

2 Splitting into Submodels

The above rather complex system (1) is split into 3 subsystems / submodels, according to the major physical / chemical processes and the numerical methods applied in their solution.

$$\frac{\partial c_s^{(1)}}{\partial t} = -\frac{\partial(uc_s^{(1)})}{\partial x} - \frac{\partial(vc_s^{(1)})}{\partial y} + \frac{\partial}{\partial x}\left(K_x \frac{\partial c_s^{(1)}}{\partial x}\right) + \frac{\partial}{\partial y}\left(K_y \frac{\partial c_s^{(1)}}{\partial y}\right) = A_1 c_s^{(1)}(t)$$

horizontal advection & diffusion

$$\frac{\partial c_s^{(2)}}{\partial t} = E_s + Q_s(c_1^{(2)}, c_2^{(2)}, \dots c_q^{(2)}) - (k_{1s} + k_{2s})c_s^{(4)} = A_2 c_s^{(2)}(t)$$

chemistry, emissions & deposition

$$\frac{\partial c_s^{(3)}}{\partial t} = -\frac{\partial(wc_s^{(3)})}{\partial z} + \frac{\partial}{\partial z}\left(K_z \frac{\partial c_s^{(3)}}{\partial z}\right) = A_3 c_s^{(3)}(t)$$

vertical transport

Various splitting schemes have been proposed and analysed in [2, 3, 4, 5, 8, 9, 13] The three splitting schemes, discussed in this paper and used in our experiments, are briefly described below.

2.1 Sequential Splitting Scheme, Used by Default in UNI-DEM

$$\left.\begin{array}{c}\frac{dc_s^{(1)}(t)}{dt} = A_1 c_s^{(1)}(t), \quad t \in ((k-1)\tau, k\tau] \\ c_s^{(1)}((k-1)\tau) = c_s^{(2)}((k-1)\tau)\end{array}\right\} \tag{2}$$

$$\left.\begin{array}{c}\frac{dc_s^{(2)}(t)}{dt} = A_2 c_s^{(2)}(t), \quad t \in ((k-1)\tau, k\tau] \\ c_s^{(2)}((k-1)\tau) = c_s^{(1)}(k\tau)\end{array}\right\} \tag{3}$$

$s = 1, 2, \ldots, q($ q – the number of chemical species).

Equations (2)–(3) describe the sequential splitting scheme in the 2-D case (without the vertical transport). The splitting error of this scheme, used by default in UNI-DEM, is $\mathcal{O}(\tau)$, where τ is the time step.

2.2 Marchuck - Strang Splitting Scheme

$$\left.\begin{array}{c}\frac{dc_s^{(1)}(t)}{dt} = A_1 c_s^{(1)}(t), \quad t \in ((k-1)\tau, (k-\frac{1}{2})\tau] \\ c_s^{(1)}((k-1)\tau) = \hat{c}^{(1)}((k-1)\tau)\end{array}\right\} \tag{4}$$

$$\left.\begin{array}{c}\frac{dc_s^{(2)}(t)}{dt} = A_2 c_s^{(2)}(t), \quad t \in ((k-1)\tau, k\tau] \\ c_s^{(2)}((k-1)\tau) = w_k^{(1)}((k-\frac{1}{2})\tau)\end{array}\right\} \tag{5}$$

$$\left.\begin{array}{c}\frac{d\hat{c}_s^{(1)}(t)}{dt} = A_1 \hat{c}_s^{(1)}(t), \quad t \in ((k-\frac{1}{2})\tau, k\tau] \\ \hat{c}_s^{(1)}((k-\frac{1}{2})\tau) = c_s^{(2)}(k\tau)\end{array}\right\} \tag{6}$$

$s = 1, 2, \ldots, q($ q – the number of chemical species).

Equations (4)–(6) describe the symmetric splitting scheme (due to Marchuck and Strang) in the 2-D case (without the vertical transport). This scheme has higher order of accuracy, $\mathcal{O}(\tau^2)$, where τ is the time step.

2.3 Weighted Sequential Splitting Scheme

The sequential splitting scheme (2)–(3) is applied twice on each step with reverse order of the two operators A_1 and A_2. The average of the two results for $c_s((k-1)\tau)$ is taken as initial value for calculations of the $c_s(k\tau)$ on the next step. This scheme has also second order of accuracy, $\mathcal{O}(\tau^2)$.

3 Parallelization Strategy and Numerical Methods, Used in the Solution of the Submodels

Although the splitting is a crucial step in the efficient numerical treatment of the model, after discretization of the large computational domain each submodel becomes itself a huge computational task. In addition, the dynamics of the chemical and photo-chemical processes requires using of small time-step to keep stability of the computations. Large parallel supercomputers must be used in order to meet the high speed and storage requirements. Moreover, development and impementation of efficient parallel algorithms is very important for improving the practical capabilities and the performance of DEM. That topic has been discussed in more detail in [10, 11, 12, 15, 16].

Distributed memory parallelization model via MPI [6] is used in the current UNI-DEM version considered in this work. For maximum portability only standard MPI routines are used in the UNI-DEM code. Parallelization is based on domain decomposition of the horizontal grid, which implies certain restrictions on the number of MPI tasks and requires communication on each time step. Improving the data locality for more efficient cache utilization is achieved by using *chunks* to group properly the small tasks in the chemistry-deposition and vertical exchange stages.

Additional **pre-processing** and **post-processing** stages are needed for scattering the input data and gathering the results. These are cheap, but their relative weight grows up with increasing the number of MPI tasks, affecting the total speed-up and scalability.

The numerical methods, used in the solution of the submodels, are given below.

- **Advection-diffusion part:** Finite elements, followed by predictor-corrector schemes with several different correctors. The native parallel tasks in this case are the calculations for a given pollutant in a given layer. There are enough parallel tasks, sometimes too big to fit into the cache (depends on the subdomain size).
- **Chemistry-deposition part:** An improved version of the QSSA (Quazi Steady-State Approximation) (Hesstvedt et al. - [7]). The native parallel tasks here are the calculations in a single grid-point. These small tasks are grouped in *chunks* for efficient cache utilization.
- **Vertical transport:** Finite elements, followed by θ-methods. The native parallel tasks here are the calculations along each vertical grid-line. The number of these tasks is large, while the tasks are relatively small. They also can be grouped in *chunks*, like those in the chemical stage.

4 Numerical Results

The experiments, presented in this section, are performed on the SUN Sunfire 6800 parallel machine (24 CPU UltraSparc-III / 750 MHz), located at the Danish Technical University (DTU) in Lyngby, Denmark. All experiments are for a

Table 1. NO_2 results (concentrations, discrepancy and correlation factor) for 1998

UNI-DEM results for NO_2 on $(96 \times 96 \times 1)$ grid, in comparison with the measurements of 35 stations in Europe										
Month	Concentration [mol/l]				‖Discrepancy %‖			Correlation		
	Obs.	**Seq.**	**Weig.**	**M.-S.**	**Seq.**	**Weig.**	**M.-S.**	**Seq.**	**Weig.**	**M.-S.**
January	2.79	3.20	3.21	2.35	13	13	19	0.67	0.66	0.62
February	3.24	4.96	4.97	3.83	35	35	15	0.61	0.61	0.59
March	2.18	1.83	1.84	1.38	19	19	58	0.69	0.69	0.75
April	1.98	1.76	1.76	1.39	12	12	42	0.59	0.59	0.49
May	1.81	2.05	2.05	1.66	12	12	9	0.64	0.64	0.66
June	1.58	2.33	2.34	1.95	32	32	19	0.56	0.56	0.47
July	1.57	1.86	1.86	1.56	15	16	1	0.50	0.50	0.49
August	1.70	1.99	1.99	1.81	15	15	6	0.59	0.59	0.57
September	1.84	2.42	2.42	2.11	24	24	13	0.65	0.65	0.56
October	1.96	2.38	2.39	1.85	18	18	6	0.57	0.57	0.53
November	3.10	3.97	3.97	3.04	22	22	2	0.65	0.65	0.70
December	3.33	4.10	4.11	3.10	19	19	8	0.70	0.69	0.71
mean	2.26	2.74	2.74	2.17	17	17	4	0.69	0.69	0.68

time period of 1 year (1998). The number of chemical species considered by the chemical submodel in the current version is $q = 35$. Chunk size equal to 48 is used in the experiments, which seems to be optimal for this problem and the target machine.

4.1 Accuracy Results

These results are obtained by experiments with real data (meteorological data sets, collected by the Norwegian Meteorological Institute; as well as emmision data) for 1998. The results are compared to the records of several stationary measuring stations throughout Europe, wich have enough measurements. It is not possible to extract only the splitting error in these experiments. Other errors, which are present in the results, are as follows:

- Input data error;
- Error of the numerical methods, used in the different submodels;
- Spatial discretization error (depending on the grid step);
- Computational (rounding) error;

and so on. In addition, the error of the measurement instruments is also included. Such accuracy results for three major pollutants: nitrogen dioxide (NO_2), sulphur dioxide (SO_2) and amonia ($NH_3 + NH_4$); are given in Tables 1 – 3 respectively.

The following abbreviations are used in the column headings of these tables:

Obs. - observed by the measurement stations;

Seq. - computed by DEM with sequenial splitting;

Weig. - computed by DEM with weighted sequenial splitting;

M.-S. - computed by DEM with Marchuck - Strang splitting.

Table 2. SO_2 results (concentrations, discrepancy and correlation factor) for 1998

UNI-DEM results for SO_2 on $(96 \times 96 \times 1)$ grid, in comparison with the measurements of 35 stations in Europe										
Month	Concentration [mol/l]				||Discrepancy %||			Correlation		
	Obs.	**Seq.**	**Weig.**	**M.-S.**	**Seq.**	**Weig.**	**M.-S.**	**Seq.**	**Weig.**	**M.-S.**
January	1.36	2.70	2.72	2.49	50	50	46	0.83	0.83	0.83
February	1.41	2.02	2.04	1.91	30	31	26	0.61	0.61	0.69
March	1.11	1.49	1.51	1.41	26	26	22	0.64	0.64	0.64
April	0.80	1.06	1.07	1.01	25	25	21	0.60	0.60	0.64
May	0.73	1.13	1.14	1.12	35	36	35	0.71	0.71	0.72
June	0.51	0.72	0.72	0.72	29	30	30	0.72	0.72	0.70
July	0.49	0.63	0.64	0.62	22	23	20	0.82	0.83	0.83
August	0.56	0.73	0.74	0.75	23	24	25	0.63	0.63	0.71
September	0.60	0.85	0.88	0.86	29	29	30	0.82	0.82	0.81
October	0.58	0.68	0.69	0.67	16	17	15	0.75	0.75	0.81
November	1.32	1.59	1.60	1.58	17	18	17	0.29	0.28	0.45
December	1.41	2.89	2.91	2.64	51	52	47	0.39	0.39	0.52
mean	0.91	1.38	1.39	1.32	34	34	31	0.71	0.71	0.77

Table 3. Amonia results (concentrations, discrepancy and correlation factor) for 1998

UNI-DEM results for $NH_3 + NH_4$ on $(96 \times 96 \times 1)$ grid, in comparison with the measurements of 24 stations in Europe										
Month	Concentration [mol/l]				||Discrepancy %||			Correlation		
	Obs.	**Seq.**	**Weig.**	**M.-S.**	**Seq.**	**Weig.**	**M.-S.**	**Seq.**	**Weig.**	**M.-S.**
January	1.02	1.27	1.37	1.04	20	26	2	0.78	0.78	0.80
February	1.51	1.43	1.51	1.18	5	0	28	0.79	0.79	0.79
March	1.19	1.42	1.52	1.16	16	22	2	0.67	0.64	0.67
April	1.36	1.29	1.35	1.12	6	0	21	0.65	0.64	0.63
May	1.46	1.67	1.75	1.31	12	17	11	0.76	0.75	0.69
June	1.20	0.99	1.02	0.93	21	18	29	0.74	0.74	0.74
July	1.09	1.00	1.03	0.97	9	6	12	0.72	0.72	0.68
August	1.01	1.17	1.18	1.06	13	15	4	0.83	0.83	0.82
September	1.39	1.86	1.96	1.53	25	29	9	0.78	0.77	0.80
October	0.73	0.89	0.93	0.76	18	21	4	0.65	0.64	0.67
November	1.19	1.82	1.95	1.59	34	39	25	0.65	0.64	0.66
December	1.13	1.70	1.83	1.46	33	38	22	0.76	0.77	0.77
mean	1.19	1.38	1.45	1.18	13	17	1	0.78	0.77	0.77

4.2 Performance Results

Some time and speed-up results, showing the performance of UNI-DEM on Sunfire 6800 computing system, are presented in Table 4.

Table 4. Time in seconds and the **(speed-up)** (given in brackets below) of UNI-DEM for running an 1-year experiment on different size grids (2-D and 3-D versions) on a Sunfire 6800 machine at DTU

Time / **(speed-up)** results of UNI-DEM on a SunFire 6800 at DTU					
Grid size / stage	**1 proc.**	**2 proc.**	**4 proc.**	**8 proc.**	**16 proc.**
[96 × 96 × 1] – total	36270	18590	10296	4788	2492
		(1.95)	**(3.52)**	**(7.58)**	**(14.55)**
– **advection**	9383	4882	2512	1540	792
		(1.92)	**(3.74)**	**(6.09)**	**(11.85)**
– **chemistry**	24428	13120	7122	3088	1544
		(1.86)	**(3.43)**	**(7.91)**	**(15.82)**
[96 × 96 × 10] – total	415442	211370	108620	52920	30821
		(1.97)	**(3.82)**	**(7.85)**	**(13.48)**
– **advection**	132660	64759	31367	15422	8612
		(2.05)	**(4.23)**	**(8.60)**	**(15.40)**
– **chemistry**	263584	133496	68138	34255	17621
		(1.97)	**(3.87)**	**(7.69)**	**(14.96)**
[288 × 288 × 1] – total	699622	352470	197811	98033	44465
		(1.98)	**(3.54)**	**(7.14)**	**(15.95)**
– **advection**	348024	169773	85812	42558	20782
		(2.05)	**(4.05)**	**(8.18)**	**(16.75)**
– **chemistry**	294155	150088	84504	37923	18498
		(1.96)	**(3.48)**	**(7.76)**	**(15.90)**

5 Conclusions and Plans for Future Work

– By using high performance parallel computers to run the variable grid-size code UNI-DEM, reliable results for a large region (whole Europe) and for a very long period (one or several years) can be obtained within a reasonable time.
– In most cases the results, obtained with the weighted sequential splitting, are quite similar to those of the sequential splitting. This could be an indication for a small commutator of the two operators.
– The splitting is not the only source of error in UNI-DEM. Nevertheless, its contribution in the total error seems to be significant. For some species (NO_2, SO_2, amonia) the Marchuck-Strang scheme gives results, closer to the measuremets than the other two splitting methods. More experiments are needed in order to investigate the consistency of such behaviour.
– The parallel code, created by using MPI standard library, appears to be highly portable and shows good efficiency and scalability. The limited size of the fast cache-memory causes superlinear speed-up in some cases.

Acknowledgments

This research was supported in part by grant I-901/99 from the Bulgarian NSF, grant ICA1-CT-2000-70016 – Center of Excellence and by NATO grant NATO ARW "Impact of future climate changes on pollution levels in Europe". A grant from the Danish Natural Sciences Research Council gave us access to all Danish supercomputers.

References

1. V. Alexandrov, A. Sameh, Y. Siddique and Z. Zlatev, Numerical integration of chemical ODE problems arising in air pollution models, Env. Modeling and Assessment, 2 (1997) 365–377.
2. M. Botchev, I. Faragó, A. Havasi, Testing weighted splitting schemes on a one-column transport-chemistry model, Int. J. Env. Pol., Vol. 22, No. 1/2.
3. P. Csomós, I. Faragó, A. Havasi, Weighted sequential splittings and their analysis, Comput. Math. Appl., (to appear)
4. I. Dimov, I. Faragó, A. Havasi, Z. Zlatev, L-commuataivity of the operators in splitting methods for air pollution models, Annales Univ. Sci. Sec. Math., 44 (2001) 127-148.
5. I. Faragó, A. Havasi, On the convergence and local splitting error of different splitting schemes, Progress in Computational Fluid Dynamics, (to appear)
6. W. Gropp, E. Lusk and A. Skjellum, Using MPI: Portable programming with the message passing interface, MIT Press, Cambridge, Massachusetts, 1994.
7. E. Hesstvedt, Ø. Hov and I. A. Isaksen, Quasi-steady-state approximations in air pollution modeling: comparison of two numerical schemes for oxidant prediction, Int. Journal of Chemical Kinetics 10 (1978), pp. 971–994.
8. G. I. Marchuk, Mathematical modeling for the problem of the environment, Studies in Mathematics and Applications, No. 16, North-Holland, Amsterdam, 1985.
9. G. J. McRae, W. R. Goodin and J. H. Seinfeld, Numerical solution of the atmospheric diffusion equations for chemically reacting flows, J. Comp. Physics 45 (1984), pp. 1–42.
10. Tz. Ostromsky, W. Owczarz, Z. Zlatev, Computational Challenges in Large-scale Air Pollution Modelling, Proc. 2001 International Conference on Supercomputing in Sorrento, ACM Press (2001), pp. 407-418
11. Tz. Ostromsky, Z. Zlatev, Parallel Implementation of a Large-scale 3-D Air Pollution Model, Large Scale Scientific Computing (S. Margenov, J. Wasniewski, P. Yalamov, Eds.), LNCS-2179, Springer, 2001, pp. 309-316.
12. Tz. Ostromsky, Z. Zlatev, Flexible Two-level Parallel Implementations of a Large Air Pollution Model, Numerical Methods and Applications (I.Dimov, I.Lirkov, S. Margenov, Z. Zlatev - eds.), LNCS-2542, Springer (2002), 545-554.
13. G. Strang, On the construction and comparison of difference schemes, SIAM J. Numer. Anal. **5**, No. 3 (1968), 506-517.
14. WEB-site of the Danish Eulerian Model, available at: *http://www.dmu.dk/AtmosphericEnvironment/DEM*
15. Z. Zlatev, Computer treatment of large air pollution models, Kluwer, 1995.
16. Z. Zlatev, I. Dimov, K. Georgiev, Three-dimensional version of the Danish Eulerian Model, Zeitschrift für Angewandte Mathematik und Mechanik, 76 (1996) pp. 473–476.

Conditioning and Error Estimation in the Numerical Solution of Matrix Riccati Equations

P. Hr. Petkov[1], M. M. Konstantinov[2], and N. D. Christov[1]

[1] Technical University of Sofia, 1000 Sofia, Bulgaria
php@tu-sofia.bg, ndchr@tu-sofia.bg
[2] University of Architecture and Civil Engineering,
1421 Sofia, Bulgaria
mmk_fte@uacg.bg

Abstract. The paper deals with the condition number estimation and the computation of residual-based forward error estimates in the numerical solution of the matrix Riccati equations. Efficient, LAPACK-based condition and error estimators are proposed involving the solution of triangular Lyapunov equations along with one-norm computation.

1 Introduction

In this paper we consider the estimation of condition numbers and the computation of residual-based forward error estimates pertaining to the numerical solution of matrix algebraic continuous-time and discrete-time Riccati equations which arise in optimal control and filtering theory.

The following notation is used in the paper: $\mathcal{R}$ – the field of real numbers; $\mathcal{R}^{m\times n}$ – the space of $m \times n$ matrices $A = [a_{ij}]$ over $\mathcal{R}$; A^T – the transpose of A; $\sigma_{\max}(A)$ and $\sigma_{\min}(A)$ – the maximum and minimum singular values of A; $\|A\|_1$ – the 1-norm of the matrix A; $\|A\|_2 = \sigma_{\max}(A)$ – the spectral norm of A; $\|A\|_F = (\sum |a_{ij}|^2)^{1/2})$ – the Frobenius norm of A; I_n – the unit $n \times n$ matrix; $A \otimes B$ – the Kronecker product of matrices A and B; $\mathrm{vec}(A)$ – the vector, obtained by stacking the columns of A in one vector; ε – the roundoff unit of the machine arithmetic.

In what follows we shall consider the continuous-time Riccati equation

$$A^TX + XA + C - XDX = 0 \tag{1}$$

and the discrete-time Riccati equation

$$A^TXA - X + C - A^TXB(R + B^TXB)^{-1}B^TXA = 0 \tag{2}$$

or its equivalent

$$X = C + A^TX(I_n + DX)^{-1}A, \;\; D = BR^{-1}B^T$$

where $A \in \mathcal{R}^{n\times n}$ and the matrices C, D, $X \in \mathcal{R}^{n\times n}$ are symmetric. We assume that there exists a non-negative definite solution X which stabilizes $A - DX$ and $(I_n + DX)^{-1}A$, respectively.

Z. Li et al. (Eds.): NAA 2004, LNCS 3401, pp. 448–455, 2005.

The numerical solution of matrix Riccati equations may face some difficulties. First of all, the corresponding equation may be *ill-conditioned*, i.e. small perturbations in the coefficient matrices A, C, D may lead to large variations in the solution. Therefore, it is necessary to have a quantitative characterization of the conditioning in order to estimate the accuracy of solution computed.

The second difficulty is connected with the stability of the numerical method and the reliability of its implementation. It is known [1] that the methods for solving the Riccati equations are generally unstable. This requires to have an estimate of the forward error in the solution computed.

The paper is organized as follows. In Section 2 we discuss the conditioning of the equations (1), (2). In Section 3 we present an efficient method for computing condition number estimates which is based on matrix norm estimator implemented in LAPACK [2]. In Section 4 we propose residual based forward error estimates which implement also the LAPACK norm estimator and may be used in conjunction with different methods for solving the corresponding equation.

2 Conditioning of Riccati Equations

Let the coefficient matrices A, C, D in (1), (2) be subject to perturbations ΔA, ΔC, ΔD, respectively, so that instead of the initial data we have the matrices $\tilde{A} = A + \Delta A$, $\tilde{C} = C + \Delta C$, $\tilde{D} = D + \Delta D$. The aim of the perturbation analysis of (1), (2) is to investigate the variation ΔX in the solution $\tilde{X} = X + \Delta X$ due to the perturbations ΔA, ΔC, ΔD. If small perturbations in the data lead to small variations in the solution we say that the corresponding equation is *well-conditioned* and if these perturbations lead to large variations in the solution this equation is *ill-conditioned.* In the perturbation analysis of the Riccati equations it is supposed that the perturbations preserve the symmetric structure of the equation, i.e. the perturbations ΔC and ΔD are symmetric. If $\|\Delta A\|$, $\|\Delta C\|$ and $\|\Delta D\|$ are sufficiently small, then the perturbed solution $\tilde{X} = X + \Delta X$ is well defined.

Consider first the Riccati equation (1). The *condition number* of the Riccati equation is defined as (see [3])

$$K = \lim_{\delta \to 0} \sup \left\{ \frac{\|\Delta X\|}{\delta \|X\|} : \|\Delta A\| \leq \delta \|A\|, \; \|\Delta C\| \leq \delta \|C\|, \; \|\Delta D\| \leq \delta \|D\| \right\}.$$

For sufficiently small δ we have (within first order terms)

$$\frac{\|\Delta X\|}{\|X\|} \leq K\delta.$$

Let $\bar{X}$ be the solution of the Riccati equation computed by a numerical method in finite arithmetic with relative precision ε. If the method is backward stable, then we can estimate the error in the solution error

$$\frac{\|\bar{X} - X\|}{\|X\|} \leq p(n)K\varepsilon$$

with some low-order polynomial $p(n)$ of n. This shows the importance of the condition number in the numerical solution of Riccati equation.

Consider the perturbed Riccati equation

$$\begin{aligned}(A+\Delta A)^T(X+\Delta X)+(X+\Delta X)(A+\Delta A)+C+\Delta C\\ -(X+\Delta X)(D+\Delta D)(X+\Delta X)=0\end{aligned} \tag{3}$$

and set $A_c = A - DX$. Subtracting (1) from (3) and neglecting the second and higher order terms in ΔX we obtain a Lyapunov equation in ΔX:

$$A_c^T\Delta X + \Delta X A_c = -\Delta C - (\Delta A^T X + X\Delta A) + X\Delta D X. \tag{4}$$

Let vec(M) denotes the vector, obtained by stacking the columns of the matrix M. Then we have that

$$\|\mathrm{vec}(M)\|_2 = \|M\|_F$$

and equation (4) can be written in the vectorized form as

$$\begin{aligned}(I_n\otimes A_c^T + A_c^T\otimes I_n)\mathrm{vec}(\Delta X) = &-\mathrm{vec}(\Delta C)\\ &-(I_n\otimes X+(X\otimes I_n)W)\mathrm{vec}(\Delta A)\\ &+(X\otimes X)\mathrm{vec}(\Delta D))\end{aligned} \tag{5}$$

where we use the representations

$$\begin{aligned}\mathrm{vec}(\Delta A^T) &= W\mathrm{vec}(\Delta A)\\ \mathrm{vec}(MZN) &= (N^T\otimes M)\mathrm{vec}(Z)\end{aligned}$$

and W is the vec-permutation matrix.

Since the matrix A_c is stable, the matrix $I_n\otimes A_c^T + A_c^T\otimes I_n$ is nonsingular and we have that

$$\begin{aligned}\mathrm{vec}(\Delta X) = &(I_n\otimes A_c^T + A_c^T\otimes I_n)^{-1}(-\mathrm{vec}(\Delta C)\\ &-(I_n\otimes X+(X\otimes I_n)W)\mathrm{vec}(\Delta A)\\ &+(X\otimes X)\mathrm{vec}(\Delta D))\end{aligned} \tag{6}$$

Equation (6) can be written as

$$\mathrm{vec}(\Delta X) = -[P^{-1},\ Q,\ -S]\begin{bmatrix}\mathrm{vec}(\Delta C)\\ \mathrm{vec}(\Delta A)\\ \mathrm{vec}(\Delta D)\end{bmatrix} \tag{7}$$

where

$$\begin{aligned}P &= I_n\otimes A_c^T + A_c^T\otimes I_n\\ Q &= P^{-1}(I_n\otimes X+(X\otimes I_n)W)\\ S &= P^{-1}(X\otimes X).\end{aligned}$$

If we set

$$\eta = \max\{\|\Delta A\|_F/\|A\|_F,\ \|\Delta C\|_F/\|C\|_F,\ \|\Delta D\|_F/\|D\|_F\}$$

then it follows from (7) that

$$\|\Delta X\|_F/\|X\|_F \leq \sqrt{3}K_F\eta,$$

where

$$K_F = \|[P^{-1},\ Q,\ S]\|_2/\|X\|_F$$

is the condition number of (1) using Frobenius norms.

The computation of K_F requires the construction and manipulation of $n^2 \times n^2$ matrices which is not practical for large n. Furthermore, the computation of the condition number of the Riccati equation involves the solution matrix X, so that the condition number can be determined only after solving the equation.

Since the computation of the exact condition number is a difficult task, it is useful to derive approximations of K that can be obtained cheaply.

Rewrite equation (4) as

$$\Delta X = -\Omega^{-1}(\Delta C) - \Theta(\Delta A) + \Pi(\Delta D) \tag{8}$$

where

$$\Omega(Z) = A_c^T Z + Z A_c$$
$$\Theta(Z) = \Omega^{-1}(Z^T X + XZ)$$
$$\Pi(Z) = \Omega^{-1}(XZX)$$

are linear operators in the space of $n \times n$ matrices, which determine the sensitivity of X with respect to the perturbations in C, A, D, respectively. Based on (8) it was suggested in [3] to use the approximate condition number

$$K_B := \frac{\|\Omega^{-1}\|\|C\| + \|\Theta\|\|A\| + \|\Pi\|\|D\|}{\|X\|} \tag{9}$$

where

$$\|\Omega^{-1}\| = \max_{Z\neq 0} \frac{\|\Omega^{-1}(Z)\|}{\|Z\|}$$

$$\|\Theta\| = \max_{Z\neq 0} \frac{\|\Theta(Z)\|}{\|Z\|}$$

$$\|\Pi\| = \max_{Z\neq 0} \frac{\|\Pi(Z)\|}{\|Z\|}$$

are the corresponding induced operator norms. Note that the quantity

$$\|\Omega^{-1}\|_F = \max_{Z\neq 0} \frac{\|Z\|_F}{\|A_c^T Z + Z A_c\|_F} = \frac{1}{\mathrm{sep}(A_c^T, -A_c)}$$

where

$$\mathrm{sep}(A_c^T, -A_c) := \min_{Z\neq 0} \frac{\|A_c^T Z + Z A_c\|_F}{\|Z\|_F} = \sigma_{\min}(I_n \otimes A_c^T + A_c^T \otimes I_n)$$

is connected to the sensitivity of the Lyapunov equation

$$A_c^T X + X A_c = -C.$$

Comparing (6) and (8) we obtain that

$$\begin{aligned} \|\Omega^{-1}\|_F &= \|P^{-1}\|_2 \\ \|\Theta\|_F &= \|Q\|_2 \\ \|\Pi\|_F &= \|S\|_2. \end{aligned} \tag{10}$$

In the case of the discrete-time Riccati equation (2) the corresponding operators are determined from

$$\begin{aligned} \Omega(Z) &= A_c^T Z A_c - Z \\ \Theta(Z) &= \Omega^{-1}(Z^T X A_c + A_c^T X Z) \\ \Pi(Z) &= \Omega^{-1}(A_c^T X Z X A_c) \end{aligned}$$

where $A_c = (I_n + DX)^{-1} A$.

3 Conditioning Estimation

The quantities $\|\Omega^{-1}\|_1$, $\|\Theta\|_1$, $\|\Pi\|_1$ arising in the sensitivity analysis of Riccati equations can be efficiently estimated by using the norm estimator, proposed in [4] which estimates the norm $\|T\|_1$ of a linear operator T, given the ability to compute Tv and $T^T w$ quickly for arbitrary v and w. This estimator is implemented in the LAPACK subroutine `xLACON` [2], which is called via a reverse communication interface, providing the products Tv and $T^T w$.

Consider for definiteness the case of continuous-time Riccati equation. With respect to the computation of

$$\|\Omega^{-1}\|_F = \|P^{-1}\|_2 = \frac{1}{\mathrm{sep}_F(A_c^T, -A_c)}$$

the use of `xLACON` means to solve the linear equations

$$Py = v, \ P^T z = v$$

where

$$P = I_n \otimes A_c^T + A_c^T \otimes I_n, \ \ P^T = I_n \otimes A_c + A_c \otimes I_n,$$

v being determined by `xLACON`. This is equivalent to the solution of the Lyapunov equations

$$\begin{aligned} A_c^T Y + Y A_c &= V \\ A_c Z + Z A_c^T &= V \end{aligned} \tag{11}$$

where $\mathrm{vec}(V) = v$, $\mathrm{vec}(Y) = y$, $\mathrm{vec}(Z) = z$.

The solution of these Lyapunov equations can be obtained in a numerically reliable way using the Bartels-Stewart algorithm [5]. Note that in (11) the matrix

V is symmetric, which allows a reduction in complexity by operating on vectors v of length $n(n+1)/2$ instead of n^2.

An estimate of $\|\Theta\|_1$ can be obtained in a similar way by solving the Lyapunov equations

$$\begin{aligned} A_c^T Y + Y A_c &= V^T X + XV \\ A_c Z + Z A_c^T &= V^T X + XV. \end{aligned} \tag{12}$$

To estimate $\|\Pi\|_1$ via `xLACON`, it is necessary to solve the equations

$$\begin{aligned} A_c^T Y + Y A_c &= XVX \\ A_c Z + Z A_c^T &= XVX \end{aligned} \tag{13}$$

where the matrix V is again symmetric and we can again work with shorter vectors.

The estimation of $\|\Omega\|_1, \|\Theta\|_1, \|\Pi\|_1$ in the case of the discrete-time Riccati equation is done in a similar way.

The accuracy of the estimates that we obtain via this approach depends on the ability of `xLACON` to find a right-hand side vector v which maximizes the ratios

$$\frac{\|y\|}{\|v\|}, \frac{\|z\|}{\|v\|}$$

when solving the equations $Py = v$, $P^T z = v$. As in the case of other condition estimators it is always possible to find special examples when the value produced by `xLACON` underestimates the true value of the corresponding norm by an arbitrary factor. Note, however, that this may happens in rare circumstances.

4 Error Estimation

A posteriori error bounds for the computed solution of the matrix equations (1), (2) can be obtained in several ways. One of the most efficient and reliable ways to get an estimate of the solution error is to use practical error bounds, similar to the case of solving linear systems of equations [6, 2] and matrix Sylvester equations [7].

Consider again the Riccati equation (1). Let

$$R = A^T \bar{X} + \bar{X} A + C - \bar{X} D \bar{X}$$

be the exact residual matrix associated with the computed solution $\bar{X}$. Setting $\bar{X} := X + \Delta X$, where X is the exact solution and ΔX is the absolute error in the solution, one obtains

$$R = (A - D\bar{X})^T \Delta X + \Delta X (A - D\bar{X}) + \Delta X D \Delta X.$$

If we neglect the second order term in ΔX, we obtain the linear system of equations

$$\bar{P}\mathrm{vec}(\Delta X) = \mathrm{vec}(R)$$

where $\bar{P} = I_n \otimes \bar{A}_c^T + \bar{A}_c^T \otimes I_n$, $\bar{A}_c = A - D\bar{X}$. In this way we have

$$\|\mathrm{vec}(X - \bar{X})\|_\infty = \|\bar{P}^{-1}\mathrm{vec}(R)\|_\infty \le \|\, |\bar{P}^{-1}|\, |\mathrm{vec}(R)|\, \|_\infty.$$

As it is known [6], this bound is optimal if we ignore the signs in the elements of $\bar{P}^{-1}$ and $\mathrm{vec}(R)$.

In order to take into account the rounding errors in forming the residual matrix, instead of R we use

$$\bar{R} = fl(C + A^T\bar{X} + \bar{X}A - \bar{X}D\bar{X}) = R + \Delta R$$

where

$$|\Delta R| \le \varepsilon(4|C| + (n+4)(|A^T|\,|\bar{X}| + |\bar{X}|\,|A|) + 2(n+1)|\bar{X}|\,|D|\,|\bar{X}|) =: R_\varepsilon$$

and fl denotes the result of a floating point computation. Here we made use of the well known error bounds for matrix addition and matrix multiplication.

In this way we have obtained the overall bound

$$\frac{\|X - \bar{X}\|_M}{\|\bar{X}\|_M} \le \frac{\|\, |P^{-1}|\,(|\mathrm{vec}(\bar{R})| + \mathrm{vec}(R_\varepsilon))\|_\infty}{\|\bar{X}\|_M} \tag{14}$$

where $\|X\|_M = \max_{i,j} |x_{ij}|$.

The numerator in the right hand side of (14) is of the form $\|\, |P^{-1}|\, r\, \|_\infty$, and as in [6, 7] we have

$$\begin{aligned} \|\, |\bar{P}^{-1}|\, r\, \|_\infty &= \|\, |\bar{P}^{-1}|\, D_R e\|_\infty = \|\, |\bar{P}^{-1}D_R|\, e\|_\infty \\ &= \|\, |\bar{P}^{-1}D_R|\, \|_\infty \ = \ \|\bar{P}^{-1}D_R\|_\infty \end{aligned}$$

where $D_R = \mathrm{diag}(r)$ and $e = [1, 1, \ldots, 1]^T$. This shows that $\|\, |P^{-1}|\, r\, \|_\infty$ can be efficiently estimated using the norm estimator xLACON in LAPACK, which estimates $\|Z\|_1$ at the cost of computing a few matrix-vector products involving Z and Z^T. This means that for $Z = \bar{P}^{-1}D_R$ we have to solve a few linear systems involving $\bar{P} = I_n \otimes \bar{A}_c^T + \bar{A}_c^T \otimes I_n$ and $\bar{P}^T = I_n \otimes \bar{A}_c + \bar{A}_c \otimes I_n$ or, in other words, we have to solve several Lyapunov equations $\bar{A}_c^T X + X\bar{A}_c = V$ and $\bar{A}_c X + X\bar{A}_c^T = W$. Note that the Schur form of $\bar{A}_c$ is already available from the condition estimation of the Riccati equation, so that the solution of the Lyapunov equations can be obtained efficiently via the Bartels-Stewart algorithm. Also, due to the symmetry of the matrices $\bar{R}$ and R_ε, we only need the upper (or lower) part of the solution of this Lyapunov equations which allows to reduce the complexity by manipulating only vectors of length $n(n+1)/2$ instead of n^2.

To avoid overflows, instead of estimating the condition number K_B an estimate of the reciprocal condition number

$$\frac{1}{\tilde{K}_B} = \frac{\widetilde{\mathrm{sep}}_1(\bar{A}_c^T, -\bar{A}_c)\|\bar{X}\|_1}{\|C\|_1 + \widetilde{\mathrm{sep}}_1(\bar{A}_c^T, -\bar{A}_c)(\|\tilde{\Theta}\|_1\|A\|_1 + \|\tilde{\Pi}\|_1\|D\|_1)}$$

is determined. Here $\bar{A}_c$ is the computed matrix A_c and the estimated quantities are denoted by tilde.

The error estimation in the solution of (2) is done in a similar way.

The software implementation of the condition and error estimates is based entirely on LAPACK and BLAS [8, 9] subroutines.

References

1. P.Hr. Petkov, N.D. Christov, M.M. Konstantinov. *Computational Methods for Linear Control Systems.* Prentice Hall, N.Y., 1991.
2. E. Anderson, Z. Bai, C. Bischof, J. Demmel, J. Dongarra, J.Du Croz, A. Greenbaum, S. Hammarling, A. McKenney, S. Ostrouchov, and D. Sorensen. *LAPACK Users' Guide.* SIAM, Philadelphia, second edition, 1995.
3. R. Byers. Numerical condition of the algebraic Riccati equation. *Contemp. Math.*, **47**, 1985, pp. 35-49.
4. N.J. Higham. FORTRAN codes for estimating the one-norm of a real or complex matrix, with applications to condition estimation (Algorithm 674). *ACM Trans. Math. Software*, **14**, 1988, pp. 381-396.
5. R.H. Bartels, G.W. Stewart. Algorithm 432: Solution of the matrix equation AX + XB = C. *Comm. ACM*, **15**, 1972, pp. 820-826.
6. M. Arioli, J.W. Demmel, I.S. Duff. Solving sparse linear systems with sparse backward error. *SIAM J. Matrix Anal. Appl.*, **10**, 1989, pp. 165-190.
7. N.J. Higham. Perturbation theory and backward error for AX - XB = C. *BIT*, **33**, 1993, pp. 124-136.
8. C.L. Lawson, R.J. Hanson, D.R. Kincaid, F.T. Krogh. Basic Linear Algebra Subprograms for FORTRAN usage. *ACM Trans. Math. Software*, **5**, 1979, pp. 308-323.
9. J.J. Dongarra, J. Du Croz, I. Duff, S. Hammarling. A set of Level 3 Basic Linear Algebra Subprograms. *ACM Trans. Math. Software*, **16**, 1990, pp. 1-17.

Numerical Modelling of the One-Phase Stefan Problem by Finite Volume Method

Nickolay Popov[1], Sonia Tabakova[1], and François Feuillebois[2]

[1] Department of Mechanics, TU - Sofia, branch Plovdiv,
4000 Plovdiv, Bulgaria
stabakova@hotmail.com
[2] Laboratoire PMMH, CNRS, ESPCI, 75231 Paris, France
feuillebois@pmmh.espci.fr

Abstract. The one-phase Stefan problem in enthalpy formulation, describing the freezing of initially supercooled droplets that impact on solid surfaces, is solved numerically by the finite volume method on a non-orthogonal body fitted coordinate system, numerically generated. The general case of third order boundary conditions on the droplet is considered. The numerical results for the simple case of a spherical droplet touching a surface at first order boundary conditions are validated well by the known 1D asymptotic solution. The proposed solution method occurs faster than another method, based on ADI implicit finite-difference scheme in cylindrical coordinates, for the same droplet shapes.

1 Introduction

The phase change process at supercooled droplet impact on a cold substrate has numerous technological applications, as well as observations at some phenomena in nature, for example ice accretion on aircrafts during flights [1 - 3]. Since a great number of parameters control the process, it is difficult to study it either experimentally and theoretically. The droplet shape initially assumed spherical deforms during impact and finally takes an uneven form as solid. The freezing process is complicated and roughly separated into two stages [4]: initial freezing (return to stable equilibrium at solidification temperature); full freezing (all material becomes solid). If the droplet is assumed undeformable after the initial freezing [1 - 3], then the full freezing problem can be considered as a one-phase Stefan problem with third or first order boundary conditions on the droplet surfaces exposed to the ambient and substrate. Although the droplet shape is supposed axisymmetrical, but generally not spherical, the Stefan problem is 2D and does not possess analytical solution. The numerical grid generation method is appropriate [5] for arbitrary shaped objects and especially for free and moving boundary problems, if the generated coordinates are "body-fitted" (BFC). Then the physical problem is solved into the transformed computational region, which is rectangular or square. The BFC can be: orthogonal, which require special techniques for complex spaces [5]; or non-orthogonal, which is more flexible, but causes complications in the physical problems [6 - 7].

Z. Li et al. (Eds.): NAA 2004, LNCS 3401, pp. 456–462, 2005.

In the present paper we study numerically the one-phase Stefan problem of a droplet of arbitrary axisymmetrical shape, which freezes at impact on a substrate. The numerical model is based on the enthalpy method in finite volume approach on a non-orthogonal BFC. The obtained results after the numerical simulations refer to the case of a supercooled water droplet. The proposed model is compared well by the 1D analytical solution described in [4] for the special case of a spherical drop and by another numerical solution in cylindrical coordinates given in [1 - 3] for the arbitrary droplet shape. The scheme based on the non-orthogonal BFC is faster than the other one, which is promising for future studies with fluid dynamics taken into account in the model.

2 Problem Formulation

We assume that at the initial moment of the full freezing stage, the droplet substance is a liquid-solid mixture with constant properties, different from those of liquid and solid phases. Then in the mixture phase Ω_m the temperature is equal to the freezing temperature $\theta = \theta_m$. The one-phase Stefan problem is given by the heat conduction equation for the solid phase Ω_s:

$$\rho_s c_s \frac{\partial \theta}{\partial t} = \kappa_s \nabla^2 \theta \tag{1}$$

and by the heat flux jump on the moving surface Γ, the interface with freezing temperature $\theta = \theta_m$:

$$\kappa_s \nabla \theta \cdot \mathbf{n} = -\rho_s L_m v_n, \tag{2}$$

where L_m is the latent heat of fusion of mixture, v_n is the interface normal velocity, ρ_s, c_s, κ_s are correspondingly the solid phase density, heat capacity and conductivity. The heat flux boundary conditions on the droplet surface facing the ambient gas and substrate are as follows:

$$-\kappa_i \nabla \theta \cdot \mathbf{n}_j = \alpha_j (\theta - \theta_j), \tag{3}$$

where the subscript (i) corresponds to mixture (m) or solid (s) and (j) to the droplet surface exposed to ambient gas (a) and to substrate (s). The heat transfer coefficient α_a or α_s is between the droplet and ambient gas or substrate. If the droplet surface is isothermal, then instead of (3), $\theta = \theta_a$. The initial temperature condition states that at impact the supercooled droplet returns to thermodynamic equilibrium at the freezing temperature:

$$\theta = \theta_m, \quad \text{at} \quad t = 0, \tag{4}$$

and the the droplet surface is the initial position of Γ.

3 Numerical Methods

3.1 Grid Generation

The cylindrical coordinate system (r, z, φ) in the physical domain, as shown in Fig.1., is connected with the coordinate system (ξ, η, φ) in the computational

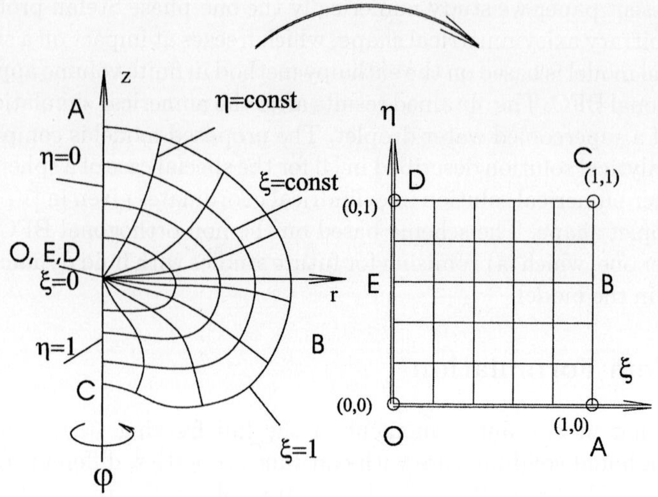

Fig. 1. The numerical mapping of the droplet physical domain (r, z, φ) to the computational domain (ξ, η, φ)

domain by the transformation $\xi = \xi(r, z)$, $\eta = \eta(r, z)$. The transformation Jacobian is given by $\sqrt{g}$, where $g = det(\mathbf{g}) = r^2(g_{11}g_{22} - g_{12}^2)$ and $\mathbf{g}$ is the covariant metric tensor in physical domain. The transformed coordinates ξ and η are solutions of the Laplace generating equations $\nabla^2\xi = 0$ and $\nabla^2\eta = 0$ with Ist order boundary conditions on the four boundaries (Fig.1). The inverse transformation $\mathbf{r} = (r(\xi, \eta), z(\xi, \eta))$ is obtained as a solution of the homogeneous quasi-liner elliptic system:

$$g_{22}\mathbf{r}_{\xi\xi} + g_{11}\mathbf{r}_{\eta\eta} - 2g_{12}\mathbf{r}_{\xi\eta} = 0, \tag{5}$$

with Ist order boundary conditions on the four boundaries (Fig.1.). The system (5) is solved numerically as in [7] by the iterative SOR method [9] with a transfinite grid as initial approximation in order to accelerate the iterative procedure.

3.2 Enthalpy Method

There exist two main groups of numerical methods for the Stefan problem [10]: front tracking methods and fixed domain methods. The most common of the latter methods is the enthalpy method and the results obtained by it in [1] for the simple case of a free spherical droplet with Ist order boundary conditions have excellent accuracy with respect to the asymptotic - analytical solution of [4]. The enthalpy reformulation of (1) - (2) includes the enthalpy function $H(\theta) = \int \rho(\theta)c(\theta)d(\theta) + L_m\rho(\theta)\eta(\theta - \theta_m)$, to reach to only one equation for the whole droplet region:

$$\frac{\partial H(\theta)}{\partial t} = \kappa_s \nabla^2\theta, \tag{6}$$

where η is the Heaviside step function $\eta(\theta-\theta_m)=\begin{cases}1 \text{ for } & \theta \geq \theta_m \\ 0 \text{ for } & \theta < \theta_m\end{cases}$ and $\rho(\theta)$ and $c(\theta)$ are the density and specific heat, such that $\rho(\theta)=\rho_m$ and $c(\theta)=c_m$ at $\theta=\theta_m$; $\rho(\theta)=\rho_s$ and $c(\theta)=c_s$ at $\theta<\theta_m$. The equation (6) has a singularity at $\theta=\theta_m$ and a smoothing over a small temperature zone is applied in the numerical scheme [10], [1 - 3].

3.3 Numerical Scheme for the Stefan Problem

The dimensionless form of (6) with (3), (4), written in BFC, is the following:

$$[St_m+\delta(T-1)]\frac{\partial T}{\partial \tau}=\frac{1}{\sqrt{g}}\left\{\frac{\partial}{\partial \xi}[J(g_{22}T_\xi-g_{12}T_\eta)]\right.$$
$$\left.+\frac{\partial}{\partial \eta}[J(g_{11}T_\eta-g_{12}T_\xi)]\right\} \quad \text{in } \Omega_m \cup \Omega_s \cup \Gamma \tag{7}$$

$$-\frac{J}{\sqrt{g_{22}}}(g_{22}T_\xi-g_{12}T_\eta)=Bi(T)T, \quad \text{on} \quad \partial\Omega_a, \tag{8a}$$

$$-\frac{J}{\sqrt{g_{11}}}(g_{11}T_\eta-g_{12}T_\xi)=Bi_t(T)T, \quad \text{on} \quad \partial\Omega_s, \tag{8b}$$

$$T=1 \quad \text{at} \quad \tau=0, \tag{9}$$

where $J=r^2/\sqrt{g}$, $St_m=c_s\triangle\theta/L_m$ is the mixture Stefan number, $\delta(T-1)$ the Dirac delta function, $T=\theta-\theta_a$ the dimensionless temperature ($\theta_a=\theta_s$ is assumed), $\tau=t/t_{fr}$ the dimensionless time, $Bi(T)=\alpha_a a_m/(\kappa_s k(T))$ and $Bi_t(T)=\alpha_s a_m/\kappa_s k(T)$ respectively the Biot numbers for the ambient gas and target substrate surface, $k(T)=1$ for $T<1$ and $k(T)=\kappa_m/\kappa_s$ at $T=1$. Here, the characteristic parameters are a_m for length, t_{fr} for freezing time, $\triangle\theta$ for temperature, ρ_s for density, κ_s for conductivity and c_s for heat capacity. If the first order boundary conditions are present on the droplet surface, their dimensionless form is: $T=0$.

The numerical scheme for the unsteady non-linear problem (7) - (9) is based on the implicit Euler method in time and finite volumes method in the computational square $0\leq\xi\leq 1, \quad 0\leq\eta\leq 1$ with constant grid step $\triangle\zeta=\triangle\xi=\triangle\eta$. At each time layer an inner iterative procedure by the SOR method with ω as relaxation parameter is performed till solution convergence is achieved ((n) is the index of time layer, (s) the index of the inner iterations):

$${}^{s+1}T_{ij}^{n+1}=\left[\frac{T_{ij}^n+A\sum_{(k,l)\neq(i,j)}C_{kl} \quad {}^sT_{kl}^{n+1}}{1-AC_{ij}}\right]\omega+(1-\omega)^sT_{ij}^{n+1}, \tag{10}$$

where $i,j=0,...,N$ are the grid points indexes in ξ and η direction, $A=\triangle\tau/\left[\triangle\zeta^2\left(St_m+\delta\left({}^sT_{ij}^{n+1}-1\right)\right)\right]$, C_{ij} the metrics coefficients in the right-hand side of (7) (not given here for sake of brevity), $\triangle\tau$ is the time step.

4 Numerical Results

Here we present the results for two droplet shapes (both correspond to the same volume): spherical and arbitrary axisymmetrical (registered from experiments of droplet impact [1 - 3]). For both of them BFC systems are generated and the obtained grids contain 4489 nodal points each, correspondent to 67 x 67 points on the unit square in the computational domain (Cf. Fig.2 for the arbitrary shape).

For the spherical case the full freezing times due to our model (BFC) are found to be in very good agreement with the analytical solution [4] (for first order boundary conditions) and another numerical solution by finite-difference ADI scheme in cylindrical coordinate system (CCS) [1 - 3] (for first and third order boundary conditions), as shown in Table 1.

Table 1. Dimensionless final freezing time for a spherical droplet

	I^{st} b.c.			IIIrd b.c., Bi = 100	
St_m	BFC	CCS [1]	Analytical [4]	BFC	CCS [3]
0.01	0.1657	0.1680	0.1682	0.1681	0.1785
0.1	0.1714	0.1811	0.1803	0.1821	0.1917
0.489	0.2168	0.2255	0.2149	0.2308	0.2389

Moreover, the temperature profile due to (BFC) possesses spherical symmetry for any ray $\eta = const.$ and is presented in Fig.3 for different times in correspondence with the analytical [4] and numerical (CCS) [1] for first order boundary conditions (droplet dimensionless radius is 0.325 and $St_m = 0.01$). A good agreement between the results in (BFC) and in (CCS) for the case of an

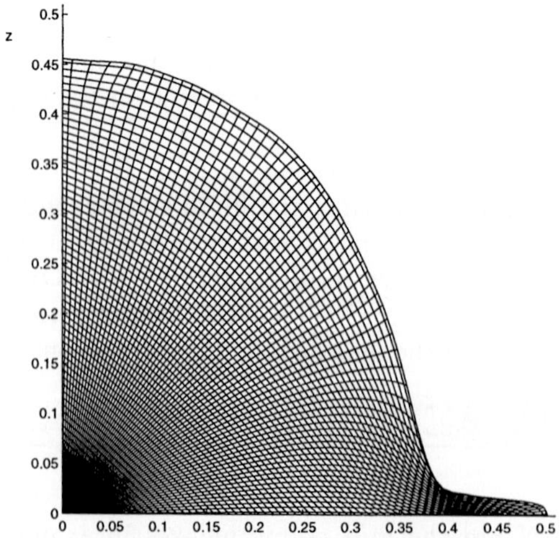

Fig. 2. Numerically generated BFC system for arbitrary droplet shape

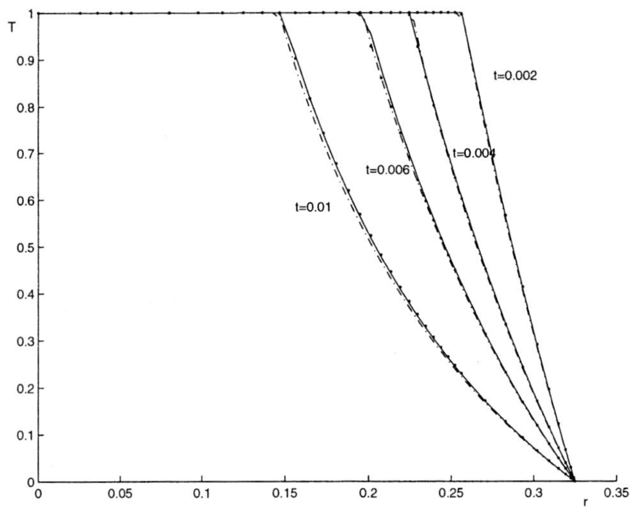

Fig. 3. Temperature profiles for spherical droplet case for $St_m = 0.01$ at different times: points represent the analytical solution [4], solid line (CCS) [1], dash-dotted line (BFC)

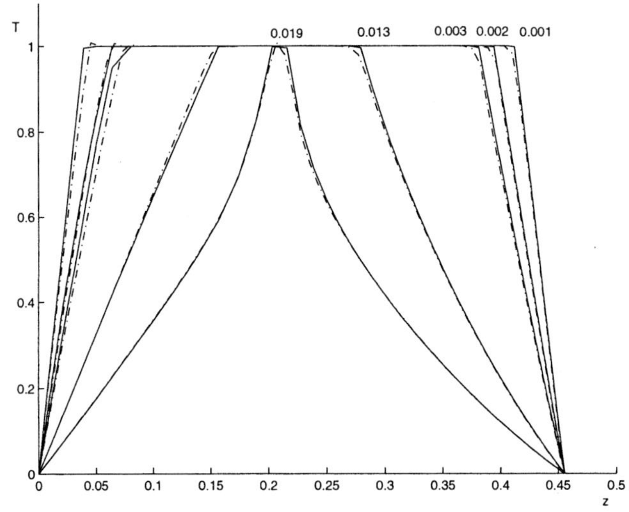

Fig. 4. Temperature profiles for arbitrary shape case at different times and $St_m = 0.489$: solid line is due to (CCS) [1], dash-dotted line due to (BFC)

arbitrary shape (as given in Fig.2) is achieved, which is presented in Fig.4 for the first order boundary conditions. Due to our numerous calculations with the proposed model in BFC and with different values of problem parameters: St_m,

Bi, Bi_t we can conclude that it is several times faster that the model (CCS) and more flexible for complicated droplet shapes.

5 Conclusions

A numerical model based on the enthalpy method in finite volume approach on a non-orthogonal BFC has been proposed to solve the one-phase Stefan problem of a droplet of arbitrary axisymmetrical shape, which freezes at impact on a substrate. It is validated for the simple case of a spherical droplet with the 1D analytical solution described in [4] and by another numerical solution in cylindrical coordinates given in [1 - 3] for an arbitrary droplet shape. The present scheme is faster than the one in [1 - 3] and on its basis our future studies, with fluid dynamics taken into account, will be developed.

Acknowledgement

This work has been financially supported by the French-Bulgarian research program "Rila", which is gratefully acknowledged by the authors.

References

1. Tabakova, S., F. Feuillebois, On the solidification of a supercooled liquid droplet posed on a surface, J. Colloid Interface Sci., 272 (2004), 225-234.
2. Feuillebois, F., Tabakova, S., Impact and freezing of subcooled water droplets, Recent Dev. Colloids Interface Res., No.1 (2003) 291-311.
3. Tabakova, S., F. Feuillebois, Numerical modeling the freezing of a supercooled droplet on a substrate at dry ice regime, Proc. 29-th Inter. Summer School AMEE, June 2003, Sozopol-Bulgaria (eds. M.Marinov, G.Venkov), Sofia 2004, 246-251.
4. Feuillebois, F, Lasek, A., Creismeas, P., Pigeonneau, F., Szaniawski, A., Freezing of a subcooled liquid droplet, J. Colloid Interface Sci., 169 (1995), 90-102.
5. Thompson, J.F., Warsi, Z., Mastin, C., Numerical Grid Generation. Foundations and Applications, Elsevier Science Publ., 1985.
6. Li, W., Yan, Y., An alternating dependent variables (ADV) method for treating slip boundary conditions of free surface flows with heat and mass transfer, Numer. Heat Trans., Part B, 41 (2002), 165-189.
7. Popov, N., Tabakova, S., Numerically generated grids for unstationary heat conduction problems, Scientific Researches of the Union of Scientists in Bulgaria-Plovdiv, ser. C. Techniques and Technology, vol.III (2004), (Scientific session held on 24.10.2003, House of Scientists - Plovdiv), 7-15.
8. Shyy, W., Udaykumar, H., Rao, M., Smith, R., Computational fluid dynamics with moving boundaries, Taylor and Francis, 1996.
9. Quarteroni, A. and Valli, A., Numerical approximation of partial differential equations, Springer, 1997.
10. Crank, J., Free and moving boundary problems, Oxford Univ. Press, 1984.

Adaptive Conjugate Smoothing of Discontinuous Fields

Minvydas Ragulskis and Violeta Kravcenkiene

Kaunas University of Technology,
Department of Mathematical Research in Systems,
Studentu st. 50, Kaunas, Lithuania

Abstract. A technique for adaptive conjugate smoothing of discontinuous fields is presented in the paper. The described technique is applicable in various engineering problems and is especially effective when hybrid numerical - experimental methodologies are used. Adaptive smoothing strategy is illustrated for a discontinuous plain stress field problem when photoelastic fringes representing the variation of stress are constructed in virtual projection plane.

Keywords: Finite element method, adaptive smoothing, visualization

1 Introduction

Adaptive smoothing of discontinuous fields is a problem of a high importance in hybrid numerical - experimental techniques when the results of experimental analysis are mimicked in virtual numerical environment [1]. Typical example is the construction of digital fringe images from finite element analysis results imitating the stress induced effect of photo-elasticity [2].

Conventional finite element analysis is based on interpolation of nodal variables (displacements) inside the domain of each element [3]. Though the field of displacements is continuous in the global domain, the field of stresses is discontinuous at inter-element boundaries due to the operation of differentiation. Construction of digital fringe images from finite element analysis results is a typical problem when discontinuous fields are to be visualized. Therefore it is important to develop numerical techniques enabling physically based smoothing applicable for visualization procedures.

The proposed strategy of smoothing parameter is based on the assumption that the larger smoothing is required in the zones where the discontinuity of the field is higher.

The finite element norm representing the residual of stress field reconstruction in the domain of the analyzed element is introduced. It can be noted that the calculation of element norms is not a straightforward procedure. First, the nodal stress values in the global domain are sought by the least square method minimizing the differences between the interpolated stress field from the nodal stress values and discontinuous stress field calculated directly from the displacement field. As the minimization is performed over the global domain and the

Z. Li et al. (Eds.): NAA 2004, LNCS 3401, pp. 463–470, 2005.

interpolations are performed over the local domains of every element, direct stiffness procedure based on Galiorkin method is developed and applied to the described problem.

When the nodal stress values are calculated, the finite element norms are calculated for each element as the average error of the field reconstruction through the interpolation of those nodal values.

It can be noted that the first step of calculation of the nodal values of stress produces a continuous stress field of stresses over the global domain. Nevertheless that field is hardly applicable for visualization procedures as the derivatives of the field are discontinuous and the plotted fringes are unsmooth. Therefore the augmented residual term is added to the previously described least squares procedure while the magnitudes of the terms for every finite element are proportional to the element norms. Explicit analysis of the smoothing procedure for the reconstruction of the stress field is presented for a one-dimensional problem.

Digital images of two-dimensional systems simulating the realistic effect of photoelasticity are presented. Those examples prove the importance of the introduced smoothing procedure for practical applications and build the ground for the development of hybrid numerical experimental techniques.

2 FEM Based Technique for Adaptive Conjugate Smoothing

Stress component σ (a continuous function) inside a finite element can be calculated in a usual way [3]:

$$\sigma = [D]\{\varepsilon\} \tag{1}$$

where $\{\varepsilon\}$ – column of generalized nodal strains of the analysed finite element; $[D]$ – vector row relating the strain component and nodal strains in the domain of the finite element. Vector $[D]$ involves the derivatives of the element's shape functions in conventional FEM models based on displacement formulation [3]. Therefore, though the displacement field is continuous in the global domain, the strain field can be discontinuous at the inter-element boundaries [3].

Techniques used for the visualization of results from FEM analysis [4] require the nodal values of the plotted parameter which are interpolated in the domain of each finite element by its shape functions. Therefore visualization of the stress field requires data on the nodal values of the stress. This is not a straightforward problem for conventional FEM formulations.

Probably the natural way for calculation of nodal values of stress would be the minimization of residual constructed as an integral of the squared difference between the interpolated and factual stress fields over the global domain:

$$\sum_{D.S.} \left(\iint \left([N]\{s\} - \sigma\right)^2 dxdy \right) \tag{2}$$

where abbreviation $D.S.$ stands for FEM direct stiffness procedure [5]; integrals are calculated over the domain of each finite element; $[N]$–row of the shape functions of finite element; $\{s\}$ – column of unknown nodal values of the component of stress in currently integrated finite element.

Though such reconstruction of stress field solves the problem of discontinuity, the derivatives of the stress field are still discontinuous at inter-element boundaries. Visualization procedures for such types of interpolated fields are highly sensitive to such discontinuities of derivatives [4]. Therefore there exists a need to design a method for smoothing the reconstructed field.

The proposed method is based on the augmentation of the residual (3) by a penalty term for fast change of the stress in any direction in the analysed plane:

$$\lambda_i\left(\left(\frac{\partial\sigma}{\partial x}\right)^2+\left(\frac{\partial\sigma}{\partial y}\right)^2\right) \tag{3}$$

where $\lambda_i > 0$ are smoothing parameters selected individually for every finite element. Keeping in mind that the stress field is interpolated by the form functions of the finite element the augmentation term can be approximately interpreted as

$$\left(\frac{\partial w}{\partial x}\right)^2+\left(\frac{\partial w}{\partial y}\right)^2 \approx \{s\}^T[C]^T[C]\{s\} \tag{4}$$

where

$$[C]=\begin{bmatrix}\frac{\partial N_1}{\partial x} & \frac{\partial N_2}{\partial x} & \cdots & \frac{\partial N_n}{\partial x}\\ \frac{\partial N_1}{\partial y} & \frac{\partial N_2}{\partial y} & \cdots & \frac{\partial N_n}{\partial y}\end{bmatrix}$$

and n–number of nodes in the analyzed element; N_i – shape functions of the $i-th$ node.

Then the nodal values of stress are found minimizing the augmented residual:

$$\frac{d\Big(\sum_{D.S.}\Big(\iint\left(([N]\{s\}-\sigma)^2+\lambda_i\{s\}^T[C]^T[C]\{s\}\right)dxdy\Big)\Big)}{ds_i}=0, \tag{5}$$

where $i = 1,\ldots,m$, s_i is an $i-th$ component ($i-th$ nodal value) of global vector of nodal values of strain $\{S\}$; m - number of nodes in the global domain. Step-by-step differentiation leads to following equality:

$$\sum_{D.S.}\left(\iint\left([N]^T[N]+[C]^T\lambda_i[C]\right)dxdy\right)\cdot\{S\}=\sum_{D.S.}\left(\iint[N]^T\sigma dxdy\right) \tag{6}$$

It can be noted that eq. 6 is a system of linear algebraic equations in respect of unknown nodal values of $\{S\}$. Moreover, this formulation conveniently involves the smoothing parameter λ into FEM formulation. $\sum_{D.S.}(\iint[N]^T[N]dxdy)$ is a positive definite matrix in conventional FEM [5], so the system matrix in eq. 6 will be also positive definite at $\lambda > 0$, what will guarantee the solution of $\{S\}$. Finally, eq. 6 is very convenient for implementation, straightforwardly falls into FEM ideology and therefore does not require extensive modifications of standard FEM codes.

3 One-Dimensional Example

The properties of conjugate smoothing can be illustrated by the following trivial example. One-dimensional system consisting from three elements and four nodes

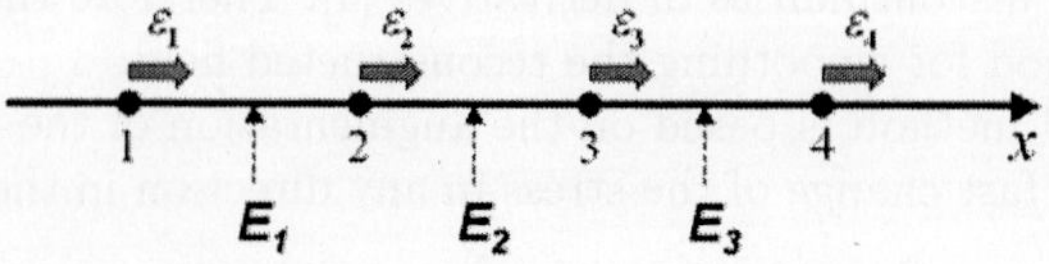

Fig. 1. One-dimensional system consisting from 3 linear finite elements and 4 nodes

is presented in Fig. 1. For simplicity it is assumed that the elements are linear, co-ordinates of the nodes x_k are:

$$x_k = k - 1, \qquad k = 1, \ldots, 4 \tag{7}$$

Then the shape functions of the $i-th$ element are:

$$\begin{aligned} N_1^{(i)}(x) &= i - x \\ N_2^{(i)}(x) &= x - i + 1, \end{aligned} \tag{8}$$

Here $i = 1, \ldots, 3$.

The nodal strains are explicitly defined as ε_k. Strain distribution in the domain of the $i-th$ finite element is approximated by appropriate shape functions:

$$\varepsilon^{(i)}(x) = N_1^{(i)}(x)\varepsilon_i + N_2^{(i)}(x)\varepsilon_{i+1} \tag{9}$$

Then the stress inside the $i-th$ element is calculated as the derivative of strain [2]:

$$\sigma^{(i)}(x) = B_1^{(i)}(x)\varepsilon_{(i)} + B_2^{(i)}\varepsilon_{i+1} = \varepsilon_{i+1} - \varepsilon_{(i)} \tag{10}$$

where

$$B_1^{(i)}(x) = \frac{\partial N_1^{(i)}(x)}{\partial x} = -1 \qquad B_2^{(i)}(x) = \frac{\partial N_2^{(i)}(x)}{\partial x} = 1 \tag{11}$$

The unknown nodal values of stress are denoted as s_k. It can be noted that $[N] = \left\lfloor N_1^{(i)}(x); N_2^{(i)}(x) \right\rfloor$; $[C] = \left\lfloor B_1^{(i)}(x); B_2^{(i)}(x) \right\rfloor$. Then, for the $i-th$ element:

$$\begin{gathered} \int\limits_{i-1}^{i} [N]^T[N]dx = \int\limits_{i-1}^{i} \begin{bmatrix} \left(N_1^{(i)}(x)\right)^2 & N_1^{(i)}(x)N_2^{(i)}(x) \\ N_1^{(i)}(x)N_2^{(i)}(x) & \left(N_2^{(i)}(x)\right)^2 \end{bmatrix} dx = \begin{bmatrix} \frac{1}{3} & \frac{1}{6} \\ \frac{1}{6} & \frac{1}{3} \end{bmatrix} \\ \int\limits_{i-1}^{i} [C]^T[C]dx = \int\limits_{i-1}^{i} \begin{bmatrix} \left(B_1^{(i)}\right)^2 & B_1^{(i)}B_2^{(i)} \\ B_1^{(i)}B_2^{(i)} & \left(B_2^{(i)}\right)^2 \end{bmatrix} dx = \begin{bmatrix} 1 & -1 \\ -1 & 1 \end{bmatrix} \\ \int\limits_{i-1}^{i} [N]^T\sigma^{(i)}(x)dx = \int\limits_{i-1}^{i} \begin{bmatrix} N_1^{(i)}(x)\sigma^{(i)}(x) \\ N_2^{(i)}(x)\sigma^{(i)}(x) \end{bmatrix} dx = \begin{bmatrix} \frac{\varepsilon_{i+1}-\varepsilon_i}{2} \\ \frac{\varepsilon_{i+1}-\varepsilon_i}{2} \end{bmatrix} \end{gathered} \tag{12}$$

Initially it is assumed that the smoothing parameter λ is the same for all elements. Direct stiffness procedure [5] over the three elements results into the following system of linear algebraic equations:

$$\begin{bmatrix} \frac{1}{3}+\lambda & \frac{1}{6}-\lambda & 0 & 0 \\ \frac{1}{6}-\lambda & \frac{2}{3}+2\lambda & \frac{1}{6}-\lambda & 0 \\ 0 & \frac{1}{6}-\lambda & \frac{2}{3}+2\lambda & \frac{1}{6}-\lambda \\ 0 & 0 & \frac{1}{6}-\lambda & \frac{1}{3}+\lambda \end{bmatrix} \cdot \begin{bmatrix} s_1 \\ s_2 \\ s_3 \\ s_4 \end{bmatrix} = \begin{bmatrix} \frac{\varepsilon_2-\varepsilon_1}{2} \\ \frac{\varepsilon_3-\varepsilon_1}{2} \\ \frac{\varepsilon_4-\varepsilon_2}{2} \\ \frac{\varepsilon_4-\varepsilon_3}{2} \end{bmatrix} \tag{13}$$

The solution of this linear system can be found using computer algebra:

$$\begin{aligned} s_1 &= \frac{36\lambda^2(\varepsilon_4-\varepsilon_1)-12\lambda(8\varepsilon_1-6\varepsilon_2-3\varepsilon_3+\varepsilon_4)-19\varepsilon_1+24\varepsilon_2-6\varepsilon_3+\varepsilon_4}{3(36\lambda^2+36\lambda+5)} \\ s_2 &= \frac{36\lambda^2(\varepsilon_4-\varepsilon_1)-6\lambda(10\varepsilon_1-3\varepsilon_2-6\varepsilon_3-\varepsilon_4)-7\varepsilon_1-3\varepsilon_2+12\varepsilon_3-2\varepsilon_4}{3(36\lambda^2+36\lambda+5)} \\ s_3 &= \frac{36\lambda^2(\varepsilon_4-\varepsilon_1)-6\lambda(\varepsilon_1+6\varepsilon_2+3\varepsilon_3-10\varepsilon_4)+2\varepsilon_1-12\varepsilon_2+3\varepsilon_3+7\varepsilon_4}{3(36\lambda^2+36\lambda+5)} \\ s_4 &= \frac{36\lambda^2(\varepsilon_4-\varepsilon_1)+12\lambda(\varepsilon_1-3\varepsilon_2-6\varepsilon_3+8\varepsilon_4)-\varepsilon_1+6\varepsilon_2-24\varepsilon_3+19\varepsilon_4}{3(36\lambda^2+36\lambda+5)} \end{aligned} \tag{14}$$

The reconstructed stress in the domain of the $i-th$ element is approximated by the shape functions of the $i-th$ element:

$$S^{(i)}(x,\lambda) = s_i(\lambda)N_1^{(i)}(x) + s_{i+1}(\lambda)N_2^{(i)}(x) \qquad i=1,\ldots,3 \tag{15}$$

It can be noted that

$$\lim_{\lambda\to\infty}\left(s_1(\lambda)\right) = \lim_{\lambda\to\infty}\left(s_2(\lambda)\right) = \lim_{\lambda\to\infty}\left(s_3(\lambda)\right) = \lim_{\lambda\to\infty}\left(s_4(\lambda)\right) = \frac{\varepsilon_4-\varepsilon_1}{3} \tag{16}$$

By the way, application of computer algebra simplifies the calculation of the following integral:

$$\sum_{i=1}^{3}\left(\int_{i-1}^{i} S^{(i)}(x,\lambda)dx\right) = \varepsilon_4-\varepsilon_1 \tag{17}$$

Remarkable is the simplicity of the result and the fact that the integral does not depend from λ . By the way,

$$\sum_{i=1}^{3}\left(\int_{i-1}^{i}\sigma^{(i)}(x)dx\right) = \sum_{i=1}^{3}\left(\int_{i-1}^{i}(\varepsilon_{i+1}-\varepsilon_i)dx\right) = \varepsilon_4-\varepsilon_1 \tag{18}$$

The produced equalities (14) and (15) enable construction of smoothed field of stresses in the analyzed domain. Further improvement of the smoothing technique is possible when the smoothing parameters are selected individually for each element. Such adaptive selection is based on the magnitude of error norms of the elements which are calculated as:

$$R_i = \sqrt{\int_{i-1}^{i}\left(N_1^{(i)}(x)s_i(0) + N_2^{(i)}(x)s_{i+1}(0) - \sigma^{(i)}(x)\right)^2 dx} \tag{19}$$

It can be noted that in determination (20) the nodal values of stress $s_i(0)$ and $s_{i+1}(0)$ require the solution of eg. (13) at $\lambda = 0$. Calculation of the error norms enables the selection of individual parameters of smoothing:

$$\lambda_i = aR_i \tag{20}$$

where a is a constant. Direct stiffness procedure will produce the following system of algebraic equations:

$$\begin{bmatrix} \frac{1}{3}+aR_1 & \frac{1}{6}-aR_1 & 0 & 0 \\ \frac{1}{6}-aR_1 & \frac{2}{3}+a(R_1+R_2) & \frac{1}{6}-aR_2 & 0 \\ 0 & \frac{1}{6}-aR_2 & \frac{2}{3}+a(R_2+R_3) & \frac{1}{6}-aR_3 \\ 0 & 0 & \frac{1}{6}-aR_3 & \frac{1}{3}+aR_3 \end{bmatrix} \cdot \begin{bmatrix} s_1 \\ s_2 \\ s_3 \\ s_4 \end{bmatrix} = \begin{bmatrix} \frac{\varepsilon_2-\varepsilon_1}{2} \\ \frac{\varepsilon_3-\varepsilon_1}{2} \\ \frac{\varepsilon_4-\varepsilon_2}{2} \\ \frac{\varepsilon_4-\varepsilon_3}{2} \end{bmatrix} \tag{21}$$

It is clear that the solution of eq.21 at $a = 0$ will coincide with the solution of eq.13 at $\lambda = 0$. Computer algebra helps to prove that

$$\lim_{a\to\infty}\left(s_1(a)\right) = \lim_{a\to\infty}\left(s_2(a)\right) = \lim_{a\to\infty}\left(s_3(a)\right) = \lim_{a\to\infty}\left(s_4(a)\right) = \frac{\varepsilon_4-\varepsilon_1}{4} \tag{22}$$

at bounded error norms R_i.

Assumption of particular values of strains - for example $\varepsilon_1 = 0.02$; $\varepsilon_2 = 0.02$; $\varepsilon_3 = 0.06$ and $\varepsilon_4 = -0.02$ enables the illustration of the reconstructed field of stress. The nodal values of strain at $\lambda = 0$ are: $s_1 = -0.083$; $s_2 = 0.045$; $s_3 = 0.021$; $s_4 = -0.131$. The error norms of the elements are: $R_1 \approx 0.0426667$; $R_2 \approx 0.0471781$; $R_3 \approx 0.0506667$.

Then the system of equations 22 can be solved at different values of a and the calculated nodal values of stress can be interpolated by the shape functions in the domain of every element. The results are visualized in Fig. 2. It can be noted that the reconstructed field of stress at $a = 0$ is continuous in the global domain, but the discontinuity of its derivatives at inter-element boundaries is an obstacle for construction of its smooth visual interpretations. Increase of parameter a helps solving this problem what is illustrated in the following section.

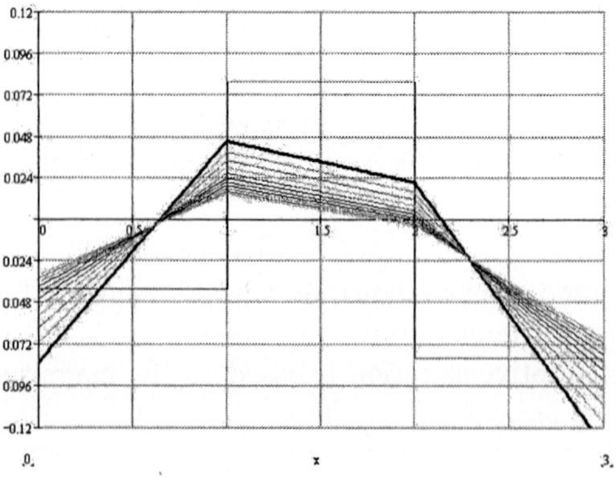

Fig. 2. Reconstructed field of stress S at $a = (i-1)*0.25$; $i = 1, \ldots, 20$; the horizontal lines represent σ

4 Adaptive Smoothing for Stress Field Visualisation in the Problem of Plane Stress

The components of stresses in the domain of the analysed finite element are calculated in the usual way [2, 5]:

$$\begin{Bmatrix} \sigma_x \\ \sigma_y \\ \tau_{xy} \end{Bmatrix} = [D][B]\{\delta_0\} \tag{23}$$

where $\{\delta_0\}$ is the vector of nodal displacements of the analysed element; $[B]$ is the matrix relating the strains with the displacements; $[D]$ is the matrix relating the stresses with the strains; σ_x, σ_y, τ_{xy} are the components of the stresses in the problem of plane stress. Conjugate smoothing of the stress field in the global domain results in the following system of equations:

$$\begin{gathered}
\left(\sum_i \iint_{e_i} ([N]^T[N] + [C]^T\lambda_i[C])dxdy\right) \cdot \{\delta_x\} = \sum_i \iint_{e_i} [N]^T\sigma_x dxdy \\
\left(\sum_i \iint_{e_i} ([N]^T[N] + [C]^T\lambda_i[C])dxdy\right) \cdot \{\delta_y\} = \sum_i \iint_{e_i} [N]^T\sigma_y dxdy \\
\left(\sum_i \iint_{e_i} ([N]^T[N] + [C]^T\lambda_i[C])dxdy\right) \cdot \{\delta_{xy}\} = \sum_i \iint_{e_i} [N]^T\tau_{xy} dxdy
\end{gathered} \tag{24}$$

where $\{\delta_x\}$ - the vector of nodal values of σ_x; $\{\delta_y\}$ - the vector of nodal values of σ_y; $\{\delta_{xy}\}$ - the vector of nodal values of τ_{xy}. Relative error norms for finite elements are calculated using the methodology described in Eq.(21) and averaging for all three components of plain stress.

5 Computational Example and Concluding Remarks

Practical applicability and usefulness of the presented technique for adaptive smoothing of discontinuous fields is illustrated by the following example where the discontinuous stress field is visualised generating digital image of isochromatics in virtual projection plane. The numerical procedure used for constructing digital fringes is presented in [4]. Three patterns of fringes corresponding to the same strain field are presented in Fig. 3. Unsmoothed fringes are presented in Fig. 3a. It can be clearly seen that the structure of fringes is not uniform at the inter-element boundaries what is especially distinct in the corners of the area. Fig. 3b presents uniformly smoothed fringes. Though the structure of the pattern is smoother than in Fig. 3a, the fringes in the corners are missing. That corresponds to an over-smoothed state and illustrates a non-physical behaviour of the system. Fig. 3c corresponds to adaptive smoothing where the value of parameter λ is selected in accordance to finite element norms as described earlier.

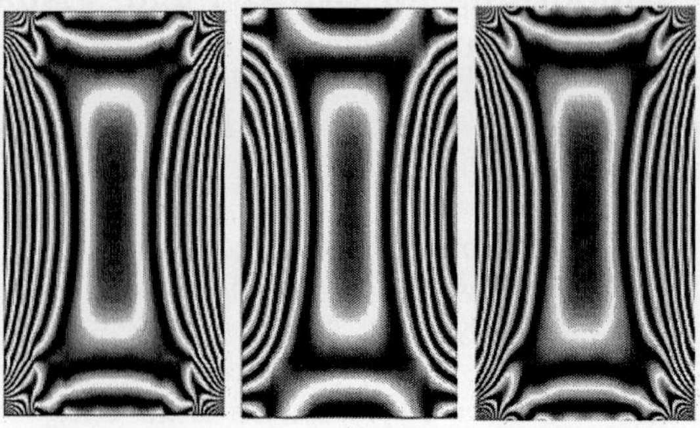

Fig. 3. Photoelastic fringes for a plane stress problem: *a*–unsmoothed fringes at $\lambda = 0$; *b*–smoothed fringes at constant $\lambda = 0,1$ for all elements; *c*–fringes after adaptive smoothing

The whole pattern of fringes is rather smooth, while the corner fringes are not eliminated.

Presented smoothing technique can be applied in different engineering applications. Especially it is useful in hybrid numerical - experimental techniques where the functionality of the systems is mimicked in virtual computational environment and the represented field of physical parameters is discontinuous.

References

1. Holstein A., Salbut L., Kujawinska M., Juptner W.: Hybrid Experimental-numerical Concept of Residual Stress Analysis in Laser Weldments. Experimental Mechanics, 41(4), 343-350 (2001).
2. Dally J.W., Riley W.F.: Experimental Stress Analysis. McGraw Hill Book Company, New York (1991).
3. Atluri S.N., Gallagher R.H., Zienkiewicz O.C., Eds.: Hybrid and Mixed Finite Element Methods. John Wiley & Sons (1983).
4. Ragulskis M., Palevicius A., Ragulskis L.: Plotting Holographic Interferograms for Visualisation of Dynamic Results From Finite Element Calculations. International Journal for Numerical Methods in Engineering, 56(11), 1647-1659 (2003).
5. Bathe K.J.: Finite element procedures in engineering analysis. New Jersey: Prentice-Hall, 1982. - p. 738.

Finite Differences Scheme for the Euler System of Equations in a Class of Discontinuous Functions

Mahir Rasulov and Turhan Karaguler

Beykent University, Department of Mathematics and Computing,
Istanbul 34900, Turkey

Abstract. In this paper, the finite difference scheme for solving the Cauchy problem for the simplified Euler system in a class of discountinuous functions, which describes irrational flow of fluid by neglecting the viscosity and temperature effects is investigated. For this purpose, firstly the Euler system is decomposed with respect to its coordinates. Then an auxiliary problem which is superiour to the main problem in terms of obtaining the solution is introduced, and shown that the solutions of this auxiliary problem are smoother than the solutions of the main problem. Additionally, the auxiliary problem provides to develop effective and efficient algorithms.

Keywords: Computational hydrodynamics, compressible and incompressible flow, Euler systems, numerical modeling, shock waves.

1 Introduction

In this paper, we will investigate the system of differential equations describing the flow of perfect fluid given as below

$$\frac{\partial \mathbf{u}}{\partial t} + (\mathbf{u}\nabla)\mathbf{u} = \mathbf{F} - \frac{\mathbf{1}}{\rho}\nabla \mathbf{p}, \tag{1.1}$$

$$\frac{\partial \rho}{\partial t} + \nabla(\rho \mathbf{u}) = 0. \tag{1.2}$$

Here, $\mathbf{u}=(u_1, u_2, u_3)$ is the velocity vector, $\mathbf{F}$ is the body force on an elementary unit volume dv; p is the surface force (pressure) on a unit surface element ds; ρ is the density; and ∇ is the nabla operator.

Further details about these equations (1.1), (1.2) can be found in [2], [3], [4], [5], [12], and etc. Also, it is a well known fact that by using Euler system of differential equations many problems of fluid dynamics can be solved. Within the limits of this paper for the simplified Euler system, the special finite differences method in a class of discontinuous functions is suggested.

If it is assumed that the problem is one-dimensional and the body force F and the surface force p are ignored, then the system of equations above become the well known Hopf's equation as $\frac{\partial u}{\partial t} + u\frac{\partial u}{\partial x} = 0$.

Z. Li et al. (Eds.): NAA 2004, LNCS 3401, pp. 471–477, 2005.

The first order nonlinear partial equations, in particular, the Hopf's equation and its exact solution are well studied in literature (see [1], [3], [6], [7], [8]).

As it is known that if the initial profile has both a negative and a positive slopes, the solution of Hopf's equation includes the first type points of discountinuity (shock waves) whose positions are unknown beforehand. Otherwise, the solution of Hopf's equation must be a discountinuous function. As to be forced to work with discountinuous functions and be able to investigate the true nature of the physical phenomena, it is required to obtain the solution of Euler system (1.1) in a class of discountinuous functions. This is why, in this paper, a special finite differences method for solving the Cauchy problem of the simplified Euler system is examined.

In order to explain the details of the suggested numerical method, we assume that the function F included in the Eq. (1.1) is zero.

We consider the system (1.1) of nonlinear partial equations with the initial condition given as

$$\mathbf{u}(\mathbf{x},\mathbf{0}) = \mathbf{u}^{(\mathbf{0})}(\mathbf{x}). \tag{1.3}$$

Here, $\mathbf{u}^{(\mathbf{0})}(\mathbf{x})$ is the given initial value of velocity vector having a compact support and a positive and a negative slopes. We introduce the potential function Φ such that

$$\mathbf{u} = \text{grad } \mathbf{\Phi} \tag{1.4}$$

and assume that the flow is irrotational, i.e,

$$curl\ \mathbf{u} = \mathbf{0}. \tag{1.5}$$

Taking account of (1.4), (1.5), and $\rho =$ constant, then and the system equation (1.1) can be rewritten as follows

$$\frac{\partial}{\partial t}\text{grad } \Phi + \text{grad}\left[\frac{1}{2}U^2 + \frac{\mathrm{p}}{\rho}\right] = 0. \tag{1.6}$$

We introduce the following notations

$$U^2 = \mathrm{u}^2 + \mathrm{v}^2 + \mathrm{w}^2, \tag{1.7}$$

and

$$\Phi = \Phi^* + \int_0^t c(\tau)d\tau. \tag{1.8}$$

In this notations, we have

$$\frac{\partial \Phi^*}{\partial t} + \frac{1}{2}U^2 + \frac{\mathrm{p}}{\rho} = 0. \tag{1.9}$$

The equation (1.9) is called Cauchy's integral for (1.1). If the function $\mathbf{u}$ independent of t, then the Eq.(1.9) can be rewritten as $\frac{1}{2}\mathbf{U^2} + \frac{p}{\rho} = 0$. This relation is called Bernoulli's integral (see, [2], [3], [5], [12]).

On the other hand, the Eq.(1.9) is the vectorial form of the first integral of the system of equations (1.1). It is obvious that, each solution of Eqs. (1.1) satisfies the equation (1.9) too, and inversely, the solutions of (1.9) are the solutions of (1.1).

In general, the arbitrary constant c, which is included in (1.8) depends on t, and therefore, the Eq.(1.6) is not defined as single-valued. Using the notations (1.4) and (1.7), the Eqs. (1.6) can be rewritten as

$$\frac{\partial \mathbf{u}}{\partial t} + \text{grad}\left[\left(\frac{U^2}{2} + \frac{p}{\rho}\right)\right] = 0. \tag{1.10}$$

Now, we consider the case that the ρ is depended on $\mathbf{x}$ and introduce the following notation

$$\Im = \int \frac{dp}{\rho}.$$

Then the system equations (1.6)) become as follows

$$\text{grad}\left[\frac{\partial \Phi}{\partial t} + \frac{u^2 + v^2 + w^2}{2} + \Im\right] = 0. \tag{1.11}$$

Now, taking account of (1.9) and (1.7), the system equation (1.11) can be rewritten as follows

$$\frac{\partial \mathbf{u}}{\partial t} + \text{grad}\left[\frac{U^2}{2} + \Im\right] = 0. \tag{1.12}$$

Thus, as seen from (1.12) these equations are the decomposed form of the system equations (1.1) with respect to its coordinates. The solution of each equation in this system becomes multivalued functions when the corresponding initial functions contain a positive and a negative slopes. Since, from a physical point of view, the multivalued functions can not be true solutions as indicated in [9], [10], [11], we are able to construct a single valued solution with the points of discontinuity of the first type instead of a multivalued solution. It is clear that the solution with these properties can be described by means of a weak solution. For that reason, the weak solution is determined as:

Definition 1. A function $\mathbf{u}(\mathbf{x}, \mathbf{t})$ *which satisfy the condition (1.3) is called a weak solution of the problem (1.1), (1.3), if the following integral relation*

$$\int_{R^4} \left\{\mathbf{u}(\mathbf{x}, t)\frac{\partial \varphi(\mathbf{x}, \mathbf{t})}{\partial \mathbf{t}} + \left(\frac{\mathbf{U^2}}{\mathbf{2}} + \frac{\mathbf{p}}{\rho}\right) grad\varphi(\mathbf{x}, \mathbf{t})\right\} dxdt$$

$$+ \int_{-\infty}^{\infty} \mathbf{u}(\mathbf{x}, \mathbf{0})\varphi(\mathbf{x}, \mathbf{0})\mathbf{dx} = 0 \tag{1.13}$$

hold for every test function $\varphi(\mathbf{x}, \mathbf{t}) \in \mathbf{H^1}(\overset{o}{\mathbf{R}}{}^4)$ *and* $\varphi(\mathbf{x}, \mathbf{T}) = \mathbf{0}$, *here,* $H^1(\overset{o}{R}{}^4)$ *is a Hilbert's space on* $\overset{o}{R}{}^4$.

2 The Auxiliary Problem

When we include the concept of a weak solution for the system (1.1), (1.3), a new problem arises that the position and time evolution of the points of discontinuities are unknown. On the other hand the existence of points of discontinuities causes many problems in the applications of well known classical numerical methods to the problem (1.1), (1.3) as the approximation of derivatives at the points of discontinuities is not expressed by finite differences formula.

To determine the weak solution of the problem (1.1), (1.3) in accordance with [9], [10], [11], we introduce the auxiliary problem as follows

$$\frac{\partial}{\partial t}\int \mathbf{u}(\mathbf{x},\mathbf{t})\mathbf{dx} + \frac{U^2}{2} + \frac{p}{\rho} = C(t), \tag{2.1}$$

$$\mathbf{u}(\mathbf{x},\mathbf{0}) = \mathbf{u}^{(0)}(\mathbf{x}). \tag{2.2}$$

Here $C = (C_1(t), C_2(t), C_3(t))$, are any constants. As it is obvious, from (2.1), (2.2) that the function $\mathbf{u}$ may be discontinuous function too. So the class of solutions of the problem (2.1), (2.2) coincide with the class of weak solutions defined by the relation (1.13).

We introduce the following notations as

$$\int \mathbf{u}(\mathbf{x},\mathbf{t})\mathbf{dx} + C(t) = \mathbf{v}(\mathbf{x},\mathbf{t}). \tag{2.3}$$

Here, $\mathbf{v}$=(v_1, v_2, v_3), and the expresion (2.3) denotes $\int u_i(x,t)dx_i + C_i(t) = v_i(x,t)$ in which C_i is an arbitrary function for each i and independent of x_i only, $(i = 1, 2, 3)$.

In these notations, the system equations (2.1), (2.2) can be rewritten as

$$\frac{\partial \mathbf{v}}{\partial t} + \frac{U^2}{2} + \frac{p}{\rho} = 0. \tag{2.4}$$

The initial condition for (2.4) is

$$\mathbf{v}(\mathbf{x},\mathbf{0}) = \mathbf{v}^{(0)}(\mathbf{x}). \tag{2.5}$$

Here, $\mathbf{v_0}(\mathbf{x})$ is any continuous solution of the following equation

$$\frac{d\mathbf{v}^{(0)}(\mathbf{x})}{dx} = \mathbf{u}^{(0)}(\mathbf{x}). \tag{2.6}$$

The auxiliary problem (2.1), (2.2) has the following advantages:

- The function vector $\mathbf{v}(\mathbf{x},\mathbf{t})$ is smoother than the vector function $\mathbf{u}(\mathbf{x},\mathbf{t})$, as their order of differentiable property is higher than the latter.
- $\mathbf{u}(\mathbf{x},\mathbf{t})$ can be determined without using the derivatives of the function $\mathbf{u}(\mathbf{x},\mathbf{t})$ with respect to x_1, x_2, x_3 and t;
- the function $\mathbf{u}(\mathbf{x},\mathbf{t})$ can be discontinuous too.

Theorem 1. *If* $\mathbf{v}(\mathbf{x},\mathbf{t})$ *is the smoother solution of the auxiliary problem (2.1), (2.2), then the function*

$$\mathbf{u}(\mathbf{x},\mathbf{t}) = \frac{\partial \mathbf{v}(\mathbf{x},\mathbf{t})}{\partial \mathbf{x}} \tag{2.7}$$

is the weak solutions of the problem (1.1), (1.3) in the sense of definition 1.

Since, the suggested auxiliary problem does not involve any derivatives of $\mathbf{u}(\mathbf{x},\mathbf{t})$ with respect to x_1, x_2, x_3 and t the numerical solution to the problem (1.1), (1.3) can be obtained with no difficulty by means of the numerical solution to the problem (2.1), (2.2).

Each equation of the (1.10) is the first order nonlinear wave equation, and furthermore, they express conservation laws. Therefore, the integral $\int \mathbf{u}(\mathbf{x},\mathbf{t})\mathbf{dx}$ independent of t. We introduce the following notation

$$E(0) = \int \mathbf{u}^{(0)}(\mathbf{x})\mathbf{dx}.$$

The number $E(0)$ is called the critical value of the function $\mathbf{v}(\mathbf{x},\mathbf{t})$.

Definition 2. The functions defined by

$$\mathbf{v}_{\mathbf{ext}}(\mathbf{x},\mathbf{t}) = \begin{cases} \mathbf{v}(\mathbf{x},\mathbf{t}),\ \mathbf{v}(\mathbf{x},\mathbf{t}) < \mathbf{E}(\mathbf{0}), \\ E(0),\quad \mathbf{v}(\mathbf{x},\mathbf{t}) \geq \mathbf{E}(\mathbf{0}) \end{cases} \tag{2.8}$$

is called the extended solutions of the problem (2.1), (2.2).

From the theorem 1, for the weak solutions of the problem (1.1), (1.3), we have

$$\mathbf{u}_{\mathbf{ext}}(\mathbf{x},\mathbf{t}) = \frac{\partial \mathbf{v}_{\mathbf{ext}}(\mathbf{x},\mathbf{t})}{\partial \mathbf{x}}. \tag{2.9}$$

Thus, the function defined by (2.9) is the extended (or weak) solution of the problem (1.1), (1.3). Thus, the geometrical location of the points at where the functions $\mathbf{v}(\mathbf{x},\mathbf{t})$ take the critical values defining the jump surface of $\mathbf{u}(\mathbf{x},\mathbf{t})$ vector. By putting another way, the jump points for the $\mathbf{u}(\mathbf{x},\mathbf{t})$ vector are such points that on the right hand side of these points, $\mathbf{u}(\mathbf{x},\mathbf{t})$ takes zero values.

Note. The weak solution for the system equations (1.12) with initial conditions (1.3) is constructed as above.

3 Numerical Algorithm in a Class of Discontinuous Functions

In order to devolope a numerical algorithm for the problem (1.1), (1.3), at first, we cover the region $R^4_{x_1,x_2,x_3,t}$ by the grid as

$$\begin{aligned} \Omega_{h_{x_1},h_{x_2},h_{x_3},\tau} = \{(x^1_i, x^2_j, x^3_l, t_k) \mid x^1_i = ih_{x_1},\ x^2_j = jh_{x_2},\ x^3_l = lh_{x_3},\ t_k = k\tau; \\ i = \ldots - M, -(M-1), \ldots, -1, 0, 1, \ldots, M, \ldots; \\ j = \ldots - N, -(N-1), \ldots, -1, 0, 1, \ldots, N, \ldots; \end{aligned}$$

$$l = \ldots - L, -(L-1), \ldots, -1, 0, 1, \ldots, L; \; k = 0, 1, 2, \ldots;$$
$$h_{x_1} > 0, \; h_{x_2} > 0, \; h_{x_3} > 0, \; \tau > 0\}.$$

Here, h_{x_1}, h_{x_2}, h_{x_3} and τ are the steps of the grid $\Omega_{h_{x_1}, h_{x_2}, h_{x_3}, \tau}$ with respect to x_1, x_2, x_3 and t variables, respectively.

At any point $(x_i^1, x_j^2, x_l^3, t_k)$ of the grid $\Omega_{h_{x_1}, h_{x_2}, h_{x_1 3}, \tau}$, we approximate the equation (2.4) by the finite difference scheme as follows

$$\mathbf{V_{k+1}} = \mathbf{V_k} - \frac{\tau}{\mathbf{2}} \overline{\mathbf{U}}_{\mathbf{k}}^{\mathbf{2}} - \frac{\wp_{\mathbf{k}}}{\rho}. \tag{3.1}$$

Here, the grid functions $\mathbf{V_k}$, $\overline{\mathbf{U}}_k^2$, and $\wp_k$ represent the approximate values of the functions $\mathbf{v}$ and p, at point $(x_i^1, x_j^2, x_l^3, t_k)$.

The finite differences analog for the initial condition (2.5) at the (x_i^1, x_j^2, x_l^3) points of the grid $\Omega_{h_{x_1}, h_{x_2}, h_{x_3}, \tau}$ is given as

$$\mathbf{V}^{(0)} = \mathbf{v}^{(0)}(\mathbf{x_i}). \tag{3.2}$$

Here, the grid functions $\mathbf{v_0}(\mathbf{x_i})$ is defined by the following equation

$$(\mathbf{v}^{(0)}(\mathbf{x_i}))_{\overline{\mathbf{x}}} = \mathbf{u}^{(0)}(\mathbf{x_i}). \tag{3.3}$$

Thus, the following theorem holds good.

Theorem 2. *If* $\mathbf{V_{k+1}}$ *is the numerical solution of the auxiliary problem (3.1), (3.2), then the relations defined by*

$$\mathbf{U_{k+1}} = (\mathbf{V}_{\overline{\mathbf{x}}})_{\mathbf{k+1}} \tag{3.4}$$

is numerical solutions of the main problem (1.1), (1.3).

The difference scheme (3.1), (3.2) is the first order with respect to τ, however, the order of it can be made higher by applying, for example, the Runge-Kutta method.

As it can be seen from (3.1),(3.2), the suggested algorithm is very effective and economic from a computational point of view.

References

1. Ames, W. F.: Nonlinear Partial Differential Equations in Engineering . Academic Press, New York, London, (1965)
2. Anderson, J. D. : Modern Compressible Flow. McGraw-Hill Book Company, New-York, (1982)
3. Anderson, D. A., Tannehill, J. C., Pletcher, R. H.:Computational Fluid Mechanics and Heat Transfer, Vol 1,2. Hemisphere Publishing Corporation, (1984)
4. Courant, R., Friedrichs, K. O. : Supersonic Flow and Shock Waves. Springer-Verlag New York Berlin Heidelberg, (1948)

5. Kochin, N. E., Kibel, I. A., Roze, N. B.: Theorical Hydrodynamics. Pub. Physical-Mathematics Sciences, Moskow, USSR, (1963), (in Russian).
6. Lax, P. D.: Weak Solutions of Nonlinear Hyperbolic Equations and Their Numerical Computations. Comm. of Pure and App. Math, **VII** (1954) 159–193
7. Lax, P. D.: Hyperbolic Systems of Conservation Laws II. Comm. of Pure and App. Math. **X** (1957) 537–566
8. Oleinik O. A.: Discontinuous Solutions of Nonlinear Differential Equations. Usp. Math. **12** (1957) 3–73
9. Rasulov, M. A.: On a Method of Solving the Cauchy Problem for a First Order Nonlinear Equation of Hyperbolic Type with a Smooth Initial Condition. Soviet Math. Dok. **43** No.1 (1991)
10. Rasulov, M. A.: Finite Difference Scheme for Solving of Some Nonlinear Problems of Mathematical Physics in a Class of Discontinuous Functions. Baku, (1996) (in Russian)
11. Rasulov, M. A., Ragimova, T. A.: A Numerical Method of the Solution of one Nonlinear Equation of a Hyperbolic Type of the First -Order. Dif. Equations, Minsk, **28** No.7 (1992) 2056–2063
12. Schlichting, H.: Boundary Layer theory. McGraw-Hill Book Company, New-York, (1968)

The λ-Error Order in Multivariate Interpolation

Dana Simian

Universitatea "Lucian Blaga" din Sibiu, Facultatea de Ştiinţe,
dr. I. Raţiu 5-7, Sibiu, România

Abstract. The aim of this article is to introduce and to study a generalization of the error order of interpolation, named λ - error order of interpolation. This generalization makes possible a deeper analysis of the error in the interpolation process. We derived the general form of the λ - error order of interpolation and then we applied it for many choices of the functional λ.

1 Preliminaries

We consider the general polynomial interpolation problem: for a given set of functionals, Λ, find a polynomial subspace, V, of the space of polynomials in "d" variables, Π^d, such that, for an arbitrary analytical function f, there is a unique polynomial $p \in V$ which matches f on the set Λ, that is

$$\lambda(p) = \lambda(f),\ \forall\ \lambda \in \Lambda;\ f \in \mathcal{A}_0 \tag{1}$$

The space V is named an interpolation space for Λ.
We associate to any functional λ the generating function, λ^ν:

$$\lambda^\nu(x) = \sum_{\alpha \in N^d} \frac{D^\alpha \lambda^\nu(0)}{\alpha!} x^\alpha = \sum_{\alpha \in N^d} \frac{\lambda(m_\alpha)}{\alpha!} x^\alpha, \tag{2}$$

$m_\alpha(x) = x^\alpha,\ x \in R^d,\ \alpha = (\alpha_1, \ldots, \alpha_d) \in N^d$.
The expression of this function can be deduced (see [4]), using the equality:

$$\lambda^\nu(z) = \lambda(e_z),\ \text{with}\ e_z(x) = e^{z \cdot x},\ z, x \in R^d. \tag{3}$$

In order to represent the action of a functional on a polynomial, we introduce the pair $< f, p >$, between an analytical function f and a polynomial p:

$$< f, p >= (p(D)f)(0) = \sum_{\alpha \in N^d} \frac{D^\alpha p(0) D^\alpha f(0)}{\alpha!}, \tag{4}$$

Next, we denote by $|\alpha| = \alpha_1 + \ldots + \alpha_d$ the length of α. If $p = \sum_{|\alpha| \le deg\ p} c_\alpha (\cdot)^\alpha$, then $p(D)$ is the differential operator with constant coefficients: $p(D) = \sum_{|\alpha| \le deg\ p} c_\alpha D^\alpha$. Obviously, the pair (4) is a genuine inner product on spaces of polynomials.

Z. Li et al. (Eds.): NAA 2004, LNCS 3401, pp. 478–485, 2005.

Thus, $\lambda(p) = \sum_{\alpha \in N^d} \frac{D^\alpha \lambda^\nu(0) D^\alpha p(0)}{\alpha!} = (p(D)\lambda^\nu)(0) =< \lambda^\nu, p >$.

For any $g \in \mathcal{A}_0$, we define the least term, $g\downarrow = T_j g$, such as the first nonzero term in the Taylor polynomial of g.

Let be

$$H_\Lambda = span\{\lambda^\nu;\ \lambda \in \Lambda\};\ H_\Lambda\downarrow = span\{g\downarrow; g \in H_\Lambda\}. \tag{5}$$

Using the results in [4] it can be proved that the polynomial subspace $H_\Lambda\downarrow$ is an interpolation space for the conditions Λ. If Λ is a set of linear independent functionals, then $\#\Lambda = dim\ H_\Lambda$ and the generating functions of the functionals in Λ form a basis for H_Λ.

There comes out from [1] the following proposition:

Proposition 1. *Let $\Lambda = \{\lambda_1, \ldots, \lambda_n\}$ be a set of linear functionals, linear independent and let be a basis, for example $p_j = \lambda_j^\nu$, $j \in \{1, \ldots, n\}$, of H_Λ. We can construct, inductive, another basis g_j, $j \in \{1, \ldots, n\}$ of H_Λ, having the "orthogonality" property: $< g_i, g_k\downarrow> \neq 0 \Leftrightarrow k = i$. The functions of the new basis are given by:*

$$g_j = p_j - \sum_{l=1}^{j-1} g_l \frac{< p_j, g_l\downarrow>}{< g_l, g_l\downarrow>} \tag{6}$$

If, in the inductive process we obtain $deg\ g_j\downarrow > deg\ g_i\downarrow$, for certain $i < j$, then we must recalculate the function g_i, in order to maintain the orthogonality property, using the formula:

$$g_i = g_i - g_j \cdot \frac{< g_i, g_j\downarrow>}{< g_j, g_j\downarrow>} \tag{7}$$

Let be

$$g_j(x) = \sum_{i=1}^{n} c_{j,i}\ \lambda_i^\nu(x),\ j \in \{1, \ldots, n\} \tag{8}$$

and $f \in \mathcal{A}_0$. We extend the inner product (4) such as

$$< g_j, f >= \sum_{i=1}^{n} c_{j,i} < \lambda_i^\nu, f >= \sum_{i=1}^{n} c_{j,i}\lambda_i(f) \tag{9}$$

with $c_{j,i}$ given in (8).

2 Properties of the Interpolation Scheme

Next, we will refer to the interpolation with respect to the conditions Λ, from the space $H_\Lambda\downarrow$. This is the "least interpolation", introduced by C. de Boor in [4]. In order to prove the main results in section 3, we need some properties of the interpolation operator and of the dual operator and we need to introduce two new notions: the λ-remainder of interpolation and the function $\varepsilon_{\Lambda,\lambda}^\nu$.

Theorem 1. *The unique element $L_\Lambda(f) \in H_\Lambda\downarrow$ which interpolates $f \in \mathcal{A}_0$ with respect to the conditions Λ is*

$$L_\Lambda(f) = \sum_{j=1}^{n} g_j\downarrow \frac{< g_j, f >}{< g_j, g_j\downarrow >}, \quad n = \#\Lambda \tag{10}$$

Proof. Using the orthogonality property of the basis g_j, $j \in \{1, \ldots n\}$ and (10), we get $< g_i, L_\Lambda(f) > = < g_i, f >$, $\forall\ f \in \mathcal{A}_0$ and hence $\lambda_k(L_\Lambda(f)) = \lambda_k(f)$, $k \in \{1, \ldots, n\}$.

The following theorem was proved in [6].

Theorem 2. *The operator*

$$L_\Lambda^*(f) = \sum_{j=1}^{n} g_j \frac{< f, g_j\downarrow >}{< g_j, g_j\downarrow >}; \ f \in \mathcal{A}_0. \tag{11}$$

has the duality property:

$$< L_\Lambda^*(g), f > = < g, L_\Lambda(f) >, \ g, \ f \in \mathcal{A}_0, \tag{12}$$

In [7] we introduced the λ-remainder, which generalizes the classical interpolation remainder.

Definition 1. *Let be the interpolation formula:*

$$f = L_\Lambda(f) + R_\Lambda(f), \tag{13}$$

with R_Λ being the remainder operator. We name λ-remainder, the value

$$R_{\Lambda,\lambda}(f) = \lambda(R_\Lambda) = \lambda[(1 - L_\Lambda)(f)]; \ f \in \mathcal{A}_0; \ \lambda \in (\Pi^d)' \tag{14}$$

We notice that for $\lambda \in \Lambda$ we get $R_{\Lambda,\lambda}(f) = 0$, $\forall f \in \mathcal{A}_0$.

Proposition 2. *The λ-remainder can be expressed such as:*

$$R_{\Lambda,\lambda}(f) = < \varepsilon^\nu_{\Lambda,\lambda}, f >, \tag{15}$$

with

$$\varepsilon^\nu_{\Lambda,\lambda} = (1 - L_\Lambda^*)(\lambda^\nu). \tag{16}$$

Proof. $R_{\Lambda,\lambda}(f) = \lambda[(1 - L_\Lambda)(f)] = < \lambda^\nu, f > - < L_\Lambda^*(\lambda^\nu), f > =$
$= < \lambda^\nu - L_\Lambda^*(\lambda^\nu), f > = < \varepsilon^\nu_{\Lambda,\lambda}, f >$.

Corollary 1. *The expression of the classical interpolation remainder is:*

$$(R_\Lambda(f))(x) = < e_x - L_\Lambda^*(e_x), f >, \ \text{with } e_x(t) = e^{x \cdot t}; \ x, t \in R^d. \tag{17}$$

Proof. $(R_\Lambda(f))(x) = R_{\Lambda,\delta_x}(f) = < \varepsilon^\nu_{\Lambda,\delta_x}, f >$ and the generating function of the evaluation functional $\lambda = \delta_x$ is $\lambda^\nu = e_x$.

Proposition 3. $\varepsilon^{\nu}_{\Lambda,\lambda}$ *satisfies the equality*

$$< \varepsilon^{\nu}_{\Lambda,\lambda}, f >= \lambda(f), \forall\ f \in kerL_{\Lambda} \tag{18}$$

Proof. $< \varepsilon^{\nu}_{\Lambda,\lambda}, f >= R_{\Lambda,\lambda}(f) =< \lambda, f > - < \lambda, L_{\Lambda}(f) >$.
But, $L_{\Lambda}(f) = 0$ for all $f \in kerL_{\Lambda}$ and hence $< \lambda, L_{\Lambda}(f) >= 0$.

Theorem 3. $\varepsilon^{\nu}_{\Lambda,\lambda} \perp H_{\Lambda}\!\downarrow$ *and every homogeneous component of* $\varepsilon^{\nu}_{\Lambda,\lambda}$ *satisfies the same orthogonality property.*

Proof. Let be $p \in H_{\Lambda}\downarrow$. Then, $< \varepsilon^{\nu}_{\Lambda,\lambda}, p >= R_{\Lambda,\lambda}(p) = 0$. Consequently, $\varepsilon^{\nu}_{\Lambda,\lambda} \perp H_{\Lambda}\!\downarrow$.

The polynomial subspace $H_{\Lambda}\!\downarrow$ is generated by homogeneous polynomials. Therefore $(\varepsilon^{\nu}_{\Lambda,\lambda})^{[k]} \perp H_{\Lambda}\!\downarrow$.

For any analytical function, $g \in \mathcal{A}_0$, we had defined the k - order homogeneous component, such as: $g^{[k]} = \sum\limits_{|\alpha|=k} D^{\alpha}g(0)(\cdot)^{\alpha}/\alpha!$

Theorem 4. *The operator* L_{Λ} *in (10) is degree reducing, that is*

$$(L_{\Lambda}(f)) \leq deg(f),\ \forall\ f \in \Pi^d$$

and the inequality is strict if and only if $f\!\uparrow\ \perp H_{\Lambda}\!\downarrow$, *with* $f\!\uparrow$ *being the leading term of the polynomial* f.

Proof. Let $f \in \Pi^d$ and $deg(f) = k$.

There comes out from (10), that $deg(L_{\Lambda}) \leq \max\limits_{j=1,\dots n} (deg(g_j\!\downarrow))$.

If $deg(g_j\!\downarrow) > k$ then $< g_j, f >= 0$. If $deg(g_j\!\downarrow) = k$, then, the following implications hold, for any $j \in \{1, \dots, n\}$:

$$f\!\uparrow \perp H_{\Lambda}\!\downarrow \Leftrightarrow < p, f\!\uparrow >= 0, \forall\ p \in H_{\Lambda}\!\downarrow \Rightarrow < g_j\!\downarrow, f\!\uparrow >= 0 \Rightarrow < g_j, f >= 0.$$

Consequently $deg(L_{\Lambda}(f)) < deg(f)$, $\forall\ f \in \Pi^d$, with $f\!\uparrow \perp H_{\Lambda}\!\downarrow$.

Let's suppose now that $deg\ L_{\Lambda}(f) < deg\ (f)$. Hence $(f - L_{\Lambda}(f))\!\uparrow = f\!\uparrow$.

We use the fact that if $p(D)$ annihilates H_{Λ}, then $p\!\uparrow (D)$ annihilates $H_{\Lambda}\!\downarrow$ (see [1]) and the following implications:

$$\lambda(p) = 0,\ \forall\ \lambda \in \Lambda \Leftrightarrow p \perp H_{\Lambda} \Rightarrow p\!\uparrow \perp H_{\Lambda}\!\downarrow .$$

Therefore:

$$\lambda(f - L_{\Lambda}(f)) = 0, \forall\ \lambda \in \Lambda \Leftrightarrow (f - L_{\Lambda}(f)) \perp H_{\Lambda} \Rightarrow (f - L_{\Lambda}(f))\!\uparrow \perp H_{\Lambda}\!\downarrow \Leftrightarrow f\!\uparrow \perp H_{\Lambda}\!\downarrow$$

Corollary 2. *The following inequality holds :* $deg\ L_{\Lambda}(\varepsilon^{\nu}_{\Lambda,\lambda}) < deg\ (\varepsilon^{\nu}_{\Lambda,\lambda})$

Proof. Taking into account theorem 3, $\varepsilon^{\nu}_{\Lambda,\lambda}\!\uparrow \perp H_{\Lambda}\!\downarrow$ and we may apply theorem 4.

3 The λ-Error Order of Interpolation

Definition 2. *(C. de Boor, [3]) Let be $L : \mathcal{A}_0 \to \Pi^d$ a polynomial interpolation operator. The error order of interpolation is the greatest integer k such that $f(x) - (L(f))(x) = 0, \forall f \in \Pi^d_{<k}$.*

We generalize this definition.

Definition 3. *For the interpolation scheme presented in section 1, we name λ-error of interpolation, the greatest integer k such that $R_{\Lambda,\lambda} = 0, \ \forall f \in \Pi^d_{<k}$, $\lambda \in (\Pi^d)'$, $R_{\Lambda,\lambda}$ being defined in (14).*

If in definition 3 we take $\lambda = \delta_x, \forall x \in R^d$, we obtain definition 2.

Theorem 5. *If $H_\Lambda\!\downarrow \neq \Pi_m$, then the λ-error order of interpolation is given by the $deg(\varepsilon^\nu_{\Lambda,\lambda}\!\downarrow)$, $\varepsilon^\nu_{\Lambda,\lambda}$ being defined in (16).*

Proof. Let $k = deg(\varepsilon^\nu_{\Lambda,\lambda}\!\downarrow)$ and $f \in \Pi^d_{<k}$. Obviously $deg(f\!\uparrow) < k$ and $< \varepsilon^\nu_{\Lambda,\lambda}, f >= 0$, hence the λ-error order of interpolation is greater or equal k.

First, let's consider that there is not any $m \in N$ such that $H_\Lambda\!\downarrow = \Pi^d_m$. Let's suppose by contradiction that $R_{\Lambda,\lambda}(f) = 0, \forall f \in \Pi^d_k$. Then $< \varepsilon^\nu_{\Lambda,\lambda}, f >= 0$ and taking into account theorem 3 we also get $< \varepsilon^\nu_{\Lambda,\lambda}\!\downarrow, f >= 0, \forall f \in \Pi^d_k$. This is a contradiction, because $\varepsilon^\nu_{\Lambda,\lambda}\!\downarrow$ is a homogeneous polynomial of degree k.

Similarly, the supposition that $R_{\Lambda,\lambda}(f) = 0, \ \forall f \in \Pi^d_{<q}$, with $q > k$ leads us to the contradiction $< \varepsilon^\nu_{\Lambda,\lambda}\!\downarrow, f^{[k]} >= 0, \forall f \in \Pi^d_{<q}$.

If $H_\Lambda\!\downarrow = \Pi^d_m$, the λ- error order of interpolation is $m+1$, be cause $\varepsilon^\nu_{\Lambda,\lambda}\!\downarrow \perp H_\Lambda\!\downarrow$.

Theorem 6. *The following equality holds:*
$deg\ \varepsilon^\nu_{\Lambda,\lambda}\!\downarrow = min\{deg\ p | p \in \Pi^d;\ \lambda(p) \neq 0;\ \lambda \in (\Pi^d)';\ \lambda \notin \Lambda;\ p \in ker(L_\Lambda)\}$.

Proof. Let $k = deg\ \varepsilon^\nu_{\Lambda,\lambda}\!\downarrow$ and
$k' = min\{deg\ p | p \in \Pi^d;\ \lambda(p) \neq 0;\ \lambda \in (\Pi^d)';\ \lambda \notin \Lambda;\ p \in ker(L_\Lambda)\}$.
Taking into account theorem 3 we get:
$p \in ker(L_\Lambda) \Rightarrow < \varepsilon^\nu_{\Lambda,\lambda}, p >= \lambda(p) \neq 0 \Rightarrow deg\ \varepsilon^\nu_{\Lambda,\lambda}\!\downarrow \leq deg\ p\!\uparrow = deg\ p \Rightarrow k \leq k'$.

On the other hand let $q = \varepsilon^\nu_{\Lambda,\lambda}\!\downarrow - L_\Lambda(\varepsilon^\nu_{\Lambda,\lambda}\!\downarrow)$. Using theorems 3 and 4, we obtain $deg\ q = deg\ \varepsilon^\nu_{\Lambda,\lambda}\!\downarrow = k$.

Much more $\lambda(q) = R_{\Lambda,\lambda}(\varepsilon^\nu_{\Lambda,\lambda}\!\downarrow) = < \varepsilon^\nu_{\Lambda,\lambda}, \varepsilon^\nu_{\Lambda,\lambda}\!\downarrow > > 0$ and $L_\Lambda(q) = 0$. But, from $q \in \Pi^d;\ \lambda(q) \neq 0$ and $q \in ker(L_\Lambda)$ we obtain $deg\ q \geq k'$, that is $k \geq k'$.

Corollary 3. *If $q \in \Pi^d_{\geq k}$, then the expression of the λ- remainder is*

$$R_{\Lambda,\lambda}(q) = < \varepsilon^\nu_{\Lambda,\lambda}, q >= \sum_{\alpha \in N^d, |\alpha| \geq k} \frac{D^\alpha \varepsilon^\nu_{\Lambda,\lambda}(0) \cdot D^\alpha q(0)}{\alpha!}$$

with $k = deg\ \varepsilon^\nu_{\Lambda,\lambda}\!\downarrow$.

Taking into account that $\left(q(D)\varepsilon^{\nu}_{\Lambda,\lambda}\right)(0) = 0, \forall\, q \in \Pi_{<k}$, we may formulate the following corollary:

Corollary 4. *$\varepsilon^{\nu}_{\Lambda,\lambda}$ vanishes to order $k = deg\ \varepsilon^{\nu}_{\Lambda,\lambda}\!\downarrow$ at 0.*

The next theorem allows us to study the classical remainder and hence to obtain the error order of interpolation.

Theorem 7. *The operator L_Λ given in (10) reproduces the monomials x^α, $x \in R^d$, $\alpha \in N^d$, if and only if $\sum\limits_{i=1}^{n} c_{k,i}\lambda_i(x^\alpha) = D^\alpha g_k(0)$, with coefficients $c_{k,i}$ given in (8).*

We proved this theorem in [5], for $d = 2$, but the proof is the same for $d > 2$.

4 Application

Let $\Lambda = \delta_\Theta$, $\Theta = \{\theta_i;\ i = 1,\dots,4\} = \{(a,0);(0,b);(-a,0);(0,-b)\}$, $a,\ b \in R_+$. We generate the basis $\{g_i;\ i = 1,\dots,4\}$ and $\{g_i\!\downarrow;\ i = 1,\dots,4\}$ of the spaces H_Λ and $H_\Lambda\!\downarrow$, using the relations (6) and (7). We start with $g_1 = e_{\theta_1}$ and obtain:

$$g_1(x,y) = \frac{1}{a^4+b^4}\left[b^4\cosh(ax) + a^4\cosh(by)\right]$$

$$g_1\!\downarrow(x,y) = 1;\ < g_1, g_1\!\downarrow> = 1$$

$$g_2(x,y) = \sinh(by) - \sinh(ax)$$

$$g_2\!\downarrow(x,y) = by - ax;\ < g_2, g_2\!\downarrow> = a^2 + b^2$$

$$g_3(x,y) = \frac{1}{(a^2+b^2)(a^4+b^4)}\cdot E(x,y)$$

$$\begin{aligned} E(x,y) = [&(a^6 - a^4 + a^4b^2 + a^2b^2 + 2b^6)\cosh(ax) - 2(a^4b^2 + b^6)e^{ax} + \\ &+(a^6 + a^4 + a^4b^2 - a^2b^2 + 2a^2b^4)\cosh(by) - 2(a^6 + a^2b^4)e^{by}] \end{aligned}$$

$$g_3\!\downarrow(x,y) = -\frac{2ab^2}{a^2+b^2}x - \frac{2a^2b}{a^2+b^2}y;\ < g_3, g_3\!\downarrow> = \frac{4a^2b^2}{a^2+b^2}$$

$$g_4(x,y) = 2[\cosh(by) - \cosh(ax)]$$

$$g_4\!\downarrow(x,y) = -a^2x^2 + b^2y^2;\ < g_4, g_4\!\downarrow> = 2(a^4+b^4).$$

The polynomials $g_2\!\downarrow$ and $g_3\!\downarrow$ are linear independent, hence, the interpolation space is: $H_\Lambda\!\downarrow = span\{g_i\!\downarrow;\ i = 1,\dots 4\} = \Pi_1 + span\{-a^2x^2 + b^2y^2\}$

The coefficients $c_{i,j}$; $i, j \in \{1 \dots 4\}$, in formula (8) are given below:

$$c_{1,1} = \frac{b^4}{2(a^4+b^4)}; \qquad c_{3,1} = k(a^6 - 3a^4b^2 - a^4 + a^2b^2 - 2b^6);$$

$$c_{1,2} = \frac{a^4}{2(a^4+b^4)}; \qquad c_{3,2} = k(-3a^6 + a^4b^2 + a^4 - a^2b^2 - 2a^2b^4);$$

$$c_{1,3} = \frac{b^4}{2(a^4+b^4)}; \qquad c_{3,3} = k(a^6 + a^4b^2 - a^4 + a^2b^2 + 2b^6);$$

$$c_{1,4} = \frac{a^4}{2(a^4+b^4)}; \qquad c_{3,4} = k(a^6 + a^4b^2 + a^4 - a^2b^2 + 2a^2b^4);$$

$$\begin{array}{ll} c_{2,1} = -1/2; & c_{4,1} = -1 \\ c_{2,2} = 1/2; & c_{4,2} = 1 \\ c_{2,3} = 1/2; & c_{4,3} = -1 \\ c_{2,4} = -1/2; & c_{4,4} = 1 \end{array}$$

We used the notation $k = \dfrac{1}{2(a^2+b^2)(a^4+b^4)}$.

Using (17), we get the expression of the remainder

$$(R_\Lambda(f))(x,y) = < e_{(x,y)}, f > - \sum_{i=1}^{4} (d_{1,i} + d_{2,i}x + d_{3,i}y + d_{4,i}x^2 + d_{5,i}y^2) \cdot f(\theta_i),$$

For $i \in \{1, \dots, 4\}$ we have:

$$d_{1,i} = c_{1,i};\ d_{2,i} = -\left(\frac{a}{a^2+b^2}c_{2,i} + \frac{1}{2a}c_{3,i}\right);\ d_{3,i} = \frac{b}{a^2+b^2}c_{2,i} - \frac{1}{2b}c_{3,i};$$

$$d_{4,i} = -\frac{a^2}{2(a^4+b^4)}c_{4,i};\ d_{5,i} = \frac{b^2}{2(a^4+b^4)}c_{4,i}$$

We will calculate and analyze the λ-error order of interpolation for various choices of functional λ.

Case 1. The case of evaluation functionals

We want to study the classical error order of interpolation, using theorem 5. In this case $\lambda = \delta_{(t_1,t_2)}$, $(t_1, t_2) \in R^2$. The generating function is $\lambda^\nu(x,y) = e^{t_1x+t_2y}$. We calculate the homogeneous components of $\varepsilon^\nu_{\Lambda,\lambda}$ and obtain: $(\varepsilon^\nu_{\Lambda,\lambda})^{[0]} = 0$;

$$(\varepsilon^\nu_{\Lambda,\lambda})^{[1]} = \frac{b(b-a)}{a^2+b^2}(t_1 - t_2)x + \frac{1}{a^2+b^2}[t_1(ab - a^2) - t_2(ab + b^2)]y$$

If $a \neq b$ then $deg(\varepsilon^\nu_{\Lambda,\lambda})\downarrow = 1$, $\forall (t_1, t_2) \in R^2$ and the λ-error order equals 1, $\forall \lambda = \delta_{(t_1,t_2)}$, that is the classical error order of interpolation is equal to 1.

If $a = b$, then for $t_2 = 0$, $(\varepsilon^\nu_{\Lambda,\lambda})^{[1]} = 0$, that is the λ-error order of interpolation is greater than 1, for $\lambda = \delta_{(t_1,0)}$.

The same result can be obtained using theorem 7.

L_Λ reproduces the constant functions, because $\sum_{i=1}^{4} c_{k,i}\delta_{\theta_i}(1) = g_k(0)$, $\forall k \in \{1,\ldots,4\}$, but it does not reproduce the polynomials of degree 1, because $\sum_{i=1}^{4} c_{3,i}\delta_{\theta_i}(m_{(1,0)}) \neq D^{(1,0)}g_3(0)$, with $m_{(1,0)}(x,y) = x$. Consequently, the error order of interpolation is equal to 1.

We notice that the analysis of λ- error order of interpolation is deeper than the analysis of classical error order of interpolation.

Case 2. The case of Birchoff type functionals

The functional $\lambda_{q,\theta}(f) = (q(D)f)(\theta)$, $q \in \Pi^d$, $\theta \in R^2$ is a generalization of the derivative of f at the point θ, that is why, it is important to study the $\lambda_{q,\theta}$ - error order of interpolation.

Let's choose $q(x,y) = m_{(1,0)}(x,y) = x$ and $\theta = (t_1,t_2) \in R^2$. The generating function is $\lambda^{\nu}_{q,\theta}(x,y) = xe^{t_1x+t_2y}$.

Next, we will denote $\lambda = \lambda_{m_{(1,0)},(t_1,t_2)}$. The results we have obtained are:

$(\varepsilon^{\nu}_{\Lambda,\lambda})^{[0]} = 0$

$(\varepsilon^{\nu}_{\Lambda,\lambda})^{[1]} = 0$

$$(\varepsilon^{\nu}_{\Lambda,\lambda})^{[2]} = x^2\left[\frac{b^4}{a^4+b^4}t_1 + \frac{a(a^6-a^4-a^4b^2+a^2b^2)}{4(a^2+b^2)(a^4+b^4)}\right] + t_2xy +$$
$$+y^2\left[\frac{a^2b^2}{a^4+b^4}t_1 + \frac{b^2(-a^6+a^4+a^4b^2-a^2b^2)}{4a(a^2+b^2)(a^4+b^4)}\right]$$

That means that $deg(\varepsilon^{\nu}_{\Lambda,\lambda}) = 2$ and the λ-error order of interpolation is equal 2. If $t_2 = 0$ and $a^6 - a^4b^2 - b^6 - b^4 = 0$ we have $deg(\varepsilon^{\nu}_{\Lambda,\lambda}) > 2$ and in this case the λ-error order is greater then 2.

References

1. de Boor C., Ron A. : On multivariate polynomial interpolation. Constr. Approx. **6** (1990) 287-302
2. de Boor C. : Polynomial interpolation in several variables. Math. Z. **210** (1992) 347-378
3. de Boor C. : On the error in multivariate polynomial interpolation. Math. Z. **220** (1992) 221-230
4. de Boor C., Ron A. : The least solution for the polynomial interpolation problem. Math.Z. **220** (1992) 347-378.
5. Simian D. : On the bivariate Polynomial Interpolation. Acta Technicum Napocensis **44** vol.1 (2001) 97-105
6. Simian D. : The dual of a polynomial interpolation operator. Proceedings of The Annual Meeting of The Romanian Society of Mathematical Sciences, Edit. Univ. Transilvania Braşov, vol.1 (2001) 289-296
7. Simian D. : The λ - remainder in minimal polynomial interpolation of multivariate functions. Proceedings of International Symposium on Numerical Analysis and Approximation Theory Cluj University Press (2002) 418-429

Computational Aspects in Spaces of Bivariate Polynomial of w-Degree n

Dana Simian[1], Corina Simian[2], and Andrei Moiceanu[1]

[1] Universitatea "Lucian Blaga" din Sibiu, Facultatea de Ştiinţe, dr. I. Raţiu 5-7, Sibiu, Româ nia
[2] Universitatea Babeş Bolyai din Cluj-Napoca, Facultatea de Matematică şi Informatică, M. Kogălniceanu 1, Cluj-Napoca, România

Abstract. Multivariate ideal interpolation schemes are deeply connected with H-bases. Both the definition of a H-basis and of an ideal interpolation space depend of the notion of degree used in the grading decomposition of the polynomial spaces. We studied, in the case of bivariate polynomials, a generalized degree, introduced by T. Sauer and named w-degree. This article give some theoretical results that allow us to construct algorithms for calculus of the dimension of the homogeneous spaces of bivariate polynomials of w - degree n. We implemented these algorithms in C++ language. The analysis of the results obtained, leads us to another theoretical conjecture which we proved in the end.

1 Introduction

The multivariate polynomial interpolation problem is a subject which is currently an active area of research, taking into account the importance of multivariate functions in modelling of real processes and phenomena. It is known that, in the univariate case, spline functions are more useful for the interpolation of large data sets, be cause the interpolation polynomials often oscillate up to a point which renders them useless. This oscillation is caused by the degree of the interpolation polynomials. In higher dimensions, the degree of polynomials increases, related to the number of conditions, lower than in univariate case. Therefore, multivariate polynomial interpolation may be a reasonable tool for a "moderate" number of conditions.

Let $\mathcal{F} \supset \Pi^d$ be a space of functions which includes polynomials in d variables and let Λ be a set of linear independent functionals. The general problem of polynomial interpolation is to find a polynomial subspace, $\mathcal{P}$, such that, for an arbitrary function $f \in \mathcal{F}$ there is an unique polynomial $p \in \mathcal{P}$ satisfying the conditions

$$\lambda(f) = \lambda(p), \ \forall \lambda \in \Lambda. \tag{1}$$

The space $\mathcal{P}$ is called an interpolation space for the conditions Λ. There are usually more interpolation spaces for a set of conditions. From a practical point of view we are interested in a such space of minimal degree.

Z. Li et al. (Eds.): NAA 2004, LNCS 3401, pp. 486–493, 2005.

On the other hand, it is known (see [1], [2]) that, in the case of an ideal interpolation scheme, that is if $ker(\Lambda)$ is a polynomial ideal, there is a strong connection between the minimal interpolation space with respect to conditions Λ and the space of reduced polynomials modulo a Gröbner basis or a H-basis of the ideal $ker(\Lambda)$.

Definition 1. *A set of polynomials, $\mathcal{H} = \{h_1, \ldots, h_s\} \subset \Pi^d \setminus \{0\}$, is a H-basis for the ideal $I = <\mathcal{H}>$ if any $p \in I$, $p \neq 0$, has an unique representation:*

$$p = \sum_{i=1}^{s} h_i g_i,\ g_i \in \Pi^d \text{ and } deg(h_i) + deg(g_i) \leq deg(p).$$

Both, in definition of interpolation spaces and in definition of H-bases, the notion of the polynomial degree is determinative.

2 The Generalized Degree

Let $(\Gamma, +)$ denote an orderer monoid, with respect to the total ordering $\prec$, such that: $\alpha \prec \beta \Rightarrow \gamma + \alpha \prec \gamma + \beta,\ \forall\ \alpha, \beta, \gamma \in \Gamma$.

Definition 2. *([3]) A direct sum*

$$\Pi = \bigoplus_{\gamma \in \Gamma} \mathcal{P}_\gamma^{(\Gamma)} \tag{2}$$

is called a grading induced by Γ, or a Γ-grading, if $\forall\ \alpha,\ \beta \in \Gamma$

$$f \in \mathcal{P}_\alpha^{(\Gamma)},\ g \in \mathcal{P}_\beta^{(\Gamma)} \Rightarrow f \cdot g \in \mathcal{P}_{\alpha+\beta}^{(\Gamma)} \tag{3}$$

The total ordering induced by Γ gives the notion of degree for the components in $\mathcal{P}_\gamma^{(\Gamma)}$. Each polynomial $f \neq 0$, has a unique representation

$$f = \sum_{i=1}^{s} f_{\gamma_i},\ f_{\gamma_i} \in \mathcal{P}_{\gamma_i}^{(\Gamma)};\quad f_{\gamma_i} \neq 0 \tag{4}$$

The terms f_{γ_i} represent the Γ- homogeneous terms of degree γ_i.

Assuming that $\gamma_1 \prec \ldots \prec \gamma_s$, the Γ- homogeneous term f_{γ_s} is called the leading term or the maximal part of f, denoted by $f^{(\Gamma)} \uparrow$.

If $\Gamma = N$ with the natural total ordering, we obtain the H-grading because of the homogeneous polynomials. In that case we have:

$$\mathcal{P}_k^{(\Gamma)} = \mathcal{P}_k^{(H)} = \Pi_k^0 \text{ and } \Pi_n = \bigoplus_{k=0}^{n} \mathcal{P}_k^{(H)} = \bigoplus_{k=0}^{n} \Pi_k^0$$

The homogeneous polynomials are: $\Pi_n^0 = \{p(x) = \sum_{|\alpha|=n} c_\alpha x^\alpha,\ x = (x_1, \ldots, x_d)\}$

Different H-gradings can be obtained, by using another total orderings.

In 1994, T. Sauer proposed, in [1], a generalization of the degree by using a weight $w \in N^d$.

Definition 3. *([1]) The w-degree of the monomial x^α is*
$\delta_w(x^\alpha) = w \cdot \alpha = (w, \alpha) = \sum_{i=1}^d w_i \cdot \alpha_i,\ \forall\ \alpha \in N^d,\ w = (w_1, \ldots, w_d) \in N^d,\ x \in R^d$.

3 The H-Grading Induced by the w-Degree

The aim of this paper is to study, in the case of two variables, the H-grading induced by the w- degree. We will denote by $\Pi_{n,w}$ the vector space of all polynomials of w- degree less than or equal to n and we will denote by $\Pi^0_{n,w}$ the vector space of all homogeneous polynomials of total w- degree exactly n:

$$\Pi_{n,w} = \left\{ \sum_{w \cdot \alpha \leq n} c_\alpha x^\alpha \mid c_\alpha \in R,\ \alpha \in N^d \right\}$$

$$\Pi^0_{n,w} = \left\{ \sum_{w \cdot \alpha = n} c_\alpha x^\alpha \mid c_\alpha \in R,\ \alpha \in N^d \right\}$$

Let consider the grading by w-degrees. In that case the polynomials in (2) are: $\mathcal{P}_k^{(\Gamma)} = \Pi^0_{k,w}$

We introduce the following notations:

$$A^0_{n,w} = \{\alpha \in N^d \mid w \cdot \alpha = n\},\ w \in (N^*)^d,\ n \in N, \text{ and } r_{n,w} = \#(A^0_{n,w})$$

The polynomial homogeneous subspace of w- degree n can be rewritten as:

$$\Pi^0_{n,w} = \left\{ \sum_{\alpha \in A^0_{n,w}} c_\alpha x^\alpha \mid c_\alpha \in R,\ \alpha \in N^d \right\}$$

We observe that $r_{n,w}$ is the dimension of the homogeneous subspace $\Pi^0_{n,w}$.

Obviously, $\delta_w(x^{\alpha+\beta}) = (\alpha_1 + \beta_1) * w_1 + (\alpha_2 + \beta_2) * w_2 = \delta_w(x^\alpha) + \delta_w(x^\beta)$ and, consequently, the equality in (3) holds. Hence

$$\Pi_{n,w} = \bigoplus_{k \in M_n} \Pi^0_{k,w}, \text{ with } M_n = \{k \in N \mid r_{k,w} > 0 \text{ and } k \leq n\}$$

The sets of multiindices $A^0_{n,w}$, is essentially for the calculus in the polynomial spaces of w- degree. We partially solved the problem of computation of the set $A^0_{n,w}$ and of the dimension $r_{n,w}$, in [4], for the case $n = 2$ and w_1, w_2 prime one to another, that is $(w_1, w_2) = 1$. We also proved in [4] the following proposition:

Proposition 1. *Let be $w = (w_1, w_2) \in (N^*)^2$ a weight and $(w_1, w_2) = p$, $p \in N$, then $\delta_w(x^\alpha) = \delta_{w'}(x^{\alpha \cdot p})$, with $w' = (w'_1, w'_2)$ and $w'_i = \frac{w_i}{p}$, $i = 1, 2$.*

The theorem 3.6 from [4] allows us to obtain an algorithm which supplies the dimension of the homogeneous polynomial subspace of w - degree n. We take again here this theorem, giving a complete proof, for arbitrary w_1, w_2.

Theorem 1. *Let $w = (w_1, w_2) \in (N^*)^2$ and let consider the functions:*

$$r : \{0, \ldots, w_1 - 1\} \to \{0, \ldots, w_1 - 1\};\ r(i) = (iw_2) \bmod w_1$$

$$\tilde{r} : \{0, \ldots, w_2 - 1\} \to \{0, \ldots, w_2 - 1\};\ \tilde{r}(i') = (i'w_1) \bmod w_2$$

Then

1. *If* $j \vdots w_1$, *that is* $j = cw_1$, *then* $r_{j,w} = \left[\frac{c}{w_2}\right] + 1$, *and* $(\alpha_1, \alpha_2) \in A^0_{j,w}$ *are given by* $\alpha_2 = kw_1$, *with* $0 \le k \le \left[\frac{c}{w_2}\right]$; $\alpha_1 = \frac{j - w_2\alpha_2}{w_1}$
2. *If* $j \vdots w_2$, *that is* $j = cw_2$, *then* $r_{j,w} = \left[\frac{c}{w_1}\right] + 1$, *and* $(\alpha_1, \alpha_2) \in A^0_{j,w}$ *are given by* $\alpha_1 = kw_2$, *with* $0 \le k \le \left[\frac{c}{w_1}\right]$; $\alpha_2 = \frac{j - w_1\alpha_1}{w_2}$
3. *If* $j \vdots w_1$ *and* $j \vdots w_2$ *then , we can use, equivalently, the results from the sentences 1 and 2 of the present theorem.*
4. *For any* j, $0 < j < \min(w_1, w_2)$, $r_{j,w} = 0$.
5. *If* $j \not\vdots w_1$ *and* $j \not\vdots w_2$, *with* $j \ge \min(w_1, w_2)$, *then*

$$r_{j,w} = \#(M_1), \text{ with } M_1 = \left\{ \left[0, \left[\frac{j - iw_2}{w_1 w_2}\right]\right] \cap N \right\}, \text{ if } \frac{j - iw_2}{w_1 w_2} \ge 0,$$

where $[\cdot]$ *is the integer part function,* $i = r^{-1}(s)$, *with* $s = j \bmod w_1$, *and* $(\alpha_1, \alpha_2) \in A^0_{j,w}$ *are given by*

$$\alpha_1 = \frac{j - (q_1 w_1 + i)w_2}{w_1}, \text{ with } q_1 \in M_1 \text{ and } \alpha_2 = \frac{i - w_1\alpha_1}{w_2}$$

6. *If* $j \not\vdots w_1$ *and* $j \not\vdots w_2$, *with* $j \ge \min(w_1, w_2)$, *then*

$$r_{j,w} = \#(M_2), \text{ with } M_2 = \left\{ \left[0, \left[\frac{j - i'w_1}{w_1 w_2}\right]\right] \cap N \right\}, \text{ if } \frac{j - i'w_1}{w_1 w_2} \ge 0,$$

where $[\cdot]$ *is the integer part function,* $i' = \tilde{r}^{-1}(p)$, *with* $p = j \bmod w_2$, *and* $(\alpha_1, \alpha_2) \in A^0_{j,w}$ *are given by*

$$\alpha_2 = \frac{j - (q_2 w_2 + i')w_1}{w_2}, \text{ with } q_2 \in M_2 \text{ and } \alpha_1 = \frac{i - w_2\alpha_2}{w_1}$$

7. *The sentences from 5 and 6 are equivalent.*
8. *If* $j \not\vdots w_1$ *and* $j \not\vdots w_2$, *with* $j \ge \min(w_1, w_2)$ *and* $\min\{j - iw_1, j - i'w_2\} < 0$, i, i' *defined in the previous statements of theorem, then* $r_{j,w} = 0$.

The proof needs the following four lemmas:

Lemma 1. *Let* $(w_1, w_2) \in (N^*)^2$ *and* $r_i = (iw_2) \bmod w_1$, $i = 0, \ldots, w_1 - 1$. *Then* $r_i \ne r_j$, $\forall\, i \ne j$.

Proof. Let $(w_1, w_2) = d$, that is there are p_1, $p_2 \in N$ such that $w_i = dp_i$, $i \in \{1, 2\}$ and $(p_1, p_2) = 1$. We suppose that there are $i, j \in \{0, \ldots, w_1 - 1\}$, $i \ne j$, such that $r_i = r_j$. Then $iw_2 = c_i w_1 + r_i$ and $jw_2 = c_j w_1 + r_j$. Hence $(i - j)w_2 = (c_i - c_j)w_1$, or equivalent $(i - j)p_2 \equiv 0 \bmod p_1$. Therefore $(i - j) \vdots p_1$, that is $i - j = cp_1$, $c \in N$. Using the fact that $i - j \le w_1 - w_2 < w_1$ we obtain $c = 0$, namely $i = j$.

Lemma 2. *Let*

$$j = k_1 w_1 + k_2 w_2,\ j \not\vdots w_1,\ j \not\vdots w_2, \tag{5}$$

$k_1, k_2, w_1, w_2 \in N^*$, *and* $r_i = (iw_2) \bmod w_1$. *Then*

1. *The function* $r : \{0, \ldots, w_1 - 1\} \to \{0, \ldots, w_1 - 1\};\ r(i) = r_i$ *is one to one.*
2. *There is an unique* $i \in \{0, \ldots, w_1 - 1\}$, *which satisfy (5) such that* $j \bmod w_1 = r_i,\ \forall\ j \in N$.
3. *If* $j \bmod w_1 = r_i$, *then* k_2 *from (5) is given by*

$$k_2 = q_1 w_1 + i, \text{ with } q_1 \in M_1 = \{[0, [a]] \cap N\}, \text{ if } [a] \geq 0$$

and there is no $k_2 \in N$ *such that (5) holds, if* $[a] < 0$, *where* $a = \dfrac{j - iw_2}{w_1 w_2}$ *and* $[\cdot]$ *denotes the integer part function.*

Lemma 3. *Let* $j \in N$, *given by (5) and* $\tilde{r}_{i'} = (i'w_1) \bmod w_2$. *Then*

1. *The function* $\tilde{r} : \{0, \ldots, w_2 - 1\} \to \{0, \ldots, w_2 - 1\};\ \tilde{r}(i) = \tilde{r}_i$ *is one to one.*
2. *There is an unique* $i' \in \{0, \ldots, w_2 - 1\}$, *which satisfy (5) such that* $j \bmod w_2 = \tilde{r}_{i'},\ \forall\ j \in N$.
3. *If* $j \bmod w_2 = \tilde{r}_{i'}$, *then* k_1 *from (5) is given by*

$$k_1 = q_2 w_2 + i', \text{ with } q_2 \in M_2,\ q_2 \in M_2 = \{[0, [a]] \cap N\}, \text{ if } [a] \geq 0$$

and there is no $k_1 \in N$ *which satisfy (5) if* $[a] < 0$, *where* $a = \dfrac{j - i'w_1}{w_1 w_2}$ *and* $[\cdot]$ *is integer part function.*

Lemma 4. *If* M_1 *and* M_2 *are the sets from lemmas 2 and 3, then* $M_1 = M_2$.

The proofs of the last three lemmas are similarly with those from [4], given for the case $(w_1, w_2) = 1$.

Corollary 1. $r_{0,w} = 1$ *and* $(0, 0) \in A^0_{0,w},\ \forall\ w \in (N^*)^2$.

4 Computational Aspects

Using the theorem 1 we obtained an algorithm which allows us to give the dimension of the w-degree homogeneous spaces and the exponents of the monomials in such a polynomial homogeneous space. The program was written in C++. It is the intention of this section to report some of the results obtained given by the implementation of the algorithm. We used both the case in which w_1 and w_2 are primes one to another, together with the proposition 1 and the theorem 1. If k satisfies both the hypothesis 1 and 2 of the theorem 1 or the hypothesis 5 and 6(like $k \vdots w_1$ and $k \vdots w_2$ or $k \not\vdots w_1$, $k \not\vdots w_2$, $k \geq \min(w_1, w_2)$) we take into account the execution time of the algorithm.

The hypothesis 1 and 2 give the same order of the execution time. If the hypothesis 5 and 6 are verified in the same time, we used the relation given in statement 5 when $w_1 < w_2$ and the relation given in statement 6 of the theorem when $w_2 < w_1$.

For the case in which w_1 and w_2 are not primes one to another, we implemented two variants of algorithms: one of them uses the proposition 1 and the other uses directly the theorem 1

We used various weights and various values of j. The results we obtained, allow us to make an analysis of the behavior of the w-homogeneous spaces.

We observe that if $(w_1, w_2) = 1$ only the homogeneous spaces of w-degree j, with $j < min(w_1, w_2)$ are missing. The case in which $(w_1, w_2) = d \neq 1$ is more unfavorable that the case in which $(w_1, w_2) = 1$, be cause there are more values of n such that the homogeneous polynomial space of w-degree n has the dimension equal 0.

Some of the results we obtained are presented in $Table$ 1, $Table$ 2 and in $Graphics 1-6$.

We observe that there is a kind of periodicity in the functional dependence

$$r_w : N \to N,\ r_w(j) = r_{w,j} \tag{6}$$

In order to characterize this periodicity, we introduce the following notions and definitions:

Definition 4. *Let $f : N \to N$ a function, $k \in N^*$ and*

$$f_i : \{i \cdot k, i \cdot k + 1, \ldots i \cdot k + k - 1\} \to N,\ f_i(x) = f(x).$$

The function $f : N \to N$ is named k-periodical on levels if

$$f_i(y) = f_{i-1}(x) + 1, \tag{7}$$

$\forall y \in \{i\cdot k, \ldots i\cdot k+k-1\}$, $\forall x \in \{(i-1)\cdot k, \ldots (i-1)\cdot k+k-1\}$, *with $x \equiv y \bmod k$.*

Theorem 2. *Let $w = (w_1, w_2) \in (N^*)^2$ a bidimensional weight with $(w_1, w_2) = 1$. The function r_w which gives the dimension of the homogeneous polynomial subspace of w - degree is $w_1 \cdot w_2$ periodical on levels.*

Proof. Let $x = j \cdot w_1 w_2 + i$, $0 \leq i \leq w_1 w_2 - 1$ and $y = (j+1)w_1 w_2 + i = x + w_1 w_2$.
Case 1: $x \,\vdots\, w_1$, that is $x = c_1 w_1$. Using the theorem 1 we obtain :

$$r_w(x) = \left[\frac{c_1}{w_2}\right] + 1 = \left[\frac{x}{w_1 w_2}\right] + 1 = j + 1$$

We observe that $y \,\vdots\, w_1$, that is $y = c_2 w_1$.

$$r_w(y) = \left[\frac{c_2}{w_2}\right] + 1 = \left[\frac{(j+1)w_1 w_2 + i}{w_1 w_2}\right] + 1 = j + 2$$

Case 2: $x \,\vdots\, w_2$ - the proof is similar with that of the case 1.

Case 3: $x \not\vdots w_1$ and $x \not\vdots w_2$. From lemma 2 we know that there is an unique $l \in \{0,\ldots,w_1-1\}$ such that $x \bmod w_1 = (lw_2) \bmod w_1$, that is, there is an unique $c \in N$ such that $x - lw_2 = cw_1$.

a) Let $x = jw_1w_2 + i \geq \min(w_1, w_2)$ and $a = \left[\frac{x-lw_2}{w_1w_2}\right] \geq 0$. Hence

$$r_w(x) = \left[\frac{x - lw_2}{w_1w_2}\right] + 1 = \left[\frac{c}{w_2}\right] + 1$$

Let $y = (j+1)w_1w_2 + i = x + w_1w_2$. There is an unique $l' \in \{0,\ldots,w_1-1\}$ such that $y \bmod w_1 = (lw_2) \bmod w_1$ or equivalent $x \equiv l'w_2 \bmod w_1$. Thus $l' = l$ and

$$r_w(y) = \left[\frac{y - lw_2}{w_1w_2}\right] + 1 = \left[\frac{x - lw_2 + w_1w_2}{w_1w_2}\right] + 1 = \left[\frac{c}{w_2}\right] + 2 = r_w(x) + 1$$

b) Let $x < min(w_1, w_2)$.

In that case $r_w(x) = 0$ and $y = x + w_1w_2 > min(w_1, w_2)$. There exists an unique $l \in \{0,\ldots,w_1-1\}$ such that $y - lw_2 \equiv 0 \bmod w_1$. Let $y - lw_2 = cw_1,\ c \in N$. We obtain

$$r_w(y) = \left[\frac{cw_1}{w_1w_2}\right] + 1 = 1 = r_w(x) + 1$$

c)Let $x \geq min(w_1, w_2)$ and $\left[\frac{x-lw_2}{w_1w_2}\right] < 0$.

In this case $r_w(x) = 0$ and $x < lw_2 < w_1w_2 - w_2$.

Let $y = x + w_1w_2$. There is an unique $l' \in \{0,\ldots,w+1-1\}$ such that $y - l'w_2 = c_1w_1$, that is $x + (w_1 - l')w_2 = c_1w_1$ and consequently $c_1 > 0$. On the other hand, $c_1 < w_2$. Therefore $r_w(y) = 1 = r_w(x) + 1$.

Theorem 3. *Let $w = (w_1, w_2) \in (N^*)^2$, $(w_1, w_2) = d > 1$ and r_w the function which gives the dimension of the homogeneous polynomial subspace of w - degree. Then $r_w(n) = 0$, $\forall\ n \in A = \{j \in N |\ j \not\vdots d\}$ and the restriction of r_w to $N \setminus A$ is $w_1 \cdot w_2$ periodical on levels.*

Proof. Let $w_1 = dp_1$, $w_2 = dp_2$, $(p_1, p_2) = 1$. The w-degree of the monomial x^α, $\alpha = (\alpha_1, \alpha_2)$ is n if and only if $w_1\alpha_1 + w_2\alpha_2 = n$, or, equivalent $d \cdot (p_1\alpha_1 + p_2\alpha_2) = n$. Consequently n can be the w-degree of a monomial if and only if $n \vdots d$. If this condition is satisfied, then $\frac{n}{d}$ is the (p_1, p_2)-degree of the monomial x^α. Using the theorem 2 the proof is complete.

The theorems 2 and 3 make possible a more efficient implementation of the algorithm which calculates the dimensions of the homogeneous polynomial spaces of w - degree. It is sufficient to compute and to store the values of $r_w(j)$ for $j \in \{0,\ldots w_1w_2-1\}$. For $j \geq w_1w_2$ we need to compute the level, $l = \left[\frac{j}{w_1w_2}\right]$ and the value $i = j \bmod w_1w_2$. The theorem 2 allows us to find $r_w(j) = r_w(i) + l$. In that way the efficiency of algorithm increases. Of course, if we want to compute the monomial basis of a homogeneous polynomial subspace of w - degree, we need to apply theorem 1.

Table 1. $w_1 = 2,\ w_2 = 5$

j	0	1	2	3	4	5	6	7	8	9
$r_w(j)$	1	0	1	0	1	1	1	1	1	1
j	10	11	12	13	14	15	16	17	18	19
r_w	2	1	2	1	2	2	2	2	2	2

Table 2. $w_1 = 2,\ w_2 = 3$

j	0	1	2	3	4	5	6	7	8	9	10	11	12
$r_w(j)$	1	0	1	1	1	1	2	1	2	2	2	2	3

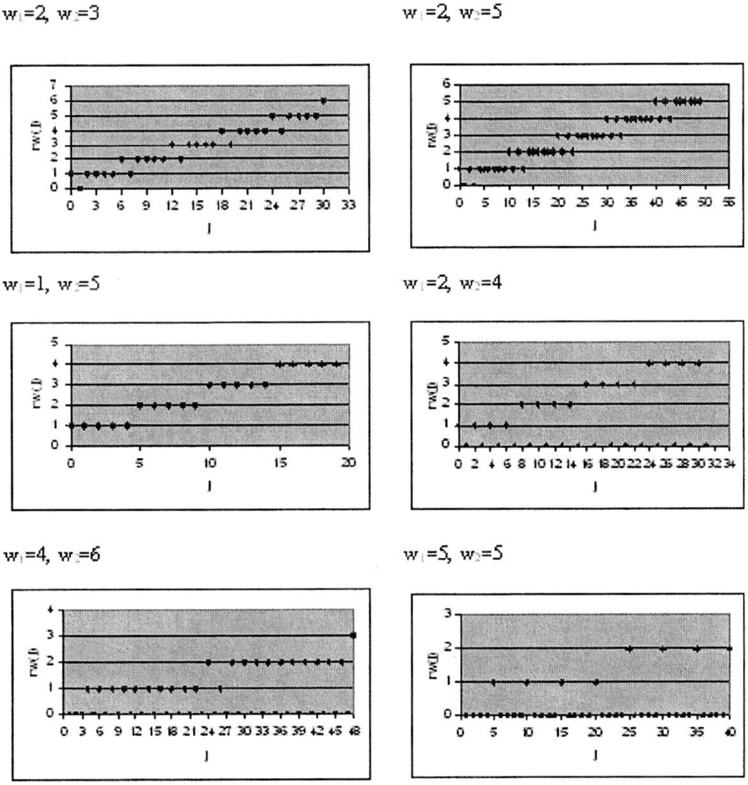

Graphics 1-6

References

1. Sauer T. : Gröbner basis, H-basis and interpolation. Transactions of the American Mathematical Society, (1994) html paper
2. Sauer T. : Polynomial interpolation of minimal degree and Gröbner bases. Gröbner Bases and Applications (Proc. of the Conf. 33 Year of Gröbner Bases) (1998) vol. **251** of London Math. Soc, Lecture Notes, 483-494
3. Möller H.M., Sauer T. : H-bases for polynomial interpolation and system solving. Advances in Computational Mathematics (1999) html paper
4. Dana Simian : Multivariate interpolation from polynomial subspaces of w-degree n. Mathematical Analysis and Approximation Theory, Proceedings of the 5-th Romanian - German Seminar on Approximation Theory and its Applications Burg-Verlag (2002) 243-254.

Restarted GMRES with Inexact Matrix–Vector Products

Gerard L.G. Sleijpen[†], Jasper van den Eshof[‡], and Martin B. van Gijzen[§]

[†]Department of Mathematics, Utrecht University,
P.O. Box 80.010, 3508 TA Utrecht, The Netherlands
sleijpen@math.uu.nl
http://www.math.uu.nl/people/sleijpen/
[‡]Department of Mathematics, University of Düsseldorf,
Universitätsstr. 1, 40224, Düsseldorf, Germany
eshof@am.uni-duesseldorf.de
http://www.am.uni-duesseldorf.de/~eshof/
[§]CERFACS, 42, avenue Gaspard Coriolis,
31057 Toulouse cedex 1, France
gijzen@cerfacs.fr
http://www.cerfacs.fr/algor

Abstract. This paper discusses how to control the accuracy of inexact matrix-vector products in restarted GMRES. We will show that the GMRES iterations can be performed with relatively low accuracy. Furthermore, we will study how to compute the residual at restart and propose suitable strategies to control the accuracy of the matrix-vector products in this computation.

1 Introduction

Iterative Krylov subspace solvers are widely used for solving large systems of linear equations. In recent years Krylov subspace methods have been used more and more for solving linear systems and eigenvalue problems in applications with full coefficient matrix where the matrix-vector products can be approximated reasonably effectively with some (iteration) method. Examples include simulations in quantum chromodynamics [9], electromagnetic applications [5], the solution of generalized eigenvalue problems [6], and of Schur complement systems [3, 8]. Relaxation strategies for controlling the accuracy of inexact matrix-vector products within Krylov methods have attracted considerable attention in the past years, see [1, 2, 3, 8, 10, 11]. These strategies allow the error in the matrix-vector product to grow as the Krylov method converges, without affecting the final accuracy too much. Relaxation strategies have been proposed for a range of different Krylov methods, and have shown to be surprisingly effective for the applications mentioned above.

In this paper we will discuss techniques for controlling the error in the matrix-vector product in the context of restarted GMRES. We will argue that the GMRES iterations at the inner level can be performed with relatively low accuracy

Z. Li et al. (Eds.): NAA 2004, LNCS 3401, pp. 494–502, 2005.

for the matrix-vector products, since only a residual reduction has to be achieved. Moreover, this low accuracy can be further reduced by applying a relaxation strategy. At restart, a suitable strategy to control the error in the matrix-vector product for the computation of the residual depends on the method by which the residual is updated. We will discuss two different methods and propose corresponding strategies to control the accuracy of the matrix-vector products in this computation. We will illustrate our ideas with two different types of Schur complement systems.

2 Restarted GMRES with Inexact Matrix-Vector Products

2.1 Preliminaries

The central problem is to find a vector $\mathbf{x}'$ that approximately satisfies the equation

$$\mathbf{A}\mathbf{x} = \mathbf{b} \quad \text{such that} \quad \|\mathbf{b} - \mathbf{A}\mathbf{x}'\|_2 < \epsilon, \tag{1}$$

for some user specified, predefined value of ϵ. For ease of presentation we will assume that the problem is scaled such that $\|\mathbf{b}\| = 1$. We will solve the above problem with the restarted GMRES-method but assume that the matrix $\mathbf{A}$ is not explicitly available. Instead, vectors of the form $\mathbf{A}\mathbf{v}$ are replaced with an approximation $\mathcal{A}_\eta(\mathbf{v})$ that has a precision η:

$$\mathcal{A}_\eta(\mathbf{v}) = \mathbf{A}\mathbf{v} + \mathbf{f} \quad \text{with} \quad \|\mathbf{f}\|_2 \leq \eta \|\mathbf{A}\|_2 \|\mathbf{v}\|_2. \tag{2}$$

2.2 Outline of the Algorithm

In restarted GMRES, in every (restart) step a correction to the current approximation is computed by at most m iterations of the GMRES method. Restarting the method is very attractive if inexact matrix-vector products are used, since in the inner loop (the actual GMRES iterations) only a residual reduction has to be achieved that can be much smaller than the final norm of the residual $\|\mathbf{b} - \mathbf{A}\mathbf{x}'\|_2 < \epsilon$ we aim for. As a consequence, lower tolerances can be used in the inner loop for the inexact matrix-vector products [11].

In Figure 1 we have summarized the general structure of the restarted GMRES method. The GMRES iterations are terminated after m steps, or if a residual reduction of ϵ_{inner} has been achieved. In the $j+1$-th GMRES iteration of the k-th step of the method we have to prescribe the tolerance $\eta_{j,k}$ for the precision of the matrix-vector product. Since we allow at most a residual reduction of ϵ_{inner} between two restarts, we can apply the relaxation strategy that was proposed in [2] to further reduce the cost of the inexact matrix-vector products. Translated in our setting, this strategy reads

$$\eta_{j,k} = \frac{\|\mathbf{r}_k\|_2}{\|\mathbf{r}_{j,k}\|_2} \epsilon_{inner}. \tag{3}$$

a. **START:**
$k = 0$, $\mathbf{x}_0 = \mathbf{0}$, $\mathbf{r}_0 = \mathbf{b}$
b. Do m steps of the GMRES method to solve $\mathbf{A}\mathbf{z}_k = \mathbf{r}_k$
(inexact matrix-vector products with $\mathcal{A}_{\eta_{j,k}}$).
Terminate early if a relative precision ϵ_{inner} is achieved.
c. **Compute update, restart or stop:**
Update solution: $\mathbf{x}_{k+1} = \mathbf{x}_k + \mathbf{z}_k$
Update residual: $\mathbf{r}_{k+1}$ (inexact matrix-vector product with $\mathcal{A}_{\eta_k}$).
Test: If $\|\mathbf{r}_{k+1}\| < \epsilon$ **STOP** else $k = k+1$ **GOTO** b.

Fig. 1. Restarted GMRES with inexact matrix-vector products

Here $\|\mathbf{r}_{j,k}\|_2$ is the norm of the residual computed in the $j+1$-th GMRES iteration of the k-th step. Notice that the limitation on the residual reduction between restarts to ϵ_{inner} is convenient in our context since it allows us to tune the accuracy of the matrix-vector multiplications within the GMRES steps using (3). In the next sections we focus on the computation of the residual at restart.

2.3 Restarting: Directly Computed Residuals

At every restart we need to compute the residual corresponding to the newly formed iterate $\mathbf{x}_{k+1}$. The usual way to compute it is directly from

$$\mathbf{r}_{k+1} = \mathbf{b} - \mathcal{A}_{\eta_k}(\mathbf{x}_{k+1}). \tag{4}$$

The question is how to determine a suitable strategy for choosing the precisions η_k. As was noted in [2] the usual strategy, where the precision is chosen high at the first restart and then decreased at subsequent restarts, does not work in this case. A suitable strategy for choosing η_k can be derived by exploiting that after the GMRES iterations we have for the true residual norm

$$\begin{aligned} \|\mathbf{b} - \mathbf{A}\mathbf{x}_{k+1}\|_2 &= \|\mathbf{b} - \mathbf{A}(\mathbf{x}_k + \mathbf{z}_k)\|_2 \\ &\leq \|\mathbf{r}_k - \mathbf{A}\mathbf{z}_k\|_2 + \|\mathbf{r}_k - (\mathbf{b} - \mathbf{A}\mathbf{x}_k)\|_2. \end{aligned} \tag{5}$$

The first term in the last expression results from the error that remains after the last GMRES iterations. The second term is the result of the error that we have made in the computation of the residual vector by (4) in the previous restart. Using this expression we find that

$$\|\mathbf{b} - \mathbf{A}\mathbf{x}_{k+1}\|_2 \leq \|\mathbf{r}_k - \mathbf{A}\mathbf{z}_k\|_2 + \eta_{k-1}\|\mathbf{A}\|_2\|\mathbf{x}_k\|_2.$$

This bound suggests to choose $\eta_{k-1} = \|\mathbf{r}_k - \mathbf{A}\mathbf{z}_k\|_2$ for the tolerances. This choice assures us that the second term is of the same order as the first term:

$$\|\mathbf{b} - \mathbf{A}\mathbf{x}_{k+1}\|_2 \leq (1 + \|\mathbf{A}\|_2\|\mathbf{x}_k\|_2)\|\mathbf{r}_k - \mathbf{A}\mathbf{z}_k\|_2.$$

As soon as $\|\mathbf{r}_k - \mathbf{A}\mathbf{z}_k\|_2$ drops below the precision ϵ we expect to have a solution that has a backward error of at most 2ϵ.

Unfortunately, $\|\mathbf{r}_{k+1} - \mathbf{A}\mathbf{z}_{k+1}\|_2$ is not known in advance, but has to be estimated. Hereto we make two realistic assumptions. The first assumption is that the residual reduction in iteration $k+1$ is at most ϵ_{inner}, which implies that

$$\epsilon_{inner}\|\mathbf{r}_k - \mathbf{A}\mathbf{z}_k\|_2 \leq \|\mathbf{r}_{k+1} - \mathbf{A}\mathbf{z}_{k+1}\|_2.$$

Secondly, we assume that the residual norm as computed by the inexact GMRES process is approximately equal to the true residual norm. This means that at the end of the k-th cycle of GMRES iterations we have that

$$\|\mathbf{r}_{m,k}\|_2 \approx \|\mathbf{r}_k - \mathbf{A}\mathbf{z}_k\|_2.$$

With these assumptions it is easy to see that a suitable choice for η_k is given by $\eta_k = \epsilon_{inner}\|\mathbf{r}_{m,k}\|_2$. The analysis of this choice gives rise to complicated formulas which we do not give here.

Notice that the accuracy with which the residual is computed at restarts is *increased* when the process comes closer to the solution. An advantage of the use of (4) is that the precision ϵ does not have to be decided a priori. Furthermore, there is no accumulation of errors (so the number of restarts does not appear in the expression for the final residual). For this reason, computation of the residuals at restart using (4) can be necessary in some applications. See e.g. [7] where the authors discuss the approximate solution of infinite dimensional systems.

Some problems require the solution of a linear system of equations for each computation of a matrix-vector products. This is the case, for example, for the Schur complement problems that we consider in Section 3. We notice that, if the restarted GMRES method converges, we have that $\mathbf{x}_k \approx \mathbf{x}_{k+1}$. We can exploit this by using the solution of the system that has to be solved for the computation of the matrix-vector product in step k as starting vector for the system that has to be solved in step $k+1$. If, in addition, the GMRES method is restarted in every iteration, then the resulting method is related to the much used Uzawa iteration method, see [4]. Restarting less frequently can be interpreted as an accelerated Uzawa type method.

2.4 Restarting: Recursively Updated Residuals

As an alternative for the computation of the residual by means of (4), we can compute the residual at restart by exploiting that $\mathbf{x}_{k+1} = \mathbf{x}_k + \mathbf{z}_k$:

$$\mathbf{r}_{k+1} = \mathbf{r}_k - \mathcal{A}_{\eta_k}(\mathbf{z}_k). \tag{6}$$

In this case we find with $\mathbf{r}_0 = \mathbf{b}$ and $\mathbf{x}_{k+1} = \sum_{j=0}^{k} \mathbf{z}_k$ that

$$\|\mathbf{r}_{k+1} - (\mathbf{b} - \mathbf{A}\mathbf{x}_{k+1})\|_2 \leq \|\mathbf{A}\|_2 \sum_{j=0}^{k} \eta_j \|\mathbf{z}_j\|_2 \tag{7}$$

Furthermore, with the estimate $\|\mathbf{z}_j\|_2 \leq \|\mathbf{A}^{-1}\|_2(\|\mathbf{r}_j\|_2 + \|\mathbf{r}_j - \mathbf{A}\mathbf{z}_j\|_2)$, we find

$$\|\mathbf{r}_{k+1} - (\mathbf{b} - \mathbf{A}\mathbf{x}_{k+1})\|_2 \leq 2\|\mathbf{A}\|_2\|\mathbf{A}^{-1}\|_2 \sum_{j=0}^{k} \eta_j \|\mathbf{r}_j\|_2. \tag{8}$$

Here, we have assumed that $\|\mathbf{r}_j - \mathbf{A}\mathbf{z}_j\|_2 \leq \|\mathbf{r}_j\|_2$ which can be shown to be true (up to a small factor). Given the relation (5) we want to achieve that the size of $\|\mathbf{r}_{k+1} - (\mathbf{b} - \mathbf{A}\mathbf{x}_{k+1})\|_2$ is of the order of $\|\mathbf{r}_{k+1} - \mathbf{A}\mathbf{z}_{k+1}\|_2$, as in the previous section. Therefore we choose the tolerance η_k equal to $\epsilon/\|\mathbf{r}_k\|_2$. This is the *relaxation strategy* proposed in [2].

We have seen that when the residuals at restart are computed recursively using (6), the precision of the matrix-vector products is *decreased* during the process, as opposed to the previous section. The advantage here is that no a priori knowledge is required about the expected residual reduction between two restarts. A disadvantage is that at termination, when $\|\mathbf{r}_{k+1}\|_2 \leq \epsilon$, the upper bound on the norm of the true residual contains the number of restarts and furthermore depends on the condition number of the matrix $\mathbf{A}$ instead of on the norm of the computed solution. This means that one must be cautious for situations where the norm of the solution is much smaller than the inverse of the smallest singular value of the matrix.

3 Numerical Experiments

3.1 Description of the Test Problem

As a model problem we consider the following set of partial differential equations on the unit square Ω:

$$-\nabla^2\psi - \frac{\partial\psi}{\partial x} - \alpha\nabla^2\zeta = f, \quad \nabla^2\psi + \zeta = 0 \quad \text{in } \Omega \tag{9}$$

plus boundary conditions $\psi = \frac{\partial\psi}{\partial n} = 0$ on Γ, the edge of the domain.

The above model problem is a simplified version of the example from oceanography that we considered in [11]. Discretisation yields the following block system

$$\begin{pmatrix} \mathbf{K} & \alpha\mathbf{L} \\ -\mathbf{L}^H & \mathbf{M} \end{pmatrix} \begin{pmatrix} \psi \\ \zeta \end{pmatrix} = \begin{pmatrix} \mathbf{f} \\ \mathbf{0} \end{pmatrix}. \tag{10}$$

The size of the test problem we use in our experiments is 16642. From (10) we can eliminate either ζ, which yields the upper Schur complement system

$$(\mathbf{K} + \alpha\mathbf{L}\mathbf{M}^{-1}\mathbf{L}^H)\psi = \mathbf{f}, \tag{11}$$

or ψ, which yields the lower Schur complement system

$$(\mathbf{M} + \alpha\mathbf{L}^H\mathbf{K}^{-1}\mathbf{L})\zeta = \mathbf{L}^H\mathbf{K}^{-1}\mathbf{f}. \tag{12}$$

These two systems have very different numerical characteristics. The upper Schur complement is a fourth order bi-harmonic operator which becomes rapidly ill conditioned if the mesh size is decreased. The lower Schur complement has the characteristics of a second order operator and hence is, for fine enough mesh size, better conditioned than the biharmonic operator. To illustrate this, the MATLAB-routine condest gives 500 for the condition number of the lower Schur complement, and a condition number of 10^6 for the upper Schur complement. Systems with $\mathbf{M}$, on the other hand, are easier to solve (since the mass matrix is a discretised unit operator) than systems with the convection-diffusion operator.

3.2 Solution Methods

For each multiplication with one of the Schur complements a linear system has to be solved, with the matrix $\mathbf{K}$ for the upper Schur complement, and with the matrix $\mathbf{M}$ for the lower Schur complement. In our experiments we solve these systems with preconditioned Bi-CGstab, with as preconditioner ILU of $\mathbf{K}$ resp. of $\mathbf{M}$), using a drop tolerance of 10^{-2}. These systems are solved up to a precision (residual reduction) η. Note that this does not mean that the matrix-vector multiplication with the Schur complements is performed with accuracy η (see also [8]), in theory an extra constant has to be taken into account.

The parameter α is taken rather small in the experiments: $\alpha = 10^{-3}$. For this reason we have applied ILU with drop tolerance 10^{-2} of $\mathbf{M}$ (of $\mathbf{K}$) as right preconditioner for the lower (resp. upper) Schur complement systems.

We consider four different methods to control the accuracy of the matrix-vector products:

- The systems to evaluate the matrix-vector products with the Schur complement are all solved to fixed accuracy $\epsilon = 10^{-8}$.
- Within GMRES the system are solved with reduced accuracy $\epsilon_{inner} = 10^{-3}$. At restart the accuracy $\epsilon = 10^{-8}$ is used.
- Within GMRES, relaxation is applied by $\eta_{j,k} = 10^{-3} \cdot (\|\mathbf{r}_k\|_2/\|\mathbf{r}_{j,k}\|_2)$. The residuals at restart are computed directly from $\mathbf{x}_{k+1}$, using precision $\eta_k = \max(10^{-3} \cdot \|\mathbf{r}_{m,k}\|_2, 10^{-8})$.
- The above relaxation strategy is used within GMRES. The residuals at restart are computed recursively using precision $\eta_k = 10^{-8}/\|\mathbf{r}_k\|_2$.

3.3 The Upper Schur Complement

Table 1 shows the numerical results for the upper Schur complement. The first column gives the method to control the accuracy. The second column gives the number of Bi-CGstab iterations. Bi-CGstab allows for two tests of the residual norm and hence may terminate halfway an iteration, which explains the fractions of two. The number of Bi-CGstab iterations gives a measure for the work in the inexact matrix-vector products. The third column gives the number of GMRES-iterations. In our experiments we restart (at least) every 20 iterations. The fourth column gives the number of restarts. The true residual norm at the end of the

Table 1. Solution of the upper Schur system: four different ways to control the accuracy of the matrix-vector products and their effect on the efficiency and on the final accuracy

Method	Iterations Bi-CGstab	Iterations GMRES	Restarts	$\|\mathbf{b} - \mathbf{A}\mathbf{x}_k\|_2$	$\|\mathbf{r}_k\|_2$
Full precision	4462.5	1700	85	$3.3 \cdot 10^{-8}$	$8.9 \cdot 10^{-9}$
Low accuracy GMRES	2317.5	2060	103	$3.3 \cdot 10^{-8}$	$8.9 \cdot 10^{-9}$
Directly computed residuals at restart	2345	2120	106	$3.3 \cdot 10^{-8}$	$9.1 \cdot 10^{-9}$
Recursively computed residuals at restart	1903.5	1800	90	$2.1 \cdot 10^{-6}$	$9.0 \cdot 10^{-9}$

Table 2. Solution of the lower Schur system: four different ways to control the accuracy of the matrix-vector products and their effect on the efficiency and on the final accuracy

Method	Iterations Bi-CGstab	Iterations GMRES	Restarts	$\|\mathbf{b} - \mathbf{A}\mathbf{x}_k\|_2$	$\|\mathbf{r}_k\|_2$
Full precision	2955	160	8	$1.0 \cdot 10^{-8}$	$9.9 \cdot 10^{-9}$
Low accuracy GMRES	489.5	180	9	$7.2 \cdot 10^{-9}$	$5.7 \cdot 10^{-9}$
Directly computed residuals at restart	464.5	200	10	$1.2 \cdot 10^{-9}$	$4.7 \cdot 10^{-9}$
Recursively computed residuals at restart	372	200	10	$2.3 \cdot 10^{-8}$	$4.4 \cdot 10^{-9}$

iterative process, which is computed using an exact matrix-vector product, is given in the fifth column. In practice the true residual norm is not available; one has at its disposal only $\|\mathbf{r}_k\|_2$ that is computed using inexact matrix-vector products. This value is given in the sixth column. Note that this value is used in the convergence test. The results tabulated in Table 1 show that the most important saving is obtained by using a lower accuracy for the GMRES iterations. No extra saving is obtained by applying the error-control strategy for directly computed residuals. The savings that are achieved in the initial iterations are lost in the extra few (costly) iterations that are needed due to the extra perturbations that are introduced in the process. The relaxation strategy at restart with recursively updated residuals, on the other hand, yields a small but significant extra saving in computational cost. Note that on average the tolerances used in the matrix-vector product are lower than for the strategy with directly computed residuals. Moreover, the number of restarts (and GMRES iterations) is less for this example. An important disadvantage of this strategy, however, is that the true residual norm stagnates at around 10^{-6}. This can be explained by the fact the Schur complement is ill conditioned, and relatively many restarts are required to solve the system. Note that both the condition number of the Schur complement and the number of restarts negatively influence the bound (8). If residuals are calculated directly, however, the true residual norm decreases close to the target 10^{-8}.

3.4 The Lower Schur Complement

The numerical results for the experiments with the lower Schur complement are tabulated in Table 2. The number of GMRES iterations and restarts are considerably less than for the upper Schur complement system. This is because this system is better conditioned. This fact is also reflected in the accuracy that is achieved using recursively updated residuals plus relaxation: the norm of the true residual stagnates around the target value 10^{-8}. The savings that are obtained for this example are very significant.

4 Conclusions

A considerable saving in computational cost can be obtained for restarted GMRES with inexact matrix-vector products by using a low accuracy for the GMRES-iterations, in combination with a high accuracy to compute the residuals at restart. The accuracy of the matrix-vector products at restart can be reduced by a strategy that depends on the way the residual is calculated. If the residual is calculated directly form the latest iterate then the precision of the matrix-vector product has to be increased as the method comes closer to the solution, whereas if the residuals are calculated recursively a relaxation strategy can be applied. This latter strategy, however, has the disadvantage that the achieved accuracy may be well above the target accuracy for ill-conditioned problems that require many restarts.

Acknowledgments. We thank Daniel Loghin for providing us with the test matrices and Françoise Chaitin-Chatelin for valuable comments.

References

1. A. Bouras and V. Frayssé, *A relaxation strategy for Arnoldi method in eigenproblems*, Technical Report TR/PA/00/16, CERFACS, France, 2000.
2. ———, *A relaxation strategy for inexact matrix-vector products for Krylov methods*, Technical Report TR/PA/00/15, CERFACS, France, 2000.
3. A. Bouras, V. Frayssé, and L. Giraud, *A relaxation strategy for inner-outer linear solvers in domain decomposition methods*, Technical Report TR/PA/00/17, CERFACS, France, 2000.
4. James H. Bramble, Joseph E. Pasciak, and Apostol T. Vassilev, *Analysis of the inexact Uzawa algorithm for saddle point problems*, SIAM J. Numer. Anal. **34** (1997), no. 3, 1072–1092.
5. B. Carpentieri, *Sparse preconditioners for dense complex linear systems in electromagnetic applications*, Ph.D. dissertation, INPT, April 2002.
6. F. Chaitin-Chatelin and T. Meškauskas, *Inner-outer iterations for mode solvers in structural mechanics: application to the Code Aster*, Contract Report TR/PA/01/85, CERFACS, Toulouse, France, 2001.
7. P. Favati, G. Lotti, O. Menchi, and F. Romani, *Solution of infinite linear systems by automatic adaptive iterations*, Linear Algebra Appl. **318** (2000), no. 1-3, 209–225.

8. V. Simoncini and D. B. Szyld, *Theory of inexact Krylov subspace methods and applications to scientific computing*, SIAM J. Sci. Comput. **25** (2003), no. 2, 454–477.
9. J. van den Eshof, A. Frommer, Th. Lippert, K. Schilling, and H.A. van der Vorst, *Numerical methods for the QCD overlap operator: I. sign-function and error bounds*, Comput. Phys. Comm. **146** (2002), 203–224.
10. J. van den Eshof and G. L. G. Sleijpen, *Inexact Krylov subspace methods for linear systems*, Preprint 1224, Dep. Math., University Utrecht, Utrecht, the Netherlands, February 2002, To appear in SIAM J. Matrix Anal. Appl. (SIMAX).
11. J. van den Eshof, G. L. G. Sleijpen, and M. B. van Gijzen, *Relaxation strategies for nested Krylov methods*, Preprint 1268, Dep. Math., University Utrecht, Utrecht, the Netherlands, March 2003.

Applications of Price Functions and Haar Type Functions to the Numerical Integration

S.S. Stoilova

Institute of Mathematics and Informatics - Bulgarian Academy of Sciences, Acad. G. Bonchev str., bl. 8, 1113 Sofia, Bulgaria
stanislavast@yahoo.com

Abstract. By analogy with the theory of good lattice points for the numerical integration of rapidly convergent Walsh series, in the present paper the author use the Price functional system and Haar type functional system, defined in the generalized number system, for numerical integration. We consider two classes of functions, whose Fourier-Price and Fourier-Haar coefficients satisfy specific conditions. For this classes we obtain the exact orders of the error of the quadrature formula with good lattice points, constructed in the generalized number system.

Keywords: Numerical integration, Good lattice points, Price functions, Haar type functions, Estimation of the error.

1 Introduction

For the numerical integration the leading role play the properties of the function, which we integrate and the set, which we use for the integration.

One of the main branches in the theory of numerical integration by number-theoretical methods is the theory of good lattice points. Good lattice points methods were introduced by Korobov [4] and are an excellent instrument for the numerical integration of functions which are represented by rapidly convergent Fourier series.

Let $s \geq 1$ be an arbitrary integer and $[0,1)^s$ is the $s-$dimensional unit cube. We denote $\mathbf{N}_0 = \mathbf{N} \cup \{0\}$. Let

$$\mathcal{T} = \left\{ e_{\mathbf{k}}(\mathbf{x}) = \prod_{i=1}^{s} \exp(2\pi \mathbf{i} k_i x_i) : \mathbf{k} \in \mathbf{Z}^s, \mathbf{k} = (k_1, \dots, k_s), \mathbf{x} = (x_1, \dots, x_s) \in \mathbf{R}^s \right\}$$

be the trigonometric functional system.

For an arbitrary vector $\mathbf{k}$ with integer coordinates and an integrable function $f : \mathbf{R}^s \to \mathbf{R}$ let ${}_{\mathcal{T}}\widehat{f}(\mathbf{k}) = \int_{[0,1)^s} f(\mathbf{x}) \overline{e}_{\mathbf{k}}(\mathbf{x}) d\mathbf{x}$ be the $\mathbf{k}-$th Fourier coefficient of f. For real constants $\alpha > 1$ and $c > 0$ Korobov introduces a functional class ${}_{\mathcal{T}}E_s^{\alpha}(c)$, composed by continuous periodical with period 1 by each their

Z. Li et al. (Eds.): NAA 2004, LNCS 3401, pp. 503–510, 2005.

arguments functions $f : \mathbf{R}^s \to \mathbf{R}$ which Fourier coefficients satisfy the estimation $\left|_T\widehat{f}(\mathbf{k})\right| \leq \frac{c}{(\overline{k}_1 \dots \overline{k}_s)^\alpha}$, for all vectors $\mathbf{k} = (k_1, \dots, k_s)$ with integer coordinates and $\overline{k}_i = \max(1, |k_i|)$, $1 \leq i \leq s$. Using the so-called lattice points set $L(n) = \left\{\left(\frac{l_1}{n}, \dots, \frac{l_s}{n}\right) : 1 \leq l_i \leq n,\ 1 \leq i \leq s\right\}$ Korobov proves that the error of the quadrature formula

$$\int_{[0,1)^s} f(\mathbf{x})d\mathbf{x} - \frac{1}{N}\sum_{l_1=1}^{n} \dots \sum_{l_s=1}^{n} f\left(\frac{l_1}{n}, \dots, \frac{l_s}{n}\right) - R$$

satisfies the estimation $R \in \mathcal{O}\left(N^{-\frac{\alpha}{s}}\right)$, where $N = n^s$.

The so-called digital nets or $(t, m, s)-$nets to the base $b, b \geq 2-$ integer, are convenient for a numerical integration. Larcher, Niederreiter and Schmid [5] obtain the order of the numerical integration error for functions, which Fourier-Walsh coefficients satisfy specific conditions, using the $(t, m, s)-$nets to the base b. Another results in this direction are obtained by Larcher and Wolf [8].

Larcher and Pirsic [6] use Walsh functions over finite abelian group for a numerical integration and they obtain the order of the error of the quadrature formula by means of these functions.

Chrestenson [1] introduces a generalization of Walsh functions. Let $b \geq 2$ be a fixed integer and $\omega = \exp\left(\frac{2\pi \mathrm{i}}{b}\right)$. The Rademacher functions $\{\phi_k(x)\}_{k\geq 0}$, $x \in \mathbf{R}$ to base b are defined as: $\phi_0(x) = \omega^a$, $\frac{a}{b} \leq x < \frac{a+1}{b}$, $a = 0, 1, \dots, b-1$, $\phi_0(x+1) = \phi_0(x)$ and for an arbitrary integer $k \geq 1$ $\phi_k(x) = \phi_0(b^k x)$.

The Walsh functions $\{\psi_k(x)\}_{k\geq 0}$, $x \in [0, 1)$ to base b are defined as: $\psi_0(x) = 1$ for every $x \in [0, 1)$ and if $k \geq 1$ has a $b-$adic representation $k = k_m b^{\alpha_m} + \dots + k_1 b^{\alpha_1}$, where $\alpha_m > \alpha_{m-1} > \dots > \alpha_1 \geq 0$ and for $1 \leq j \leq m$ $k_j \in \{1, 2, \dots, b-1\}$, then the $k-$th Walsh function to base b is defined as $\psi_k(x) = \phi_{\alpha_m}^{k_m}(x) \dots \phi_{\alpha_1}^{k_1}(x)$ for every $x \in [0, 1)$.

For a vector $\mathbf{k} = (k_1, \dots, k_s)$ with non-negative integer coordinates the function $\psi_{\mathbf{k}}(\mathbf{x})$ is defined as $\psi_{\mathbf{k}}(\mathbf{x}) = \prod_{j=1}^{s} \psi_{k_j}(x_j)$, $\mathbf{x} = (x_1, \dots, x_s) \in [0, 1)^s$. We denote by $\psi(b) = \{\psi_{\mathbf{k}}(\mathbf{x})\}_{\mathbf{k}\in\mathbf{N}_0{}^s}$, $\mathbf{x} \in [0, 1)^s$. For an arbitrary vector $\mathbf{k}$ with non-negative integer coordinates and for an integrable function $f : \mathbf{R}^s \to \mathbf{R}$ let ${}_{\psi(b)}\widehat{f}(\mathbf{k}) = \int_{[0,1)^s} f(\mathbf{x})\overline{\psi}_{\mathbf{k}}(\mathbf{x})d\mathbf{x}$ be the $\mathbf{k}-$th Fourier-Walsh coefficient of the function f.

Larcher and Traunfellner [7] showed that the good lattice points methods can be used as a means for the numerical integration of functions which are represented by rapidly convergent Walsh series to base b, $b \geq 2$. They consider functions $f : \mathbf{R}^s \to \mathbf{R}$ for which the estimation $|{}_{\psi(b)}\widehat{f}(\mathbf{k})| \leq \frac{c}{(\overline{k}_1 \dots \overline{k}_s)^\alpha}$ where $\alpha > 1$ and $c > 0$ are fixed real constants, holds for each vector $\mathbf{k} = (k_1, \dots, k_s)$ with non-negative integer coordinates and for $1 \leq i \leq s$ $k_i = \max(1, k_i)$. They use the net $L(b^n) = \left\{ \left(\frac{l_1}{b^n}, \ldots, \frac{l_s}{b^n} \right) : 0 \le l_i < b^n, \quad 1 \le i \le s \right\}$ for an arbitrary integer $n \ge 1$ and prove that the error of the quadrature formula satisfies the inequality $R_N(f) \in \mathcal{O}(N^{-\frac{\alpha}{s}})$, where $N = b^{sn}$.

Here, we will use the good lattice points methods as a means for numerical integration of functions which are represented by rapidly convergent Fourier-Price series and Fourier-Haar type series, defined in generalized number system.

Let $\mathcal{B} = \{b_1, b_2, \ldots, b_j, \ldots, \quad b_j \ge 2, \quad j \ge 1\}$ be a fixed sequence of integers. Using the sequence $\mathcal{B}$, we define the sequence of the generalized powers $\{B_j\}_{j\ge 0}$ as $B_0 = 1$ and for $j \ge 1$ $B_j = \prod_{i=1}^{j} b_i$. For each $j \ge 1$ we define $\omega_j = \exp\left(\frac{2\pi \mathrm{i}}{b_j}\right)$. We will recall the definition of Price [9] functions.

Definition 1. *(i) For real $x \in [0,1)$ in the $\mathcal{B}-$adic form $x = \sum_{j=1}^{\infty} x_j B_j^{-1}, \quad x_j \in \{0, 1, \ldots, b_j - 1\}$ and each integer $j \ge 0$ Price define the functions $\chi_{B_j}(x) = \omega_{j+1}^{x_{j+1}}$.*

(ii) For each integer $k \ge 0$ in the $\mathcal{B}-$adic form $k = \sum_{j=0}^{n} k_{j+1} B_j, \quad k_{j+1} \in \{0, 1, \ldots, b_{j+1} - 1\}, \quad k_{n+1} \ne 0$ and real $x \in [0,1)$ the $k-$th function of Price $\chi_k(x)$ is defined as

$$\chi_k(x) = \prod_{j=0}^{n} \left[\chi_{B_j}(x)\right]^{k_{j+1}}.$$

For a vector $\mathbf{k} = (k_1, \ldots, k_s)$ with non-negative integer coordinates the function $\chi_{\mathbf{k}}(\mathbf{x})$ is defined as $\chi_{\mathbf{k}}(\mathbf{x}) = \prod_{i=1}^{s} \chi_{k_i}(x_i), \quad \mathbf{x} = (x_1, \ldots, x_s) \in [0,1)^s$. The system of functions $\chi(\mathcal{B}) = \{\chi_{\mathbf{k}}(\mathbf{x})\}_{\mathbf{k} \in \mathbf{N}_0^s}, \quad \mathbf{x} \in [0,1)^s$ is a complete orthonormal system in $L_2([0,1)^s)$ and it is usually called Price system or Vilenkin [11] system.

A class of Haar type functions is defined in Schipp, Wade, Simon [10]. The author and Grozdanov [2] use an another class of the so-called Haar type functions for an investigation of the uniform distribution of sequences. We will recall the definition of this functional system.

Definition 2. *We put $h_0(x) = 1$ for every $x \in [0,1)$. Let $k \ge 1$ be an arbitrary integer. We choose the unique integer $g \ge 0$ such that $B_g \le k < B_{g+1}$ and for k use the $\mathcal{B}-$adic representation $k = k_g B_g + p$, with $k_g \in \{1, \ldots, b_{g+1} - 1\}$ and $0 \le p \le B_g - 1$. Then the $k-$th Haar type function $h_k(x), \quad x \in [0,1)$ is defined as*

$$h_k(x) = \begin{cases} \sqrt{B_g} \;\; \omega_{g+1}^{k_g a}, & \text{if } \frac{p b_{g+1} + a}{B_{g+1}} \le x < \frac{p b_{g+1} + a + 1}{B_{g+1}}, \quad a = 0, 1, \ldots, b_{g+1} - 1, \\ 0, & \text{otherwise.} \end{cases}$$

For a vector $\mathbf{k} = (k_1, \ldots, k_s)$ with non-negative integer coordinates the function $h_{\mathbf{k}}(\mathbf{x})$ is defined as $h_{\mathbf{k}}(\mathbf{x}) = \prod_{i=1}^{s} h_{k_i}(x_i), \quad \mathbf{x} = (x_1, \ldots, x_s) \in [0,1)^s$. The functional system $h(\mathcal{B}) = \{h_{\mathbf{k}}(\mathbf{x})\}_{\mathbf{k}\in\mathbf{N}_0^s}$, $\mathbf{x} \in [0,1)^s$ is complete and orthonormal system in $L_2([0,1)^s)$.

When in the sequence $\mathcal{B}$, all $b_j = 2$, then from Definition 2 the original system of Haar [3] is obtained.

Let the function $f : \mathbf{R}^s \to \mathbf{R}$ be a continuous, periodical function with period 1 of each its arguments. For a vector $\mathbf{k}$ with non-negative integer coordinates we signify by ${}_{\chi(\mathcal{B})}\widehat{f}(\mathbf{k})$ and ${}_{h(\mathcal{B})}\widehat{f}(\mathbf{k})$ respectively Fourier-Price and Fourier-Haar coefficients of this function

$$ {}_{\chi(\mathcal{B})}\widehat{f}(\mathbf{k}) = \int_{[0,1)^s} f(\mathbf{x})\overline{\chi}_{\mathbf{k}}(\mathbf{x})d\mathbf{x} \ \ and \ \ {}_{h(\mathcal{B})}\widehat{f}(\mathbf{k}) = \int_{[0,1)^s} f(\mathbf{x})\overline{h}_{\mathbf{k}}(\mathbf{x})d\mathbf{x}. $$

In the next two definitions, we introduce two classes of functions, whose Fourier-Price and Fourier-Haar coefficients satisfy the conditions respectively.

Definition 3. *For fixed real constants $\alpha > 1$ and $c > 0$ we will call the function $f : \mathbf{R}^s \to \mathbf{R}$ belongs to the class ${}_{\chi(\mathcal{B})}E_s^{\alpha}(c)$, if the estimation $|{}_{\chi(\mathcal{B})}\widehat{f}(\mathbf{k})| \leq \dfrac{c}{(\overline{k}_1 \ldots \overline{k}_s)^{\alpha}}$, holds for each vector $\mathbf{k} = (k_1, \ldots, k_s)$ with non-negative integer coordinates and for $1 \leq i \leq s$ $\overline{k}_i = \max(1, k_i)$.*

For an arbitrary integer $k \geq 0$ we define $\tilde{k}$ as

$$ \tilde{k} = \begin{cases} 1, & if \ k = 0 \\ B_g, & if \ \ B_g \leq k < B_{g+1}, \ \ g \geq 0, \ \ g \in \mathbf{Z}. \end{cases} $$

Definition 4. *For fixed real constants $\alpha > 1$ and $c > 0$ we will call the function $f : \mathbf{R}^s \to \mathbf{R}$ belongs to the class ${}_{h(\mathcal{B})}E_s^{\alpha}(c)$, if the estimation $|{}_{h(\mathcal{B})}\widehat{f}(\mathbf{k})| \leq \dfrac{c}{(\tilde{k}_1 \ldots \tilde{k}_s)^{\alpha+\frac{1}{2}}}$, holds for each vector $\mathbf{k} = (k_1, \ldots, k_s)$ with non-negative integer coordinates.*

Let $N \geq 1$ be an arbitrary integer and $\xi_N = \{\mathbf{x}_1, \ldots, \mathbf{x}_N\}$ is a set of N points in $[0,1)^s$. For an arbitrary integrable in Riemann sense function $f(\mathbf{x}), \mathbf{x} \in [0,1)^s$ we signify by $R(\xi_N; f)$ the error of the quadrature formula

$$ R(\xi_N; f) = \frac{1}{N}\sum_{j=1}^{N} f(\mathbf{x}_j) - \int_{[0,1)^s} f(\mathbf{x})d\mathbf{x}. \tag{1} $$

For an arbitrary fixed integer $n \geq 1$ using the sequence $\mathcal{B}$, let $\mathcal{B}(n) = \{b_1, \ldots, b_n\}$. The set $L(\mathcal{B}(n)) = \left\{\left(\dfrac{l_1}{B_n}, \ldots, \dfrac{l_s}{B_n}\right) : 0 \leq l_i < B_n, 1 \leq i \leq s\right\}$, composed of $N = B_n^s$ points in $[0,1)^s$, we will call a lattice points set.

In the present paper using the lattice points set $L(\mathcal{B}(n))$, defined in a generalized number system, we obtain the exact order $\mathcal{O}\left(N^{-\frac{\alpha}{s}}\right)$ of the error of the quadrature formula (1) for the functions from ${}_{\chi(\mathcal{B})}E_s^{\alpha}(c)$. The obtained result (see Theorem 1) is a generalization of the result of Larcher and Traunfellner [7, Theorem 1].

For the introduced class ${}_{h(\mathcal{B})}E_s^{\alpha}(c)$ we obtain an order $\mathcal{O}\left(N^{-\frac{\alpha}{s}}\right)$ of the error of the quadrature formula (1), constructed with respect to the lattice point $L(\mathcal{B}(n))$, (see Theorem 2).

2 Statement of the Results

In the next two theorems we will represent the main results.

Theorem 1. *For an arbitrary function $f \in {}_{\chi(\mathcal{B})}E_s^{\alpha}(c)$, the error of the quadrature formula (1), constructed for this function with respect to the lattice points set $L(\mathcal{B}(n))$ satisfies the estimation*

$$|R(L(\mathcal{B}(n)); f)| \leq C_1(\alpha, c, s) N^{-\frac{\alpha}{s}} + \mathcal{O}(N^{-\frac{2\alpha}{s}}),$$

with $C_1(\alpha, c, s) = \dfrac{c\alpha s}{\alpha - 1}$ *and* $N = B_n^s$.

The obtained estimation is exact, in a sense that a function $f_1(\mathbf{x}) \in {}_{\chi(\mathcal{B})}E_s^{\alpha}(c)$ exists, such that

$$|R(L(\mathcal{B}(n)); f_1)| > cN^{-\frac{\alpha}{s}}.$$

Remark: In the special case when in the sequence $\mathcal{B}$ all bases $b_j = b$ $j \geq 1$ from Theorem 1, we obtain the result of Larcher and Traunfellner [7, Theorem 1].

Theorem 2. *Let an absolute constant $B > 0$ exists, so that for all $j \geq 1$ $b_j \leq B$ and $b = \min\{b_j : j \geq 1\}$. For an arbitrary function $f \in {}_{h(\mathcal{B})}E_s^{\alpha}(c)$, the error of the quadrature formula (1), constructed for this function with respect to the lattice points set $L(\mathcal{B}(n))$ satisfies the estimation*

$$|R(L(\mathcal{B}(n)); f)| \leq C_2(\alpha, c, s, \mathcal{B}) N^{-\frac{\alpha}{s}} + \mathcal{O}(N^{-\frac{2\alpha}{s}})$$

with $C_2(\alpha, c, s, \mathcal{B}) = \dfrac{cs(B-1)b^{\alpha-1}}{b^{\alpha-1}-1}$, *and* $N = B_n^s$.

3 Preliminary Statements

To prove the main results, we need some preliminary results.

If the function $f \in {}_{\chi(\mathcal{B})}E_s^{\alpha}(c)$, then Fourier-Price series of this function is absolutely convergent. Really, for each vector $\mathbf{k}$ and each $\mathbf{x} \in [0,1)^s$ from $|\chi_{\mathbf{k}}(\mathbf{x})| = 1$ and Definition 3 we have

$$\sum_{\mathbf{k}=\mathbf{0}}^{\infty} \left| {}_{\chi(\mathcal{B})}\widehat{f}(\mathbf{k}) \chi_{\mathbf{k}}(\mathbf{x}) \right| = \sum_{\mathbf{k}=\mathbf{0}}^{\infty} \left| {}_{\chi(\mathcal{B})}\widehat{f}(\mathbf{k}) \right| \leq c \sum_{k_1, \ldots, k_s = 0}^{\infty} \frac{1}{(\overline{k}_1 \ldots \overline{k}_s)^{\alpha}} = c \left(2 + \frac{1}{\alpha - 1} \right)^s.$$

By analogy we will prove that if the function $f \in {}_{h(\mathcal{B})}E_s^{\alpha}(c)$, then Fourier-Haar series of this function is absolutely convergent. For each vector $\mathbf{k} = (k_1, \ldots, k_s)$ and each $\mathbf{x} = (x_1, \ldots, x_s) \in [0,1)^s$ from definition of the function $h_{\mathbf{k}}(\mathbf{x})$, definition of $\widetilde{k}$ and Definition 4 we have

$$\sum_{\mathbf{k}=\mathbf{0}}^{\infty} \left| {}_{h(\mathcal{B})}\widehat{f}(\mathbf{k}) h_{\mathbf{k}}(\mathbf{x}) \right| = 1 + \sum_{\mathbf{k} \neq \mathbf{0}}^{\infty} \left| {}_{h(\mathcal{B})}\widehat{f}(\mathbf{k}) h_{\mathbf{k}}(\mathbf{x}) \right| \leq 1 + c(B-1)^s \left(1 + \frac{b^{\alpha-1}}{b^{\alpha-1}-1} \right)^s.$$

Lemma 1. *Let $\xi_N = \{\mathbf{x}_1, \dots, \mathbf{x}_N\}$ be a set of $N \geq 1$ points in $[0,1)^s$.*

(i) If the function $f(\mathbf{x})$, $\mathbf{x} \in [0,1)^s$ is given by absolutely convergent series

$$f(\mathbf{x}) = \sum_{\mathbf{k}=\mathbf{0}}^{\infty} {}_{\chi(\mathcal{B})}\widehat{f}(\mathbf{k})\chi_{\mathbf{k}}(\mathbf{x}),$$

then the error of the quadrature formula (1) $R(\xi_N; f)$ for the function $f(\mathbf{x})$, constructed with respect to the set ξ_N, satisfies the equation

$$R(\xi_N; f) = \frac{1}{N} \sum_{\mathbf{k} \neq \mathbf{0}} {}_{\chi(\mathcal{B})}\widehat{f}(\mathbf{k}) S(\chi(\mathcal{B}); \xi_N; \mathbf{k}), \quad where \quad S(\chi(\mathcal{B}); \xi_N; \mathbf{k}) = \sum_{j=1}^{N} \chi_{\mathbf{k}}(\mathbf{x}_j);$$

(ii) If the function $f(\mathbf{x})$, $\mathbf{x} \in [0,1)^s$ is given by absolutely convergent series

$$f(\mathbf{x}) = \sum_{\mathbf{k}=\mathbf{0}}^{\infty} {}_{h(\mathcal{B})}\widehat{f}(\mathbf{k}) h_{\mathbf{k}}(\mathbf{x}),$$

then the error of the quadrature formula (1) $R(\xi_N; f)$ for the function $f(\mathbf{x})$, constructed with respect to the set ξ_N, satisfies the equation

$$R(\xi_N; f) = \frac{1}{N} \sum_{\mathbf{k} \neq \mathbf{0}} {}_{h(\mathcal{B})}\widehat{f}(\mathbf{k}) S(h(\mathcal{B}); \xi_N; \mathbf{k}), \quad where \quad S(h(\mathcal{B}); \xi_N; \mathbf{k}) = \sum_{j=1}^{N} h_{\mathbf{k}}(\mathbf{x}_j).$$

Corollary 1. *Let $\xi_N = \{\mathbf{x}_1, \dots, \mathbf{x}_N\}$ be a set of $N \geq 1$ points in $[0,1)^s$.*

(i) If the function $f(\mathbf{x}) \in {}_{\chi(\mathcal{B})}E_s^{\alpha}(c)$, then the error of the quadrature formula (1) $R(\xi_N; f)$ for function $f(\mathbf{x})$, constructed with respect to the set ξ_N, satisfies the inequality

$$|R(\xi_N; f)| \leq \frac{c}{N} \sum_{\mathbf{k} \neq \mathbf{0}} \frac{|S(\chi(\mathcal{B}); \xi_N; \mathbf{k})|}{(\overline{k}_1 \dots \overline{k}_s)^{\alpha}}, \quad \mathbf{k} = (k_1, \dots, k_s);$$

(ii) If the function $f(\mathbf{x}) \in {}_{h(\mathcal{B})}E_s^{\alpha}(c)$, then the error of the quadrature formula (1) $R(\xi_N; f)$ for function $f(\mathbf{x})$, constructed with respect to the set ξ_N, satisfies the inequality

$$|R(\xi_N; f)| \leq \frac{c}{N} \sum_{\mathbf{k} \neq \mathbf{0}} \frac{|S(h(\mathcal{B}); \xi_N; \mathbf{k})|}{(\tilde{k}_1 \dots \tilde{k}_s)^{\alpha + \frac{1}{2}}}, \quad \mathbf{k} = (k_1, \dots, k_s).$$

For an arbitrary integer $k \geq 0$ we define the function $\delta_{B_n}(k)$ as

$$\delta_{B_n}(k) = \begin{cases} 1, \ if \ \ k \equiv 0 (mod \ \ B_n) \\ 0, \ if \ \ k \not\equiv 0 (mod \ \ B_n). \end{cases}$$

Lemma 2. *For each vector $\mathbf{k} = (k_1, \dots, k_s)$ with non-negative integer coordinates the equality holds*

$$\sum_{l_1, \dots, l_s = 0}^{B_n - 1} \chi_{\mathbf{k}} \left(\frac{l_1}{B_n}, \dots, \frac{l_s}{B_n} \right) = B_n^s \prod_{i=1}^{s} \delta_{B_n}(k_i).$$

Lemma 3. *Let $k \geq 1$ be an arbitrary integer and $g \geq 0$ is the unique integer, such that $B_g \leq k < B_{g+1}$. Then we have*

$$\textit{(i)}\quad \textit{If } g < n, \quad \textit{then} \quad \sum_{l=0}^{B_n-1} h_k\left(\frac{l}{B_n}\right) = 0;$$

$$\textit{(ii)}\quad \textit{If } g = n, \quad \textit{then} \quad \sum_{l=0}^{B_n-1} h_k\left(\frac{l}{B_n}\right) = \sqrt{B_g};$$

$$\textit{(iii)}\quad \textit{If } g > n, \quad \textit{then} \quad \left|\sum_{l=0}^{B_n-1} h_k\left(\frac{l}{B_n}\right)\right| \leq \sqrt{B_g}.$$

4 Proof of Theorem 1

From Corollary 1 (i) and Lemma 2 for the lattice points set $L(\mathcal{B}(n))$ we obtain

$$|R(L(\mathcal{B}(n));f)| \leq \frac{c}{N}\sum_{\mathbf{k}\neq \mathbf{0}} \frac{|S(\chi(\mathcal{B}); L(\mathcal{B}(n));\mathbf{k})|}{(\overline{k}_1\ldots\overline{k}_s)^\alpha} = c\sum_{\mathbf{k}\neq \mathbf{0}} \frac{\delta_{B_n}(k_1)\ldots\delta_{B_n}(k_s)}{(\overline{k}_1\ldots\overline{k}_s)^\alpha}. \tag{2}$$

From the definition of $\delta_{B_n}(k)$, we obtain that when $k_i \equiv 0(mod\ B_n)$, $\delta_{B_n}(k_i) = 1$ for each i, $1 \leq i \leq s$. For $1 \leq i \leq s$ we put $k_i = B_n t_i$, for an integer $t_i \geq 0$. From here and (2) we obtain

$$|R(L(\mathcal{B}(n));f)| \leq c \sum_{t_1,\ldots,t_s=0}^{\infty}{}' \frac{1}{(\overline{B_n t_1}\ldots\overline{B_n t_s})^\alpha}, \tag{3}$$

where $\sum'$ denotes summing up of all vectors $(t_1,\ldots,t_s) \neq (0,\ldots,0)$. From Definition 3 and (3) we obtain

$$|R(L(\mathcal{B}(n));f)| \leq c\left[\left(1+\frac{1}{B_n^\alpha}\frac{\alpha}{\alpha-1}\right)^s - 1\right]. \tag{4}$$

We have that $B_n^\alpha = N^{\frac{\alpha}{s}}$ and from (4) we obtain

$$|R(L(\mathcal{B}(n));f)| \leq \frac{c\alpha s}{\alpha-1}N^{-\frac{\alpha}{s}} + \mathcal{O}(N^{-\frac{2\alpha}{s}}),$$

so, the first part of Theorem 1 is proved with the constant $C_1(\alpha,c,s) = \dfrac{c\alpha s}{\alpha-1}$.

It is easy to prove the second part of Theorem 1.

5 Proof of Theorem 2

According to Corollary 1 (ii) and Lemma 3 we have the inequality

$$|R(L(\mathcal{B}(n));f)| \leq \frac{c}{N}\sum_{\mathbf{k}\neq \mathbf{0}} \frac{\left|\sum_{l_1=0}^{B_n-1} h_{k_1}\left(\frac{l_1}{B_n}\right)\ldots\sum_{l_s=0}^{B_n-1} h_{k_s}\left(\frac{l_s}{B_n}\right)\right|}{(\tilde{k}_1\ldots\tilde{k}_s)^{\alpha+\frac{1}{2}}}. \tag{5}$$

For each vector $\mathbf{k} = (k_1, \ldots, k_s)$ with non-negative integer coordinates and $\mathbf{k} \neq \mathbf{0}$, without restriction to community of consideration, we will propose that for some integer r, $1 \leq r \leq s$, $k_1 \neq 0, \ldots, k_r \neq 0, k_{r+1} = \ldots = k_s = 0$. From (5) and Definition 4 we obtain

$$|R(L(\mathcal{B}(n)); f)| \leq c \sum_{r=1}^{s} B_n^{-r} \binom{s}{r} \prod_{i=1}^{r} \left[\sum_{k_i=1}^{\infty} \frac{\left| \sum_{l_i=0}^{B_n-1} h_{k_i}\left(\frac{l_i}{B_n}\right) \right|}{\tilde{k}_i^{\alpha+\frac{1}{2}}} \right] \leq$$

$$\leq c \sum_{r=1}^{s} B_n^{-r} \binom{s}{r} \left[\frac{(B-1)b^{\alpha-1}}{b^{\alpha-1}-1} \frac{1}{B_n^{\alpha-1}} \right]^r = \frac{cs(B-1)b^{\alpha-1}}{b^{\alpha-1}-1} \frac{1}{B_n^{\alpha}} + \mathcal{O}\left(\frac{1}{B_n^{2\alpha}}\right). \quad (6)$$

Because $N = B_n^s$, then from (6) the next estimation is finally obtained

$$|R(L(\mathcal{B}(n)); f)| \leq C_2(\alpha, c, s, \mathcal{B}) N^{-\frac{\alpha}{s}} + \mathcal{O}(N^{-\frac{2\alpha}{s}}),$$

where $C_2(\alpha, c, s, \mathcal{B}) = \dfrac{cs(B-1)b^{\alpha-1}}{b^{\alpha-1}-1}$. The proof of Theorem 2 is completed.

References

1. Chrestenson, H.E.: A class of generalized Walsh functions. Pacific J. Math. **5** (1955) 17–31
2. Grozdanov, V.S., Stoilova, S.S.: Haar type functions and uniform distribution of sequences. Comp. ren. Acad. Bul. Sci. **54** No 12 (2001) 23–28
3. Haar, A.: Zür Theorie der orthogonalen Functionensysteme. Math. Ann. **69** (1910) 331–371
4. Korobov, N.M.: Number-theoretical methods in approximate analysis. Fizmatgiz, Moscow, (1963) (in Russian)
5. Larcher,G., Niederreiter, H., Schmid, W.: Digital nets and sequences constructed over finite rings and their application to quasi-Monte Carlo integration.Monatsh. Math. **121,** (1996) 231–253
6. Larcher, G., Pirsic, G.: Base change problems for generalized Walsh series and multivariate numerical integration. Pac. J. Math. **189** No 1 (1999) 75–105
7. Larcher, G., Traunfellner, C.: On the numerical integration of Walsh series by number-theoretic methods. Math. Comp. **63** No 207 (1994) 277–291
8. Larcher, G., Wolf, R.: Nets constructed over finite fields and the numerical integration of multivariate Walsh series. Finite Fields Appl. **2** (1996) 241–273
9. Price, J.J.: Certain groups of orthonormal step functions. Canad. J. Math. **9** No 3 (1957) 413–425
10. Schipp, F., Wade, W.R., Simon, P.: Walsh Series. An Introduction to Dyadic Harmonic Analysis. Adam Hilger, Bristol and New York (1990)
11. Vilenkin, N.Ja.: On a class of complete orthogonal system. Izv. AN SSSR, series math. **11** (1947) 363–400

Numerical Modelling of the Free Film Dynamics and Heat Transfer Under the van der Waals Forces Action

Sonia Tabakova[1] and Galina Gromyko[2]

[1] Department of Mechanics, TU - Sofia, branch Plovdiv,
4000 Plovdiv, Bulgaria
stabakova@hotmail.com

[2] Institute of Mathematics of NAS of Belarus,
220072 Minsk, Belarus
grom@im.bas-net.by

Abstract. In the present work a numerical model of the heat transfer of a hot free thin viscous film attached on a rectangular colder frame is proposed. If the film is cooled down to its solidification temperature the Stefan boundary condition for the heat flux jump is introduced and a part of the liquid film is transformed into a rigid one. The film is assumed to be under the action of the capillary forces and attractive intermolecular van der Waals forces and to be symmetric to a middle plane. Taking the film thickness as a small parameter, the thermal-dynamic problem in its one-dimensional form is solved numerically by a conservative finite difference scheme on a staggered grid.

1 Introduction

Previous investigations on free film dynamics are mainly on periodic or semi-infinite films [5], while in [2] we focus on the film confined laterally by solid boundaries. In these works the van der Waals forces action has a profound effect on film behavior, eventually leading to its rupture. Their influence on the film we studied in [6], where we found that they destabilize also the temperature field. The solidification of a free thin film in its thermostatic approach (without dynamics) is considered in [4] and the interface evolution in time is found for different boundary temperature regimes and process parameters. The current work is an extension of [1], [2] and [6] to include the solidification to the the heat transfer and dynamics of a free thin film attached on a rectangular frame surrounded by an ambient gas. The film is assumed initially hot and cooling by conduction with the colder frame, convection due to the film liquid dynamics and radiation with the colder ambient gas. The one-dimensional form of the thermal-dynamic problem is solved numerically by a finite difference scheme. The numerical results for the film shape, longitudinal velocity and temperature are obtained for different Reynolds numbers, dimensionless Hamaker constants and radiation numbers.

Z. Li et al. (Eds.): NAA 2004, LNCS 3401, pp. 511–518, 2005.

2 Formulation of the Problem

The fluid is assumed Newtonian viscous with constant physical properties: density ρ, dynamic viscosity μ, thermal conductivity κ, heat capacity c and radiation β. The film is thin enough to be under the action of the intermolecular van der Waals forces, while the gravity is neglected. It is supposed symmetrically attached on a rectangular horizontal frame with a stable center plane $z = 0$ in a Cartesian coordinate system (x, y, z) connected with the frame $x = \pm a$, $y = \pm a$. The mean film thickness is εa, such that $\varepsilon << 1$. The free symmetrical surfaces are defined as $z = \pm h/2$, where $h(x, y, t) = O(\varepsilon)$ represents the film shape. The heat transfer is supposed to be due to conduction, convection and radiation on film interfaces with ambient gas, whose temperature is θ_a.

The uncoupled thermal-dynamic system for the film thickness, longitudinal velocity and temperature of order $O(\epsilon)$ is given by [6]:

$$\frac{\partial h}{\partial t} + \nabla_s \cdot (h\mathbf{v}_s) = 0, \tag{2.1}$$

$$\rho\{\frac{\partial \mathbf{v}_s}{\partial t} + \mathbf{v}_s \cdot \nabla_s \mathbf{v}_s\} = \frac{1}{h}\nabla_s \cdot \hat{\mathbf{T}}, \tag{2.2}$$

$$\rho c\{\frac{\partial \theta}{\partial t} + \mathbf{v}_s \cdot \nabla_s \theta\} = \frac{\kappa}{h}\nabla_s(h\nabla_s \theta) + \frac{2\beta}{h}(\theta_a^4 - \theta^4), \tag{2.3}$$

where $\mathbf{v}_s = (u, v)$ is the surface film velocity vector, ∇_s is the surface gradient, θ is the temperature, $\hat{\mathbf{T}} = -\mathbf{P} + \mathbf{T}$ is the surface film stress tensor, $\mathbf{P} = -0.5\sigma\left[h\nabla_s^2 h\mathbf{I_s} + 0.5(\nabla_s h)^2\mathbf{I_s} - \nabla_s h \otimes \nabla_s h\right] + 1.5h\phi$ is the pressure tensor with $\mathbf{I_s}$ standing for the identical surface tensor, $\phi = = A'h^{-3}/(6\pi\rho)$ is the potential function of van der Waals forces (A' is the Hamaker constant) and the viscous stress tensor is given as $\mathbf{T} = 2\mu h\left\{(\nabla_s \cdot \mathbf{v_s})\mathbf{I_s} + 0.5\left[\nabla_s \mathbf{v} + (\nabla_s \mathbf{v})^T\right]\right\}$.

The obtained system (2.1) - (2.3) in dimensionless form is simplified for the one-dimensional case, when the effects in y direction are negligible, to:

$$\frac{\partial h}{\partial t} + \frac{\partial}{\partial x}(uh) = 0, \tag{2.4}$$

$$\frac{\partial u}{\partial t} + u\frac{\partial u}{\partial x} = \frac{\varepsilon}{We}\frac{\partial^3 h}{\partial x^3} + \frac{4}{Re\, h}\frac{\partial}{\partial x}(h\frac{\partial u}{\partial x}) + \frac{A}{h^4}\frac{\partial h}{\partial x}, \tag{2.5}$$

$$\frac{\partial T}{\partial t} + u\frac{\partial T}{\partial x} = \frac{1}{Pe\, h}\frac{\partial}{\partial x}(h\frac{\partial T}{\partial x}) + \frac{Ra}{Pe\, h}(T_a^4 - T^4), \tag{2.6}$$

where $Re = \rho a U/\mu$ is the Reynolds number, $We = ReCa = 2\rho a U^2/\sigma$ – the Weber number, Ca – the capillary number, $A = A'/(2\pi\rho U^2 a^3 \varepsilon^3)$ – the dimensionless Hamaker constant, $Pe = RePr = \rho c a U/\kappa$ – the Peclet number and $Ra = 2\beta a \theta_m^3/\varepsilon\,\kappa$ – the radiation number, θ_m – the solidification temperature.

The system (2.4) - (2.6) is of order $O(\varepsilon)$, then $Re \leq \varepsilon^{-1}$, $We \leq 1$, $A \geq \varepsilon$, $Pe \leq \varepsilon^{-1}$ and $Ra \geq \epsilon Pe$. The boundary and initial conditions for h, u and T are:

$$u(0, t) = u(1, t) = 0, \tag{2.7}$$

$$\frac{\partial T}{\partial x}(0,t) = 0, \qquad T(1,t) = T_1, \tag{2.8}$$

$$\frac{\partial h}{\partial x}(0,t) = 0, \qquad \frac{\partial h}{\partial x}(1,t) = \tan\alpha, \tag{2.9}$$

$$h(x,0) = 1, \qquad u(x,0) = 0, \qquad T(x,0) = T_0, \tag{2.10}$$

where T_1 is the dimensionless frame temperature, $\pi/2 - \alpha$ is the wetting angle with the frame and T_0 is the dimensionless initial temperature of the film. The mass conservation condition of the film

$$\int_0^1 (h-1)dx = 0 \tag{2.11}$$

is satisfied directly, if (2.4) is integrated and (2.7) is considered.

If the liquid film is cooled down to its solidification temperature, then the Stefan boundary condition on the solid/liquid interface point $x_{sl} = \xi(t)$, such that $T(\xi,t) = 1$, is added:

$$c_2\frac{\partial \mathrm{T}}{\partial x}|_s - \frac{\partial \mathrm{T}}{\partial x}|_l = \frac{\mathrm{Pe}}{\mathrm{St}}\cdot\frac{d\xi}{dt}, \tag{2.12}$$

where $\frac{d\xi}{dt}$ is the interface normal velocity, $c_2 = \frac{\kappa_s}{\kappa}$, κ_s is the solid thermal conductivity, $St = \frac{c\theta_m}{L_m}$ is the Stefan number, with L_m latent heat (indices (l) and (s) correspond to solid and liquid phase). In the solid phase the velocity $u = 0$, the film shape does not modify and remains as that of the liquid film before solidification, while the temperature changes by conduction and radiation similarly to (2.6):

$$c_1\frac{\partial T}{\partial t} = \frac{c_2}{Pe\, h}\frac{\partial}{\partial x}(h\frac{\partial T}{\partial x}) + \frac{Ra}{Pe\, h}(T_a^4 - T^4), \tag{2.13}$$

where $c_1 = \frac{\rho_s c_s}{\rho c}$, ρ_s is the solid density and c_s - the solid heat capacity. We assume that the solution of the system (2.4) - (2.13) possesses the required smoothness $u, h, T \subset C^4(\Omega)$ in the problem domain $\Omega = \{0 \le x \le 1\}$ for $t > 0$.

3 Numerical Scheme

The control volumes method is applied as it conserves exactly mass, momentum, heat flux on each group of control volumes and therefore on the whole problem domain. A temporary grid with a variable time step Δt_j is used: $\overline{\Omega_t} = \{t_{j+1} = t_j + +\Delta t_j,\ \Delta t_j > 0,\ \ t_0 = 0\}$. A staggered grid is exploited in space, such that for the functions u and T the grid points are with integer indexes $\overline{\Omega_x^{u,T}} = \{x_i = (i-1)\Delta x,\ \ i = 1, ..., N+1;\ \ x_{N+1} = 1\}$, while for the function h the grid points are with half integer indexes $\overline{\Omega_x^h} = \{x_{i-0.5} = (i-0.5)\Delta x,\ \ i = 1, ..., N+1;\ \ x_{N+1} = 1 - 0.5\Delta x\}$.

Integrating (2.4) on the control volume $[x_{i-1}, x_i]$ we have [2]:

$$h_t + (\Delta M - \Delta M_-)/\Delta x = 0, \quad i = \overline{2, N}, \tag{3.1}$$

where $\Delta M = \Delta M_i =< h >_i< u >_i$. The notation $<>$, as in [3], means that the values of h and u are on the current cell border.

For big Re and Pe corresponding to strong convection dominance, we reach to singular problems, characterized with regions (boundary layers) of strong changes of solution. Therefore, for such cases particular difference schemes are developed as in [2] and [6], which we utilize for velocity and temperature in the liquid film:

$$u_{\bar{t}} + 0.5(u_{+0.5}u)_{\bar{x}} = (\tilde{\mu}_{+0.5}\, u_x)_{\bar{x}} + (\varepsilon/We)h_{\bar{x}x\dot{x}} - (A/3)(h^{-3})_{\dot{x}}\,, \tag{3.2}$$

$$(hT)_{\bar{t}} + (h_{+1}u_{+0.5}T)_{\bar{x}} = (\tilde{\varphi}_{+0.5}\, T_x)_{\bar{x}} + \frac{Ra}{Pe}(T_a^4 - T^4), \tag{3.3}$$

where $\tilde{\mu}_{+0.5} = \mu_{+0.5}F_B(R_{+0.5})$, $\mu_{+0.5} = 4h_{+1}/(Re\,\bar{h})$, $\bar{h} = 0.5(h+h_{+1})$, $R_{+0.5} = 0.125\;Re\;\Delta x\;u_{+0.5}\bar{h}/h$ and the function F_B is a piecewise linear approximation of $R/(\exp R - 1)$, $\tilde{\varphi}_{+0.5} = h_{+1}\;F_B(S_{+0.5})/Pe$, $S_{+0.5} = \Delta x\;u_{+0.5}\;Pe$.

The boundary condition (2.8) on $x = 0$ is approximated as:

$$(hT)_{\bar{t},0} + 2h_1u_{0.5}T_0/\Delta x = 2\tilde{\varphi}_{0.5}\;T_{\bar{x},1}/\Delta x + \frac{Ra}{Pe}(T_a^4 - T_0^4). \tag{3.4}$$

For the solid film in the domain $[x_{sl}^1, x_{sl}^2]$, in which $u = 0$, we exploited an implicit difference scheme:

$$c_1(hT)_{\bar{t}} = \frac{c_2}{Pe}(h_{+1}T_x)_{\bar{x}} + \frac{Ra}{Pe}(T_a^4 - T^4). \tag{3.5}$$

The boundaries x_{sl}^1 and x_{sl}^2 are found from (2.12):

$$x_{sl}^1 = \check{x}_{sl}^1 - \frac{St}{Pe}(c_2T_x - T_{\bar{x}})|_{x_{sl}^1}, x_{sl}^2 = \check{x}_{sl}^2 - \frac{St}{Pe}(c_2T_{\bar{x}} - T_x)|_{x_{sl}^2}. \tag{3.6}$$

Iteration procedures based on Thomas algorithm are performed for h, u and T as described in details in [2] and [6] for the liquid film. If we denote the iteration number with s, then the iteration process for interface position and solid film temperature distribution is:

$$c_1(\overset{s+1}{h}\overset{s+1}{T})_{\bar{t}} = \frac{c_2}{Pe}(\overset{s+1}{h_{+1}}\overset{s+1}{T_x})_{\bar{x}} + \frac{Ra}{Pe}(T_a^4 + 3\overset{s}{T}{}^4 - 4\overset{s}{T}{}^3\overset{s+1}{T}) \tag{3.7}$$

$$\overset{s+1}{x}{}^1_{sl} = \check{x}_{sl}^1 - \frac{St}{Pe}(c_2\overset{s}{T}_x - \overset{s}{T}_{\bar{x}})|_{\overset{s}{x}{}^1_{sl}}, \overset{s+1}{x}{}^2_{sl} = \check{x}_{sl}^2 - \frac{St}{Pe}(c_2\overset{s}{T}_{\bar{x}} - \overset{s}{T}_x)|_{\overset{s}{x}{}^2_{sl}}. \tag{3.8}$$

As an initial approximation we take the solution from the preceding time level: $\overset{0}{T} = \check{T}$ and $\overset{0}{x}{}^k_{sl} = \check{x}_{sl}^k$, $k = 1, 2$. The iteration process is performed till reaching some convergence criterium.

Since the cooling process starts when the film is fluid, the interface arises when the the temperature becomes $T_{i^*} \le T_m$ in some point x_{i^*}. Depending on the process parameters, i.e., on Re, the problem can possess 1 or 2 interface points.

The iteratiion process is finished when we reach steady solutions for u and T or when the film thickness $h \to h_{cri}$, where $h_{cri} > 0$ is a critical thickness, at which the actual rupture of the film occurs and in its region the problem has a singularity.

4 Numerical Results

Since in the present work we examine the influence of the inertial and van der Waals forces on the heat transfer and solidification, we performed our numerical calculations based on the described numerical scheme for a large range of Re ($1 \leq Re \leq 100$) and dimensionless Hamaker constant, A ($0.1 \geq A \geq 0$), radiation numbers Ra ($0 \leq Ra \leq 100$) and constant capillary number Ca, i.e., Weber number, We depends only on Re ($We = CaRe$). The presented numerical results are found at constant values of $Ca = \varepsilon = 0.01$, $\alpha = 1.37$, $Pr = 1$, $St = 2.8$, $c_1 = 1.024$ and $c_2 = 1.072$ (the data are due to [4]). Then $Pe = Re$, since $Pe = Re.Pr$. In our calculations we used $\Delta x = 0.02$, $\Delta t = 10^{-5}$ and $h_{cri} = 10^{-5}$.

In Fig.1 the evolution in space and time of the film thickness h, longitudinal velocity u and temperature T are given at $A = 0.1, Re = Pe = 1, We = 0.01, T_0 = 1.19, T_1 = 1, T_a = 0.9, Ra = 1$. The same case has been studied only

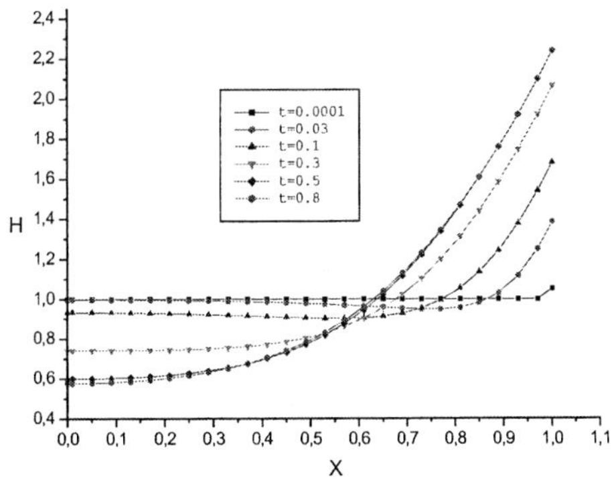

Fig. 1. Evolution of h, u and T at $A = 0.1$, $Re = Pe = 1$, $We = 0.01$, $St = 2.8$, $T_0 = 1.19$, $T_1 = 1, T_a = 0.9, Ra = 1$, $\alpha = 1,37$: a) h(x,t)

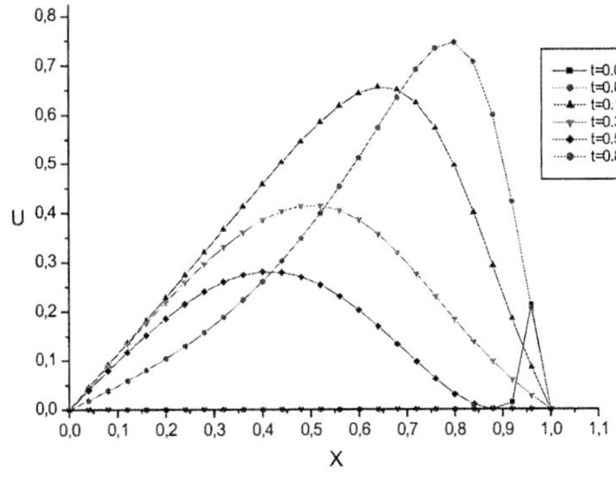

Fig. 1. b) u(x,t)

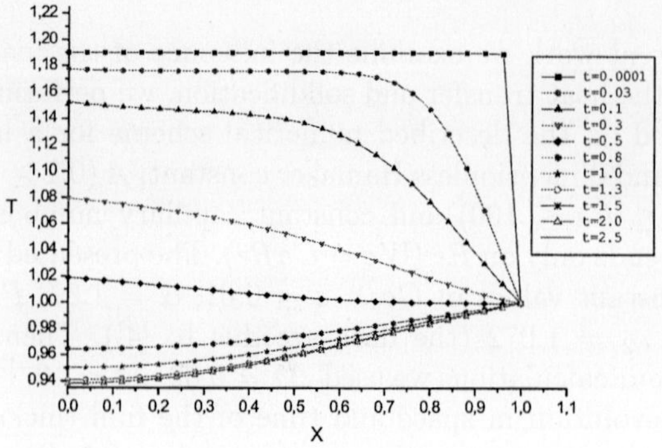

Fig. 1. c) T(x,t)

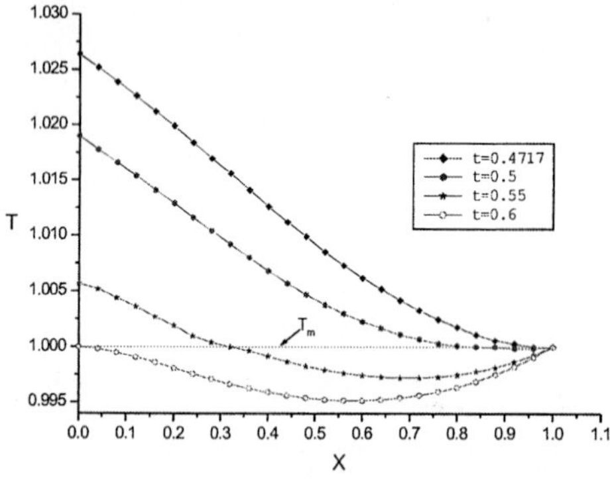

Fig. 1. d) T(x,t) only for solid/liquid film

dynamically in [2], where the rupture is observed at $x = 0$ for a finite time $t = 2.4$. As it is seen from Fig.1, the solidification starts at $t = 0.4717$ and at $t = 0.6$ the whole film is solid, but if continues to cool down till reaching a steady temperature distribution at $t = 2.5$. For bigger Ra the solidification mechanism is similar, but more rapid, e.g., for $Ra = 10$ the solidification starts at $t = 0.051$ and at $t = 0.079$ the whole film is solid, but cools till reaching a steady temperature distribution at $t = 0.5$.

For bigger Re ($Re > 1$) the velocity and thickness solutions have big amplitudes and the rupture point moves from the center $x = 0$ towards the frame point $x = 1$. In the case of $Re = 100$, $We = 1$ and $A = 0.1$ the rupture is achieved for time $t = 1.757$ in the point $x = 0.53$ [2]. As in the previous case, the solidification starts before film rupture and for example at $T_0 = 1.19, T_1 = 1, T_a = 0.9$ and $Ra = 100$, this is at $t = 0.65$ and finishes at $t = 0.9$, while the steady temperature regime is achieved at $t = 1.33$.

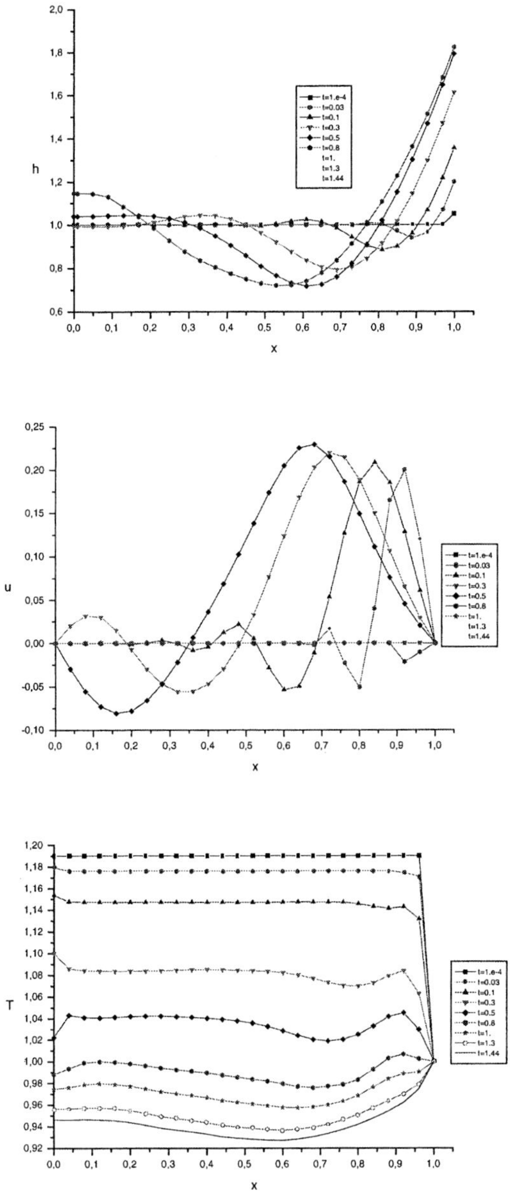

Fig. 2. Evolution of h, u and T at $A = 0.1$, $\alpha = 1.37$, $Re = Pe = 100$, $We = 1.$, $St = 2.8$, $T_0 = 1.19$, $T_1 = 1, T_a = 0.9, Ra = 1$: a) h(x,t); b) u(x,t) and c) T(x,t)

5 Conclusions

This work is a continuation of authors series of papers [1], [2] and [6], dealing with the dynamics and heat transfer of a free thin film attached on a rectangular frame, and take into the account the solidification problem, i.e., Stefan problem. The film dynamics is significantly influenced by the van der Waals forces, which make the problem singular when the film is near to rupture. The 1D nonlinear thermal - dynamic problem is solved numerically by a conservative difference scheme on a staggered space grid.

Numerical results for the film thickness, velocity and temperature evolution are obtained for different Reynolds numbers, Re ($1 \leq Re \leq 100$), dimensionless Hamaker constant, A ($0 \leq A \leq 0.1$) and radiation numbers, Ra ($1 \leq Ra \leq 10$). At $Re = O(1)$ only one interface point is observed that moves in time from the frame towards center. However, at higher Re, since the solidification is affected highly by the film dynamics, two interface points are observed.

References

1. L.Popova, G.Gromyko, S. Tabakova, Numerical modeling of free thin film dynamics, Mathematical Modelling and Analysis 8 (1), 2003, 291 – 311.
2. L.Popova, G.Gromyko, S. Tabakova, Numerical Solution of a Nonlinear One-Dimensional System of Differential Equations Describing the Motion of a Free Thin Film, Differential Equations 39 (7), 2003, 1037 – 1043.
3. Yu.I. Shokin, N.N. Yanenko Method differential approximation, Novosibirsk, Nauka, 1985.
4. S. Tabakova, L. Carotenuto, Numerical modeling of the solidification of free thin films, Microgravity Quarterly 4 (1), 1994, 55 – 61.
5. D. Vaynblat, J. R. Lister, T. P. Witelski, Rupture of thin viscous films by van der Waals forces: Evolution and self-similarity, Phys. Fluids 13 (5), 2001, 1130 – 1140.
6. G.G. Gromyko, S. Tabakova, L. Popova, On the cooling of a free thin film at the presence of the van der Waals forces, Mathematical Modelling and Analysis (accepted).

Two Resultant Based Methods Computing the Greatest Common Divisor of Two Polynomials

D. Triantafyllou and M. Mitrouli

Department of Mathematics, University of Athens,
Panepistimiopolis 15784, Athens, Greece
dtriant@math.uoa.gr, mmitroul@cc.uoa.gr

Abstract. In this paper we develop two resultant based methods for the computation of the Greatest Common Divisor (GCD) of two polynomials. Let S be the resultant Sylvester matrix of the two polynomials. We modified matrix S to S^*, such that the rows with non-zero elements under the main diagonal, at every column, to be gathered together. We constructed modified versions of the LU and QR procedures which require only the $\frac{1}{3}$ of floating point operations than the operations performed in the general LU and QR algorithms. Finally, we give a bound for the error matrix which arises if we perform Gaussian elimination with partial pivoting to S^*. Both methods are tested for several sets of polynomials and tables summarizing all the achieved results are given.

Keywords: Greater Common Divisor, Sylvester matrix, Gaussian elimination, QR factorization.

1 Introduction

The computation of the greatest common divisor (GCD) of two or more polynomials is one of the most frequent problems in several fields such as numerical analysis, linear and numerical linear algebra, control theory, matrix theory, statistics etc. Many numerical algorithms have been created to solve this problem [1],[5]. We can classify these algorithms in two categories [4] :

- Numerical methods that are based on Euclid's algorithm.
- Numerical methods that are based on procedures involving matrices.

A resultant based computation of the greatest common divisor of two polynomials belongs to the second of the above categories. We can use either non-orthogonal or orthogonal algorithms in order to find the GCD. Non-orthogonal algorithms are faster but we can not prove their stability in contrast to orthogonal algorithms that are slower but stable. In practice there are only few cases in which non-orthogonal algorithms give wrong results. Except for the previous dilemma there is another one : We know that during numerical operations on a floating point arithmetic rounding off errors (catastrophic cancellation) are

Z. Li et al. (Eds.): NAA 2004, LNCS 3401, pp. 519–526, 2005.

caused. On this account we introduce a numerical accuracy as we will see below. Finally there are also methods which compute the approximate GCD.

Let R[s] be the ring of real polynomials [3], $\Re^{m \times n}$ the set of all mxn real matrices, r(A) the rank of the matrix $A \epsilon \Re^{m \times n}$ and ϑ\{f(s)\} the degree of a polynomial. Consider two polynomials a(s),b(s)$\epsilon R[s]$, ϑ\{a(s)\}=m, where a(s) is a monic polynomial and ϑ\{b(s)\}=n, with $n \leq$m,where [1]

$$\begin{aligned} a(s) &= s^m + a_{m-1}s^{m-1} + \ldots + a_1 s + a_0 \\ b(s) &= b_n s^n + b_{n-1}s^{n-1} + \ldots + b_1 s + b_0 \end{aligned}$$

We define the resultant S(a,b) of the two polynomials by

$$S(a,b) = \begin{bmatrix} 1 & a_{m-1} & a_{m-2} & \cdots & \cdots & \cdots & \cdots & a_0 & 0 & \cdots & \cdots & 0 & 0 \\ 0 & 1 & a_{m-1} & a_{m-2} & \cdots & \cdots & \cdots & \cdots & a_0 & 0 & \cdots & 0 & 0 \\ . & . & . & \cdots & \cdots & \cdots & \cdots & \cdots & . & \cdots & \cdots & . & . \\ 0 & 0 & 0 & \cdots & 1 & a_{m-1} & . & . & . & . & \cdots & a_1 & a_0 \\ - & - & - & - & - & - & - & - & - & - & - & - & - \\ b_n & b_{n-1} & b_{n-2} & \cdots & \cdots & b_0 & 0 & \cdots & \cdots & \cdots & \cdots & 0 & 0 \\ 0 & b_n & b_{n-1} & \cdots & \cdots & \cdots & b_0 & 0 & \cdots & \cdots & \cdots & 0 & 0 \\ . & . & . & \cdots & \cdots & \cdots & \cdots & \cdots & \cdots & \cdots & \cdots & . & . \\ 0 & 0 & 0 & \cdots & \cdots & \cdots & \cdots & b_n & b_{n-1} & \cdots & \cdots & b_1 & b_0 \end{bmatrix}$$

Obviously $S(a,b) \epsilon \Re^{(n+m)\times(n+m)}$. Applying elementary row operations with scalars from $\Re$ to matrix S(a,b) we transform it to an upper triangular form (this can be managed to using e.g. gaussian or orthogonal transformations). The existence, the degree and the coefficients of the GCD are arising from the following theorem :

Theorem 1. *[3] Let a(s),b(s)$\in$R[s], ϑ\{a(s)\}=m, ϑ\{b(s)\}=n,m $\geq$n and let $z(s) = s^r + z_{r-1}s^{r-1} + \ldots + z_1 s + z_0$ be the GCD. The following properties hold true:*

(i) (a(s),b(s)) are coprime, if and only if rank(S(a,b))=n+m.
(ii) r=ϑ\{z(s)\}=m+n-rank(S(a,b)).
(iii) If $S_H(a,b)$ is the row echelon form of S(a,b), then the last non-vanishing row gives the coefficients of GCD of (a(s),b(s)). □

In the previous matrix S we observe that its first column has only two non-zero elements. We interchange the second and the (n+1)-th row in order to collect the two non-zero elements of the first column in the two first rows of S. Afterwards only three elements of the second column of S are non-zero. Interchanging the corresponding rows, we collect them to the three first rows of S. The number of non-zero elements under the diagonal will be increasing per one at every column until the deg\{b(x)\}-1 column. From the next column and until the m+n-deg\{b(x)\} column the number of non-zero elements is constant and equal to deg\{b(x)\}+1. Finally at the remaining deg\{b(x)\}-1 columns the number of non-zero elements will be decreasing per one. With the corresponding row interchanges we collect at every column all those rows that have non-zero entries under the diagonal.

The modified Sylvester matrix will have the following form:

$$S^*(a,b) = \begin{bmatrix} 1 & a_{m-1} & a_{m-2} & a_{m-3} & \cdots & a_{m-n} & a_{m-n-1} & a_{m-n-2} & \cdots & a_0 & 0 & 0 & \cdots & 0 \\ b_n & b_{n-1} & b_{n-2} & b_{n-3} & \cdots & b_0 & 0 & 0 & \cdots & 0 & 0 & 0 & \cdots & 0 \\ 0 & 1 & a_{m-1} & a_{m-2} & \cdots & a_{m-n+1} & a_{m-n} & a_{m-n-1} & \cdots & a_1 & a_0 & 0 & \cdots & 0 \\ 0 & b_n & b_{n-1} & b_{n-2} & \cdots & b_1 & b_0 & 0 & \cdots & 0 & 0 & 0 & \cdots & 0 \\ 0 & 0 & 1 & a_{m-1} & \cdots & a_{m-n+2} & a_{m-n+1} & a_{m-n} & \cdots & a_2 & a_1 & a_0 & \cdots & 0 \\ 0 & 0 & b_n & b_{n-1} & \cdots & b_2 & b_1 & b_0 & \cdots & 0 & 0 & 0 & \cdots & 0 \\ . & . & . & . & \cdots & . & . & . & \cdots & . & . & . & \cdots & . \\ 0 & 0 & 0 & 0 & \cdots & 0 & 1 & a_{m-1} & \cdots & a_{m-n+1} & a_{m-n} & a_{m-n-1} & \cdots & a_0 \\ 0 & 0 & 0 & 0 & \cdots & 0 & b_n & b_{n-1} & \cdots & b_1 & b_0 & 0 & \cdots & 0 \\ 0 & 0 & 0 & 0 & \cdots & 0 & 0 & b_n & \cdots & b_2 & b_1 & b_0 & \cdots & 0 \\ . & . & . & . & \cdots & . & . & . & \cdots & . & . & . & \cdots & . \\ . & . & . & . & \cdots & . & . & . & \cdots & . & . & . & \cdots & . \\ 0 & 0 & 0 & 0 & \cdots & 0 & 0 & 0 & \cdots & 0 & 0 & b_n & \cdots & b_0 \end{bmatrix}$$

2 The Resultant LU Method

Corollary 1. *Let $S(a,b)$ be the resultant matrix of the pair of polynomials $(a(s), b(s))$, ρ=rank($S(a,b)$) and let $S_1(a,b)$ denote the upper triangular form of $S(a,b)$ obtained under the Gauss row transformations [3], i.e.*

$$S_1(a,b) = \begin{bmatrix} x & x & \cdots & \cdots & x & \cdots & \cdots & x \\ 0 & x & \cdots & \cdots & x & \cdots & \cdots & x \\ . & . & \cdots & \cdots & . & \cdots & \cdots & . \\ . & . & \cdots & \cdots & . & \cdots & \cdots & . \\ . & . & \cdots & \cdots & . & \cdots & \cdots & . \\ 0 & 0 & \cdots & \cdots & x & \cdots & \cdots & x \\ 0 & 0 & \cdots & \cdots & 0 & \cdots & \cdots & 0 \\ . & . & \cdots & \cdots & . & \cdots & \cdots & . \\ . & . & \cdots & \cdots & . & \cdots & \cdots & . \\ 0 & 0 & \cdots & \cdots & 0 & \cdots & \cdots & 0 \end{bmatrix}$$

where the x leading element of each nonzero row is also nonzero. The nonzero elements of the last nonzero row of $S_1(a,b)$ define the coefficients of GCD in reverse order. □

These results hold for the modified matrix S^* too.

As we have seen, the Sylvester matrix S has a specific form. We can take advantage of this form, during the procedure of triangulation. More particular, because the Sylvester matrix has many zeros in many columns, we can zero only the non-zero elements of S. After we produce the Sylvester matrix, we know exactly how many non-zero elements has each column of S. So, we transform the Sylvester matrix S to the modified matrix S^*. Then the nullification of the elements at k step in column k, is not necessary to take place to the whole sub-column (i=k+1:m+n), but only to the first s elements (i=k+1:k+s) since the other elements are already zero (where s is the number of non-zero elements under the diagonal)(i=r:s denotes i=r,..., s). The required row interchanges do not affect the final result, the greater common divisor. After we zero the elements under the diagonal at stage k, we must update the elements in submatrix $S^*_{k+1:m+n,k+1:m+n}$. But the big advantage is that at row k the non-zero elements are until the column $[\frac{k+1}{2}]$+deg{a(x)} for the first 2·deg{b(x)} rows and until the deg{a(x)}+deg{b(x)} column for the last ones. So the part of the matrix that we must update is only the submatrix k+1:s,k+1:P_k at k-stage, where the element P_k of matrix P is the number of the column, that after it we have only zeros at row k.

The number of non-zero elements under the diagonal will be increasing per one at every step until the deg{b(x)}-1 step. From the next step and until deg{a(x)} step the number of non-zero elements is constant and equal to deg{b(x)}. Finally at the remaining deg{b(x)}-1 steps the number of non-zero elements will be decreasing per one at every step. We mention that because of the special form of S^* only few elements (which are adjoining) are needed to be zero at every step. Also not the entire (m+n-k+1)x(m+n-k+1) submatrix at stage k must be updated. These two observations will decrease a lot the complexity.

So, we can also modify the LU Factorization with considerable reduction of flops. It follows a non-orthogonal algorithm using Gaussian transformation (with partial pivoting to improve numerical stability) :

Algorithm Modified LU Factorization

m=deg{a(x)}+1
n=deg{b(x)}+1
Construct matrix P
Comment : S is of order m+n
Comment : p is the number of non zero elements which are under the diagonal and which we will zero
p=0
for k=1:deg{b(x)}-1
  p=p+1
  Find r : $|S_{r,k}| = max_{k+1 \leq i \leq k+p}\{|S_{i,k}|\}$
  Interchange rows k and r
  $m_{ik} = s_{ik}/s_{kk}$, i=k+1:k+p
  $s_{ij} = s_{ij} - m_{ik}s_{kj}$, i=k+1:k+p, j=k+1:$P_k$
  Set $s_{i,j} = 0$ if $|s_{i,j}| \leq$ accuracy , i=k:m+n , j=k:m+n

Repeat the previous procedure for k=deg{b(x)} : *deg*{a(x)} and for constant elements p=deg{b(x)} (under the diagonal which we must zero).

Repeat the previous procedure for k=deg{a(x)}+1:deg{a(x)}+deg{b(x)}-1 and for decreasing per one in every step elements p, starting from p=deg{b(x)} (the elements under the diagonal which we must zero).

Complexity
The complexity of the previous algorithm is $O\ (\frac{5}{6}n^3)$ flops which is much fewer than the $O\ (\frac{8}{3}n^3)$ flops [2] of the general case for a $2n \times 2n$ matrix (we have taken the worse case : m=n=max$\{m, n\}$.

Error Analysis
As we have mentioned we could have catastrophical results during the numerical operations in a floating point arithmetic. Is the previous algorithm stable? The following theorem refers to the stability of this method.

Theorem 2. *Let S(a,b) be a given matrix of order k. If we perform Gaussian elimination with partial pivoting using floating point arithmetic with unit round off u and $S^{'}(a,b)$ is the resultant upper triangular matrix, the following relation holds :*

$$L \cdot S^{'}(a,b)=S(a,b)+E,\ \|E\|_\infty \leq (n^2+2n)\cdot \rho u \cdot \|S(a,b)\|_\infty + (\tfrac{n^2}{2}+\tfrac{n}{2})\varepsilon_G$$

where L a lower triangular matrix with units on the diagonal, $\rho = \frac{max_{i,j,k}|s_{ij}^{(k)}|}{max_{i,j}|s_{ij}|}$ is the growth factor of the Gaussian elimination and ε_G is the accuracy of gaussian elimination (elements with absolute value less than ε_G will be set equal to zero during the gaussian elimination). □

So, $L\cdot S^{'}=S+E$, where $E = E^{(1)} + E^{(2)} + \ldots E^{(2n-1)}$ is the sum of the errors at every stage and it is bounded from the previous quantity.

We can use B-scaling at every stage after the first step of the LU factorization in order to keep the absolute values of the elements less or equal to 1, to avoid numerical errors. So the growth factor ρ will be less or equal to 1 and the final error will be :

$$\|E\|_\infty \leq (n^2+2n)u\|S\|_\infty + (\tfrac{n^2}{2}+\tfrac{n}{2})\varepsilon_G.$$

Remark1
Adjusting the error analysis of Gauss to the specific form of matrix S(a,b) we attain as error bound n^2 instead of $4n^2$ that appear if we directly applied the formulas of the general error analysis to an 2n×2n matrix. □

Remark2
Using the above error analysis it can be proved that instead of computing the GCD of a(x) and b(x) we compute the GCD of $a(x)^{'}$ and $b(x)^{'}$, where

$$a(x)^{'} = x^n + a^{'}_{n-1}x^{n-1} + a^{'}_{n-2}x^{n-2} + \ldots + a^{'}_3x^3 + a^{'}_2x^2 + a^{'}_1x + a^{'}_0$$

$$b(x)^{'} = b^{'}_nx^n + b^{'}_{n-1}x^{n-1} + b^{'}_{n-2}x^{n-2} + \ldots + b^{'}_3x^3 + b^{'}_2x^2 + b^{'}_1x + b^{'}_0$$

where $a^{'}_i = a_i + \varepsilon_i$, with $|\varepsilon_i| \leq$(2n-2)gu+(n-1)ε_G , $b^{'}_i = b_i + \varepsilon^{'}_i$, with $|\varepsilon^{'}_i| \leq$2ngu +nε_G. □

3 The Resultant QR-Method

We can accommodate the QR factorization to the previous special form of S^*, reducing thereby the number of the needed flops sufficiently. Every row has some nonzero elements after the diagonal. After these elements there are only zeros. Therefore we can update only the nonzero elements, which are less or equal to m+1 or n+1 in every row. The number of the elements which we must zero at every stage is the same as in the LU-case. After these observations the complexity of the triangulation of S is decreasing a lot. It follows the algorithm of the modified QR factorization:

Algorithm Modified QR Factorization

m=deg{a(x)}+1 , n=deg{b(x)}+1
Construct matrix P
Comment : S is of order m+n and q is the number of non zero elements which are under the diagonal and which we will zero
q=0
for k=1:deg{b(x)}-1
  q=q+1
  $u_{k:m+n} = zeros_{1,m+n-k+1}$
  [u,t]=house1($S_{k:k+q-1,k}$)
  w=$u_{k:m+n}$
  s_{kk}=t
  $r = \sum_{i=1}^{q} w_i^2$
  $b = \frac{2}{r}$
  sum=$\sum_{i=k}^{k+q-1} u_i \cdot s_{i,j}$, j=k+1:k+P_k
  r=b*sum
  $s_{i,j} = s_{i,j} - r \cdot u_i$, i=k:k+q-1, j=k+1:k+P_k
  Set $s_{i,j} = 0$ if $|s_{i,j}| \leq$ accuracy , i=k:m+n , j=k:m+n

Repeat the previous procedure for k=deg{b(x)}:n-deg{v(x)} and for constant elements q=deg{b(x)}+1 (under the diagonal which we must zero).

Repeat the previous procedure for k=n-deg{b(x)}+1:n-1 and for decreasing per 1 elements in every step, starting from q=deg{b(x)}+1 (the elements under the diagonal which we must zero).

where house1 [2] zeros the entries of the vector x after the first element :

m=max$|x_i|$, i=1,. . .,n
$x_i \equiv u_i = \frac{x_i}{m}$, i=1,. . .,n
t=sign$(u_1)\sqrt{u_1^2 + u_2^2 + \ldots + u_n^2}$
$x_1 \equiv u_1 = u_1 + t$
t=-m·t

Complexity
The previous algorithm consists of three QR-parts. The only difference of each one is the number of elements that is needed to zero. In first part, the number of non-zero elements under the diagonal at step 1 is equal to 1 and it increases until it catches the deg{b(x)}-1. In second part, the number of non-zero elements under the diagonal is constant and equal to deg{b(x)}-1 for n-2deg{b(x)}+1 steps and in third part, the number of non-zero elements under the diagonal is equal to deg{b(x)}-1 and decreases until the triangulation has been completed.

Finally : flops(part1)+flops(part2)+flops(part3)=$O\left(\frac{5}{3}n^3\right)$ flops, which is less enough than the $O(\frac{16}{3}(n)^3)$ flops that requires the classical QR factorization of an $2n \times 2n$ (we have taken the worse case : m=n=max$\{m, n\}$). The housel algorithm requires only $O(4(n+1))$ flops.

4 Numerical Results

Numerical results, which were arised applying the classical LU and QR factorization and the modified versions of them to polynomials, are given next. In the tables we compare the four methods. Let ε_{ac} be the accuracy of the Gaussian or QR factorization. This means that values less than ε_{ac} are supposed to be zero.

Example 1

$a(x) = x^7 - 28x^6 + 322x^5 - 1960x^4 + 6769x^3 - 13132x^2 + 13068x - 5040$
$b(x) = x^4 - 28x^3 + 269x^2 - 962x + 720$
$GCD = x - 1$

Method	Relative Error	ε_{ac}	flops
LU	0	10^{-12}	1321
modified LU	0	10^{-12}	561
QR	0	10^{-10}	2298
modified QR	0	10^{-10}	1218

Example 2 [5]

$a(x) = x^{16} - 30x^{15} + 435x^{14} - 4060x^{13} + 27337x^{12} - 140790x^{11} + 573105x^{10} - 1877980x^9 + 4997798x^8 - 10819380x^7 + 18959460x^6 - 26570960x^5 + 29153864x^4 - 24178800x^3 + 14280000x^2 - 5360000x + 960000$
$b(x) = x^{14} - 140x^{12} + 7462x^{10} - 191620x^8 + 2475473x^6 - 15291640x^4 + 38402064x^2 - 25401600$
$GCD = x^4 - 10x^3 + 35x^2 - 50x + 24$

Method	Relative Error	ε_{ac}	flops
LU	$7.6194 \cdot 10^{-9}$	10^{-1}	26341
modified LU	$7.6194 \cdot 10^{-9}$	10^{-1}	6797
QR	$8.0689 \cdot 10^{-7}$	10^{-3}	39278
modified QR	$8.0689 \cdot 10^{-7}$	10^{-3}	17235

Example 3

$a(x) = x^{16} - 13.6x^{15} + 85x^{14} - 323.68x^{13} + 839.402x^{12} - 1569.52x^{11} + 2185.03x^{10} - 2305.72x^9 + 1859.53x^8 - 1146.9x^7 + 537.452x^6 - 188.616x^5 + 48.366x^4 - 8.70777x^3 + 1.02992x^2 - 0.0707343x + 0.00209228$
$b(x) = x - 0.1$
$GCD = x - 0.1$

Method	Relative Error	ε_{ac}	flops
LU	0	10^{-4}	4801
modified LU	0	10^{-4}	496
QR	0	10^{-4}	7724
modified QR	0	10^{-4}	1568

Example 4

$a(x) = x^{16} - 13.6x^{15} + 85x^{14} - 323.68x^{13} + 839.402x^{12} - 1569.52x^{11} + 2185.03x^{10} - 2305.72x^9 + 1859.53x^8 - 1146.9x^7 + 537.452x^6 - 188.616x^5 + 48.366x^4 - 8.70777x^3 + 1.02992x^2 - 0.0707343x + 0.00209228$
$b(x) = x^4 - 2.6x^3 + 1.31x^2 - 0.226x + 0.012$
$GCD = x^3 - 0.6x^2 + 0.11x + 0.006$

Method	Relative Error	ε_{ac}	flops
LU	$2.5663 \cdot 10^{-6}$	10^{-6}	7772
modified LU	$2.5663 \cdot 10^{-6}$	10^{-6}	2096
QR	$2.8106 \cdot 10^{-6}$	10^{-6}	12237
modified QR	$2.9524 \cdot 10^{-6}$	10^{-6}	4078

References

1. S. Barnet, Greatest Common Divisor from Generalized Sylvester Resultant Matrices, *Linear and Multilinear Algebra*, 8 (1980), 271-279.
2. B.N. Datta, Numerical Linear Algebra and Applications, Second Edition, Brooks/Ccole Publishing Company, United States of America, 1995.
3. N. Karcanias, M. Mitrouli and S. Fatouros, A resultant based computation of the GCD of two polynomials, *Proc. of 11th IEEE Mediteranean Conf. on Control and Automation, Rodos Palace Hotel, MED'03, June 18-20, Rhodes, Greece.*
4. M. Mitrouli, N. Karcanias and C. Koukouvinos, Further numerical aspects of ERES algorithm for the computation of the GCD of polynomials and comparison with other existing methodologies, *Utilitas Mathematica*, 50 (1996), 65-84.
5. I.S. Pace and S. Barnet, Comparison of algorithms for calculation of GCD of polynomials, *International Jornal of System Science*, 4 (1973), 211-226.

Conservative Difference Scheme for Summary Frequency Generation of Femtosecond Pulse

Vyacheslav A. Trofimov, Abdolla Borhanifar, and Alexey G. Volkov

Lomonosov Moscow State University,
Department of Computational Mathematics & Cybernetics,
Leninskye Gory, Moscow 119992, Russia
vatro@cs.msu.su

Abstract. A Conservative difference scheme is proposed for a problem of three femtosecond pulses interaction in medium with quadratic nonlinearity. Second order dispersion and dispersion of nonlinear response are taking into account under the consideration of femtosecond pulses propagation in nonlinear optical fiber.

1 Introduction

An analysis of femtosecond pulse propagation in nonlinear medium represents the great interest for various scientific and applied problems [1-3] in connection with unique properties of such pulse: extremely short duration and possibility to achieve laser light intensity, which on some orders greater than intensity of intrinsic electric field of atom. As it is known, nonlinear propagation of femtosecond pulses is described by so-called combined nonlinear Schrodinger equation (CNLSE). It differs from usual nonlinear Schrodinger equation (NLSE) by presence of derivative on time from nonlinear response of medium (dispersion of the nonlinear response). Presence of this derivative leads, in particular, to formation of optical shock waves, restriction of maximal intensity, light pulse compression under the propagation in medium with Kerr nonlinearity [4].

It is necessary to emphasize, that the proved difference schemes for the considering problem are practically absent in the literature . It has been caused, in particular, by absence of invariants (conservation laws) for CNLSE till recent time. The given blank in consideration of the equation has been filled in [4-6]. Obtained invariants allow to use a principle of conservatism [7] for construction of difference schemes with reference to this class of problems. So, in [8-9] comparison of various approaches to description of three pulses interaction without consideration into account a second order dispersion is carried out. In the present report we consider the similar problem, but taking into account the second derivative on time from complex amplitudes of interacting pulses.

2 Basic Equations

Light pulse propagation in optical fiber is described by the following wave equation $[1-3]$

Z. Li et al. (Eds.): NAA 2004, LNCS 3401, pp. 527–534, 2005.

$$\frac{\partial^2 E(x,t)}{\partial x^2} - \frac{1}{c^2}\frac{\partial^2 E(x,t)}{\partial t^2} = \frac{4\pi}{c^2}\frac{\partial^2 P(x,t)}{\partial t^2},\ 0 < t < L_t,\ 0 < x < L_x, \quad (1)$$

with corresponding initial and boundary conditions. Here $E(x,t)$ is an electric field strength, $x-$coordinate along which the light pulse is propagated, $t-$time, L_t and L_x are the time interval and medium length correspondingly on which the propagation of waves is investigated, c - light velocity, P describes medium polarization. For the medium with quadratic nonlinear response its polarization can be written as:

$$P = \chi(\omega)E + \chi^{(2)}(\omega,\omega)E^2.$$

Above $\chi(\omega)$ and $\chi^{(2)}(\omega,\omega)$ are correspondingly linear and quadratic susceptibilities of medium on frequency ω.

To write the equation for slowly varying amplitude we'll represent electric field strength in the following way

$$E(x,t) = \frac{1}{2}[A_1 e^{i(\omega_1 t - k_1 x)} + A_2 e^{i(\omega_2 t - k_2 x)} + A_3 e^{i(\omega_3 t - k_3 x)}] + c.c., \quad (2)$$

where k_j, ω_j (j = 1, 2, 3) accordingly wave number and wave frequency, A_j- slowly varying complex amplitude. Letters c.c. mean a complex conjugation. Substituting (2) in (1), without consideration the second derivatives on longitudinal coordinate and carrying out averaging on the period of each from interacting waves, one can write down the following dimensionless system of equations

$$\frac{\partial A_j}{\partial x} + \nu_j \frac{\partial A_j}{\partial t} + iD_j \frac{\partial^2 A_j}{\partial t^2} + F_j = 0, 0 < x, 0 < t < L_t, j = 1,2,3 \quad (3)$$

with initial and boundary conditions

$$A_j|_{x=0} = A_{j0}(t),\ A_j|_{t=0,L_t} = \frac{\partial A_j}{\partial t}|_{t=0,L_t} = 0,\ j = 1,2,3. \quad (4)$$

It should be stressed, for simplicity we don't introduce new notations of variables. Here $A_j(x,t)$- complex amplitude that is normalized on the maximal value of one from interacting waves, x is normalized on medium length L_x, $t-$time is normalized on characteristic pulse duration, coefficient D_j describes second order dispersion, $\nu_j = u_j^{-1} - u_1^{-1}$ characterizes group velocity mismatch, $i-$imaginary unit, u_j product of pulse group velocity for characteristic time in units of which it is measured. L_t-dimensionless time interval during which the problem is analyzed.

Taking into account an exchanging of energy between waves, the relation between frequencies of interacting waves $\omega_3 = \omega_1 + \omega_2$ and phase mismatching $\Delta\bar{k} = k_1 + k_2 - k_3$, one can write functions F_j in the form

$$F_j = \begin{cases} \gamma_1(iA_3A_2^* + \frac{1}{\bar{\omega}_1}\frac{\partial}{\partial t}(A_3A_2^*))e^{i\Delta kx},\ j = 1, \\ \gamma_2(iA_3A_1^* + \frac{1}{\bar{\omega}_2}\frac{\partial}{\partial t}(A_3A_1^*)e^{i\Delta kx},\ j = 2, \\ \gamma_3(iA_1A_2 + \frac{1}{\bar{\omega}_3}\frac{\partial}{\partial t}(A_1A_2))e^{-i\Delta kx},\ j = 3, \end{cases}$$

where $\bar{\omega}_j$ - product of the half frequency of light pulse for characteristic pulse duration, $\bar{\omega}_3 = \bar{\omega}_1 + \bar{\omega}_2$, γ_j - dimensionless factors of nonlinear coupling of interacting waves, $\Delta k = \Delta \bar{k} L_x$.

3 Transformation of the Equations and Invariants of Problem

During the interaction an energy of three waves preserves

$$I(x) = \sum_{j=1}^{3} \frac{\bar{\omega}_j}{\gamma_j} \int_0^{L_t} |A_j|^2 d\eta = const. \tag{5}$$

To write others invariants, it is necessary to transform the equations (1) to ones which are more convenient for numerical modelling [5]. With this purpose we'll introduce new functions by a rule

$$E_j = \int_0^t A_j e^{i\bar{\omega}_j(\eta - t)} d\eta, \text{ j} = 1{,}2{,}3,$$

which as well satisfy to relaxation equations

$$\frac{\partial E_j}{\partial t} + i\bar{\omega}_j E_j = A_j, \text{ j} = 1{,}2{,}3. \tag{6}$$

In new variables, the equations (1) are transformed to form being not to contain derivatives on time from the nonlinear response

$$\frac{\partial E_j}{\partial x} + v_j \frac{\partial E_j}{\partial t} + iD_j \frac{\partial^2 E_j}{\partial t^2} + \overline{F_j} = 0, \; 0 < x, \; 0 < t < L_t, \; j = 1,2,3 \tag{7}$$

and functions $\overline{F_j}$ look as follows

$$\overline{F_j} = \begin{cases} \frac{\gamma_1}{\bar{\omega}_1} A_3 A_2^* e^{i\Delta k x}, \; j = 1, \\ \frac{\gamma_2}{\bar{\omega}_2} A_3 A_1^* e^{i\Delta k x}, \; j = 2, \\ \frac{\gamma_3}{\bar{\omega}_3} A_1 A_2 e^{-i\Delta k x}, \; j = 3. \end{cases}$$

Initial and boundary conditions for new functions are

$$E_j|_{x=0} = E_{j0}(t), \; j = 1,2,3,$$

$$E_j|_{t=0} = \frac{\partial E_j}{\partial t}|_{t=0} = (\frac{\partial E_j}{\partial t} + i\bar{\omega}_j E_j)|_{t=L_t} = (\frac{\partial E_j}{\partial x} - i(\bar{\omega}_j^2 D_j + \bar{\omega}_j \nu_j) E_j)|_{t=L_t} = 0.$$

It is necessary to note the initial distribution of functions E_j, corresponding to input complex amplitudes, is calculated using the formula (5).

Conservation laws for problem under the consideration are written as

$$I_1(x) = \int_0^{L_t} \sum_{j=1}^{3} \bar{\omega}_j (i\bar{\omega}_j |E_j|^2 - E_j \frac{\partial E_j^*}{\partial t}) dt = const, \tag{8}$$

$$I_2(x) = \int_0^{L_t} \sum_{j=1}^{3} \bar{\omega}_j E_j A_j^* dt = const\ , \tag{9}$$

$$I_3(x) = \int_0^{L_t} \sum_{j=1}^{3} (E_j \frac{\partial A_j^*}{\partial t} + E_j^* \frac{\partial A_j}{\partial t}) dt = const. \tag{10}$$

It should be noticed that invariants (8) - (10) haven't physical meaning with regard to well known concepts (mass, energy, impulse).

4 Construction and Realization of a Conservative Scheme

Let's introduce into area $G = \{(x,t) : 0 \le x \le L_x, 0 \le t \le L_t\}$ uniform grids on x and t coordinates

$$\omega_x = \{x_n = nh,\ n = 0, 1, ..., N_x,\ h = L_x/N_x\},$$
$$\omega_t = \{t_k = k\tau,\ k = 0, 1, ..., N_t,\ \tau = L_t/N_t\},\ \omega = \omega_x \times \omega_t.$$

Let's define mesh functions A_h, E_h on ω using without-index notations

$$\begin{aligned} &A_j = A_j(x_n, t_k), \hat{A}_j = A_j(x_n + h, t_k), (A_j)_{\pm l} = A_j(x_n, t_k \pm l\tau), \\ &\overset{0.5}{A_j} = 0.5(\hat{A}_j + A_j), \hat{E}_j = E_j(x_n + h, t_k), (\hat{E}_j)_{\pm l} = E_j(x_n + h, t_k \pm l\tau), \\ &E_j = E_j(x_n, t_k), (E_j)_{\pm l} = E_j(x_n, t_k \pm l\tau), \overset{0.5}{E_j} = 0.5(\hat{E}_j + E_j). \end{aligned} \tag{11}$$

Further for simplicity an index h at mesh functions we'll omit. For the problem (5), (6) we propose the following nonlinear two-layer difference scheme which is written below in internal units of the grid

$$(\hat{E}_j - E_j)/h + v_j((\overset{0.5}{E_j})_{+1} - (\overset{0.5}{E_j})_{-1})/\tau + iD_j((\overset{0.5}{E_j})_{+1} - 2\overset{0.5}{E_j} + (\overset{0.5}{E_j})_{-1})/\tau^2 + \overline{\overline{F_j}} = 0,$$

$$\overline{\overline{F_j}} = \begin{cases} \frac{\gamma_1}{2\bar{\omega}_1} \overset{0.5}{A_3} \overset{0.5}{A_2^*} (e^{i\Delta k(x+h)} + e^{i\Delta k x}),\ j = 1, \\ \frac{\gamma_2}{2\bar{\omega}_2} \overset{0.5}{A_3} \overset{0.5}{A_1^*} (e^{i\Delta k(x+h)} + e^{i\Delta k x}),\ j = 2, \\ \frac{\gamma_3}{2\bar{\omega}_3} \overset{0.5}{A_1} \overset{0.5}{A_2} (e^{-i\Delta k(x+h)} + e^{-i\Delta k x}),\ j = 3. \end{cases} \tag{12}$$

The complex amplitude A_j is defined in the same mesh points as a solution of equation

$$\overset{0.5}{A}_j = ((\overset{0.5}{E}_j)_{+1} - (\overset{0.5}{E}_j)_{-1})/(2\tau) + i\bar{\omega}_j \overset{0.5}{E}_j,\ j = 1,2,3. \tag{13}$$

At boundary mesh points there are conditions

$$\overset{0.5}{E}_{j0} = \overset{0.5}{E}_{j1} = 0, \tag{14}$$

$$\tau(\hat{E}_{jN_t} - E_{jN_t})/h - i(D_j + v_j/\bar{\omega}_j)((\overset{0.5}{E}_{jN_t} - \overset{0.5}{E}_{jN_t-1})/\tau + i\bar{\omega}_j \overset{0.5}{E}_{jN_t}) = 0, j = 1,2,3.$$

It is necessary to emphasize, that only the suggested approximation of the right domain condition guarantees a validity of conservation laws for the constructed difference scheme.

Since the constructed difference scheme is nonlinear, for it's removability we use an iterative process

$$(\overset{s+1}{\hat{E}_j} - E_j)/h + v_j((\overset{s+1}{\overset{0.5}{E_j}})_{+1} - (\overset{s+1}{\overset{0.5}{E_j}})_{-1})/\tau + iD_j((\overset{s+1}{\overset{0.5}{E_j}})_{+1} - 2\overset{s+1}{\overset{0.5}{E_j}} + (\overset{s+1}{\overset{0.5}{E_j}})_{-1})/\tau^2 + \overline{\overline{F_j}} = 0,$$

$$\overset{s+1}{\overset{0.5}{A_j}} = ((\overset{s+1}{\overset{0.5}{E_j}})_{+1} - (\overset{s+1}{\overset{0.5}{E_j}})_{-1})/(2\tau) + i\bar{\omega}_j \overset{s+1}{\overset{0.5}{E_j}},\ j = 1,2,3, \tag{15}$$

$$\overline{\overline{F_j}} = \begin{cases} \frac{\gamma_1}{2\bar{\omega}_1} \overset{s}{\overset{0.5}{A_3}}\, \overset{s}{\overset{0.5}{A_2^*}}(e^{i\Delta k(x+h)} + e^{i\Delta kx}),\ j = 1, \\ \frac{\gamma_2}{2\bar{\omega}_2} \overset{s}{\overset{0.5}{A_3}}\, \overset{s}{\overset{0.5}{A_1^*}}(e^{i\Delta k(x+h)} + e^{i\Delta kx}),\ j = 2, \\ \frac{\gamma_3}{2\bar{\omega}_3} \overset{s}{\overset{0.5}{A_1}}\, \overset{s}{\overset{0.5}{A_2}}(e^{-i\Delta k(x+h)} + e^{-i\Delta kx}),\ j = 3. \end{cases}$$

In boundary points accordingly we have the equations

$$\overset{s+1}{\overset{0.5}{E_{j0}}} = \overset{s+1}{\overset{0.5}{E_{j1}}} = 0, \tag{16}$$

$$\tau(\overset{s+1}{\hat{E}_{jN_t}} - E_{jN_t})/h - i(D_j + v_j/\bar{\omega}_j)((\overset{s+1}{\overset{0.5}{E_{jN_t}}} - \overset{s+1}{\overset{0.5}{E_{jN_t-1}}})/\tau + i\bar{\omega}_j \overset{s+1}{\overset{0.5}{E_{jN_t}}}) = 0, j = 1,2,3.$$

Above index s=0,1,2,3... denotes iteration number. As initial approach for iteration procedure (values of mesh functions on the high layer on zero iteration) the values of mesh functions from the previous layer on longitudinal coordinate are chose

$$\overset{s=0}{\hat{A}_j} = A_j, \ \overset{s=0}{\hat{E}_j} = E_j, \ j = 1, 2, 3.$$

The iterative process stops, if the following conditions are executed

$$\max_{t_k} | \overset{s+1}{\hat{E}_j} - \overset{s}{\hat{E}_j} | < \varepsilon_1 \max_{t_k} |\overset{s}{\hat{E}_j}| + \varepsilon_2, \ \varepsilon_1, \varepsilon_2 > 0, \ j = 1, 2, 3.$$

Constructed difference scheme approximates difference equations with second order on time and spatial coordinate with regard to a point (x+h/2, t). However, the right domain condition in point L_t is approximated with the first order on time.

5 Computer Simulation

For instance on Fig. 1 we present shape of pulses which computed using the conservative difference scheme (12)-(14) and nonconservative one. Computing was made for initial Gaussian shape of first and second pulses

$$A_{1,0}(t) = A_{2,0}(t) = exp(-(t - L_t/2)^2)$$
$$A_{30}(t) = 0,$$

and parameters:

$$D_{1,2,3} = 0.5, \nu_{1,2,3} = 0, \Delta k = 0, \gamma_1 = 5, \gamma_2 = 10, \omega_1 = 1, \omega_2 = 2$$

and mesh steps:

$$\tau = 0.005, h = 0.0002.$$

The nonconservative difference scheme preserves only the energy of interacting waves but other invariants (8)-(10) are changeable and their values depends on longitudinal coordinate. We can see with increasing of longitudinal coordinate the shape of pulses distinguishes (Fig.1). The biggest distinguishing takes place between intensities distributions of second and third waves. It increases, dramatically for our case of the interaction parameters after section z=0,1. It should be emphasized that invariant of energy (5) is constant for both schemes. But invariant (8) varies not constant for nonconservative scheme (Fig. 2).

The other advantage of conservative difference scheme concludes in possibility of an essential increasing of the mesh steeps in comparison with nonconservative method to obtain a solution with requiring accuracy on given distance.

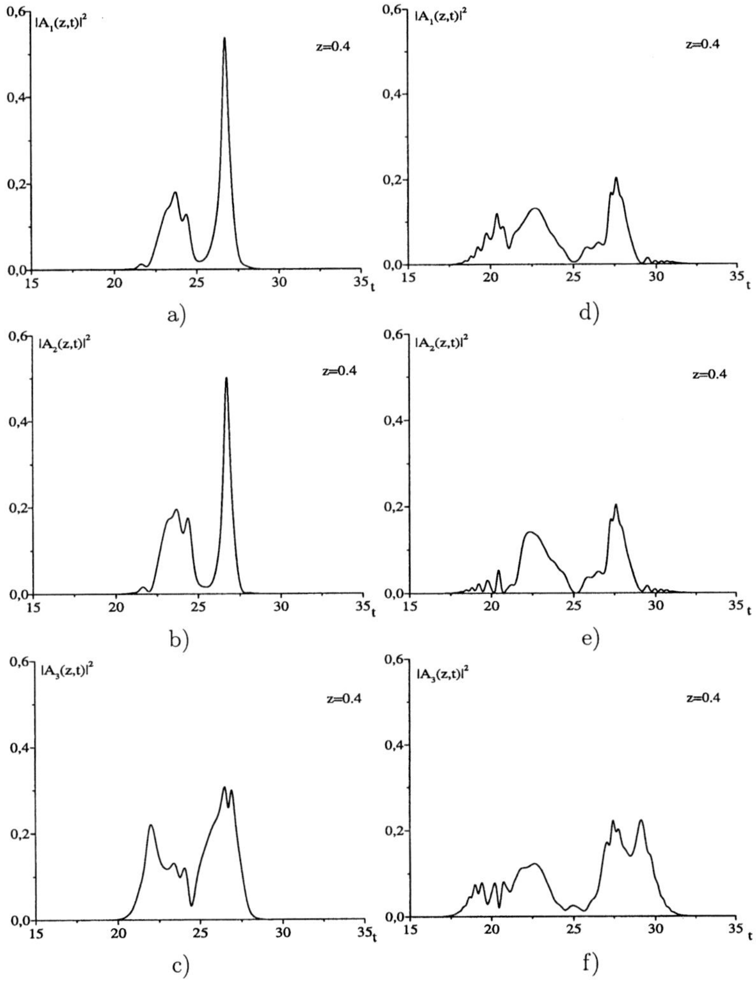

Fig. 1. Shape of pulses for three interacting waves, which is computed on the base of conservative difference scheme (a,b,c) and nonconservative one (d,e,f)

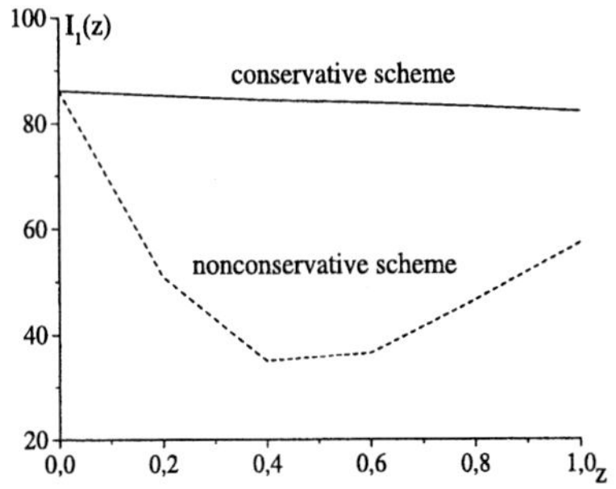

Fig. 2. Evolution of $I_1(z)$ for conservative (solid) and nonconservative (dash line) difference schemes

6 Conclusions

Thus, in the present paper the conservative difference scheme for system of CNLSE describing three femtosecond pulses interaction in medium with quadratic nonlinearity in view of derivative on time from the nonlinear medium response is proposed. For its realization the method of simplest iteration in combination with matrix running algorithm for the solution of corresponding system of linear equations can be used.

Aknowledgement. This paper was supported partly by RFBR (grant N 02-01-727).

References

1. Ahmanov S.A., Vyslouh V.A., Chirkin A.S. Optics of femtosecond laser pulses Moscow: *Nauka* (1988) (In Russian)
2. Agraval G. Nonlinear fiber optics. Translated from English. Moscow: *Mir* (1996) (In Russian)
3. Sukhorukov A.P. Nonlinear wave interactions in optics and radio-physics Moscow: *Nauka* (1988) (In Russian)
4. Trofimov V.A., Volkov A.G., Varentsova S.A. Influence of dispersion of nonlinear response on self-focusing of femtosecond pulse propagation in optical fiber In "Laser Physics and Photonics, Spectroscopy and Modeling II " / Eds. Derbov V.L. et. al. Proceedings of SPIE. **4706** (2002) 88-97
5. Varentsova S.A., Trofimov V.A. Invariants of nonlinear interaction femtosecond laser pulses in view of dispersion of the third order. J. of Calculation Math. & Math. Physics. **42** (2002) 709-717 (In Russian)
6. Trofimov V.A. On new approach to modelling of nonlinear propagation of super-short laser pulses. J. of Calculation Math. & Math. Physics. **38** (1998) 835-839 (In Russian)
7. Samarskii A.A. Theory of difference schemes. Moscow: *Nauka* (1989) (In Russian)
8. Borhanifar A., Trofimov V.A. Comparison of efficiency of various approaches to computer simulation of nonlinear interaction of three femtosecond pulses in an optical fiber: Vestnik Moscow University. Series of calculations math. and cybernetics. **2** (2004) 20-27 (In Russian)
9. Trofimov V.A., Borhanifar A. Conservative difference schemes for three waves interaction of femtosecond pulses in optical fiber: International Conference "Mathematical Modeling and Analysis". Trokai, Lituania. (2003) 76

Comparison of Some Difference Schemes for Problem of Femtosecond Pulse Interaction with Semiconductor in the Case of Nonlinear Mobility Coefficient

Vyacheslav A. Trofimov and Maria M. Loginova

Lomonosov Moscow State University,
Department of Computational Mathematics & Cybernetics,
Leninskye Gory, Moscow 119992, Russia
vatro@cs.msu.su

Abstract. A difference scheme for problem of femtosecond pulse interaction with semiconductor is proposed in the case of nonlinear dependencies of light energy absorption coefficient and mobility of free electrons from electric field strength. Comparison of some difference schemes efficiency for computation of various regimes of laser pulse interaction with semiconductor is carried out. It is shown that approach developed by us allows to analyze regimes for which difference schemes known earlier are unsuitable.

1 Introduction

As is known, interaction of high-intensity femtosecond pulse with semiconductor is accompanied by various nonlinear effects [1]. Among them we'll note the phenomenon of power levels shift under the action of such pulse [1,2]. On the basis of this phenomenon in [3] the optical bistability scheme is proposed. It takes into account the change of band gap owing to shift of power levels under the action of high-intensity laser femtosecond pulse. However at carrying out of computer simulation it was revealed that conservative scheme [3], known in the literature, for some sets of parameters prove to be inapplicable: iterative process doesn't converge or the difference scheme loses property of conservatism. In this connection in [4] the difference scheme based on transformation of initial equations has been created. It allowed to remove the difficulties described above for the case of constant electrons mobility. It is necessary to emphasize, that application of this scheme in the case of nonlinear dependence of electrons mobility on electric field strength prove to be not effective. In the present work the generalization of this difference scheme for the case of nonlinear electrons mobility is executed.

2 Problem Statement

Femtosecond laser pulse interaction with semiconductor within the framework of optical thin layer is described by the following system of dimensionless differential equations [5,6]:

Z. Li et al. (Eds.): NAA 2004, LNCS 3401, pp. 535–542, 2005.

$$\frac{\partial^2 \varphi}{\partial\, x^2} = \gamma(n - N), \tag{1}$$

$$\frac{\partial n}{\partial t} = D\frac{\partial}{\partial x}\left(\frac{\partial n}{\partial x} - \mu n\frac{\partial \varphi}{\partial x}\right) + G(n, N, \varphi) - R(n, N),$$

$$\frac{\partial N}{\partial t} = G(n, N, \varphi) - R(n, N),\ 0 < x < L_x = 1,\ t > 0$$

with boundary and initial conditions

$$\left.\frac{\partial \varphi}{\partial x}\right|_{x=0,L_x} = 0, \qquad \left.\frac{\partial n}{\partial x}\right|_{x=0,L_x} = 0, \qquad n|_{t=0} = N|_{t=0} = n_0 \tag{2}$$

which correspond to absence of external electric field and current through semiconductor surface. It is supposed, that at the initial time moment semiconductor is electrically neutral. Above functions G and R, describing generation and recombination of semiconductor free charges particles correspondingly, we'll determine as follows:

$$G = q_0 q_1(x) q_2(t) \delta(n, N, \varphi), \qquad R = \frac{nN - n_0^2}{\tau_p}. \tag{3}$$

Electrons mobility is set by function

$$\mu(x,t) = \frac{\mu_0}{1 + |E|/E_k}, \tag{4}$$

where μ_0 non-negative constant, $E_k > 0$ characterizes nonlinearity of mobility. Light energy absorption coefficient $\delta(n, N, \varphi)$ is approximated by one of the following functions

$$\delta(n, N, \varphi) = (1 - N)\left\{e^{-(\alpha-\beta|\varphi|)}, \quad ch(\beta E), \quad e^{-\psi(1-\xi n)} e^{-(\alpha-\beta|\varphi|)}\right\}, \tag{5}$$

$$E = -\frac{\partial \varphi}{\partial x}.$$

In the system of equations (1) - (5) the following variables are introduced: x - dimensionless cross-section coordinate normalized on the input optical beam radius, t - time, that is measured in units of relaxation time, $n(x,t)$ and $N(x,t)$ - concentrations of free electrons in conductivity zone of semiconductor and ionized donors , which are normalized on their maximum possible value in the given conditions. Function $\varphi(x,t)$ - dimensionless electric field potential, D - coefficient of electrons diffusion. Parameter γ depends, in particular, on the maximum possible concentration of free charge carriers, n_0 - equilibrium value of free electrons concentration and the ionized donors one, τ_p characterizes recombination time. Parameters α, β, ψ, ξ - non-negative constants. Functions $q_1(x)$, $q_2(t)$ describe intensity profile and time shape of optical pulse correspondingly, q_0 - maximum light intensity.

Under the computer simulation experiments the action of gauss beam on semiconductor is considered

$$q_1(x)q_2(t) = exp\{-(\frac{1-0.5L_x}{0.1L_x})^2\}(1-e^{-10t}). \tag{6}$$

As is known, for interaction of optical radiation with semiconductor the law of charge preservation takes place

$$Q(t) = \int_0^{L_x} (n(t,\varsigma) - N(t,\varsigma))\,d\varsigma = 0. \tag{7}$$

This is necessary to taking into account at computer simulation of light beam interaction with semiconductor as conservation law.

3 Construction of Differences Schemes

For definiteness we'll consider absorption coefficient dependence on electric field potential of a kind as $\delta(n,N,\varphi) = (1-N)e^{-(\alpha-\beta|\varphi|)}$. Differences schemes construction for its other dependencies is similarly. With the purpose of differences schemes creation we introduce into area $\bar{G} = \{0 \le x \le L_x\} \times \{0 \le t \le L_t\}$ an uniform grid $\Omega = \omega_x \times \omega_t$:

$$\omega_x = \left\{x_i = ih,\ i = \overline{0,N_x},\ h = {}^{L_x}\!/_{N_x}\right\}, \tag{8}$$

$$\omega_t = \left\{t_j = j\tau,\ j = \overline{0,N_t}, \tau = {}^{L_t}\!/_{N_t}\right\}.$$

On this grid let define mesh functions $n_h,\ N_h,\ \varphi_h,\ E_h$ as follows:

$$n = n_h(x_i,t_j), \quad N = N_h(x_i,t_j), \quad \varphi = \varphi_h(x_i,t_j), \quad E = E_h(x_i,t_j), \tag{9}$$

$$\hat{n} = n_h(x_i,t_{j+1}), \quad \hat{N} = N_h(x_i,t_{j+1}), \quad \hat{\varphi} = \varphi_h(x_i,t_{j+1}), \quad \hat{E} = E_h(x_i,t_{j+1}).$$

Below for briefly index h at functions we'll lower.One of the possible conservative difference scheme for problem (1) - (3) is described in [4] (we entitle it as the **Scheme 1**). It approximates initial system of differential equations with accuracy $O\left(h^2+\tau^2\right)$. Boundary conditions are approximated with the first order on x-coordinate. It is caused by necessity of scheme conservatism realization. However, in some cases this scheme loses convergence in view of nonlinear dependence of electrons mobility on electric field induced by laser light. Therefore below new **Scheme 2** is considered.

Scheme 2. Let introduce new function $\overline{n}(x,t)$ as follows

$$n(x,t) = \overline{n}(x,t)e^{\mu_0 F(x,t)}. \tag{10}$$

In the case of constant mobility $F = \varphi$ and similar transformation was made, for example, in [7]. So initial system of equations (1) can be written down as

$$\frac{\partial E}{\partial t} = -\gamma D e^{\mu_0 F} \frac{\partial \bar{n}}{\partial x}, \quad \frac{\partial \varphi}{\partial x} = -E, \quad \frac{\partial N}{\partial t} = G - R, \tag{11}$$

$$n = \frac{1}{\gamma}\frac{\partial^2 \varphi}{\partial x^2} + N, \quad \frac{\partial F}{\partial x} = -\frac{E}{1+|E|/E_k}, \quad 0 < x < L_x, \quad t > 0$$

with boundary and initial conditions

$$E|_{x=0,L_x} = \left.\frac{\partial n}{\partial x}\right|_{x=0,L_x} = 0, \quad n|_{t=0} = N|_{t=0} = n_0, \quad E|_{t=0} = 0. \tag{12}$$

Let's construct the difference scheme for problem (10) - (12). With this aim we define on the grid Ω mesh function $\bar{n} = \bar{n}_h(x_i, t_j)$.This difference scheme is nonlinear so for its solvability the iteration method is used. Proposed scheme with iteration method is written in following manner

$$\frac{\overset{s+1}{\hat{E}} - E}{\tau} = -\frac{1}{4}\gamma D \left(\overset{s}{\hat{\bar{n}}}_{\overset{\circ}{x}} + \bar{n}_{\overset{\circ}{x}} \right) \left(e^{\mu_0 \overset{s}{\hat{F}}} + e^{\mu_0 F} \right), \quad \overset{s+1}{\hat{\varphi}}_x = -\overset{s+1}{\hat{E}}, \tag{13}$$

$$\overset{s+1}{\hat{F}}_x = -\frac{\overset{s+1}{\hat{E}}}{1 + \left|\overset{s+1}{\hat{E}}\right| / E_k}, \quad \overset{s+1}{\hat{n}} = \overset{s+1}{\hat{N}} - \frac{1}{\gamma} \overset{s+1}{\hat{E}}_x, \quad \overset{s+1}{\hat{\bar{n}}} = \frac{\overset{s+1}{\hat{n}}}{e^{\mu_0 \overset{s+1}{\hat{F}}}},$$

$$\frac{\overset{s+1}{\hat{N}} - N}{\tau} = \frac{1}{2} q(x) \left(e^{-\left(\alpha - \beta \left|\overset{s+1}{\hat{\varphi}}\right|\right)} (1 - \overset{s+1}{\hat{N}}) + e^{-(\alpha - \beta|\varphi|)}(1 - N) \right) -$$

$$-\frac{1}{2}\left(\frac{\overset{s}{\hat{n}} \overset{s+1}{\hat{N}} - n_0^2}{\tau_p} + \frac{nN - n_0^2}{\tau_p} \right), \quad \overset{s+1}{\underset{0.5}{n}}_{x,0} = \overset{s+1}{\underset{0.5}{n}}_{\bar{x},N_x} = 0, \quad \overset{s+1}{\hat{E}}_0 = \overset{s+1}{\hat{E}}_{N_x} = 0,$$

$$n_i|_{j=0} = N_i|_{j=0} = n_0, \quad E_i|_{j=0} = 0, \quad i = \overline{0, N_x}.$$

The difference scheme (13) approximates initial system of differential equations with accuracy $O\left(h^2 + \tau^2\right)$. Boundary conditions are approximated with the first order on spatial coordinate. As initial approach for iterative process values of functions on the previous time layer is chosen:

$$\overset{s=0}{\hat{n}}(x_i, t_{j+1}) = n(x_i, t_j), \quad \overset{s=0}{\hat{N}}(x_i, t_{j+1}) = N(x_i, t_j), \tag{14}$$

$$\overset{s=0}{\hat{\varphi}}(x_i, t_{j+1}) = \varphi(x_i, t_j), \quad \overset{s=0}{\hat{E}}(x_i, t_{j+1}) = E(x_i, t_j).$$

Iteration process terminates when the following conditions are satisfied

$$\left|\overset{s+1}{\hat{n}} - \overset{s}{\hat{n}}\right| \le \varepsilon_1 \left|\overset{s}{\hat{n}}\right| + \varepsilon_2, \quad \left|\overset{s+1}{\hat{N}} - \overset{s}{\hat{N}}\right| \le \varepsilon_1 \left|\overset{s}{\hat{N}}\right| + \varepsilon_2, \tag{15}$$

$$\left| \overset{s+1}{\hat{\varphi}} - \overset{s}{\hat{\varphi}} \right| \le \varepsilon_1 \left| \overset{s}{\hat{\varphi}} \right| + \varepsilon_2, \qquad \varepsilon_1, \varepsilon_2 > 0.$$

It is necessary to emphasize, that constructed difference **Scheme 2** is monotonous. Disadvantage of the given scheme is necessity of finding of function F and calculations of exponents from it on each iteration. That leads to increasing of operations number. To increase an accuracy and on the big time intervals the Fourier fast discrete transformation method is used for computation function F, potential of electric field and function E derivation. We found out some regimes of femtosecond pulse interaction with semiconductor, for which computation using the **Scheme 1** is unsuitable.

4 Comparison of Difference Schemes Efficiency

For comparison of the **Scheme 1** and the **Scheme 2** efficiency the computer simulations of various regimes, determined by parameters and absorption coefficient, were carried out.

As an example on Fig. 1 distributions of electrons concentration, computed using the **Scheme 1** (Fig. 1a, c, e) and the **Scheme 2** (Fig. 1b, d, f) with the same steps of difference grid are represented. It is necessary to emphasize, that using of the **Scheme 1** for this regime the computation is possible only till some time moment *($t \le 64$)* as further the iterations number quickly increases

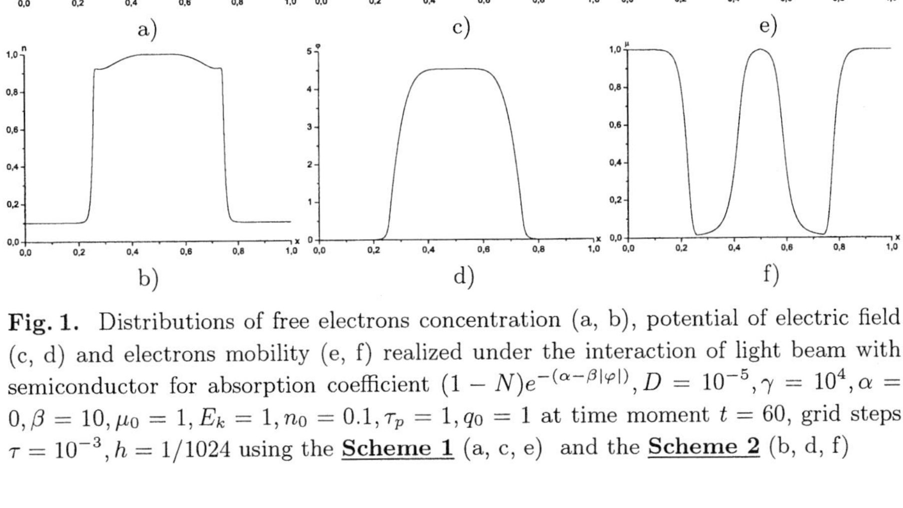

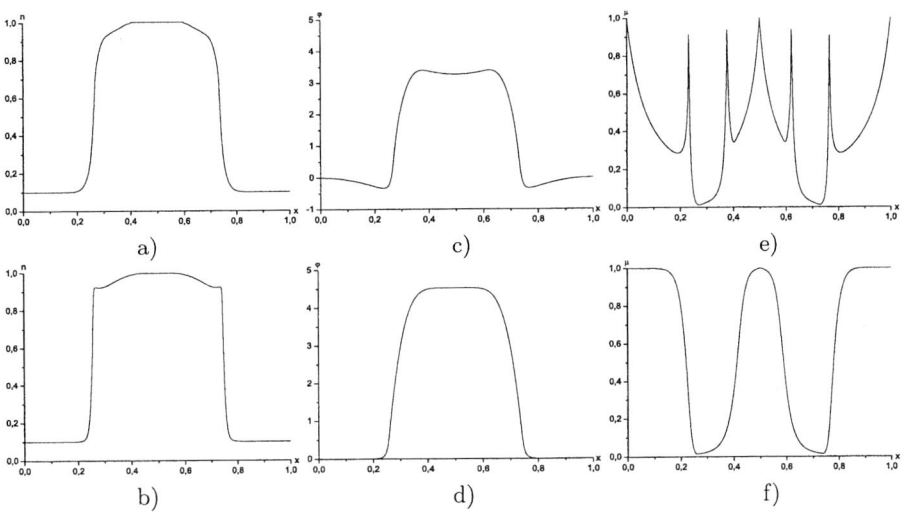

Fig. 1. Distributions of free electrons concentration (a, b), potential of electric field (c, d) and electrons mobility (e, f) realized under the interaction of light beam with semiconductor for absorption coefficient $(1 - N)e^{-(\alpha - \beta|\varphi|)}, D = 10^{-5}, \gamma = 10^4, \alpha = 0, \beta = 10, \mu_0 = 1, E_k = 1, n_0 = 0.1, \tau_p = 1, q_0 = 1$ at time moment $t = 60$, grid steps $\tau = 10^{-3}, h = 1/1024$ using the **Scheme 1** (a, c, e) and the **Scheme 2** (b, d, f)

Table 1. Comparison of accuracy of invariant $Q(t)$ preservation by difference schemes. $D=10^{-3}$, $\gamma=10^3$, $\alpha=0$, $n_0=0.1$, $\tau_p=1$

	Scheme 1		**Scheme 2**	
	$h=1/256$, $\tau=10^{-2}$	$h=1/1024$, $\tau=10^{-3}$	$h=1/256$, $\tau=10^{-2}$	$h=1/1024$, $\tau=10^{-3}$
Absorption coefficient	$\delta(n,N,\varphi)=(1-N)e^{-(\alpha-\beta\|\varphi\|)}$			
Parameters	$\beta=1$, $\mu_0=1$, $E_k=2$, $q_0=10$			
$Q(25)$	0.00105166	0.00025889	0.00012645	0.00001136
$Q(250)$	0.11952817	0.02929271	0.00049349	0.00004645
$Q(500)$	0.26601535	0.06430590	0.00049349	0.00004645
Absorption coefficient	$\delta(n,N,\varphi)=(1-N)ch\beta E$			
Parameters	$\beta=1$, $\mu_0=1$, $E_k=1$, $q_0=2$			
$Q(25)$	0.00004463	0.00001037	0.00001019	0.00000236
$Q(50)$	0.03870489	0.00930603	0.00023420	0.00004869
$Q(100)$	0.08324952	0.02005396	0.00025810	0.00007989
Absorption coefficient	$\delta(n,N,\varphi)=(1-N)e^{-\psi(1-\xi n)}e^{-(\alpha-\beta\|\varphi\|)}$			
Parameters	$\beta=10$, $\psi=2.553$, $\xi=5/\psi$, $\mu_0=10$, $E_k=1$, $q_0=1$			
$Q(25)$	-0.00000001	0	-0.00000040	0
$Q(50)$	-0.01883673	-0.00378525	-0.00003634	0
$Q(100)$	-0.06697757	-0.01330692	-0.00002418	-0.00000868

up to several thousand. At use the <u>**Scheme 2**</u> it doesn't occur: computations can be carried out with large grid steps and on unlimited time interval. Also it is important to note, that the results got on the base of the <u>**Scheme 1**</u> till moment of abrupt increase of iterations number, qualitatively differ from the corresponding ones got under using the <u>**Scheme 2**</u>. It is necessary to pay an attention as well that similar properties of the schemes were observed with zero electrons mobility (Fig. 2).

One more difference of analyzed schemes consists in existence of regimes of femtosecond pulse interaction with semiconductor which computations using the <u>**Scheme 1**</u> doesn't cause the difficulties described above (iterative process converges quickly), but because of mistakes summation and boundary conditions of the second type it loses conservatism: invariant (8) doesn't hold out (Fig. 3). At use of the <u>**Scheme 2**</u> it doesn't occur (Table). As example on Fig. 3a we can see uniform growth of electrons concentration n in time under preservation of high concentration domain form. Using the <u>**Scheme 2**</u> similar effects are not observed: the system reaches the steady state (Fig. 3c).

It is important to emphasize, that the specified growth of free electrons concentration is slowed down at decreasing of grid steps in computations using the

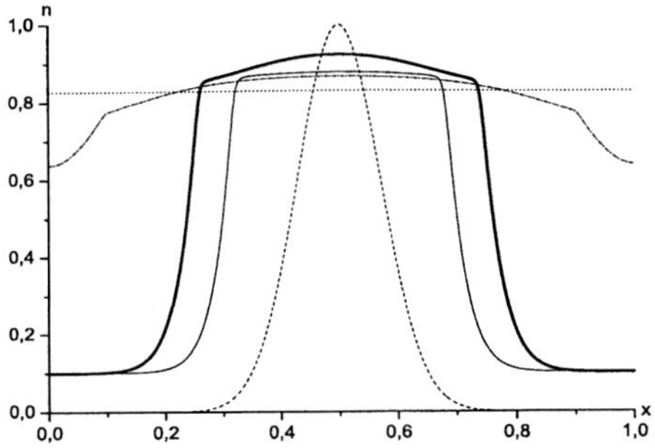

Fig. 2. Distributions of electrons concentration, realized under the interaction of light beam with semiconductor for absorption coefficient $(1-N)e^{-(\alpha-\beta|\varphi|)}e^{-\psi(1-\xi n)}$, $D = 10^{-3}, \gamma = 10^3, \alpha = 0, \beta = 5, \psi = 2.553, \xi = 5/\psi, \mu_0 = 0, E_k = 1, n_0 = 0.1, \tau_p = 1, q_0 = 1$, grid steps $\tau = 10^{-3}, h = 1/1024$ using the **Scheme 1** (solid fat line) and the **Scheme 2** (solid thin, dot-dash and dot lines) at the time moment $t = 5$ (solid lines), $t = 25$ (dot-dash line), 100 (dot line). The dash line represents initial distribution of entrance intensity

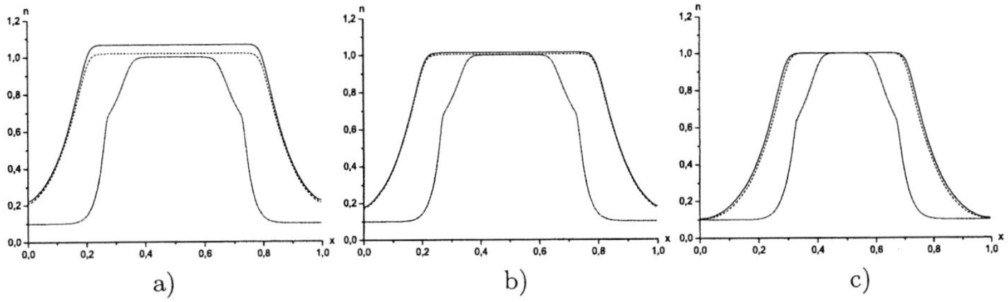

Fig. 3. Distributions of electrons concentration, realized under the interaction of light beam with semiconductor for absorption coefficient $(1-N)e^{-(\alpha-\beta|\varphi|)}e^{-\psi(1-\xi n)}$, $D = 10^{-3}, \gamma = 10^3, \alpha = 0, \beta = 10, \psi = 2.553, \xi = 5/\psi, \mu_0 = 10, E_k = 1, n_0 = 0.1, \tau_p = 1, q_0 = 1$, grid steps $\tau = 10^{-2}, h = 1/256$ (a, c), $\tau = 10^{-3}, h = 1/1024$ (b) at the time moment $t = 5$ (dot-dash line), 50 (dash line), 100 (solid line) computed using the **Scheme 1** (a, b), **Scheme 2** (c)

Scheme 1 (Fig. 3b). In this case the results of computations come nearer to ones received using the **Scheme 2**.

5 Conclusions

The monotonous scheme constructed in the present work allows to simulate regimes of femtosecond pulse interaction with semiconductor for which known before deference schemes are unsuitable because of iterations number fast growth at transition to following time layer or because of the loosing conservatism property. Nevertheless, it is necessary to emphasize, that on the certain time interval and at using grids with small steps both schemes one can receive identical results. Regimes of optical radiation interaction with semiconductor when monotonous Scheme 2 proves to be inefficient weren't revealed.

Acknowledgment

This paper was supported partly by RFBR (grant N 02-01-727).

References

1. Delone N.B., Kraynov V.P.: Nonlinear ionization of atoms by laser radiation. Moscow: Phismatlit. (2001) 310 (in Russian)
2. Varentsova S.A., Trofimov V.A.: Influence of strong light field on the shift of hydrogen-like atom spectrum. In Book "Numerical methods of mathematical physics." Proceeding of department CMC MSU. Editors Kostomarov D.P., Dmitriev V.I. Moscow: Pub. MSU. (1998) 67–75 (in Russian)
3. Varentsova S.A., Loginova M.M., Trofimov V.A. Mathematical model and the difference scheme for the optical bistability problem on the base of dependence of semiconductor band gap from electric field. Vestnik moskovskogo universiteta. Ser. calculation mathematics and cybernetics. **1** (2003) 20–27 (in Russian)
4. Loginova M.M., Trofimov V.A.: Comparison of some difference schemes for a problem of femtosecond pulse action on the semiconductor. Vestnik moskovskogo universiteta. Ser. computation mathematics and cybernetics. **1** (2004) (in Russian)
5. Smith R. Semiconductors.: Transl. from English. Moscow: Mir. (1982)560 (in Russian)
6. Bonch-Bruevich V.L., Kalashnikov S.G.: Physics of semiconductors. Moscow: Nauka. (1990) 685 (in Russian)
7. Il'in V.P.: Difference methods for electrophysics problems. Novosibirsk: Nauka. (1985) 335 (in Russian)

Soliton-Like Regime of Femtosecond Laser Pulse Propogation in Bulk Media Under the Conditions of SHG

Vyacheslav A. Trofimov and Tatiana M. Lysak

Lomonosov Moscow State University,
Department of Computational Mathematics and Cybernetics,
Leninskye Gory, Moscow 119992, Russia
Fax: +7 (095) 939-2596; phone: +7 (095) 939-5255;
vatro@cs.msu.su

Abstract. Results of numerical simulation of soliton-like propagation of two interacting femtosecond laser pulses in bulk media with radial symmetry are presented. Propagation is analysed under conditions of SHG with strong self-action and remarkable phase mismatch for nonzero input amplitudes of both waves. Two NLSE in two spatial coordinates and time with quadratic and cubic nonlinearities govern the process. Numerical simulation was made on the base of conservative difference scheme, taking into account the conservation laws of the problem. Long-time approximation of flat beam profiles was used to reveal the conditions of soliton-like propagation.

1 Introduction

Investigation of various regimes of SHG by high intensive laser femtosecond pulses is still a very important task, which attracts attention of many authors [1-11]. Some investigators concentrate their efforts at the achievement of high generation efficiency [10,11]. Others are interested in the formation of special regimes of propagation, e.g. solitons and spatiotemporal solitons – solutions that are localized in space and time [7-9].

Formation of solitons and soliton-like regimes of high intense femtosecond pulses under the SHG is promoted by self-action (cubic non-linearity) that under certain conditions becomes comparable with quadratic non-linearity for pulses in femtosecond diapason. The influence of self-action on laser light propagation increases dramatically with the pulse intensity growth. As a result, violation of the optimal phase conditions due to cubic non-linearity causes dramatic distortion of pulses' shapes. The pulse is divided into several narrow pulses, number of which grows with propagation length, and the intensities in the center of pulses grow, causing the destruction of the media. To avoid these problems special conditions of propagation should be set, that could guarantee the preservation of the main characteristics of pulse.

Z. Li et al. (Eds.): NAA 2004, LNCS 3401, pp. 543–550, 2005.

In this paper, using numerical stimulation, we offer possibility of soliton-like regime propagation of laser beam in bulk media, which is characterized by constant intensity in the central part of the high intense femtosecond pulses. For simplicity we assume the radial symmetry of media and input beams. Both pulses preserve their width. The pulse of the basic wave propagates without any noticeable distortions, while the pulse on doubling frequency propagates with constant central part and oscillating front at both radial and time coordinates. Conditions of appearence of such regime are derived in the framework of long-time approximation for flat beam profiles. We conducted our numerical simulation on the base of conservative difference scheme, taking into account the conservation laws of the problem.

2 Basic Equations

The process of SHG by femtosecond pulses in the framework of slowly varying envelope and under the condition of pulse self-action is governed by the pair of dimensionless nonlinear Shrodinger equarions (NLSE)

$$\frac{\partial A_1}{\partial z} + iD_1\frac{\partial^2 A_1}{\partial \eta^2} + iD_\perp \Delta_\perp A_1 + i\gamma A_1^* A_2 e^{-i\Delta kz} + i\alpha_1 A_1\left(|A_1|^2 + 2|A_2|^2\right) = 0,$$

$$\frac{\partial A_2}{\partial\, z} + \nu\frac{\partial A_2}{\partial \eta} + iD_2\frac{\partial^2 A_2}{\partial\, \eta^2} + i\frac{D_\perp}{2}\Delta_\perp A_2 + i\gamma A_1^2 e^{i\Delta kz} + i\alpha_2 A_2\left(2|A_1|^2 + |A_2|^2\right) = 0, \tag{1}$$

$$0 < z \le L_z, \quad \Delta_\perp = \frac{1}{r}\frac{\partial}{\partial r}\left(r\frac{\partial}{\partial r}\right), \quad \alpha_2 = 2\alpha_1 = 2\alpha.$$

Here η - dimensionless time in the system of coordinates moving with the basic wave pulse, z - normalized longitudinal coordinate, r - radial coordinate normalized on initial beam radius of the first wave , $D_j \sim -0.5\frac{\partial^{\,2}\bar{k}_j}{\partial\,\bar{\omega}_j^2}$ - coefficients that characterize dispersion of group velocity, $\bar{k}_j$, $\bar{\omega}_j$ - scaled wave number and frequency of j-th wave correspondingly, $D_\perp$ - diffraction coefficient, γ - coefficient of nonlinear coupling of interaction waves, $\Delta k = k_2 - 2k_1$ - dimensionless mismatching of their wave numbers, α_j - coefficients of self-action of waves, A_j - complex amplitudes ($j = 1, 2$), normalized on the maximal amplitude of first harmonic in the input section of medium ($z = 0$). Parameter ν is proportional to the difference between inverse values of group velocities of the second harmonic wave and the basic wave, L_z - nonlinear medium length.

At the input section the initial distributions of FH (fundamental harmonic) basic pulse and SH (second harmonic) pulse are defined

$$A_1(z = 0, \eta, r) = A^0{}_1(\eta, r)\exp(is_{10}),$$

$$A_2(z = 0, \eta, r) = A^0{}_2(\eta, r)\exp(is_{20}) \quad 0 \le \eta \le L_t \;, \quad 0 \le r \le L_r \tag{2}$$

$$A_i^0(\eta, r) = A_{i0} \exp\left(-((\eta - L_t)/\tau)^{m_t}/2\right) \exp\left(-\left(r/R\right)\right)^{m_r}/2\right), \tag{3}$$

where A_{i0} - dimensionless amplitudes, L_t - dimensionless time, L_r - dimensionless domain on transverse coordinate. Maximum input intensities of interacting waves are normalized at the sum of their intensities:

$$|A_{10}|^2 + |A_{20}|^2 = 1.$$

SHG process has the following invariants

$$I_1 = \int_0^{L_t}\int_0^{L_r} r(|A_1|^2 + |A_2|^2) dr d\eta \quad I_2 = \int_0^{L_t}\int_0^{L_r} r(A_1 \frac{\partial A_1^*}{\partial \eta} + A_2 \frac{\partial A_2^*}{\partial \eta}) dr d\eta,$$

$$I_3 = \int_0^{L_t}\int_0^{L_r} r \left[-2D_1 \left|\frac{\partial A_1}{\partial \eta}\right|^2 - D_2 \left|\frac{\partial A_2}{\partial \eta}\right|^2 - 2D_\perp \left|\frac{\partial A_1}{\partial r}\right|^2 - \frac{D_\perp}{2} \left|\frac{\partial A_2}{\partial r}\right|^2 \right.$$

$$\left. - \nu \left(A_2 \frac{\partial A_2^*}{\partial \eta} \right) \right] dr d\eta + \int_0^{L_t}\int_0^{L_r} r\gamma \left[A_2 A_1^{*2} e^{-i\Delta k z} + A_2^* A_1^2 e^{i\Delta k z} \right] dr d\eta -$$

$$- \int_0^{L_t}\int_0^{L_r} r \left[\Delta k \left(2|A_1|^2 + |A_2|^2 \right) + \alpha \left(|A_1|^4 + |A_2|^4 + 4|A_1|^2 |A_2|^2 \right) \right] dr d\eta. \tag{4}$$

Conservative difference schemes that preserve difference analog of these invariants were used with the aim of controlling of computer simulation results. Below we supposed group velocities matching for numerical calculation. It results in zero value of parameter ν (group synchronism).

3 Conditions for Soliton-Like Regime Formation

Formation of soliton-like regime depends on the problem parameters, particularly on the ratio of quadratic and cubic nonlinearities, phase mismatch and input characteristics of both waves. Conditions of soliton-like regime appearance can be written in the framework of long-time approximation for flat beam profiles.

Using standard writing for complex amplitudes of FH and SH waves

$$A_j = a_j e^{i\varphi_j}, \quad j = 1, 2$$

equations (1) can be rewritten in the form

$$\frac{da_1}{dz} = \gamma\, a_1 a_2 \sin\varphi, \quad \frac{da_2}{dz} = -\gamma\, a_1^2 \sin\varphi \quad ,$$

$$\frac{d\varphi}{dz} = 2\alpha\left(a_2^2 - a_1^2\right) - \gamma\cos\varphi\left(\frac{a_1^2}{a_2} - 2a_2\right) + \Delta k, \quad \varphi = \varphi_2 - 2\varphi_1 \qquad (5)$$

with the initial conditions

$$a_2\big|_{z=0} = a_{20}, \quad a_1\big|_{z=0} = a_{10}, \quad \varphi\big|_{z=0} = \varphi_0.$$

Invariants (4) in this case are the following

$$I_1 = a_1^2 + a_2^2, \quad I_3 = 2\gamma\, a_2 a_1^2 \cos\varphi + \alpha\left(|a_1|^4 + |a_2|^4 + 4|a_1|^2|a_2|^2\right) - \Delta k a_1^2.$$

For the further analysis is useful to introduce modified third integral $\bar{I}_3$

$$\bar{I}_3 = \left(I_3 - \alpha\, I_1^2\right)/2\alpha = \left(1 - a_2^2\right)\left(a_2^2 + \ a_2 q\ \cos\varphi - p\right).$$

Here we have denoted $q = \alpha/\gamma$, $p = \Delta k/(2\alpha)$ and used relation

$$I_1 = a_1^2 + a_2^2 = 1.$$

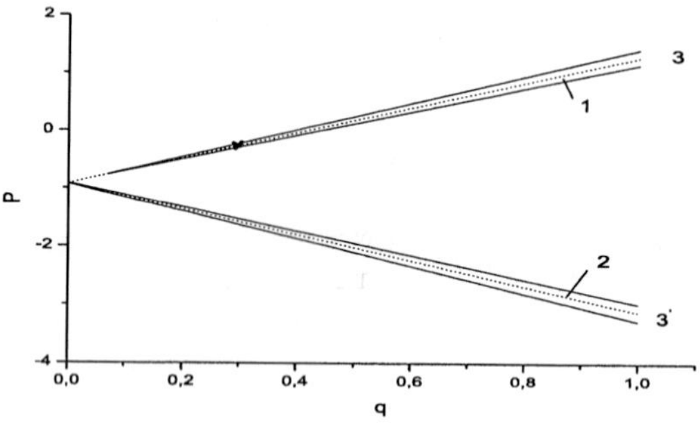

Fig. 1. Areas of χ - stability in the plain ($q = \alpha/\gamma$, $p = \Delta k/(2\alpha)$) for $a_{20} = 0.2, a_{10} = \sqrt{1 - a_{20}^2}$ and $\varphi_0 = \pi$ (area 1), $\varphi_0 = 0$ (area 2). Lines 3 and $3'$ correspond to soliton-like regimes for $\varphi_0 = \pi$ (3), $\varphi_0 = 0$ ($3'$)

Amplitudes a_1 and a_2, such as phase difference φ, remain constant during propagation if the following conditions take place:

$$\left.\frac{\partial \bar{I}_3}{\partial a_2}\right|_{a_2=a_{20},\varphi=\varphi_0} = 0 \ , \quad \left.\frac{\partial \bar{I}_3}{\partial \varphi}\right|_{a_2=a_{20},\varphi=\varphi_0} = 0$$

for input amplitudes and phase. This gives the following equation for parameters p and q

$$-4a_{20}^3 - 3a_{20}^2 q\cos\varphi_0 + 2a_{20}(1+p) + q\cos\varphi_0 = 0, \qquad (6)$$

where input phase difference φ_0 can take values 0 or π. For the given values of input amplitudes equation (6) determines values of parameters p and q for which amplitudes of FH and SH waves remain constant. Fig. 1 shows lines on the plain (p, q), determined by equation (6) for $a_{20} = 0.2$ and $\varphi_0 = 0$ (line 3') and $\varphi_0 = \pi$ (line 3). Areas 1 and 2 are the ones of χ – stability of such solutions for $\chi = 0.1$. For values (p, q) from these areas the relative change of SH amplitude does not exceed χ:

$$\frac{|a_2 - a_{20}|}{a_{20}} < \chi.$$

Propagation of flat beams of FH and SH waves under conditions (6) was numerically investigated in [12] for Gaussian (m_t=2) and hyper Gaussian (m_t=6) shapes of input pulses and various values of dispersion coefficients under their interaction in optical fiber. It was shown that hyper Gaussian pulses leads to its better preservation. Strong self-action results in considerable change of SH pulse shape at the front and the back of the pulse, where essential generation of SH takes place. Increasing of second order dispersion influence results in reduction of the medium length for which undistorted propagation of pulses takes place.

Here we present the results of numerical simulation of SH and FH pulses propagation in the bulk media under conditions (6).

4 Results of Modelling

At our numerical simulations we have chosen the following parameters: $\alpha = 10$, $\gamma = 3$, $\Delta k = -5.2$, $A_{20} = 0.2$, $A_{10} = \sqrt{1 - A_{20}^2}$, $s_{10} = 0$, $s_{20} = \pi$. Corresponding values of parameters $p$and q are shown in Fig.1 by the cross. Input pulses had hyper Gaussian shape and profile ($m_t = 6$, $m_r = 6$). Hyper Gaussian shape has been chosen because unchanging of intensity in the pulse center are better pronounced for this shape. Results of numerical simulation are shown in Fig. 2-4.

It should be noted that propagation length, at which pulses have soliton-like shape and profile, depends on dispersion and diffraction coefficients. Numerical calculations have shown that for small dispersion and diffraction ($\sim 10^{-6}$) this length exceeds 4 dimensionless units. For dispersion and diffraction 10 times larger ($\sim 10^{-5}$) it declines to 2 dimensionless units. Dispersion $\sim 10^{-1}$ reduces the length of soliton-like propagation up to 0.5 dimensionless units, while anomalous dispersion $\sim -10^{-1}$supports soliton-like propagation so that the length exceeds 1 dimensionless unit in our notation.

While propagating as soliton-like solution, FH pulse preserves its shape (Fig.2) with slight disturbances out the central part of pulse (Fig3a) far from beam axis (Fig. 3b). Slight oscillating of the central plateau part of both pulses is due to the perturbation in setting of input amplitudes – corresponding point in the plain (p, q)(Fig.1) is not precisely in the line 3, though it is in the area 1 of χ-stability.

Propagation of soliton-like SH pulse differs from FH pulse propagation (Fig.2). While the central plateau part remains constant, intensity out of this part of pulse undergoes high amplitude oscillations (Fig.3c,d). Maximum of such oscillations is 10 times greater than intensity at the plateau part of profile. As a

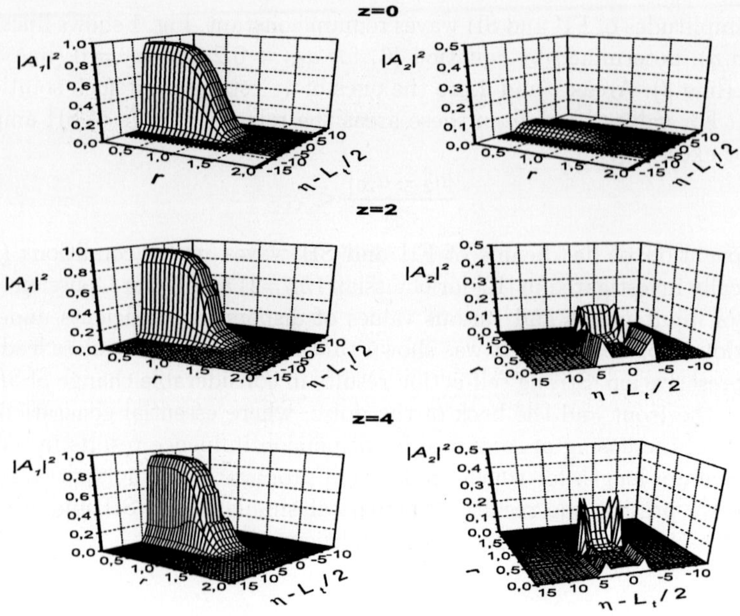

Fig. 2. Spatio-temporal distributions of intensities at various cross sections along the longitudinal coordinate for FH (left column) and SH (right column). $L_t = 30$, $\alpha = 10$, $\gamma = 3$, $\Delta k = -5.2$, $D_1 = D_2 = D_\perp = 10^{-6}$

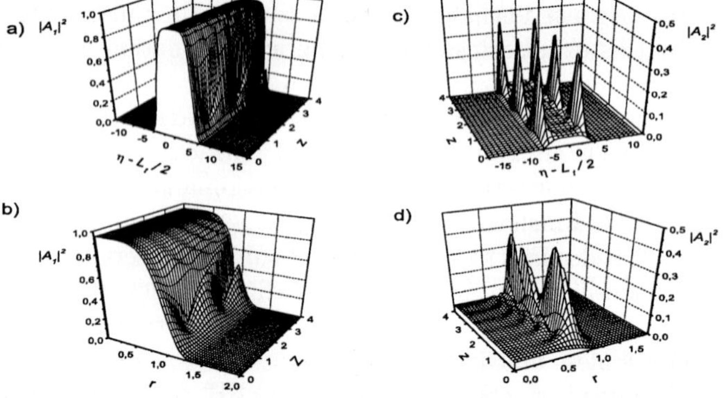

Fig. 3. Evolution of the central cross sections of the FH pulse (a,b) and SH pulse (c,d). Pictures a, c correspond to the radial cross section at $r = 0$, pictures b,d - to the time cross sections at $\eta = L_t/2$. $L_t = 30$, $\alpha = 10$,$\gamma = 3$, $\Delta k = -5.2$, $D_1 = D_2 = D_\perp = 10^{-6}$

result transvers distribution in the center of SH pulse consists of circle plateau of input intensity surrounded by the narrow ring of higher intensity (Fig.4). The intensity of the surrounding can be 2 to 10 times higher than the intensity of plateau part of the beam, depending on output section. So effective SH takes

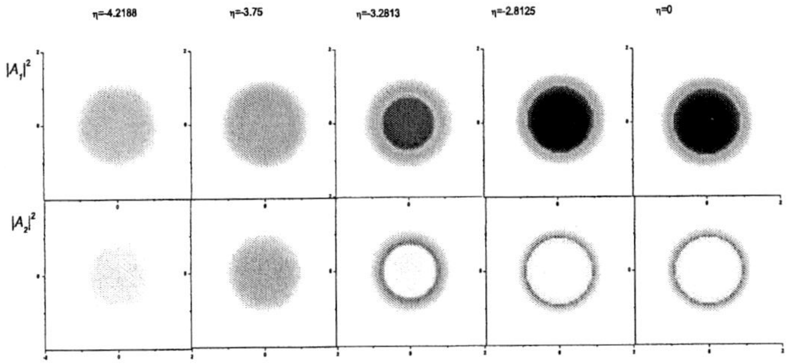

Fig. 4. A sequence of time snapshots for FH (upper set) and SH (lower set)pulses at $z = 1$. Higher intensities correspond to darker areas. Values of η given in pictures should be added to $\eta = L_t/2$ to get instances of time. $L_t = 30$, $\alpha = 10$, $\gamma = 3$, $\Delta k = -5.2$, $D_1 = D_2 = D_\perp = 10^{-6}$

place in the narrow ring around the plateau intensity domain, thus increasing total efficiency of generation up to 6 times.

5 Conclusions

Nonlinear interaction between the basic and doubling frequency waves can result in soliton-like propagation regime of both waves, even when self-action is strong. The shape and profile intensity of basic wave during soliton-like regime remains constant with slight disturbances at the pulse fronts. The doubling frequency pulse undergoes stronger distortion at the fronts, though central part of the pulse propagates without any changes. Pulse distortions increase with the growth of strong-action and results in oscillation of intensity of SH wave in the narrow ring around center plateau part of pulse. Effective SHG takes place in this ring. Its efficiency depends on the cross section.

Special conditions have to be fulfilled to get such behavior. In particular, input waves amplitudes should be nonzero and remarkable phase mismatch is required under the self-focusing. The length of soliton-like propagation depends on dispersion and diffraction of the beam. So that the stronger are dispersion and diffraction, the shorter is this length. At the same time, anomalous dispersion strongly promotes soliton-like regime. As numerical simulation has shown, anomalous dispersion increases the length by several times compared to normal dispersion of the same order.

Acknowledgement

This paper was supported in part by RFBR (grant N 02-01-727).

References

1. Steinmeyer G., Sutte P.H., Gallmann L. et al: Frontiers in ultrashort pulse generation: pushing the limits in linear and nonlinear optics. Science. **286** (1999) 1507–1560
2. Razumihina T.B., Teleguin L.S., Cholodnych A.I. et al: Three frequences interaction of intense light waves in media with quadratic and qubic nonlinearities. Quantovaya electronica (in Russian). **11** (1984) 2026–2030
3. Suchorukov. A.P. Nonlinear wave interactions in optics and radiophysics. M.: Nauka. 1988.
4. Ditmire T., Rubenchik A.M., Eimerl A. et al: The effects of cubic nonlinearity on the frequency doubling of high power laser pulses. JOSA B. **13** (1996) 649–652.
5. Arbore M.A., Fejer M.M., Ferman M. et al: Frequancy doubling of femtosecond erbium-fiber soliton lasers in periodically poled lithium niobate. Optic. Lett. **22** (1997) 13.
6. Arbore M.A., Galvanouskas A. , Harter D. et al: Egineerable compression of ultrashort pulses by use of second-harmonic generation in chirped-period-poled-lithium niobate. Optic. Lett.**22** (1997) 1341.
7. Liu X., Beckwitt K., Wise F.W.: Two-dimensional optical spatiotemporal solitons in quadratic media. Phys. Rev. **62** (2000) 1328–1340.
8. Liu X., Beckwitt K., Wise F.W.: Transverse instability of optical spatiotemporal solitons in quadratic media. Phys. Rev. Lett. **85** (2000) 1871–1874.
9. Zhang J.Y., Huang J.Y., Wang H. et al: Second-harmonic generation from regeneratively amplified femtosecond laser pulses in BBO and LBO crystals. JOSA B. **15** (1998) 200–209.
10. Wise F., Trapani P.: The hunt for light bullets - spatiotemporal solitons. Opt. Photon. News. **13** (2002) 27–30.
11. Krylov V., Rebane A., Kalintsev A.G. et al: Second-harmonic generation of amplified femtosecond Ti:sapphire laser pulses. Optic. Lett. **20** (1995) 198–200.
12. Lysak T.M., Trofimov V.A.: On the possibility of soliton-like regime for the two-wave femtosecond pulse propagation in optic fiber under conditions of SH generation. Optica & Spectroscopia (in Russian). **94** (2003) 808–814.

Computational Method for Finding of Soliton Solutions of a Nonlinear Shrödinger Equation

Vyacheslav A. Trofimov and Svetlana A. Varentsova

Lomonosov Moscow State University,
Department of Computational Mathematics and Cybernetics,
Leninskye Gory, Moscow 119992, Russia
vatro@cs.msu.su

Abstract. An effective difference method is proposed for finding of eigenvalues and eigenfunctions of a nonlinear Shrödinger equation (NLSE) with a cubic nonlinearity, describing the light beam propagation in an optical fiber. The methodical recommendations are given for the NLSE eigenfunctions construction depending on the nonlinearity coefficient and the transverse size of a waveguide.

1 Introduction

The problem of finding of spatial or temporal or spatiotemporal solitons of a light propagation in the nonlinear medium is very important for many applications. This fact causes the great interest to such problems [1]–[11]. It is well-known, the soliton solution of the corresponding nonlinear optics equation doesn't change either in time or along the coordinate of an optical radiation propagation in the nonlinear medium. If the medium is an optical fiber then the soliton preserves its shape in time. In this case the problem of information transmission by optical methods is the most essential application of solitons.

Let's note, there isn't a general method for finding of soliton solutions of corresponding nonlinear equations. We can mention the method of inverse scattering problem [1], allowing to calculate the spectrum of NLSE in the case of a weak nonlinearity. However it is very difficult to find the soliton shape (or profile) on its base. Another shortcoming of this method — it is impossible to generalize it on n-dimension problems. From our point of view there aren't other sufficiently effective and universal methods for constructing of finite on space (or time) solitons with limited energy. For this reason the soliton, which is well-known in the literature, is one that has a $\cosh^{-2}(t)$–shape, where t is, for example, a normalized time. Finite solitons of a higher order (with a more complicated shape) are practically unknown. We don't discuss below the trivial case of a $\cosh^{-2}(t)$–shape solitons composition.

Let's emphasize that in contrast to papers mentioned above we consider the problem of finding of NLSE solitons on a limited domain along the transverse coordinate. From the mathematical point of view in this case the solitons are the eigenfunctions of the corresponding NLSE with given boundary conditions.

Z. Li et al. (Eds.): NAA 2004, LNCS 3401, pp. 551–557, 2005.

Thus, it is evident to try to generalize the well-known computational methods for solving of the linear eigenvalue problem on this class of nonlinear equations. However, it turned out that many recommendations for finding of eigenfunctions and eigenvalues in the linear case aren't applied for this nonlinear case. In particular, for linear eigenvalue problems considered on the unlimited (or semi-unlimited) interval, under the computer simulation it is necessary to increase the interval on which the computational solution is looking for in order to calculate the eigenvalues more exactly. But in the case of NLSE this way isn't effective even for little values of a nonlinearity coefficient α. Therefore, the main aim of the present paper consists in finding the rule of constructing of arbitrary k-order eigenfunctions for NLSE depending on a nonlinearity coefficient not only for a weak nonlinearity but for $\alpha >> 1$ as well.

It should be noticed the difference between the eigenfunctions and the corresponding NLSE solitons. The eigenfunction hasn't to preserve the property of being the NLSE soliton in general. The reason consists in the formulation of the problem: we look for the finite solutions on a limited in time or space area. When we use the eigenfunction as the initial condition of NLSE, it can change along the propagation coordinate, if the transverse coordinate is enlarged (or shortened). However we don't consider the case of enlargement in the present paper.

2 Basic Equations

The optical beam propagation in a plane optical waveguide with a cubic nonlinear medium response can be described by a dimensionless NLSE:

$$\frac{\partial A}{\partial z} + i\frac{\partial^2 A}{\partial x^2} + i\alpha|A|^2 A = 0, \qquad z > 0, \qquad 0 < x < L \tag{1}$$

with initial and boundary conditions

$$A\Big|_{z=0} = A_0(x), \quad A\Big|_{x=0,L} = 0.$$

Above $A(x,z)$ is a complex amplitude, normalized to its maximum, x is a dimensionless transverse coordinate, z is a longitudinal coordinate along which an optical beam propagates, α is a coefficient characterizing the propagation nonlinearity, L is a transverse size of a plane waveguide. The meaning $\alpha > 0$ corresponds to the process of beam self-focusing, $\alpha < 0$ — to its defocusing.

Soliton solutions of (1) are given by

$$A(x,z) = \psi(x)\, e^{-i\lambda z},$$

with real functions $\psi(x)$ and real eigenvalues λ. Let's reduce (1) to the problem which is nonlinear on the eigenfunction:

$$\frac{d^2\psi(x)}{dx^2} + \alpha\psi^3(x) = \lambda\psi(x), \quad 0 < x < L, \tag{2}$$

$$\psi(0) = \psi(L) = 0.$$

It is easy to see that for $\alpha = 0$ the problem (2) is linear with a well-known solution:

$$\psi_k^d(x) = \sin\left(\frac{\pi k x}{L}\right), \qquad \lambda_k^d = -\left(\frac{\pi k}{L}\right)^2, \qquad k = 1, 2, \dots \tag{3}$$

In some papers (see for example [9]) the authors propose to find the soliton solutions in the form:

$$A(x, z) = \frac{1}{\sqrt{\alpha}} \psi(x)\, e^{-i\lambda z},$$

with a fixed coefficient at the nonlinear term in (2). From our point of view, it isn't always suitable because such substitution doesn't allow to study how the properties and the behavior of eigenfunctions and eigenvalues of (2) depends on α if, for example, α is close to zero.

3 Difference Scheme

For a numerical calculation of the problem (2) we introduce the regular grid on the segment $[0,\ L]$:

$$\omega_h = \{x_i = ih,\ i = 0, \dots, N,\ h = L/N\}$$

and write a difference scheme for (2) with the following iterative process:

$$\psi_{\bar{x}x,i}^{s+1} + \alpha(\psi_i^s)^2 \psi_i^{s+1} = \lambda^{s+1}\psi_i^{s+1}, \qquad i = 1, \dots, N-1, \tag{4}$$

$$\psi_0^{s+1} = \psi_N^{s+1} = 0, \qquad s = 0, 1, \dots$$

It easy to show the scheme (4) is of second-order accuracy: $\Psi = O(h^2)$. At each iteration step s the difference problem (4) is a linear eigenvalue one

$$\mathbf{L}^s\, \psi^{s+1} = \lambda^{s+1}\psi^{s+1}$$

with a three-diagonal symmetric matrix of the operator $\mathbf{L}^s$, defined as:

$$\mathbf{L}^s = \Lambda + \alpha\, I\, (\psi^s)^2.$$

Here $\Lambda y_i = y_{\bar{x}x,i}$, I is an unit operator, $(\psi_1^s, \psi_2^s, \dots, \psi_N^s) = \psi^s$. At each iteration step s the problem (4) is solved by the bisection method. The corresponding eigenvector is found by means of a three-diagonal algorithm.

Function ψ at the initial iteration $s = 0$ is built in special way. In the case of a weak nonlinearity ($|\alpha| \le 1$), it is chosen as a corresponding eigenfunction of the linear ($\alpha = 0$) difference problem:

$$\psi_{k,i}^0 = \sin\left(\frac{\pi k x_i}{L}\right).$$

However, this case isn't of great interest both for practice and for computational mathematics since for a little $|\alpha|$ the problem (4) is quasi-linear one: the influence of α on its solution is weak and the solution of (4) can be found analytically by the method of a scattering inverse problem. The case $|\alpha| \geq 1$ is of interest for us.

The iterations are terminated if the following condition is fulfilled:

$$||\psi^{s+1} - \psi^s||_C < \varepsilon_1 ||\psi^s||_C + \varepsilon_2, \qquad \varepsilon_1, \varepsilon_2 > 0.$$

The absolute solution error Ψ_k in the C-norm is checked too:

$$\Psi_k = \left|\left| \frac{d^2\psi_k(x)}{dx^2} + \alpha|\psi_k(x)|^2 \psi_k(x) - \lambda_k \psi_k(x) \right|\right|_C .$$

As we mentioned above, the amplitude $A(x, z)$ is normalized to its maximum, so we'll solve the difference problem (4) with $||\psi_k(x)||_C = 1$.

4 Computational Results

Calculations show that for all considered values of the domain length $L = 0.5 \div 20$ and a nonlinearity parameter α the problem (2) has a discrete spectrum. The iterative process converges for any examined eigenvalue numbers $k \leq 50$. The comparison of few first eigenvalues on two grids with a decreasing mesh step shows that λ_k^h are calculated with a second order accuracy. To test our method we used the first eigenfunction of (2), which can be written analytically (see for example [1–3], [8]). The second test consists in finding of eigenvalues and eigenfunctions in the case of a weak nonlinearity ($|\alpha| \leq 0.01$). For small $|\alpha|$ eigenvalues of the nonlinear difference scheme (4) are negative and close to the corresponding ones of the linear ($\alpha = 0$) case.

4.1 Case of Self-Focusing: $\alpha > 0$

We begin computer simulations for the case of self-focusing, $\alpha > 0$. For the practice of eigenfunction computations it is very important to study how the convergence and the kind of spectrum depends on a domain size L. Computer simulations show that for the length $L = 0.5 \div 2.0$ the iterative process (4) converges for any α, which is smaller than 100, and eigennumbers $k \leq 50$, for example with the step $h = 0.01$. Let's note, eigenvalues λ_k increase with increasing of α. Moreover, the value of λ_1 tends to the value $\alpha/2$ with increasing of the domain length L. For example, in the Table α_{cr} is presented, for which the following condition is fulfilled:

$$|\lambda_1 - \alpha/2| \leq 10^{-2}.$$

We can see that α_{cr} decreases very quickly with a growth of L.

For domain lengths $2 < L < 20$ the values of λ_k for $k = 2 \div 10$ with a growth of α tend to $\alpha/2$ too and form the group of very closely located eigenvalues. It causes the essential difficulties with the computation of eigenvalues and eigenfunctions.

Table 1

L	2	5	10	20
α_{cr}	36	6	0.9	0.2

It is important, if the group of the first eigenvalues λ_k are close to $\alpha/2$, the use of the exact solution of the linear ($\alpha = 0$) difference scheme (4) as the initial approximation is impossible because the iterative process doesn't converge to the required eigenvalue and eigenfunction. We use another way in such case, and construct the initial approximation giving the convergence of the iterative process. It consists in the calculation of the first eigenfunction ψ_1 on the segment with a less domain length L (for example $L = 1$) and a large value of α, for which the first k eigenvalues λ_k don't yet form the group close to $\alpha/2$. Let's note, for large meanings of α the shape of the eigenfunction ψ_1 is concave and narrow enough. Then this function ψ_1 is iterated k times on the segment with a more length L and imitates the corresponding eigenfunction ψ_k with required number of oscillations. The function constructed in such way is used as the initial approximation to find ψ_k on the segment length $L = 2, 5, 10, 20$ for those α under which the group of eigenvalues λ_k is close to $\alpha/2$.

The iterative process converges for the constructed initial approximation ψ_k^0 if the following condition takes place:

$$||\psi_k^0 - \psi_k||_C \leq \varepsilon, \tag{5}$$

where $\varepsilon = \varepsilon(L, k)$. The condition (5) means the constructed approximation must be close enough to the solution of (4). Computer simulations show, for $L = 5$ and $k = 2$ the value of ε must satisfy the condition $\varepsilon \leq 0.05$, meanwhile for $L \geq 10$ and the same k the value of ε must be less than 0.0001. It should be noted, the choice of initial α (when the length of the initial approximation segment is fixed, $L = 1$) depends on the length under consideration and on the eigennumber which are defined. In this case one can estimate the width of one separated initial approximation and choose the corresponding α.

Thus, if the first k eigenvalues form the group close to $\alpha/2$, it is possible to separate them using the specially constructed initial approximation. Unfortunately we didn't succeed to construct such approximation in any case of closely located eigenvalues.

With increasing of the domain length L up to $L = 10$ and 20 it is difficult to compute even the first eigenvalue λ_1 because in this case the first k eigenvalues tend to $\alpha/2$ much more rapidly than for the segment length $L < 10$. However, there is a possibility to improve the convergence by increasing the mesh step h and then to use the obtained eigenfunction ψ_1 as an initial approximation for the grid function with a greater number of steps. Indeed, the number of grid eigenfunctions decreases with a growth of a mesh step h and this fact simplifies the separation of eigenvalues.

There aren't difficulties with the calculation of eigenvalues and eigenfunctions with numbers $10 \leq k \leq 50$ in large intervals on α using this algorithm. For example, on the segment length $L = 10$ the values λ_k, ψ_k with $20 \leq k \leq 50$ can be calculated for $\alpha \leq 100$. The eigenvalues and eigenfunctions corresponding to $10 \leq k < 20$ are computed for $\alpha \leq 40$. If $L = 20$ it is possible to calculate λ_k, ψ_k for $30 \leq k \leq 50$ if $\alpha \leq 40$. It should be stressed, the profile of obtained eigenfunctions differs dramatically from the set of consecutive functions of the $\cosh^{-2}(x)$–type.

In conclusion of this section one can mention that in all computer experiments carried out the absolute solution error Ψ_k in the C-norm depending on a grid step h didn't exceed 10^{-4} for separate eigenvalues and 0.05 for eigenvalues close to $\alpha/2$.

4.2 Case of Defocusing: $\alpha < 0$

Computer simulations show that the case $\alpha < 0$ is similar to the previous one. Namely, the first eigenvalues and eigenfunctions can be obtained without difficulties on segments with a small length $L \leq 2$. For the fixed segment length L with a growth of the eigennumber k the interval on α for which the eigenvalue λ_k can be obtained, increases as well. For example, if $L = 1$ the eigenvalue λ_1 is computed for $|\alpha| \leq 50$, λ_2 — for $|\alpha| \leq 119$, the values of λ_{10} are obtained for $|\alpha| \leq 840$. The profile of the corresponding eigenfunction in contrast to the previous case is convex and for the fixed eigennumber k its width increases with a growth of the absolute value of α.

5 Conclusions

The results obtained above allow to give some recommendations for calculation of eigenvalues and eigenfunctions of NLSE on a limited segment length L for the practically important case $|\alpha| > 1$ (including $|\alpha| \gg 1$).

First, it is necessary to decrease the segment length to separate the eigenvalues in opposite to the case of linear Shrödinger equation considered on an unlimited interval. The reason is that any eigenvalue λ_k for $\alpha > \alpha_{cr}(L)$ tends to the $\alpha/2$ and $\alpha_{cr}(L)$ decreases dramatically with a growth of a segment length L.

Second, with a growth of the eigennumber k (for which the corresponding λ_k is close to $\alpha/2$) it is necessary to choose the initial approximating eigenfunction, which is close enough to the corresponding eigenfunction. This provides the convergence of the iterative process. We proposed one of the possible ways of such initial approximation construction based on the property of a first eigenfunction in the case of a self-focusing: the eigenfunction width decreases with a growth of a nonlinearity coefficient.

Third, it is useful to take into account the well-known property of difference schemes: the maximum number of calculated eigenfunction depends on a grid step. So, to limit the number of eigenfunctions with eigenvalues, which are close to $\alpha/2$, it is necessary to use rough grids. In this case, one can find the first k eigenfunctions. Then it is necessary to decrease the grid step and to use the eigenfunctions computed on a rough grid as an initial approximation to obtain more exact results.

Thus, the computation method for finding of NLSE eigenvalues and eigenfunctions proposed in the paper is effective for the practically interesting case $|\alpha| \in [1, 100]$. Using it the eigenfunctions ψ_k, $k \leq 50$ are found. Eigenfunctions ψ_k were used as the initial condition $A(x, 0) = \psi_k(x)$ for NLSE (1) to proof obtained results (in the point of view if they are really solitons). We considered their evolution along the coordinate z belonging to the interval $1 \leq L_z \leq 20$ with the same transverse length L. The calculations confirmed, eigenfunctions preserved their profile along z ($|A(x,0)|^2 = |A(x,z)|^2$) for the length L_z much more than the diffraction length of a separate subbeam. However, if the transverse size of an area is increased, the profile of the eigenfunctions propagating along z with $\lambda_k < \alpha/2$ doesn't preserve.

It should be noticed, the method proposed in the paper can be generalized on the case of two coordinates. The iteration method for solving of the nonlinear on eigenfunction problem (4) can be also applied to the case of a $f(\psi)\psi$–type nonlinear term in (4) and can be used on adaptive grids.

Acknowledgement. This paper was supported partly by RFBR (grant N 02-01-727).

References

1. Zaharov V.E., Manakov S.V., Novikov S.P., Pitaevskii L.P. Theory of Solitons: Inverse Problem Method. Moscow *Nauka* (1980) (in Russian)
2. Ablowitz M.Y., Segur H.: Solitons and the Inverse Scattering Transformation. SIAM Philadelphia (1981)
3. Kivshar Yu. S., Pelinovsky D.E.: Self-focusing and transverse instabilities of solitary waves. Phys. Reports. **33** (2000) 117–195
4. Maymistov A.I. Propagation of the supershort electromagnetic pulses in the nonlinear medium. Some models: Quantum electronics. **30** (2000) 287–304 (in Russian)
5. Lysak T.M., Trofimov V.A. About the possibility of the soliton-like regime of the femtosecond pulses propagation in the optical waveguide under SHG: Optics and Spectroscopy. **94** (2003) 808–814 (in Russian)
6. Liu X., Beckwitt K., Wise F.W. Two-dimensional optical spatiotemporal solitons in quadratic media: Phys. Rev. E. **1** (2000) 1328–1340
7. Liu X., Beckwitt K., Wise F.W. Transverse instability of optical spatiotemporal solitons in quadratic media: Phys. Rev. Lett. **85** (2000) 1871–1874
8. Akhmediev N.N., Mitzkevich N.V. Extremely high degree of N-soliton pulse compression in an optical fiber: IEEE J. Quant. Electron. **37** (1991) 849–857
9. Ankiewicz A., Krolikowski W., Akhmediev N.N. Partially coherent solitons of variable shape in a slow Kerr-like medium: Exact solutions: Phys. Rev. E. **59** (1999) 6079–6087
10. Gromov E.M., Talanov V.I. High order approximations of the nonlinear waves theory in the homogeneous and nonhomogeneous media: Izvestiya RAS. Physics Ser. **60** (1996) 16–28 (in Russian)
11. Montesinos G.D., Perez-Garsia V.M., Torres P.J. Stabilization of solitons of the multidimensional nonlinear Shrödinger equation: matter-wave breathers: Physica D. **191** (2004) 193–210

Convergence Analysis for Eigenvalue Approximations on Triangular Finite Element Meshes

Todor D. Todorov

Department of Mathematics, Technical University,
5300 Gabrovo, Bulgaria
paralaxview@yahoo.com

Abstract. The paper is devoted to the eigenvalue problem for a second order strongly elliptic operator. The problem is considered on curved domains, which require interpolated boundary conditions in approximating finite element formulation. The necessary triangulations for solving the eigenvalue problem consists of isoparametric elements of degree n, where n is any integer greater than two.

An approximating numerical quadrature eigenvalue problem is investigated. The considered convergence analysis is a crucial point for estimating of the error in approximating eigenvalues. An isoparametric approach is the basic tool for proving the convergence.

Keywords: eigenvalue problem, isoparametric FEM, numerical integration.

1 Introduction and Setting of the Problem

The interest in the second order eigenvalue problems consist of their considerable practical importance. Different results are obtained with respect to the finite element discretizations, convexity of the domains, smoothness of the exact solutions and boundary, type of the boundary conditions etc.

Most of the authors use piecewise linear trial functions which gives the lowest rate of convergence. Armentano and Duran [5] analyze the effect of mass-lumping in the linear triangular finite element approximation of second order elliptic eigenvalue problems. The original domain in their case is a plane polygon. The authors prove that in the case of convex domain the eigenvalue obtained by using mass-lumping is always below the one obtained with exact integration. For singular eigenfunctions, as those arising in non convex polygons, they show that the eigenvalue obtained by mass-lumping is above the exact eigenvalue when the mesh size is small enough.

Hernández and Rodriguez [11] consider the spectral problem for the Laplace equation with Neumann boundary conditions on a curved non convex domain. They prove convergence and optimal order error estimates for the eingenpairs introducing nonconforming finite element method. Unfortunately their result is

Z. Li et al. (Eds.): NAA 2004, LNCS 3401, pp. 558–565, 2005.

valid for standard piecewise linear continuous elements and polygonal computational domain only. Hernández and Rodriguez extend their results for spectral acoustic problems on curved domains in [12]. There the optimal rate of convergence is proved by straight Raviart-Thomas elements. Duran, Padra and Rodriguez present results for 3D-eigenvalue problem on polyhedral domains in [10].

The convergence of the piecewise linear eigenvalue approximations on a non convex domains are considered also by Vanmaele and Ženišek [14]. They investigate second order eigenvalue problem with Dirichlet boundary conditions. The results in [14] have been further extended by the same authors to include multiple eigenvalues [15] and numerical integration effects [16] (thus allowing to considering non constant coefficients).

General result on polygonal domains are obtained on arbitrary Lagrangian rectangular triangulations by Andreev, Kascieva and Vanmaele [1]. Note that in this case the computational domain is the original domain.

Lebaud analyzed in [13] an eigenvalue problem for second order elliptic operator with variable coefficients on domain with Lipschitz-continuous boundary. She consider simple eigenvalues and Dirichlet boundary conditions, assuming exact integration. Lebaud construct the so-called "good" approximation of the boundary by isoparametric triangular finite elements of degree $k \in \mathbf{N}$. She prove optimal rate of convergence for the eigenpairs assuming that the exact eigenfunctions are smooth enough. Lebaud does not investigate the effect of numerical integration.

Andreev and Todorov [2] consider lumped mass method for the problem introduced by Lebaud [13]. To lumping of the mass matrix they use the 7-node isoparametric element and corresponding quadrature formula. Proving the convergence they suppose that the bilinear a-form is computed exactly.

The paper deals with a second order eigenvalue problem for selfadjoint elliptic operator defined on domain with complex geometry. The problem for obtaining the rate of convergence for eigenpair approximations obtained by triangular meshes with elements of arbitrary degree is still open. The solutions of such a type problems consists of three important steps:

(i) Computing of the total quadrature error on the corresponding triangulations;
(ii) Proving of the convergence of the eigenvalue approximations;
(iii) Determining of the rate of convergence of the approximate eigenpairs;

The step (i) is already done and the results are published by Andreev and Todorov in [4]. The problem (ii) is considered in the present paper for triangulations constructed by arbitrary triangular Lagrangian finite elements of degree $k \geq 2,\ k \in \mathbf{N}$. The effect of the numerical integration is studied in the most complicated case when both sides of the eigenvalue problem are approximated by quadrature formula. For this purpose a pure isoparametric approach is used.

The major difficulty solving the problem (iii) consists of the fact that the computational domain is not the original domain. The precisely estimates of the remainders are necessary for completing (iii). The solution of (ii) is a crucial point for realizing (iii). The problem (iii) is an object of further investigations.

Let Ω be a bounded domain in $\mathbf{R}^2$, with Lipschitz-continuous boundary. We consider the case when the piecewise C^{n+1} (n is any integer greater than two) boundary $\Gamma = \partial\Omega$ is curved. Denote the space of generalized functions on Ω by $\mathcal{D}'(\Omega)$. Define the operator

$$Lu = -\sum_{i,j=1}^{2} \frac{\partial}{\partial x_j}\left(a_{ij}\frac{\partial u}{\partial x_i}\right) + a_0 u, \quad u \in \mathcal{D}'(\Omega),$$

where a_{ij} belong to $C^1(\Omega)$, $a_{ij} = a_{ji}$, $i, j = 1, 2$ and $a_0 \in C(\Omega)$, $a_0(x) \geq \underline{a} > 0$, $\forall x \in \Omega$, $\underline{a} = \text{const}$. Assume that L is strongly elliptic, i.e. there exists a constant $\alpha_0 > 0$ such that

$$\sum_{i,j=1}^{2} a_{ij}(x)\xi_i\xi_j \geq \alpha_0 \sum_{i=1}^{2} \xi_i^2, \quad \forall \xi, x \in \mathbf{R}^2.$$

We study the eigenvalue problem

$$\mathcal{P} \; : \quad \begin{cases} \text{find } \lambda \in \mathbf{R} \text{ and a nontrivial function } u \in \mathcal{D}'(\Omega) \text{ such that} \\ Lu = \lambda u \text{ in } \Omega, \\ u = 0 \text{ on } \Gamma, \end{cases}$$

for the operator L.

Let $H^m(\Omega)$, $m \in \mathbf{N}$ be the real Sobolev space provided with the seminorms $|\cdot|_{m,\Omega}$ and norms $\|\cdot\|_{m,\Omega}$ [7].

Define the space $\mathbf{V} = \{v \in H^1(\Omega) \mid v = 0 \text{ on } \Gamma\}$. Consider the bilinear form

$$a(u,v) = \int_\Omega \sum_{i,j=1}^{2} a_{ij}(x)\frac{\partial u}{\partial x_i}\frac{\partial v}{\partial x_j}\, dx + \int_\Omega a_0(x)uv\, dx$$

on $\mathbf{V} \times \mathbf{V}$. Write the weak formulation of the problem $\mathcal{P}$ by

$$\mathcal{P}_W \; : \quad \begin{cases} \text{find } \lambda \in \mathbf{R} \text{ and a nontrivial function } u \in \mathbf{V} \text{ such that} \\ a(u,v) = \lambda(u,v), \quad \forall v \in \mathbf{V}. \end{cases}$$

2 Preliminaries

We use isoparametric triangular Lagrangian finite elements of degree $n \geq 2$. The elements of higher degree (quartic and quintic elements) are considered by Brenner and Scott [6–p. 72]. Introduce a finite element discretization for solving the problem $\mathcal{P}_W$.

We suppose that all finite elements in the triangulation τ_h of the domain Ω are isoparametric equivalent to one finite element $(\hat{K}, \hat{P}, \hat{\Sigma})$ called finite element of reference:

$\hat{K} = \{(\hat{x}_1, \hat{x}_2) \mid \hat{x}_1 \geq 0,\ \hat{x}_2 \geq 0,\ \hat{x}_1 + \hat{x}_2 \leq 1\}$ is the canonical 2-simplex;

$\hat{P} = P_n(\hat{K})$, where P_n is the space of all polynomials of degree, not exceeding n;

$\hat{\Sigma} = \{\hat{x} = (\hat{x}_1, \hat{x}_2) \mid \hat{x}_1 = \frac{i}{n},\ \hat{x}_2 = \frac{j}{n};\ i+j \leq n;\ i,j \in \mathbf{N}\bigcup\{0\}\}$ is the set of all Lagrangian interpolation nodes.

An arbitrary finite element $K \in \tau_h$ is defined by $K = F_K(\hat{K})$, where $F_K \in \hat{P}^2$ is an invertible transformation. Let

$$h_K = \mathrm{diam}(K), \quad h = \max_{K \in \tau_h} h_K, \quad \forall K \in \tau_h.$$

We use not only straight elements but also isoparametric elements with one curved side for getting good approximation of the boundary Γ. Thus we obtain a perturbed domain $\Omega_h = \bigcup_{K \in \tau_h} K$ of the domain Ω.

The notations $C,\ C_1,\ C_2, \ldots$, are reserved for generic positive constants, which may vary with the context.

We define a finite element space associated with a triangulation τ_h by

$$\mathbf{V}_h = \{v \in C(\mathbf{R}^2) \mid v(x) = 0 \text{ if } x \notin \Omega_h;\ v_{|K} \in P_K, K \in \tau_h\},$$

where $P_K = \{p : K \to \mathbf{R} \mid p = \hat{p} \circ F_K^{-1},\ \hat{p} \in \hat{P}\}$. The finite element space V_h is constructed on the basis of $(n+1)(n+2)/2$ - node isoparametric triangular elements.

The boundaries Γ and Γ_h are close by a little h, then there exists a bounded open set $\widetilde{\Omega}$, which satisfies $\Omega \subset \widetilde{\Omega},\ \Omega_h \subset \widetilde{\Omega}$ for all considered triangulations τ_h. We suppose that every function from $\mathbf{V}$ is extended by zero outside of Ω to $\mathbf{R}^2$ in a continuous way.

Further we shall use the space $\mathbf{W}_h = \mathbf{V} + \mathbf{V}_h$ (see [6–p. 198]). We define the approximating bilinear form

$$A_h(u,v) = \int_{\widetilde{\Omega}} \sum_{i,j=1}^{2} \widetilde{a}_{ij}(x) \frac{\partial u}{\partial x_i} \frac{\partial v}{\partial x_j}\, dx + \int_{\widetilde{\Omega}} \widetilde{a}_0(x) uv\, dx, \quad u, v \in \mathbf{W}_h, \tag{1}$$

where $\widetilde{a}_{ij},\ \widetilde{a}_0 \in L^\infty(\widetilde{\Omega})$ are continuous extensions of the coefficients $a_{ij},\ a_0$ over $\widetilde{\Omega}$. We shall write the scalar product in the spaces $L^2(\Omega)$, $L^2(\Omega_h)$ and $L^2(\widetilde{\Omega})$ by one and the same denotation $(\cdot,\cdot)$. Suppose that the bilinear forms (1) are uniformly $\mathbf{W}_h$-elliptic, i.e. there exists a constant $\widetilde{\beta}$ independent of the spaces $\mathbf{W}_h$ such that for all h sufficiently small and $\forall v \in \mathbf{W}_h$

$$\widetilde{\beta} \|v\|^2_{1,\widetilde{\Omega}} \leq A_h(v,v).$$

Thus we obtain the consistent mass eigenvalue problem

$$\mathcal{P}_C \quad : \quad \begin{cases} \text{find } \lambda^h \in \mathbf{R} \text{ and a nontrivial function } u^h \in \mathbf{V}_h \text{ such that} \\ A_h(u^h, v) = \lambda^h (u^h, v), \quad \forall v \in \mathbf{V}_h. \end{cases}$$

Further we present some results concerning isoparametric numerical integration.

To evaluate integrals over the finite element of reference $\hat{K}$ numerically, we use the quadrature formula

$$\int_{\hat{K}} \hat{\psi}(\hat{x})\, d\hat{x} \cong \hat{I}(\hat{\psi}) = \sum_{i=1}^{\hat{N}} \hat{\omega}_i \hat{\psi}(\hat{a}_i), \tag{2}$$

where $\hat{a}_i$ are the nodes of the quadrature formula and $\hat{\omega}_i > 0$.

Define a quadrature formula $I_K(\psi)$ over the finite element K for continuous ψ by

$$\int_K \psi(x)\, dx \cong I_K(\psi) = \hat{I}(J_{F_K}\hat{\psi}), \tag{3}$$

where J_{F_K} is the Jacobian of transformation F_K. The integrals over Ω_h will be computed element by element using (3).

Denoting the error functionals

$$\hat{E}(\hat{\psi}) = \int_{\hat{K}} \hat{\psi}\, d\hat{x} - \hat{I}(\hat{\psi}), \quad \forall \hat{\psi} \in C(\hat{K}),$$

$$E_K(\psi) = \int_K \psi\, dx - I_K(\psi), \quad \forall \psi \in C(K),$$

$$E(u,v) = \sum_{K\in\tau_h} E_K(uv), \quad \forall u,v \in \mathbf{W}_h,$$

$$E^a(u,v) = \sum_{K\in\tau_h} E^a_K(uv)$$

$$= \sum_{K\in\tau_h} E_K\left(\widetilde{a}_{ij}\frac{\partial u}{\partial x_i}\frac{\partial v}{\partial x_j} + \widetilde{a}_0 uv\right), \quad \forall u,v \in \mathbf{W}_h.$$

Further, we need the following hypotheses.

(H1) The Jacobian J_{F_K} of isoparametric transformation F_K, $K \in \tau_h$ satisfies the relations:

$$|J_{F_K}|_{m,\infty,\hat{K}} = Ch^{m+2}, \quad m = 0,1,\ldots,n-1, \quad J_{F_K} \in P_{n-1}(\hat{K}). \qquad \square$$

(H2) The triangulation τ_h is n-regular in the sense of Ciarlet and Raviart [8]. $\square$

(H3) We assume that the quadrature formula (2) is exact for all polynomials of $P_{2n-1}(\hat{K})$, $\hat{\omega}_i > 0$, $\forall i = 1,2,...,\hat{N}$ and the union $\bigcup_{i=1}^{\hat{N}} \hat{\omega}_i$ contains $\hat{P}$-unisolvent subset. $\square$

The estimates

$$|E(u_h,v_h)| \le Ch^{2s}\|u_h\|_{s,\Omega_h}\|v_h\|_{s,\Omega_h}, \quad s = 0,1, \tag{4}$$

$$|E^a(u_h,v_h)| \le Ch^{2s}\|u_h\|_{s,\Omega_h}\|v_h\|_{s,\Omega_h}, \quad s = 0,1, \tag{5}$$

$\forall u_h, v_h \in \mathbf{V}_h$, $\widetilde{a}_{ij}$, $\widetilde{a}_0 \in W^{2n,\infty}(\widetilde{\Omega})$ were proved as a corollary of hypotheses (H1)-(H3) in [4].

3 Numerical Quadrature Isoparametric Problem

Introduce the scalar product and norm in V_h by

$$(u,v)_h = \sum_{K\in\tau_h} I_K(uv), \quad \|v\|_h = \sqrt{(v,v)_h}, \quad \forall u,v \in \mathbf{V}_h.$$

Taking into account the fact that Hypothesis (H3) holds, it is easy to see that the norms $\|\cdot\|_h$ and $\|\cdot\|_{0,\Omega_h}$ are uniformly equivalent on the space V_h.

Define the bilinear form

$$a_h(u,v) = \sum_{K\in\tau_h} I_K\left(\sum_{i,j=1}^{2} \widetilde{a}_{ij}\frac{\partial u}{\partial x_i}\frac{\partial v}{\partial x_j} + \widetilde{a}_0 uv\right).$$

Introduce an approximate eigenvalue problem

$$\mathcal{P}_{NQ} \quad : \quad \begin{cases} \text{find } \widetilde{\lambda}^h \in \mathbf{R} \text{ and a nontrivial function } \widetilde{u}^h \in \mathbf{V}_h \text{ such that} \\ a_h(\widetilde{u}^h, v) = \widetilde{\lambda}^h(\widetilde{u}^h, v)_h, \quad \forall v \in \mathbf{V}_h, \end{cases}$$

corresponding to $\mathcal{P}_W$.

Denote the approximate eigenpairs by $(\widetilde{\lambda}^h_k, \widetilde{u}^h_k)$, $1 \le k \le N(h) = \dim V_h$. We have a set of positive eigenvalues

$$0 < \widetilde{\lambda}^h_1 \le \widetilde{\lambda}^h_2 \le \dots \le \widetilde{\lambda}^h_{N(h)}$$

and normalize the eigenfunctions by

$$(\widetilde{u}^h_i, \widetilde{u}^h_j)_h = \delta_{ij}, \quad 1 \le i,j \le N(h).$$

4 Convergence of the Eigenvalue Approximations

The hypotheses (H2) and (H3) assure that the bilinear form $a_h(\cdot,\cdot)$ are uniformly $\mathbf{V}_h$-elliptic. Define the Rayleigh quotients by

$$R(u,v) = \frac{a(u,v)}{(u,v)}, \quad u,v,\in \mathbf{V},$$

$$R_h(u_h,v_h) = \frac{A_h(u_h,v_h)}{(u_h,v_h)}, \quad \widetilde{R}_h(u_h,v_h) = \frac{a_h(u_h,v_h)}{(u_h,v_h)_h}, \quad \forall u,v,\in \mathbf{V}_h.$$

The next theorem contains the main result of the paper.

Theorem 1. *Let λ_k and $\widetilde{\lambda}^h_k$ be simple eigenvalues of problems $\mathcal{P}_W$ and $\mathcal{P}_{NQ}$ respectively. Let also hypotheses (H1)-(H3) hold, then $\widetilde{\lambda}^h_k \to \lambda_k$ when $h \to 0$.*

Proof. It is well known fact that if λ_k is a simple exact eigenvalue of problem $\mathcal{P}_W$ then the following inf-sup condition is valid (see Courant and Hilbert [9])

$$\lambda_k = \inf_{E^k\subset \mathbf{V}} \sup_{v\in E^k} R(v,v),$$

where E^k denotes a k-dimensional subspace of $\mathbf{V}$ ($k \le \dim \mathbf{V}_h$). Then for the consistent mass approximations and numerical quadrature formula approximations we have

$$\lambda_k^h = \inf_{E_h^k \subseteq \mathbf{V}_h} \sup_{v_h \in E_h^k} R_h(v_h, v_h)$$

and

$$\widetilde{\lambda}_k^h = \inf_{E_h^k \subseteq \mathbf{V}_h} \sup_{v_h \in E_h^k} \widetilde{R}_h(v_h, v_h).$$

The set E_h^k is a k-dimensional subspace of $\mathbf{V}_h$.

We estimate the difference between the Rayleigh quotients by the error functionals

$$\left|\widetilde{R}_h^{-1}(v_h, v_h) - R_h^{-1}(v_h, v_h)\right| \le \left|\frac{(v_h, v_h)_h}{a_h(v_h, v_h)} - \frac{(v_h, v_h)}{A_h(v_h, v_h)}\right|$$

$$\le \left|\frac{(v_h, v_h)_h}{a_h(v_h, v_h)} - \frac{(v_h, v_h)}{a_h(v_h, v_h)}\right| + \left|\frac{(v_h, v_h)}{a_h(v_h, v_h)} - \frac{(v_h, v_h)}{A_h(v_h, v_h)}\right|$$

$$= \frac{|E(v_h, v_h)|}{a_h(v_h, v_h)} + \frac{(v_h, v_h)}{A_h(v_h, v_h)} \left|\frac{A_h(v_h, v_h) - a_h(v_h, v_h)}{a_h(v_h, v_h)}\right|$$

$$= \frac{1}{a_h(v_h, v_h)} \left(|E(v_h, v_h)| + R_h^{-1}(v_h, v_h)\,|E^a(v_h, v_h)|\right).$$

We obtain

$$\left|\widetilde{R}_h^{-1}(v_h, v_h) - R_h^{-1}(v_h, v_h)\right| \le Ch^2 \left(1 + R_h^{-1}(v_h, v_h)\right)$$

by the $\mathbf{V}_h$-ellipticity of bilinear form $a_h(.,.)$ and inequalities (4) and (5). Then

$$(1 - Ch^2)R_h^{-1}(v_h, v_h) - Ch^2 \le \widetilde{R}_h^{-1}(v_h, v_h) \le (1 + Ch^2)R_h^{-1}(v_h, v_h) + Ch^2.$$

Taking inf-sup in the latter two inequalities we have

$$(1 - Ch^2)\left(\lambda_k^h\right)^{-1} - Ch^2 \le \left(\widetilde{\lambda}_k^h\right)^{-1} \le (1 + Ch^2)\left(\lambda_k^h\right)^{-1} + Ch^2$$

for h sufficiently small.

Since $\lambda_k^h \to \lambda_k$ when $h \to 0$ (see Lebaud [13]) we get that

$$\lim_{h \to 0} \left(\widetilde{\lambda}_k^h\right)^{-1} = (\lambda_k)^{-1},$$

which completes the proof. □

Theorem 1 enable us to obtain optimal order estimates for the error in solutions of problem $\mathcal{P}_{NQ}$, when triangular elements of arbitrary degree are used. On the other hand this theorem could be used for computing of the rate of convergence of the lumped mass approximations.

References

1. A. B. ANDREEV, V. A. KASCIEVA, M. VANMAELE, *Some results in lumped mass finite element approximation of eigenvalue problems using numerical quadrature formulas,* J. Comp. Appl. Math., vol. 43, 1992, pp. 291-311.
2. A. B. ANDREEV, T. D. TODOROV, *Lumped mass approximation for an isoparametric finite element eigenvalue problem,* Sib. J. of Numerical Mathematics, $N^{\underline{o}}4$, vol. 2, 1999, pp. 295-308.
3. A. B. ANDREEV, T. D. TODOROV, *Lumped mass error estimates for an isoparametric finite element eigenvalue problem,* Sib. J. of Numerical Mathematics, $N^{\underline{o}}3$, vol. 3, 2000, pp. 215-228.
4. A. B. ANDREEV, T. D. TODOROV, *Isoparametric Numerical Integration on Triangular Finite Element Meshes,* Comptes rendus de l'Academie bulgare des Sciences, v 57, 7, 2004.
5. M. G. ARMENTANO AND R. G. DURAN, *Mass lumping or not mass lumping for eigenvalue problems,* Numer. Meth. PDE's 19(5), 2003, pp. 653-664.
6. S. C. BRENNER AND L. R. SCOTT, *The mathematical theory of finite element methods,* Texts in Appl. Math., Springer-Verlag, 1994.
7. P. G. CIARLET, *The finite element method for elliptic problems,* North-Holland, Amsterdam, 1978.
8. P. G. CIARLET AND P. A. RAVIART, *The combined effect of curved boundaries and numerical integration in isoparametric finite element method,* Math. Foundation of the FEM with Applications to PDE, A. K. Aziz, Ed., New York, Academic Press, 1972, pp. 409-474.
9. R. COURANT AND HILBERT, *Methods of Mathematical Physics,* vol. II, Interscience, New York, 1962.
10. R. DURAN, C. PADRA, R. RODRIGUEZ, *A posteriori error estimates for the finite element approximation of eigenvalue problems*, Mathematical Models and Methods in Applied Sciences (M3AS), 13, 2003, pp. 1219-1229.
11. E. HERNÁNDEZ; R. RODRIGUEZ, *Finite element approximation of spectral problems with Neumann boundary conditions on curved domains,* Math. Comp. 72, 2003, pp. 1099-1115.
12. E. HERNÁNDEZ, R. RODRIGUEZ, *Finite element approximation of spectral acoustic problems on curved domains,* Numerische Mathematik, 97, 2004, pp. 131-158.
13. M. P. LEBAUD, *Error estimate in an isoparametric finite element eigenvalue problem,* Math. of Comp. vol. 63, 1994, pp. 19-40.
14. M. VANMAELE AND ŽENIŠEK, *External finite element approximations of eigenvalue problems.* R.A.I.R.O., Modélisation Math. Anal. Numér. 27, (1993), 565-589.
15. M. VANMAELE AND ŽENIŠEK, *External finite-element approximations of eigenfunctions in the case of multiple eigenvalues*, J. Comput. Appl. Math., 50, 1994, 51-66.
16. M. VANMAELE AND ŽENIŠEK, *The combined effect of numerical integration and approximaion of the boundary in the finite element method for eigenvalue problems,* Numer. Math. 71 (1995), 253-273.

A Standard CGA Implementation

Algorithm CGS(**in** A, adr, c, y;**out** x) (* CGA *Axelsson's variant without preconditioning* *)
Auxiliary: double $d[1,\ldots,n], p[1,\ldots,n], r[1,\ldots,n], nom, nory, denom, \alpha$;
(1) $nory = 0$;
(2) **for** $i = 1$ **to** n **do**
(3) $\{ \; nory+ = y[i] * y[i]; \;\; r[i] = -y[i]; \;\; x[i] = 0.0; \;\; d[i] = y[i]; \; \}$
(4) $nom = nory$;
(5) **while** (residual vector is "large") **do** {
(6) **call** MVM($A, adr, c, d; p$);
(7) **call** Dot_product($d, p; denom$); $\alpha = nom/denom$;
(8) **for** $i = 1$ **to** n **do** $\{ \; x[i]+ = \alpha * d[i]; \;\; r[i]+ = \alpha * p[i]; \; \}$
(9) $denom = nom$; **call** Dot_product($r, r; nom$); $\alpha = nom/denom$;
(10) **for** $i = 1$ **to** n **do** $\{ \; d[i] = \alpha * d[i] - r[i]; \; \}$

This code has a serious drawback. If the data cache size is less than the total memory requirements for storing all input, output, and auxiliary arrays $(A, adr, c, d, p, r, x, y)$, then due to thrashing misses, part or all of these arrays are flushed out of the cache and must be reloaded during the next iteration of the **while** loop at codelines (5)-(10). This inefficiency can be reduced by application of the *loop reversal* [3] and *loop fusion* [3].

An Improved CGA Implementation Based on Loop Restructuring

Algorithm CGM(**in** A, adr, c, y;**out** x) (* *modified implementation* of CGS*)
Auxiliary: double $d[1,\ldots,n], p[1,\ldots,n], r[1,\ldots,n], nom, nory, denom, \alpha$;
(1) $nory = 0$;
(2) **for** $i = 1$ **to** n **do**
(3) $\{nory+ = y[i] * y[i]; \;\; r[i] = -y[i]; \;\; x[i] = 0.0; \;\; d[i] = y[i];\}$
(4) $nom = nory$;
(5) **while** (residual vector is "large") **do** {
(6') **Loop fusion** (MVM($A, adr, c, d; p$), Dot_product($d, p; denom$));
(8') $\alpha = nom/denom; \;\; nory = 0.0$;
(8'') **for** $i = 1$ **to** n **do** $\{x[i]+ = \alpha * d[i]; \;\; r[i]+ = \alpha * p[i]; \;\; nory+ = r[i] * r[i];\}$
(9') $denom = nom; \;\; nom = nory; \;\; \alpha = nom/denom$;
(10') **for** $i = n$ **downto** 1 **do** $\{ \; d[i] = \alpha * d[i] - r[i]; \; \}$

CGM code has been obtained from CGS code by applying 3 transformations:

1. Codelines (6, 7) in CGS are grouped together by *loop fusion* and this allows to reuse immediately the new computed values of array p.
2. Similarly, codelines (8, 9) in CGS are grouped together by loop fusion and this allows to reuse immediately the new computed values of array r (see codelines (8"), (9') in CGM).

3. The loop on codeline (10) in `CGS` is reversed by *loop reversal* so that the last elements of arrays d and r that remain in the cache from the loop on codelines (8, 9) can be reused.

The `CGM` code has better *temporal locality* of data and in Section 4.2 we perform its quantitative analysis.

4 Probabilistic Analysis of the Cache Behavior

4.1 Sparse Matrix-Vector Multiplication

Algorithms `MVM_CSR` and `MVM_SSS` (see Section 2.1) produce the same sequence of memory access instructions and therefore, the analytical model is the same for both. It is based on the following simplified assumptions (same as in [1]):

1. There is no correlation among mappings of arrays A, c, and x into cache blocks. Hence, we can view load operations as mutually independent events.
2. We consider *thrashing misses* only for loads from arrays A, c, and x.
3. We assume that the whole cache size is used for data of $SpM{\times}V$.
4. We assume that each execution of $SpM{\times}V$ starts with the empty cache.

We use the following notation.

- $P(Z[i])$ denotes the probability of a thrashing miss of the cache block containing element i of array Z.
- N_{CM} denotes the number of cache misses during one execution of $SpM{\times}V$.
- d denotes the average number of iterations of the innermost loop of `MVM_CSR` at codeline (4) between 2 reuses of the cache block containing some element of array $x[1,\ldots,n]$.

We distinguish 3 relevant types of sparse matrices to estimate the value of d. (1) A symmetric sparse matrix $\mathcal{A}$ with bandwidth w_{B} and with *uniform* distribution of nonzero elements on rows. Then d can be approximated by $d = w_{\mathrm{B}}$ [2]. (2) A symmetric sparse banded matrix $\mathcal{A}$ with *similar row structure*. Two rows i and $i+1$ are said to be *similar* if row i contains nonzero elements $\cdots, \mathcal{A}[i][i-\Delta], \mathcal{A}[i][i], \mathcal{A}[i][i+\Delta], \cdots$, whereas row $i+1$ contains nonzero elements $\cdots, \mathcal{A}[i+1][i+1-\Delta], \mathcal{A}[i+1][i+1], \mathcal{A}[i+1][i+1+\Delta], \cdots$, where Δ is a constant. In other words, we assume that the indexes of nonzero elements of row $i+1$ are only "shifted" by one position with respect to row i. For A with structurally similar rows, the cache block containing an element $x[i]$ is reused with high probability during loading $x[i+1]$ after n_{zpr} iterations[1]. Hence, $d = \min(w_{\mathrm{B}}, n_{\mathrm{zpr}}) = n_{\mathrm{zpr}}$.

---

[1] For simplicity, we assume that all rows corresponding to discretization of internal mesh nodes are similar. But in real applications, boundary mesh nodes produce equations with a slightly different structure. This simplification should not have a significant impact, since the number of boundary nodes is of order of square root of all mesh nodes.

(3) A sparse banded matrix A where $\frac{B}{zpr} \approx 1$. Then $x[i]$ is reused with high probability during loading $x[i+1]$ in the next iteration, and therefore, $d = 1$.

Due to the assumption that each new execution of $SpM\times V$ starts with the empty cache, all n_Z elements of arrays A and c and all n elements of arrays x, adr, y must be loaded into the cache once and the number of *compulsory misses* is $N_{CM}^{C} = \frac{z\ (\ D+\ I)+\ \ (2\cdot\ D+\ I)}{S}$.

Since caches have always limited size, thrashing misses occur: Data loaded into a cache set may cause replacement of other data that will be needed later. Hence, the total number of cache misses is $N_{CM} = N_{CM}^{C} + N_{CM}^{T}$.

Symmetric SpM×V. The same assumptions are valid even for symmetric case. We consider the SSS format (see 2),let denote n_Z' the number of nonzero elements in strictly lower triangular submatrix. Then the number of *compulsory misses* is ${N_{CM}^{C}}' = \frac{z'\ (\ D+\ I)+\ \ (3\cdot\ D+\ I)}{S}$, ${N_{CM}}' = {N_{CM}^{C}}' + {N_{CM}^{T}}'$.

Direct Mapped Cache ($s = 1$). The innermost loop of the `MVM_CSR` algorithm has n_Z iterations. Each element of arrays A, c, and x in each iteration is either reused or replaced due to thrashing. Under our assumption of independence of these replacements for all 3 arrays, the total number of thrashing misses can be approximated by formula: $N_{CM}^{T} = n_Z\,(P(A[j]) + P(c[j]) + P(x[k])) \quad ; \forall j, k$.

The probability that 2 randomly chosen cache blocks from the cache are distinct is $1 - \frac{s}{S} = 1 - h^{-1}$. Hence, $P(c[j]) = P(A[j]) = 1 - (1 - h^{-1})^2$.

Arrays A and c are accessed linearly, their indexes are incremented in each iteration of the innermost loop, and in a given moment, only one cache block is actively used unless thrashing occurs. The access pattern for array x is more complicated due to indirect addressing. In the worst case, an element of x, after loading into the cache, is reused only after d iterations of the innermost loop, since it is used at each row of matrix A only once (said otherwise, array x actively uses d cache blocks). Every load during this time can cause cache thrashing. Hence,$P(x[k]) = 1 - (1 - h^{-1})^3$.

Symmetric SpM×V. We can make similar assumptions. Vector $\boldsymbol{y}$ is accessed using the same pattern as vector $\boldsymbol{x}$, so $P(c[j]) = P(A[j]) = 1-(1-h^{-1})^3 \quad P(x[k]) = P(y[k]) = 1 - (1 - h^{-1})^4 \quad {N_{CM}^{T}}' = n_Z'\,(P(A[j]) + P(c[j]) + P(x[k])) \quad ; \forall j, k$.

s-Way Set-Associative Cache, Random Replacement Strategy
Standard SpM×V. The probability can be derived as in the previous section, only the thrashing occurs with probability $\frac{1}{s}$. So, one of s cache blocks containing elements of A or c can be replaced by both loads, which are assumed independent, with probability $P(A[j]) = P(c[j]) = \frac{1}{s}(1 - (1 - h^{-1})^2)$ and $P(x[k]) = \frac{1}{s}(1 - (1 - h^{-1})^3\ )$.

Symmetric SpM×V. We can make similar assumptions. Vector $\boldsymbol{y}$ is accessed with same pattern like vector $\boldsymbol{x}$, so $P(A[j]) = P(c[j]) = \frac{1}{s}(1 - (1 - h^{-1})^3)$ and $P(x[k]) = P(y[k]) = \frac{1}{s}(1 - (1 - h^{-1})^4\ )$.

s-Way Set-Associative Cache, LRU Replacement Strategy
In the LRU case, cache block can be replaced only if at least s immediately preceding loads accessed this block. Hence,

$$P(A[j]) = P(c[j]) = \begin{cases} 0 & \text{if } s > 2, \\ h^{-2} & \text{if } s = 2, \\ 1-(1-h^{-1})^2 & \text{if } s = 1. \end{cases} \tag{1}$$

Arrays A and c are accessed in the linear order of indexes as before and during d iterations, they completely fill $\varrho = \lfloor \frac{(D+1)}{S} \rfloor$ cache sets. We distinguish 2 cases of sparsity of matrix A:

- $\frac{B}{zpr} \approx 1$. This corresponds to a sparse banded matrix $\mathcal{A}$ of type (3) in Section 4.1 where cache sets are almost completely filled with array x. Then memory access pattern is the same as for arrays A and c, $P(x[k]) = P(A[j]) = P(c[j])$.
- $\frac{B}{zpr} \geq n_{\mathrm{dpb}}$. This corresponds to a sparse matrix $\mathcal{A}$ of type (1) or (2) in Section 4.1 such that every cache block contains at most one element of array x during one execution of $SpM{\times}V$. If the load operations for arrays A or c replace a cache block containing array x, one thrashing miss occurs, $P(x[k]) = 1 - (1 - h^{-1})$.

Symmetric $SpM{\times}V$. Similarly,

$$P(A[j]) = P(c[j]) = \begin{cases} 0 & \text{if } s > 3, \\ h^{-3} & \text{if } s = 3, \\ 3h^{-2} & \text{if } s = 2, \\ 1-(1-h^{-1})^3 & \text{if } s = 1. \end{cases} \tag{2}$$

Vector $\boldsymbol{y}$ is accessed with the same pattern like vector $\boldsymbol{x}$, so $P(y[k]) = P(x[k])$.

4.2 CGA

We consider the same data structures A, adr, c, x, y as in Algorithms `CGS` and `CGM`. Let us further assume the following simplified conditions.

1. There is no correlation among mappings of arrays into cache blocks.
2. Thrashing misses occur only within the subroutines for $SpM{\times}V$. Hence, we consider only *compulsory* cache misses.
3. We again assume that the whole cache size is used for input, output, and auxiliary arrays.
4. We assume that each iteration of the **while** loop of the `CGA` starts with an empty cache.

Define

NS = the predicted number of cache misses that occur on codeline (l) of the `CGS` algorithm.
NM = the predicted number of cache misses that occur on codeline (l) of the `CGM` algorithm.

The total number of cache misses for $SpM{\times}V$ on codeline (6) in CGS and on codeline (6') in CGM was evaluated in Section 4.1, $NS_{(6)} = NM_{(6')} = N_{\mathrm{CM}}$. The number of compulsory misses for the dot product of 2 vectors on codeline (7) in CGS is the number of cache blocks needed for $2n$ doubles. Whereas in CGM, the dot product is computed directly during the multiplication due to the loop fusion and all elements of arrays d and p are reused. Hence, $NS_{(7)} = \frac{2}{\mathrm{dpb}}$ and $NM_{(7')} = 0$.

The codeline (8) contains 2 linear operations on 4 vectors and the same holds for CGM. Therefore,$NS_{(8)} = NM_{(8'+8'')} = \frac{4}{\mathrm{dpb}}$. The codeline (9) contains a dot product of vector r, whereas in CGM on the codeline (8") the same dot product is computed directly during those linear vector operations and all elements of array r are reused. So,$NS_9 = \frac{}{\mathrm{dpb}}$ and $NS_{9'} = 0$.

The codeline (10) contains linear operations on 2 vectors. For large n, we can assume that after finishing the loop on the codeline (9), the whole cache is filled only with elements of arrays d, p, r, and x. In CGM, the loop on codeline (10') is reversed and so, in the best case, the last $\frac{\mathrm{S}}{4\ \mathrm{D}}$ elements of array r (similarly for array d) can be reused. Therefore, $NS_{10} = \frac{2}{\mathrm{dpb}}$ and $NM_{10'} = \frac{2}{\mathrm{dpb}} - \frac{\mathrm{S}}{2\ \mathrm{S}}$.

4.3 Evaluation of the Probabilistic Model

Figure 1 gives performance numbers for Pentium Celeron 1GHz, 256 MB, running W2000 Professional with the following cache parameters: L1 cache is data cache with $B_{\mathrm{S}} = 32$, $C_{\mathrm{S}} = 16K$, $s = 4$, $h = 128$, and LRU replacement strategy. L2 cache is unified with $B_{\mathrm{S}} = 32$, $C_{\mathrm{S}} = 128K$, $s = 4$, $h = 1024$, and LRU strategy. The parameters $R_1 = \frac{\mathrm{CM}(\ 1)}{\mathrm{CM}(\ 1)}$ and $R_2 = \frac{\mathrm{CM}(\ 2)}{\mathrm{CM}(\ 2)}$ denote the ratios of the estimated and real numbers of misses for L1 and L2 caches, respectively. They represent the accuracy of the probabilistic model.

Figure 1 illustrates that the real number of cache misses was predicted with average accuracy 97% in case of algorithm MVM_CSR and 95% in case of algorithm MVM_SSS. Figure 1 also shows that the accuracy of the probabilistic cache model

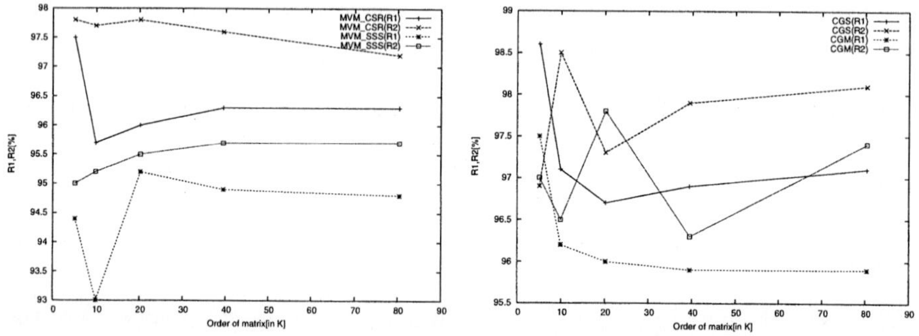

Fig. 1. The accuracy of the model (a) for algorithms MVM_CSR,MVM_SSS, (b) for algorithms CGS,CGM

is around 96% of algorithms CGS and CGM. The accuracy of an analytical model is influenced by the following assumptions.

1. Successive loads of items of array x are assumed mutually independent from the view of mapping into cache sets and that happens if the structure of matrix A is random, but this is not true in real cases.
2. In the $SpM\times V$, we consider only arrays A, c, and x, but the algorithms MVM_CSR and MVM_SSS (see Section 2.1) also load arrays adr and y to caches.
3. We assume that both L1 and L2 caches are data caches. In the Intel architecture, this assumption holds only for the L1 cache, whereas the L2 cache is unified and it is used also for storing instructions. This fact is not taken into account in our formulas, but the error is small due to small code sizes. Similarly, a small part of the error is due to system task codes in L2.
4. In the CGAs, we assume that every iteration is independent. This assumption is valid for $C_{\mathrm{S}} \leq n(5 \cdot S_{\mathrm{D}} + S_{\mathrm{I}}) + n_{\mathrm{zpr}}(S_{\mathrm{D}} + S_{\mathrm{I}})$ (all memory requirements for storing the arrays A, adr, c, d, p, r, x, y in the CGAs).

5 Conclusions

Our analytical probabilistic model for predicting cache behavior is similar to the model in [2], but we differ in 2 aspects. We have explicitly developed and verified a model for s-way set associative caches with LRU block replacement strategy and obtained average accuracy of predicted numbers of cache misses equal to about 97% for both $SpM\times V$ and CGA. We have derived models for both general and symmetric matrices. In contrast to [2], (1) our results indicate that cache miss ratios for these 2 applications are sensitive to the replacement strategy used, (2) we consider both compulsory and thrashing misses for arrays A and c.

Acknowledgements

This research has been supported by grant GA AV ČR IBS 3086102, by IGA CTU FEE under CTU0409313, and by MŠMT under research program #J04/98: 212300014.

References

1. P. Tvrdík and I. Šimeček: Analytical modeling of sparse linear code. PPAM **12** (2003) 617-629 Czestochova, Poland
2. Olivier Temam and William Jalby: Characterizing the Behavior of Sparse Algorithms on Caches. Supercomputing (1992) 578-587
3. K. R. Wadleigh and I. L. Crawford: Software optimization for high performance computing. Hewlett-Packard professional books (2000)

Modeling and Simulating Waves in Anisotropic Elastic Solids*

Valery G. Yakhno and Hakan K. Akmaz

Dokuz Eylul University, Fen-Edebiyat,
Department of Mathematics, 35160, Buca, Izmir, Turkey
{valery.yakhno, hakan.akmaz}@deu.edu.tr

Abstract. An iterative procedure of finding a solution of the initial value problem for a linear anisotropic system of elasticity with polynomial data is described. 3-D images and animated movies of elastic wave propagations in different anisotropic crystals are generated. These images are collected in the library of images.

Keywords: 3-D simulation, anisotropic solids, initial value problem, wave propagation.

1 Introduction

Search and development of new materials with specific properties are needed for different industries such as chemistry, microelectronics, etc. When new materials are created we must be able to have the possibility to model and study their properties. Mathematical models of physical processes can provide cutaway views that let you see aspects of something that would be invisible in the real artifact but computer models can also provide visualization tools.

Our current activity includes mathematical modeling and simulating the wave propagation in anisotropic solids and crystals with different structure of anisotropy. Dynamic mathematical models of elastic wave propagations in anisotropic media are described by the system of partial differential equations [3], [4], [7] .

The initial value problem for a linear system of anisotropic elasticity is considered in this paper. We describe an iterative procedure of finding a solution of this initial value problem with polynomial data. Wave fields for different anisotropic materials are simulated by this procedure. We have used Mathematica 4.0 to generate 3-D images and animated movies of elastic wave propagations in crystals. These images are collected in the library of images. Our library can serve as a collection of patterns and samples when we analyze the structure of anisotropic materials or evaluate the performance of numerical methods.

The structure of our paper is as follows. The linear system of anisotropic elasticity and an initial value problem of this system are described in Section 2.

* This research has been supported by Dokuz Eylul University of Turkey under the research grant 03.KB.FEN.077.

Z. Li et al. (Eds.): NAA 2004, LNCS 3401, pp. 574–582, 2005.

An iterative procedure of solving the considered initial value problem is given in Section 3. Visualization examples of dynamic wave propagations in different anisotropic solids are presented in Section 4. General remarks are at the end of the paper.

2 Statement of the Problem and Properties of Its Solution

Let $x = (x_1, x_2, x_3) \in \mathbb{R}^3$. We assume that $\mathbb{R}^3$ is an elastic medium, whose small amplitude vibrations

$$\mathbf{u}(x,t) = (u_1(x,t), u_2(x,t), u_3(x,t)) \tag{1}$$

are governed by the system of partial differential equations and initial conditions (see, [3], [4], [7])

$$\rho \frac{\partial^2 u_j}{\partial t^2} = \sum_{k=1}^{3} \frac{\partial \tau_{jk}}{\partial x_k} + f_j(x,t), \quad x \in \mathbb{R}^3, \quad t > 0, \quad j = 1,2,3, \tag{2}$$

$$u_j(x,0) = \varphi_j(x), \quad \left. \frac{\partial u_j(x,t)}{\partial t} \right|_{t=0} = \psi_j(x), \quad j = 1,2,3. \tag{3}$$

Here ρ is the density of the medium,

$$\tau_{jk} = \sum_{l,m=1}^{3} C_{jklm} \sigma_{lm}, \qquad j,k = 1,2,3 \tag{4}$$

is the stress tensor,

$$\sigma_{lm} = \frac{1}{2}\left(\frac{\partial u_l}{\partial x_m} + \frac{\partial u_m}{\partial x_l} \right), \qquad l,\ m = 1,2,3 \tag{5}$$

is the strain tensor, and $\{C_{jklm}\}_{j,k,l,m=1}^{3}$ are the elastic moduli of the medium, $\varphi_j(x)$, $\psi_j(x)$, $f_j(x)$, $j = 1,2,3$ are smooth functions. We assume that ρ and C_{jklm} are constants.

It is convenient and customary to describe the elastic moduli in terms of a 6×6 matrix according to the following conventions relating a pair (j,k) of indices $j,k = 1,2,3$ to a single index $\alpha = 1,\ldots,6$ (see, [3], [4]):

$$\begin{array}{lll} (1,1) \longleftrightarrow 1, & (2,2) \longleftrightarrow 2, & (3,3) \longleftrightarrow 3, \\ (2,3),(3,2) \longleftrightarrow 4, & (1,3),(3,1) \longleftrightarrow 5, & (1,2),(2,1) \longleftrightarrow 6. \end{array} \tag{6}$$

This correspondence is possible due to the symmetry properties $C_{jklm} = C_{kjlm} = C_{jkml}$. The additional symmetry property $C_{jklm} = C_{lmjk}$ implies that the matrix

$$C = (C_{\alpha\beta})_{6\times 6} \tag{7}$$

of all moduli where $\alpha = (jk)$, $\beta = (lm)$, is symmetric. We will assume also that $\rho > 0$ and the matrix $(C_{\alpha\beta})_{6\times 6}$ is positive definite.

In this paper we analyze relations (2), (3) as the Cauchy problem for the second order hyperbolic equations system with the polynomial initial data. Using the reasoning similar to the work [8] the system (2) can be written as symmetric hyperbolic system of the first order. Applying the symmetric hyperbolic system theory (see, [1], [5]) to obtained system we get the following proposition.

Proposition 1. *Let T be fixed positive, p be non-negative numbers; $x^0 = (0,0,0)$, $Q = (x^0, T)$ be a point from four dimensional space, $\Gamma(Q)$ be the conoid of dependence (the characteristic conoid) of the hyperbolic system (2) at the point Q. Suppose also initial data and non-homogeneous term of (2) are given in the following polynomial form*

$$\varphi_j(x) = \sum_{l=0}^{p}\sum_{m=0}^{p} \varphi_j^{l,m}(x_3) x_1^m x_2^l, \qquad j = 1,\, 2,\, 3, \tag{8}$$

$$\psi_j(x) = \sum_{l=0}^{p}\sum_{m=0}^{p} \psi_j^{l,m}(x_3) x_1^m x_2^l, \qquad j = 1,\, 2,\, 3, \tag{9}$$

$$f_j(x,t) = \sum_{l=0}^{p}\sum_{m=0}^{p} F_j^{l,m}(x_3,t) x_1^m x_2^l, \qquad j = 1,\, 2,\, 3, \tag{10}$$

$$\begin{aligned} &\varphi_j^{l,m}(x_3) \in C^2,\ \psi_j^{l,m}(x_3) \in C^1,\ F_j^{l,m}(x_3,t) \in C^1(\overline{\Delta(T)}), \\ &\Delta(T) = \{(x_3,t) \in \mathbb{R}^2 : (0,0,x_3,t) \in \Gamma(Q)\}. \end{aligned} \tag{11}$$

Then the solution $\mathbf{u} = (u_1, u_2, u_3)$ of (2), (3) has the following structure in the conoid $\Gamma(Q)$

$$u_j(x_1,x_2,x_3,t) = \sum_{k=0}^{p}\sum_{s=0}^{p} U_j^{s,k}(x_3,t) x_1^s x_2^k, \tag{12}$$

where $U_j^{s,k}(x_3,t) \in C^2(\overline{\Delta(T)})$, $j = 1,2,3;\ s,k = 0,1,2,\ldots,p$.

We note that the problem (2), (3) may be written in the form

$$\rho\frac{\partial^2 \mathbf{u}}{\partial t^2} = G\frac{\partial^2 \mathbf{u}}{\partial x_1^2} + H\frac{\partial^2 \mathbf{u}}{\partial x_2^2} + B\frac{\partial^2 \mathbf{u}}{\partial x_3^2} + F\frac{\partial^2 \mathbf{u}}{\partial x_1 \partial x_2} + D\frac{\partial^2 \mathbf{u}}{\partial x_1 \partial x_3} + E\frac{\partial^2 \mathbf{u}}{\partial x_2 \partial x_3} + \mathbf{f}, \tag{13}$$

$$u_j(x,0) = \varphi_j(x), \qquad \left.\frac{\partial u_j(x,t)}{\partial t}\right|_{t=0} = \psi_j(x), \quad j = 1,2,3, \tag{14}$$

where the components of vector functions $\boldsymbol{\varphi} = (\varphi_1,\varphi_2,\varphi_3)$, $\boldsymbol{\psi} = (\psi_1,\psi_2,\psi_3)$, $\mathbf{f} = (f_1,f_2,f_3)$ satisfy (8), (9), (10);

$$G = \begin{bmatrix} c_{11}\ c_{16}\ c_{15} \\ c_{16}\ c_{66}\ c_{56} \\ c_{15}\ c_{56}\ c_{55} \end{bmatrix},\ F = \begin{bmatrix} c_{16}+c_{16} & c_{12}+c_{66} & c_{14}+c_{56} \\ c_{66}+c_{12} & c_{26}+c_{26} & c_{46}+c_{25} \\ c_{56}+c_{14} & c_{25}+c_{46} & c_{45}+c_{45} \end{bmatrix}, \tag{15}$$

$$H = \begin{bmatrix} c_{66} & c_{26} & c_{46} \\ c_{26} & c_{22} & c_{24} \\ c_{46} & c_{24} & c_{44} \end{bmatrix}, \quad D = \begin{bmatrix} c_{15}+c_{15} & c_{14}+c_{56} & c_{13}+c_{55} \\ c_{56}+c_{14} & c_{46}+c_{46} & c_{36}+c_{45} \\ c_{55}+c_{13} & c_{45}+c_{36} & c_{35}+c_{35} \end{bmatrix}, \tag{16}$$

$$B = \begin{bmatrix} c_{55} & c_{45} & c_{35} \\ c_{45} & c_{44} & c_{34} \\ c_{35} & c_{34} & c_{33} \end{bmatrix}, \quad E = \begin{bmatrix} c_{56}+c_{56} & c_{46}+c_{25} & c_{36}+c_{45} \\ c_{25}+c_{46} & c_{24}+c_{24} & c_{23}+c_{44} \\ c_{45}+c_{36} & c_{44}+c_{23} & c_{34}+c_{34} \end{bmatrix}. \tag{17}$$

3 Procedure of Solving (13), (14)

In this section we describe a procedure of finding recurrence relations for terms $U_j^{s,k}(x_3,t)$, $j=1,2,3$; s, $k=0,1,2,\dots,p$ from the equation (12). The starting step is to find $U^{p,p}(x_3,t) = (U_1^{p,p}(x_3,t), U_2^{p,p}(x_3,t), U_3^{p,p}(x_3,t))$. Differentiating (13), (14) p times with respect to x_1 and then p times with respect to x_2, using (10), (12) we find

$$\rho \frac{\partial^2 \mathbf{U}^{p,p}}{\partial t^2} = B \frac{\partial^2 \mathbf{U}^{p,p}}{\partial x_3^2} + \mathbf{F}^{p,p}, \qquad (x_3,t) \in \Delta(T), \tag{18}$$

with initial conditions

$$\mathbf{U}^{p,p}\big|_{t=0} = \varphi^{p,p}(x_3), \qquad \left.\frac{\partial \mathbf{U}^{p,p}}{\partial t}\right|_{t=0} = \psi^{p,p}(x_3), \qquad x_3 \in O(T), \tag{19}$$

where

$$O(T) = \{x_3 \in \mathbb{R} : (x_3, 0) \in \Delta(T)\}, \tag{20}$$

B is defined by (17) and $\varphi^{p,p}(x_3)$, $\psi^{p,p}(x_3)$, $\mathbf{F}^{p,p}(x_3,t)$ are given coefficients of polynomial expansions (8), (9), (10) of vector functions φ, ψ, $\mathbf{f}$. Since C defined by (7) is a real symmetric positive definite matrix then B defined by (17) is a real symmetric positive definite matrix also. Hence B is congruent to a diagonal matrix of its eigenvalues. That is, there exists an orthogonal matrix Z such that

$$Z^{-1} B Z = \Lambda = \begin{bmatrix} \lambda_1 & 0 & 0 \\ 0 & \lambda_2 & 0 \\ 0 & 0 & \lambda_3 \end{bmatrix}. \tag{21}$$

Because B is positive definite, real and symmetric then its eigenvalues λ_i, $i = 1,2,3$ are real and positive.

Setting

$$\mathbf{U}^{p,p} = Z\mathbf{Y}^{p,p} \tag{22}$$

in (18) we get

$$\rho \frac{\partial^2 Z\mathbf{Y}^{p,p}}{\partial t^2} = B \frac{\partial^2 Z\mathbf{Y}^{p,p}}{\partial x_3^2} + \mathbf{F}^{p,p}, \qquad (x_3,t) \in \Delta(T). \tag{23}$$

We multiply left hand side of (23) by Z^{-1} to obtain

$$\rho\frac{\partial^2\mathbf{Y}^{p,p}}{\partial t^2} = \Lambda\frac{\partial^2\mathbf{Y}^{p,p}}{\partial x_3^2} + Z^{-1}\mathbf{F}^{p,p}, \qquad (x_3,t)\in\Delta(T). \tag{24}$$

Denoting $\nu_j = \left(\frac{\lambda_j}{\rho}\right)^{\frac{1}{2}}$, $j=1,2,3$ and using d'Alembert formula we can solve the Cauchy problem for (24) with the following initial data

$$\mathbf{Y}^{p,p}|_{t=0} = Z^{-1}\varphi^{p,p}(x_3), \qquad \left.\frac{\partial\mathbf{Y}^{p,p}}{\partial t}\right|_{t=0} = Z^{-1}\psi^{p,p}(x_3), \quad x_3\in O(T). \tag{25}$$

Using this fact and (22) we can solve the Cauchy problem (18), (19) in $\Delta(T)$. The solution of this problem is given by

$$U_i^{p,p}(x_3,t) = \sum_{j=1}^{3} Z_{ij}\left\{\frac{1}{2}\left[(Z^{-1}\varphi^{p,p})_j(x_3-\nu_j t) + (Z^{-1}\varphi^{p,p})_j(x_3+\nu_j t)\right]\right.$$
$$\left. + \frac{1}{2\rho\nu_j}\int_0^t\int_{x_3-\nu_j(t-\tau)}^{x_3+\nu_j(t-\tau)} \left[Z^{-1}\mathbf{F}^{p,p}\right]_j(\sigma,\tau)d\sigma d\tau\right\}$$
$$+\frac{1}{2\nu_j}\int_{x_3-\nu_j t}^{x_3+\nu_j t}(Z^{-1}\psi^{p,p})_j(\sigma)d\sigma, \qquad i=1,2,3, \quad (x_3,t)\in\Delta(T), \tag{26}$$

where Z_{ij}, $i,j=1,2,3$ are elements of the matrix Z.

The next step consists of finding $U^{p-1,p}(x_3,t)$ and $U^{p,p-1}(x_3,t)$ if we suppose that $U_j^{p,p}(x_3,t)$, $j=1,2,3$ were found by (26).

Differentiating (13), (14) $p-1$ times with respect to x_1 and p times with respect to x_2 and repeating reasoning which we did before we find

$$U_i^{p-1,p} = \sum_{j=1}^{3} Z_{ij}\left\{\frac{1}{2}\left[(Z^{-1}\varphi^{p-1,p})_j(x_3-\nu_j t) + (Z^{-1}\varphi^{p-1,p})_j(x_3+\nu_j t)\right]\right.$$
$$+\frac{1}{2\rho\nu_j}\int_0^t\int_{x_3-\nu_j(t-\tau)}^{x_3+\nu_j(t-\tau)}\left(p\tilde{D}\left[Z^{-1}\frac{\partial\mathbf{U}^{p,p}}{\partial x_3}\right] + Z^{-1}\mathbf{F}^{p-1,p}\right)_j(\sigma,\tau)d\sigma d\tau$$
$$\left.+\frac{1}{2\nu_j}\int_{x_3-\nu_j t}^{x_3+\nu_j t}(Z^{-1}\psi^{p-1,p})_j(\sigma)d\sigma\right\}. \tag{27}$$

Differentiating (13), (14) p times with respect to x_1 and $p-1$ times with respect to x_2 in the similar way we get the following formula

$$U_i^{p,p-1} = \sum_{j=1}^{3} Z_{ij}\left\{\frac{1}{2}\left[(Z^{-1}\varphi^{p,p-1})_j(x_3-\nu_j t) + (Z^{-1}\varphi^{p,p-1})_j(x_3+\nu_j t)\right]\right.$$
$$+\frac{1}{2\rho\nu_j}\int_0^t\int_{x_3-\nu_j(t-\tau)}^{x_3+\nu_j(t-\tau)}\left(p\tilde{E}\left[Z^{-1}\frac{\partial\mathbf{U}^{p,p}}{\partial x_3}\right] + Z^{-1}\mathbf{F}^{p,p-1}\right)_j(\sigma,\tau)d\sigma d\tau$$
$$\left.+\frac{1}{2\nu_j}\int_{x_3-\nu_j t}^{x_3+\nu_j t}(Z^{-1}\psi^{p,p-1})_j(\sigma)d\sigma\right\}. \tag{28}$$

Using analogous reasoning we find formulas for $U^{p-1,p-1}(x_3,t)$, $U^{p-m,p}(x_3,t)$, $m = 2,\ldots,p$, $U^{p,p-m}(x_3,t)$, $m = 2,\ldots,p$; $U^{p-m,p-n}(x_3,t)$, $m,n = 2,\ldots,p$ as follows

$$U_i^{p-1,p-1} = \sum_{j=1}^{3} Z_{ij}\Bigg\{ \frac{1}{2\rho\nu_j}\int_0^t \int_{x_3-\nu_j(t-\tau)}^{x_3+\nu_j(t-\tau)} \Bigg(p\tilde{D}\left[Z^{-1}\frac{\partial \mathbf{U}^{p,p-1}}{\partial x_3}\right]$$
$$+p\tilde{E}\left[Z^{-1}\frac{\partial \mathbf{U}^{p-1,p}}{\partial x_3}\right] + p^2\tilde{F}\left[Z^{-1}\mathbf{U}^{p,p}\right] + Z^{-1}\mathbf{F}^{p-1,p-1}\Bigg)_j (\sigma,\tau)d\sigma d\tau$$
$$+\frac{1}{2}\left[(Z^{-1}\varphi^{p-1,p-1})_j(x_3-\nu_j t) + (Z^{-1}\varphi^{p-1,p-1})_j(x_3+\nu_j t)\right]$$
$$+\frac{1}{2\nu_j}\int_{x_3-\nu_j t}^{x_3+\nu_j t} (Z^{-1}\psi^{p-1,p-1})_j(\sigma)d\sigma\Bigg\}, \tag{29}$$

$$U_i^{p-m,p}(x_3,t) = \sum_{j=1}^{3} Z_{ij}\Bigg\{ \frac{1}{2\nu_j}\int_{x_3-\nu_j t}^{x_3+\nu_j t} (Z^{-1}\psi^{p-m,p})_j(\sigma)d\sigma$$
$$+\frac{1}{2\rho\nu_j}\int_0^t \int_{x_3-\nu_j(t-\tau)}^{x_3+\nu_j(t-\tau)} \Bigg((p-m+1)\tilde{D}\left[Z^{-1}\frac{\partial \mathbf{U}^{p-m+1,p}}{\partial x_3}\right]$$
$$+(p-m+1)(p-m+2)\tilde{G}\left[Z^{-1}\mathbf{U}^{p-m+2,p}\right] + Z^{-1}\mathbf{F}^{p-m,p}\Bigg)_j (\sigma,\tau)d\sigma d\tau$$
$$+\frac{1}{2}\left[(Z^{-1}\varphi^{p-m,p})_j(x_3-\nu_j t) + (Z^{-1}\varphi^{p-m,p})_j(x_3+\nu_j t)\right]\Bigg\}, m=2,\ldots,p, \tag{30}$$

$$U_i^{p,p-m}(x_3,t) = \sum_{j=1}^{3} Z_{ij}\Bigg\{ \frac{1}{2\nu_j}\int_{x_3-\nu_j t}^{x_3+\nu_j t} (Z^{-1}\psi^{p,p-m})_j(\sigma)d\sigma$$
$$+\frac{1}{2\rho\nu_j}\int_0^t \int_{x_3-\nu_j(t-\tau)}^{x_3+\nu_j(t-\tau)} \Bigg((p-m+1)\tilde{E}\left[Z^{-1}\frac{\partial \mathbf{U}^{p,p-m+1}}{\partial x_3}\right]$$
$$+(p-m+1)(p-m+2)\tilde{H}\left[Z^{-1}\mathbf{U}^{p,p-m+2}\right] + Z^{-1}\mathbf{F}^{p,p-m}\Bigg)_j (\sigma,\tau)d\sigma d\tau$$
$$+\frac{1}{2}\left[(Z^{-1}\varphi^{p,p-m})_j(x_3-\nu_j t) + (Z^{-1}\varphi^{p,p-m})_j(x_3+\nu_j t)\right]\Bigg\}, m=2,\ldots,p, \tag{31}$$

$$U_i^{p-m,p-n}(x_3,t) = \sum_{j=1}^{3} Z_{ij}\Bigg\{ \frac{1}{2\rho\nu_j}\int_0^t \int_{x_3-\nu_j(t-\tau)}^{x_3+\nu_j(t-\tau)} \Bigg(Z^{-1}\mathbf{F}^{p-m,p-n}$$
$$+(p-m+1)\tilde{D}\left[Z^{-1}\frac{\partial \mathbf{U}^{p-m+1,p-n}}{\partial x_3}\right] + (p-n+1)\tilde{E}\left[Z^{-1}\frac{\partial \mathbf{U}^{p-m,p-n+1}}{\partial x_3}\right]$$

$$
\begin{aligned}
&+(p-m+1)(p-n+1)\tilde{F}\left[Z^{-1}\mathbf{U}^{p-m+1,p-n+1}\right] \\
&+(p-m+1)(p-m+2)\tilde{G}\left[Z^{-1}\mathbf{U}^{p-m+2,p-n}\right] \\
&+(p-n+1)(p-n+2)\tilde{H}\left[Z^{-1}\mathbf{U}^{p-m,p-n+2}\right]\Bigg)_j (\sigma,\tau)d\sigma d\tau \\
&+\frac{1}{2}\left[(Z^{-1}\varphi^{p-m,p-n})_j(x_3-\nu_j t)+(Z^{-1}\varphi^{p-m,p-n})_j(x_3+\nu_j t)\right] \\
&\quad +\frac{1}{2\nu_j}\int_{x_3-\nu_j t}^{x_3+\nu_j t}(Z^{-1}\psi^{p-m,p-n})_j(\sigma)d\sigma\Bigg\}, \qquad m,n=2,\ldots,p, \qquad (32)
\end{aligned}
$$

where $(x_3,t)\in\Delta(T)$ and

$$
\tilde{D}=Z^{-1}DZ,\ \tilde{E}=Z^{-1}EZ,\ \tilde{F}=Z^{-1}FZ,\ \tilde{G}=Z^{-1}GZ,\ \tilde{H}=Z^{-1}HZ. \qquad (33)
$$

Using mentioned arguments all terms $U_j^{s,k}(x_3,t)$, $j=1,2,3$; $s,k=0,1,2,\ldots,p$ of (12) may be found successively in the order which we fixed in this section.

4 Visualization Examples of Wave Propagations

In this section we present the images of the wave propagations in three crystals belonging to different crystal systems. These pictures are obtained by fixing one of the space variables in the solution which we get by the method explained in the previous section.

Consider (13), (14) with the following initial conditions

$$
\begin{aligned}
&\boldsymbol{\varphi}=(\varphi_1,\varphi_2,\varphi_3), \qquad \boldsymbol{\psi}=(\psi_1,\psi_2,\psi_3), \\
&\varphi_j=p(x_1)p(x_2)p(x_3), \qquad \psi_j\equiv 0, \qquad j=1,2,3,
\end{aligned}
\qquad (34)
$$

where the function $p(z)$ is the interpolating polynomial of

$$
f(z)=\frac{1}{z}\sin\left(\frac{5z}{2}\right) \qquad (35)
$$

in the interval $[-5,5]$ with 16 points.

Let $(x_1,x_2,x_3)\in\mathbb{R}^3$ be space variables and one of these variables be fixed, for example $x_2=0$. The three-dimensional graph of each function $\varphi_j(x)$ has a hillock shape which is shown in Figure 1. The horizontal axes here are x_1, x_3, the vertical axis is φ_j for $x_2=0$. In Figure 1 level plots of the same surface is shown. The different colors correspond to different levels of the surface.

We consider this problem for selenium, tin and gold which are trigonal, tetragonal and cubic crystals, respectively. The elastic constants in units of 10^{12}dyn/cm^2 and densities (gr/cm^3) of these crystals are as follows:

For selenium,

$$
\begin{aligned}
&C_{11}=C_{22}=0.1870, \quad C_{13}=C_{23}=0.2620, \quad C_{44}=C_{55}=0.1490, \\
&C_{12}=0.0710, \quad C_{33}=0.7410, \quad C_{66}=(C_{11}-C_{12})/2, \\
&C_{14}=-C_{15}=C_{56}=0.0620, \quad \rho=4.838,
\end{aligned}
\qquad (36)
$$

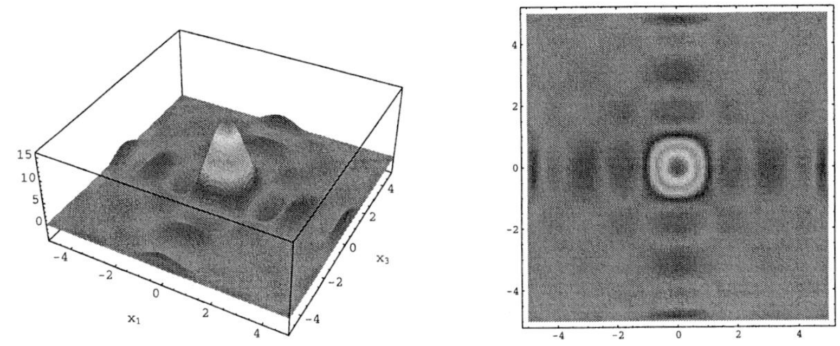

(a) 3–D level plot of $\varphi_j(x_1, 0, x_3)$ (b) 2–D level plot of $\varphi_j(x_1, 0, x_3)$

Fig. 1. The j – th component of initial vector function

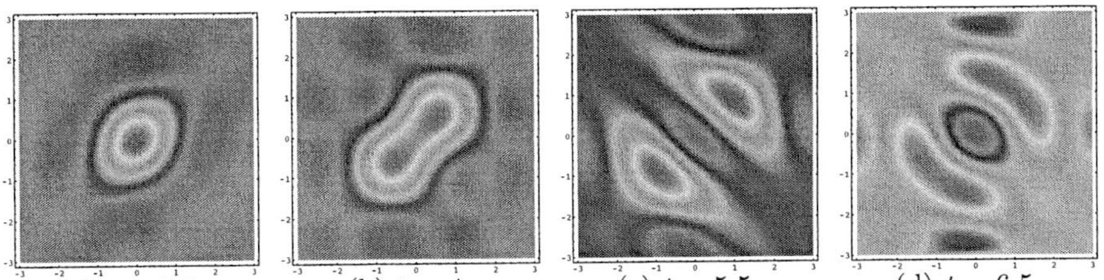

(a) $t = 2.5$ (b) $t = 4$ (c) $t = 5.5$ (d) $t = 6.5$

Fig. 2. $u_2(x_1, 0, x_3, t)$ for selenium (trigonal media)

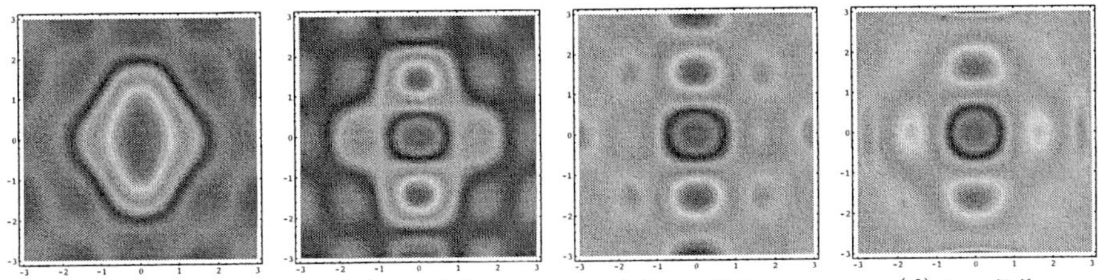

(a) $t = 3$ (b) $t = 3.5$ (c) $t = 5.5$ (d) $t = 7.5$

Fig. 3. $u_2(x_1, 0, x_3, t)$ for tin (tetragonal media)

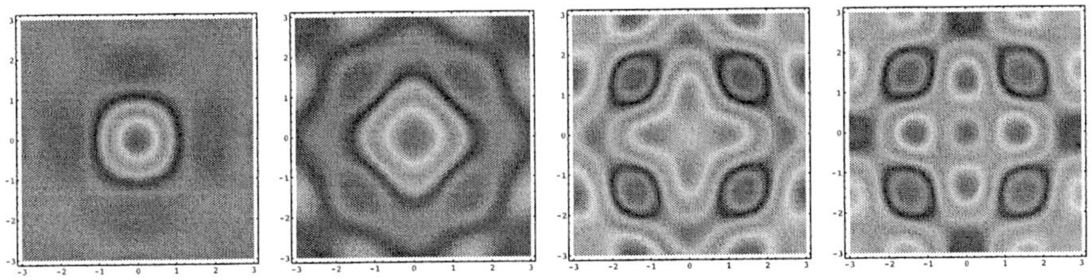

(a) $t = 2.5$ (b) $t = 4$ (c) $t = 4.5$ (d) $t = 5.5$

Fig. 4. $u_2(x_1, 0, x_3, t)$ for gold (cubic media)

for tin,

$$\begin{aligned} &C_{11} = C_{22} = 0.7529, \quad C_{13} = C_{23} = 0.4400, \quad C_{44} = C_{55} = 0.2193, \\ &C_{12} = 0.6156, \quad C_{33} = 0.9552, \quad C_{66} = 0.2336, \quad \rho = 7.29, \end{aligned} \tag{37}$$

for gold,

$$\begin{aligned} &C_{11} = C_{22} = C_{33} = 1.9244, \quad C_{12} = C_{13} = C_{23} = 1.6298, \\ &C_{44} = C_{55} = C_{66} = 0.4200, \quad \rho = 19.283, \end{aligned} \tag{38}$$

and other components are 0. The Figures 2 – 4 contain four screen shots of the wave propagations in crystals selenium, tin and gold. These figures are 2-D level plots of $u_2(x_1, 0, x_3, t)$ for time in units of 10^{-6} sec.

5 General Remarks

On the base of the formulas described in this paper a library of images and animated movies of elastic wave propagations in different crystals was created. This library can serve as a collection of patterns and samples for the analysis of the structure of anisotropic crystals or evaluation of the numerical methods performance. The results of this paper and [8] were used to generalize formulas and the described procedure to the case of vertical non-homogeneous anisotropic media in which the elastic moduli and density are functions of one space variable.

References

1. Courant, R. and Hilbert, D.: Methods of Mathematical Physics, Vol. 2, Interscience, New York, 1962.
2. CRC Handbook of Chemistry and Physics, 80th Edition (Ed. D.R. Lide), CRC Press, 1999.
3. Dieulesaint, E. and Royer, D.: Elastic Waves in Solids, John Wiley & Sons, Chichester, 1980.
4. Fedorov, F.I.: Theory of Elastic Waves in Crystals, Plenum Press, New York, 1968.
5. Mizohata, S.: The Theory of Partial Differential Equations, Cambridge University Press, 1973.
6. Sacks, P.E., Yakhno, V.G.: The inverse problem for a layered anisotropic half space, Journal of Mathematical Analysis and Applications **228** (1998) 377–398.
7. Ting, T.C.T., Barnett, D.M. Wu, J.J.: Modern Theory of Anisotropic Elasticity and Applications, SIAM Philadelphia, 1990.
8. Yakhno, V.G.: Inverse problem for differential equations system of electromagnetoelasticity in linear approximation. Inverse Problems, Tomography and Image Processing, Newark, DE, 1997 (Ed. A.G. Ramm),p. 211–240, Plenum, New York, 1998.

Numerical Method for a Chemical Nonlinear Reaction Boundary Value Problem*

A.I. Zadorin and O.V. Kharina

Omsk Branch of Institute of Mathematics,
Siberian branch of Russian Academy of Sciences, Russia
zadorin@iitam.omsk.net.ru
harina@iitam.omsk.net.ru

Abstract. A second order singularly perturbed differential equation with nonlinear source term modelling chemical processes on semi-infinite interval is considered. The question of transformation of the boundary conditions to a finite interval is studied. We proved that the problem posed on a finite interval does not involve a distinct boundary layer and can be solved with the use of an upwind difference scheme that is uniformly convergent with respect to the small parameter. Numerical experiments are discussed.

1 Introduction

A boundary value problem for a nonlinear equation of second order on a semi-infinite interval is considered. Nonlinearity in the equation corresponds to modelling of a chemical reaction, when a speed of reaction depends on the temperature according to Arrenius law. The essential feature of the problem consists in the unboundeness of the domain and in presence of a boundary layer of solution corresponding to the zone of the chemical reaction.

Consider a boundary value problem:

$$\varepsilon u'' - au' + F(u) = 0, \quad u(0) = A, \quad \lim_{x \to \infty} u(x) = B, \tag{1}$$

where $\varepsilon > 0,\ a > 0,\ F(u) \geq 0,\ B > A > 0$. At modelling of chemical reaction the function F is given in a form

$$F(u) = K(B - u) \exp\left(- E/u \right),$$

where $K > 0,\ E > 0$, u is a temperature, ε is a coefficient of diffusion, a is a speed of expansion of a flame, K is a constant of reaction speed, E is an energy of activation, $F(u)$ corresponds to a thermal emission according to Arrenius law.

* Supported by the Russian Foundation of Basic Research under Grant 02-01-01166, 04-01-00578.

Z. Li et al. (Eds.): NAA 2004, LNCS 3401, pp. 583–589, 2005.

Expressing a derivative in an obvious form, it is possible to show, that $u(x)$ is increasing function, $A \leq u(x) \leq B$.

Some phenomena in different areas of science, by example in chemical reactions, are models of singularly perturbed problems of reaction-diffusion-convection type, for which the diffusion coefficient can be very small with respect to the reaction terms. The presence of boundary (interior) layers makes necessary the use of robust numerical methods, which give effective solutions for any value of the diffusion parameter. There are many publications dealing with equation (1) on finite interval [2]-[6].

Many authors have been developed the fitted mesh method, i.e., classical schemes defined on special meshes condensing the mesh points in the boundary layer regions. Most popular are Shishkin's and Bakhvalov's meshes [2]-[5].

In this paper, we first discuss the reduction of the problem (1) to a suitable boundary value problem on finite interval. We analyze in Section 2 the transfer the limit boundary condition from infinity. The new problem posed on a finite interval we solve using an upwind difference scheme, Section 3. Uniform convergence of discrete solution to the continuous one is also proved. Numerical experiments are discussed in Section 4.

2 Reduction of the Problem to a Finite Interval

Consider the question, how to transfer a problem on infinite interval to a problem for a finite interval. We shall do this, following [1], [6] on a base of extraction of all set of solutions satisfying the boundary condition at infinity, by means of the first order equation:

$$u'(x) = \gamma(u(x) - B) + \beta(u(x)). \tag{2}$$

where γ is a negative root of the equation

$$\varepsilon\gamma^2 - a\gamma - K\exp(-E/B) = 0, \tag{3}$$

$\beta(u)$ is solution of the singular Cauchy problem

$$\varepsilon\beta'(u)(\gamma(u-B) + \beta(u)) + (\varepsilon\gamma - a)\beta(u) =$$

$$= K(u-B)\Big(\exp(-E/u) - \exp(-E/B)\Big), \ \beta(B) = 0. \tag{4}$$

Using the set (2), we get a problem on a finite interval:

$$\varepsilon u'' - au' + F(u) = 0, \tag{5}$$

$$u(0) = A, \quad u'(L) = \gamma(u(L) - B) + \beta(u(L)).$$

The solutions of problems (1) and (5) coincide for all $0 \leq x \leq L$. It follows from uniqueness of solutions of these problems and from a way of extraction of a steady manifold, when on the extracted variety the initial equation has form of identity. Note, that the reduced problem is formulated on a finite interval, but the function $\beta(u)$ in the boundary condition is solution of the corresponding singular Cauchy problem. The solution of this problem can be found approximately as asymptotic series:

$$\tilde{\beta}_m(u) = \sum_{k=0}^{m} \beta_k(u)\varepsilon^k.$$

Substituting this expansion in (4), and expanding γ in a series on ε, we receive recurrent formulas for $\beta_k(u)$

$$\beta_k = \frac{1}{a}\sum_{i=0}^{k-1} \beta_i' \beta_{k-1-i} + \sum_{i=1}^{k} (-1)^i \frac{2^{i-1} P_i\, K^i \exp(-iE/B)}{i!\, a^{2i}} (\beta_{k-i}'(u-B) + \beta_{k-i}),$$

$$\beta_0 = \frac{1}{a} K(u-B)\Big(\exp(-E/B) - \exp(-E/u)\Big),\ P_i = \prod_{j=0}^{i-2}(2j+1),\ P_1 = 1, \quad (6)$$

and $\|\beta - \tilde{\beta}_m\| \le C\varepsilon^{m+1}$.

Therefore the coefficients in the boundary condition of the reduced problem can be found with given accuracy. We can show, that problem (5) is stable to perturbation of this coefficients. The next problem is how to find solution of a problem (5).

3 Construction of the Difference Scheme

Now we construct a difference scheme for the problem (5). We consider spline method of construction of difference scheme. Feature of a method that at construction of difference scheme it is taken into account boundary layer growth of the solution, though boundary layer function is not used.

Let Ω is a grid on an interval $[0, L]$, $\Delta_n = (x_{n-1}, x_n]$, $n = 1, 2, \ldots, N$. Consider the next problem instead of (5):

$$\varepsilon\tilde{u}'' - a\tilde{u}' + \tilde{F}(\tilde{u}) = 0,$$

$$\tilde{u}(0) = A, \quad \tilde{u}'(L) + g(\tilde{u}(L)) = 0, \quad (7)$$

where $\tilde{F}(\tilde{u}(x)) = F(\tilde{u}(x_n))$ at $x \in \Delta_n$, $\ g(u) = -\gamma(u-B) - \beta(u)$.

On the segment $\Delta_n = (x_{n-1}, x_n]$ this problem has the form:

$$\varepsilon\tilde{u}'' - a\tilde{u}' + F(\tilde{u}_n) = 0,$$

$$\tilde{u}(x_{n-1}) = u_{n-1}^h,\ \tilde{u}(x_n) = u_n^h.$$

The solution of this equation has the explicit form:

$$\tilde{u}(x) = \frac{\tilde{F}(\tilde{u}_n)}{a}x + C_n^1 + C_n^2 \exp\left(\frac{a}{\varepsilon}(x - x_n)\right).$$

Finding constants C_n^1, C_n^2 from boundary conditions and taking into account continuity of the derivative on the boundary of the adjoining intervals

$$\lim_{x \to x_n - 0} \tilde{u}'(x) = \lim_{x \to x_n + 0} \tilde{u}'(x),$$

we get the monotonous difference scheme:

$$\frac{\Phi_{n+1}\exp(-\tau_{n+1})}{h_{n+1}} u_{n+1}^h - \left(\frac{\Phi_{n+1}\exp(-\tau_{n+1})}{h_{n+1}} + \frac{\Phi_n}{h_n}\right) u_n^h + \frac{\Phi_n}{h_n} u_{n-1}^h =$$

$$= \frac{F(u_n^h)}{a}(1 - \Phi_n) - \frac{F(u_{n+1}^h)}{a}\left(1 - \Phi_{n+1}\exp(-\tau_{n+1})\right),$$

$$u_0^h = A, \quad \Phi_N \frac{u_N^h - u_{N-1}^h}{h_N} = \frac{F(u_N^h)}{a}(\Phi_N - 1) - g(u_N^h), \tag{8}$$

where $\tau_n = ah_n/\varepsilon$, $\Phi_n = \tau_n/(1 - \exp(-\tau_n))$.

Lemma 1. *Let* $F'(u) \le -\beta_0 < 0, \; g'(u) \ge 0$. *Then*

$$|u(x) - \tilde{u}(x)| \le \frac{1}{\beta_0} \max_x |\tilde{F}(\tilde{u}) - F(\tilde{u})|.$$

Proof. Let $z = u - \tilde{u}$. Then

$$\varepsilon z'' - az' + F(u) - \tilde{F}(\tilde{u}) = 0,$$

$$z(0) = 0, \; z'(L) + g'(s)z(L) = 0,$$

which implies

$$\varepsilon z'' - az' + \frac{F(u) - F(\tilde{u})}{u - \tilde{u}} z = \tilde{F}(\tilde{u}) - F(\tilde{u}),$$

$$z(0) = 0, \; z'(L) + g'(s)z(L) = 0.$$

An application of the maximum principle leads to the assertion.

Corollary 1. *Let* $F(u)$ *is a monotonous function on each grid interval. If for all* $n = 1, 2, \ldots, N$

$$|F(u_n^h) - F(u_{n-1}^h)| \le \Delta,$$

then

$$\|u^h - [u]\| \le \frac{1}{\beta_0}\Delta, \quad \|u^h - [u]\| = \max_n |u_n^h - u(x_n)|.$$

Thus, condensing a grid and solving a boundary value problem on series of meshes, we can achieve the necessary accuracy of the difference scheme.

If we replace the problem (5) by the problem:

$$\varepsilon\tilde{u}'' - a\tilde{u}' + K(B - \tilde{u})f(u) = 0,$$

$$\tilde{u}(0) = A, \quad \tilde{u}'(L) = \gamma(\tilde{u}(L) - B) + \beta(\tilde{u}(L)), \tag{9}$$

where $f(u(x)) = \exp(-E/u(x_n))$ at $x \in \Delta_n$, then we obtain the difference scheme:

$$\frac{\tau_{n+1}\exp(-\lambda^1_{n+1}h_{n+1})}{\phi_{n+1}-1}u^h_{n+1} - \left(\frac{\lambda^2_n\phi_n - \lambda^1_n}{\phi_n - 1} + \frac{\lambda^2_{n+1} - \lambda^1_{n+1}\phi_{n+1}}{\phi_{n+1}-1}\right)u^h_n +$$

$$+\frac{\tau_n\exp(\lambda^2_n h_n)}{\phi_n - 1}u^h_{n-1} = B\left(\frac{\lambda^1_n + \tau_n\exp(\lambda^2_n h_n) - \lambda^2_n\phi_n}{\phi_n - 1} +\right.$$

$$\left. + \frac{\lambda^1_{n+1}\phi_{n+1} + \tau_{n+1}\exp(-\lambda^1_{n+1}h_{n+1}) - \lambda^2_{n+1}}{\phi_{n+1}-1}\right),$$

$$u^h_0 = A, \quad \frac{(\lambda^2_N\phi_N - \lambda^1_N)u^h_N - \tau_N\exp(\lambda^2_N h_N)u^h_{N-1}}{\phi_N - 1} = \gamma(u^h_N - B) +$$

$$+B\frac{\lambda^2_N\phi_N - \lambda^1_N - \tau_N\exp(\lambda^2_N h_N)}{\phi_N - 1} + \beta(u^h_N),$$

where $\tau_n = \lambda^2_n - \lambda^1_n$, $\phi_n = \exp(h_n\tau_n)$.

Now we turn to investigate the accuracy of the upwind difference scheme for problem (5):

$$L^h_n u^h = 2\varepsilon\frac{h_n(u^h_{n+1} - u^h_n) - h_{n+1}(u^h_n - u^h_{n-1})}{h_n h_{n+1}(h_n + h_{n+1})} - a\frac{u^h_n - u^h_{n-1}}{h_n} + F(u^h_n) = 0,$$

$$u^h_0 = A, \quad \frac{u^h_N - u^h_{N-1}}{h_N} + g(u^h_N) = 0. \tag{10}$$

Let $z^h = u^h - [u]$. Using results of [3], we obtain

$$|L^h_n z^h| \le C\int_{x_{n-1}}^{x_{n+1}} [\varepsilon|u'''(s)| + |u''(s)|]\,ds.$$

Assuming that for n

$$\varepsilon h_n \max_{x\in\Delta_n}|u'''(s)| + h_n\max_{x\in\Delta_n}|u''(s)| \le \Delta, \tag{11}$$

we find $|L^h_n z^h| \le C\Delta$. Next using the maximum principle, we can prove, that in this case $|u^h_n - u(x_n)| \le C\Delta$ for every n.

Thus, by condensing of the grid Ω it is possible to provide the required accuracy of the difference scheme. To fulfill the condition (11) it is necessary to calculate approximately u''_n, u'''_n in terms of u^h_n. At modelling of chemical reactions the function $u(x)$ has great gradients on an internal boundary layer and for calculation of these derivatives we can not use difference approximations. We can find derivatives u''_n, u'''_n from degenerate equation $au' = F(u)$. Then the inequality (11) has the form:

$$\frac{\varepsilon h_n}{a^3}|F''_n F^2_n + F'^2_n F_n| + \frac{h_n}{a^2}|F'_n F_n| \le \Delta.$$

Consider the iterative method for the nonlinear difference scheme (10):

$$\varepsilon\frac{h_n(u^k_{n+1}-u^k_n)-h_{n+1}(u^k_n-u^k_{n-1})}{h_nh_{n+1}(h_n+h_{n+1})}-a\frac{u^k_n-u^k_{n-1}}{h_n}-Mu^k_n=-F(u^{k-1}_n)-Mu^{k-1}_n,$$

$$u^k_0=A,\quad \frac{u^k_N-u^k_{N-1}}{h_N}+M_1u^k_N=M_1u^{k-1}_N-g(u^{k-1}_N).$$

Lemma 2. *Let*

$$-B_0\le F'(u)\le-\beta_0<0,\quad 0<\beta_1\le g'(u)\le B_1, \tag{12}$$

$$M=\frac{B_0+\beta_0}{2},\; M_1=\frac{B_1+\beta_1}{2},$$

then

$$\|u^h-u^k\|\le q\|u^h-u^{k-1}\|,\quad q=\max_{i=0,1}\left\{\frac{B_i-\beta_i}{B_i+\beta_i}\right\}.$$

This method has the property of monotone convergence, even if conditions (12) are not fulfilled, but in this case we can not estimate the number of iterations.

4 Numerical Experiments

Consider the iterative method for the upwind difference scheme on an uniform grid:

$$\varepsilon\frac{u^k_{n+1}-2u^k_n+u^k_{n-1}}{h^2}-a\frac{u^k_n-u^k_{n-1}}{h}+K(B-u^k_n)\exp(-E/u^{k-1}_n)=0,$$

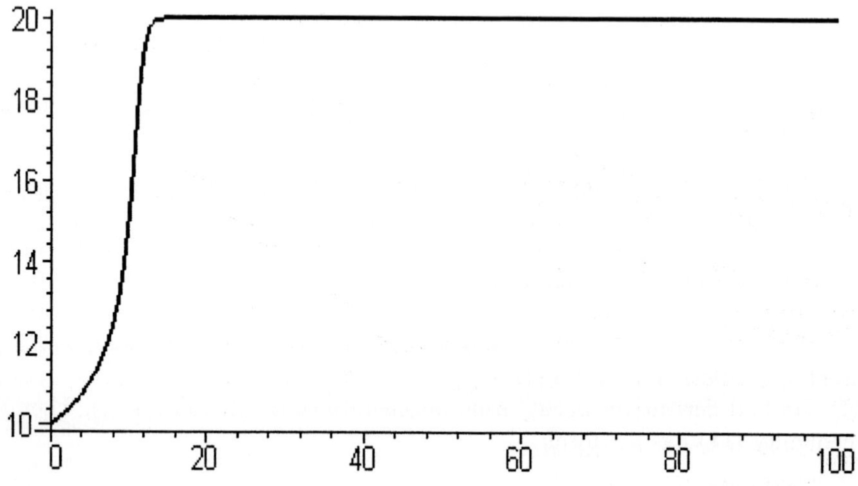

Fig. 1

Table 1

L	boundary conditions			
	Dirichlet	Neumann	$m = 0$	$m = 1$
5	8.975	0.034	0.040	0.0016
10	5.023	0.211	0.024	0.0009
20	0	0	0	0

$$u_0^k = A, \quad \frac{u_N^k - u_{N-1}^k}{h} - \gamma u_N^k = -B\gamma + \beta(u_N^{k-1}).$$

On Figure 1 the graph of the solution of the problem (5) is displayed at $\varepsilon = 0.001, A = 10, B = 20, K = 300, E = 100, h = 0.01$.

The accuracy of a boundary condition transfer from infinity was compared on the base of conditions Dirichlet $u(L) = 0$ and Neumann $u'(L) = 0$ and on the base of extraction of steady manifold with transition to a problem (5). In the last case was used asymptotic expansions with $m = 0$ $m = 1$. As the solution of the scheme on the infinite interval was considered a solution of scheme on a large enough interval with length $L_0 = 100$. Comparisons were performed at $L = 5, L = 10, L = 20$. Let u_∞^h is the projected solution from large interval. In Table 1 the error $\delta = \|u^h - u_\infty^h\|$ is displayed depending on the boundary conditions.

The results of computations confirm the advantage of the approach presented.

References

1. Abramov A.A., Balla K., Konyukhova N.B.: Transfer of the boundary conditions from special points for systems of linear ordinary differential equations. Messages on calculative mathematics. — Moscow: Computing center of Academy of Sciences USSR, 1981.
2. Farrell P.A., Hegarty A.F., J.J.H. Miller, O'Riordan E., Shishkin G.I.: Robust Computational Methods for Boundary Layers, Chapman and Hall / CRC, Boca Raton, 2000.
3. Kellogg R.B., Tsan A.: Analysis of some difference approximations for a singular perturbation problems without turning points. Math. Comput. (1978) **32** No.144. 1025–1039.
4. Miller J.J.H., O'Riordan E., Shishkin G.I.: Solution of Singularly Perturbed Problems with ε - uniform Numerical Methods, World Scientific, Singapore, 1996.
5. Roos H.G., Stynes M., Tobiska L.: Numerical Methods for Singularly Perturbed Differential Equations, Springer-Verlag, Berlin, 1996.
6. Zadorin A.I.: The transfer of the boundary condition from the infinity for the numerical solution to the second order equations with a small parameter. Siberian Journal of Numerical Mathematics (1999) **2** No.1. 21–36.

Parametrically Driven Dark Solitons: A Numerical Study

E.V. Zemlyanaya[1], I.V. Barashenkov[2], and S.R. Woodford[2]

[1] Laboratory for Information Technologies,
Joint Institute for Nuclear Research, Dubna 141980, Russia
elena@jinr.ru
[2] Department of Applied Mathematics, University of Cape Town,
Rondebosch 7701, South Africa
igor@cenerentola.mth.uct.ac.za
woodford@maths.uct.ac.za

Abstract. We show that unlike the bright solitons, the parametrically driven kinks of the nonlinear Schrödinger equation are immune from instabilities for all damping and forcing amplitudes; they can also form stable bound states. In the undamped case, the two types of kinks and their complexes can stably travel with nonzero velocities. The bistability of the Bloch and Néel walls within the NLS contrasts the properties of these solutions within the Ginzburg-Landau equation, where they cannot stably coexist.

1 Introduction

We consider the parametrically driven, damped nonlinear Schrödinger (NLS) equation, with the "defocusing" cubic nonlinearity

$$i\psi_t + \frac{1}{2}\psi_{xx} - |\psi|^2\psi + \psi = h\overline{\psi} - i\gamma\psi, \tag{1}$$

where γ is the damping coefficient, and h is the amplitude of parametric driving that should compensate dissipative losses of localized states in various media. (Without loss of generality, one can consider $h, \gamma \geq 0$.)

Eq.(1) has a number of applications in physics. In fluid dynamics, the "defocusing" parametrically driven NLS governs the amplitude of the water surface oscillations in a vibrated channel with a large width-to depth ratio [1, 2]. The same equation arises as an amplitude equation for the upper cutoff mode in the parametrically driven damped nonlinear lattices [3]. In the optical context, it was derived for the doubly resonant $\chi^{(2)}$ optical parametric oscillator in the limit of large second-harmonic detuning [4]. Finally, Eq.(1) describes inhomogeneities of the magnetisation of an easy-plane ferromagnet in an external field [5], and also arises in the anisotropic XY-model [5, 6, 7].

The localized solutions forming in the defocusing media are domain walls, or kinks, also known as "dark solitons" in the context of nonlinear optics. Our purpose was to numerically explore the stability and bifurcations of dark solitons of Eq.(1) and their bound states.

Z. Li et al. (Eds.): NAA 2004, LNCS 3401, pp. 590–597, 2005.

2 Formulation of Problem; Numerical Techniques

Numerical study of Eq.(1) included three stages. First of all, we made numerical continuation of stationary solutions of Eq.(1) in a wide range of parameters. Next, we examined the stability properties of the stationary solutions applying a corresponding eigenvalue problem for the linearized operator. Finally, we performed direct numerical simulation of time-dependent equation (1) for some values of parameters.

For steadily travelling solutions we assumed the form $\psi(x - Vt)$, in which case Eq.(1) reduces to an ODE

$$\frac{1}{2}\psi_{xx} - iV\psi_x - |\psi|^2\psi + \psi = h\overline{\psi}, \tag{2}$$

where velocity V plays a role of additional parameter.

For the numerical solution of Eq.(2) we used the Newtonian iteration with the fourth-order Numerov's finite-difference approximation. The calculations were performed on the interval (-100,100), with the stepsize $\Delta x = 5 \cdot 10^{-3}$. The numerical continuation procedure that we applied is described in [8]. As two bifurcation measures of stationary solutions, we calculated the energy and momentum integrals:

$$E = \mathrm{Re} \int \left(\frac{|\psi_x|^2}{2} + \frac{|\psi|^4}{2} - |\psi|^2 + h\psi^2 + \frac{|\psi^{(0)}|^4}{2} \right) dx, \tag{3}$$

$$P = \frac{i}{2} \int (\overline{\psi_x}\psi - \psi_x\overline{\psi})\, dx. \tag{4}$$

In the undamped case, integrals (3) and (4) of Eq.(1) are conserved. Stability of stationary solutions was examined numerically, by computing eigenvalues of

$$\mathcal{H}\varphi = \lambda J \varphi, \tag{5}$$

where the column $\varphi = (u, v)^T$; the operator $\mathcal{H}$ has the form

$$\mathcal{H} = -\frac{I}{2}\partial_x^2 + \begin{pmatrix} 3\mathcal{R}^2 + \mathcal{I}^2 + h & 2\mathcal{R}\mathcal{I} - V\partial_x + \gamma \\ 2\mathcal{R}\mathcal{I} + V\partial_x - \gamma & \mathcal{R}^2 + 3\mathcal{I}^2 - h \end{pmatrix}$$

and J is the skew-symmetric matrix:

$$J = \begin{pmatrix} 0 & -1 \\ 1 & 0 \end{pmatrix}.$$

The eigenvalue problem (5) is obtained by linearising eq.(1) about $\psi = \mathcal{R} + i\mathcal{I}$ in the co-moving frame, and letting $\delta\psi = (u + iv)e^{\lambda t}$. The solution is stable when there are no eigenvalues with positive real part.

To solve (5), we used the Fourier expansion of eigenfunctions u and v:

$$u(x) = \sum_{m=-N/2}^{N/2} u^m \exp(-i\omega_m x), \quad v(x) = \sum_{m=-N/2}^{N/2} v^m \exp(-i\omega_m x), \tag{6}$$

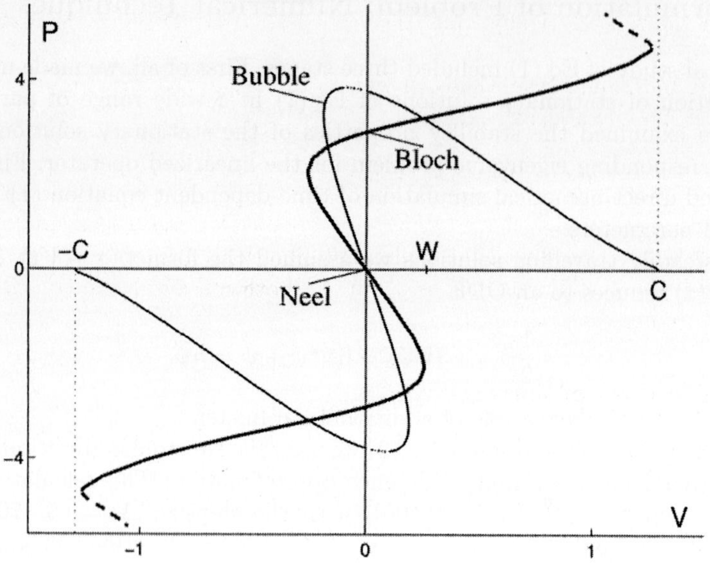

Fig. 1. The momentum of the travelling Bloch and Néel walls (thick line) and their nonoscillatory bubble-like complex (thin line). For $|V|$ close to but slightly lower than c, the wall attaches a small-amplitude bubble on each flank; this accounts for the turn of the thick curve near $|V| = c$. The dotted segments of the continuous branches indicate unstable solutions. This figure pertains to $h = 1/15$

where $\omega_m = 2\pi m/L$. Substituting Eqs.(6) into Eq.(5) gives us a matrix eigenvalue problem that was solved with the help of the standard EISPACK code. Typically, we chose $N = 600$ modes on the interval $(-L, L)$, with $L = 30$.

Finally, the direct numerical simulation of the time-dependent equation (1) was performed with the help of the pseudospectral code developed in [9, 10].

3 Results of Numerical Study

Some stationary solutions of Eq.(1), localized and nonlocalized, are known explicitly. First, there is a pair of flat solutions. One of them is always unstable; the other one is stable for all $h \geq \gamma \geq 0$. It reads

$$\psi^{(0)} = iAe^{i\theta}, \tag{7}$$

where

$$A = \sqrt{1 + \sqrt{h^2 - \gamma^2}}, \quad \theta = \frac{1}{2} \arcsin \frac{\gamma}{h}. \tag{8}$$

Next, one nonhomogeneous solution is known in literature as the Néel wall [1, 2, 4]:

$$\psi_N(x) = iA \tanh(Ax) e^{i\theta}. \tag{9}$$

One can show, analytically, that the Néel wall is stable for all h and γ (with $h \geq \gamma \geq 0$) [5].

In the undamped case, the Néel wall (9) coexists with another solution that is also available analytically [11]:

$$\psi_B(x) = iA\tanh(2\sqrt{h}x) \pm \sqrt{1-3h}\,\mathrm{sech}(2\sqrt{h}x). \tag{10}$$

This solution is usually referred to as the Bloch wall; numerical stability analysis reveals that the Bloch wall is stable throughout its existence region (which is $h < \frac{1}{3}$ and $\gamma = 0$). Other solutions of Eq.(1), both quiescent and travelling, were obtained numerically.

3.1 The Case of Zero Damping

In the case $\gamma = 0$, we have found that the Bloch and Néel walls can travel with nonzero velocity. The corresponding dependence $P(V)$ [where P is the momentum defined by eq.(4)] is shown in the diagram Fig.1. The distinction between the Bloch and Néel walls becomes less visible as the velocity V grows. When $V = w$, the two branches merge. This scenario is similar to the case of easy-axis ferromagnets [12], where w corresponds to Walker's velocity.

Our numerical analysis of eq.(5) shows that the entire Bloch-Néel branch of travelling walls is stable. This coexistence of two stable kinks is in contrast with the relativistic

$$\frac{1}{2}(\psi_{tt} - \psi_{xx}) + |\psi|^2\psi - \psi + h\overline{\psi} = 0 \tag{11}$$

and diffusive

$$\psi_t = \frac{1}{2}\psi_{xx} - |\psi|^2\psi + \psi - h\overline{\psi} \tag{12}$$

counterparts of eq.(1) — in both eq.(11) and eq.(12), the Bloch wall is the only stable solution in the region $h < \frac{1}{3}$, where the Bloch and Néel walls coexist [7, 13, 14].

We also found other travelling solutions, namely bubble-like bound states of a Bloch wall and a Néel wall. The corresponding $P(V)$ curve is also shown in Fig.1. The point of special interest of this curve is $V = 0$. At this point, similar to the bright solitons case [15], there is a one-parameter family of bound states, with the parameter z characterising the separation distance between the two walls. Members of this family are generally unstable, nonsymmetric complexes of Bloch and Néel walls. However, there is a particular separation $z = \zeta$ for which the bubble is symmetric: $\overline{\psi}_\zeta(-x) = -\psi_\zeta(x)$. For $h = 1/15$, the analytical form of the symmetric bubble is available:

$$\psi_\zeta(x) = iA\left[1 - (3/4)\,\mathrm{sech}^2\left(Ax/4 \mp i\pi/4\right)\right]. \tag{13}$$

Our numerical study showed that the symmetric bubble has the largest momentum over bubbles with various z; at the same time, ζ is the smallest possible separation: $z \geq \zeta$. The symmetric bubble is the *only* stable bubble. Furthermore, for each h, only the symmetric, stable bubble can be continued to nonzero velocity.

As $|V| \to c = \sqrt{1 + 2h + \sqrt{4h(1+h)}}$ (which is the minimum phase velocity of linear waves), the bubble degenerates into the flat solution (7). On the other hand, when $V, P \to 0$, the bubble transforms into a pair of Néel walls with the separation $z \to \infty$. The entire branch of moving bubbles is stable, with the exception of a small region between $V = 0$ and the point of maximum $|P|$, inside which a real pair $\pm\lambda$ occurs (Fig.1). The change of stability at points where $dP/dV = 0$ is explained in [15].

3.2 The Dissipative Case

A natural question is which parts of the bifurcation diagram Fig.1 persist for the case of nonzero γ. This problem was studied in [16] for bright solitons. When $\gamma \neq 0$, the energy and momentum are, in general, changing with time:

$$\dot{P} = -2\gamma P,$$

$$\dot{E} = \gamma \int (|\psi^{(0)}|^4 - |\psi|^4) dx - 2\gamma E,$$

and therefore in order to be continued to nonzero γ, a steadily travelling soliton has to satisfy

$$P = 0$$

and

$$E = \frac{1}{2} \int (|\psi^{(0)}|^4 - |\psi|^4) dx.$$

In fact, only the first condition needs to be ensured for the continuability. Indeed, for $\psi = \psi(x - Vt)$, Eq.(1) with $\gamma = 0$ yields the identity

$$VP = E - \frac{1}{2} \int (|\psi^{(0)}|^4 - |\psi|^4) dx,$$

and hence if $P = 0$,

$$E = \frac{1}{2} \int (|\psi^{(0)}|^4 - |\psi|^4) dx$$

immediately follows. Since $P = 0$ only at the stationary Néel wall, we conclude that no other solutions can be continued to small nonzero γ.

In the case $\gamma \neq 0$, we obtained, numerically, a new solution with $V = 0$. This is the stationary bubble-like complex of two Néel walls. The typical shape of this solution is shown in Fig.2a. Our simulations revealed a corridor of h values, $h_1(\gamma) < h < h_2(\gamma)$, where two Néel walls attract and form this stable, stationary bubble. This corridor is shown in Fig.2b (shaded region). The inset depicts the energy (4) of the bubble as it is continued in h for fixed $\gamma = 0.35$. Note that this solution is *not* continuable to $\gamma = 0$.

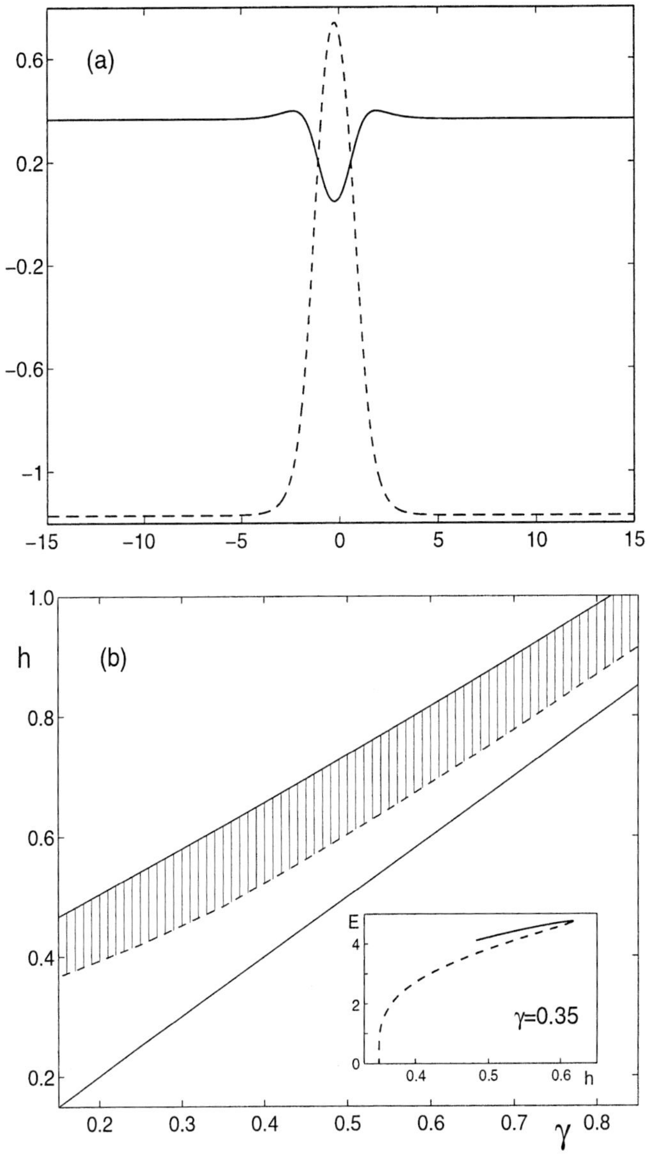

Fig. 2. (a) The stable, stationary bubble-like solution in the case of nonzero damping. Here $h = 0.617$, $\gamma = 0.35$. (Solid line: real part; dashed line: imaginary part.) (b) Main panel: the existence and stability diagram of the damped bubble on the (γ, h)-plane. Solid lines bound the band-like region of the bubble's existence; its shaded subregion gives the stability corridor.
Inset: the energy of the bubble as a function of h for fixed $\gamma = 0.35$. Here, the solid and dashed branches represent stable and unstable solutions, respectively. The solid line terminates at $h \approx \sqrt{\frac{1}{9} + \gamma^2}$, at which point the distance between the bound walls becomes infinite. The dashed line terminates at $h = \gamma$, where the bubble degenerates into the flat solution (7)

4 Concluding Remarks

In summary, we have shown that:

- Unlike the bright solitons, the parametrically driven kinks are immune from instabilities for all dampings and forcing amplitudes; they can also form stable bound states.
- In the case of zero damping, the two types of stable kinks (Bloch and Néel walls) and their complexes can move with nonzero velocities. The moving solitons are also stable.
- Of all the undamped solutions, only the stationary Néel wall can be continued to nonzero γ.
- For nonzero γ, two Néel walls can form a stationary bound state. This bound state exists for a wide range of values of h and γ, but cannot be continued to $\gamma = 0$.

The remarkable stability of the damped-driven kinks and their bound states is in sharp contrast with stability properties of the bright solitons, which become unstable when the parametric driving strength exceeds a certain (rather low) threshold [10, 15, 17, 18]. Finally, the stable coexistence of two types of domain walls is also worth emphasising; this multistability is not observed in the parametrically driven Klein-Gordon and Ginzburg-Landau equations [eqs.(11) and (12)].

We thank Nora Alexeeva for writing a pseudospectral code for the time-dependent NLS (1). IB and SW were supported by the NRF of South Africa under grant 2053723; EZ by the Russian Fund for Basic Research (grant 03-01-00657).

References

1. C. Elphick and E. Meron, Phys. Rev. A **40**, 3226 (1989)
2. B. Denardo *et al*, Phys. Rev. Lett. **64**, 1518 (1990)
3. B. Denardo *et al*, Phys. Rev. Lett. **68**, 1730 (1992); G. Huang, S.-Y. Lou, and M. Velarde, Int. J. Bif. Chaos **6**, 1775 (1996)
4. S. Trillo, M. Haelterman and A. Sheppard, Opt. Lett. **22**, 970 (1997)
5. I.V. Barashenkov, S.R. Woodford and E.V. Zemlyanaya. Phys. Rev. Lett. **90**, 054103 (2003)
6. L.N. Bulaevskii and V.L. Ginzburg, Sov. Phys. JETP, **18**, 530 (1964); J. Lajzerowicz and J.J. Niez, J. de Phys. **40**, L165 (1979)
7. P. Coullet *et al*, Phys. Rev. Lett. **65**, 1352 (1990); P. Coullet, J. Lega and Y. Pomeau, Europhys. Lett. **15**, 221 (1991); D.V. Skryabin *et al*, Phys. Rev. E **64**, 056618 (2001)
8. E.V. Zemlyanaya and I.V. Barashenkov. JINR Preprint P11-2004-17, Dubna (2004); Math. Modelling **67** (2004) (in press)
9. M. Bondila, I.V. Barashenkov and M.M. Bogdan, Physica D**87**, 314 (1995)
10. N.V. Alexeeva, I.V. Barashenkov and D.E. Pelinovsky, Nonlinearity **12**, 103 (1999)
11. S. Sarker, S.E. Trullinger and A.R. Bishop, Phys. Lett. A **59** (1976) 255

12. A.M. Kosevich, B.A. Ivanov and A.S. Kovalev, Phys. Rep. **194**, 117 (1990)
13. P. Hawrylak, K.R. Subbaswamy and S.E. Trullinger, Phys. Rev. D **29**, 1154 (1984)
14. B.A. Ivanov, A.N. Kichizhiev and Yu.N. Mitsai, Sov. Phys. JETP **75**, 329 (1992)
15. I.V. Barashenkov, E.V. Zemlyanaya and M. Bär, Phys. Rev. E **64**, 016603 (2001)
16. I.V. Barashenkov and E.V. Zemlyanaya, SIAM J. Appl. Maths. **64** 3, 800 (2004)
17. I.V. Barashenkov, M.M. Bogdan and V.I. Korobov, Europhys. Lett. **15**, 113 (1991)
18. I.V. Barashenkov and E.V. Zemlyanaya, Phys. Rev. Lett. **83**, 2568 (1999)

Numerical Investigation of the Viscous Flow in a Bioreactor with a Mixer

Ivanka Zheleva[1] and Anna Lecheva[2]

[1] University of Rousse,
Technology College – Razgrad,
BG – 7200 Razgrad, POB 110, Bulgaria
izheleva@ru.acad.bg,
[2] University of Rousse, Silistra branch
milen.anna@infotel.bg

Abstract. Tank reactors with different mixers are very often used for many chemical and biological processes. For their effective use is necessary to know in details hydrodynamics and mass transfer which take place in them. For many practically important cases the experimental study of these processes is very expensive, very difficult or impossible. This is why recently mathematical modelling of the complex swirling flows in such reactors becomes an effective method for investigation of the behavior of the fluids in these reactors.

This paper presents mathematical model and numerical results for viscous swirling flow in a cylindrical tank reactor with a mixer.

The model is based on the Navier-Stokes equations in cylindrical coordinate system. These equations are written in terms of the stream function, vorticity and the momentum of the tangential component of the velocity. The flow is supposed to be stationary and axis-symmetric. A special attention is devoted to the correct formulation of the boundary conditions.

A numerical algorithm for studying this motion of the fluid is proposed in [6]. The grid for the discretization of the equations is nonuniform. Difference scheme of Alternating Direction Implicit Method for solving Navier-Stokes equations is used.

Numerical results for the stream function, the velocity field and the momentum of the tangential velocity for different Reynolds numbers are obtained by this numerical algorithm. The results are presented graphically. They are discussed and compared with results of other authors. It is observed a very good agreement between them.

Keywords: Stirred tank reactor, Navier – Stokes equations; axis-symmetrical swirling viscous flow, numerical investigation.

1 Introduction

The industrial biotechnologies for production of foods, drinks, antibiotics etc. are based on interdependent and very complex mass transfer processes, which are usually conducted in apparatus with intensive mixing.

Z. Li et al. (Eds.): NAA 2004, LNCS 3401, pp. 598–605, 2005.

For example the fermentation medium is a complex multiphase system with inhomogeneous and dynamically changing characteristics during the processes.

Such mediums consist usually of three phases – solid, liquid and gaseous. The liquid phase includes water, dissolved components, cell products, dispersed particles, oil, etc. The solid phase consists of cells, maze flour and other similar particles. The gaseous phase includes air bubbles with different concentration of oxygen (O_2), carbon dioxide (CO_2) and nitrogen (N). In the gaseous phase could be also volatile fermentation components. These three phases have to be well homogenized through mechanical mixing and aeration. When the biomass is aerated by air babbles, because of the low solubility of oxygen in water, the oxygen is only partially absorbed and its concentration in the different fermentation mediums varies within the limits $4 - 7\,mg/l$. Such a small concentration of oxygen allows microorganisms to live only for several minutes, which leads to the necessity of continuous aeration in the volume [1]. That is why the oxygen concentration is considered as a most important factor for the biosynthesis. The aeration conditions and mechanical mixing in biosynthesis processes to a considerable extent define the characteristics of the final products. For this reason, the role of the hydrodynamics of these processes is very important.

Because of the very high complexity of the biosynthesis, the literature does not provide sufficient data to predict all the regimes and to effectively manage the work of the equipment. The experimental optimization of the apparatuses where such complex processes are conducted is very expensive and time-consuming procedure. The only alternative to natural experiments is the computational one, which allows a considerable reduction of costs and time for optimization of the existing technologies or for creation of new ones.

A common practice for computational experiments is to use a hierarchy of models. This method allows to evaluate the role of each parameter and to estimate its influence on the technology effectiveness. Except that, if necessary, it is possible to increase the model's complexity in order to examine the process more precisely.

In this paper, following such a scheme, we deal with one of the most important factors for all biosynthesis – the hydrodynamics in apparatuses with mechanical mixing and aeration.

2 Mathematical Model of the Problem

2.1 Geometrical Domain

The reactor is a cylinder with radius R and height Z. It is filled up by incompressible viscous fluid. The mixer is attached to the axis of the cylinder on the height H_1 and it is rotating with constant angle velocity Ω. We assume in this paper that the mixer is a solid disk with a given radius R_1 and thickness L_1 and it is attached in the middle of the cylinder axis. The motion of the viscous fluid in the domain is due to rotating of the disk and we assume that it is axis-symmetrical. The tank axis is a symmetry line for the investigated fluid motion.

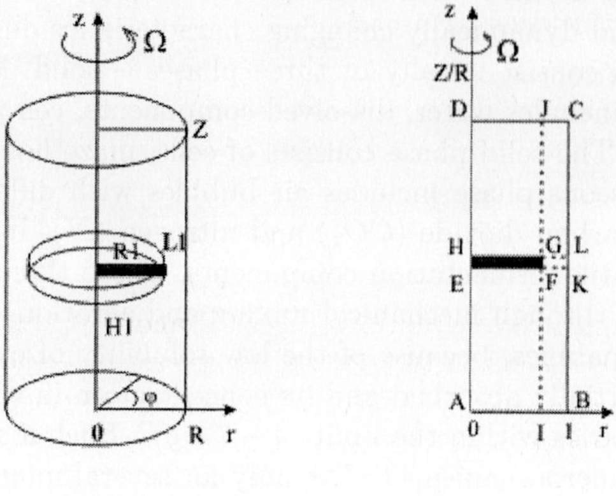

Fig. 1. Scheme of the tank reactor with a mixer (a) and the computational area (b)

It is convenient to introduce a cylindrical coordinate system (r, φ, z). The first coordinate r is along the tank radius direction, the third coordinate z is along the tank height direction and φ is an angle coordinate (see Fig. 1).

2.2 Basic Equations

We consider Navier – Stokes equations and continuity equation for incompressible viscous steady state flow written in cylindrical coordinates (r, φ, z). We assume that the fluid motion is axis-symmetrical which means that every unknown function depends only on r and z but do not depend on the angle variable φ. In this case the first derivative with respect to the angle φ is equal to zero for all unknown functions, i.e. $\frac{\partial}{\partial \varphi} = 0$. We will examine the flow in the reactor long time after it has started rotating and we will look only for a stationary solution, therefore the unknown functions do not depend on the time. Thus, the Navier – Stokes equations written in the terms of stream function, vorticity and momentum of the tangential velocity in dimensionless form are [3]:

$$\frac{\partial}{\partial r}\left(\frac{1}{r}\frac{\partial \psi}{\partial r}\right) + \frac{\partial}{\partial z}\left(\frac{1}{r}\frac{\partial \psi}{\partial z}\right) = -\omega \tag{1}$$

$$\frac{1}{r}\frac{\partial rUM}{\partial r} + \frac{\partial WM}{\partial z} = \frac{1}{Re}\left[r\frac{\partial}{\partial r}\left(\frac{1}{r}\frac{\partial M}{\partial r}\right) + \frac{\partial^2 M}{\partial z^2}\right], \tag{2}$$

$$\frac{\partial U\omega}{\partial r} + \frac{\partial W\omega}{\partial z} - \frac{1}{r^3}\frac{\partial M^2}{\partial z} = \frac{1}{Re}\left[\frac{\partial}{\partial r}\left(\frac{1}{r}\frac{\partial r\omega}{\partial r}\right) + \frac{\partial^2 \omega}{\partial z^2}\right], \tag{3}$$

$$\frac{\partial (rU)}{\partial r} + \frac{\partial (rW)}{\partial z} = 0, \tag{4}$$

where $\vec{V}\,(U, V, W)$ is the velocity vector with its components in the cylindrical coordinate system, Re is the Reynolds number $Re = \frac{\Omega R^2}{\nu}$, R and ΩR are the length and velocity scales, respectively, ψ is the stream function, $\vec{\omega} = rot\vec{V} = (0, \omega, 0)$ is the vorticity vector and $M = rV$ is the magnitude of the momentum of the tangential velocity.

The relation between components U and W of the velocity vector and the stream function ψ in cylindrical coordinate system is as follows:

$$U = -\frac{1}{r}\frac{\partial\psi}{\partial z}, \quad W = \frac{1}{r}\frac{\partial\psi}{\partial r}. \tag{5}$$

As a result we have five equations (1) – (5) and five unknown functions: the stream function ψ, the magnitude of the momentum of the tangential velocity M, the second component ω of the vorticity vector $\vec{\omega}$, the first U and the third W component of the velocity vector.

2.3 Boundary Conditions

For the correct mathematical formulation of the problem it is necessary to define appropriate boundary conditions for all unknown functions.

On all solid walls we have: $\vec{V}_{fluid} = \vec{V}_{wall}$ and $\psi = const$. The boundary condition for the momentum of the tangential velocity is $M = 0$ except on the solid disk, where $M = r^2/R^2$. The boundary conditions for the vorticity ω are defined on the base of values of the stream function. Only on the symmetry line we set up the boundary condition $\omega = 0$, [3, 5, 6].

Thus, we have Dirichlet boundary conditions for all unknown functions. The problem is to find out a solution of the system (1) – (5) which satisfies the defined boundary conditions.

3 Numerical Algorithm

A finite difference method for solving the Dirichlet boundary value problem is developed in [5, 6]. It is based on a construction of a special non-uniform grid with condensation close to the solid boundaries of the domain. The equations (1) – (5) are approximated by finite differences on the non-uniform grid by a 5-point star stencil.

The ADI difference schemes are based on introducing a fictitious time t into the equation (2) for the momentum of the tangential velocity M an into the equation (3) for the vorticity ω. It is well known [4] that if the boundary conditions of the unknown functions does not dependent on the time then at $t \to \infty$ the solution of the parabolic differential equations system will converge to the solution of the stationary system (2), (3). In the ADI methods the step size τ of the fictitious time t is realized into two time layers – $k + 1/2$ and $k + 1$. So that, for each equation the coefficient matrices on each time layer are tri-diagonal, [3]. More details for ADI scheme, which is used here, can be found in [3, 4].

The equation (1) for the stream function ψ can be solved by different ways, [3, 4]. One of them is to introduce its own fictitious time t_1 different from the time t. The ADI scheme for the stream function is described in [3].

The equations of the momentum and of the vorticity are integrated together. The stationary solution of the equation of the stream function have to be obtained on each iteration for M and ω.

The solution is considered to be convergent if

$$P^f(k) = \frac{\max_{i,j} \left| f_{i,j}^{k+1} - f_{i,j}^{k} \right|}{\max_{i,j} \left| f_{i,j}^{k+1} \right|} \leq \varepsilon_f, \tag{6}$$

where f is any of functions: the vorticity ω, the momentum M and the stream function ψ, $k+1$ is the current iteration with respect to fictitious time t, ε_f is the accuracy constant.

3.1 Numerical Results and Discussion

All numerical results are obtained on the computational area, given on Fig. 1 (b) with following values of geometrical parameters: $Z/R = 3$; $H_1 = 1.5$; $R_1 = 0.4$; $L_1 = 0.125$. The value of the accuracy constant of the stream function ψ is $\varepsilon_\psi = 10^{-6}$ for all Reynolds numbers. All results presented in this section are obtained using accuracy constants $\varepsilon_\omega = 10^{-4}$ and $\varepsilon_M = 10^{-4}$.

The results for the stream function ψ, obtained for different Reynolds number in the interval $50 \leq Re \leq 1000$ are presented on Fig. 2.

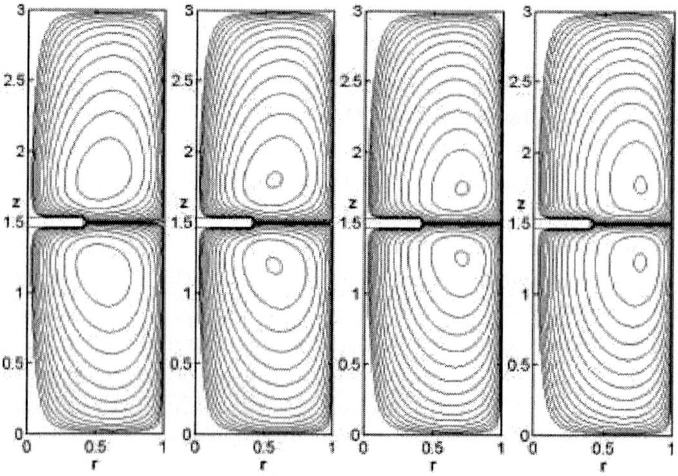

(a) (b) (c) (d)

Fig. 2. Stream function lines for different Reynolds numbers: $Re = 50, 100, 500, 1000$

Values of lines are: $\pm 2.1069e-005$; $\pm 3.1432e-005$; $\pm 4.6891e-005$; $\pm 609953e-005$; ± 0.00010436; ± 0.00015568; ± 0.00023225; ± 0.00034648; ± 0.00051689; ± 0.0007711; ± 0.0011503; ± 0.0017161; ± 0.0025602; ± 0.0038193; ± 0.0056977; ± 0.0085

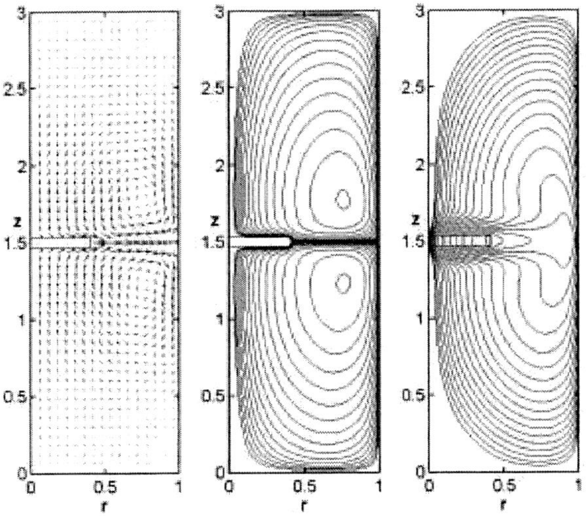

(a) (b) (c)

Fig. 3. Results for $Re = 1000$, $\varepsilon_\omega = 10^{-4}$; (a) Velocity field; (b) Stream-function ψ; (c) Momentum M

Let us mentioned that for all unknown functions zero initial conditions are used. It is observed in all cases that the dominant flow structure in the system is two ring vortices. First of them is formed below the mixer and the second one is formed above the mixer. Their positions and size are found to be depended on the Reynolds number and on the position of the mixer. The values of the stream function are positive numbers above the mixer and the values of the stream function are negative numbers below the mixer.

From Fig. 2 easily can be seen that when the Reynolds number increases the centers of the two observed vortices becomes closer to the mixer and to the wall of the reactor. These results are in good agreement with the results, received by Lamberto et al. [2].

The numerical results for the velocity field $\vec{V}$, the stream function ψ and the momentum M are graphically presented on Fig. 3 for $Re = 1000$.

As it was mentioned above Lamberto et al (see [2]) have studied similar problems. Some results from an experimental study and from computational simulations can be found in [2]. Now we will present some comparison results of our computer simulations and their experiments. The following values of geometrical parameters are used in [2]: $Z/R = 2.2$; $H_1 = 0.7\,R$; $R_1 = 0.38\,R$ and the values of the Reynolds number are $8 < Re < 70$.

We provide our computations for the same values for the parameters as in [2]. But the number of points in our non-uniform grid is approximately 12 000 and our simulation requires 3 hours CPU time for PC computer while approximately 150 000 points and 1 000 hours CPU time for SUN SPARC 20 workstation are reported in [2].

Stability Analysis of a Nonlinear Model of Wastewater Treatment Processes

Plamena Zlateva[1] and Neli Dimitrova[2]

[1] Institute of Control and System Research,
Bulgarian Academy of Sciences
Acad. G. Bonchev Str. Blok 2,
P. O. Box 79, 1113 Sofia, Bulgaria
plamzlateva@icsr.bas.bg

[2] Institute of Mathematics and Informatics,
Bulgarian Academy of Sciences
Acad. G. Bonchev Str. Blok 8, 1113 Sofia, Bulgaria
nelid@bio.bas.bg

Abstract. For a nonlinear model of a wastewater treatment process the input-output static caracteristics are studied with respect to uncertainties in process parameters. Computer simulations and visualizations are performed in *Maple* to show how these uncertainties influence the efficiency of the bioreactor.

1 Introduction

Clean water is essential for health, recreation and life protection among other human activities. The activated sludge processes are most widely used biological systems in wastewater treatment. These processes are very complex due to their nonlinear dynamics, large uncertainty in uncontrolled inputs and in the model parameters and structure [5].

The activated sludge wastewater treatment process is carried out in a system, which consists of an aeration tank and a secondary settler (see Figure 1). It is assumed that the hydraulic characteristics of aeration tank are those of a continuously stirred tank bioreactor with cell recycle. Ideal conditions are assumed to prevail in the settler [2].

Using mass and energy balance equations, the dynamical model of the process in the bioreactor is described by the following two nonlinear ordinary differential equations [3]

$$\frac{dx}{dt} = \mu x + rDx_r - (1+r)Dx \tag{1}$$

$$\frac{ds}{dt} = -k\mu x + D(s_{in} - s) \tag{2}$$

[1] This research has been partially supported by the Bulgarian National Science Fund under contract No. MM-1104/01.

Z. Li et al. (Eds.): NAA 2004, LNCS 3401, pp. 606–612, 2005.

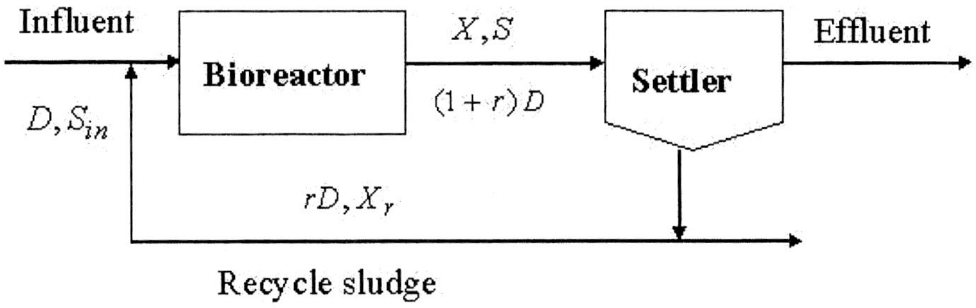

Fig. 1. Schematic diagram of the sludge wastewater treatment process

where $x = x(t)$ and $s = s(t)$ are state variables,

- x is biomass concentration (activated sludge) [mg/l],
- s is substrate concentration (biological oxygen demand) [mg/l],
- D is dilution rate [1/h],
- s_{in} is influent substrate concentration [mg/l],
- x_r is recycle biomass concentration [mg/l],
- μ is specific growth rate [1/h],
- r is sludge recycle ratio
- k is yield coefficient.

The specific growth rate $\mu = \mu(s)$ is presented by the Monod law

$$\mu(s) = \frac{\mu_m \cdot s}{k_s + s},$$

where μ_m and k_s are kinetic process parameters.

It is assumed that the control input is $u = D(t)$ and the output is $y = s(t)$ or $y = x(t)$.

The main difficulty in controlling the activated sludge wastewater treatment processes comes from the variation of the process parameters and the influent waste load. These variations induce process state changes that may lead to a reduction of the water treatment efficiency, unless the plant operation is continuously adjusted.

Therefore, to overcome this problem, the steady state analysis has been performed. The activated sludge process steady state may critically depend on the process parameters and on the value of the influent waste load.

Practical investigations and computer simulations show that the parameters s_{in}, x_r and r in the model (1)–(2) are unknown but bounded. Assume now that instead of numerical values for s_{in}, x_r and r we are given intervals $[s_{in}]$, $[x_r]$ and $[r]$ respectively.

The aim of this paper is to study the input-output static characteristics $y_s(u) = s(u)$ and $y_x(u) = x(u)$ involving intervals in the above mentioned parameters. Using advanced techniques from recently developed interval analysis we compute the so called interval static characteristic which presents the set of all functions $y_s(u)$, $y_x(u)$ for any values of the coefficients in the given intervals.

The paper is organized as follows. In Section 2 the interval input-output static characteristics of the dynamic process are obtained. Some numerical results and graphic presentations in the computer algebra system *Maple* are reported in Section 3.

2 Interval Static Characteristics

The static model of the process is delivered from (1)–(2) by setting the right-hand side functions equal to zero. Thus we obtain the nonlinear algebraic system with respect to s and x

$$\mu(s) \cdot x + rux_r - (1+r)ux = 0$$
$$k \cdot \mu(s) \cdot x - u(s_{in} - s) = 0$$

or more precisely,

$$\left(\frac{\mu_m s}{k_s + s} - (1+r)u\right) x + rux_r = 0 \tag{3}$$

$$k\frac{\mu_m s}{k_s + s} x - u(s_{in} - s) = 0. \tag{4}$$

By expressing $x = x(s, u)$ from the second equation (4) and then substituting it in the first one we obtain the quadratic equation with respect to s

$$as^2 + bs - c = 0$$

where

$$a = (1+r)u - \mu_m, \quad b = \mu_m krx_r - s_{in}((1+r)u - \mu_m) + (1+r)uk_s,$$
$$c = (1+r)uk_s s_{in}.$$

Its discriminant is given by

$$\begin{aligned}\Delta(u) &= b^2 + 4ac \\ &= (\mu_m krx_r - s_{in}((1+r)u - \mu_m) - (1+r)uk_s)^2 + 4\mu_m krx_r(1+r)uk_s.\end{aligned}$$

Obviously, $\Delta(u) > 0$ for any $u \geq 0$ is valid. The two roots, $s_1(u)$ and $s_2(u)$ are then

$$s_1(u) = \frac{-b + \sqrt{\Delta(u)}}{2a} = \frac{2c}{\sqrt{\Delta(u)} + b}, \quad s_2(u) = \frac{-b - \sqrt{\Delta(u)}}{2a}.$$

Taking into account the biotechnological restriction $0 < s < s_{in}$, the second root $s_2(u)$ is excluded from further consideration. Moreover, $s_1(u)$ satisfies

$$s_1(u) < s_{in} \quad \text{and} \quad \lim_{u \to \infty} s_1(u) = s_{in}.$$

Thus the (biologically reasonable) steady state is $s(u) = s_1(u)$, that is

$$s(u) = \frac{2k_s(1+r)us_{in}}{\sqrt{\Delta(u)} + \mu_m krx_r - s_{in}((1+r)u - \mu_m) + (1+r)uk_s}. \tag{5}$$

Then (4) implies

$$x(u) = \frac{u(s_{in} - s(u))}{k \cdot \mu(s(u))}. \tag{6}$$

The functions

$$y_s(u) = s(u), \quad y_x(u) = x(u) \tag{7}$$

are called input-output static characteristics of the dynamic activated sludge wastewater treatment process.

Assume now that the coefficients s_{in}, x_r and r in the model (3)–(4) are enclosed by the intervals $[s_{in}] = [s_{in}^-, s_{in}^+]$, $[x_r] = [x_r^-, x_r^+]$ and $[r] = [r^-, r^+]$ respectively.

For arbitrary but fixed $u > 0$ consider $y_s(u) = y_s(u; s_{in}, x_r, r)$ and $y_x(u) = y_x(u; s_{in}, x_r, r)$ from (7) as functions of the variables s_{in}, x_r and r defined on the vector with intervals in the components (interval vector) $[z] = ([s_{in}], [x_r], [r])$. As a next step we shall compute the ranges of $y_s(u; s_{in}, x_r, r)$ and $y_x(u; s_{in}, x_r, r)$ on $[z]$. The main difficulty here is the strong dependence (repeatability) on the uncertain parameters in the expressions for $y_s(u)$ and $y_x(u)$ [4]. To overcome it we shall use monotonicity properties of the functions with respect to s_{in}, x_r and r. The gradients of y_s and y_x,

$$grad(y_s) = \left(\frac{\partial y_s}{\partial s_{in}}, \frac{\partial y_s}{\partial x_r}, \frac{\partial y_s}{\partial r}\right)$$
$$grad(y_x) = \left(\frac{\partial y_x}{\partial s_{in}}, \frac{\partial y_x}{\partial x_r}, \frac{\partial y_x}{\partial r}\right)$$

have constant signs with respect to each one of the components on $[z]$, that is

$$\frac{\partial y_s}{\partial s_{in}} > 0, \quad \frac{\partial y_s}{\partial x_r} < 0, \quad \frac{\partial y_s}{\partial r} < 0;$$
$$\frac{\partial y_x}{\partial s_{in}} > 0, \quad \frac{\partial y_x}{\partial x_r} > 0, \quad \frac{\partial y_x}{\partial r} > 0.$$

Then the range of $y_s(u) = y_s(u; s_{in}, x_r, r)$ and $y_x(u) = y_x(u; s_{in}, x_r, r)$ on the interval vector $[z] = ([s_{in}], [x_r], [r])$ is presented (see [1]) by

$$[y_s](u) = [y_s^-(u),\, y_s^+(u)] = [y_s(u; s_{in}^-,\, x_r^+,\, r^+),\, y_s(u; s_{in}^+,\, x_r^-,\, r^-)],$$
$$[y_x](u) = [y_x^-(u),\, y_x^+(u)] = [y_x(u; s_{in}^-,\, x_r^-,\, r^-),\, y_x(u; s_{in}^+,\, x_r^+,\, r^+)].$$

The functions $[y_s](u)$ and $[y_x](u)$ are interval-valued functions of the real variable u. They are called interval input-output static characteristics of the process (1)–(2) with respect to the outputs s and x respectively. They present the hull of all static characteristics $y_s(u) = y_s(u; s_{in}, x_r, r)$ and $y_x(u) = y_x(u; s_{in}, x_r, r)$ when the coefficients s_{in}, x_r and r vary in the prescribed intervals.

The interval function $[y_s](u)$ is uniquely defined by its boundary functions $y_s^-(u)$ and $y_s^+(u)$ that is $[y_s](u) = [y_s^-(u), y_s^+(u)]$ with $y_s^-(u) \le y_s^-(u)$ for any $u > 0$. We shall give explicit expressions for $y_s^-(u)$ and $y_s^-(u)$ now. Denote first

$$\begin{aligned}\Delta_1(u) = &(\mu_m k r^+ x_r^+ - s_{in}^-((1+r^+)u - \mu_m) - (1+r^+)uk_s)^2 \\ &+ 4\mu_m k r^+ x_r^+ (1+r^+)uk_s,\end{aligned}$$

$$\begin{aligned}\Delta_2(u) = &(\mu_m k r^- x_r^- - s_{in}^+((1+r^-)u - \mu_m) - (1+r^-)uk_s)^2 \\ &+ 4\mu_m k r^- x_r^- (1+r^-)uk_s.\end{aligned}$$

Then the boundary functions are presented by

$$y_s^-(u) = \frac{2k_s(1+r^+)us_{in}^-}{\sqrt{\Delta_1(u)} + \mu_m k r^+ x_r^+ - s_{in}^-((1+r^+)u - \mu_m) + (1+r^+)uk_s},$$

$$y_s^+(u) = \frac{2k_s(1+r^-)us_{in}^+}{\sqrt{\Delta_2(u)} + \mu_m k r^- x_r^- - s_{in}^+((1+r^-)u - \mu_m) + (1+r^-)uk_s}.$$

The boundary functions $y_x^-(u)$ and $y_x^+(u)$ for $[y_x](u)$ can be computed similarly by substituting $s_{in} = s_{in}^-$, $x_r = x_r^-$, $r = r^-$ and $s_{in} = s_{in}^+$, $x_r = x_r^+$, $r = r^+$ in $y_x(u)$ respectively.

3 Numerical Experiments

From the literature and from practical experiments some average values for the coefficients in the model (1)–(2) are known [3]:

$$\begin{aligned}\mu_m &= 0.35;\ \ k_s = 100;\ \ k = 2; \\ s_{in} &= 250;\ \ x_r = 2000;\ \ r = 0.2.\end{aligned}$$

In our computer experiments the above values for the coefficients are considered as centers of the intervals. The radii are given by $\varrho_\alpha \cdot \alpha$, $\alpha \in \{s_{in}, x_r, r\}$ with $0 < \varrho_\alpha < 1$; ϱ_α is called deviation from α.

Thus

$$\begin{aligned}[s_{in}] &= [s_{in}(1 - \varrho_{s_{in}}),\ s_{in}(1 + \varrho_{s_{in}})], \\ [x_r] &= [x_r(1 - \varrho_{x_r}),\ x_r(1 + \varrho_{x_r})], \\ [r] &= [r(1 - \varrho_r),\ r(1 + \varrho_r)].\end{aligned}$$

Giving different values to ϱ_α, intervals $[s_{in}]$, $[x_r]$ and $[r]$ with different radii (widths) are obtained.

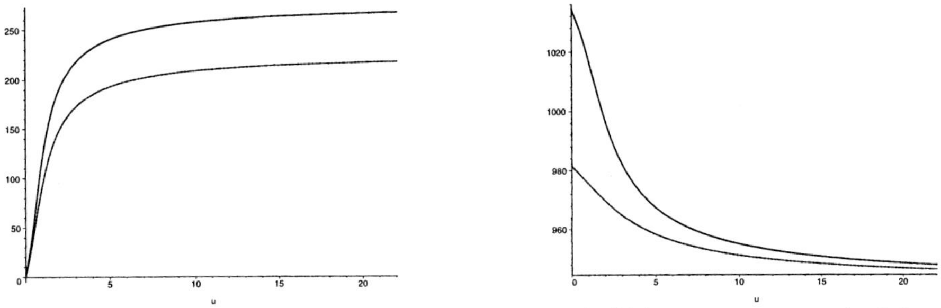

Fig. 2. Plots of $[y_s](u)$ with $\varrho_{s_{in}} = 0.1$ (left) and $[y_x](u)$ with $\varrho_{s_{in}} = 0.4$ (right)

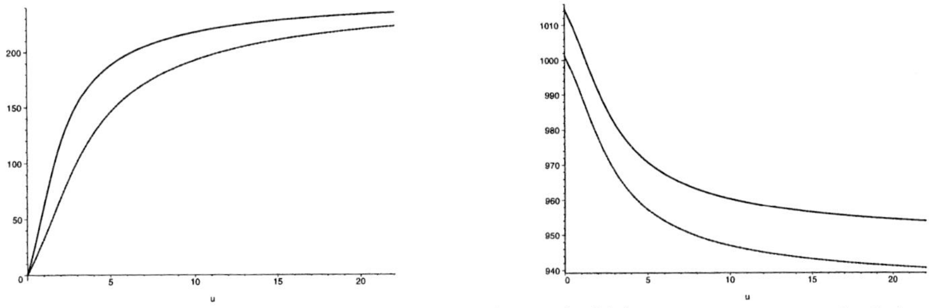

Fig. 3. Plots of $[y_s](u)$ with $\varrho_{x_r} = 0.3$ (left) and $[y_x](u)$ with $\varrho_{x_r} = 0.007$ (right)

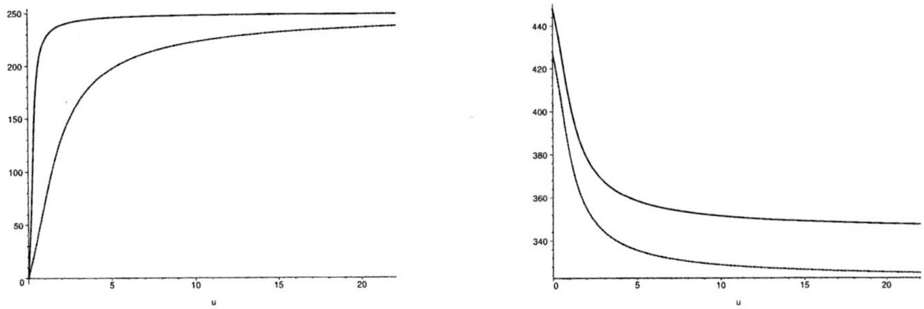

Fig. 4. Plots of $[y_s](u)$ with $\varrho_r = 0.9$ (left) and $[y_x](u)$ with $\varrho_r = 0.04$ (right)

All computations and visualizations are performed in the computer algebra system *Maple*.

Assume first that only s_{in} is uncertain, $s_{in} \in [s_{in}]$, and the other two coefficients x_r and r are exactly known, that is $\varrho_{x_r} = \varrho_r = 0$ (or equivalently

$[x_r] = x_r$, $[r] = r$). Figure 2 shows the boundary functions of $[y_s]_{s_{in}}(u)$ and $[y_x]_{s_{in}}(u)$ with different deviations.

Let be $x_r \in [x_r]$ and $[s_{in}] = s_{in}$, $[r] = r$. Figure 3 visualizes the interval static characteristics $[y_s]_{x_r}(u)$ and $[y_x]_{x_r}(u)$.

For $r \in [r]$ and $[s_{in}] = s_{in}$, $[x_r] = x_r$ Figure 4 presents the boundary functions of $[y_s]_r(u)$ and $[y_x]_r(u)$.

4 Discussion and Conclusion

This work presents qualitative analysis of the input-output static characteristics of the activated sludge wastewater treatment process involving uncertainties in the inflow parameters. This analysis is necessary in order to design and operate such system stable and efficiently.

The considered mathematical model of the process does not require special bounds on the dilution rate u. In practice, an operating admissible upper bound $\hat{u}$ for u always exists and depends on the technical caracteristic of the bioreactor.

Figure 2 shows that the large increase of u leads to strong decrease of the activated sludge concentration x, so that x may become smaller than a technologically given minimal concentration $x_{\min}$. This may cause instability and inefficiency of the bioreactor. The left picture presents the strong influence of s_{in} on $y_s(u)$ for small values of u, whereas large values for u retain the width $y_s^+(u) - y_s^-(u)$ of $[y_s](u)$ near to the width $s_{in}^+ - s_{in}^-$ of $[s_{in}]$.

Figure 3 shows strong dependence of the activated sludge concentration x with respect to the recycled biomass concentration x_r. The left picture shows that the deviations in x_r have strong impact on $s(u)$ in the first (exponential) phase, but for large rates u it becomes negligible. Similar effects are observed on Figure 4 as well.

The qualitative analysis is widely used in optimal control design of an activated sludge wastewater treatment process in the bioreactor.

References

1. N. Dimitrova: On a Numerical Approach for Solving a Class of Nonlinear Systems, in: Scientific Computing and Validated Numerics, G. Alefeld, A. Frommer, B. Lang (eds.), Akademie Verlag, Berlin, 147–153 (1996).
2. I. Dunn, E. Heinzle, J. Ingham, J. Prenosil: Biological Reaction Engineering, VCH, Weiheim, Germany (1992).
3. M. Puteh, K. Minekawa, N. Hashimoto, Y. Kawase: Modeling of Activated Sludge Wastewater Treatment Process, Bioprocess Engineering 21, 249–254 (1999).
4. H. Ratschek, J. Rockne: Computer Methods for the Range of Functions. Ellis Horwood Publ., Halsted Press, New York (1984).
5. S. Tzonkov, D. Filev, I. Simeonov, L. Vaklev: Control of the biotechnological processes. Technica, Sofia (in Bulgarian) (1992)

Numerical Search for the States with Minimal Dispersion in Quantum Mechanics with Non–negative Quantum Distribution Function

Alexander V. Zorin, Leonid A. Sevastianov, and Gregory A. Belomestny

Peoples' Friendship University of Russia
Moscow, 117198, Russia
sevast@sci.pfu.edu.ru
http://mathmod.sci.pfu.edu.ru

Abstract. We consider problems of quantum mechanics of Kuryshkin which pass to eigenvalue problem of conventional quantum mechanics when passing to the limit. From the demand of experimental confirmation of the theory's results are derived linearized equations for eigenstates of observables. The method of solving derived equations is illustrated on an example of hydrogen-like atom, for which were constructed matrices $O_{ij}(H)$ and $O_{ij}(H^2)$. An example of the solution is presented.

1 Quantum and Classical Mechanics in the Phase-Space Representation

In the paper of Moyal [1], caused by the preceding paper of Wigner [2], was made an attempt to interpret the conventional quantum mechanics as a statistical theory. The author proposed to represent the averages of an observables A, which take the form [3]:

$$\langle A\rangle_\Psi = (\hat{A}\Psi, \Psi) \tag{1.1}$$

in the CQM, in the form of the integral by the phase-space with so-called quantum distribution function (QDF) $F_\Psi(q,p)$:

$$\langle A\rangle_\Psi = \iint dqdpA(q,p)F_\Psi(q,p). \tag{1.2}$$

Reconciliation of the expressions (1.1) and (1.2) leaded the author of [1] to the conclusion, that the use of the QDF of Wigner from the paper [2] in the capacity of F_Ψ with the necessity requires to represent the operators $\hat{A}(\hat{q},\hat{p})$ in the form, proposed by Weyl [4].

In his paper [5] Shirokov has formulated the model (the concept) of the generalized algebra of observables combining in itself both model of classical mechanics and the model of quantum mechanics in the form of two pairs of operations σ_0, π_0 and σ_h, π_h on the common set of observables, realized by functions $A(q,p)$ on the phase-space. As a realization of his model the author [5] demonstrated the Weyl quantization rule and considered it's passage to the limit $\sigma_h \to \sigma_0$ and $\pi_h \to \pi_0$

Z. Li et al. (Eds.): NAA 2004, LNCS 3401, pp. 613–620, 2005.

while $\hbar \to 0$. In this connection Shirokov commented the papers of Moyal [1] and Wigner [2] with their statistical interpretation of the Weyl–Wigner–Moyal's quantum mechanics.

His reasoning about the passage to the limit of the statistical construction of the quantum mechanics into the classical statistical mechanics does not take into account nonclassical character of the Wigner's QDF (which is a generalized function, satisfying to a pseudo-differential evolution equations). These singularities were marked in the papers [6, 7] devoted to applications of QDF to different concrete problems. In these papers was marked the non-classical character of Wigner's QDF as the authors of [2] himself wrote.

At the same time the problem of the validity of the passage to the limit while $\hbar \to 0$ of the statistical (probabilistic) construction of the quantum mechanics was considered into the papers [8, 9]. This problem was completed in the papers of Kuryshkin [10, 11, 12] (see also the review [13]).

2 States with Minimal Dispersion in the Quantum Mechanics of Kuryshkin

In the quantum mechanics of Kuryskin with non-negative QDF the operators are defined [10, 11, 12, 13, 14] up to an arbitrary set of square-integrable on q functions of mocking configuration space and time $\{\varphi_k(q,t)\}$ normed by the condition

$$\sum_k \int |\varphi_k(q,t)|^2 dq = 1. \tag{2.1a}$$

By virtue of their square integrability the functions φ_k permit the Fourier transformation

$$\tilde{\varphi}(p,t) = (2\pi\hbar)^{-n/2} \int \exp\left\{-\frac{\mathrm{i}}{\hbar}(q,p)\right\} \varphi_k(q,t)dq. \tag{2.1b}$$

Further is introduced the function of the mocking phase space and time $\Phi(q,p,t)$ constructed with the help of auxiliary functions $\{\varphi_k\}$ in the form

$$\Phi(q,p,t) = (2\pi\hbar)^{-n/2} \exp\left\{-\frac{\mathrm{i}}{\hbar}(q,p)\right\} \sum_k \varphi_k(q,t)\tilde{\varphi}_k^*(p,t). \tag{2.2}$$

The correspondence rule of the constructing the operators in the quantum mechanics of Kuryshkin is formulated as follows: to the classical function $A(q,p,t)$ corresponds the linear operator $O(A)$ whose action on an arbitrary Fourier transformable function $\Psi(q,p)$ is defined by the relation

$$[O(A)\Psi](q,t) = (2\pi\hbar)^{-n} \int \Phi(\xi - q, \eta - p) A(\xi,\eta,t) \times \\ \exp\left\{\frac{\mathrm{i}}{\hbar}\Big((q-q'),p\Big),\right\} \Psi(q')d\xi d\eta dq' dp. \tag{2.3}$$

Then are valid two equivalent ways of evaluation the averages (experimentally measured) $\langle A \rangle$ of an arbitrary observable quantity A :

a) the quantum-mechanical average

$$\langle A\rangle_\Psi = \int \Psi^*(q,t)[O(A)\Psi](q,t)dq \tag{2.4a}$$

and

b) the quantum-statistical average

$$\langle A\rangle_\Psi = \int A(q,p,t)F_\Psi(q,p,t)dqdp \tag{2.4b}$$

so that the relation

$$F_\Psi(q,p,t) = (2\pi\hbar)^{-n}\sum_k \left|\int \tilde{\varphi}_k^*(q-\xi,t)\Psi(\xi,t)\exp\left\{-\frac{\mathrm{i}}{\hbar}(\xi,p)\right\}d\xi\right|^2 \tag{2.5}$$

determines the correspondence between the wave function and the probability distribution in the phase-space.

As in every statistical theory in the quantum mechanics of Kuryshkin[1] an important role plays the degree of incertainty of an average $\langle A\rangle_\Psi$ the quantitative measure of which is it's dispersion:

$$\langle(\Delta A)^2\rangle_\Psi = \big\langle O\big((A-\langle A\rangle_\Psi)^2\big)\big\rangle_\Psi\,. \tag{2.6}$$

The main task of the conventional quantum mechanics is to find the states of the physical system described by the eigenwavefunctions of the operators of the observable A :

$$\hat{A}\Psi_j^0 = a_j^0\Psi_j^0\,. \tag{2.7}$$

The dispersion of this observable in such states equal to

$$\left((\hat{A}-\langle A\rangle_j)^2\Psi_j^0,\Psi_j^0\right) = \left((\hat{A}-a_j^0)^2\Psi_j^0,\Psi_j^0\right) \tag{2.8}$$

is identically equal to zero.

In the quantum mechanics of Kuryshkin in analogous states Ψ_j satisfying the equation

$$O(A)\Psi_j = a_j\Psi_j \tag{2.9}$$

the dispersion of observable A equal to

$$\begin{aligned}\left(O\left((A-\langle A\rangle_j)^2\right)\Psi_j,\Psi_j\right) = \left(O(A^2)\Psi_j,\Psi_j\right) - 2a_j^2 + a_j^2 = \\ \left(O(A^2)\Psi_j,\Psi_j\right) - \left(O(A)\Psi_j,\Psi_j\right)^2\end{aligned} \tag{2.10}$$

is not identically equal to zero.

---

[1] The same concerns any quantization rule except for the Neumann's one but the authors do not know that these quantization rules would consider from the point of view of dispersions of averages.

In papers [11, 14] is shown that the quantum mechanics with non-negative QDF on the subset of observables M_0 goes over to the CQM when

$$\sum_k |\varphi_k(q,t)|^2 \rightarrow \delta(q) \Leftrightarrow \sum_k |\tilde{\varphi}_k(p,t)|^2 \rightarrow \delta(p). \tag{2.11}$$

In the cases of auxiliary functions $\{\varphi_k\}$ the relations (2.9), (2.10) differs from the relations (2.7), (2.8).

In the quantum mechanics of Kuryshkin (QMK) takes place the eigenvalue and eigenvector problem (2.9) the variational formulation of which has the form

$$\frac{(O(A)\Psi, \Psi)}{(\Psi, \Psi)} \rightarrow \min_{\Psi \neq O} . \tag{2.9a}$$

The solutions of the problem (2.9a) differs from the solutions of the corresponding problem in CQM

$$\frac{\left(\hat{A}\Psi^0, \Psi^0\right)}{(\Psi^0, \Psi^0)} \rightarrow \min_{\Psi^0 \neq O} . \tag{2.7a}$$

At the same time one may formulate the problem of statistical nature to find the states with minimal dispersion

$$\frac{\left(O\left((A - \langle A\rangle_\Psi)^2\right)\Psi, \Psi\right)}{(\Psi, \Psi)} \rightarrow \min_{\Psi \neq O} \tag{2.10a}$$

The analogous problem in CQM coincides identically with the problem (2.7a) and in the eigenstates of the observable A the dispersion of A is identically equal to zero and thus minimal. The solutions of the problem (2.10a) also differ from the solutions of the problem (2.7a)[2].

Let us note that similarly to the problems (2.7a), (2.9a) if Ψ is the solution of the problem (2.10a) then $c \cdot \Psi$ also is it's solution for $\forall c$. Thus the solutions of the problem (2.10a) are the one-dimensional subspaces (and their sums) so that the problems (2.7a), (2.9a), (2.10a) may be reformulated in the equivalent forms

$$\left(\hat{A}\Psi^0, \Psi^0\right) \rightarrow \min_{\|\Psi^0\|=1} , \tag{2.7b}$$

$$(O(A)\Psi, \Psi) \rightarrow \min_{\|\Psi\|=1} , \tag{2.9b}$$

$$\left(O\left((A - (O(A)\Psi, \Psi))^2\right)\Psi, \Psi\right) \rightarrow \min_{\|\Psi\|=1} . \tag{2.10b}$$

[2] All the three variational problems (2.7a), (2.9a), (2.10a) are solving in the same Hilbert space $L_2(Q)$.

3 An Approximate Evaluation of the Eigenstates and of the States with Minimal Dispersion

So we are considering three variational problems

$$\left(\hat{A}\Psi^0, \Psi^0\right) \to \min_{\|\Psi^0\|=1}, \tag{3.1}$$

$$(O(A)\Psi, \Psi) \to \min_{\|\Psi\|=1}, \tag{3.2}$$

$$\left(O\left((A - (O(A)\Psi, \Psi))^2\right)\Psi, \Psi\right) \to \min_{\|\Psi\|=1}, \tag{3.3}$$

the first order necessary conditions of which (named Euler's equations and constraints) take the form

$$\hat{A}\Psi_j^0 = a_j^0\Psi_j^0, \qquad \|\Psi_j^0\| = 1, \tag{3.4}$$

$$O(A)\Psi_j = a_j\Psi_j, \qquad \|\Psi\| = 1, \tag{3.5}$$

$$O\left(A^2\right)\Psi - 2\left(O(A)\Psi, \Psi\right)O(A)\Psi = 0, \qquad \|\Psi\| = 1. \tag{3.6}$$

Both the solutions of (3.2) and (3.5) and the solutions of (3.3) and (3.6) differ from the solutions of (3.1) and (3.4). And while the dispersions of the solutions of (3.1) and (3.4) are equal to zero, the dispersions both of the solutions of (3.2) and (3.5) and the solutions of (3.3) and (3.6) are not identically equal to zero. Because of the fact that the predictions of the CQM are confirmed by experimental data, distinctions of the QMK would be confirmed by experimental data in case of $\left(|a_j^1 - a_j^0| / |a_j^0|\right) \ll 1$ and $\left(|a_j^2 - a_j^0| / |a_j^0|\right) \ll 1$ where $a_j^2 = (O(A)\Psi_j, \Psi_j)$ – for (3.3) and (3.6). In just the same way the dispersions of the corresponding solutions must satisfy the estimates

$$\left(\delta_j^1 / |a_j^0|\right) \ll 1, \quad \left(\delta_j^2 / |a_j^0|\right) \ll 1.$$

Consequently the local minimums of the problems (3.2) and (3.5) and the problems (3.3) and (3.6) must be located in small neighborhoods of the local minimums of the problems (3.1) and (3.4).

We shall look for the solutions of (3.2) and (3.3) in the form $\Psi = \sum c_j\Psi_j^0$, where $\left\{\Psi_j^0\right\}$ is the full system of the solutions of the problem (3.1). Then in coordinate form problems (3.2) and (3.3) take the form

$$\sum_{i,j} O_{ij}(A)c_ic_j \to \min_{\sum_i c_i^2=1}, \tag{3.7}$$

$$\sum_{i,j} O_{ij}\left(\left(A - \left(\sum_{i,j} O_{ij}(A)c_ic_j\right)\right)^2\right)c_ic_j \to \min_{\sum c_i^2=1}, \tag{3.8}$$

where $O_{ij}(A) = \left(O(A)\Psi_i^o, \Psi_j^0\right)$.

We look for the solution Ψ_p of the problem (3.2) near the solution Ψ_p^0 of the problem (3.1) in the form

$$\Psi_p = \sum_k (\delta_{pk} + \delta c_k)\Psi_k^0. \tag{3.9}$$

That means that we look for the solution of the problem (3.7) in the form

$$\{c_k^p\} = \{\delta_{pk} + \delta c_k\}. \tag{3.10}$$

The first order necessary conditions of the problem (3.7) has the form

$$\sum_j O_{ij}(A)c_j^p - \left(\sum_{jk} O_{jk}(A)c_j^p c_k^p\right) c_i^p = 0; \quad i = 1, 2, \ldots, \tag{3.11a}$$

$$\sum_i \left(c_i^p\right)^2 = 1. \tag{3.11b}$$

Because of the norming $\sum_k \left(c_k^p\right)^2 = 1$ we have $\delta c_p = 0$ so that

$$\{c_k^p\} = \left\{\delta c_1^p, \delta c_2^p, \ldots, \delta c_{p-1}^p, \delta_{pk}, \delta_{p+1}^p, \ldots\right\}. \tag{3.12}$$

The first term of the sum (3.11) thus has the form

$$O_{ip}(A) + \sum_{j \neq p} O_{ij}(A)\delta c_j.$$

The first factor of the second term of (3.11) has the form

$$\sum_{j\neq p}\left(\sum_{k\neq p} O_{jk}(A)\delta c_j \delta c_k + O_{jp}(A)\delta c_j\right) + \sum_{k\neq p} O_{pk}(A)\delta c_k + O_{pp}(A).$$

The second factor of the second term of (3.11) has the form (3.12) consequently all the second term of the (3.11) has the form (up to the small's of the first order)

$$O_{pp}(A)\delta c_i + 2\sum_{j\neq p} O_{pj}(A)\delta c_j \delta_{pi}.$$

Thus all the equation up to the small's of the first order has the form

$$\sum_{j\neq p} \left(O_{ij}(A) - O_{pp}(A)\delta_{ij}\right)\delta c_{ij} = -O_{ip}(A). \tag{3.13}$$

We look for the solution $\tilde{\Psi}_p$ of the problem (3.3) near the solution Ψ_p^0 of the problem (3.1) in the form

$$\tilde{\Psi}_p = \sum_k \left(\delta_{pk} + \delta \tilde{c}_k\right)\Psi_k^0.$$

That means that we look for the solution of the problem (3.8) in the form $\{\tilde{c}_k^p\} = \{\delta_{pk} + \delta\tilde{c}_k\}$.

The first order necessary conditions of the problem (3.8) has the form

$$\sum_j O_{ij}\left(A^2\right)\tilde{c}_j^p - 2\left(\sum_{i,j} O_{ij}(A)\tilde{c}_i^p\tilde{c}_j^p\right)\sum_j O_{ij}(A)\tilde{c}_j^p = 0, \tag{3.14a}$$

$$\sum\left(\tilde{c}_i^p\right)^2 = 1. \tag{3.14b}$$

As in case of (3.11b) we have from (3.14b)

$$\{\tilde{c}_k^p\} = \left\{\delta\tilde{c}_1^p, \delta\tilde{c}_2^p, \ldots, \delta\tilde{c}_{p-1}^p, 1, \delta\tilde{c}_{p+1}^p, \ldots\right\}. \tag{3.15}$$

In just the same way from (3.14a) with regard to (3.15) we obtain up to the small's of the first order

$$\begin{gathered}\sum_{k\neq p}\Big(O_{jk}(A^2) - O_{pp}(A)O_{jk}(A) - 2O_{jp}(A)O_{pk}(A)\Big)\delta\tilde{c}_k^p = \\ O_{jp}(A)O_{pp}(A) - O_{jp}(A^2); \quad j = 1, 2, \ldots.\end{gathered} \tag{3.16}$$

As an example we consider Hamiltonian $H(\boldsymbol{r},\boldsymbol{p}) = \left(\boldsymbol{p}^2/2\mu\right) - \left(Ze^2/|\boldsymbol{r}|\right)$ of the hydrogen-like atom in the quantum mechanics of Kuryskin. In papers [15, 16, 17] is shown, that the infinite-dimensional matrix $O_{jk}(H)$ is stably approximated by finite-dimensional matrices $\{O_{jk}(H)\}_{j,k=1}^N$, where $N_n = \sum_{k=1}^n k^2 : N_1 = 1, N_2 = 5, N_3 = 14$. In just the same way one may prove that the infinite-dimensional matrix $O_{jk}(H^2)$ is stably approximated by finite-dimensional matrices $\{O_{jk}(H^2)\}_{j,k=1}^N$. For $N_3 = 14$ we solve the systems of linear algebraic equations (3.13) and (3.16) with matrices (see [17]) $O_{ij}(H)$, and $O_{ij}(H^2)$. Let us present for example in case $N_2 = 5$ two matrices M_k received with the help of two auxiliary functions φ_k so that $O_{ij}(H) = \sum_k a_k^2 M_k$:

$$M_1 = \left|\begin{array}{ccccc} -\frac{(1.25\ Z\,b-1.)}{b^2} & -\frac{0.18\ Z}{b} & 0. & 0. & 0. \\ -\frac{0.18\ Z}{b} & -\frac{(0.42\ Z\,b-1.)}{b^2} & 0. & 0. & 0. \\ 0. & 0. & -\frac{(-1.+0.49\ Z\,b)}{b^2} & 0. & 0. \\ 0. & 0. & 0. & -\frac{(-1.+0.49\ Z\,b)}{b^2} & 0. \\ 0. & 0. & 0. & 0. & -\frac{(-1.+0.49\ Z\,b)}{b^2} \end{array}\right|,$$

$$M_2 = \left|\begin{array}{ccccc} -\frac{(0.42\ Z\,b-0.25)}{b^2} & -\frac{0.018\ Z}{b} & 0. & 0. & 0. \\ -\frac{0.018\ Z}{b} & -\frac{(-0.33\ Z\,b-0.25)}{b^2} & 0. & 0. & 0. \\ 0. & 0. & -\frac{(0.33\ Z\,b-0.25)}{b^2} & 0. & 0. \\ 0. & 0. & 0. & -\frac{(0.33\ Z\,b-0.25)}{b^2} & 0. \\ 0. & 0. & 0. & 0. & -\frac{(0.33\ Z\,b-0.25)}{b^2} \end{array}\right|.$$

Matrices M_k^2 for $O_{ij}(H^2) = \sum_k a_k^2 M_k^2$ have more complicated structure, so we present here only the result of calculation.

lems along coordinate directions. The simple and convenient scalar technique based on the solution of three-point difference equations by Tri-Diagonal Matrix Algorithm (TDMA) has received a wide recognition of researchers and propagation in practical calculations.

This methodology lies in a basis of line-by-line method (the terminology [8]). For desired grid function from other coordinate directions the Seidel's approach is additional used. An experience of practical using [9, 10] have shown, that on detailed grids this method frequently has a low convergence rate. Therefore, the problem of increasing its efficiency is appeared.

A Modified Line-by-Line Method (MLLM) with conservation of initial technology were offered in [9, 10]. A block variant of this method was considered in [11]. The considerable improvement in convergence is related to the increasing of algorithm ellipticity. It is reached by the implicit consideration of difference flux from other grid direction, and by the diminution a norm of iterative expression.

An accumulated experience in methods of incomplete factorization [2, 6] together with a diagonal compensation (the terminology [6]) with constant value of iterative parameter have been used in [10] for this purpose. The parametric analyses carried out in [10] have shown that the optimal value of compensation parameter (nearly to 1) in the concrete problem allows sharp decreasing a necessary quantity of iterations. It indicates the necessity of giving it a certain degree of freedom, as the iteration situation strongly differs from a node to a node in calculation domain. As it is mentioned in [2], the practical experience of a variable compensation parameter application in incomplete factorization methods is absent in the literature and a problem of its best value determination still waits for the solution. Therefore the first empirical attempts in this direction, made in given paper for increasing of MLLM effectiveness, represent certain interest. It is create a basis for the further theoretical investigations and generalizations.

2 Mathematical Formulation of Problem

We consider a general stationary differential equation of convection-diffusion

$$\frac{\partial \rho u \Phi}{\partial x} + \frac{\partial \rho v \Phi}{\partial y} = \frac{\partial}{\partial x}\left(\Gamma \frac{\partial \Phi}{\partial x}\right) + \frac{\partial}{\partial y}\left(\Gamma \frac{\partial \Phi}{\partial y}\right) + S_P \Phi + S_C, \tag{1}$$

in two-dimensional rectangular domain $G(x, y)$ supplemented with general boundary conditions on the boundary G_k in the form

$$q_{1k}\frac{\partial \Phi}{\partial n} + q_{2k}\Phi = q_{3k}, \tag{2}$$

where n is the outward normal to the boundary G_k, Φ is the desired function, ρ is the density, (u, v) are the components of velocity field, S_C, S_P are the sources of Φ. The coefficients Γ, S_P, S_C, q are piecewise continuous functions of (x, y) and $\Gamma > 0$, $S_P \le 0$. Equation (1) is the conservation law for the scalar variable and is widely used in solving problems of convective heat transfer.

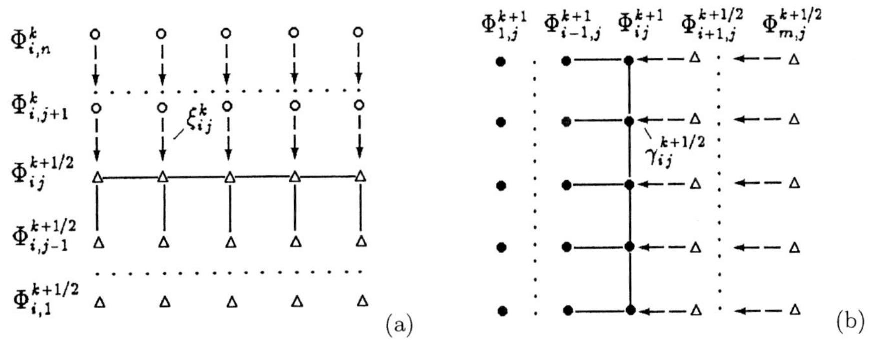

Fig. 1. The iterative algorithm scheme: ○ — k, △ — k+1/2, ● — k+1 stages of iteration step

$$i = \overline{1, m}, \quad j = \overline{n, 2};$$

$$a'_P = (a_P - \theta_{i,j}(a_E + a_W)), \quad b' = b + a_E\Phi_{i+1,j} + a_W\Phi_{i-1,j} - \theta_{i,j}(a_E + a_W)\Phi_{i,j}.$$

The processes of transfer in i-direction in the equation (3) are selected similarly. After usage of diagonal compensation

$$f_{i,j}(\Phi, \theta) = -\theta_{i,j}(a_S + a_N)\Phi_{i,j}$$

the equation (3) becomes

$$-a_W\Phi_{i-1,j} + [a_P - \theta_{i,j}(a_S + a_N)]\Phi_{i,j} - a_E\Phi_{i+1,j} \tag{11}$$

$$= b + a_S\Phi_{i,j-1} + a_N\Phi_{i,j+1} + f_{i,j}(\Phi).$$

By supposing in (11), that the iterated expression is equal to zero

$$F_{i,j}(\Phi, \theta) = a_S\Phi_{i,j-1} + a_N\Phi_{i,j+1} - \theta_{i,j}(a_S + a_N)\Phi_{i,j},$$

we receive the analogous (9) formula for calculation $\theta_{i,j}$:

$$\theta_{i,j} = \begin{cases} 0, & \theta_{NS}^{calc} \le 0, \\ \theta_{NS}^{calc}, & 0 < \theta_{NS}^{calc} < 1, \\ 1, & \theta_{NS}^{calc} \ge 1 \end{cases} \qquad \theta_{NS}^{calc} = \frac{a_N\Phi_{i,j+1} + a_S\Phi_{i,j-1}}{(a_N + a_S)\Phi_{i,j}}. \tag{12}$$

Solving the equation (11) by left-hand TDMA, we receive the necessary coefficients of the two-point recurrent relation $(\gamma, d)_{i,j}$ in i-direction.

The two-stage scheme of the iteration algorithm is represented on fig. 1. The light circles denote the grid function $\Phi_{i,j}^k$ computed on the preceding k-iteration step. The triangles and black circles designate the grid function on the first, $(k+1/2)$, and the second, $(k+1)$ stages of current iteration step. The algorithm includes the following stages.

Stage 1. At the known field $\Phi_{i,j}^k$, using (9) for calculation $\theta_{i,j}$, we found $(\xi, \eta)_{i,j}$ from equation (8), by applying the formulas (10) for left-hand TDMA [1]. Solving further the equations (6) for all lines $j = \overline{1,n}$, we obtain the field $\Phi_{i,j}^{k+1/2}$ (see fig. 1a).

Stage 2. Using obtained $\Phi_{i,j}^{k+1/2}$ and formula (12) for calculation $\theta_{i,j}$, from equation (11) we found $(\gamma, d)_{i,j}^{k+1/2}$, by applying the left-hand TDMA [1]. Solving the equations (7) for all lines $i = \overline{1,m}$, we obtain the field $\Phi_{i,j}^{k+1}$ (see fig. 1b).

4 Computational Experiment

The estimation of practical convergence rate and effectiveness of the offered iteration method for the SLAE solution (3), and also a comparison with other algorithms, were carried out on the Dirichlet's problem for a Poisson equation with variable coefficients in a unit square (see [1]):

$$\frac{\partial}{\partial x}\left(a_1\frac{\partial \Phi}{\partial x}\right) + \frac{\partial}{\partial y}\left(a_2\frac{\partial \Phi}{\partial y}\right) = -\varphi, \quad (x,y) \in [0,1], \quad \Phi|_{G_k} = 0 \qquad (13)$$

$$a_1 = 1 + C[(x-0.5)^2 + (y-0.5)^2], \quad a_2 = 1 + C[0.5 - (x-0.5)^2 - (y-0.5)^2],$$

$$c_1 = 1 \le a_1, a_2(x,y) \le c_2 = 1 + 0.5C, \quad C = \text{const}.$$

Right-hand side $\varphi(x,y)$ of (13) was chosen so, that the exact solution of equation (13) corresponds to the test function $\Phi(x,y) = x(1-x)y(1-y)$. The usual five-point difference scheme (3) of [1] was used for approximation of (13) on the uniform grid $\omega = \{(x_i, y_j) = (ih, jh),\ 0 \le i,j \le N,\ h = 1/N\}$. The initial approximation was $\Phi_{i,j}^0 = 1$. Varying the constant C, we change the ratio c_2/c_1 and the region of coefficients a_1, a_2 in (13) (c_2, c_1 are their maximum and minimum values).

The iteration process was monitored by the relative change in the norms of the residual $\|r^k\|/\|r^0\|$ and error $\|z^k\|/\|z^0\|$ vectors:

$$z_{i,j} = (\Phi_{i,j} - \Phi_{i,j}^*), \quad r_{i,j} = a_P\Phi_{i,j} - \sum a_{nb}\Phi_{nb} - b, \quad \|r\| = (r,r)^{1/2},$$

where k is the number of current iteration, nb is a neighboring stencil's nodes, $*$ notes the exact numerical solution computed in advance with high accuracy.

5 Numerical Results and Their Analysis

Figure 2 shows trajectories of the residual $\|r^k\|$ and error $\|z^k\|$ vectors during iteration process. The curve 1 corresponds to the initial line-by-line method [8], the curves 2, 3 are the modified method with constant parameter $\theta = 0$ [9] and $\theta = 1$ [10] respectively, the curve 4 is the given paper with the variable compensation parameter $\theta_{i,j}$. The curve of the Stone's Strongly Implicit Method (SIP) [13] (5) and the curve of Seidel's method (6) are given for comparison. The initial parameters of the problem are $c_2/c_1 = 32$, $N = 32$, $\|r^0\| = 197$.

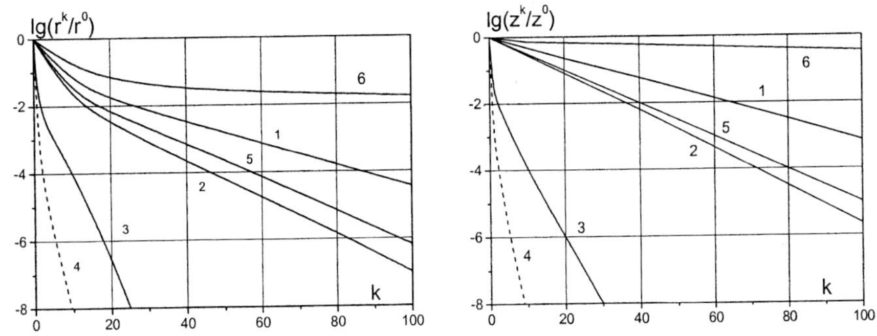

Fig. 2. The dynamics of the residual r^k and error z^k vectors in iterative process. 1 — LLM, 2 — LLM ($\theta = 0$), 3 — MLLM ($\theta = 1$), 4 — LLMVC (the given paper, θ =var), 5 — SIP [13], 6 — Seidel's method. $(N \times N) = (32 \times 32)$, $c_2/c_1 = 32$

Table 1. A quantity of the required iteration steps for accuracy $\|r^k\|/\|r^0\| = 10^{-4}$ in test problem

$N \times N$	c_2/c_1	SIP [13]	LLM [8]	MLLM		LLMVC
				$\theta = 0$ [9]	$\theta = 1$ [10]	$\theta_{i,j}$ =var
32×32	2	88	149	80	12	2
	32	55	87	49	9	3
	512	51	80	45	9	3
64×64	2	290	490	264	20	3
	32	181	285	159	16	3
	512	166	261	146	16	3
128×128	2	916	1544	833	32	3
	32	576	907	505	27	3
	512	525	833	454	27	3

The vectors of residual and error are shown by dotted curves 4 (the given work). They are decreasing very rapidly (up to 4 orders) on the first 3–4 iterations, in contrast to other curves. On a smooth approximation their decreasing more than two orders are observed already after the first iteration step. A comparison with curve 3 ($\theta = 1$) shows that such effectiveness is entirely connected with the application of the variable compensation parameter. Using a sequence of grids, we examined the behavior of the iterative characteristics depending on the coefficients a_1 and a_2 in (13) and the quantity of grid nodes. Table 1 shows the quantity k of iterations for reducing the initial residual by a factor of 10^4 (i. e., $\|r^k\|/\|r^0\| = 10^{-4}$). An analysis of results shows that the offered algorithm practically does not depend on the ratio of coefficients c_2/c_1 in comparison with the line-by-line method [8]. The convergence rate depends from a grid dimension very weakly. Therefore an application of iteration method is most effective on detailed grids.

Thus, the results obtained confirm a high effectiveness of the proposed iteration algorithm and are emphasized the necessity of implicit registration of the difference fluxes (or their analogues) from all the grid directions. The integral form of the equation (3) represents a balance of the fluxes and sources for desired variable, therefore the iteration method should reflect this nature. The select of compensation parameter can be carried out by the different ways and here there is a wide field of activity for future investigations. The results obtained are shown the perspectives of using a variable compensation parameter in the solution of difference equations.

6 Conclusions

The new iteration Line-by-Line Method with Variable Compensation parameter (LLMVC) is offered for solving the difference equations which arise at the numerical approximation of the two-dimensional elliptic and parabolic differential equations. The scalar technology of an alternating direction method is retained in algorithm. Its convergence rate is not sensitive to a variation of coefficients at highest derivatives in differential equation and weakly depends on a quantity of grid nodes. Test computations showed that the offered iteration method on detailed grids allows reducing the quantity of required iteration steps by two orders of magnitude in comparison with the initial line-by-line method [8] and almost on the order in relation to its modified variant [10].

References

1. Samaskii, A.A., Nikolaev, E.S.: Metody resheniya setochnykh uravnenii (Methods for Solving Difference Equations). Moscow: Nauka. (1978)
2. Il'in, V.P.: Incomplete factorization methods. Singapore: World Sci. Publ. Co. (1992)
3. Marchuk G.I.: Metody vychislitel'noi matematiki (Methods of Computational Mathematics). Moscow: Nauka. (1989)
4. Hageman, L.A., Young, D.M.: Applied Iterative Methods. New York: Academic. (1981)
5. Samaskii, A.A., Gulin, A.V.: Chislennye metody (Numerical Methods). Moscow: Nauka. (1989)
6. Buleev, N.I.: Prosranstvennaya model' turbulentnogo obmena (A Three-Dimensional Model of Turbulent Transfer). Moscow: Nauka. (1989)
7. Chetverushkin, B.N.: Iterative Algorithm for Solving Difference Equations. Zh. Vychisl. Mat. Mat. Fiz. **16** (2) (1976) 519–524
8. Patankar, S.V.: Numerical Heat Transfer and Fluid Flow. Washington: Hemisphere. (1980)
9. Zverev, V.G., Zvereva, L.V.: Modifitsirovannyi polineinyi metod reshenija dvumernykh setochnykh uravnenii (A Modified Line-by-Line Method for Two-Dimensional Difference Equations). Available from VINITI. (N76-B92) (1992) 21
10. Zverev, V.G.: Modified Line-by-Line Method for Difference Elliptic Equations. Comput. Math. Math. Physics. **38** (9) (1998) 1490–1498
11. Zverev, V.G.: Implicit Block Iterative Method for solving Two-Dimensional Elliptic Equations. Comput. Math. Math. Physics. **40** (4) (2000) 562–569
12. Il'in, V.P.: Metody konechnykh raznostei i konechnykh ob'emov dlya ellipticheskikh uravnenii. Novosibirsk: Inst. Matematiki. (2000)
13. Stone, H. L.: Iterative solution of implicit approximation of multidimensional partial differential equations. SIAM J. Numer. Analys. **5** (1968) 530–538

Author Index

Lecture Notes in Computer Science

For information about Vols. 1–3299

please contact your bookseller or Springer

Vol. 3418: U. Brandes, T. Erlebach (Eds.), Network Analysis. XII, 471 pages. 2005.

Vol. 3412: X. Franch, D. Port (Eds.), COTS-Based Software Systems. XVI, 312 pages. 2005.

Vol. 3410: C.A. Coello Coello, A. Hernández Aguirre, E. Zitzler (Eds.), Evolutionary Multi-Criterion Optimization. XVI, 912 pages. 2005.

Vol. 3406: A. Gelbukh (Ed.), Computational Linguistics and Intelligent Text Processing. XVII, 829 pages. 2005.

Vol. 3404: V. Diekert, B. Durand (Eds.), STACS 2005. XVI, 706 pages. 2005.

Vol. 3403: B. Ganter, R. Godin (Eds.), Formal Concept Analysis. XI, 419 pages. 2005. (Subseries LNAI).

Vol. 3401: Z. Li, L. Vulkov, J. Waśniewski (Eds.), Numerical Analysis and Its Applications. XIII, 630 pages. 2005.

Vol. 3398: D.-K. Baik (Ed.), Systems Modeling and Simulation: Theory and Applications. XIV, 733 pages. 2005. (Subseries LNAI).

Vol. 3397: T.G. Kim (Ed.), Artificial Intelligence and Simulation. XV, 711 pages. 2005. (Subseries LNAI).

Vol. 3393: H.-J. Kreowski, U. Montanari, F. Orejas, G. Rozenberg, G. Taentzer (Eds.), Formal Methods in Software and Systems Modeling. XXVII, 413 pages. 2005.

Vol. 3391: C. Kim (Ed.), Information Networking. XVII, 936 pages. 2005.

Vol. 3388: J. Lagergren (Ed.), Comparative Genomics. VIII, 133 pages. 2005. (Subseries LNBI).

Vol. 3387: J. Cardoso, A. Sheth (Eds.), Semantic Web Services and Web Process Composition. VIII, 147 pages. 2005.

Vol. 3386: S. Vaudenay (Ed.), Public Key Cryptography - PKC 2005. IX, 436 pages. 2005.

Vol. 3385: R. Cousot (Ed.), Verification, Model Checking, and Abstract Interpretation. XII, 483 pages. 2005.

Vol. 3382: J. Odell, P. Giorgini, J.P. Müller (Eds.), Agent-Oriented Software Engineering V. X, 239 pages. 2005.

Vol. 3381: P. Vojtáš, M. Bieliková, B. Charron-Bost, O. Sýkora (Eds.), SOFSEM 2005: Theory and Practice of Computer Science. XV, 448 pages. 2005.

Vol. 3379: M. Hemmje, C. Niederee, T. Risse (Eds.), From Integrated Publication and Information Systems to Information and Knowledge Environments. XXIV, 321 pages. 2005.

Vol. 3378: J. Kilian (Ed.), Theory of Cryptography. XII, 621 pages. 2005.

Vol. 3376: A. Menezes (Ed.), Topics in Cryptology – CT-RSA 2005. X, 385 pages. 2004.

Vol. 3375: M.A. Marsan, G. Bianchi, M. Listanti, M. Meo (Eds.), Quality of Service in Multiservice IP Networks. XIII, 656 pages. 2005.

Vol. 3374: D. Weyns, H.V.D. Parunak, F. Michel (Eds.), Environments for Multi-Agent Systems. X, 279 pages. 2005. (Subseries LNAI).

Vol. 3372: C. Bussler, V. Tannen, I. Fundulaki (Eds.), Semantic Web and Databases. X, 227 pages. 2005.

Vol. 3368: L. Paletta, J.K. Tsotsos, E. Rome, G.W. Humphreys (Eds.), Attention and Performance in Computational Vision. VIII, 231 pages. 2005.

Vol. 3366: I. Rahwan, P. Moraitis, C. Reed (Eds.), Argumentation in Multi-Agent Systems. XII, 263 pages. 2005. (Subseries LNAI).

Vol. 3363: T. Eiter, L. Libkin (Eds.), Database Theory - ICDT 2005. XI, 413 pages. 2004.

Vol. 3362: G. Barthe, L. Burdy, M. Huisman, J.-L. Lanet, T. Muntean (Eds.), Construction and Analysis of Safe, Secure, and Interoperable Smart Devices. IX, 257 pages. 2005.

Vol. 3361: S. Bengio, H. Bourlard (Eds.), Machine Learning for Multimodal Interaction. XII, 362 pages. 2005.

Vol. 3360: S. Spaccapietra, E. Bertino, S. Jajodia, R. King, D. McLeod, M.E. Orlowska, L. Strous (Eds.), Journal on Data Semantics II. XI, 223 pages. 2004.

Vol. 3359: G. Grieser, Y. Tanaka (Eds.), Intuitive Human Interfaces for Organizing and Accessing Intellectual Assets. XIV, 257 pages. 2005. (Subseries LNAI).

Vol. 3358: J. Cao, L.T. Yang, M. Guo, F. Lau (Eds.), Parallel and Distributed Processing and Applications. XXIV, 1058 pages. 2004.

Vol. 3357: H. Handschuh, M.A. Hasan (Eds.), Selected Areas in Cryptography. XI, 354 pages. 2004.

Vol. 3356: G. Das, V.P. Gulati (Eds.), Intelligent Information Technology. XII, 428 pages. 2004.

Vol. 3355: R. Murray-Smith, R. Shorten (Eds.), Switching and Learning in Feedback Systems. X, 343 pages. 2005.

Vol. 3353: J. Hromkovič, M. Nagl, B. Westfechtel (Eds.), Graph-Theoretic Concepts in Computer Science. XI, 404 pages. 2004.

Vol. 3352: C. Blundo, S. Cimato (Eds.), Security in Communication Networks. XI, 381 pages. 2005.

Vol. 3350: M. Hermenegildo, D. Cabeza (Eds.), Practical Aspects of Declarative Languages. VIII, 269 pages. 2005.

Vol. 3349: B.M. Chapman (Ed.), Shared Memory Parallel Programming with Open MP. X, 149 pages. 2005.

Vol. 3348: A. Canteaut, K. Viswanathan (Eds.), Progress in Cryptology - INDOCRYPT 2004. XIV, 431 pages. 2004.

Vol. 3347: R.K. Ghosh, H. Mohanty (Eds.), Distributed Computing and Internet Technology. XX, 472 pages. 2004.

Vol. 3346: R.H. Bordini, M. Dastani, J. Dix, A.E.F. Seghrouchni (Eds.), Programming Multi-Agent Systems. XIV, 249 pages. 2005. (Subseries LNAI).

Vol. 3345: Y. Cai (Ed.), Ambient Intelligence for Scientific Discovery. XII, 311 pages. 2005. (Subseries LNAI).

Vol. 3344: J. Malenfant, B.M. Østvold (Eds.), Object-Oriented Technology. ECOOP 2004 Workshop Reader. VIII, 215 pages. 2005.

Vol. 3342: E. Şahin, W.M. Spears (Eds.), Swarm Robotics. IX, 175 pages. 2005.

Vol. 3341: R. Fleischer, G. Trippen (Eds.), Algorithms and Computation. XVII, 935 pages. 2004.

Vol. 3340: C.S. Calude, E. Calude, M.J. Dinneen (Eds.), Developments in Language Theory. XI, 431 pages. 2004.

Vol. 3339: G.I. Webb, X. Yu (Eds.), AI 2004: Advances in Artificial Intelligence. XXII, 1272 pages. 2004. (Subseries LNAI).

Vol. 3338: S.Z. Li, J. Lai, T. Tan, G. Feng, Y. Wang (Eds.), Advances in Biometric Person Authentication. XVIII, 699 pages. 2004.

Vol. 3337: J.M. Barreiro, F. Martin-Sanchez, V. Maojo, F. Sanz (Eds.), Biological and Medical Data Analysis. XI, 508 pages. 2004.

Vol. 3336: D. Karagiannis, U. Reimer (Eds.), Practical Aspects of Knowledge Management. X, 523 pages. 2004. (Subseries LNAI).

Vol. 3335: M. Malek, M. Reitenspieß, J. Kaiser (Eds.), Service Availability. X, 213 pages. 2005.

Vol. 3334: Z. Chen, H. Chen, Q. Miao, Y. Fu, E. Fox, E.-p. Lim (Eds.), Digital Libraries: International Collaboration and Cross-Fertilization. XX, 690 pages. 2004.

Vol. 3333: K. Aizawa, Y. Nakamura, S. Satoh (Eds.), Advances in Multimedia Information Processing - PCM 2004, Part III. XXXV, 785 pages. 2004.

Vol. 3332: K. Aizawa, Y. Nakamura, S. Satoh (Eds.), Advances in Multimedia Information Processing - PCM 2004, Part II. XXXVI, 1051 pages. 2004.

Vol. 3331: K. Aizawa, Y. Nakamura, S. Satoh (Eds.), Advances in Multimedia Information Processing - PCM 2004, Part I. XXXVI, 667 pages. 2004.

Vol. 3330: J. Akiyama, E.T. Baskoro, M. Kano (Eds.), Combinatorial Geometry and Graph Theory. VIII, 227 pages. 2005.

Vol. 3329: P.J. Lee (Ed.), Advances in Cryptology - ASIACRYPT 2004. XVI, 546 pages. 2004.

Vol. 3328: K. Lodaya, M. Mahajan (Eds.), FSTTCS 2004: Foundations of Software Technology and Theoretical Computer Science. XVI, 532 pages. 2004.

Vol. 3327: Y. Shi, W. Xu, Z. Chen (Eds.), Data Mining and Knowledge Management. XIII, 263 pages. 2005. (Subseries LNAI).

Vol. 3326: A. Sen, N. Das, S.K. Das, B.P. Sinha (Eds.), Distributed Computing - IWDC 2004. XIX, 546 pages. 2004.

Vol. 3325: C.H. Lim, M. Yung (Eds.), Information Security Applications. XI, 472 pages. 2005.

Vol. 3323: G. Antoniou, H. Boley (Eds.), Rules and Rule Markup Languages for the Semantic Web. X, 215 pages. 2004.

Vol. 3322: R. Klette, J. Žunić (Eds.), Combinatorial Image Analysis. XII, 760 pages. 2004.

Vol. 3321: M.J. Maher (Ed.), Advances in Computer Science - ASIAN 2004. XII, 510 pages. 2004.

Vol. 3320: K.-M. Liew, H. Shen, S. See, W. Cai (Eds.), Parallel and Distributed Computing: Applications and Technologies. XXIV, 891 pages. 2004.

Vol. 3319: D. Amyot, A.W. Williams (Eds.), Telecommunications and beyond: Modeling and Analysis of Reactive, Distributed, and Real-Time Systems. XII, 301 pages. 2005.

Vol. 3318: E. Eskin, C. Workman (Eds.), Regulatory Genomics. VIII, 115 pages. 2005. (Subseries LNBI).

Vol. 3317: M. Domaratzki, A. Okhotin, K. Salomaa, S. Yu (Eds.), Implementation and Application of Automata. XII, 336 pages. 2005.

Vol. 3316: N.R. Pal, N.K. Kasabov, R.K. Mudi, S. Pal, S.K. Parui (Eds.), Neural Information Processing. XXX, 1368 pages. 2004.

Vol. 3315: C. Lemaître, C.A. Reyes, J.A. González (Eds.), Advances in Artificial Intelligence – IBERAMIA 2004. XX, 987 pages. 2004. (Subseries LNAI).

Vol. 3314: J. Zhang, J.-H. He, Y. Fu (Eds.), Computational and Information Science. XXIV, 1259 pages. 2004.

Vol. 3313: C. Castelluccia, H. Hartenstein, C. Paar, D. Westhoff (Eds.), Security in Ad-hoc and Sensor Networks. VIII, 231 pages. 2005.

Vol. 3312: A.J. Hu, A.K. Martin (Eds.), Formal Methods in Computer-Aided Design. XI, 445 pages. 2004.

Vol. 3311: V. Roca, F. Rousseau (Eds.), Interactive Multimedia and Next Generation Networks. XIII, 287 pages. 2004.

Vol. 3310: U.K. Wiil (Ed.), Computer Music Modeling and Retrieval. XI, 371 pages. 2005.

Vol. 3309: C.-H. Chi, K.-Y. Lam (Eds.), Content Computing. XII, 510 pages. 2004.

Vol. 3308: J. Davies, W. Schulte, M. Barnett (Eds.), Formal Methods and Software Engineering. XIII, 500 pages. 2004.

Vol. 3307: C. Bussler, S.-k. Hong, W. Jun, R. Kaschek, D.. Kinshuk, S. Krishnaswamy, S.W. Loke, D. Oberle, D. Richards, A. Sharma, Y. Sure, B. Thalheim (Eds.), Web Information Systems – WISE 2004 Workshops. XV, 277 pages. 2004.

Vol. 3306: X. Zhou, S. Su, M.P. Papazoglou, M.E. Orlowska, K.G. Jeffery (Eds.), Web Information Systems – WISE 2004. XVII, 745 pages. 2004.

Vol. 3305: P.M.A. Sloot, B. Chopard, A.G. Hoekstra (Eds.), Cellular Automata. XV, 883 pages. 2004.

Vol. 3303: J.A. López, E. Benfenati, W. Dubitzky (Eds.), Knowledge Exploration in Life Science Informatics. X, 249 pages. 2004. (Subseries LNAI).

Vol. 3302: W.-N. Chin (Ed.), Programming Languages and Systems. XIII, 453 pages. 2004.

Vol. 3300: L. Bertossi, A. Hunter, T. Schaub (Eds.), Inconsistency Tolerance. VII, 295 pages. 2005.

Lecture Notes in Computer Science 3401

Commenced Publication in 1973
Founding and Former Series Editors:
Gerhard Goos, Juris Hartmanis, and Jan van Leeuwen

Editorial Board

David Hutchison
Lancaster University, UK
Takeo Kanade
Carnegie Mellon University, Pittsburgh, PA, USA
Josef Kittler
University of Surrey, Guildford, UK
Jon M. Kleinberg
Cornell University, Ithaca, NY, USA
Friedemann Mattern
ETH Zurich, Switzerland
John C. Mitchell
Stanford University, CA, USA
Moni Naor
Weizmann Institute of Science, Rehovot, Israel
Oscar Nierstrasz
University of Bern, Switzerland
C. Pandu Rangan
Indian Institute of Technology, Madras, India
Bernhard Steffen
University of Dortmund, Germany
Madhu Sudan
Massachusetts Institute of Technology, MA, USA
Demetri Terzopoulos
New York University, NY, USA
Doug Tygar
University of California, Berkeley, CA, USA
Moshe Y. Vardi
Rice University, Houston, TX, USA
Gerhard Weikum
Max-Planck Institute of Computer Science, Saarbruecken, Germany